Springer Proceedings in Mathematics & Statistics

Volume 464

This book series features volumes composed of selected contributions from workshops and conferences in all areas of current research in mathematics and statistics, including data science, operations research and optimization. In addition to an overall evaluation of the interest, scientific quality, and timeliness of each proposal at the hands of the publisher, individual contributions are all refereed to the high quality standards of leading journals in the field. Thus, this series provides the research community with well-edited, authoritative reports on developments in the most exciting areas of mathematical and statistical research today.

Melvyn B. Nathanson
Editor

Combinatorial and Additive Number Theory VI

CANT, New York, USA, 2022 and 2023

Springer

Editor
Melvyn B. Nathanson
Department of Mathematics
Lehman College (CUNY)
Bronx, NY, USA

ISSN 2194-1009 ISSN 2194-1017 (electronic)
Springer Proceedings in Mathematics & Statistics
ISBN 978-3-031-65063-5 ISBN 978-3-031-65064-2 (eBook)
https://doi.org/10.1007/978-3-031-65064-2

Mathematics Subject Classification: 11B13, 11B34, 11B05, 11B75, 11P99, 20M14

This Springer imprint is published by the registered company Springer Nature Switzerland AG
The registered company address is: Gewerbestrasse 11, 6330 Cham, Switzerland

Preface

Workshops on Combinatorial and Additive Number Theory (CANT) have been organized at the CUNY Graduate Center in New York every year since 2003. We would usually meet for three or four days in May, in the week immediately preceding Memorial Day. These in-person meetings have become a fixed point in the number theory calendar, and persisted even during the COVID pandemic when it was impossible to meet in person and we met on Zoom. In 2022, the 20th annual meeting again took place on Zoom, but the 21st annual meeting in 2023 reconvened at the CUNY Graduate Center. This meeting was hybrid, with some talks given on Zoom by lecturers in many different countries, and others talks presented on campus and simultaneously broadcast on Zoom. Future CANT conferences will continue in this hybrid Zoom/in person form.

I am grateful to Springer for making possible the publication of the proceedings of the CANT 2022 and CANT 2023 conferences. I thank *Journal of Number Theory* for its continuing financial support of these meetings.

Previous volumes in this series are listed below.

Bronx, USA Melvyn B. Nathanson

References

1. M. B. Nathanson, editor, *Combinatorial and Additive Number Theory–CANT 2011 and 2012*, Springer Proc. Math. Stat., vol. 101, Springer, New York, 2014.
2. M. B. Nathanson, editor, *Combinatorial and Additive Number Theory II–CANT 2015 and 2016*, Springer Proc. Math. Stat., vol. 220, Springer, New York, 2017.
3. M. B. Nathanson, editor, *Combinatorial and Additive Number Theory III–CANT 2017 and 2018*, Springer Proc. Math. Stat., vol. 297, Springer, New York, 2020.
4. M. B. Nathanson, editor, *Combinatorial and Additive Number Theory IV–CANT 2019 and 2020*, Springer Proc. Math. Stat., vol. 347, Springer, New York, 2021.
5. M. B. Nathanson, editor, *Combinatorial and Additive Number Theory V–CANT 2021*, Springer Proc. Math. Stat., vol. 395, Springer, New York, 2022.

Contents

Chessboard Domination: Introduction of Some New Problems

S. D. Adhikari, Shruti Hegde, and Md Ibrahim Molla

Abstract For a positive integer n, considering the $n \times n$ chessboard, for a chess piece P, one is interested in the graph P_n by taking the squares of the chessboard as its vertices where two squares are *adjacent* if the piece P situated at one of the squares is able to move directly to the other. After recalling some known results and open questions on the domination number $\gamma(P_n)$ for the square board P_n, we here introduce some new questions.

1 Introduction

The problems of covering chessboards with minimal number of chess pieces date back to more than hundred and fifty years ago (see Cockayne [3], for instance). Though these questions are mainly combinatorial, they find a place among problems in number theory (see Guy [9]). For references to a connection with a problem in number theory and the problem of placing n queens on an $n \times n$ chessboard so that no two queens attack, one may look into [10]. Similarly, covering an $n \times n$ chessboard with a minimum number of queens on a major diagonal is related to the function $r_3(n)$, which is the largest cardinality of a subset of $\{1, ..., n\}$ which contains no 3-term arithmetic progression (see [6]).

For any two positive integers m, n, for a chess piece P, one considers the graph $P_{m\times n}$ by taking the squares of an $m \times n$ chessboard as its vertices and two squares are *adjacent* if piece P situated at one of the squares is able to move directly to the

---

S. D. Adhikari—Formerly at Harish-Chandra Research Institute.

S. D. Adhikari · S. Hegde (✉) · M. I. Molla
Department of Mathematics, Ramakrishna Mission Vivekananda Educational and Research Institute, Belur 711202, India
e-mail: shrutihg2112@gmail.com

S. D. Adhikari
e-mail: adhikarisukumar@gmail.com

M. I. Molla
e-mail: mdibimya@gmail.com

M. B. Nathanson (ed.), *Combinatorial and Additive Number Theory VI*,
Springer Proceedings in Mathematics & Statistics 464,
https://doi.org/10.1007/978-3-031-65064-2_1

other. When $m = n$, one simply writes P_n for $P_{n\times n}$. Corresponding to the queen, we have the *queen's graph* $Q_{m\times n}$ where two squares are *adjacent* if they are in the same row, column or diagonal.

A set D of squares of P_n is said to be a *dominating set* for P_n if every square of P_n is either in D or adjacent to some square in D. The minimum size of a dominating set of P_n is called the *domination number* for P_n and it is denoted by $\gamma(P_n)$.

In this article, we concern ourselves with problems related to the square boards; for some discussions related to the queen's graph for rectangular boards, one may look into the articles [1], [18] (Sect. 6) and [19].

In the next section, we go through some known results and open questions on the domination number $\gamma(P_n)$ for the square board P_n. In the later sections, we introduce some new questions and dwell on them.

2 Some Known Domination Problems

For the queen's graph, the following result on the domination number $\gamma(Q_n)$ was observed independently by Raghavan and Venketesan [12] and Spencer (see [3]). A proof is available in the survey article [3].

Theorem 1 *For a positive integer n, we have*

$$\gamma(Q_n) \geq (n-1)/2.$$

Weakley [16] showed that $\gamma(Q_n) = (n-1)/2$ implies $n \equiv 3 \pmod 4$, and hence for a positive integer $n \equiv 1 \pmod 4$, we have $\gamma(Q_n) \geq (n+1)/2$. While it is easy to observe that $\gamma(Q_3) = 1$ (see [18]), a dominating set given in [14, 15] (see also [18]) shows that $\gamma(Q_{11}) = 5$. In [16], it was observed that for $n \equiv 3 \pmod 4$, $\gamma(Q_n) = (n-1)/2$ holds only for $n = 3, 11$.

For small values of n, $\gamma(Q_n)$ is known and one is interested in knowing (see [18]) for which integers n, we have $\gamma(Q_n) = (n+1)/2$.

Welch (see [3]) established the following upper bound for $\gamma(Q_n)$.

Theorem 2 *For a positive integer n, writing* $n = 3q + r$, $0 \leq r < 3$, *we have*

$$\gamma(Q_n) \leq 2q + r.$$

A proof of the above has been given in [3].

Successive improvements [2, 11, 17] of the upper bound in Theorem 2 followed; in [11], it has been proved that for any positive integer n, we have $\gamma(Q_n) \leq 69n/133 + O(1)$.

We place the board in such a way that its sides are parallel to the coordinate axes and centre of each square has integer coordinates. The square with its centre at (x, y) is in *column* x and *row* y. The square (x, y) is said to be *even* if $x + y$ is even and *odd* if $x + y$ is odd.

Corresponding to a chess piece P, a set of squares of P_n is said to be *monochromatic* if all its squares have the same parity; otherwise the set is called *bichromatic*. By $\gamma_m(P_n)$ we mean the minimum size of a monochromatic dominating set of P_n.

While the columns and rows are referred to as *orthogonals*, the *difference diagonal* (respectively *sum diagonal*) through the square (x, y) is the set of all squares with centres on the line of slope $+1$ (respectively -1)passing through the point (x, y). The orthogonals and the diagonals are collectively called the *lines*. The value of $y - x$ is the same for each square (x, y) on a difference diagonal and sometimes one refers to the diagonal by this value. Similarly, $y + x$ is the same for each square (x, y) on a sum diagonal and one associates this value with this sum diagonal.

The following result has been recently proved in [19].

Theorem 3 *Let n be a positive integer and D a dominating set of Q_n of size $\lfloor n/2 \rfloor$. Then D is monochromatic if and only if it is possible to place the board so that the squares of D include exactly one from every even-indexed orthogonal.*

Another result in [19] is the following:

Theorem 4 *Among the positive integers n, $n \equiv 0 \pmod 4$, only for $n = 4, 12$ does Q_n have a monochromatic dominating set of size $n/2$.*

It has been conjectured in [19] that for each even integer $n > 4$, if $\gamma(Q_n) = n/2$, then every minimum dominating set of Q_n is monochromatic.

For more results on monochromatic dominating sets, we refer to [19].

Suppose that the squares of the chessboard are labelled (i, j), so that black and white squares have $(i + j)$ even and odd, respectively.

A subset K of $N = \{1, \ldots, n\}$ is called a *diagonal dominating set* if queens placed in positions $\{(k, k) : k \in K\}$ on the black major (or main) difference diagonal, dominate the entire board. One defines

$$\operatorname{diag}(n) = \min\{|K| : K \text{ is a diagonal dominating set}\}.$$

The following theorem [6] relates $\operatorname{diag}(n)$ to the well known number-theoretic function $r_3(n)$, which is the largest cardinality of a subset of $\{1, \ldots, n\}$ which contains no 3-term arithmetic progression.

Theorem 5 *For any n, we have*

$$\operatorname{diag}(n) = n - r_3\left(\left\lceil \frac{n}{2} \right\rceil\right).$$

From the above theorem and the result $lim_{n\to\infty}(r_3(n)/n) = 0$ of Roth [13], one gets

$$lim_{n\to\infty}(\operatorname{diag}(n)/n) = 1.$$

Defining $\operatorname{col}(n)$ by the minimum number of queens on any single column which are required to dominate the entire $n \times n$ chessboard, $\operatorname{col}(n)$ has been found [4] related to the above mentioned arithmetic function $r_3(n)$.

Corresponding to a chess piece P, a set I of squares of P_n is called *independent* if no two squares of I are adjacent. The minimum size of an independent dominating set of P_n is called the *independent domination number* for P_n and is denoted by $i(P_n)$. Clearly, $\gamma(P_n) \leq i(P_n)$.

For a positive integer n, for the rook's graph R_n and the king's graph K_n, it was shown in [8] that

$$\gamma(R_n) = i(R_n) = n \text{and}$$

$$\gamma(K_n) = i(K_n) = \lfloor \frac{n+2}{3} \rfloor^2.$$

For the Bishop's graph too, from [5] we have

$$\gamma(B_n) = i(B_n) = n.$$

For the queen's graph, it was proved in [7] that

$$i(Q_n) < 0.705n + 0.895.$$

3 New Domination Problems

We start with some observations.

Observation 1 (i) A queen on any line can cover at most three squares on any other parallel line and it covers all the squares on the line it occupies.

(ii) A bishop on any orthogonal can cover at most two squares on any other parallel orthogonal and it covers exactly one square on the orthogonal it occupies. Also, a bishop on any black (or white) diagonal can cover at most one square on any other parallel black (or white) diagonal and it cannot cover any square on a parallel white (or black) diagonal.

(iii) A rook on any orthogonal covers exactly one square on any other parallel orthogonal. Also, a rook on any diagonal can cover at most two squares on any other parallel diagonal and exactly one square on the diagonal it occupies. □

Now we proceed to introduce some new chessboard domination problems involving two types of chess pieces. Let X and Y be two types of chess pieces. For any positive integer n, for a fixed $1 \leq k < \gamma(X_n)$, let $\eta(X^kY)_n$ denote the minimum number of chess pieces of type Y required, along with k number of chess pieces of type X, to dominate the whole $n \times n$ chessboard. We would like to call $\eta(X^kY)_n$ the *mixed domination number of type* X^kY. We note that when we use different chess pieces, then whether a square (x, y) dominates another is determined after putting a particular chess piece at (x, y); therefore the situation here does not have a graph theoretic interpretation.

We call a square *X-needy* if it cannot be covered by any chess piece of type Y already placed on the board. We call a line *Y-empty* if it is not occupied by any chess piece of type Y.

We name the rows from bottom to top as $r_1, r_2, \ldots, r_n$ and columns from left to right as $c_1, c_2, \ldots, c_n$. A square (c_i, r_j) (sometimes written simply as (i, j)) is the square at which the row r_j and the column c_i intersect (see Fig. 1).

We name the sum diagonals from the left bottom corner of the board through the right top corner sequentially as $s_1, s_2, \ldots, s_{2n-1}$. Similarly, we name the difference diagonals from the right bottom corner through the left top corner sequentially as $d_1, d_2, \ldots, d_{2n-1}$ (see Fig. 2a, b).

Note that for any $1 \leq u \leq n$, each of the diagonals s_u and s_{2n-u} contains u number of squares and similarly, each of d_v and d_{2n-v} contains v number of squares for any $1 \leq v \leq n$.

In this article, we will refer to a diagonal all whose squares have even parity (black squares) as a black diagonal and a diagonal all whose squares have odd parity (white squares) as a white diagonal.

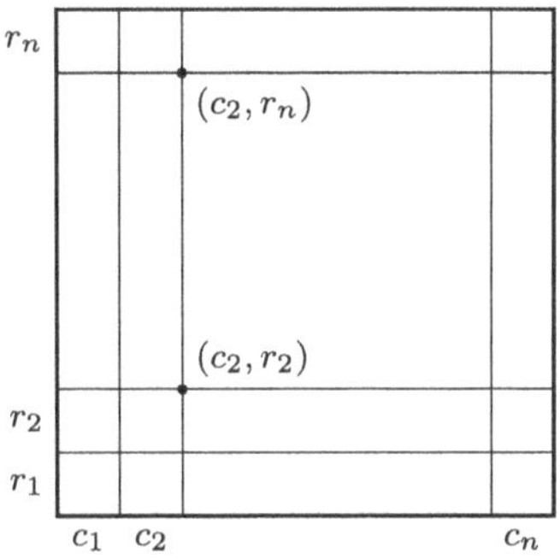

Fig. 1 Rows and columns in $n \times n$ chessboard

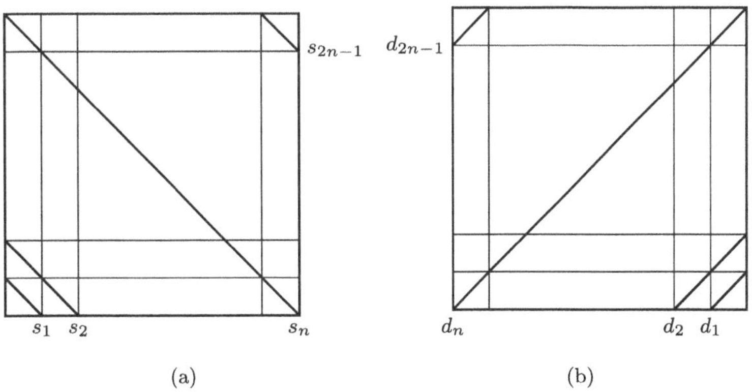

Fig. 2 Diagonals in $n \times n$ chessboard

Observation 2 When n is odd, the sum diagonals $s_1, s_3, \ldots, s_{2n-1}$ and the difference diagonals $d_1, d_3, \ldots, d_{2n-1}$ are all black diagonals; the diagonals $s_2, s_4, \ldots, s_{2n-2}$ and $d_2, d_4, \ldots, d_{2n-2}$ are all white diagonals. In particular, the main diagonals s_n and d_n are both black diagonals.

When n is even, the sum diagonals $s_1, s_3, \ldots, s_{2n-1}$ and the difference diagonals $d_2, d_4, \ldots, d_{2n-2}$ are all black diagonals; the diagonals $s_2, s_4, \ldots, s_{2n-2}$ and $d_1, d_3, \ldots, d_{2n-1}$ are all white diagonals. In particular, the main sum diagonal s_n is a white diagonal and the main difference diagonal d_n is a black diagonal.

A bishop occupying a black (or even) square is called a black bishop and a bishop occupying a white (or odd) square is called a white bishop.

Observation 3 If s_u (resp. d_u) is a *bishop-empty* white sum (resp. difference) diagonal containing u number of squares and m is the number of white bishops on the board, then at most m number of squares on s_u (resp. d_u) are covered by the m white bishops and hence at least $u - m$ number of squares on s_u (resp. d_u) are not covered. A similar statement can be made about a *bishop-empty* black diagonal containing v number of squares.

Observation 4 Let a bishop B be placed on an $n \times n$ board at the square (c, r). Then one can easily observe that

Number of squares covered by B, except (c, r), is $\leq$
min{ *number of rows above r, number of columns right to c*}
+ min{ *number of rows above r, number of columns left to c*}
+ min{ *number of rows below r, number of columns right to c*}
+ min{ *number of rows below r, number of columns left to c*}. □

4 New Domination Problems: Some Theorems

In this section, we see some results on the mixed domination numbers.

Theorem 6 *For any positive integer $n \geq 4$,*

$$\eta(B^1R)_n = n - 2.$$

Proof Let $n \geq 4$. Assume that $n - 3$ rooks and a bishop can cover the whole $n \times n$ board, if possible.

Clearly, $n - 3$ rooks occupy at most $n - 3$ rows and at most $n - 3$ columns, leaving at least 3 rows and at least 3 columns *rook-empty*. Let $c_{i_1}, c_{i_2}, c_{i_3}$ be three *rook-empty* columns and $r_{j_1}, r_{j_2}, r_{j_3}$ be three *rook-empty* rows (see Fig. 3a).

Then the squares in $\mathcal{A} = \{(c_{i_p}, r_{j_q}) : 1 \leq p, q \leq 3\}$ are not covered by the rooks and have to be covered by the only bishop available. That is, the squares in $\mathcal{A}$ are *bishop-needy*.

By Observation 1, a bishop on an orthogonal can cover at most two squares on any other parallel orthogonal. But, there are three or more *bishop-needy* squares on each c_{i_p}, $1 \leq p \leq 3$. Therefore, there will be at least one *bishop-needy* square on each of the columns c_{i_p}, $1 \leq p \leq 3$, which are not covered by the bishop.

This shows that we need at least $n - 2$ rooks along with a bishop to dominate the whole board and hence

$$\eta(B^1R)_n \geq n - 2.$$

On the other hand, place a bishop at the square (2, 2), a rook at each of the squares (2, 4), (4, 2) and (i, i), for all $5 \leq i \leq n$ (which exist unless $n = 4$), as shown in Fig. 3b.

Now, all the columns c_i where $5 \leq i \leq n$ and all the rows r_j where $5 \leq j \leq n$ are covered by the $n - 4$ rooks placed on the main difference diagonal. The 4×4 sub-board formed by the rows r_j, $1 \leq j \leq 4$ and columns c_i, $1 \leq i \leq 4$, is dominated by the bishop and the rooks lying here.

Therefore, we have

$$\eta(B^1R)_n = n - 2.$$

Remark 1 Now, let $n = 3$. Note that by placing a bishop at the square (2, 2) and a rook at each (2, 3) and (3, 2), we can dominate the whole 3×3 board (see Fig. 4a).

If only one rook is placed on a 3×3 board anywhere except at the square (2, 2), there will be at least one black *bishop-needy* square and at least one white *bishop-needy* square and only one bishop cannot cover them both.

If the rook is placed at (2, 2), then a bishop can cover all of the *bishop-needy* squares only if it can occupy (2, 2) which is already occupied by the rook (see Fig. 4b).

So, we get

$$\eta(B^1R)_3 = 2.$$

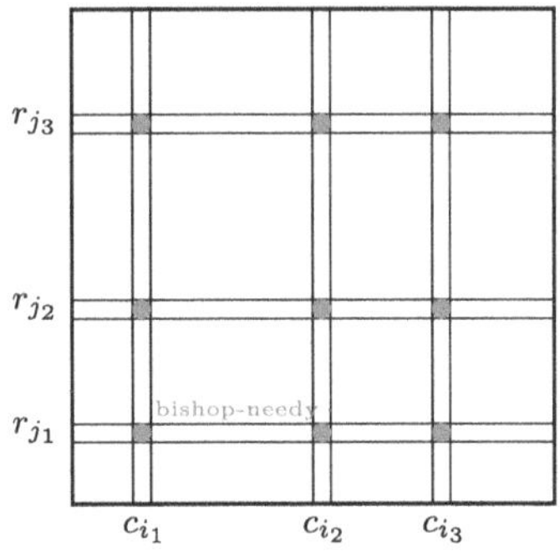

(a)

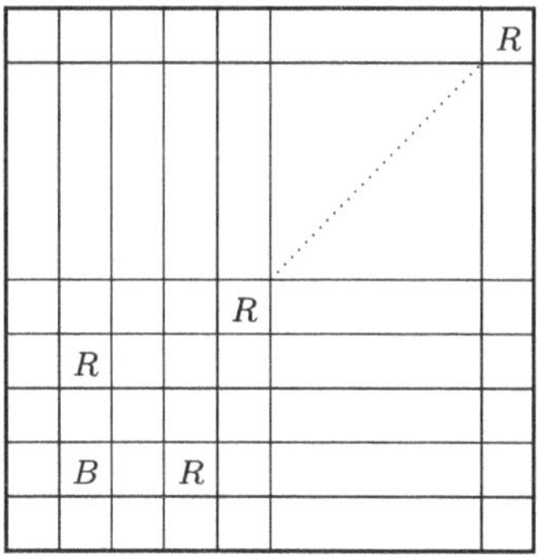

(b)

Fig. 3 For $\eta(B^1R)_n$

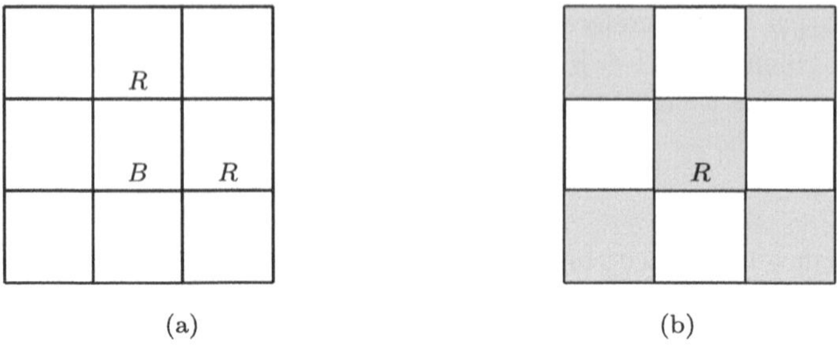

Fig. 4 For $\eta(B^1R)_3$

Theorem 7 *For any positive integer $n \geq 4$,*

$$\eta(Q^1R)_n = n - 3.$$

Proof Assume that $n-4$ rooks and a queen can cover the whole $n \times n$ board, if possible.

As in the proof of Theorem 6, $n-4$ rooks occupy at most $n-4$ rows and at most $n-4$ columns. Let c_{i_p}, $1 \leq p \leq 4$ be four *rook-empty* columns and r_{j_q}, $1 \leq q \leq 4$ be four *rook-empty* rows (see Fig. 5a). Then, all the squares in $\mathcal{A} = \{(c_{i_p}, r_{j_q}) : 1 \leq p, q \leq 4\}$ are *queen-needy*.

By Observation 1, a queen on an orthogonal can cover at most three squares on any other parallel orthogonal. But, we have four or more *queen-needy* squares on each c_{i_p}, $\forall 1 \leq p \leq 4$. Therefore, there will be at least three of the columns c_{i_p}, $1 \leq p \leq 4$ containing a *queen-needy* square that cannot be dominated by the queen.

This is a contradiction and we get

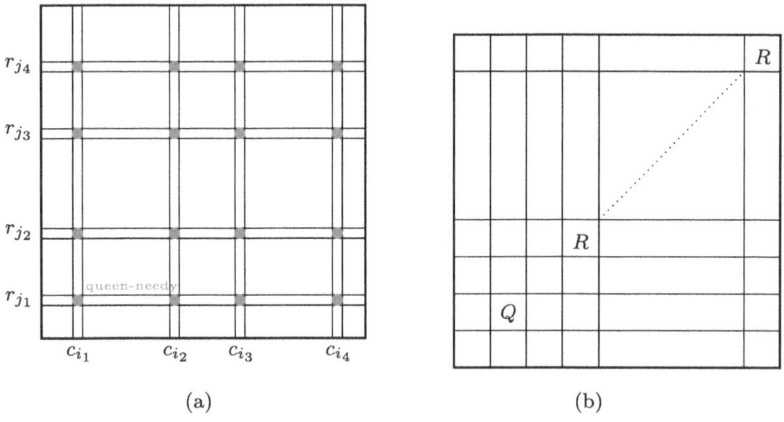

Fig. 5 For $\eta(Q^1R)_n$

$$\eta(Q^1 R)_n \geq n - 3.$$

Next, as shown in Fig. 5b, place a queen at the square (2, 2) and $n-3$ rooks on the main difference diagonal, that is, one each at the squares (i, i) for $4 \leq i \leq n$.

The columns c_i, $\forall 4 \leq i \leq n$ and the rows r_j, $\forall 4 \leq j \leq n$ are covered by the $n-3$ rooks placed on the main difference diagonal. The 3×3 sub-board formed by the rows r_j, $1 \leq j \leq 3$ and columns c_i, $1 \leq i \leq 3$, is dominated by the queen occupying (2, 2).

Thus, the whole board is dominated by a queen and $n-3$ rooks if they are placed on the board in the above manner.

Therefore, we have

$$\eta(Q^1 R)_n = n - 3.$$

Theorem 8 *For any positive integer $n \geq 3$,*

$$\eta(B^2 R)_n = n - 2.$$

Proof Suppose $n-3$ rooks and two bishops dominate the $n \times n$ board. Then, there are three columns, say $c_{i_1}, c_{i_2}, c_{i_3}$, and three rows, say $r_{j_1}, r_{j_2}, r_{j_3}$, which are *rook-empty*. As a result, the squares (c_{i_p}, r_{j_q}), $1 \leq p, q \leq 3$, will be *bishop-needy* squares. Assume, without loss of generality, that $1 \leq i_1 < i_2 < i_3 \leq n$ and $1 \leq j_1 < j_2 < j_3 \leq n$.

Since we have at least nine *bishop-needy* squares and two bishops have to cover them, by Pigeonhole principle, one of the bishops has to cover at least five *bishop-needy* squares. Let B_1 denote the position of the bishop that covers five or more *bishop-needy* squares.

If B_1 lies on or to the left of column c_{i_1}, then the number of *rook-empty* columns lying on the right side of the column of B_1 is at most 3 and that on the left side is zero. Also, the numbers of *rook-empty* rows above and below the row of B_1 add up to 3 or less.

Now, by Observation 4, B_1 can cover a maximum of three *bishop-needy* squares. As that is not desirable, due to symmetry, we can assume that B_1 lies between columns c_{i_1} and c_{i_3} and between rows r_{j_1} and r_{j_3}.

If B_1 lies on a *rook-empty* row (or column) and on a rook occupied column (or row), then it does not cover any *bishop-needy* square on its row (or column), which follows from Observation 1. Again by Observation 1, the second bishop cannot cover all the three *bishop-needy* squares on the row (or column) of B_1.

Therefore, B_1 has to occupy either a *bishop-needy* square or a square at the intersection of a rook occupied row with a rook occupied column.

Now, it is enough to consider the following two cases due to symmetry.

Case 1: B_1 lies at (c_{i_2}, r_{j_2}).

As B_1 has to cover at least five *bishop-needy* squares, both of its diagonals have to meet the two *rook-empty* columns c_{i_1} and c_{i_3} at *bishop-needy* squares. Thus, B_1 necessarily covers the squares (c_{i_1}, r_{j_1}), (c_{i_1}, r_{j_3}), (c_{i_3}, r_{j_1}) and (c_{i_3}, r_{j_3}) (see Fig. 6a).

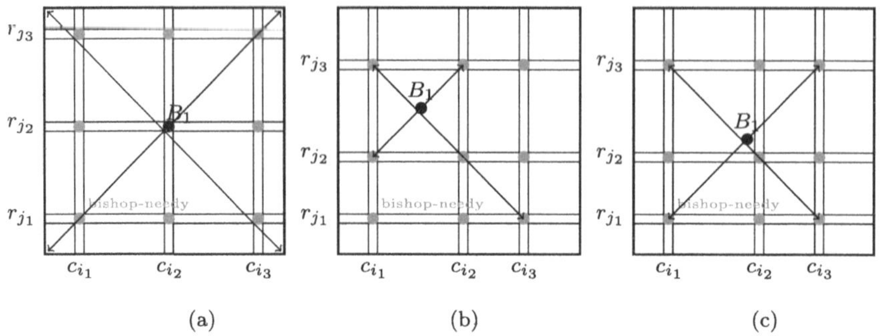

Fig. 6 For $\eta(Q^1R)_n$

Note that no three of the remaining *bishop-needy* squares (c_{i_1}, r_{j_2}), (c_{i_2}, r_{j_1}), (c_{i_2}, r_{j_3}), (c_{i_3}, r_{j_2}) lie on the same diagonal, so that both the diagonals of the second bishop B_2 have to cover two of these remaining squares each. But, for any two of these squares lying on the same diagonal d, the other two are on at least two diagonals perpendicular to d. Therefore, it is impossible for B_2 to cover all the four remaining *bishop-needy* squares.

Case 2: B_1 lies between rows r_{j_2} and r_{j_3} and columns c_{i_1} and c_{i_2}.

Again by the choice of B_1, both of its diagonals have to meet two *rook-empty* columns and one of its diagonals has to meet the third *rook-empty* column, all at *bishop-needy* squares. Thus, we have only two possibilities:

If B_1 covers (c_{i_1}, r_{j_2}) and (c_{i_1}, r_{j_3}) on c_{i_1}, then B_1 covers all the squares in

$$\{(c_{i_1}, r_{j_2}), (c_{i_1}, r_{j_3}), (c_{i_2}, r_{j_2}), (c_{i_2}, r_{j_3}), (c_{i_3}, r_{j_1})\}$$

and if B_1 covers (c_{i_1}, r_{j_1}) and (c_{i_1}, r_{j_3}) on c_{i_1}, then it covers all of the squares in

$$\{(c_{i_1}, r_{j_1}), (c_{i_1}, r_{j_3}), (c_{i_3}, r_{j_1}), (c_{i_3}, r_{j_3}), (c_{i_2}, r_{j_2})\}$$

(see Fig. 6b, c).

In either case, it is not difficult to see that among the remaining *bishop-needy* squares, for any two squares on the same diagonal d, the other squares lie on two or more diagonals perpendicular to d. So, as in Case 1, the second bishop cannot cover all the remaining *bishop-needy* squares.

This shows that two bishops and $n-3$ rooks cannot cover the whole board. Therefore, we have

$$\eta(B^2R)_n \geq n-2.$$

In the other direction, by Theorem 6,

$$\eta(B^2R)_n \leq \eta(B^1R)_n = n-2.$$

Thus, we have

$$\eta(B^2R)_n = n - 2.$$

Lemma 1 *When $n = 4k + 1$ for some positive integer k, then along with a queen, at least $n - 3$ bishops are needed to dominate the whole $n \times n$ board.*

Proof Assume $n - 4$ bishops and a queen dominate the whole $n \times n$ board, if possible. Let m be the number of white bishops and l be the number of black bishops so that

$$m + l = n - 4. \tag{1}$$

We claim that the queen has to lie on a sum diagonal (see Fig. 7a) in $\mathcal{S} = \{s_{\frac{n+3}{2}}, s_{\frac{n+5}{2}}, \ldots, s_{2n-\frac{n+3}{2}}\}$ and on a difference diagonal in $\mathcal{D} = \{d_{\frac{n+3}{2}}, d_{\frac{n+5}{2}}, \ldots, d_{2n-\frac{n+3}{2}}\}$.

If possible, assume that the queen lies on a sum diagonal not in $\mathcal{S}$. By Observation 2, note that $s_{\frac{n+3}{2}}$ and $s_{2n-\frac{n+3}{2}}$ are white diagonals and $\mathcal{S}$ contains $2k = \frac{n-1}{2}$ white diagonals and $2k - 1 = \frac{n-3}{2}$ black diagonals.

If $m \leq \frac{n-5}{2}$, then by Observation 3, on any *bishop-empty* white sum diagonal $s_u \in \mathcal{S}$, with $u \in [\frac{n+3}{2}, 2n - \frac{n+3}{2}]$, there are at least $u - m \geq \frac{n+3}{2} - \frac{n-5}{2} = 4$ squares that are not covered by the bishops. These four or more *queen-needy* squares cannot be covered by the queen that is lying on a sum diagonal outside $\mathcal{S}$ by Observation 1. Since $m \leq \frac{n-5}{2}$ and there are $\frac{n-1}{2}$ white diagonals in $\mathcal{S}$, there will be a *bishop-empty* white sum diagonal in $\mathcal{S}$ that will contain four or more *queen-needy* squares.

Therefore, $m \geq \frac{n-3}{2}$. By Eq. (1), we have $l \leq \frac{n-5}{2}$.

Similarly, there will be a *bishop-empty* black sum diagonal s_u with $u \in [\frac{n+3}{2}, 2n - \frac{n+3}{2}]$, which will contain at least $u - l \geq \frac{n+3}{2} - \frac{n-5}{2} = 4$ *queen-needy* squares. As seen above, a queen lying on a sum diagonal outside $\mathcal{S}$ cannot cover these four or more *queen-needy* squares.

Thus, the queen has to lie on a sum diagonal in $\mathcal{S}$.

By merely imitating the above arguments for difference diagonals, we can show that the queen must lie on a difference diagonal in

$$\mathcal{D} = \{d_{\frac{n+3}{2}}, d_{\frac{n+5}{2}}, \ldots, d_{2n-\frac{n+3}{2}}\}.$$

This proves our claim.

Clearly, on the board, the region covered by the sum diagonals from $s_{\frac{n+3}{2}}$ to $s_{2n-\frac{n+3}{2}}$ and the difference diagonals from $d_{\frac{n+3}{2}}$ to $d_{2n-\frac{n+3}{2}}$ does not contain any edge square (see Fig. 7b). Therefore the queen does not lie on any edge square.

However, on an $n \times n$ board, there are $(4n - 4)$ edge squares. The $(n - 4)$ bishops can cover a maximum of $4(n - 4)$ edge squares and the queen not lying on any edge square can cover at most 8 edge squares. Thus, there are at least

$$4n - 4 - 4(n - 4) - 8 = 4$$

edge squares which are not covered by the bishops and the queen.

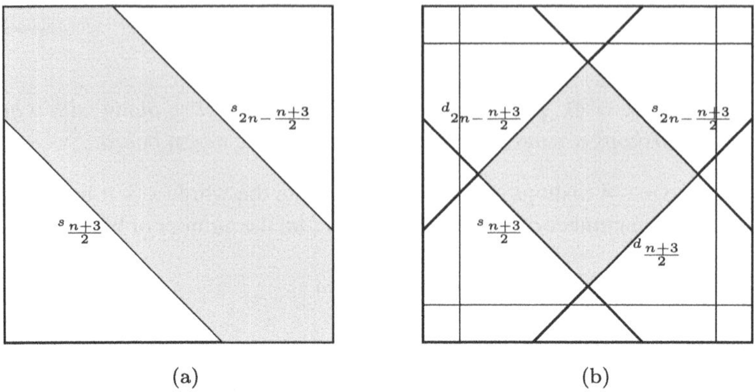

Fig. 7 For Lemma 1

This shows that our initial assumption that $n-4$ bishops and a queen can dominate the whole board is wrong.

Lemma 2 *When $n = 4k$ for some positive integer k, then along with a queen, at least $n-2$ bishops are needed to dominate the whole $n \times n$ board.*

Proof Let $n = 4k$ for some positive integer $k \geq 2$. Suppose that $n-3$ bishops and a queen dominate the whole board. Let m be the number of white bishops and l be the number of black bishops on the board so that

$$m + l = n - 3. \tag{2}$$

Let $\mathcal{S}' = \{s_{\frac{n+4}{2}}, s_{\frac{n+6}{2}}, \ldots, s_{2n-\frac{n+4}{2}}\}$ (see Fig. 8a). By Observation 2, there are $2k - 1 = \frac{n-2}{2}$ white and $2k - 2 = \frac{n-4}{2}$ black diagonals in $\mathcal{S}'$.

If $m \leq \frac{n-6}{2}$, then a *bishop-empty* white sum diagonal s_u, for an even $u \in [\frac{n+4}{2}, 2n - \frac{n+4}{2}]$, has at least $u - m \geq \frac{n+4}{2} - \frac{n-6}{2} = 5$ squares that are not covered by the bishops. Since $m \leq \frac{n-6}{2}$ and there are $\frac{n-2}{2}$ white diagonals in $\mathcal{S}'$, there are two *bishop-empty* white sum diagonals in $\mathcal{S}'$. Now, only one queen cannot cover five or more *queen-needy* squares on both of these diagonals by Observation 1.

This contradiction shows that $m \geq \frac{n-4}{2}$. By Eq. (2), $l \leq \frac{n-2}{2}$.

Now, let $\mathcal{D}' = \{d_{\frac{n+4}{2}}, d_{\frac{n+6}{2}}, \ldots, d_{2n-\frac{n+4}{2}}\}$. By Observation 2, there are $2k - 1 = \frac{n-2}{2}$ black and $2k - 2 = \frac{n-4}{2}$ white diagonals in $\mathcal{D}'$. Again, if $l \leq \frac{n-6}{2}$, then as seen above, we get two *bishop-empty* black difference diagonals each containing at least five *queen-needy* squares and a single queen cannot cover them. Therefore, $l \geq \frac{n-4}{2}$.

Thus, by Eq. (2), we have

$$\frac{n-4}{2} \leq m, l \leq \frac{n-2}{2} \text{ and } m + l = n - 3. \tag{3}$$

For $n = 4$, (3) holds trivially.

Without loss of generality, we assume $m = \frac{n-4}{2}$ and $l = \frac{n-2}{2}$. (If we assume $l = \frac{n-4}{2}$ and $m = \frac{n-2}{2}$, then the roles of sum and difference diagonals have to be interchanged in the following arguments.)

Claim: There is exactly one sum diagonal s_u in $\mathcal{S}'$ containing four or more *queen-needy* squares and this s_u is a *bishop-empty* white sum diagonal.

Now, by Observations 2 and 3, any *bishop-empty* diagonal in $\mathcal{S}'$ would have at least four *queen-needy* squares. Since there are $m = \frac{n-4}{2}$ white bishops and $l = \frac{n-2}{2}$ black bishops and there are $\frac{n-2}{2}$ white diagonals and $\frac{n-4}{2}$ black diagonals in $\mathcal{S}'$, all the diagonals in $\mathcal{S}'$, except one white sum diagonal s_u, have to be occupied by bishops. This proves our claim.

Since $l = \frac{n-2}{2}$ and we have exhausted $\frac{n-4}{2}$ black bishops in $\mathcal{S}'$, at least one of the black diagonals $s_{\frac{n+2}{2}}$ and $s_{2n-\frac{n+2}{2}}$ must be *bishop-empty*. Due to symmetry, we can assume that $s_{\frac{n+2}{2}}$ is *bishop-empty*.

For $s_{\frac{n+2}{2}}$, $\frac{n+2}{2} - l = \frac{n+2}{2} - \frac{n-2}{2} = 2$. Therefore by Observation 3, there are at least two *queen-needy* squares on $s_{\frac{n+2}{2}}$.

Case 1: If one of the *queen-needy* squares on $s_{\frac{n+2}{2}}$ is not an edge square, then the black difference diagonal d_v, on which this *queen-needy* square lies, has to be *bishop-empty*. Moreover, d_v must be in $\mathcal{D}'$ which follows easily from Fig. 8b. So, d_v contains at least $\frac{n+4}{2}$ squares.

But, $v - l \geq \frac{n+4}{2} - \frac{n-2}{2} = 3$ implies that the *bishop-empty* black diagonal d_v contains at least three *queen-needy* squares. Now, a queen at any square on the board cannot cover all (three or more) *queen-needy* squares on the black diagonal d_v and all (four or more) *queen-needy* squares on the white diagonal s_u. This follows from Observation 1.

Case 2: If both the *queen-needy* squares on $s_{\frac{n+2}{2}}$ are edge squares, then these squares are in particular $(1, \frac{n+2}{2})$ and $(\frac{n+2}{2}, 1)$ (see Fig. 8b). Again by Observation 1, a queen anywhere on the board cannot cover both $(1, \frac{n+2}{2})$ and $(\frac{n+2}{2}, 1)$ and all (four or more) *queen-needy* squares on the white diagonal s_u.

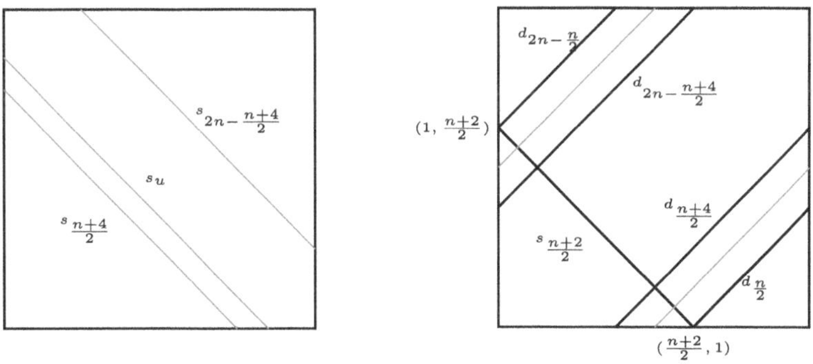

(a) (b)

Fig. 8 For Lemma 2

This shows that our initial assumption is wrong and we need at least $n-2$ bishops along with a queen to dominate the whole board.

Theorem 9 *Let $n \geq 4$ be an integer.*

$$\eta(Q^1 B)_n = \begin{cases} n-3 & \text{when n is odd} \\ n-2 & \text{when n is even.} \end{cases}$$

Proof Let $n \geq 5$ be an odd integer. If $n \equiv 1 \pmod 4$, then by Lemma 1, we get $\eta(Q^1 B)_n \geq n-3$.

When $n \equiv 3 \pmod 4$, let

$$\mathcal{S}_1 = \{s_{\frac{n+3}{2}}, s_{\frac{n+5}{2}}, \ldots, s_{2n-\frac{n+3}{2}}\} \text{ and } \mathcal{D}_1 = \{d_{\frac{n+3}{2}}, d_{\frac{n+5}{2}}, \ldots, d_{2n-\frac{n+3}{2}}\}.$$

Clearly, both $\mathcal{S}_1$ and $\mathcal{D}_1$ contain $\frac{n-1}{2}$ black diagonals and $\frac{n-3}{2}$ white diagonals. A mere imitation of the proof of Lemma 1 by interchanging the roles of black and white bishops will give $\eta(Q^1 B)_n \geq n-3$ in this case also.

Next, place the queen at the square $(\frac{n+1}{2}, \frac{n+1}{2})$; place $\frac{n-3}{2}$ bishops above and $\frac{n-3}{2}$ bishops below the queen on the column $c_{\frac{n+1}{2}}$ as shown in Fig. 9a.

So, all the sum diagonals except

$$s_1, s_2, \ldots, s_{\frac{n+1}{2}} \text{ and } s_{2n-\frac{n+1}{2}}, s_{2n-\frac{n-1}{2}}, \ldots, s_{2n-1}$$

are occupied by $n-3$ bishops and a queen. Similarly, all the difference diagonals except $d_1, d_2, \ldots, d_{\frac{n+1}{2}}$ and $d_{2n-\frac{n+1}{2}}, d_{2n-\frac{n-1}{2}}, \ldots, d_{2n-1}$ are occupied.

One can easily see that the diagonals of these chess pieces cover all squares on the board except $(1, \frac{n+1}{2})$, $(\frac{n+1}{2}, 1)$, $(n, \frac{n+1}{2})$ and $(\frac{n+1}{2}, n)$ and these squares are covered by the orthogonals of the queen. Thus the whole board is dominated and this shows that

$$\eta(Q^1 B)_n = n-3,$$

when n is odd and $n \geq 5$.

Now, let $n \geq 4$ be an even integer. When $n \equiv 0 \pmod 4$, from Lemma 2, we get $\eta(Q^1 B)_n \geq n-2$.

Let $n \equiv 2 \pmod 4$. Let

$$\mathcal{S}_2 = \{s_{\frac{n+4}{2}}, s_{\frac{n+6}{2}}, \ldots, s_{2n-\frac{n+4}{2}}\} \text{ and } \mathcal{D}_2 = \{d_{\frac{n+4}{2}}, d_{\frac{n+6}{2}}, \ldots, d_{2n-\frac{n+4}{2}}\}.$$

Then, there are $\frac{n-2}{2}$ black and $\frac{n-4}{2}$ white diagonals in $\mathcal{S}_2$ and there are $\frac{n-2}{2}$ white and $\frac{n-4}{2}$ black diagonals in $\mathcal{D}_2$. Again, using the arguments in the proof of Lemma 2 with the roles of sum and difference diagonals interchanged, we get $\eta(Q^1 B)_n \geq n-2$.

Thus, $\eta(Q^1 B)_n \geq n-2$ when $n \geq 4$ is an even integer.

On the other hand, place a queen at $(\frac{n}{2}, \frac{n}{2})$; place $\frac{n-2}{2}$ bishops below and $\frac{n-2}{2}$ bishops above the queen on the column $c_{\frac{n}{2}}$ as shown in Fig. 9b.

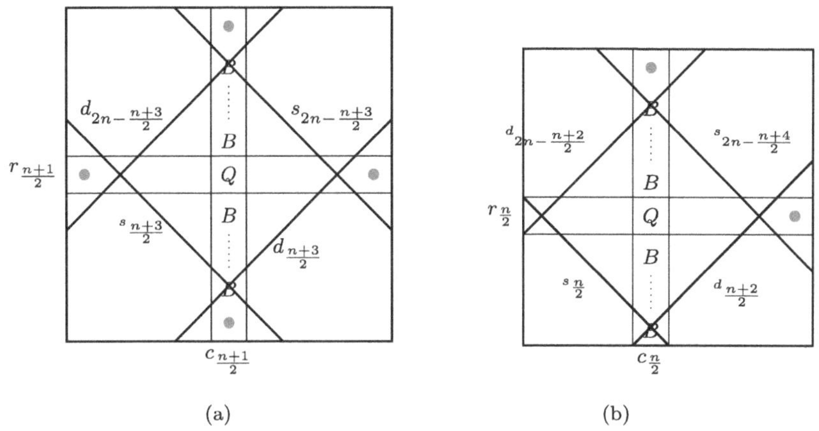

Fig. 9 For $\eta(Q^1B)_n$

Clearly, all the sum diagonals except $s_1, s_2, \ldots, s_{\frac{n-2}{2}}$ and $s_{2n-\frac{n+2}{2}}, s_{2n-\frac{n}{2}}, \ldots, s_{2n-1}$ and all the difference diagonals except $d_1, d_2, \ldots, d_{\frac{n}{2}}$ and $d_{2n-\frac{n}{2}}, d_{2n-\frac{n-2}{2}}, \ldots, d_{2n-1}$ are occupied by $n-2$ bishops and the queen. Clearly, all the squares on these diagonals except $(n, \frac{n}{2})$ and $(\frac{n}{2}, n)$ are covered by the diagonals of these chess pieces. The squares $(n, \frac{n}{2})$ and $(\frac{n}{2}, n)$ are covered by the queen orthogonally.

This shows that

$$\eta(Q^1B)_n = n - 2,$$

when n is even.

This completes the proof.

Lemma 3 *For any integer $n = 4k + 1$ for some $k \in \mathbb{Z}^+$, at least $n - 2$ bishops are needed, along with a rook, to dominate the whole $n \times n$ board.*

Proof Suppose $n - 3$ bishops and a rook dominate the whole board. Let m be the number of white bishops and l be the number of black bishops so that

$$m + l = n - 3. \tag{4}$$

Let $\mathcal{S} = \{s_{\frac{n+1}{2}}, s_{\frac{n+3}{2}} \ldots, s_{2n-\frac{n+1}{2}}\}$. Clearly $\mathcal{S}$ contains $2k + 1 = \frac{n+1}{2}$ black diagonals and $2k = \frac{n-1}{2}$ white diagonals.

Suppose $m \le \frac{n-3}{2}$, if possible. Since $m \le \frac{n-3}{2}$ and there are $\frac{n-1}{2}$ white diagonals in $\mathcal{S}$, there will necessarily be a *bishop-empty* white diagonal in $\mathcal{S}$. Note that $s_{\frac{n+1}{2}}$ and $s_{2n-\frac{n+1}{2}}$ are black diagonals by Observation 2. For any *bishop-empty* white diagonal $s_u \in \mathcal{S}$, we have $u - m \ge \frac{n+3}{2} - \frac{n-3}{2} = 3$, that is, s_u will contain at least three *rook-needy* squares by Observation 3. But, a rook cannot cover more than two squares on any diagonal, by Observation 1.

Therefore, $m \ge \frac{n-1}{2}$. By Eq. (4), $l \le \frac{n-5}{2}$.

For any *bishop-empty* black diagonal $s_u \in \mathcal{S}$, $u - l \geq \frac{n+1}{2} - \frac{n-5}{2} = 3$ and hence s_u will have at least three *rook-needy* squares. Also, since $l \leq \frac{n-5}{2}$ and there are $\frac{n+1}{2}$ black diagonals in $\mathcal{S}$, there will be a *bishop-empty* black sum diagonal in $\mathcal{S}$. This is again a contradiction by Observation 1.

Thus, $n - 3$ bishops and a rook cannot cover the whole board.

Lemma 4 *For any integer $n = 4k$ for some $k \in \mathbb{Z}^+$, at least $n - 1$ bishops are needed, along with a rook, to dominate the whole $n \times n$ board.*

Proof Assume, if possible, that $n - 2$ bishops and a rook dominate the whole board. Let m be the number of white bishops and l be the number of black bishops so that

$$m + l = n - 2. \tag{5}$$

Let $\mathcal{S}' = \{s_{\frac{n+4}{2}}, s_{\frac{n+6}{2}} \dots, s_{2n-\frac{n+4}{2}}\}$ and $\mathcal{D}' = \{d_{\frac{n+4}{2}}, d_{\frac{n+6}{2}} \dots, d_{2n-\frac{n+4}{2}}\}$.

Clearly, $\mathcal{S}'$ contains $2k - 1 = \frac{n-2}{2}$ white diagonals and $2k - 2 = \frac{n-4}{2}$ black diagonals.

If $m \leq \frac{n-4}{2}$, then for any *bishop-empty* white sum diagonal $s_u \in \mathcal{S}'$, $u - m \geq \frac{n+4}{2} - \frac{n-4}{2} = 4$, that is, there are at least four *rook-needy* squares on s_u. Since there are $\frac{n-2}{2}$ white diagonals in $\mathcal{S}'$ and $m \leq \frac{n-4}{2}$, there will be a *bishop-empty* white sum diagonal having four or more *rook-needy* squares. We get a contradiction by Observation 1.

Therefore, we must have $m \geq \frac{n-2}{2}$ which implies $l \leq \frac{n-2}{2}$ by Eq. (5).

Now, $\mathcal{D}'$ contains $2k - 1 = \frac{n-2}{2}$ black diagonals and $2k - 2 = \frac{n-4}{2}$ white diagonals.

If $l \leq \frac{n-4}{2}$, then for any *bishop-empty* black difference diagonal $d_v \in \mathcal{D}'$, $v - l \geq \frac{n+4}{2} - \frac{n-4}{2} = 4$. And since there are $\frac{n-2}{2}$ black diagonals in $\mathcal{D}'$, there will be a *bishop-empty* black difference diagonal having four or more *rook-needy* squares. We again get a contradiction by Observation 1.

Therefore, $l \geq \frac{n-2}{2}$ and by Eq. (5), we get $m = l = \frac{n-2}{2}$.

Note that by Observation 3, any *bishop-empty* diagonal in $\mathcal{S}'$ would contain at least three *rook-needy* squares. Therefore, by Observation 1, we do not want any *bishop-empty* diagonal in $\mathcal{S}'$. So, all the $n - 3$ sum diagonals in $\mathcal{S}'$ must be occupied by $\frac{n-2}{2}$ white bishops and at least $\frac{n-4}{2}$ black bishops.

Then, at least one of the black sum diagonals $s_{\frac{n+2}{2}}$ and $s_{2n-\frac{n+2}{2}}$ has to be *bishop-empty*. Due to symmetry, we can assume $s_{\frac{n+2}{2}}$ is *bishop-empty* without loss of generality. As $u - l = \frac{n+2}{2} - \frac{n-2}{2} = 2$, it follows from Observation 3 that there are at least two *rook-needy* squares on $s_{\frac{n+2}{2}}$.

If one of the *rook-needy* squares on $s_{\frac{n+2}{2}}$ is not an edge square, then the black difference diagonal d_v on which this square lies has to be *bishop-empty*. Further, as seen in Fig. 8b, $d_v \in \mathcal{D}'$. By Observation 3, this *bishop-empty* black diagonal d_v will contain three or more *rook-needy* squares. This is a contradiction by Observation 1. Therefore, both the *rook-needy* squares on $s_{\frac{n+2}{2}}$ have to be edge squares.

Now we repeat the above arguments for difference diagonals and we get one of the white difference diagonals $d_{\frac{n+2}{2}}$ and $d_{2n-\frac{n+2}{2}}$ *bishop-empty* and the two edge squares on it are *rook-needy*.

So, depending on whether $d_{\frac{n+2}{2}}$ or $d_{2n-\frac{n+2}{2}}$ is *bishop-empty*, the rook has to cover all the squares in either $\{(1, \frac{n+2}{2}), (\frac{n+2}{2}, 1), (n, \frac{n+2}{2}), (\frac{n}{2}, 1)\}$ or in $\{(1, \frac{n+2}{2}), (\frac{n+2}{2}, 1), (1, \frac{n}{2}), (\frac{n+2}{2}, n)\}$ (see Fig. 10a, b). Clearly a rook at any square on the board cannot cover all the squares of either of these sets. This contradiction shows that the assumption we made in the beginning is wrong.

Theorem 10 *For any integer* $n \geq 6$, $\eta(R^1B)_n = \begin{cases} n-2 & \text{when } n \text{ is odd,} \\ n-1 & \text{when } n \text{ is even.} \end{cases}$

Proof Let $n \geq 7$ be an odd integer. When $n \equiv 1 \pmod 4$, from Lemma 3, it follows that $\eta(R^1B)_n \geq n-2$.

When $n \equiv 3 \pmod 4$, we interchange the roles of white and black bishops in the proof of Lemma 3 to get $\eta(R^1B)_n \geq n-2$.

Let $\mathcal{S}'' = \{s_{\frac{n+3}{2}}, s_{\frac{n+5}{2}} \ldots, s_{2n-\frac{n+3}{2}}\}$ and $\mathcal{D}'' = \{d_{\frac{n+3}{2}}, d_{\frac{n+5}{2}} \ldots, d_{2n-\frac{n+3}{2}}\}$. Observe that there are $n-2$ sum diagonals in $\mathcal{S}''$ and $n-2$ difference diagonals in $\mathcal{D}''$.

Now, place $n-2$ bishops in such a way that there is exactly one bishop on each sum diagonal in $\mathcal{S}''$ and on each difference diagonal in $\mathcal{D}''$. This can be achieved by placing the bishops one each at $(\frac{n+1}{2}, i)$ $\forall i \in [2, n-1] \setminus \{\frac{n+1}{2}, \frac{n+5}{2}\}$ and $(\frac{n-1}{2}, \frac{n+3}{2}), (\frac{n+3}{2}, \frac{n+3}{2})$. Now place a rook at the square $(\frac{n+1}{2}, \frac{n+1}{2})$.

As shown in Fig. 11a, all the squares on the board, except the squares

$$(1, \frac{n+1}{2}), (\frac{n+1}{2}, 1), (\frac{n+1}{2}, n) \text{ and } (n, \frac{n+1}{2}),$$

are covered by the diagonals of the bishops; these squares are covered by the rook.

Thus the whole board can be dominated by a rook and $n-2$ bishops when $n \geq 7$ is odd. That is, $\eta(R^1B)_n = n-2$ when $n \geq 7$ is odd.

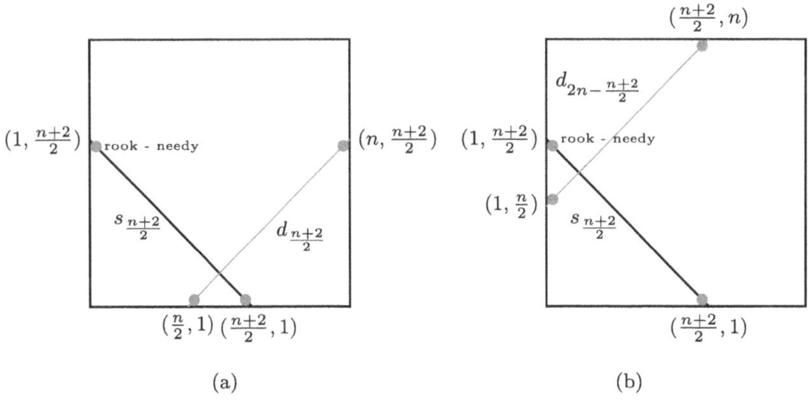

Fig. 10 For Lemma 4

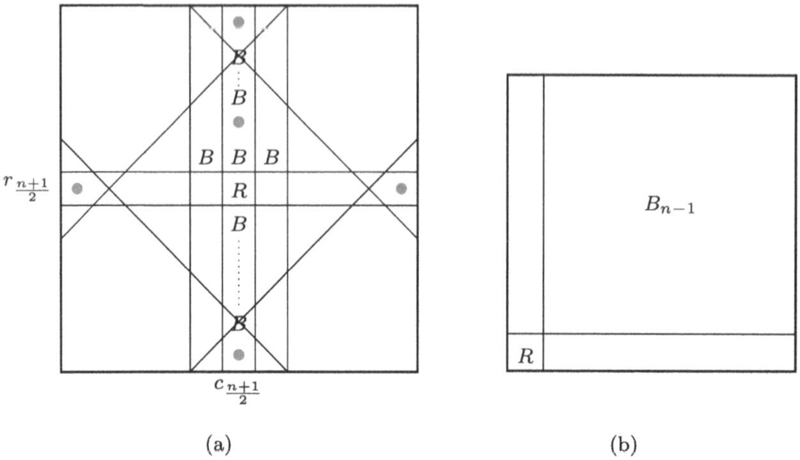

Fig. 11 For $\eta(R^1 B)_n$

Now, let $n \geq 6$ be an even integer. If $n \equiv 0 \pmod 4$, from Lemma 4, we get $\eta(R^1 B)_n \geq n - 1$.

When $n \equiv 2 \pmod 4$, we imitate the proof of Lemma 4 with the roles of sum and difference diagonals interchanged and we get $\eta(R^1 B)_n \geq n - 1$.

On the other hand, place a rook at the square $(1, 1)$ (see Fig. 11b). Now the $(n - 1) \times (n - 1)$ sub-board formed by the rows r_j, $2 \leq j \leq n$ and columns c_i, $2 \leq i \leq n$ can be dominated by $n - 1$ bishops as $\gamma(B_{n-1}) = n - 1$. Thus, $\eta(R^1 B)_n = n - 1$ when n is even.

This completes the proof.

Remark 2 Using the fact that $\gamma(B_{n-1}) = n - 1$, we can easily say that $\eta(R^1 B)_n \leq n - 1$, when $n = 4, 5$.

Since by Lemma 4, $\eta(R^1 B)_4 \geq 3$, we have $\eta(R^1 B)_4 = 3$.

Let $n = 5$. Suppose, if possible, that three bishops and a rook cover the whole 5×5 board. Clearly, s_4 and s_6 are white diagonals with four squares each and s_5 is a black diagonal with five squares.

If $m \leq 1$, then one of s_4 and s_6 is *bishop-empty* and that diagonal will contain three or more *rook-needy* squares, a contradiction by Observation 1. Therefore, we must have $m = 2$ for otherwise, s_5 will be a *bishop-empty* diagonal containing five *rook-needy* squares.

Thus, we have $m = 2, l = 1$. Now, if the black bishop does not lie on both d_5 and s_5, then d_5 will also be a *bishop-empty* diagonal containing five *rook-needy* squares, again a contradiction. Therefore, the black bishop has to occupy $(3, 3)$. Also, on *bishop-empty* black diagonal s_3, there will be two *rook-needy* squares, namely $(3, 1)$ and $(1, 3)$. To cover these squares, the rook has to occupy $(3, 3)$ which is not available.

Therefore, $\eta(R^1 B)_5 \geq 4$ and hence $\eta(R^1 B)_5 = 4$.

5 Concluding Remarks

There are still several intriguing questions in the type of problems we considered in the previous section. We give a few instances here.

If a rook is placed at the square (1, 1) of an $n \times n$ chessboard, it is clear that the $(n-1) \times (n-1)$ subboard obtained by deleting the first row and the first column can be dominated by $\gamma(Q_{n-1})$ queens placed suitably on this subboard. Therefore, we have

$$\eta(R^1 Q)_n \leq \gamma(Q_{n-1}). \tag{6}$$

We are naturally tempted to expect that equality should hold in (6).

It is already known that $\gamma(Q_4) = 2$ (see [1]). By appropriately placing two queens on the 4×4 subboard formed by c_i, $1 \leq i \leq 4$ and r_j, $1 \leq j \leq 4$, this subboard can be dominated by the two queens. Now placing a rook at each of the squares $(i, i), 5 \leq i \leq n$, one can easily see that the whole board is dominated. Therefore,

$$\eta(Q^2 R)_n \leq n - 4.$$

Again, we expect that $\eta(Q^2 R)_n = n - 4$ should be true.

Also, determining the value of $\eta(B^1 Q)_n$ will be an interesting question.

Acknowledgements The second and the third authors would like to acknowledge CSIR, Govt. of India, for research fellowships.

References

1. S. Bozóki, P. Gál, I. Marosi, and W. D. Weakley, Domination of the rectangular queen's graph, *Electron. J. Combin.* 26(4) (2019), #P4.45
2. A. P. Burger, E. J. Cockayne and C. M. Mynhardt, Domination and irredundance in the queen's graph, *Discrete Math.* 163 (1997), no. 1–3, 47-66.
3. E. J. Cockayne, Chessboard domination problems, *Discrete Math.* 86 (1990), no. 1–3, 13–20.
4. E. J. Cockayne, B. Gamble and B. Shepherd, Domination of chessboards by queens on a column, *Ars Combin.* 19 (1985), 105–118.
5. E. J. Cockayne, B. Gamble and B. Shepherd, Domination parameters for the bishops graph, *Discrete Math.* 58 (1986), no. 3, 221–227.
6. E. J. Cockayne and S. T. Hedetniemi, On the diagonal queens domination problem, *J. Combin. Theory Ser. A* 42 (1986), no. 1, 137–139.
7. E. J. Cockayne and P. H. Spencer, On the independent queens covering problem, *Graphs Combin.* 4 (1988), no. 2, 101–110.
8. G. Fricke, S. M. Hedetniemi, S. T. Hedetniemi, A. McRae, C. K. Wallis, M. S. Jacobson, H. W. Martin, and W. D. Weakley. Combinatorial problems on chessboards: a brief survey. In Graph Theory, Combinatorics, and Algorithms: *Proc. 7th Quadrennial Internat. Conf. on the Theory and Applications of Graphs* (Y. Alavi and A. Schwenk, eds.), volume 1, pages 507–528, 1995.
9. R. K. Guy, Unsolved Problems in Number Theory, *Springer*, 3rd edition, 2004.
10. D. A. Klarner, Klarner's review of G. B. Huff's paper 'On pairings of the first $2n$ natural numbers' (Acta Arith. 23 (1973), 117–126), AMS Math. Reviews MR0332522.

11. P. R. J. Östergård and W. D. Weakley, Values of domination numbers of the queen's graph, *Electron. J. Combin.*, 8, (2001), no. 1, Research paper 29, 19pp.
12. V. Raghavan, S. M. Venketesan, On bounds for a covering problem. *Inf. Process. Lett.* 25, 281–284 (1987).
13. K. F. Roth, Sur quelques ensembles d'entiers, *C.R. Acad. Sci. Paris* 234 (1952) 388–390.
14. K. von Szily, Das Minimalproblem der Damen, in *Deutsche Schachzeitung*, vol 57 (1902), pp. 326–328.
15. K. von Szily, Das Minimalproblem der Damen, in *Deutsche Schachzeitung*, vol 58 (1903), pp. 65–68.
16. W. D. Weakley, Domination in the queen's graph, in *Graph Theory, Combinatorics, and Algorithms, Vol. 2 (Kalamazoo, MI 1992)*, ed. by Y. Alavi, A. Schwenk (Wiley, New York, 1995), pp. 1223–1232.
17. W. D. Weakley, Upper bounds for domination numbers of the queen's graph, *Discrete Math.* 242 (2002), no. 1–3, 229–243.
18. W. D. Weakley, Queens around the world in twenty-five years, *Graph theory - favorite conjectures and open problems. 2*, 43–54, Probl. Books in Math., Springer, Cham, 2018.
19. W. D. Weakley, Queen Domination of Even Square Boards, *Electron. J. Combin.* 29(2) (2022), #P2.50

Bounds on Distinct and Repeated Dot Product Trees

Aaron Autry, Slade Gunter, Christopher Housholder, and Steven Senger

Abstract We study questions inspired by Erdős' celebrated distance problems with dot products in lieu of distances, and for more than a single pair of points. In particular, we study point configurations present in large finite point sets in the plane that are described by weighted trees. We give new lower bounds on the number of distinct sets of dot products serving as weights for a given type of tree in any large finite point set. We also as demonstrate the existence of many repetitions of some special sets of dot products occurring in a given type of tree in different constructions, narrowing gap between the best known upper and lower bounds on these configurations.

1 Introduction

1.1 Background

In 1946, in [10], Paul Erdős posed a problem that continues to inspire mathematicians: In a large finite set of points in the plane, what is the upper bound of the number pairs can determine the same distance? This is often called the unit distance problem. A related problem also described in [10] is the so-called distinct distances problem, which asks for a lower bound on the number of distinct distances that must be determined by any large finite set of points in the plane.

For the unit distance problem, after some initial activity, general progress seems to have halted in 1984, at the upper bound of a constant times $n^{\frac{4}{3}}$, given by Szemerédi and Trotter in [24]. In the case of the distinct distances problem, there was much activity and incremental progress up until 2010, when Guth and Katz essentially resolved the problem in [13]. For a more detailed look, see [12, 13], and the references contained therein.

Though Erdős originally asked about pairs of points determining a fixed distance, there are a number of well-studied variants involving k-tuples of points determining

A. Autry · S. Gunter · C. Housholder · S. Senger (✉)
Missouri State University, 901 S. National Ave., Springfield, MO 65897, USA
e-mail: stevensenger@missouristate.edu

M. B. Nathanson (ed.), *Combinatorial and Additive Number Theory VI*,
Springer Proceedings in Mathematics & Statistics 464,
https://doi.org/10.1007/978-3-031-65064-2_2

other values, such as direction, angle, or dot product. Moreover, there are many different settings besides finite point sets in $\mathbb{R}^2$, such fractal sets, higher dimensions, or even different types of vector spaces and modules over finite fields and rings. For some examples and discussions, see [6, 9, 16, 18, 28] and the references contained therein. Also, we will assume that the dot products discussed below are all nonzero, unless otherwise specified, as zero dot products can exhibit quite distinct geometric structure. See [14, 18, 20] for more on the unique geometric properties of zero dot products.

We highlight for the moment on the specific problem of the number of distinct dot products determined by a large finite point set in the plane. Given a set of n points in the plane, a straightforward argument using the celebrated Szemerédi-Trotter theorem (from [27]) guarantees the existence of at least a constant times $n^{\frac{2}{3}}$ distinct dot products. We prove this below as Corollary 1. This was the best-known bound for many years until the recent developments by Hanson, Roche-Newton, and the fourth listed author in [15], where an exponent of $\frac{2}{3} + \frac{1}{2739}$ was achieved by relying on tools from additive combinatorics.

1.2 Setting

In this note, our primary focus will be trees. We now describe the setup in detail. For a given large finite point set $E \subseteq \mathbb{R}^d$, with $d \geq 2$, we consider G_E the complete edge-weighted graph whose vertices are points from E and whose edge weights are determined by the dot product between the associated vertices. Now, given a tree T, we estimate two things: how many different sequences of edge weights correspond to instances of T as a subgraph in G_E, and how many times a specific sequence of edge weights occurs as the weights for a copy of T in G_E. The former is analogous to the distinct distances problem, while the latter is akin to the unit distance problem.

To this end, we introduce some notation. Given a tree T with vertices $(v_1, \ldots, v_{k+1})$ we denote its k edges by $\{\{v_{i_1}, v_{i_2}\}, \{v_{i_3}, v_{i_4}\}, \ldots, \{v_{i_{2k-1}}, v_{i_{2k}}\}\}$, where the indices for each pair of vertices corresponding to an edge, $\{v_{i_{2j-1}}, v_{i_{2j}}\}$, are indexed in lexicographical order. Now, given a sequence of real number weights $\mathbf{w} := (\alpha_1, \ldots, \alpha_k)$, we can consider the weighted tree $T_{\mathbf{w}}$ where the edge connecting $v_{i_{2j-1}}$ to $v_{i_{2j}}$ has weight α_j, for $j = 1, \ldots, k$. In what follows, we will usually be concerned with edge weights equal to the dot product determined by the pair of points corresponding to the pair of vertices.

There is an abundance of literature on the case that our tree T is a path of length k, sometimes called a k-chain. See for example [2, 11, 22] for distance chain bounds, and see [3] and [20] for results related to dot product chains. More general trees have also been widely studied, but we direct special attention to [21], by Ou and Taylor, and later, [23], by Dung, Pham, and the fourth listed author, for motivating the results here.

Throughout this paper, we assume that k is like a constant compared to the size of any subset E. To ease exposition, we use the asymptotic symbols $X \lesssim Y$ if $X =$

$O(Y)$, and $X \approx Y$ when $X = \Theta(Y)$. Moreover, we write $X \gtrapprox Y$ when for every $\epsilon > 0$, there exists a constant $C_\epsilon > 0$ such that $X \gtrsim C_\epsilon q^\epsilon Y$.

1.3 Main Results

Our main result follows the ideas in [21]. It gives lower bounds for the number of distinct sets of dot products much arise as edge weights for trees in a large finite point set in the plane.

Theorem 1 *Given a large finite set of points $E \subseteq \mathbb{R}^2$, and a tree T on k edges, the number of distinct k-tuples of dot products, $\mathbf{w}$, for which the tree $T_\mathbf{w}$ is realized in E is at least $\gtrsim n^{\frac{2k}{3}}$.*

The key to proving this result is a pinned dot product result between two sets, stated precisely below as Theorem 6. While this result does not immediately extend to to higher dimensions, we can still obtain a nontrivial pinned estimate.

Theorem 2 *Given a large finite set of n points $E \subseteq \mathbb{R}^d$, we have that there exists a point $x \in E$ for which*

$$|\{x \cdot y : y \in E\}| \gtrsim n^{\frac{2}{2d-1}}.$$

The other result we have gives lower bounds on how often a particular set of dot product edge weights can occur in some specific point set constructions. It is based on two different constructions which have different strengths and weaknesses.

Theorem 3 *Given a tree T on k edges, and a large finite natural number n, there exists a k-tuple of dot products $\mathbf{w}$ and set of n points in $\mathbb{R}^d$ that exhibits*

$$\gtrsim \max\left\{n^{1+\frac{k(d-1)}{(d+1)}}, n^{\lceil \frac{k+1}{2} \rceil}\right\}$$

copies of $T_\mathbf{w}$.

We can see that depending on k and the ambient dimension, one or the other bound may dominate. To compare this with known results, we can consider a special case. Suppose T is a perfect binary tree of height h. It is known that this tree has $2^{h+1} - 1$ vertices, meaning it has $2^{h+1} - 2$ edges. So if we set $k = 2^{h+1} - 2$, Theorem 3 guarantees the existence of a set of n points and a k-tuple of dot products $\mathbf{w}$ in the plane that has $\gtrsim n^{2^h}$ copies of $T_\mathbf{w}$. This agrees with the power of n of the upper bounds on the same, given in [14], as the following result will show, showing that both bounds are tight. We note here that in [14], Theorems 1.12 and 1.13 are correct for c-ary trees when $c = 2$, but have some superficial typos when $c \neq 2$.

Theorem 4 (Theorem 1.13 from [14]) *Given a large finite set of n points $E \subseteq \mathbb{R}^2$, a complete balanced binary tree T of height h, and a $(2^{h+1} - 2)$-tuple of dot products $\mathbf{w}$, there cannot be more than $\lesssim n^{2^h}$ copies of $T_\mathbf{w}$ determined by points in E.*

The structure of the rest of the paper is as follows. We first acquaint the reader with our primary tools, then prove the main technical theorem, which is a pinned estimate. We then show how to use this to prove our first main result, which is a lower bound on distinct k-tuples of dot products determined by trees isomorphic to a given tree. Finally, we conclude with two constructions that together establish our second main result, a lower bound on how often a particular k-tuple of dot products determined by trees isomorphic to a given tree can occur.

2 Proofs

2.1 Tools

Given a point $p = (p_1, p_2) \in \mathbb{R}^2 \setminus \{(0, 0)\}$, the set of points $q \in \mathbb{R}^2$ that determine dot product α with p form a line whose slope is perpendicular to the line through both p and the origin. To see this, we simply recall the definition of dot product and rearrange the equation $p_1q_1 + p_2q_2 = \alpha$. We sometimes call this the α-line of p, and denote it $\ell_\alpha(p)$. A direct consequence of this definition is that $p \neq q$ implies $\ell_\alpha(p) \neq \ell_\alpha(q)$. We now recall the Szemerédi-Trotter theorem from [27]. Given a point p and a line ℓ, we call the pair (p, ℓ) an incidence if the point p is on the line ℓ.

Theorem 5 *Given a set of n points and m lines in the plane, the number of incidences between the points and lines is at most*

$$\lesssim (mn)^{\frac{2}{3}} + n + m.$$

We now give the following well-known corollary alluded to above to illustrate some of the basic ideas to come.

Corollary 1 *Given a large finite set of n points $E \subseteq \mathbb{R}^2$, the point pairs determine at least $n^{\frac{2}{3}}$ distinct dot products.*

Proof Fix any $\alpha \neq 0$, construct the set of α-lines $\ell_\alpha(p)$ for every $p \in E$. Any time a point q is on an α-line $\ell_\alpha(p)$, we have that $p \cdot q = \alpha$, and vice versa. So the number of times that the dot product α occurs between pairs of points in the set E is exactly the number of incidences between points and α-lines. Since there were n points, and each determines a unique α-line, we have n lines. So by Theorem 5, we have that α can occur no more than $\lesssim \left(n^2\right)^{\frac{2}{3}} = n^{\frac{4}{3}}$ times. Since this is true for any $\alpha \neq 0$, and there are n^2 distinct pairs of points, we get that the number of distinct dot products must be at least

$$\frac{n^2}{n^{\frac{4}{3}}} \gtrsim n^{\frac{2}{3}}.$$

Corollary 1 shows that there are many distinct dot products, but it does not show that any particular point determines many distinct dot products. For this, we introduce another concept. Given a point p and a set of points E, we define the *pinned* dot product set $\Pi_p(E)$, to be the set of dot products determined by p and the points in the set E. In this case, we call the point p a *pin*. Specifically,

$$\Pi_p(E) := \{p \cdot q : q \in E\}.$$

2.2 The Pinned Result

We now state a pinned version of Corollary 1.

Theorem 6 *Given two subsets E and F of $\mathbb{R}^2$ consisting of n points each, where E has no more than $\gtrsim n^{\frac{2}{3}}$ points on any line through the origin, we have that there is a subset $E' \subset E$ of size $|E'| \geq \frac{1}{2}n$ with the property that for any $x \in E'$, the number of distinct dot products determined by x and the points in F is at least $\gtrsim n^{\frac{2}{3}}$. Moreover, if $E = F$, we get the same results without the hypothesis about the number of points on any line through the origin.*

The proof we give essentially comes from Sect. 9.2 in [12], but Theorem 6 highlights the pinned nature of the result, which is not explicitly stated in the original text. Moreover, we slightly modify the proof to get a result for two sets of points instead of one, which is what we will need later. Unfortunately, because we are specifying which set the good pin must come from, this brings in the possibility of pathological pairs of point sets. For example, if E is a set of n points on a line through the origin and F is a set of n points on a line perpendicular to the line containing E, then we would have that each point $x \in E$ we have $|\Pi_x(F)| = 1$. However, if $E = F$ above, we will see in the proof that we can drop the hypothesis that no line through the origin has more than $\gtrsim n^{\frac{2}{3}}$ points of E.

In order to prove this result, we will rely on a version of the so-called crossing number lemma. In order to state the lemma, we remind the reader of two concepts. First, a multigraph is a generalization of a graph allowing for multiple distinct edges connecting pairs of vertices. We refer to the number of distinct edges connecting a given pair of vertices as the edge multiplicity of that pair of vertices. Second, given a graph G, its crossing number, denoted $cr(G)$ is the minimum number of pairs of edges that cross taken over any drawing of G in the plane. So, a planar graph has crossing number zero, but any graph that is not planar has a positive crossing number. We now state a multigraph version of the crossing number lemma, from Székely in [26].

Lemma 1 *Given a multigraph G with v vertices, e edges, and a maximum edge multiplicity of m, satisfying $e > 5mv$, we have that*

$$cr(G) \gtrsim \frac{e^3}{mv^2}.$$

With this in tow, we prove Theorem 6.

Proof We will also assume that the origin is not in E or F, as it can only determine the dot product zero anyway. Construct a graph, G, whose vertex set is the point set $E \cup F$. Note, although we defined a graph G_E related to a point set E above, this graph will have very different edges than G_E. For every point $p \in E$, draw the α-lines for each $\alpha \in \Pi_p(F)$. By definition, for each p, the associated lines drawn will be parallel to one another. Now, for every such $\ell_\alpha(p)$ with more than one point on it, we connect vertices by an edge if their corresponding points are consecutive along that line. Notice that we will disregard such lines with exactly one point on them. However, if p and q lie on the same radial line, it is possible that $\ell_\alpha(p) = \ell_\beta(q)$. In this case, the construction of the graph H could lead to multiple edges between some pairs of vertices. Any time that there are multiple edges between a pair of points, r and s, draw each edge as a smooth curve, $\mathscr{C}$, connecting the appropriate points close enough to the line segment, $\mathscr{L}$, connecting those points so that there are no other points from E in the region bounded by $\mathscr{C}$, $\mathscr{L}$, and the points r and s.

We now deal with a possible degenerate case. Suppose there is a point p whose α-lines contribute much fewer than n edges. In this case, this means that for most of the values of $\alpha \in \Pi_p(F)$, we have that $\ell_\alpha(p)$ only has one point. That would imply that p determines $\gtrsim n$ distinct dot products with points in F, and we would be done. So we can assume that it takes about n edges for each point in E. Therefore, there must be about n^2 edges in H.

Define t to be the maximum number of distinct dot products determined by any point in $p \in E$ with points in F. That is, we define

$$t := \max_{p \in E}\{p \cdot q : q \in F\}.$$

The the bulk of rest of the proof is devoted to showing that $t \gtrsim n^{\frac{2}{3}}$. For any fixed radial line, ℓ, the vertices of consecutive points along each of the parallel lines perpendicular to ℓ will be connected by as many edges as there are points on ℓ. Because of our hypothesis on the set E, we know ℓ has no more than $n^{\frac{2}{3}}$ points from E on it. Thus, in our graph, the maximum edge multiplicity will be $m \lesssim n^{\frac{2}{3}}$.

We pause here to note a slight change that applies to the case where $E = F$ to handle the 'moreover' part of the theorem. Namely, if we were to have $E = F$, we would not need the hypothesis on the number of points on a radial line, as if there were more than $n^{\frac{2}{3}}$ points on a radial line, any of those points would determine at least as many distinct dot products as are on the radial line, with the points on the radial line in question, so each such point would automatically satisfy the conclusion of the theorem statement.

Now that we know the number of vertices, edges, and maximum edge multiplicity, we can try to apply Lemma 1. Recall that we need $e > 5mv$ to apply the lemma. If this condition were to fail, we would have $e \leq 5mv$, but plugging in our values of e, m, and v would give us $n^2 \lesssim 5n^{\frac{2}{3}}n$ which is a contradiction. Therefore, we can proceed assuming that this condition holds, and by plugging in values of e, m, and v, we are left with

$$cr(G) \gtrsim \frac{e^3}{mv^2} \gtrsim \frac{n^6}{n^{\frac{2}{3}}n^2} \gtrsim n^{\frac{10}{3}}.$$

Next, we will pair this with an upper bound for the crossing number of G. Any crossing between edges only occurs when a line perpendicular to the radial line of one point, Say p, crosses a line perpendicular to the radial line of another point, say q. Since each point has no more than t such associated parallel lines (its α-lines), each pair of points can contribute at most t^2 crossings. Because we know that there are about n^2 different pairs of points, we are guaranteed that the total number of crossings is bounded above by $\lesssim n^2t^2$. Now, while we don't know what the minimum number of crossings in any drawing of G in the plane must be, we can be sure that it is no more than n^2t^2, because we have found an explicit drawing with no more than n^2t^2 crossings. Putting the upper and lower bounds for the crossing number together:

$$n^{\frac{10}{3}} \lesssim cr(G) \lesssim n^2t^2.$$

This tells us that $t \gtrsim n^{\frac{2}{3}}$, as claimed. Now, we have shown the existence of one point $x \in E$ that determines $\gtrsim n^{\frac{2}{3}}$ distinct dot products with points in F. We call this point a *good pin*. To finish the proof, we just need to show that there are many good pins. To see this, define the set $E_1 := E \setminus \{x\}$, and set F_1 to be any subset of F with $n-1$ points. We can run the same argument to get that there is a good pin $x_1 \in E_1$, that determines many distinct dot products with the points in F_1. Repeat this process until we have at least $m = \lceil n/2 \rceil$ such good pins, and obtain the set E_m. To finish, set $E' = E_m$, which is of size $\geq n/2$, and consists entirely of good pins.

2.3 *Proof of Theorem 2*

With a simple pigeonholing argument, we can use Theorem 6 to get a pinned result in higher dimensions in the case that $E = F$.

Proof Suppose t_1 is the maximum number of parallel $(d-1)$-hyperplanes supported by points in E that are the level sets of dot products determined by points in E. That is, these are the higher dimensional analogs of α-lines. Let x be such a point determining t_1 distinct dot products with points in E. Pick one of these parallel $(d-1)$-hyperplanes, with as many or more points than any of the other parallel $(d-1)$-hyperplanes and call it P_1. By the pigeonhole principle, we can see that it has at least $\gtrsim n/t_1$ points of E. Now we repeat this process within P_1. Specifically, we select $x_2 \in P_1$ be a point determining t_2 distinct dot products, so that no other points in P_1 determine more dot products with points in P_1. As before, pick a $(d-2)$-hyperplane P_2, that contains greater than $(n/t_1)/t_2$ points of E. We continue doing this, getting successive hyperplanes of one dimension lower until P_{d-2} is a plane with $(n)/(t_1t_2...t_{d-2})$ points. By appealing to Theorem 6, we see that this plane has a point x_{d-2} that determines $(n/(t_1t_2...t_{d-2}))^{\frac{2}{3}}$ distinct dot products.

Now we collect what we know about each of the points $x_1, \ldots, x_{d-1}$. Namely, x_1 determines t_1 distinct dot products with the points of E, x_2 determines t_2, distinct dot products with the points of E, and for all other $j < d-1$, we have that x_j determines t_j distinct dot products with the points of E. Finally, x_{d-1} determines is at least $\gtrsim ((n)/(t_1 t_2 ... t_{d-1}))^{\frac{2}{3}}$ distinct dot products with the points in E. Putting all of this together, we have

$$\max_{x \in E} |\Pi_x(E)| \gtrsim \max \left\{ t_1, t_2, \ldots, t_{d-2}, ((n)/(t_1 t_2 ... t_{d-2}))^{\frac{2}{3}} \right\}. \tag{1}$$

Now we consider the case when all of the components are approximately equal, namely if

$$t_1 \approx t_2 \approx \cdots \approx ((n)/(t_1 t_2 ... t_{d-2}))^{\frac{2}{3}}.$$

In this case, we get that

$$t_1 \approx \left(\frac{n}{t_1^{d-2}} \right)^{\frac{2}{3}},$$

which gives us

$$\max_{x \in E} |\Pi_x(E)| \approx t_1 \gtrsim n^{\frac{2}{2d-1}}.$$

Now we deal with the case that one of the t_j is significantly bigger one of the others. It is straightforward to see that this would only increase the lower bound for $|\Pi_x(E)|$. The same is true if the final term, $((n)/(t_1 t_2 ... t_{d-2}))^{\frac{2}{3}}$, is significantly larger than the t_j. To prove this rigorously, one can set up a Lagrange multiplier argument, or take logarithms of each term in base n and use linear programming.

To summarize, we are able to pigeonhole our previous argument onto higher dimensions by running the algorithm on each hyperplane determined by our points. In general, our lower bound for higher dimensions is:

$$\max_{x \in E} |\Pi_x(E)| \gtrsim n^{\frac{2}{2d-1}}.$$

2.4 *Proof of Theorem 1*

This argument closely follows the arguments in [23], by Dung, Pham, and the fourth listed author, which was inspired by the work of Ou and Taylor in [21]. The proof of Theorem 1 will essentially follow by applying this more technical result. For brevity, we often conflate a point in $\mathbb{R}^2$ with its associated vertex in a graph. Given a tree T with a vertex v, we will often refer to the pair of them by (T, v).

In essence, what the theorem says is that if we fix a tree T and isolate a vertex v, then there are many points x that serve as the vertex v in many trees T', isomorphic to T, and that these isomorphic copies T' have a wide variation of dot product edge weights.

Theorem 7 *Given two subsets E and F of $\mathbb{R}^2$ consisting of n points each, where E has no more than $n^{\frac{2}{3}}$ points on any line through the origin, a tree T with k edges, and an arbitrary vertex v of T, we have that there is a subset $E' \subset E$ of size $|E'| \geq 2^{-k}n - o(n)$ with the property that for any $x \in E'$, the number of distinct edge weight k-tuples, $\mathbf{w}$, from trees $T'_{\mathbf{w}}$ with (T, v) isomorphic to (T', x), and $T_{\mathbf{w}} \setminus \{x\} \subset F$ is at least $\gtrsim n^{\frac{2k}{3}}$.*

Proof We now prove Theorem 7 by induction on k. The base case is when $k = 1$, which holds by applying Theorem 6. We assume that the conclusion holds for any tree on k vertices, and then show that it also holds for any tree on $k+1$ vertices. We will do this by splitting into two cases: the case where v is a leaf of T (only adjacent to one other vertex of T) and the case where v is not a leaf of T.

If v is a leaf of T, then it has a unique adjacent vertex, which we call u. Notice that the tree $S := T \setminus v$ has k edges, so we can apply the induction hypothesis to get that there is a subset $E' \subset E$ with size at least $2^{-k}n$ such that for all $x \in E'$, there are $\gtrsim n^{\frac{2k}{3}}$ distinct dot product edge weight k-tuples $\mathbf{w}$ for trees $S'_{\mathbf{w}}$ with x as a vertex so that $(S'_{\mathbf{w}}, x)$ is isomorphic to (S, u). Recall that for tree isomorphism here, we are not concerned with the edge weights. We are showing that there are many different trees with the same general shape, but different edge weights. Next, we can apply Theorem 6 to the set E' and any subset of $F' \subseteq F$ of size $|E'|$ to get that there a subset $E'' \subseteq E'$ of size at least $|E'|/2$, so that every point $x \in E''$ determines at least $\gtrsim n^{\frac{2}{3}}$ distinct dot products with points in F'. Putting this all together, we have that there is a set, E'' of size at least $|E'|/2 \geq 2^{-(k+1)}n$ so that any point $x \in E''$ satisfies the conclusion of the theorem. This handles the case that v is a leaf (Fig. 1).

We now turn our attention to the case that v is not a leaf. In this case, we will break up T into two trees that share only the vertex v. So T_1 is a tree with at least one edge and v as a vertex, T_2 is also a tree with at least one edge with v as a vertex, and the union of the two trees is T. Suppose the number of edges in T_1 is k_1, and the number of edges in T_2 is k_2. As k is the total number of edges in T, and they are all accounted for without repeating by considering both T_1 and T_2, we can see that $k = k_1 + k_2$.

Next, we arbitrarily partition E into two subsets of equal size (either both $n/2$ or $(n+1)/2$ and $(n-1)/2$), so we have

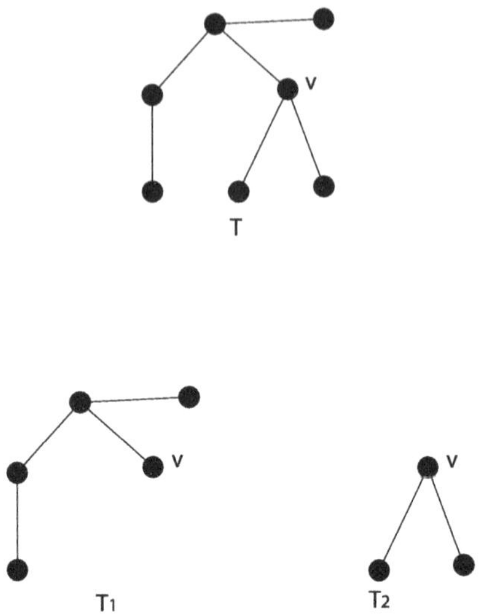

Fig. 1 Here we can see the tree T with vertex v decomposed into T_1 and T_2, both of which also have vertex v

$$E = E_1 \sqcup E_2.$$

Next, we similarly partition F into two subsets of nearly equal size, to get F_1 and F_2. We will now perform the following steps for both E_1 and E_2 separately. In what follows, we will often need various sets to be the same size, but they could be off by one. If this is the case, we merely delete a point from the larger set as needed providing a negligible error considering the fact that n is large.

Because $k_1 \leq k$, we can apply the induction hypothesis to E_i and F_1 with (T_1, v) to find a subset $E_i' \subseteq E_i$ of size at least $2^{-k_1}|E_i|$ with the property that for every $x \in E_i'$, there are at least $\gtrsim n^{\frac{2k_1}{3}}$ distinct dot product k_1-tuples $\mathbf{w}_1$ corresponding to trees isomorphic to T_1 with x serving as the vertex v and the other vertices coming from points in F_1. Now, we find an arbitrary subset $F_2' \subseteq F_2$ of size $|E_1'|$. Because $k_2 \leq k$, we can again appeal to the induction hypothesis for the sets E_i' and F_2' with the tree T_2 and vertex v. This guarantees the existence of a set $E_i'' \subseteq E_i'$ whose size is at least

$$|E_i''| \geq 2^{-k_2}|E_i'| \geq 2^{-k_2-k_1}|E_i| = 2^{-k}|E_i| \geq 2^{-k-1}n - 1, \tag{2}$$

with the property that for every $y \in E_i''$, there are at least $\gtrsim n^{\frac{2k_2}{3}}$ distinct dot product k_2-tuples $\mathbf{w}_2$ corresponding to trees isomorphic to T_2 with y serving as the vertex v and the other vertices coming from points in F_2'. Because $E_i'' \subseteq E_i'$, each of these points $y \in E_i''$ also determines at least $\gtrsim n^{\frac{2k_1}{3}}$ distinct dot product k_1-tuples $\mathbf{w}_1$ corresponding to trees isomorphic to T_1 with y serving as the vertex v. Because we chose F_1 disjoint from F_2, we know that the points in F_1 that serve as vertices for isomorphic copies of T_1 are disjoint from the points from F_2 that serve as vertices for the isomorphic copies of T_2.

We now generalize a term from before. Here, a *good pin* is a choice of v determining sufficiently many distinct dot product tuples from their respective dot product trees, in analogy to the same term applied to single dot products in the proof of Theorem 6. Recall that we run the above argument for both E_1 and E_2, which were disjoint. So we can add the good pins v coming from E_1 to the good pins v coming from E_2 and appeal to (2) to get a lower bound on the number good pins to be

$$|E'| \geq |E_1'' \sqcup E_2''| \geq (2^{-k-1}n - 1) + (2^{-k-1}n - 1) = 2^{-k}n - 2 = 2^{-k}n - o(1),$$

as claimed.

To finish, notice that by definition, the dot product k-tuple $\mathbf{w}$ is what we get by combining $\mathbf{w}_1$ and $\mathbf{w}_2$. Moreover, F_1 and F_2 were disjoint, so the total number of distinct dot product k-tuples $\mathbf{w}$ determined by isomorphic copies of T with v coming from E and the other vertices coming from F is at least the product of the number of dot product k_1-tuples $\mathbf{w}_1$ associated to isomorphic copies of T_1 from F_1' times the number of dot product k_1-tuples $\mathbf{w}_2$ associated to isomorphic copies of T_2 coming from points in F_2'.

$$\gtrsim n^{\frac{2k_1}{3}} \cdot n^{\frac{2nk_2}{3}} = n^{\frac{2(k_1+k_2)}{3}} = n^{\frac{2k}{3}}, \tag{3}$$

finishing the induction step, and therefore the proof of Theorem 7.

To complete the proof of Theorem 1, we need to look at lines through the origin. Now, if there is no radial line with more than $n^{\frac{2}{3}}$ points on it, we can repeatedly apply Theorem 7 with $F = E$ to get the desired result. If there are radial lines with too many points, we have a decision to make. If half of the points in E lie on radial lines with fewer than $n^{\frac{2}{3}}$ points, then we just apply Theorem 7 with E and F consisting of these points. So all that remains is the case that at least half of the points are lying on radial lines with more than $n^{\frac{2}{3}}$ points each.

In this case, fix a radial line ℓ with more than $n^{\frac{2}{3}}$ points on it, then partition $E \cap \ell$ into k sets of roughly the same size, called E_i, and we can see that there are at least $n^{\frac{2k}{3}}$ different trees by selecting distinct choices of x_i from $E_i \cap \ell$. Without loss of generality, suppose ℓ is the x-axis. Then the points in E_i have coordinates $(a_{i,j}, 0)$ where $j = 1, \ldots, n^{\frac{2}{3}}/k$, and the $a_{i,j}$ are all distinct. Define

$$A_i := \{a_{i,j} : j = 1, \ldots, n^{\frac{2}{3}}/k\}.$$

Now we can see that the different edge weights of our trees correspond to the elements of the Cartesian product of product sets of the form

$$A_i A_{i'} := \{a_{i,j} a_{i',j} : a_{i,j} \in A_i, a_{i',j'} \in A_{i'}\}.$$

We now proceed by induction on k. It is trivial to see that $|A_1 A_2| \geq |A_1|$, as one need only consider the products of a fixed element of one set with each of the distinct elements of the other. So the claim is true for $k = 1$. Now, suppose that we have $\gtrsim n^{\frac{2(k-1)}{3}}$ distinct $(k-1)$-tuples coming from the product sets $A_i A_{i'}$. We just need to show that these will lead to at least $\gtrsim n^{\frac{2k}{3}}$ distinct k-tuples when we factor in the final product set $A_{i''} A_{k+1}$, where the vertex numbered $k+1$ is a leaf adjacent to the vertex of index i'' that has already been dealt with. Now, for every choice of $a_{i'',j}$, there are $\gtrsim n^{\frac{2}{3}}/(k+1)$ distinct choices of $a_{k+1,j'}$, each leading to a different product, meaning there are at least $n^{\frac{2}{3}}/(k+1)$ distinct products for any of the $\gtrsim n^{\frac{2(k-1)}{3}}$ trees already determined, and we are done.

2.5 *Proof of Theorem 3*

Theorem 3 will follow from the following two constructions, which we state separately as propositions. The first one is adapted from a construction given in [20].

Proposition 1 *Given a tree T on k edges, there exists a set of n points in $\mathbb{R}^d$ and a k-tuple of dot product edge weights $\mathbf{w}$ so that there are $\gtrsim n^{\lceil (k+1)/2 \rceil}$ isomorphic copies of T in the set whose edge weights are $\mathbf{w}$.*

The second proposition is adapted from a construction given in [19].

Proposition 2 *Given a tree T on k edges, there exists a set of n points in $\mathbb{R}^d$ so that there are $\gtrsim n^{1+k(d-1)/(d+1)}$ isomorphic copies of T in the set whose edge weights are 1.*

Combining these two yields Theorem 3. We now prove each of the propositions.

2.5.1 Proof of Proposition 1

This construction is a modification of a chains construction in [20]. It can be easily modified any number of ways, but we restrict to one way that is relatively easy to communicate, while still showing the essential components that one might want to modify. We begin by 2-coloring the vertices tree T, which we can do because the chromatic number of any tree is at most 2. Call the two bipartition sets U and V, and suppose without loss of generality that $|U| \geq |V|$. Now let k_1 denote the size of U and let k_2 denote the size of $|V|$. So $k_1 + k_2 = k + 1$. Recall that by definition, we know $k_1 = |U| \geq \lceil k/2 \rceil$. Now we proceed with the following algorithm. Pick any vertex in U and call it u_1.

We will now put points on the standard xy-coordinate plane. Place $\lfloor (n - k_2)/k_1 \rfloor$ points along the line $x = 1$ in the plane. Now suppose that u_1 has a_1 neighbors in T called $v_{1,1}$ through v_{1,a_1}. For each such neighbor $v_{1,j}$, place a point at $(1 + j, 0)$. So any of the points along the line $x = 1$ have dot product $1 + j$ with the point at $(1 + j, 0)$. Now, call the b neighbors of $v_{1,1}$ (other than u_1) $u_2, u_3, \ldots, u_b$. For each such neighbor u_j, place $\lfloor (n-k_2)k_1 \rfloor$ points on the line $x = 1 + a_1 + j$. Continue by placing single points along the x-axis for vertices from V, and columns of $\lfloor (n-k_2)/k_1 \rfloor$ points for each vertex from U, keeping track of the relative dot products. After all of the vertices in U and V have been processed, place any other remaining points in such a way that they do not overlap with previously placed points, until you have a total of n points, and the construction is finished.

2.5.2 Proof of Proposition 2

This construction is essentially the discrete part of the construction in [19], though with some slight modifications as the constants chosen there may lead to some confusion. The new piece here is how to embed the trees. Also, in Proposition 2, we stress that the dot products are always 1, as opposed to the previous construction, for Proposition 1, where our choices of dot product edge weights might have to be different. The basic idea is to construct two sets, E and F, where every point in either set makes the dot product 1 with many points in the other set. The set E will be a kind of lattice, so it will have many points on many hyperplanes. The set F will be the points that make dot product 1 with many of the hyperplanes. We now sketch the modified construction.

For a large natural number n, fix $q \approx dn^{\frac{1}{d+1}}$. Now define two sets of numbers:

$$A := \left\{ \frac{q+i}{dq} : i = 1, \ldots, q \right\},$$

and

$$B := \left\{ \frac{q^2+j}{d^2q^2} : j = 1, \ldots, q^2 \right\}.$$

To build our lattice, E, we will take the $(d-1)$-fold Cartesian product of A with itself, and then take the Cartesian product of this set with B. By construction, we can see that $|E| = q^{d-1} \cdot q^2 \approx n$. Next, we focus on the point set F, which will also be built using the sets of numbers A and B.

$$F := \left\{ \left(\frac{-c_1}{b}, \frac{-c_2}{b}, \ldots, \frac{-c_{d-1}}{b}, \frac{1}{b} \right) : c_j \in A, b \in B \right\}.$$

Now we want to get a handle on point pairs whose dot product is 1. Dot products are constant on hyperplanes, so given a d-tuple, $(c_1, c_2, \ldots, c_{d-1}, b)$, we can define the hyperplane

$$h(c_1, c_2, \ldots, c_{d-1}, b) := \left\{ (x_1, x_2, \ldots, x_d) \in \mathbb{R}^d : x_d = \left(\sum_{j=1}^{d-1} c_j x_j \right) + b \right\}.$$

We will now focus on the following set of hyperplanes

$$H := \{h(c_1, c_2, \ldots, c_{d-1}, b) : c_j \in A, b \in B\}.$$

These hyperplanes are the sets of points that determine dot product 1 with points in F. To prove this, consider a d-tuple, $(c_1, c_2, \ldots, c_{d-1}, b) \in A^{d-1} \times B$, and compute the dot product of the associated element of F and any point on the hyperplane associated to the same d-tuple. To be precise, consider the point $f \in F$ given by

$$f = \left(\frac{-c_1}{b}, \frac{-c_2}{b}, \ldots, \frac{-c_{d-1}}{b}, \frac{1}{b}\right) \text{ to be associated to } h_f := h(c_1, \ldots, c_{d-1}, b),$$

a hyperplane in H, and compute its dot product with an arbitrary $x \in h_f$. Their dot product will be

$$\begin{aligned} f \cdot x &= \left(\frac{-c_1}{b}, \frac{-c_2}{b}, \ldots, \frac{-c_{d-1}}{b}, \frac{1}{b}\right) \cdot \left(x_1, x_2, \ldots, x_{d-1}, \left(\sum_{j=1}^{d-1} c_j x_j\right) + b\right) \\ &= \left(\sum_{j=1}^{d-1} \frac{-c_j x_j}{b}\right) + \left(\sum_{j=1}^{d-1} \frac{c_j x_j}{b}\right) + \frac{b}{b} = 1. \end{aligned}$$

We now consider any $f \in F$. The associated hyperplane, $h_f \in H$, consists of points that have dot product 1 with f. By construction, for most[1] $f \in F$, the associated hyperplane h_f will have $\approx q^{d-1}$ points from E. Moreover, the dual relation also holds. That is, for most $e \in E$, there are about q^{d-1} points $f \in F$ such that $e \cdot f = 1$.

Now we embed our trees as in the proof of the previous result. For any tree T, we 2-color the vertices to obtain disjoint vertex sets U and V. Fix some vertex from U to be u_1. Let any point $f_1 \in F$ be a choice for u_1. We consider the set of q^{d-1} choices of $e \in h_{f_1}$ that have dot product 1 with f_1. Suppose that u_1 has a_1 neighbors in T, and designate these by $v_{1,1}, v_{1,2}, \ldots, v_{1,a_1}$. For each of these, we select a different point in h_{f_1}. Notice that there are q^{d-1} choices for each of them. Now consider the neighbors of $v_{1,1}$ in T, and for each representative point $e_{1,1} \in h_{f_1}$ of each of them, choose from the q^{d-1} points of F that make dot product 1 with $e_{1,1}$. Continuing in this way, each new vertex that is processed in the tree has $\approx q^{d-1}$ possible choices. Since there were $\approx n$ choices for u_1, and each of the k subsequent points had $q^{d-1} \approx n^{\frac{d-1}{d+1}}$ choices, we end up with a total of $\gtrsim n^{1+k(d-1)/(d+1)}$ isomorphic copies of T in the set whose edge weights are 1, as claimed.

Remark 1 We pause to note that the higher dimensional results may appear unnecessary in light of a slight modification of the construction in Remark 2 of [20], which

[1] Here we notice that there could be some members of f whose hyperplanes cut through the edge of E, and therefore hit significantly fewer than $\approx q^{d-1}$ points, but this will happen for $\lesssim |F|$ points. As this is a negligible loss, we ignore it and similar instances of it.

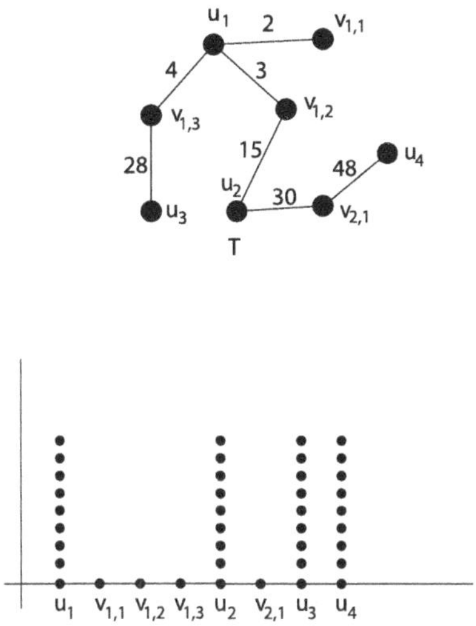

Fig. 2 Here we can see the tree T with vertices broken up into U and V, along with a set of points in $\mathbb{R}^2$ that correspond to it. Note for example, the point at (6, 0) will always a choice for $v_{2,1}$, and its dot product with any point in the column of points with $x = 8$, each of which is a possible choice for u_4, will have dot product 48, so we have the dot product edge weight of 48 on the edge connecting $v_{2,1}$ and u_4

can give n^k trees for certain k-tuples of edge weights in dimensions $d \geq 3$. However, the constructions given here are not concentrated on lines, and therefore are s-adaptable for a wider range of s. See [19, 20] for more on s-adaptability, and how it connects to constructing fractal sets. For completeness, we give the construction for k-paths, and describe how to modify it for general trees.

Given n and k, arrange $\lfloor n/(k+1) \rfloor$ points on each of the following lines perpendicular to the x-axis: $\{(1, y, 0)\}, \{2, 0, z\}, \{(3, y, 0)\}, \{(4, 0, z)\} \ldots$ alternating whether y or z is the free variable on every other line. We then 2-color the path, and map the vertices of the (potentially) larger color class to the points on the lines with odd x coordinates and map the vertices of the other color to the points on the lines with even x coordinate. To modify this for general trees, we begin by 2-coloring the tree, then make sure we have a y-line for every vertex of one color and a z-line for every vertex of the other color. So the tree given in Fig. 2 could be given by $\lfloor n/8 \rfloor$ points on each of the following lines:

$$\{(1, y, 0)\}, \{2, 0, z\}, \{(3, 0, z)\}, \{(4, 0, z)\}, \{(5, y, 0)\}, \{6, 0, z\}, \{(7, y, 0)\}, \{(8, y, 0)\}.$$

Acknowledgements We would like to thank Alex Iosevich, Ethan Lynch, and Caleb Marshall for valuable discussions that improved the results here and the exposition thereof.

References

1. D. Barker and S. Senger, *Upper bounds on pairs of dot products*, Journal of Combinatorial Mathematics and Combinatorial Computing, Volume 103, November, 2017, pp. 211–224.
2. M. Bennett, A. Iosevich, and K. Taylor, *Finite chains inside thin subsets of* $\mathbb{R}^d$, Analysis and PDE, volume 9, no. 3, (2016).
3. V. Blevins, D. Crosby, E. Lynch, and S. Senger, *On the Number of Dot Product Chains in Finite Fields and Rings*, Nathanson M. (eds) Combinatorial and Additive Number Theory V. CANT 2021. pp. 1–20. Springer Proceedings in Mathematics & Statistics. Springer, Cham.
4. J. Bourgain, N. Katz, and T. Tao, *A sum-product estimate in finite fields, and applications*, Geom. Funct. Anal. **14** (2004), pp. 27–57.
5. P. Brass, W. Moser, and J. Pach, *Research Problems in Discrete Geometry*, Springer (2000), 499 pp.
6. J. Chapman, B. Erdoğan, D. Hart, A. Iosevich, D. Koh, *Pinned distance sets, k-simplices, Wolff's exponent in finite fields and sum-product estimates*, Math. Z. 271 (2012), no. 1-2, 63–93.
7. D. Covert and S. Senger, *Pairs of dot products in finite fields and rings*, Nathanson M. (eds) Combinatorial and Additive Number Theory II. CANT 2015, CANT 2016. Springer Proceedings in Mathematics & Statistics, vol 220. Springer, Cham. (Appeared 14 Jan. 2018)
8. D. Covert, A. Iosevich, J. Pakianathan, *Geometric configurations in the ring of integers modulo* p^ℓ, Indiana University Mathematics Journal, 61 (2012), 1949–1969.
9. D. Covert, D. Hart, A. Iosevich, S. Senger, and I. Uriarte-Tuero, *An analog of the Furstenberg-Katznelson-Weiss theorem on triangles in sets of positive density in finite field geometries*, Discrete Math. 311, no. 6, 423–430, (2011).
10. P. Erdős, *On sets of distances of n points*, Amer. Math. Monthly **53** (1946) 248–250.
11. N. Frankl and A. Kupavskii, *Almost sharp bounds on the number of discrete chains in the plane*, Combinatorica 42 (Suppl 1), 1119–1143 (2022).
12. J. Garibaldi, A. Iosevich, and S. Senger, *Erdős distance problem*, AMS Student Library Series, 56, (2011).
13. L. Guth and N. H. Katz, *On the Erdős distinct distance problem in the plane*, Annals of Math., Pages 155–190, Volume 181 (2015), Issue 1.
14. S. Gunter, E. Palsson, B. Rhodes, and S. Senger, *Bounds on point configurations determined by distances and dot products*, Nathanson M. (eds) Combinatorial and Additive Number Theory IV. CANT 2019, CANT 2020. Springer Proceedings in Mathematics & Statistics, vol 347. Springer, Cham.
15. B. Hanson, O. Roche-Newton, and S. Senger, *Convexity, superquadratic growth, and dot products*, Journal of the London Mathematical Society, Volume 107, Issue 5, May 2023, pp. 1900–1923.
16. D. Hart, A. Iosevich D. Koh, M. Rudnev, *Averages over hyperplanes, sum-product theory in vector spaces over finite fields and the Erdős-Falconer distance conjecture*, Trans. Amer. Math. Soc. 363 (2011), no. 6, 3255–3275.
17. A. Iosevich and M. Rudnev, *Erdős-Falconer distance problem in vector spaces over finite fields*, Trans. Amer. Math. Soc., 359 (2007), no. 12, 6127–6142.
18. A. Iosevich and S. Senger, *Orthogonal systems in vector spaces over finite fields*, Electronic J. of Combinatorics, Volume 15, December (2008).
19. A. Iosevich and S. Senger, *Falconer-type estimates for dot products*, Bulletin of the Hellenic Mathematical Society **64** 98–110 (2020).
20. S. Kilmer, C. Marshall, and S. Senger, *Dot product chains*, (accepted) arXiv:2006.11467 (2020).
21. Y. Ou and K. Taylor, *Finite point configurations and the regular value theorem in a fractal setting*, Indiana Univ. Math. J. 71 (2022), no. 4, 1707–1761.
22. E. Palsson, S. Senger, and A. Sheffer, *On the number of discrete chains*, (to appear in the Proceedings of the American Mathematical Society).
23. T. Pham, S. Senger, and D. Tran, *Distribution of pinned distance trees in the plane* $\mathbb{F}_p^2$, Discrete Mathematics, Volume 346, Issue 12, 2023, 113613.

24. J. Spencer, E. Szemerédi, and W. T. Trotter, *Unit distances in the Euclidean plane* B. Bollobás, editor, "Graph Theory and Combinatorics", pages 293–303, Academic Press, New York, NY, (1984).
25. J. M. Steele, *The Cauchy-Schwarz Master Class ICM Edition: An Introduction to the Art of Mathematical Inequalities*, Cambridge University Press, (2010).
26. L.A. Székely, *Crossing numbers and hard Erdős problems in discrete geometry*, Combin. Probab. Comput. **6** (1997), no. 3, 353–358.
27. E. Szemerédi and W. T. Trotter, *Extremal problems in discrete geometry*, *Combinatorica* **3** (1983), 381–392.
28. T. V. Pham and L. A. Vinh *Orthogonal Systems in Vector Spaces over Finite Rings*, Electronic J. of Combinatorics, Volume 19, Issue 2 (2012).

Fractal Dimension, Approximation and Data Sets

L. Betti, I. Chio, J. Fleischman, A. Iosevich, F. Iulianelli, S. Kirila, M. Martino, A. Mayeli, S. Pack, Z. Sheng, C. Taliancic, A. Thomas, N. Whybra, E. Wyman, U. Yildirim, and K. Zhao

Abstract The purpose of this paper is to study the fractal phenomena in large data sets and associated questions of dimension reduction. We examine situations where the classical Principal Component Analysis is not effective in identifying the salient underlying fractal features of the data set. Instead, we employ the discrete energy, a technique borrowed from geometric measure theory, to limit the number of points of a given data set that lie near a k-dimensional hyperplane or, more generally, near a set of a given upper Minkowski dimension. Concrete motivations stemming from naturally arising data sets are described, and future directions are outlined.

1 Introduction

One of the basic questions in data science, which is to determine the "effective" dimension of a large data set. If a data set has 10^6 points in 1000-dimensional space, it is extremely useful to be able to detect if a significant proportion of these points live on a lower-dimensional plane, or a more complicated surface, reflecting the hidden relationships between the features present in the data set. One of the main tools in this area is Principal Component Analysis (PCA) (see e.g. [5]). This method has been and will remain a fundamental tool in the study of dimensionality of data sets.

A. Iosevich—The work of the fourth listed author was supported in part by the NSF HDR Tripods 1934962 grant and the NSF DMS 2154232 grant.
E. Wyman—The work of the fourteenth listed author was supported in part by the NSF DMS 2204397 grant.

L. Betti · I. Chio · J. Fleischman · A. Iosevich (✉) · F. Iulianelli · S. Kirila · M. Martino · S. Pack · Z. Sheng · C. Taliancic · A. Thomas · N. Whybra · U. Yildirim · K. Zhao
Department of Mathematics, University of Rochester, Rochester, NY 14627, USA
e-mail: iosevich@gmail.com

A. Mayeli
CUNY, New York, USA

E. Wyman
SUNY, Binghamton, USA

M. B. Nathanson (ed.), *Combinatorial and Additive Number Theory VI*,
Springer Proceedings in Mathematics & Statistics 464,
https://doi.org/10.1007/978-3-031-65064-2_3

We propose to complement this method with a set of tools that capture important dimensionality phenomena that PCA does not see.

As a simple synthetic example, let us use the stages of construction of a Cantor-type subset of [0, 1] consisting of real numbers that have only 0s and 2s in their base 4 expansions. More precisely, divide [0, 1] in four equal segments, and remove the segments (1/4, 1/2) and (3/4, 1]. We keep the endpoints of the remaining segments, obtaining {0, 1/4, 1/2, 3/4}. Repeating the same procedure with the remaining two intervals of length 1/4, we obtain 16 points at the next stage, and so on. Let $C_{2^{k+1}}$ denote the resulting set and note that it has 2^{k+1} points. Let $n = 2^{2(k+1)}$ and define $P_n = C_{2^{k+1}} \times C_{2^{k+1}}$. From the point of view of PCA, this set is two-dimensional, and yet the fractal pattern that is present may be of considerable practical significance. We are going to explore this idea from a numerical point of view below.

In order to illustrate the ubiquity of fractional dimension phenomena in large data sets, let's consider the following hypothetical example. Suppose that we have a time series representing sales of a retail store going back 40 years. Suppose that one wished to look at the times when the sales were in the top 5% of all sales in a given year, and it turned out that this happened every July, December and April, and that during those months it happened during the first week, and during that week it happened on Fridays and Saturdays, and that on those days, the sales peaked in the mornings. As the reader can see, this structure is highly reminiscent of the Cantor type construction from the previous paragraph. While this type of phenomenon has been extensively studied in terms of seasonality, we believe that a fractal perspective can be of considerable value in view of the fact that if the specific months, weeks, days, and times in the hypothetical above change, while their relative number remains roughly the same, the seasonality considerations no longer apply, while the fractal dimension analysis, as we shall see, is still valid and effective.

Another manifestation of the fractional dimensional phenomena in large data sets comes from the stock price data (see, e.g., [11]). It has been noted by several authors that fractional Brownian motion can be used to model stock price volatility. The fractional Brownian path has the upper Minkowski dimension > 1, reflecting the volatility of the data (see also [2]). Combined with the discussion in the previous paragraph about the variants of seasonality, we arrive at a very interesting situation where we have a set of effective dimensions < 1 on the time axis, combined with the volatile data modeled by a function whose graph has dimension > 1. A proper understanding of a situation of this type calls for advanced tools and perspectives from geometric measure theory, frame theory and harmonic analysis.

The fractal phenomenon in data sets has been studied before. We have been particularly influenced by the investigations by Smaller Jr., Turcotte and others in [13–15].

1.1 Structure of the Paper

In order to describe our results, we need to set up the notion of fractional dimension of finite points. In Sect. 2, we develop the notion of Hausdorff dimension for families of finite point sets living in the d-dimensional unit cube $[0, 1]^d$, $d \geq 2$. We place particular focus on point sets that are given as graphs of a function from $[0, 1]^{d-1}$ to $[0, 1]$, with the idea of modeling real-life data sets where output, such as sales figures, is viewed as a function of the inputs that may include, the date, location, inflation rate, and other figures.

Theoretical results: After developing the notion of Hausdorff dimension of point sets by introducing the discrete s-energy

$$I_s(P_n) = n^{-2} \sum_{p \neq p'; p, p' \in P_n; |P_n| = n} |p - p'|^{-s},$$

we show (Theorem 2) that if the discrete s-energy of a point set P_n is suitably small, then P_n cannot contain too many points on k-dimensional surfaces, $k < s$, thus showing that small s energy limits the extent to which an effective dimension reduction can be implemented on a given data set. Note that $|p - p'|$ above denotes the Euclidean distance between p and p', whereas $|P_n|$ denotes the size of the finite set P_n.

For a comprehensive treatment of discrete energy, see [3]. We also note that discrete energy has previously been studied in the context of geometric combinatorics and geometric measure theory. See, e.g. [8, 9].

Numerical experiments: In order to illustrate the utility of the ideas described above, we apply the standard PCA (Principal Component Analysis) to a discretized version of the Cartesian product of two Cantor sets and show that PCA does not fundamentally distinguish between this sparse example and a scale integer grid. We then show that the discrete dimensionality of the same set is quickly and efficiently estimated using a Python implementation of the discrete energy function described above. These results are described in Sect. 2.2 below and carried out later in the paper.

Future directions: After identifying discrete energy as an effective tool for determining the concentration of points of a numerical data set, we shall turn our attention towards understanding how the fractal nature of a data set can be exploited to make accurate forecasts using neural network models. More precisely, we shall ask how the architecture of a neural network should be influenced by the complexity of the data set as measured by discrete energy and other analytic tools. These ideas will be explored thoroughly in a sequel.

2 Discrete Fractional Dimension and Concentration of Data Sets

2.1 Definitions and Core Results

In this subsection, we develop some basic definitions of fractional dimension in a discrete setting and state some results that will be proven later in the paper.

Let $P_n \subset [0,1]^d$, $d \geq 2$ be a finite set of size n. We consider a family of such sets,

$$\mathcal{P} = \{P_n\},$$

where n ranges over some subset of the positive integers. For $s \in [0,d]$, define the *discrete s-energy* of P_n as

$$I_s(P_n) = n^{-2} \sum_{p \neq p'; p, p' \in P_n} |p - p'|^{-s}, \tag{1}$$

where here the sum is over pairs (p, p') with $p, p' \in P_n$ and $p \neq p'$.

Definition 1 Let $\mathcal{P}$ be as above. Define the discrete Hausdorff dimension of $\mathcal{P}$ as

$$dim_{\mathcal{H}}(\mathcal{P}) = \sup \left\{ s \in [0,d] : \sup_n I_s(P_n) < \infty \right\}.$$

We shall sometimes restrict our attention to point sets of the form

$$G_n(f) = \left\{ (j/q, f(j/q)) : j \in \mathbb{Z}^{d-1} \cap [0,q)^{d-1} \right\}, \qquad n = q^{d-1}, \tag{2}$$

for some $f : [0,1]^{d-1} \to [0,1]$. We shall refer to these as point set graphs. However, whenever possible, we shall establish results for the more general point sets defined above.

Before we start stating and proving our results, we describe the practical motivation for the point sets considered above. Suppose that a company X has d branches around the country, and we wish to describe the sales total for each branch on each of n dates. The resulting point set is a 2-dimensional array

$$\{p_{k,i}\}_{1 \leq k \leq n; 1 \leq i \leq d}.$$

We can regard this point set in two equivalent ways. We can fix a branch, labeled by i, and consider the time series array

$$\{p_{1i}, p_{2i}, \ldots, p_{ni}\}. \tag{3}$$

We have d such time series, so in this way we may regard our point set as d vectors in $\mathbb{R}^n$. Alternatively, we may fix a date and consider a vector of sales figures at the different branches on a given day:

$$\{p_{k1}, p_{k2}, \ldots, p_{kd}\}. \tag{4}$$

In this way, we may regard the point set as a collection of n vectors in $\mathbb{R}^d$, as we have it set up above.

Each of the two perspectives described above has its practical advantages from the point of view of the results described below. We are going to show that if the discrete s-energy of a point set is suitably small, then the point set cannot be too concentrated on a lower dimensional hyper-plane determining a certain linear relationship between the data points. If we look at this from the point of view of the time series in (3), the linear relationship signifies the possibility that sales figures for different branches are strongly related to one another across the board in an easy-to-describe fashion. From the point of view of (4), a concentration of data on a hyper-plane would say that for a significant number of different dates, there is a simple linear relationship between the sales figures for the various branches.

Since the number of dates is likely to be considerably larger than the number of branches of a company, the analysis in each case is of a different nature, as we shall see below.

We now turn our attention to the technical results. We begin with considering n points in d-dimensional space, as we set up at the beginning of the section, but we shall describe the "flipped" perspective afterward.

Lemma 1 *Let* $\mathcal{P} = \{G_n\}$ *be a family of point set graphs as in* (2) *above. Then* $\dim_{\mathcal{H}}(\mathcal{P}) \geq d - 1$.

We want to be able to describe the support of a family $\mathcal{P} = \{P_n\}$ of point sets of a given discrete Hausdorff dimension. To do so, we introduce a little structure. Take a compact set $E \subset \mathbb{R}^d$ and a positive number $\delta \leq 1$. We let $N_E(\delta)$ denote the minimal number of closed balls of radius δ needed to cover E. Recall

$$\underline{\dim}_{\mathcal{M}}(E) = \liminf_{\delta \to 0} \frac{\log(N_E(\delta))}{\log(1/\delta)} \quad \text{and} \quad \overline{\dim}_{\mathcal{M}}(E) = \limsup_{\delta \to 0} \frac{\log(N_E(\delta))}{\log(1/\delta)}$$

are the lower and upper Minkowski dimension of E, respectively. When the lower and the upper Minkowski dimensions agree, we call their common value the Minkowski dimension of E.

We will typically assume that, for some dimension k (which need not be an integer),

$$N_E(\delta) \leq \max(C_E \delta^{-k}, 1) \qquad \text{for } \delta > 0 \tag{5}$$

for some constant C_E. Here, C_E depends on E, but also on k as well. If such a constant exists, we must have $\overline{\dim}_{\mathcal{M}}(E) \leq k$. Not only this, but E must have finite k-dimensional upper Minkowski content. The condition (5) includes a wide range of essential examples, such as planar regions, piecewise-smooth manifolds, and a range of fractal sets including all of the Cantor sets we discuss in later sections.

We are ready for our core statement about the relationship between the discrete Hausdorff dimension and the Minkowski dimension of the set on which it is supported.

Theorem 1 *Let P_n be a point set of size n contained in a compact subset $E \subset \mathbb{R}^d$ satisfying* (5) *with $k > 0$. If $s > k$, we have the lower bound*

$$I_s(P_n) \geq \frac{k}{s-k} C_E^{-\frac{s}{k}} (n^{\frac{s}{k}-1} - 1). \tag{6}$$

Consequently, if $\mathcal{P} = \{P_n\}$ is a family of point sets contained in E of upper Minkowski dimension $\overline{\dim}_{\mathcal{M}}(E) \leq k$, then $\dim_{\mathcal{H}} \mathcal{P} \leq k$.

We now turn our attention to the question of dimension reduction. Suppose that P_n is as above and $\dim_{\mathcal{H}}(\mathcal{P}) = s$. The question we ask is whether it is possible that the points of $\mathcal{P}$ concentrate on k-dimensional hyper-planes or, more generally, on smooth k-dimensional surfaces. Our next result says that the fractal dimension places significant limitations on such possibilities.

Theorem 2 *Let E be a compact subset of $\mathbb{R}^d$ satisfying* (5)*, and let P_n be a point set in $\mathbb{R}^d$ of size n. Then, for $s > k$,*

$$|P_n \cap E| \leq \left(1 + C_E^{\frac{s}{k}} \left(\frac{s}{k} - 1\right) I_s(P_n)\right)^{\frac{1}{\frac{s}{k}+1}} n^{\frac{2}{\frac{s}{k}+1}}. \tag{7}$$

Moreover, the exponent $\frac{2}{1+\frac{s}{k}}$ is, in general, best possible.

Remark 1 Since P_n is finite, so is $I_s(P_n)$. Thus, in view of Theorem 2, it is imperative to minimize the quantity

$$\left(1 + C_E^{\frac{s}{k}} \left(\frac{s}{k} - 1\right) I_s(P_n)\right)^{\frac{1}{\frac{s}{k}+1}} n^{\frac{2}{\frac{s}{k}+1}}.$$

To see the interplay between the various quantities above, we assume first that P_n has a diameter not exceeding 1. Then, observe that for a fixed n, $I_s(P_n)$ is an increasing function of s, as is

$$\left(\frac{s-k}{k}\right)^{\frac{1}{1+\frac{s}{k}}}.$$

On the other hand, $n^{\frac{2}{1+\frac{s}{k}}}$ is a decreasing function of s (with k fixed). Ultimately, an effective algorithm is needed to choose s given k and the point set P_n.

It is important, in practice, that we introduce a bit of "wiggle room" into Theorems 1 and 2, and instead consider point sets that are in some ϵ-neighborhood of E. Thankfully, both theorems have thickened versions with the same exponents. This flexibility is quite important for potential applications.

Theorem 3 *Let E be a compact subset of $\mathbb{R}^d$ satisfying* (5) *with $k > 0$. Let P_n be a point set of size n such that each point is within a distance of $\epsilon > 0$ from E. Then, if $s > k$, we have the discrete energy bound*

$$I_s(P_n) \geq \frac{k}{s-k} C_E^{-\frac{s}{k}} \left(\left(1 - \frac{\epsilon A_{s,k}}{C_E^{\frac{1}{k}} n^{-\frac{1}{k}}} \right) n^{\frac{s}{k}-1} - 1 \right)$$

with constant

$$A_{s,k} = \frac{s(s+1)(s-k)}{k(s-k+1)}.$$

In particular

$$I_s(P_n) \geq \frac{k}{s-k} C_E^{-\frac{s}{k}} \left(\frac{1}{2} n^{\frac{s}{k}-1} - 1 \right) \quad \text{if } \epsilon \leq \frac{C_E^{\frac{1}{k}}}{2A_{s,k}} n^{-\frac{1}{k}}.$$

Theorem 4 *Let P_n be a point set in $\mathbb{R}^d$ and let E be a compact subset of $\mathbb{R}^d$ satisfying* (5). *Then, if E^ϵ is the ϵ-thickening of E with*

$$\epsilon = \frac{C_E^{\frac{1}{k}}}{2A_{s,k}} n^{-\frac{1}{k}},$$

then

$$|P_n \cap E^\epsilon| \leq 2^{\frac{1}{\frac{s}{k}+1}} \left(1 + C_E^{\frac{s}{k}} \left(\frac{s}{k} - 1 \right) I_s(P_n) \right)^{\frac{1}{\frac{s}{k}+1}} n^{\frac{2}{\frac{s}{k}+1}}.$$

We will turn our attention to the case where our family of point sets lies along the graph of a function as above. To apply the results from above, we will need to connect the regularity of the function to the dimension of its graph. Recall, we say that $f : [0, 1]^{d-1} \to \mathbb{R}$ is Hölder continuous of order $\alpha \in (0, 1]$ if there exists a constant $\rho > 0$ such that

$$|f(x) - f(y)| \leq \rho |x - y|^\alpha$$

for all $x, y \in [0, 1]^{d-1}$.

Lemma 2 *Let G_n be as above, and consider the corresponding $\mathcal{P}$. Suppose that $f : [0, 1]^{d-1} \to [0, 1]$ is Hölder continuous of order $\alpha \in (0, 1]$. Then*

$$\dim_{\mathcal{H}}(\mathcal{P}) \leq d - \alpha.$$

Our last result of this section shows that every possible discrete dimension in $[d-1, d]$ is possible for the point set $G_n(f)$.

Theorem 5 *For each* $s \in [d-1, d]$ *there exists* $f : [0, 1]^{d-1} \to [0, 1]$, *such that if* $\mathcal{P} = \{G_n(f)\}$ *(with* $G_n(f)$ *defined as above), then* $dim_{\mathcal{H}}(\mathcal{P}) = s$.

2.2 Computational Results and Examples

2.2.1 Discrete Dimension of Cartesian Products of Cantor-Type Sets

Start with the interval $[0, 1]$ and positive integers $m, n \in \mathbb{Z}^+$ s.t. $m < n$. Divide $[0, 1]$ into n subintervals of equal length, and choose m of those intervals. Let $C^1_{m,n}$ be the set of endpoints of the intervals chosen. At the k-th step, split each of the remaining intervals into n equal subintervals, choose m of those intervals, and let $C^k_{m,n}$ be the set of endpoints of the intervals chosen. At each step, we pick the same m subintervals from the remaining intervals. Since the m subintervals we choose to keep are arbitrary, the set $C^k_{m,n}$ is not unique and merely denotes one set satisfying these conditions.

Theorem 6 *Let* $d \in \mathbb{Z}^+$. *Let* $C^k_{m_1,n_1}, \ldots, C^k_{m_d,n_d}$ *be discrete Cantor sets. Set* $A_k = \prod_{i=1}^{d} C^k_{m_i,n_i}$. *Then*

$$dim_{\mathcal{H}}(A_k) = \frac{\ln(m_1)}{\ln(n_1)} + \cdots + \frac{\ln(m_d)}{\ln(n_d)}.$$

Now that we have this method for computing the dimension of discrete Cantor set products, we can compute the discrete s-energy of the sets in these families to better observe their behavior. For instance, if we compute the discrete s-energy of the discrete Cantor set products from Fig. 6 below and other Cantor-type data distribution, where for each set, s is the dimension of that set computed by Theorem 6, we can see in Figs. 1, 3, 4 and 5 that the discrete s-energy of the sets in

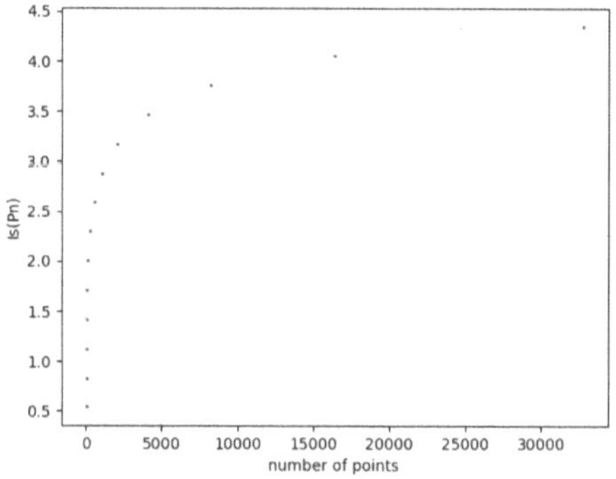

Fig. 1 Discrete s-energy of $C^k_{2,4}$ for $1 \le k \le 15$ with $s = \frac{1}{2}$

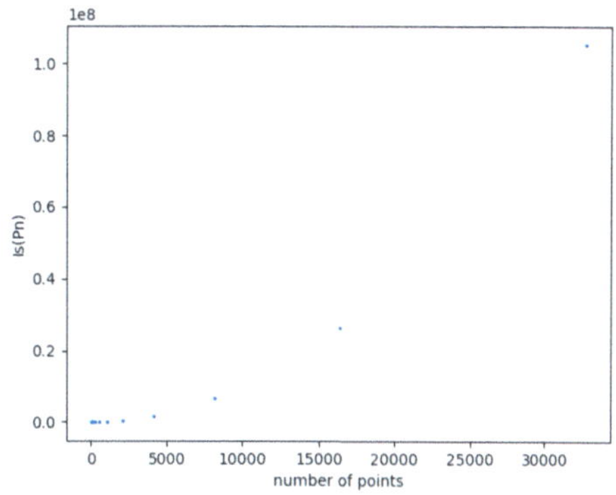

Fig. 2 Discrete s-energy of $C^k_{2,4}$ with 2^{k+1} points for $1 \leq k \leq 15$ with s = 1.5

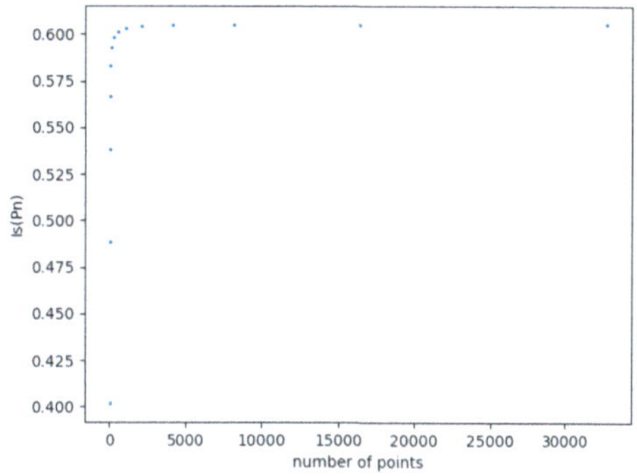

Fig. 3 Discrete s-energy of $C^k_{2,4}$ with 2^{k+1} points for $1 \leq k \leq 15$ with s = 0.1

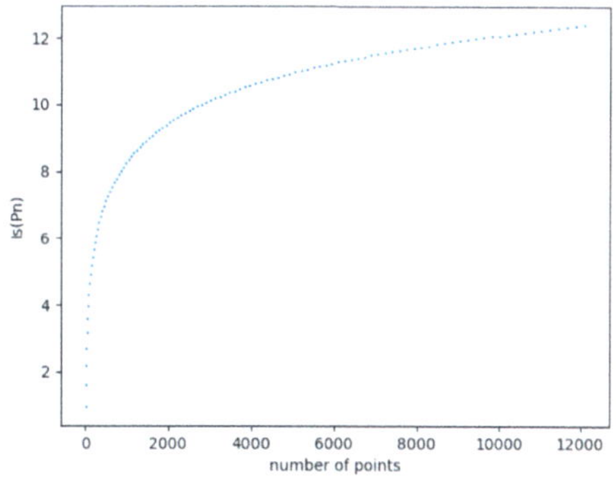

Fig. 4 Discrete s-energy of 2-dimensional lattice with $(k+1)^2$ points for $1 \leq k \leq 110$, $s = 2$

each case increases slowly. This is what we would expect given that for each family of sets, s is the critical value, and for any $s' < s$, the discrete s'-energy of the sets are all bounded by a single constant (Fig. 2).

Meanwhile, if we compute the discrete 2-energy of a family of lattices in $[0, 1]^2$, then we can see from Fig. 4 that the discrete 2-energy of the lattices increases slowly as well, possibly suggesting that the family of lattices is 2-adaptable. This would make sense intuitively because we would expect the family of lattices to be 2-dimensional,

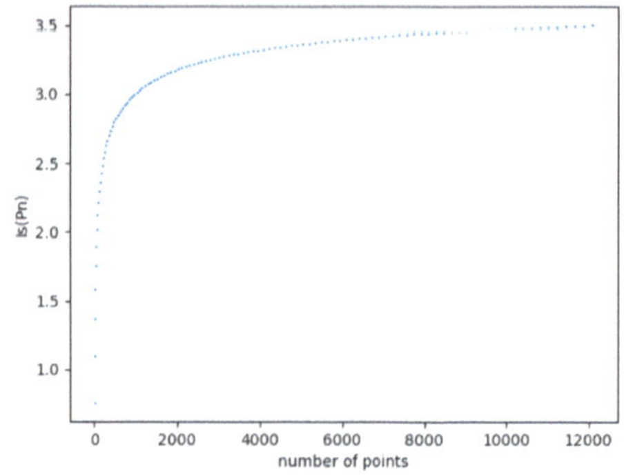

Fig. 5 Discrete s-energy of 2-dimensional lattice with $(k+1)^2$ points for $1 \leq k \leq 110, s = 1.5$

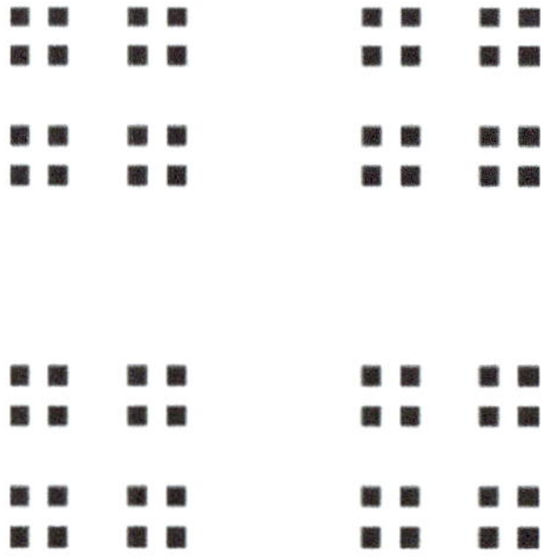

Fig. 6 The Cartesian product of two middle-half Cantor sets (otherwise known as the Garnett set)

suggesting that this new notion of dimension may align with our general notion of dimension.

2.2.2 PCA Analysis of Fractal and Non-Fractal Objects

As we denote in the introduction, the classical PCA method does not distinguish some key types of data sets. For example, it is not great at distinguishing a grid from, say, a Cartesian product of two Cantor sets, as we shall discuss below.

Recall the process of PCA: suppose there is a data set after preprocessing with the mean subtraction and standardization: $X = \{x_1, x_2, x_3, \ldots, x_n\}, x_i \in \mathbb{R}^D$. Mean subtraction will compute the mean of the dataset and subtract it from every single data point. Standardization will divide the data points by the standard deviation for every dimension [5]. To reduce set X's dimension to a lower one without losing too much information, find the data covariance matrix XX^T, calculating the m-th largest eigenvalues of that covariance matrix. Name the m eigenvectors $w_1, w_2, \ldots, w_m$ as m principle components, corresponding to the m largest eigenvalues. After generating a new matrix $W = \{w_1, w_2, w_3, \ldots.w_m\}, w_i \in R^D$, the data can be compressed to $y_i = W^T x_i$ for each x_i. The dimension of the original data set is reduced and represented as $X' = \{y_1, y_2, y_3, \ldots.y_n\}, y_i \in \mathbb{R}^M$, where $M < D$.

To illustrate some limitations of PCA, we test it with the grid and Cartesian product of two Cantor sets. The PCA analysis described above shows that the covariance matrix of the grid and Cartesian product of two Cantor sets are diagonal, and the eigenvalues for both data distributions are identical. Such results indicate that the two corresponding eigenvectors explain the same weight of the variance of the data set, meaning that there is no need to remove any components or further reduce the dimensions. Therefore, PCA regards the 2-dimensional lattice and Cartesian product of two Cantor sets as 2-dimensional sets.

Nevertheless, the discrete s-energy remains robust to identify the dimension of a cantor set is, in fact, less than 1. For instance, our codes and output figures 1-3 show that if choosing s such that $s < \frac{ln(2)}{ln(4)}$, the discrete s-energy of $C_{2,4}^k$ converges to a relatively small number as the number of points (depend on k) increases. Otherwise, discrete s-energy diverges. The results are consistent with Theorem 6.

3 Proof of Results from Sect. 2

3.1 *Proof of Lemma 1*

We will show the s-discrete energy of G_n is uniformly bounded if $0 \leq s < d-1$. As usual, we let $n = q^{d-1}$ for a positive integer q. Let $p = (j/q, f(j/q)) \in G_n$, where here $j \in \mathbb{Z}^{d-1} \cap [0, q)^{d-1}$. Note, for each such pair p, p',

$$|p - p'|^{-s} \leq |j/q - j'/q|^{-s} = q^s |j - j'|^{-s}.$$

Hence, we have

$$I_s(P_n) = n^{-2} \sum_{p \neq p'} |p - p'|^{-s} \leq q^{-2(d-1)+s} \sum_{j \neq j'} |j - j'|^{-s},$$

where here j and j' range over $\mathbb{Z}^{d-1} \cap [0, q)^{d-1}$. By a change of variables, the above is bounded by

$$q^{-(d-1)+s} \sum_{\substack{j \in \mathbb{Z}^{d-1} \cap (-q,q)^{d-1} \\ j \neq 0}} |j|^{-s} \leq Cq^{-(d-1)+s} q^{d-1-s}$$

by the integral test, where here C is a constant depending on d and s only. The conclusion of Lemma 1 follows.

3.2 Proofs of Theorems 1 and 3

We start by only assuming that E is a compact subset of $[0, 1]^d$ and P_n is any set of n points in E. We will not introduce the regularity condition (5) until a bit later.

We write

$$|p - p'|^{-s} = s \int_0^\infty \mathbf{1}_{[0,\infty)}(r - |p - p'|) r^{-s-1}\, dr,$$

so that

$$\begin{aligned} I_s(P_n) &= n^{-2} \sum_{p \neq p'} |p - p'|^{-s} \\ &= sn^{-2} \sum_{p \neq p'} \int_0^\infty \mathbf{1}_{[0,\infty)}(r - |p - p'|) r^{-s-1}\, dr \\ &= sn^{-2} \int_0^\infty \left(|\{(p, p') : |p - p'| \leq r\}| - n \right) r^{-s-1}\, dr. \end{aligned}$$

The subtracted n is here to eliminate the contribution to $|\{(p, p') : |p - p'| \leq r\}|$ of the pairs (p, p') with $p = p'$. To estimate $I_s(P_n)$, we will need a lower bound on the size of this set.

Lemma 3 *Let E be a compact subset of $\mathbb{R}^d$ and let $P_n \subset E$ be a point set in E of size n. Then,*

$$|\{(p, p') \in P_n \times P_n : |p - p'| \leq r\}| \geq \frac{n^2}{N_E(r)}.$$

Proof Take a minimal cover of E by $N_E(r)$ balls of radius r. Select a partition of E into $N_E(r)$ pairwise disjoint sets

$$E = E_1 \cup \cdots \cup E_{N_E(r)}$$

where each part E_i is contained in its respective ball in the cover. Note,

$$|\{(p, p') : |p - p'| \leq r\}| \geq \sum_j |E_j \cap P_n|^2 \geq \frac{1}{N_E(r)} \left(\sum_j |E_j \cap P_n| \right)^2 = \frac{n^2}{N_E(r)},$$

where the second inequality is an application of Cauchy-Schwarz. This concludes the proof of the lemma.

Remark 2 The Cauchy-Schwarz inequality used above is optimal when $|E_j \cap P_n|$ is constant across j, i.e. when P_n is well-distributed in E. If, say, P_n is contained in a plane E in $[0, 1]^d$, then the energy $I_s(P_n)$ is minimized (or nearly minimized) if P_n is a lattice in E.

We now resume the proof of Theorem 1. Take δ to be any number with $0 < \delta \leq 1$. Then, the lemma gives

$$|\{(p, p') : |p - p'| \leq r\}| - n \geq \begin{cases} 0 & r < \delta \\ N_E(r)^{-1}n^2 - n & r \geq \delta. \end{cases}$$

It now follows that

$$I_s(P_n) \geq sn^{-2} \int_{\delta}^{\infty} \left(N_E(r)^{-1}n^2 - n\right) r^{-s-1}\, dr.$$

After some minor calculations, it follows:

Lemma 4 *Let E be a compact subset of $[0, 1]^d$ and P_n a set of n points in E. Then,*

$$I_s(P_n) \geq s \left(\int_{\delta}^{\infty} N_E(r)^{-1} r^{-s-1}\, dr \right) - n^{-1}\delta^{-s}$$

where $0 < \delta \leq 1$.

We are now ready to prove Theorem 1.

Proof (Theorem 1) We now can assume (5) from which we have

$$N_E(r)^{-1} \geq \min(C_E^{-1} r^k, 1).$$

Hence by the lemma, we have

$$\begin{aligned} I_s(P_n) &\geq C_E^{-1} s \int_{\delta}^{C_E^{\frac{1}{k}}} r^{k-s-1}\, dr + s \int_{C_E^{\frac{1}{k}}}^{\infty} r^{-s-1}\, dr - n^{-1}\delta^{-s} \\ &= \frac{C_E^{-1} s}{k-s} \left(C_E^{1-\frac{s}{k}} - \delta^{k-s} \right) + C_E^{-\frac{s}{k}} - n^{-1}\delta^{-s}. \end{aligned}$$

Taking $\delta = (C_E^{-1} n)^{-\frac{1}{k}}$ yields the bound, after some simplification.

Theorem 3 is also now within easy reach.

Proof (Theorem 3) Let E^{ϵ} denote the ϵ-thickening of E. Now, if E satisfies (5), then

$$N_{E^{\epsilon}}(r) \leq N_E(r - \epsilon) \leq \max(C_E(r - \epsilon)^{-k}, 1).$$

By the lemma, we have

$$\begin{aligned} I_s(P_n) &\geq s \int_{\delta+\epsilon}^{\infty} (N_E(r - \epsilon))^{-1} r^{-s-1}\, dr - n^{-1}(\delta + \epsilon)^{-s} \\ &\geq s \int_{\delta}^{\infty} N_E(r)^{-1}(r + \epsilon)^{-s-1}\, dr - n^{-1}\delta^{-s}. \end{aligned}$$

Since r^{-s-1} is convex, we have

$$(r+\epsilon)^{-s-1} \geq r^{-s-1} - (s+1)\epsilon r^{-s-2} \qquad \text{for } r > 0.$$

Hence,

$$I_s(P_n) \geq \left(s \int_{\delta}^{\infty} N_E(r)^{-1} r^{-s-1}\, dr - n^{-1}\delta^{-s} \right) - s(s+1)\epsilon \int_{\delta}^{\infty} N_E(r)^{-1} r^{-s-2}\, dr.$$

Taking $\delta = C_E^{\frac{1}{k}} n^{-\frac{1}{k}}$ and repeating the argument in the proof of Theorem 1 yields

$$\begin{aligned} I_s(P_n) \geq{}& \frac{k}{s-k} C_E^{-\frac{s}{k}} (n^{\frac{s}{k}-1} - 1) \\ & -\epsilon s(s+1) \left(C_E^{-1} \int_{C_E^{\frac{1}{k}} n^{-\frac{1}{k}}}^{C_E^{\frac{1}{k}}} r^{k-s-2}\, dr + \int_{C_E^{\frac{1}{k}}}^{\infty} r^{-s-2}\, dr \right). \end{aligned}$$

By change of variables, we have

$$C_E^{-1} \int_{C_E^{\frac{1}{k}} n^{-\frac{1}{k}}}^{C_E^{\frac{1}{k}}} r^{k-s-2}\, dr = \frac{C_E^{-\frac{s+1}{k}}}{s-k+1} \left(n^{-1+\frac{s+1}{k}} - 1 \right)$$

and

$$\int_{C_E^{\frac{1}{k}}}^{\infty} r^{-s-2}\, dr = \frac{C_E^{-\frac{s+1}{k}}}{s+1}.$$

Their sum is then bounded above by $\frac{C_E^{-\frac{s+1}{k}}}{s-k+1} n^{-1+\frac{s+1}{k}}$, and hence we have the lower bound

$$I_s(P_n) \geq \frac{k}{s-k} C_E^{-\frac{s}{k}} (n^{\frac{s}{k}-1} - 1) - \frac{\epsilon}{C_E^{\frac{1}{k}} n^{-\frac{1}{k}}} \frac{C_E^{-\frac{s}{k}} s(s+1)}{s-k+1} n^{\frac{s}{k}-1}.$$

The bound in the theorem follows.

3.3 *Proofs of Theorems 2 and 4*

Proof (Theorem 2) For fixed n, let $m = |P_n \cap E|$ and $P'_m = P_n \cap E$. On one hand, we have an upper bound

$$I_s(P'_m) = m^{-2} \sum_{\substack{p \neq p' \\ p, p' \in P_{m'}}} |p - p'|^{-s} \leq m^{-2} \sum_{p \neq p';\, p, p' \in P_n} |p - p'|^{-s} = m^{-2} n^2 I_s(P_n)$$

on $I_s(P'_m)$. At the same time, since $P'_m \subset E$, (6) gives us a lower bound

$$I_s(P'_m) \geq \frac{k}{s-k} C_E^{-\frac{s}{k}} (m^{\frac{s}{k}-1} - 1).$$

By comparing the upper and lower bounds, we obtain

$$(m^{\frac{s}{k}-1} - 1)m^2 \leq C_E^{\frac{s}{k}} \left(\frac{s}{k} - 1\right) I_s(P_n) n^2.$$

The left side is bounded below by $m^{\frac{s}{k}+1} - n^2$, and hence we have

$$m^{\frac{s}{k}+1} \leq \left(1 + C_E^{\frac{s}{k}} \left(\frac{s}{k} - 1\right) I_s(P_n)\right) n^2.$$

The bound in the theorem follows.

Theorem 2 is, in general, best possible. To see this, we start with a family $\mathcal{P}$ of lattices

$$P_n = \left\{ \frac{j}{n^{\frac{1}{d}}} : j \in \mathbb{Z}^d, 0 \leq j_i \leq n^{\frac{1}{d}}, 1 \leq i \leq d \right\}.$$

It is not difficult to check that $I_s(P_n) \leq C$ independently of n if $s < d$, from which we have $\dim_{\mathcal{H}} \mathcal{P} = d$. Now let us modify P_n by replacing the $n^{\frac{k}{d}}$ points in the plane $x_1 = \cdots = x_k = 0$ with a lattice $Z := m^{-1/k}\mathbb{Z}^k \cap [0, m^{1/k})^k$ of m points. We will take $m \gg n^{\frac{k}{d}}$, which effectively increases the number of points on the plane. However, we will not add so many as to disturb the bounds on the s-energy of $\mathcal{P}$. To this end, we require that if $s < k$, there is a constant C for which

$$I_s(P_n) \leq C \qquad \text{for each } n.$$

The left side reads as a bounded term plus

$$n^{-2} \sum_{\substack{u,u' \in Z \\ u \neq u'}} \left| \frac{u - u'}{m^{\frac{1}{k}}} \right|^{-s} \approx n^{-2} m^{\frac{s}{k}+1} \sum_{w \in Z} |w|^{-s} \approx n^{-2} m^{\frac{s}{k}+1},$$

and hence the desired bounds hold if and only if $m^{\frac{s}{k}+1} \leq Cn^2$, i.e. $m \leq Cn^{\frac{2}{1+\frac{s}{k}}}$. Here and throughout, $X \approx Y$ means that there exists a uniform constant C such that $C^{-1}Y \leq X \leq CY$.

Next, we proceed with the proof of Theorem 4, the thickened version of Theorem 2.

Proof (Theorem 4) We follow the proof of Theorem 2 above, except we take $P'_m = P_n \cap E^\epsilon$ where here

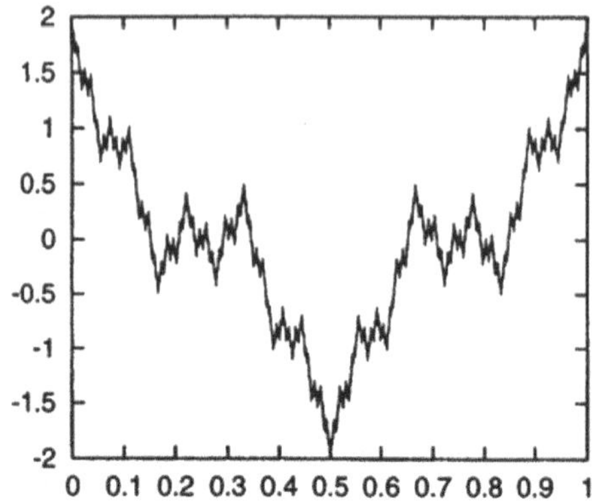

Fig. 7 Graph of $f(x)$ with $a = 0.5$ and $b = 3$ and $\theta_n = 0$ for any n [17]

$$\epsilon = \frac{C_E^{\frac{1}{k}}}{2A_{s,k}} n^{-\frac{1}{k}} \leq \frac{C_E^{\frac{1}{k}}}{2A_{s,k}} m^{-\frac{1}{k}}$$

with notation as in Theorem 3. The same theorem then tells us

$$I_s(P'_m) \geq \frac{k}{s-k} C_E^{-\frac{s}{k}} \left(\frac{1}{2} m^{\frac{s}{k}-1} - 1 \right).$$

We similarly have an upper bound

$$I_s(P'_m) \leq m^{-2} n^2 I_s(P_n).$$

Proceeding as before, we obtain

$$\frac{1}{2} m^{\frac{s}{k}+1} - n^2 \leq \frac{1}{2} m^{\frac{s}{k}+1} - m^2 \leq C_E^{\frac{s}{k}} \left(\frac{s}{k} - 1 \right) I_s(P_n) n^2.$$

It follows

$$m \leq 2^{\frac{1}{\frac{s}{k}+1}} \left(1 + C_E^{\frac{s}{k}} \left(\frac{s}{k} - 1 \right) I_s(P_n) \right)^{\frac{1}{\frac{s}{k}+1}} n^{\frac{2}{\frac{s}{k}+1}}$$

as needed (Fig. 7).

3.4 *Proof of Lemma 2*

We claim the graph of f satisfies (5) for $k = d - \alpha$; afterward, Theorem 1 completes the proof. We do this instead by covering the graph of f with cubes of length δ. Passing from cubes to balls will only affect the constant in (5).

For fixed q, let Q_j denote the cube $\prod_{i=1}^{d-1}[\frac{j_i}{q}, \frac{j_i+1}{q}]$ where here j ranges over $\mathbb{Z}^{d-1} \cap [0, q)^{d-1}$. Since f is Hölder continuous of order α, $f(Q_j)$ has diameter at most $\rho(\sqrt{d-1}q^{-1})^{\alpha}$. Hence, we can cover the graph of f over Q_j by $\rho\sqrt{d-1}q^{1-\alpha}$ d-dimensional cubes of side length q^{-1}. Repeating for each of the q^{d-1} cubes Q_j, we cover the whole graph of f by $q^{d-1} \cdot \rho\sqrt{d-1}q^{1-\alpha} = \rho\sqrt{d-1}q^{d-\alpha}$ cubes of length q^{-1}, and our claim is proved after taking $\delta = q^{-1}$.

4 Proof of Theorem 5

4.1 2-Dimensional Case

Proof In order to prove the case for $d = 2$, the construction used by B.Hunt is followed [17]. In such construction, we consider Weierstrass functions $f_\theta : [0, 1] \to [0, 1]$ with random phases of the form:

$$f_\theta(x) = \sum_{n=0}^{\infty} a^n cos\left(2\pi(b^n x + \theta_n)\right) \tag{8}$$

where $0 < a < 1 < b$, $ab < 1$, and $\theta = (\theta_1, \theta_2, \dots) \in [0, 1]^\infty = H$ is randomly chosen by sampling each of its entries according to the uniform distribution on $[0, 1]$.

Note that, for any $\theta \in H$, f_θ is an Holder continuous function with exponent $\alpha = -\frac{\log(a)}{\log(b)}$. In fact, for any $x, y \in [0, 1]$, substituting $a = b^{-\alpha}$:

$$\begin{aligned}
|f_\theta(x) - f_\theta(y)| &= \left|\sum_{n=0}^{\infty} b^{-n\alpha}\cos(2\pi(b^n x + \theta_n)) - \sum_{n=0}^{\infty} b^{-n\alpha}\cos(2\pi(b^n y + \theta_n))\right| \\
&\le \sum_{n=0}^{\infty} b^{-n\alpha}\min(2, 2\pi b^n |x-y|) \\
&\le \sum_{n=0}^{m-1} 2\pi b^{n(1-\alpha)}|x-y| + \sum_{n=m}^{\infty} 2b^{-n\alpha} \\
&= 2\pi\frac{b^{(1-\alpha)m} - 1}{b^{(1-\alpha)} - 1}|x-y| + 2\frac{b^{-\alpha m}}{1-b^{-\alpha}}
\end{aligned}$$

for any integer $m > 0$. The first inequality followed, thanks to the fact that the cosine function is Lipschitz (an immediate consequence of the mean value theorem). Set m to be the positive integer such that $b^{-m} < |x-y| \le b^{-(m-1)}$. Then, resuming the estimate:

$$
\begin{aligned}
|f_\theta(x) - f_\theta(y)| &\leq 2\pi \frac{\left(\frac{b}{|x-y|}\right)^{(1-\alpha)}}{b^{(1-\alpha)} - 1} |x - y| + 2\frac{|x - y|^\alpha}{1 - b^{-\alpha}} \\
&\leq \left(\frac{2\pi b^{(1-\alpha)}}{b^{(1-\alpha)} - 1} + \frac{2}{1 - b^{-\alpha}} \right) |x - y|^\alpha .
\end{aligned}
\tag{9}
$$

Thus, combining (9) and Lemma 2, we have that $dim_{\mathcal{H}}(\mathcal{P}) \geq 2 - \alpha$. To prove the other inequality, the random phases are very useful as they allow us to estimate the energy integral indirectly. In fact, the key point of the proof is to show that, for any $s < 2 - \alpha$, $\mathbb{E}_H(I_s(G_n)) < \infty$ independently of n (i.e., independently of q). Consequently, this implies that for almost every random sequence of phases $\theta \in H$, f_θ satisfies $dim_{\mathcal{H}}(\mathcal{P}) \leq 2 - \alpha$ (completing the proof for $d = 2$). To perform the estimate by linearity of expectation:

$$
\mathbb{E}_H(I_s(G_n)) = q^{-2} \sum_{\substack{j \neq j' \\ j, j' \in [0,q) \cap \mathbb{Z}}} \mathbb{E}_H \left(\left| \left(\frac{j}{q}, f_\theta \left(\frac{j}{q} \right) \right) - \left(\frac{j'}{q}, f_\theta \left(\frac{j'}{q} \right) \right) \right|^{-s} \right).
$$

Next, the following lemma (proved by B. Hunt [17]) provides an upper bound for the expectations inside the sum when $\frac{j}{q}$ and $\frac{j}{q}$ are close enough.

Proposition 1 *For any x, y and any such that $|x - y| < 1/2b^2$:*

$$
\mathbb{E}_H \left(|(x, f_\theta(x)) - (y, f_\theta(y))|^{-s} \right) \leq C\, |x - y|^{(1-\alpha-s)} \tag{10}
$$

for any $s \in (1, 2 - \alpha)$

Then, resuming the proof, we can decompose the sum into two terms:

$$
\mathbb{E}_H(I_s(G_n)) = I + II,
$$

where

$$
I = q^{-2} \sum_{\substack{j \neq j' \\ j, j' \in [0,q) \cap \mathbb{Z} \\ |j - j'| \leq \frac{q}{2b^2}}} \mathbb{E}_H \left(\left| \left(\frac{j}{q}, f_\theta \left(\frac{j}{q} \right) \right) - \left(\frac{j'}{q}, f_\theta \left(\frac{j'}{q} \right) \right) \right|^{-s} \right),
$$

$$
II = q^{-2} \sum_{\substack{j \neq j' \\ j, j' \in [0,q) \cap \mathbb{Z} \\ |j - j'| > \frac{q}{2b^2}}} \mathbb{E}_H \left(\left| \left(\frac{j}{q}, f_\theta \left(\frac{j}{q} \right) \right) - \left(\frac{j'}{q}, f_\theta \left(\frac{j'}{q} \right) \right) \right|^{-s} \right).
$$

The second term is easy to be bounded as:

$$II \leq q^{-2} \sum_{\substack{j \neq j' \\ j,j' \in [0,q) \cap \mathbb{Z} \\ |j-j'| > \frac{q}{2b^2}}} \left| \frac{j}{q} - \frac{j'}{q} \right|^{-s} \leq q^{-2} q^2 (2b^2)^s = (2b^2)^s.$$

For the first term, using proposition (1):

$$I \lesssim q^{-2} \sum_{\substack{j \neq j' \\ j,j' \in [0,q) \cap \mathbb{Z} \\ |j-j'| \leq \frac{q}{2b^2}}} \left| \frac{j}{q} - \frac{j'}{q} \right|^{1-\alpha-s} \lesssim q^{-3+\alpha+s} \sum_{\substack{j \neq j' \\ j,j' \in [0,q) \cap \mathbb{Z}}} |j - j'|^{1-\alpha-s}$$

$$\lesssim q^{-2+\alpha+s} \sum_{\substack{j \in (-q,q) \cap \mathbb{Z} \\ j \neq 0}} |j|^{1-\alpha-s} \lesssim q^{-2+\alpha+s} q^{2-\alpha-s} \lesssim 1.$$

completing the proof. Note that the third and fourth inequality are, respectively, due to the change of variable $j - j'$ to j and due to the integral test (where the integral converges for values of s strictly smaller than $2 - \alpha$, as desired). The 2-dimensional case is fully proved since for any given value $x \in (0, 1]$, we can find values of a and b such that $\alpha = x$. This is a consequence of the fact that $\alpha = -\frac{\log(a)}{\log b}$ and $0 < a < 1 < b$ with $ab \geq 1$. In fact, fixing a positive $a < 1$, $b = \frac{1}{a^x} \geq \frac{1}{a}$ for any value of $x \in (0, 1]$ and $\alpha = x$.

4.2 Higher Dimensional Case for $d \geq 3$

Proof To generalize to higher dimensions (i.e. $d \geq 3$), we consider the following functions $g_\theta : [0, 1]^{d-1} \to [0, 1]$ of the form:

$$g_\theta(x) = g_\theta(x_1, \ldots, x_{d-1}) = f_\theta(x_1) \tag{11}$$

where the functions f_θ are defined as in (8).

For any $\theta \in H$, g_θ is again Holder continuous with exponent α. This is because, using (9) and (11), for any $x, y \in [0, 1]^{d-1}$:

$$\begin{aligned} |g_\theta(x) - g_\theta(y)| = |f_\theta(x_1) - f_\theta(y_1)| &\leq \left(\frac{2\pi b^{(1-\alpha)}}{b^{(1-\alpha)} - 1} + \frac{2}{1 - b^{-\alpha}} \right) |x_1 - y_1|^\alpha \\ &\leq \left(\frac{2\pi b^{(1-\alpha)}}{b^{(1-\alpha)} - 1} + \frac{2}{1 - b^{-\alpha}} \right) |x - y|^\alpha . \end{aligned} \tag{12}$$

Thus, combining (12) and Lemma 2, $dim_{\mathcal{H}}(\mathcal{P}) \geq d - \alpha$.

To prove the other inequality, looking at the expectation over all random phases as in the case $d = 2$, we can perform the following decomposition:

$$\begin{aligned}\mathbb{E}_H(I_s(G_n)) &= q^{-2(d-1)}\,\mathbb{E}_H\left(\sum_{\substack{j\neq j'\\ j,j'\in[0,q)^{d-1}\cap\mathbb{Z}^{d-1}}}\left|\left(\frac{j}{q}, g_\theta\left(\frac{j}{q}\right)\right)-\left(\frac{j'}{q}, g_\theta\left(\frac{j'}{q}\right)\right)\right|^{-s}\right)\\ &= q^{-2(d-1)}\,\mathbb{E}_H\left(\sum_{\substack{j\neq j'\\ j,j'\in[0,q)^{d-1}\cap\mathbb{Z}^{d-1}}}\left|\left(\frac{j}{q}, f_\theta\left(\frac{j_1}{q}\right)\right)-\left(\frac{j'}{q}, f_\theta\left(\frac{j_1'}{q}\right)\right)\right|^{-s}\right)\\ &= I + II\end{aligned}$$

where

$$I = q^{-2(d-1)}\sum_{\substack{j_1\neq j_1'\\ j_1,j_1'\in[0,q)\cap\mathbb{Z}}}\mathbb{E}_H\left(\sum_{\bar{j},\bar{j}'\in[0,q)^{d-2}\cap\mathbb{Z}^{d-2}}\left(\left|\frac{j_1}{q}-\frac{j_1'}{q}\right|^2+\left|f_\theta\left(\frac{j_1}{q}\right)-f_\theta\left(\frac{j_1'}{q}\right)\right|^2+\left|\frac{\bar{j}}{q}-\frac{\bar{j}'}{q}\right|^2\right)^{-\frac{s}{2}}\right)$$

$$II = q^{-2(d-1)}\sum_{\substack{j_1= j_1'\\ j_1,j_1'\in[0,q)\cap\mathbb{Z}}}\mathbb{E}_H\left(\sum_{\substack{\bar{j}\neq\bar{j}'\\ \bar{j},\bar{j}'\in[0,q)^{d-2}\cap\mathbb{Z}^{d-2}}}\left(\left|\frac{j_1}{q}-\frac{j_1'}{q}\right|^2+\left|f_\theta\left(\frac{j_1}{q}\right)-f_\theta\left(\frac{j_1'}{q}\right)\right|^2+\left|\frac{\bar{j}}{q}-\frac{\bar{j}'}{q}\right|^2\right)^{-\frac{s}{2}}\right)$$

where $\bar{j} = (j_2, ..., j_{d-1})$. We just need to worry about bounding term I since it is greater than term II. To see this, note that any addend in the inner sum of II is comparable to an addend of the inner sum of II but, at the same time, the outer sum of I contains way more terms than the outer sum of II (due to the difference between the conditions $j_1 = j_1'$ and $j_1 \neq j_1'$). To bound the first term, we can further decompose it into two pieces $I = III + IV$, where

$$III = q^{-2(d-1)}\sum_{\substack{j_1\neq j_1'\\ j_1,j_1'\in[0,q)\cap\mathbb{Z}\\ |j_1-j_1'|\leq\frac{q}{2b^2}}}\mathbb{E}_H\left(\sum_{\bar{j},\bar{j}'\in[0,q)^{d-2}\cap\mathbb{Z}^{d-2}}\left(\left|\frac{j_1}{q}-\frac{j_1'}{q}\right|^2+\left|f_\theta\left(\frac{j_1}{q}\right)-f_\theta\left(\frac{j_1'}{q}\right)\right|^2+\left|\frac{\bar{j}}{q}-\frac{\bar{j}'}{q}\right|^2\right)^{-\frac{s}{2}}\right)$$

$$IV = q^{-2(d-1)}\sum_{\substack{j_1\neq j_1'\\ j_1,j_1'\in[0,q)\cap\mathbb{Z}\\ |j_1-j_1'|>\frac{q}{2b^2}}}\mathbb{E}_H\left(\sum_{\bar{j},\bar{j}'\in[0,q)^{d-2}\cap\mathbb{Z}^{d-2}}\left(\left|\frac{j_1}{q}-\frac{j_1'}{q}\right|^2+\left|f_\theta\left(\frac{j_1}{q}\right)-f_\theta\left(\frac{j_1'}{q}\right)\right|^2+\left|\frac{\bar{j}}{q}-\frac{\bar{j}'}{q}\right|^2\right)^{-\frac{s}{2}}\right)$$

The last term is bounded as follows:

$$
\begin{aligned}
IV &\le q^{-2(d-1)} \sum_{\substack{j_1 \neq j_1' \\ j_1, j_1' \in [0,q)\cap\mathbb{Z} \\ |j_1 - j_1'| > \frac{q}{2b^2}}} \sum_{\bar{j}, \bar{j}' \in [0,q)^{d-2}\cap\mathbb{Z}^{d-2}} \left| \frac{j_1}{q} - \frac{j_1'}{q} \right|^{-s} \\
&\le q^{-2(d-1)} q^2 q^{2(d-2)} (2b^2)^s = (2b^2)^s .
\end{aligned}
$$

In the inner sum of III, the quantity $\left|\frac{j_1}{q} - \frac{j_1'}{q}\right|^2 + \left|f_\theta(\frac{j_1}{q}) - f_\theta(\frac{j_1'}{q})\right|^2$ is constant (i.e. does not depend on $\bar{j}, \bar{j}'$). For the sake of clarity, set $\beta^2 = \left|\frac{j_1}{q} - \frac{j_1'}{q}\right|^2 + \left|f_\theta(\frac{j_1}{q}) - f_\theta(\frac{j_1'}{q})\right|^2$. We can then find an upper bound for such inner sum:

$$
\begin{aligned}
&\sum_{\bar{j}, \bar{j}' \in [0,q)^{d-2}\cap\mathbb{Z}^{d-2}} \left(\beta^2 + \left| \frac{\bar{j}}{q} - \frac{\bar{j}'}{q} \right|^2 \right)^{-\frac{s}{2}} \\
&\qquad = q^{d-2}\beta^{-s} + \sum_{\substack{\bar{j} \neq \bar{j}' \\ \bar{j}, \bar{j}' \in [0,q)^{d-2}\cap\mathbb{Z}^{d-2}}} \left(\beta^2 + \left| \frac{\bar{j}}{q} - \frac{\bar{j}'}{q} \right|^2 \right)^{-\frac{s}{2}} \\
&\qquad \le q^{d-2}\beta^{-s} + q^{d-2+s} \sum_{\substack{\bar{j} \neq \bar{j}' \\ \bar{j} \in (-q,q)^{d-2}\cap\mathbb{Z}^{d-2}}} \left((q\beta)^2 + |\bar{j}|^2 \right)^{-\frac{s}{2}} \qquad (13) \\
&\qquad \lesssim q^{d-2}\beta^{-s} + q^{d-2+s} \int_{x \in [0,q)^{d-2}} \left((q\beta)^2 + |x|^2 \right)^{-\frac{s}{2}} dx \\
&\qquad \lesssim q^{d-2}\beta^{-s} + q^{d-2+s} (q\beta)^{d-2-s} \int_{x \in \mathbb{R}^{d-2}} \left(1 + |x|^2 \right)^{-\frac{s}{2}} dx \\
&\qquad \lesssim q^{d-2}\beta^{-s} + q^{d-2+s} (q\beta)^{d-2-s} \\
&\qquad \lesssim q^{2(d-2)} \beta^{d-2-s} .
\end{aligned}
$$

In evaluating the integral, the change of variable x to $(q\beta)x$ is performed. Note that the above integral is convergent for any value $s > d-2$ (which is fine since we consider s such that $d-1 < s < d-\alpha$). The first inequality follows by the change of variable $\bar{j} - \bar{j}'$ to $\bar{j}$ and the second one by integral test. Additionally, since we consider $j_1 \neq j_1'$ and then $\beta^2 \ge q^{-2}$, the last inequality follows because

$$
\begin{aligned}
q^{d-2+s}(q\beta)^{d-2-s} &= \beta^{-s} q^{d-2+s} (q)^{d-2-s} (\beta)^{d-2} \\
&= \beta^{-s} q^{2(d-2)} (\beta)^{d-2} \ge \beta^{-s} q^{2(d-2)} (q^{-1})^{d-2} \\
&= \beta^{-s} q^{(d-2)}
\end{aligned}
$$

for $d \ge 2$.

Consequently, using (13) and by substituting back for β:

$$III \lesssim q^{-2(d-1)} q^{2(d-2)} \sum_{\substack{j_1 \neq j_1' \\ j_1, j_1' \in [0,q) \cap \mathbb{Z} \\ |j_1 - j_1'| \leq \frac{q}{2b^2}}} \mathbb{E}_H \left(\left(\left| \frac{j_1}{q} - \frac{j_1'}{q} \right|^2 + \left| f_\theta\left(\frac{j_1}{q}\right) - f_\theta\left(\frac{j_1'}{q}\right) \right|^2 \right)^{\frac{d-2-s}{2}} \right)$$

$$= q^{-2} \sum_{\substack{j_1 \neq j_1' \\ j_1, j_1' \in [0,q) \cap \mathbb{Z} \\ |j_1 - j_1'| \leq \frac{q}{2b^2}}} \mathbb{E}_H \left(\left| \left(\frac{j_1}{q}, f_\theta\left(\frac{j_1}{q}\right) \right) - \left(\frac{j_1'}{q}, f_\theta\left(\frac{j_1'}{q}\right) \right) \right|^{d-2-s} \right)$$

$$\lesssim q^{-2} q^{-1+\alpha-d+2+s} \sum_{\substack{j_1 \neq j_1' \\ j_1, j_1' \in [0,q) \cap \mathbb{Z}}} |j_1 - j_1'|^{1-\alpha+d-2-s}$$

$$\lesssim q^{-1+\alpha-d+s} q \sum_{j_1 \in [0,q) \cap \mathbb{Z}} |j_1|^{\alpha+d-1-s} \lesssim q^{-d+\alpha+s} q^{d-\alpha-s} \lesssim 1$$

completing the proof also for the case $d \geq 3$. The second inequality follows thanks to Proposition 1, the third one by the change of variable $j_1 - j_1'$ to j_1 and the fourth one by integral test (where the integral converges for values s strictly smaller than $d - \alpha$, as desired). Note that we could use proposition 1 successfully since $1 = -d + (d-1) + 2 < -d + s + 2 < -d + (d - \alpha) + 2 = 2 - \alpha$. Again, the proof is complete since for any given value $x \in (0, 1]$, we can find values of a and b such that $\alpha = x$. This is a consequence of the fact that $\alpha = -\frac{\log(a)}{\log b}$ and $0 < a < 1 < b$ with $ab \geq 1$. In fact, fixing a positive $a < 1$, $b = \frac{1}{a^x} \geq \frac{1}{a}$ for any value of $x \in (0, 1]$ and $\alpha = x$.

4.3 Proof of Theorem 6

We will proceed by approximating the sum $|A_k|^{-2} \sum_{a \neq a'} |a - a'|^{-s}$ by using sets centered at each $a' \in A_k$ such that for each point a in a given set, $|a_i - a_i'| \approx n_i^{-j_i}$ for all i, where $j_1, \ldots, j_d$ all range from 1 to k. In this case, $|a_i - a_i'| \approx n_i{}^{-j_i}$ means that $n_i{}^{-j_i-1} < |a_i - a_i'| \leq n_i{}^{-j_i}$. We do not need to consider side lengths with dimensions $n_i^{-j_i}$ for $j_i > k$ because by construction, the i-th discrete Cantor set $C^k_{m_i,n_i}$ does not have any distinct points whose distance is less than n_i^{-k}. Approximating the sum this way gives us

$$|A_k|^{-2} \sum_{a \neq a'} |a - a'|^{-s} \approx |A_k|^{-2} \sum_{j_1, \ldots, j_d = 1}^{k} \sum_{\substack{a \neq a' \\ |a_1 - a_1'| \approx n_1^{-j_1} \\ \vdots \\ |a_d - a_d'| \approx n_d^{-j_d}}} |a - a'|^{-s}. \tag{14}$$

Next, note that for any $a \neq a'$,

$$|a - a'|^{-s} \approx (|a_1 - a_1'| + \cdots + |a_d - a_d'|)^{-s} \leq \max_{1 \leq i \leq n} \{|a_i - a_i'|\}^{-s}. \tag{15}$$

Putting (14) and (15) together, we get

$$\begin{aligned} &|A_k|^{-2} \sum_{j_1, \ldots, j_d = 1}^{k} \sum_{\substack{a \neq a' \\ |a_1 - a_1'| \approx n_1^{-j_1} \\ \vdots \\ |a_d - a_d'| \approx n_d^{-j_d}}} |a - a'|^{-s} \\ &\approx |A_k|^{-2} \sum_{j_1, \ldots, j_d = 1}^{k} \sum_{\substack{a \neq a' \\ |a_1 - a_1'| \approx n_1^{-j_1} \\ \vdots \\ |a_d - a_d'| \approx n_d^{-j_d}}} \max_{1 \leq i \leq n} \{|a_i - a_i'|\}^{-s} \\ &= |A_k|^{-2} \sum_{j_1, \ldots, j_d = 1}^{k} \sum_{\substack{a \neq a' \\ |a_1 - a_1'| \approx n_1^{-j_1} \\ \vdots \\ |a_d - a_d'| \approx n_d^{-j_d}}} \max_{1 \leq i \leq n} \{n_i^{-j_i}\}^{-s}. \end{aligned} \tag{16}$$

Temporarily fix $a' \in A_k$. For any i, the number of elements $a_i \in C^k_{m_i, n_i}$ such that $|a_i - a_i'| \approx n_i^{-j_i}$ is approximately $\frac{|C^k_{m_i, n_i}|}{m_i^{j_i}}$. Therefore, it follows that the number of $a \in A_k$ such that $|a_i - a_i'| \approx n_i^{-j_i}$ for all i is

$$\frac{|C^k_{m_1, n_1}|}{m_1^{j_1}} \cdots \frac{|C^k_{m_d, n_d}|}{m_d^{j_d}} = \frac{|A_k|}{m_1^{j_1} \cdots m_d^{j_d}}.$$

Finally, since there are $|A_k|$ possible choices for a', we can see that

$$\begin{aligned}
&|A_k|^{-2} \sum_{j_1,\dots,j_d=1}^{k} \sum_{\substack{a \neq a' \\ +|a_1-a_1'| \approx n_1^{-j_1} \\ \vdots \\ |a_d-a_d'| \approx n_d^{-j_d}}} \max_{1\leq i\leq n} \{n_i^{-j_i}\}^{-s} \\
&= |A_k|^{-2} \sum_{j_1,\dots,j_d=1}^{k} \max_{1\leq i\leq n} \{n_i^{-j_i}\}^{-s} \sum_{\substack{a \neq a' \\ +|a_1-a_1'| \approx n_1^{-j_1} \\ \vdots \\ |a_d-a_d'| \approx n_d^{-j_d}}} 1 \qquad (17) \\
&\approx |A_k|^{-2}|A_k||A_k| \sum_{j_1,\dots,j_d=1}^{k} m_1^{-j_1} \cdots m_d^{-j_d} \max_{1\leq i\leq n} \{n_i^{-j_i}\}^{-s} \\
&= \sum_{j_1,\dots,j_d=1}^{k} m_1^{-j_1} \cdots m_d^{-j_d} \max_{1\leq i\leq n} \{n_i^{-j_i}\}^{-s}
\end{aligned}$$

Now, consider the case where $\max_{1\leq i\leq n}\{n_i^{-j_i}\} = n_1^{-j_1}$. This would mean that for all i,

$$n_1^{-j_1} \geq n_i^{-j_i} \implies n_1^{j_1} \leq n_i^{j_i} \implies j_1 \frac{\ln(n_1)}{\ln(n_i)} \leq j_i.$$

Note that the inequalities in (17) also hold for $n_i^{-j_i}$ when it is maximal. Therefore, we can further split the inner sum in (16) by considering the different cases in which each $n_i^{-j_i}$ is maximal to get

$$\begin{aligned}
\sum_{j_1,\dots,j_d=1}^{k} m_1^{-j_1} \cdots m_d^{-j_d} \max_{1\leq i\leq n} \{n_i^{-j_i}\}^{-s} &= \sum_{j_1 \frac{\ln(n_1)}{\ln(n_i)} \leq j_i \leq k} m_1^{-j_1} \cdots m_d^{-j_d} n_1^{j_1 s} + \cdots \\
&+ \sum_{j_d \frac{\ln(n_d)}{\ln(n_i)} \leq j_i \leq k} m_1^{-j_1} \cdots m_d^{-j_d} n_d^{j_1 s}.
\end{aligned}$$

We restrict our focus to the first term of the above sum in order to determine for which s the sum converges. For $2 \leq i \leq n$, the term $n_i^{-j_i}$ in the sum is a geometric series starting at $j_1 \frac{\ln(n_1)}{\ln(n_i)}$, so we can approximate each of these terms by $m_i^{j_1 \frac{\ln(n_1)}{\ln(n_i)}}$. This gives us that

$$\sum_{j_1 \frac{\ln(n_1)}{\ln(n_i)} \leq j_i \leq k} m_1^{-j_1} \cdots m_d^{-j_d} n_1^{j_1 s} \approx \sum_{j_1=1}^{n} m_1^{-j_1} \cdots m_d^{-j_1 \frac{\ln(n_1)}{\ln(n_d)}} n_1^{j_1 s} = \sum_{j_1=1}^{n} \left(m_1^{-1} \cdots m_d^{-\frac{\ln(n_1)}{\ln(n_d)}} n_1^s \right)^{j_1}.$$

Now we have a geometric series, so we know that for this series to converge as $n \to \infty$, we must have that

$$m_1^{-1} \cdots m_d^{-\frac{\ln(n_1)}{\ln(n_d)}} n_1^s < 1 \iff -\left(\ln(m_1) + \ln(m_2)\frac{\ln(n_1)}{\ln(n_2)} + \cdots + \ln(m_d)\frac{\ln(n_1)}{\ln(n_d)} \right) + s\ln(n_1) < 0$$

$$\iff s < \frac{\ln(m_1)}{\ln(n_1)} + \cdots + \frac{\ln(m_d)}{\ln(n_d)}.$$

We can use an identical argument to show that all of the other sums

$$\sum_{j_l \frac{\ln(n_l)}{\ln(n_i)} \leq j_i \leq k} m_1^{-j_1} \cdots m_d^{-j_d} n_l^{j_l s}$$

also converge exactly when

$$s < \frac{\ln(m_1)}{\ln(n_1)} + \cdots + \frac{\ln(m_d)}{\ln(n_d)}.$$

Furthermore, since

$$I_s(A_k) = |A_k|^{-2} \sum_{a \neq a'} |a - a'|^{-s} \approx \sum_{j_1 \frac{\ln(n_1)}{\ln(n_i)} \leq j_i \leq k} m_1^{-j_1} \cdots m_d^{-j_d} n_1^{j_1 s} + \cdots + \sum_{j_d \frac{\ln(n_d)}{\ln(n_i)} \leq j_i \leq k} m_1^{-j_1} \cdots m_d^{-j_d} n_d^{j_1 s},$$

this implies that $\{I_s(A_k)\}_{k \geq 1}$ converges exactly when

$$s < \frac{\ln(m_1)}{\ln(n_1)} + \cdots + \frac{\ln(m_d)}{\ln(n_d)}.$$

Hence, $\{A_k\}_{k \geq 1}$ is s-adaptable for exactly these values of s, so by definition, we may conclude that

$$s_{critical} = \frac{\ln(m_1)}{\ln(n_1)} + \cdots + \frac{\ln(m_d)}{\ln(n_d)}$$

as desired.

With this result, if we revisit the discrete Cantor set products from before, then we get that the dimension of $\{C_{2,3}^k \times C_{2,3}^k\}_{k \geq 1}$ is $\frac{\ln(2)}{\ln(3)} + \frac{\ln(2)}{\ln(3)} = 2\frac{\ln(2)}{\ln(3)}$, the dimension of $\{C_{2,4}^k \times C_{2,4}^k\}_{k \geq 1}$ is $\frac{\ln(2)}{\ln(4)} + \frac{\ln(2)}{\ln(4)} = 1$ and the dimension of $\{C_{2,4}^k \times C_{2,3}^k \times C_{2,3}^k\}_{k \geq 1}$ is $\frac{\ln(2)}{\ln(4)} + \frac{\ln(2)}{\ln(3)} + \frac{\ln(2)}{\ln(3)} = \frac{1}{2} + 2\frac{\ln(2)}{\ln(3)}$, which makes sense given how sparse these sets are. This demonstrates that our alternate notion of dimension is better at capturing the dimensionality of discrete Cantor set products than PCA.

References

1. Alexander Bain, *Mind and Body: The Theories of Their Relation*, New York: D. Appleton and Company (1873).
2. E. Alos and J. Lion, *An Intuitive Introduction to Fractional and Rough Volatilities*, MDPI, (2021).
3. A. Borodachov, D. Hardin, and E. Saff, *Discrete energy on rectifiable sets*, Springer Monographs in Mathematics. Springer, New York, (2019).
4. H. Cohn and N. Elkies, *New upper bounds on sphere packings* I. Ann. of Math. (2) **157** (2003), no. 2, 689-714.
5. M. Deisenroth, A. Faisal, and C. Ong, *Mathematics for machine learning*, Cambridge University Press, (2020).
6. G. Folland, *Fourier Analysis and Its Applications*, Pure and Applied undergraduate text, American Mathematical Society 1992.
7. S. Shalev-Shwartz and S. Ben-David, *Understanding Machine Learning: From Theory to Algorithms*, Cambridge University Press, (2014).
8. L. Guth, A. Iosevich, Y. Ou and H. Wang, *On Falconer's distance problem in the plane*, arXiv:1808.09346 Invent. Math. 219 (2020), no. 3, 779-830.
9. A. Iosevich, M. Rudnev, and I. Uriarte-Tuero, *Theory of dimension for large discrete sets and applications* (English summary) Math. Model. Nat. Phenom. **9** (2014), no. 5, 148-169.
10. W. James, *The Principles of Psychology*, New York: H. Holt and Company, (1890).
11. B. Mandelbrot, *How fractals can explain what's wrong with Wall Street*, Scientific American, September 15, (2008).
12. F. Rosenblatt, *The Perceptron: A Probalistic Model For Information Storage And Organization In The Brain*, Psychological Review. **65** (6): 386-408, (1958).
13. R. Smalley Jr, J. Chatelain, D. Turcotte, and R. Prevot, *A fractal approach to the clustering of earthquakes: applications to the seismicity of the new hebrides*, Bulletin of the Seismological Society of America, Vol. **77**, No. 4, pp. 1368-1381, (1987).
14. D. Turcotte, *Fractals and fragmentation*, J. Geophys. Res. **91**, 1921-1926, (1986).
15. D. Turcotte, *A fractal model for crustal deformation*, Tectonophysics **132**, 261-269, (1987).
16. P. Werbos, *Beyond regression: New tools for prediction and analysis in the behavioral sciences*, Thesis (Ph.D.)- Harvard University, (1975).
17. Brian R. Hunt, *The Hausdorff dimension of graphs of Weierstrass functions*, Proceedings of the American Mathematical Society, (1998)

Towards the Gaussianity of Random Zeckendorf Games

Justin Cheigh, Guilherme Zeus Dantas e Moura, Ryan Jeong, Jacob Lehmann Duke, Wyatt Milgrim, Steven J. Miller, and Prakod Ngamlamai

Abstract Zeckendorf proved that any positive integer has a unique decomposition as a sum of non-consecutive Fibonacci numbers, here indexed by $F_1 = 1, F_2 = 2, F_{n+1} = F_n + F_{n-1}$. Motivated by this result, Baird et al. [3] defined the two-player Zeckendorf game, in which two players take turns acting on a multiset of Fibonacci numbers that always sums to N. The game terminates when no possible moves remain, which importantly always happens, and the final player to perform a move wins. Notably, Baird et al. [3] empirically studied the setting of random games, in the sense that the game proceeds by always choosing an available move uniformly at random, and conjecture that as the input $N \to \infty$, the distribution of random game lengths converges to a Gaussian. We study various combinatorial questions concerning the Zeckendorf game. We found that the sum of the number of times certain moves are performed is constant. We prove that the number of shortest games on input N is at least $\prod_{k=1}^{n-2} \mathrm{Cat}(F_k)$, where n denotes the index of

J. Cheigh · J. L. Duke · S. J. Miller (✉)
Department of Mathematics and Statistics, Williams College, Williamstown, MA 01267, USA
e-mail: sjm1@williams.edu; Steven.Miller.MC.96@aya.yale.edu

J. Cheigh
e-mail: jhc5@williams.edu

J. L. Duke
e-mail: jl34@williams.edu

G. Z. Dantas e Moura
Department of Mathematics and Statistics, Haverford College, Haverford, PA 19041, USA
e-mail: gdantasemo@haverford.edu; zeusdanmou@gmail.com

R. Jeong
Department of Mathematics, University of California, Los Angeles, CA 90095, USA
e-mail: rsjeong@sas.upenn.edu

W. Milgrim
Department of Mathematics, University of Pennsylvania, Philadelphia, PA 19104, USA
e-mail: wmilgrim@math.ucla.edu

P. Ngamlamai
Department of Mathematics, The Ohio State University, Columbus, OH 43210, USA
e-mail: ngamlamai.1@osu.edu

M. B. Nathanson (ed.), *Combinatorial and Additive Number Theory VI*,
Springer Proceedings in Mathematics & Statistics 464,
https://doi.org/10.1007/978-3-031-65064-2_4

the largest Fibonacci number in the Zeckendorf decomposition of N and $\mathrm{Cat}(F_k)$ is the F_kth Catalan number. The works of Baird, Epstein, Flint, and Miller [3] and Cuzensa et al. [5] determined how to play in order to achieve the shortest and longest possible Zeckendorf game on a given input N, respectively: we improve the current understanding of achievable game lengths by establishing that for any input N, the range of possible game lengths constitutes an interval of natural numbers; in other words, for every input N, every game length between the shortest and longest game lengths can be achieved by some Zeckendorf game. Motivated towards the resolution of the Gaussianity conjecture, we also further the study of probabilistic aspects of random Zeckendorf games. In particular, we study two probability measures on the space of all Zeckendorf games on input N: the uniform measure, and the measure induced by choosing moves uniformly at random at any given configuration. We show under both measures that in the limit $N \to \infty$, both players win with probability $1/2$ when playing under the random game setting. We also find natural partitions of the collection of all Zeckendorf games of a fixed input N, on which we observe weak convergence to a Gaussian in the limit $N \to \infty$. We conclude the work with many open problems.

1 Introduction

The Fibonacci numbers are widely considered to be the most interesting and well-known recursive sequence in mathematics. In this article, we shall index the Fibonacci numbers by $F_1 = 1$, $F_2 = 2$, and for general $n \geq 3$, $F_n = F_{n-1} + F_{n-2}$. Zeckendorf proved the following fundamental theorem: the decomposition in the proceeding theorem is referred to as the **Zeckendorf decomposition** of the positive integer N.

Theorem 1 ([9]) *Every positive integer N can be decomposed uniquely into a sum of distinct, non-consecutive Fibonacci numbers.*

Inspired by this result, the authors of [3] constructed the **two-player Zeckendorf game**.

Definition 1 ([3]) Given input $N \in \mathbb{N}$, the Zeckendorf game is played on a multiset of Fibonacci numbers, initialized at $\mathcal{S} = \{F_1^N\}$. On each turn, a player can act on the multiset by performing one of the following moves if it is available.

1. If we have two consecutive Fibonacci numbers F_{k-1}, F_k for some $k \geq 2$, then we can replace them by F_{k+1}, denoted $F_{k-1} \wedge F_k \to F_{k+1}$.
2. If we have two instances of the same Fibonacci number F_k, then

   a. If $k = 1$, we can play $F_1 \wedge F_1 \to F_2$.
   b. If $k = 2$, we can play $F_2 \wedge F_2 \to F_1 \wedge F_3$.
   c. If $k \geq 3$, we can play $F_k \wedge F_k \to F_{k-2} \wedge F_{k+1}$.

The two players alternate turns until no playable moves remain. The last player to move wins the game.

Observe that the moves of the game are consistent with the Fibonacci recurrence: we either *combine* two consecutive terms, or *split* terms with multiple instances. Perhaps more intuitively, we can understand the game as acting on a row of bins, with bin k corresponding to the Fibonacci number F_k and its height being the multiplicity of F_k in the multiset.

1.1 Prior Work

The article [3] introduces the two-player Zeckendorf game, determine upper and lower bounds on the length of a game on input N (showing in particular that the game always terminates), and shows non-constructively that Player 2 has the winning strategy for all $N \geq 2$. In particular, they provide the following explicit formula for the length of the shortest Zeckendorf game on input N, achieved by only playing combine moves.

Theorem 2 ([3]) *The number of combine moves in any Zeckendorf game on input N is is $N - Z(N)$. Furthermore, any such shortest game terminates in $N - Z(N)$ moves, where $Z(N)$ is the number of terms in the Zeckendorf decomposition of N.*

The works [7] and [5] successively improved the upper bound of [3] on the length of a Zeckendorf game on fixed input N; the former article finds a deterministic game which has longest possible length for input N, while the latter generalizes this paradigm. We frequently make use of the following two results from [5] in our arguments. We note that although the following result provides a strategy to achieve a longest game, finding a convenient closed form for the length of the longest game for non-Fibonacci input N remains open (with the case of Fibonacci input treated by [5]).

Theorem 3 ([5]) *The longest game on any N is achieved by applying split moves or combine 1s (in any order) whenever possible, and, if there is no split or combine 1 move available, combine consecutive indices from smallest to largest.*

Theorem 4 ([5]) *A Zeckendorf game on input N can be played with strictly splitting and combine 1 moves if and only if $N = F_k - 1$ for some $k \geq 2$.*

Finally, we remark that analogous two-player games have been developed for other recurrences: the work [2] extends [3] by defining and studying such games for recursive sequences defined by linear recurrence relations of form $G_n = \sum_{i=1}^{k} cG_{n-i}$ ($c = k - 1 = 1$ yielding the Fibonacci numbers), again giving lower and upper bounds on game lengths (including showing termination) and showing non-constructively that Player 2 has a winning strategy, while [4] similarly studies recurrences of form $a_{n+1} = na_n + a_{n-1}$.

1.2 Notation and Conventions

We let C_1 denote the combine move $F_1 \wedge F_1 \to F_2$, and (for $k \geq 2$) let C_k denote the combine move $F_{k-1} \wedge F_k \to F_{k+1}$. Let S_2 denote the splitting move $2F_2 \to F_1 \wedge F_3$, and (for $k \geq 3$) let S_k denote the splitting move $2F_k \to F_{k-2} \wedge F_{k+1}$. We prefix a particular type of move with M to denote the number of such moves (e.g. MC_1 denotes the number of C_1's in a game).

We let $Z(N)$ denote the number of terms in the Zeckendorf decomposition of N. We loosely refer to the number of instances of F_k as the **height** of bin k, denoted h_k; it will usually be clear from context at which point in the game the quantity h_k refers to. When discussing the height of bin k after a specific number m of moves in the game, we notate this by $h_k(m)$. For $\lambda \in \mathbb{N}$, we shall also occasionally use the shorthand $[\lambda] = \{1, 2, \dots, \lambda\}$.

In this work, we shall generally work under the assumption that $F_n \leq N < F_{n+1}$ for some $n \in \mathbb{N}$ (i.e., n is the index of the largest Fibonacci number that is no larger than N).[1] While proving Theorem 7, we occasionally refer to moves $C_1, S_2, \dots, S_{n-1}$ as **Type A moves**, and all other moves (namely, moves C_k for $k \geq 2$) as **Type B moves**. The work [5] also achieved an understanding of precisely when playing strictly Type A moves throughout the whole game is possible.

Finally, the present article furthers the study of random Zeckendorf games. Here, we let Ω_N denotes the (finite) collection of all Zeckendorf games on input N, with $\mathcal{F}_N = 2^{\Omega_N}$ the associated σ-algebra, and express a given Zeckendorf game $\mathcal{G} \in \Omega_N$ as a (finite) sequence of λ moves, written as $\mathcal{G} = (M_1, M_2, \dots, M_\lambda)$. We study two probability measures to complete the space $(\Omega_N, \mathcal{F}_N)$: the uniform measure μ_N, defined by

$$\mu_N(\mathcal{G}) = \frac{1}{|\Omega_N|} \text{ for all } \mathcal{G} \in \Omega_N$$

and the probability measure $\mathbb{P}_N$ induced by choosing, at every configuration along a given game, uniformly at random among available moves, defined by

$$\mathbb{P}_N(\mathcal{G}) = \prod_{k=1}^{\lambda} \frac{1}{\text{num. playable moves after } (M_1, \dots, M_{k-1})}$$

for $\mathcal{G} = (M_1, \dots, M_\lambda) \in \Omega_N$. All of the results we derive in this context apply to both probability spaces.

[1] This is why we have elected to deviate from notation traditionally used in papers concerning this game.

1.3 Main Results

The work [5] determines an upper bound on the length of a game on input N. We improve this upper bound using similar techniques as in the work [3].

Theorem 5 *The length of a Zeckendorf game on input N is upper-bounded by*

$$\lfloor \varphi^2 N - Z_I(N) - 2Z(N) + (\varphi - 1) \rfloor$$

where $Z_1(N)$ represents the index sum of the Zeckendorf decomposition of N. Furthermore, the bound is sharp for infinitely many N.

Much of our work was inspired by the following conjecture (the only one still unresolved in the paper it was introduced in), initially posed by [3], which concerns distributional properties of the length of random Zeckendorf games on input N in the limit $N \to \infty$.

Conjecture 1 ([3, 7]) In the limit $N \to \infty$, the distribution of the number of moves in a random Zeckendorf game on input N converges to a Gaussian, with expectation and variance approximately $0.215N$.[2]

As such, many of our main results have largely arisen from attempting to understand those aspects of Zeckendorf games which may potentially aid in resolving the aforementioned conjecture (and in striving to determine what such aspects are). First, we have the following lower bound on the number of shortest Zeckendorf games of length N. Intuitively, if the distribution of random game lengths were indeed Gaussian, this should be an extreme underestimate compared to the number of ways to achieve other game lengths (shortest games involve the fewest number of decisions, so one might naturally expect that the probability of achieving one via a random game is larger than longer games), yet it still explodes in N.

Theorem 6 *Let $F_n \leq N < F_{n+1}$. Then the number of shortest Zeckendorf games with input N is at least $\prod_{k=1}^{n-2} Cat(F_k)$, where $Cat(F_k)$ denotes the F_kth Catalan number.*

The shortest game and longest game were studied in [3] and [5], respectively. It is natural to ask whether every game length between the shortest and longest game length is achievable: we resolve this in the affirmative.

Theorem 7 *For any input N to the Zeckendorf game, let M denote the length of the longest Zeckendorf game with input N. Then for any m satisfying $N - Z(N) \leq m \leq M$, there exists a Zeckendorf game of length m on input N. In other words, the set of achievable game lengths constitutes an interval in the natural numbers.*

---

[2] The authors of [3] posed this conjecture based on numerical data gathered from 9,999 simulations of a random game with $n = 18$. The authors of [7] gathered further numerical evidence with a sample of 1,000 games with $n = 1,000,000$. We ran a brute force enumeration over all possible games for $n \leq 18$ and found the distribution of lengths appeared to be Gaussian. This is a slightly different problem than the random game though is closely related.

We also study the winning odds of players in the limit $N \to \infty$ of infinite input when studying random Zeckendorf games, for which one might expect that both players win with probability $1/2$ in the limit if if Conjecture 1 holds as the variance of the conjectured Gaussian grows with N. We establish that this is indeed true by proving a much more general result: we can understand Theorem 8 as saying that in the limit of infinite input, a Z-player random Zeckendorf game is fair, in the sense that all Z players have the same probability of winning.

Theorem 8 *For any integer $Z \geq 1$ and $z \in \{0, 1, \ldots, M-1\}$, we have that*

$$\lim_{N\to\infty} \mu_N(\textit{Game length} \equiv z \bmod Z) = \lim_{N\to\infty} \mathbb{P}_N(\textit{Game length} \equiv z \bmod Z) = \frac{1}{Z}.$$

Taking $Z = 2$ in Theorem 8 above yields the following result for the classical two-player Zeckendorf game.

Theorem 9 *For the two-player Zeckendorf game, in the limit $N \to \infty$ under both probability measures μ_N and $\mathbb{P}_N$, Player 1 and Player 2 are equally likely to win. Explicitly,*

$$\lim_{N\to\infty} \mu_N(\textit{Player 1 wins}) = \lim_{N\to\infty} \mu_N(\textit{Player 2 wins}) = \frac{1}{2},$$
$$\lim_{N\to\infty} \mathbb{P}_N(\textit{Player 1 wins}) = \lim_{N\to\infty} \mathbb{P}_N(\textit{Player 2 wins}) = \frac{1}{2}.$$

Finally, we establish that there exist natural ways to partition the collection of Zeckendorf games Ω_N on input N so that the distribution of game lengths over the corresponding classes are nearly Gaussian with high probability in the limit $N \to \infty$. The construction of the subsets $\mathcal{R}_N^{\mathcal{P}} \subset \Omega_N$ and $\mathcal{R}_N^{\mathcal{S}} \subset \Omega_N$, and the sets $\mathcal{A}_N(\mathcal{R})$, is elaborated in Propositions 5 and 6.

Theorem 10 *For $\mathcal{R} \in \mathcal{R}_N^{\mathcal{P}}$, let $F_N^{\mathcal{R}}(x) : \mathbb{R} \to [0,1]$ denote the distribution function corresponding to game lengths in $\mathcal{A}_N(\mathcal{R})$ over the conditional distribution induced by $\mathbb{P}_N$, normalized to have expectation 0 and variance 1. Let $\Phi : \mathbb{R} \to [0,1]$ denote the distribution function of the standard normal. Then for any $\epsilon > 0$,*

$$\lim_{N\to\infty} \mathbb{P}_N\left(\sup_{x\in\mathbb{R}} \left|F_N^{\mathcal{R}}(x) - \Phi(x)\right| \geq \epsilon\right) = 0.$$

Similarly, for $\mathcal{R} \in \mathcal{R}_N^{\mathcal{S}}$, let $F_N^{\mathcal{R}}(x) : \mathbb{R} \to [0,1]$ denote the distribution function corresponding to game lengths in $\mathcal{A}_N(\mathcal{R})$ over the conditional distribution induced by $\mathbb{P}_N$, normalized to have expectation 0 and variance 1. Then for any $\epsilon > 0$,

$$\lim_{N\to\infty} \mathbb{P}_N\left(\sup_{x\in\mathbb{R}} \left|F_N^{\mathcal{R}}(x) - \Phi(x)\right| \geq \epsilon\right) = 0.$$

The analogous results hold for the uniform measure μ_N.

2 Structural Results

In this section, we include some straightforward, but fundamental results concerning the nature of the Zeckendorf game; some of these will be invoked in proofs of deeper theorems.

2.1 Combinatorial Observations

We begin by exploring some basic properties of the Zeckendorf game, observable by studying deterministic subroutines of moves. The following simple result mirrors techniques in [3].

Proposition 1 *Consider any decomposition of N into a sum of (possibly non-distinct, non-consecutive) Fibonacci numbers: this decomposition can be achieved via a sequence of combine moves from the starting configuration of the Zeckendorf game.*

Proof We "play the game in reverse": consider the configuration corresponding to this decomposition, and construct a sequence of moves by always taking the game piece not in the first bin (i.e., F_1) and farthest out, and replace it as the result of a combine move. Specifically, if $k \geq 3$, then replace F_k by $\{F_{k-2}, F_{k-1}\}$; if $k = 2$, replace with $2F_1$. Then reverse all the moves to get a Zeckendorf game from the initial state to this state.

The following is an easy consequence of Theorem 3 above, which states that the longest game paradigm from the starting position extends to intermediate game positions on input N which are given by converting $F_n - 1$ instances of $F_1 = 1$ into the Zeckendorf decomposition of $F_n - 1$, where n denotes the index of the largest Fibonacci number in the Zeckendorf decomposition of the input N.

Lemma 1 *Let n denote the index of the largest Fibonacci number in the Zeckendorf decomposition of the input N. A longest Zeckendorf game from an intermediate configuration given by converting $F_n - 1$ instances of 1 into the Zeckendorf decomposition of $F_n - 1$ is given by greedily playing any Type A move whenever possible, and if no such Type A move can be played, play the available Type B move with the smallest index.*

Proof If a game achieved by playing Type A moves whenever possible from this configuration were not maximal (i.e., there existed a Zeckendorf game of strictly larger length), then by initially playing the longest game on input $F_n - 1$ via all Type A moves (possible by Theorem 4), we can play the game exactly according to Theorem 3 but fail to achieve a game of maximal length, contradicting Theorem 3.

Using similar techniques as in [3], we derive the following results in order to improve the upper bound found in [5].

Lemma 2 *Let n be the largest summand in the Zeckendorf decomposition of N, we get that for any $2 \leq k \leq n-1$, the following sum is constant:*

$$MS_k + MC_k + MC_{k+1} + \cdots + MC_{n-1}.$$

Proof Consider the following relabeling of the board:

F_1	$\cdots$	F_k	$F_{k+1}-1$	$F_{k+2}-2$	$F_{k+3}-4$	$F_{k+4}-7$	$\cdots$

where after the kth bin, the value of a bin is equal to one less than the sum of the values of the two bins which precede it. We get that only the moves $S_k, C_k, C_{k+1}, \ldots$ can change the weighted sum of the tokens by the relabeled values, and each of these moves reduce the sum by 1. Since we have a fixed initial sum N and a fixed ending sum depending on the Zeckendorf Decomposition of N, we get that the sum of those moves must be constant regardless of how the game is played.

Lemma 3 *For any Zeckendorf game starting with N tokens,*

$$MC_1 - MS_2 \approx (2-\varphi)N$$

with approximation error $\leq \varphi - 1$.

Proof Similarly, we prove this with a relabeling of the board

2	3	5	$\cdots$	F_{k+1}	$\cdots$

and observing that the sum of token values goes from $2N$ to $\sim \varphi N$ with the sum decreasing by 1 only by performing C_1 and increasing by 1 only by performing S_2. Note that the final sum is equal to shifting each Zeckendorf summands of N forward by one, which is approximated by multiplying each by φ. By Binet's formula, we have $F_k = \frac{\varphi^{k+1}-(-1/\varphi)^{k+1}}{\sqrt{5}}$ due to how we index the Fibonacci sequence. Thus, the error of approximating the summand F_{k+1} with φF_k is

$$\left|\frac{(\varphi^{k+1}-(-1/\varphi)^{k+1})\varphi - (\varphi^{k+2}-(-1/\varphi)^{k+2})}{\sqrt{5}}\right| = \frac{\varphi^2+1}{\varphi^{k+2}\sqrt{5}}$$

The largest error happens when $N = F_1 + F_3 + F_5 + \cdots$ with error at most

$$\frac{\varphi^2+1}{\varphi^3\sqrt{5}}\left(\sum_{i=0}^{\infty}\frac{1}{\varphi^{2i}}\right) = \frac{\varphi^2+1}{\varphi^3(1-1/\varphi^2)\sqrt{5}} = \frac{1}{\varphi} = \varphi - 1$$

which yields the desired.

As a corollary, we prove Theorem 5

Theorem 5 *The length of a Zeckendorf game on input N is upper-bounded by*

$$\lfloor \varphi^2 N - Z_I(N) - 2Z(N) + (\varphi - 1) \rfloor$$

where $Z_I(N)$ represents the index sum of the Zeckendorf decomposition of N. Furthermore, the bound is sharp for infinitely many N.

Proof Using the relabeling of the board given by

3	4	5	...	$k+2$	...

we get that

$$2MC_1 + 2MC_2 + 3MC_3 + 4MC_4 + \cdots = 3N - 2Z(N) - Z_I(N)$$
$$+MS_3 + MS_4 + \cdots$$

Applying Lemma 3, we get

$$MC_1 + 2MC_2 + 3MC_3 + 4MC_4 + \cdots \leq 3N - 2Z(N) - Z_I(N)$$
$$+(2 - \varphi)N - (\varphi - 1) + MS_2 + MS_3 + MS_4 + \cdots$$

Thus, if we subtract the excess $MC_2 + 2MC_3 + 3MC_4 + \cdots$ from the left hand side, we get the upper bound

$$MC_1 + MC_2 + MC_3 + MC_4 + \cdots \leq (1 + \varphi)N - 2Z(N) - Z_I(N) + (\varphi - 1)$$
$$+MS_2 + MS_3 + MS_4 + \cdots$$

which we round down as the number of all moves performed is an integer. Note that if we consider $N = F_n - 1$ then we get that there are no C_2, C_3, performed within the longest game. Furthermore, if we consider $N = F_{2n} - 1$, then our approximation error is less than 1 and thus must be sharp after rounding down.

Recall that [3] proved non-constructively that Player 2 always has a winning strategy for any $N \geq 3$: finding such a winning strategy remains open. It is generally believed [8] that the key to such a strategy lies in understanding "parity swaps": distinct sequences of moves of differing length which yield the same effect on the board. The following definition follows from the easy observation that whenever playable, the sequences of moves

$$S_k \to S_{k-1} \to \cdots \to S_{k-\ell+1} \to C_{k-\ell} \quad C_{k-\ell} \to S_{k-\ell+1} \to \cdots \to S_{k-1} \to S_k$$

both have the same effect on the board as the move C_k, for some $k \geq 2$.

Definition 2 For any $\ell \geq 0$ and $k \geq 2$, call a sequence of moves of form $S_k \to S_{k-1} \to \cdots \to S_{k-\ell+1} \to C_{k-\ell}$ an *(ℓ, k)-prefix*, and a sequence of moves of form

$C_{k-\ell} \to S_{k-\ell+1} \to S_{k-\ell+2} \to \cdots \to S_k$ an *(ℓ, k)-suffix*. We call an (ℓ, k)-prefix a *prefix of C_k*, and an (ℓ, k)-suffix a *suffix of C_k*.

It should be emphasized that an (ℓ, k)-prefix or (ℓ, k)-suffix corresponds to an equivalent action as the combine move C_k, for all lengths $\ell \geq 0$ and $k \geq 2$; the combine move is thus "expanded" via a sequence of ℓ splitting moves with contiguous indices. The next result captures the intuition that the variation in game lengths is entirely due to the parity swaps described in Definition 2, namely by describing arbitrary Zeckendorf games via *permutations of suffixes.*

Proposition 2 *Any Zeckendorf game on input N can be achieved by taking a shortest game, expanding combine moves via suffixes, then shifting the splitting moves.*

Proof For an arbitrary Zeckendorf game, greedily take the earliest split move of a game, move it back to be played as early as possible, and compress it into a combine move. Proceed similarly until we only have combine moves, from which we achieve the original game by reversing the compressions and playing split moves later.

We can interpret the statement of Proposition 2 as saying that we can greedily embed an arbitrary Zeckendorf game on input N into a shortest game on input N in a natural way.

2.2 Shortest Games

Proposition 2 suggests that a study of shortest Zeckendorf games might be fruitful, as any particular Zeckendorf game can be understood as an extension and permutation of a particular shortest game. We first observe the following.

Proposition 3 *Shortest games are exactly those games which strictly use combine moves. Furthermore, such a game exists for any input N, and the multiset of combine moves for any such shortest game is unique.*

Proof A move decreases the number of pieces by at most one, so $N - Z(N)$ lower-bounds the number of moves necessary, achieved exactly by those games using strictly combine moves; such games exist by Proposition 1. To establish uniqueness of the multiset of combine moves for any such game, say $F_n \leq N < F_{n+1}$, and study bin k for $1 \leq k \leq n$. Moves affecting h_k are known precisely: letting the Zeckendorf decomposition of N be denoted $(z_1, z_2, \ldots, z_n)$ (where $z_i \in \{0, 1\}$), this yields the system

$$\begin{aligned} N - 2MC_1 - MC_2 &= z_1 \\ MC_1 - MC_2 - MC_3 &= z_2 \\ &\cdots \\ MC_{n-3} - MC_{n-2} - MC_{n-1} &= z_{n-2} \\ MC_{n-2} - MC_{n-1} &= z_{n-1} \\ MC_{n-1} &= z_n = 1 \end{aligned}$$

from which it easily follows that this system must have a unique solution.

Proposition 4 *For $n = n(N)$ and any $\delta \in (0, 1)$,*

$$\lim_{N\to\infty} \max_{\mathcal{G}\in\Omega_N} \frac{\#\{C_k \in \mathcal{G}, k > \delta n\}}{\#\{\text{Combine moves in } \mathcal{G}\}} = \lim_{N\to\infty} \max_{\mathcal{G}\in\Omega_N} \frac{\#\{C_k \in \mathcal{G}, k > \delta n\}}{N-1} = 0.$$

Proof For $\mathcal{G} \in \Omega_N$, consider all combine moves C_k with $k > \delta n$; each such combine move corresponds to a particular token "jumped," for which we consider all combine moves with index $\leq \delta n$ that led this token to land in this position, of which there must be at least $\lfloor \delta n/2 \rfloor$. The sets of combine moves with index $\leq \delta n$ corresponding to distinct combine moves with index $> \delta n$ are observed to be disjoint, so

$$\frac{\#\{C_k \in \mathcal{G}, k > \delta n\}}{N-1} \leq \frac{1}{\lfloor \delta n/2 \rfloor + 1}$$

which vanishes as $N \to \infty$.

Proposition 3 yields the following interesting lower bound[3] on the number of shortest Zeckendorf games with input N.

Theorem 6 *Let $F_n \leq N < F_{n+1}$. Then the number of shortest Zeckendorf games with input N is at least $\prod_{k=1}^{n-2} Cat(F_k)$, where $Cat(F_k)$ denotes the F_kth Catalan number.*

Proof It suffices to study $N = F_n$, since the number of distinct shortest Zeckendorf games is increasing in N: for input $N \neq F_n$, one can first play a shortest Zeckendorf game on input F_n, then proceed by always playing the rightmost available combine move to achieve a shortest game on input N by Proposition 3. (See Lemma 4, where we understand this as incrementing instances of F_1s after playing a shortest game on F_n.) The Zeckendorf decomposition is $(z_1, z_2, \ldots, z_{n-1}, z_n) = (0, 0, \ldots, 0, 1)$, where z_i denotes the number of instances of F_i in the decomposition. By Proposition 3, there exists a unique multiset of combine moves constituting a shortest game: by solving the system above, this multiset is defined by

$$(MC_1, MC_2, \ldots, MC_{n-1}) = (F_{n-2}, F_{n-3}, F_{n-4}, \ldots, F_3, F_2, F_1, F_0)$$

where we let $F_0 = 1$. A permutation of these moves constitutes a game if and only if every move is valid, i.e., no move would force the height of any bin to become negative. Specifically, at any intermediate point in the sequence, the number of C_1's played is no less than the sum of the number of C_2's and C_3's played (bin 2 is nonnegative), the number of C_2's is no less than the number of C_3's and C_4's played (bin 3 is nonnegative), and so on, to the number or C_{n-2}'s being no less than the number of C_{n-1}'s played (bin $n-1$ is nonnegative). (We need not study bin 1 or bin n, which will necessarily always have nonnegative height.)

[3] We suspect the lower bound of Theorem 6 to be somewhat loose, as much is lost when crudely pursuing the interweaving of the Dyck paths $\pi_2, \pi_3, \ldots, \pi_{n-1}$ (see the proof for details).

Let us restrict our attention to the moves C_1, C_2, C_3 (moves affecting the height of bin 2): the number of permutations of these moves such that the number of C_1's played being no less than the sum of the number of C_2's and C_3's performed holds at any point in the game is in bijective correspondence with (up-down) Dyck paths on $j = F_{n-2}$ ($C_1 \leftrightarrow U$; $C_2, C_3 \leftrightarrow D$), the number of which is $\text{Cat}(F_{n-2})$. Similarly, by studying moves affecting bin $k \geq 2$, we achieve bijective correspondences with Dyck paths on $j = F_{n-k}$.

For any choice of Dyck paths $\pi_2, \pi_3, \ldots, \pi_{n-1}$ on $j = F_{n-2}, F_{n-3}, \ldots, F_1$, respectively, there exists a shortest Zeckendorf game on input N where the ordering of the relevant moves is consistent with the bijections described above. To construct such a game, begin by placing $2F_{n-2}$ moves along a line, labeling F_{n-2} of them as C_1 in a manner consistent with π_2 ($C_1 \leftrightarrow U$). Among the $F_{n-2} = F_{n-3} + F_{n-4}$ unlabeled moves, label F_{n-3} of them as C_2 (importantly, including the first D move) in a manner consistent with π_3 (not too many unlabeled moves between consecutive instances of C_2) and the other F_{n-4} as C_3. Now add F_{n-5} instances of C_4 along this line to complete π_3 (include all missing $C_4 \leftrightarrow D$ moves) while respecting π_4 (not too many unlabeled moves between consecutive instances of C_3). Specifically, construct a labeling of the $F_{n-3} = F_{n-4} + F_{n-5}$ D moves in π_3 with F_{n-4} C_3's and F_{n-5} C_4's such that the first D is labeled C_3, and there are not too many unlabeled D moves between consecutive instances of C_3 (with respect to π_4). Insert C_4's to be adjacent to established instances of C_2 and C_3 to be consistent with this labeling.

Continue by similarly adding, for $k \geq 5$, F_{n-k-1} instances of C_k, completing π_{k-1} while respecting π_k, until we add $F_0 = 1$ instance of C_{n-1} such that we complete π_{n-1}. This results in a shortest Zeckendorf game with the ordering of the relevant moves being consistent with $\pi_2, \pi_3, \ldots, \pi_{n-1}$.

3 The Set of Possible Game Lengths Constitute an Interval

In this section, we prove Theorem 7, which we restate below.

Theorem 7 *For any input N to the Zeckendorf game, let M denote the length of the longest Zeckendorf game with input N. Then for any m satisfying $N - Z(N) \leq m \leq M$, there exists a Zeckendorf game of length m on input N. In other words, the set of achievable game lengths constitutes an interval in the natural numbers.*

We begin by establishing some intermediate results that we shall invoke in the proof of the main theorem. In the first lemma, in discussing the position given by the Zeckendorf decomposition of $N - 1$, we refer to the terminal position of the Zeckendorf game when played on input $N - 1$.

Lemma 4 *Consider the Zeckendorf game on input N, satisfying $F_n \leq N < F_{n+1}$, from the position given by the Zeckendorf decomposition on $N - 1$ (as specified above) with an additional instance of 1. There is a unique sequence of moves from this configuration to the Zeckendorf decomposition of N, all of which are combine* *moves. Furthermore, the number of such combine moves performed is bounded by* $\lfloor n/2 \rfloor$.

Proof Since no moves can be played from the Zeckendorf decomposition on $N-1$, any playable move from this position (on input N) is necessarily either C_1 or C_2, possible if and only if $F_1 = 1$ or $F_2 = 2$ is in the Zeckendorf decomposition of $N-1$, respectively (such cases are disjoint, since the Zeckendorf decomposition on $N-1$ does not contain consecutive Fibonacci numbers); otherwise, no moves can be played. We study both cases.

- If the Zeckendorf decomposition of $N-1$ contains an instance of F_1, then after playing C_1, it is easy to see that the only possible move is C_3 (iff the decomposition of $N-1$ contains an F_3), then C_5 (iff the decomposition of $N-1$ contains an F_5), and so on, until we exhaust all playable moves.
- If the Zeckendorf decomposition of $N-1$ contains an instance of F_2, then after playing C_2, it is easy to see that the only possible move is C_4 (iff the decomposition of $N-1$ contains an F_4), then C_6 (iff the decomposition of $N-1$ contains an F_6), and so on, until we exhaust all playable moves.

In both cases, it is straightforward to confirm that we cannot play strictly more than $\lfloor n/2 \rfloor$ such combine moves, as otherwise there must be an instance of F_k for k either $n+1$ or $n+2$ after completing this sequence of moves, a contradiction on $N < F_{n+1}$.

We shall also frequently use the following lemma. Intuitively, this states that if we have isolated the Zeckendorf game to a suffix of bins all of height 0 or 1, and this suffix is separated from earlier bins by a bin of height 0 (say bin k), then we can ignore bins $1, \ldots, k$ for the remainder of the game.

Lemma 5 *For some* $1 \leq k \leq n$, *let* $(x_1, x_2, \ldots, x_k, x_{k+1}, \ldots, x_n)$ *denote the heights of bins* $1, \ldots, n$ *at some point during a Zeckendorf game, with* x_i *denoting the height of bin* i. *Assume* $x_i \in \{0, 1\}$ *for* $k \leq i \leq n$, $x_k = 0$, *no playable moves involving bins* $1, \ldots, k$ *exist, and we play according to Theorem 3. Then heights* $x_1, \ldots, x_k$ *remain fixed for the rest of the game.*

Proof We prove this on $k = n - j$ by induction on $0 \leq j \leq n-2$. The statement trivially holds if $j \in \{0, 1, 2\}$; assuming it for all values less than $j \geq 3$, if $x_{k+1} = 0$ or $x_{k+2} = 0$ we can apply the induction hypothesis to $j-1$ or $j-2$, respectively, so assume $x_{k+1} = x_{k+2} = 1$. Take $r \leq n$ to be largest possible such that $x_{k+1} = \cdots = x_r = 1$: the game proceeds by playing according to Theorem 3, i.e., by playing the sequence of moves

$$(C_{k+1} \to S_{k+3} \to S_{k+4} \to \cdots \to S_r) \to (C_{k+1} \to S_{k+3} \to \cdots \to S_{r-2}) \to \cdots$$

where the final subsequence of moves, either C_{k+1} or $C_{k+1} \to S_{k+3}$, depends on the parity of $r-k$. It follows immediately by studying the moves involved that $x_1, \ldots, x_k$

remain unchanged during this sequence. Following this sequence, we have $x_r = 0$: invoke the induction hypothesis on $k = r$ (i.e., $n - r < j$) afterwards to complete the proof.

We now proceed with the proof of Theorem 7.

Proof *(Proof of Theorem 7)* We have confirmed this statement for $N \leq F_6 - 1 = 12$ via a computer check. Thus, for $N \geq F_6 = 13$, assume the statement holds for all input sizes at most $N - 1$: we aim to show the result holds for N. We explicitly specify that $F_n \leq N < F_{n+1}$.

Let I_1' denote the interval (by induction hypothesis) of possible Zeckendorf game lengths for input $N - 1$: by Lemma 4, if we include an additional instance of F_1 to the Zeckendorf decomposition of $N - 1$, there is a sequence of combine moves from the resulting configuration to the Zeckendorf decomposition of N. On input N, consider initially playing the Zeckendorf game (to completion) as if the input were $N - 1$, then executing this sequence of moves to terminate the game. Let $I_1 = [L_1, R_1]$ denote the interval I_1' shifted by the length of this sequence: by the preceding description, it follows that every game length in I_1 is achievable for input N. Furthermore, $L_1 = N - Z(N)$, since the game length L_1 as studied above results from playing strictly combine moves (see Lemma 4), which necessarily yields a shortest game by Proposition 3, and for which the length is $N - Z(N)$ by Theorem 2.

By Theorem 4, it is possible to play the Zeckendorf game on input $F_n - 1$ strictly using Type A moves. Let I_2' denote the interval (by induction hypothesis) of possible Zeckendorf game lengths for input $F_n - 1 \leq N - 1$: on input N, consider initially playing the Zeckendorf game (to completion) as if the input were $F_n - 1$, then executing the longest possible sequence of moves from the resulting position to terminate the game. Let $I_2 = [L_2, R_2]$ denote the interval I_2' shifted by the number of moves of this longest sequence. By Theorem 3, R_2 is necessarily the length of the longest Zeckendorf game on input N, since the above approach is consistent with playing Type A moves whenever possible.

Thus, it suffices to show that $L_2 \leq R_1$ to complete the induction and establish the theorem, as this yields that $I_1 \cup I_2 = [L_1, R_2]$ is an interval of achievable game lengths on input N, with the endpoints being the shortest and longest possible game lengths for input N.

The known game with length R_1 corresponds to playing the longest game on input $N - 1$, then performing the unique sequence of combine moves to achieve the Zeckendorf decomposition of N; by Theorems 3 and 4, we can take the first phase of this game (longest game on input $N - 1$) as playing the longest game on input $F_n - 1$, then playing the longest remaining game (on input $N - 1$). The known game with length L_2 corresponds to playing the shortest game on input $F_n - 1$, then playing the longest remaining game until we achieve the Zeckendorf decomposition of N. By Lemma 1, we can take the second phase of this game (playing the longest remaining game) as playing consistent to Theorem 3 until achieving the Zeckendorf decomposition.

We henceforth denote these games with length R_1 and L_2 by $\mathcal{G}_1$ and $\mathcal{G}_2$, respectively, which we depict as follows; "longest on k" indicates that during this phase, we think of the game as being played on input k, and leaving the appropriate number of instances of $F_1 = 1$ in the first bin fixed.

$$\mathcal{G}_1 : \big[(\text{Longest on } F_n - 1) \rightarrow (\text{longest remaining on } N - 1)\big] \rightarrow (\text{combine on } N)$$
$$\mathcal{G}_2 : (\text{Shortest on } F_n - 1) \rightarrow (\text{longest remaining on } N)$$

Denoting the difference between the lengths of the longest and shortest games on input $F_n - 1$ by $\ell(n)$, game $\mathcal{G}_1$ took exactly $\ell(n)$ more moves than game $\mathcal{G}_2$ on input $F_n - 1$. However, $\mathcal{G}_2$ may take longer afterwards to finish the game on input N: we aim to show that the discrepancy in the game lengths after initially playing on input $F_n - 1$ is dominated by $\ell(n)$, from which we conclude that $\mathcal{G}_2$ is no longer than $\mathcal{G}_1$, i.e., that $L_2 \leq R_1$. In particular, it certainly suffices to show that the first two segments of $\mathcal{G}_1$ involve at least as many moves as the first two segments of $\mathcal{G}_2$ to establish the result: this is how we shall proceed.

By Theorem 3 and Lemma 1, we can choose how we would like to play a longest game after playing the game on input $F_n - 1$ (specifically, we can fix an ordering on Type A moves which determines what we play when given multiple Type A moves): until the two games diverge,[4] pursue a longest remaining game by always playing the rightmost Type A move whenever a Type A move is playable. Following this ordering on Type A moves, we study the first move on which games $\mathcal{G}_1$ and $\mathcal{G}_2$ deviate. This move is necessarily either C_1 or C_2 in game $\mathcal{G}_2$ (the move must not have been playable in game $\mathcal{G}_1$, and thus must involve bin 1), and by the ordering on Type A moves established before, bins $2, 3, \ldots, n$ are either 0 or 1 when it is played. We perform casework on which move the two games deviate on.

Case 1: Move is C_1. In game $\mathcal{G}_1$, this configuration can be represented by the vector $(1, x_2, x_3, \ldots, x_n)$, where $x_i \in \{0, 1\}$ for $i \geq 2$ denotes the height of the ith bin (the first entry would be a 2 for game $\mathcal{G}_2$). Let us first study the setting $x_2 = 0$, and consider what happens after game $\mathcal{G}_2$ plays C_1. If $x_3 = 0$, then by Lemma 5 applied to $k = 3$, both games are consistent on bins $4, \ldots, n$ so that game $\mathcal{G}_2$ takes one more move than $\mathcal{G}_1$ to finish. Otherwise (i.e., $x_3 = 1$), Lemma 5 on $k = 2$ yields that game $\mathcal{G}_1$ works strictly over bins $3, \ldots, n$, and Lemma 5 on $k = 1$ yields that game $\mathcal{G}_2$ works strictly over the bins $2, \ldots, n$ (i.e., bin 1 becomes irrelevant). Thus, the resulting setting corresponds exactly to Case 2 over the $n - 1$ bins $2, \ldots, n$; here, we have an upper bound of $n - 1$ for the number of additional moves $\mathcal{G}_2$ takes, for a total upper bound of $(n - 1) + 1 = n$ (with the C_1 in $\mathcal{G}_2$) for the number of additional moves $\mathcal{G}_2$ takes.

Thus, assume $x_2 = 1$, and let position $k + 1$, with $k \geq 2$, denote the first index that is 0. Study the length-k prefix $(1, 1, 1, \ldots, 1)$ (in game $\mathcal{G}_1$; first entry is 2 in

[4] We shall assume this does happen, as otherwise the lengths of the second segments of $\mathcal{G}_1$ and $\mathcal{G}_2$ are equal, and thus the inequality $L_2 \leq R_1$ is immediate.

game $\mathcal{G}_2$) with no zero entries; we can exactly describe how the two games proceed when playing according to the longest game paradigm of Theorem 3.

- Game $\mathcal{G}_1$: We play the sequence of moves

$$(C_2 \to S_3 \to S_4 \to \cdots \to S_k) \to (C_2 \to S_3 \to \cdots \to S_{k-2}) \to \cdots$$

  continuing similarly until one of the first two bins is empty (so we cannot play C_2). In general, shorten the contiguous 1s down two indices and repeat until one of the first two bins is empty (the bin which ends up empty depends on the parity of k).
- Game $\mathcal{G}_2$: We play the sequence of moves

$$(C_1) \to (S_2) \to (S_3 \to C_1) \to (S_4 \to S_2) \to \cdots$$

  continuing similarly until we play all splitting moves S_j for $j \leq k$. In general, when first playing S_j, we play the sequence $S_j \to S_{j-2} \to \cdots$ (final move is S_2 or C_1, depending on parity of j), and this continues until position k.

After these subroutines, it is straightforward to confirm that bin k is empty, bins $k+1, \ldots, n-1$ are all 0 or 1 with heights agreeing between $\mathcal{G}_1$ and $\mathcal{G}_2$, and no playable moves involving bins $1, \ldots, k$ exist. Thus, both games proceed strictly over bins $k+1$ to $n-1$ by Lemma 5, and perform the same sequence of moves; we need only compute the difference in the lengths of these two subroutines on this prefix of length k.

- The length of the sequence of moves in the game of length L_2 is

$$\begin{cases} 2\sum_{j=1}^{(k-1)/2} j + \frac{k+1}{2} & k \text{ odd}, \\ 2\sum_{j=1}^{k/2} j & k \text{ even}. \end{cases}$$

- The length of the sequence of moves in the game of length R_1 is

$$\begin{cases} \sum_{j=1}^{(k-1)/2}(k-2j+1) = 2\sum_{j=1}^{(k-1)/2} j & k \text{ odd}, \\ \sum_{j=1}^{k/2}(k-2j+1) = \sum_{j=1}^{k/2}(2j-1) = 2(\sum_{j=1}^{k/2} j) - \frac{k}{2} & k \text{ even}. \end{cases}$$

We can thus study this difference exactly: the difference is given by

$$\begin{cases} \left[2\sum_{j=1}^{(k-1)/2} j + \frac{k+1}{2}\right] - 2\sum_{j=1}^{(k-1)/2} j = \frac{k+1}{2} - \lceil k/2 \rceil & k \text{ odd} \\ 2\sum_{j=1}^{k/2} j - \sum_{j=1}^{k/2}(2j-1) = \frac{k}{2} & k \text{ even} \end{cases}$$

so in general, the difference is bounded by $\lceil n/2 \rceil \leq n$.

Case 2: Move is C_2. In game $\mathcal{G}_1$, this configuration can be represented by the vector $(0, x_2, x_3, \ldots, x_n)$, where $x_i \in \{0, 1\}$ for $i \geq 2$ denotes the height of the ith bin (the first entry would be a 1 for game $\mathcal{G}_2$) and $x_2 = 1$ (since otherwise C_2 cannot be played in game $\mathcal{G}_2$). Let us first study the setting $x_3 = 0$, and consider what happens after game $\mathcal{G}_2$ plays C_2. Applying Lemma 5 on $k = 3$ for $\mathcal{G}_1$ and $k = 2$ for $\mathcal{G}_2$ yield that both games work strictly over bins $3, \ldots, n$, and this reduces to the same setting on the suffix of bins $3, \ldots, n$. Say we reduce the problem to a suffix with length reduced by 2 in this manner m times, so we study the case where $x_3 = 1$ over $n - 2m$ bins: by extracting the bound in the following argument (i.e., the $x_3 = 1$ case) for the number of additional moves $\mathcal{G}_2$ takes, this yields a bound of $m + (n - 2m) \leq n$.

Thus, assume $x_3 = 1$, and let position $k + 1$, with $k \geq 3$, denote the first index that is 0. Study the length-k prefix $(0, 1, 1, \ldots, 1)$ (in game $\mathcal{G}_1$; first entry is 1 in game $\mathcal{G}_2$): we can explicitly describe how the two games necessarily proceed when playing according to the longest game paradigm.

- Game $\mathcal{G}_1$: We play the sequence of moves

$$(C_3 \to S_4 \to S_5 \to \cdots \to S_k) \to (C_3 \to S_4 \to \cdots \to S_{k-2}) \to \cdots$$

  continuing similarly until either bin 2 or bin 3 is empty (so we cannot play C_3). In general, shorten the contiguous 1s down two indices and repeat until either bin 2 or bin 3 is empty (the bin which ends up empty depends on the parity of k).
- Game $\mathcal{G}_2$: This is exactly the same as game $\mathcal{G}_1$ in Case 1.

After these subroutines, it is straightforward to confirm that bin k is empty, bins $k + 1, \ldots, n - 1$ are all 0 or 1 with heights agreeing between $\mathcal{G}_1$ and $\mathcal{G}_2$, and no playable moves involving bins $1, \ldots, k$ exist. Thus, both games proceed strictly over bins $k + 1$ to $n - 1$ by Lemma 5, and perform the same sequence of moves; we need only compute the difference in the lengths of these two subroutines on this prefix of length k.

The length of the sequence of moves in game $\mathcal{G}_2$ was computed in Case 1, while the length of the sequence of moves in game $\mathcal{G}_1$ is

$$\begin{cases} \sum_{j=1}^{(k-1)/2}(k - 2j) = \sum_{j=1}^{(k-1)/2}(2j - 1) = 2(\sum_{j=1}^{(k-1)/2} j) - \frac{k-1}{2} & k \text{ odd} \\ \sum_{j=1}^{(k-2)/2}(k - 2j) = \sum_{j=1}^{(k-2)/2}(2j - 1) = 2(\sum_{j=1}^{(k-2)/2} j) - \frac{k-2}{2} & k \text{ even.} \end{cases}$$

so the difference is given by

$$\begin{cases} 2\sum_{j=1}^{(k-1)/2} j - \left[2(\sum_{j=1}^{(k-1)/2} j) - \frac{k-1}{2}\right] = \frac{k-1}{2} = \lfloor k/2 \rfloor & k \text{ odd} \\ \left[2(\sum_{j=1}^{k/2} j) - \frac{k}{2}\right] - \left[2(\sum_{j=1}^{(k-2)/2} j) - \frac{k-2}{2}\right] = k - 1 & k \text{ even} \end{cases}$$

so in general, the difference is bounded by $\max\{\lfloor k/2 \rfloor, k - 1\} \leq n - 1 \leq n$.

We observe that in both cases, the difference in the lengths of the second segments of these games is bounded by n. We now show that the difference $\ell(n)$ between the lengths of the longest and shortest games on input $F_n - 1$ is at least n for all $n \geq 6$. One can confirm, by playing a longest game according to Theorem 3 on input $N = F_6 - 1 = 12$, that $\ell(6) = 17 - (12 - Z(12)) = 17 - 9 \geq 6$; it is similarly easy to confirm that $\ell(4), \ell(5) \geq 1$. Since we have $F_{n+1} - 1 = (F_n - 1) + F_{n-1} \geq (F_n - 1) + (F_{n-1} - 1)$, one can pursue a game on input $F_{n+1} - 1$ by first playing a game on input $F_n - 1$, then a game on input $F_{n-1} - 1$, and finally performing some fixed sequence of moves to completion. By combining respective shortest games and longest games on input $F_n - 1$ and input $F_{n-1} - 1$, we observe that $\ell(n+1) - \ell(n) \geq 1$ for any $n \geq 6$, and thus $\ell(n) \geq n$ for all $n \geq 6$ (recall $\ell(6) \geq 6$).

Therefore, we have that for all $N \geq F_6 = 13$,

$$\begin{aligned} R_1 - L_2 \geq & \ell(n) - \{(\text{len. longest remaining on } N) \\ & - (\text{len. longest remaining on } N - 1)\} \\ \geq & \ell(n) - n \geq \ell(n) - \ell(n) = 0 \end{aligned}$$

from which we conclude that $L_2 \leq R_1$.

4 Winning Odds in the Limit $N \to \infty$

We dedicate this section to establishing Theorem 9, which follows as an immediate corollary of Theorem 8: recall that Theorem 8 is given as follows.

Theorem 8 *For any integer $Z \geq 1$ and $z \in \{0, 1, \ldots, M - 1\}$, we have that*

$$\lim_{N\to\infty} \mu_N(\textit{Game length} \equiv z \bmod Z) = \lim_{N\to\infty} \mathbb{P}_N(\textit{Game length} \equiv z \bmod Z) = \frac{1}{Z}.$$

The case $Z = 2$ gives Theorem 9.

Theorem 9 *For the two-player Zeckendorf game, in the limit $N \to \infty$ under both probability measures μ_N and $\mathbb{P}_N$, Player 1 and Player 2 are equally likely to win. Explicitly,*

$$\begin{aligned} \lim_{N\to\infty} \mu_N(\textit{Player 1 wins}) &= \lim_{N\to\infty} \mu_N(\textit{Player 2 wins}) = \frac{1}{2}, \\ \lim_{N\to\infty} \mathbb{P}_N(\textit{Player 1 wins}) &= \lim_{N\to\infty} \mathbb{P}_N(\textit{Player 2 wins}) = \frac{1}{2}. \end{aligned}$$

4.1 Overview

We provide a brief sketch of the proof of Theorem 9. We specifically focus on the number of $(1, k)$-prefixes which occur over the course of a game, recalling that whenever applicable, the sequence of moves (S_k, C_{k-1}) has the same effect on the game as C_k. We partition the collection of games Ω_N into subsets of games which differ only via such $(1, k)$-prefixes, describe the conditional distribution induced by any such subset (under both μ_N and $\mathbb{P}_N$), and argue that in the limit of infinite input, the sizes of the subsets induced by this partition grow so fast that the probability of Player 1 or Player 2 winning can be reduced to the outcome of a binomial random variable with exploding variance.

4.2 Partitioning the Collection of Possible Games

As observed in Definition 2, for $k \geq 2$, the sequence[5] (S_k, C_{k-1}) is a $(1, k)$-prefix. We establish some notation to use later: define $\mathcal{R}_N \in \mathcal{F}_N$ to be the collection[6] of all Zeckendorf games on input N such that (for all $k \geq 2$) any instance of the sequence (S_k, C_{k-1}) is immediately preceded by S_{k+1}; in other words, $\mathcal{R}_N$ is the collection of all Zeckendorf games on input N such that there are no $(1, k)$-prefixes for any $k \geq 2$. We can express this collection as

$$\begin{aligned}\mathcal{R}_N =&\{(M_1, \dots, M_\lambda) \in \Omega_N : (M_i, M_{i+1}) = (S_k, C_{k-1}) \implies M_{i-1} = S_{k+1} \\ &\forall\, i \in [\lambda - 1], k \geq 2\}.\end{aligned}$$

For a game $\mathcal{R} = (M_1, M_2, \dots, M_\lambda) \in \mathcal{R}_N$, construct the subset of indices $\mathcal{I}_N(\mathcal{R})$ to denote all combine moves in $\mathcal{R}$, not involved in a sequence (S_k, C_{k-1}) for $k \geq 2$, for which the latter bin has height at least 2 (i.e., combine moves C_k for $k \geq 2$ replaceable with a $(1, k)$-prefix):

$$\mathcal{I}_N(\mathcal{R}) = \{i \in [\lambda] : M_i = C_k,\ M_{i-1} \neq S_{k+1},\ h_k(i) \geq 2 \text{ for some } k \geq 2\}.$$

Now, construct the *formal* sequence of moves $\mathcal{M}(\mathcal{R}) = (\tilde{M}_1, \tilde{M}_2, \dots, \tilde{M}_\lambda)$ by replacing M_i by a symbol $\mathcal{E}_k$ (the subscript being the corresponding $k \geq 2$) for all $i \in \mathcal{I}_N(\mathcal{R})$; call $\mathcal{M}(\mathcal{R})$ the **base sequence** of $\mathcal{R} \in \mathcal{R}_N$. Let $\mathcal{A}_N(\mathcal{R}) \in \mathcal{F}_N$ denote the collection of all Zeckendorf games resulting from replacing each instance of $\mathcal{E}_k$ in $\mathcal{M}$ by either C_k or the 1-prefix (S_k, C_{k-1}) (for every $k \geq 2$). We establish the following important result, which makes it clear why we have pursued this construction in the manner that we did.

---

[5] We deviate from notation earlier in the paper and write move sequences as tuples.

[6] We elect to use the notation $\mathcal{R}_N$ as we think of these games on input N as *representatives* of the corresponding classes $\mathcal{A}_N(\mathcal{R})$ that we define later in this discussion.

Lemma 6 *The sets* $\{\mathcal{A}_N(\mathcal{R}) : \mathcal{R} \in \mathcal{R}_N\}$ *partition* Ω_N.

Proof We first show that the sets $\mathcal{A}_N(\mathcal{R})$ are disjoint. Take distinct $\mathcal{R}_1 = (M_1^1, \ldots, M_\lambda^1), \mathcal{R}_2 = (M_1^2, \ldots, M_{\lambda'}^2) \in \mathcal{R}_N$ with base sequences $\mathcal{M}_1(\mathcal{R}_1) = (\tilde{M}_1^1, \tilde{M}_2^1, \ldots, \tilde{M}_\lambda^1)$ and $\mathcal{M}_2(\mathcal{R}_2) = (\tilde{M}_1^2, \tilde{M}_2^2, \ldots, \tilde{M}_{\lambda'}^2)$, respectively. Since $\mathcal{R}_1 \neq \mathcal{R}_2$, take smallest i for which $M_i^1 \neq M_i^2 \implies \tilde{M}_i^1 \neq \tilde{M}_i^2$, and study the construction of any two games $\mathcal{G}_1 \in \mathcal{A}_N(\mathcal{R}_1)$ and $\mathcal{G}_2 \in \mathcal{A}_N(\mathcal{R}_2)$ as described above. If $\mathcal{G}_1$ and $\mathcal{G}_2$ are consistent prior to $\tilde{M}_i^1$ and $\tilde{M}_i^2$ (i.e., for $k \geq 2$, all $\mathcal{E}_k$ are replaced by the same choice of C_k or (S_k, C_{k-1})), which is certainly the only way the two games remain equal up to this point of their construction, then we necessarily produce a difference in the two games on $\tilde{M}_i^1$ and $\tilde{M}_i^2$ if $M_i^1 = \tilde{M}_i^1$ and $M_i^2 = \tilde{M}_i^2$ $(M_i^1 \neq M_i^2)$ or $M_i^1 \neq \tilde{M}_i^1$ and $M_i^2 \neq \tilde{M}_i^2$ $(\tilde{M}_i^1 = \mathcal{E}_{k_1}, \tilde{M}_i^2 = \mathcal{E}_{k_2}, k_1 \neq k_2)$. If $M_i^1 = \tilde{M}_i^1$ and $M_i^2 \neq \tilde{M}_i^2$ (say $\tilde{M}_i^2 = \mathcal{E}_k$, so necessarily $M_i^2 = C_k$ and $M_{i-1}^2 \neq S_{k+1}$ for some $k \geq 2$), then game $\mathcal{G}_2$ is next filled with either $C_k \neq M_i^1$ $(M_i^1 = M_i^2$ if $C_k = M_i^1)$ or (S_k, C_{k-1}). In the latter case, if the two games were equal after this, then necessarily $(M_{i-1}^1, M_i^1, M_{i+1}^1) = (S_{k+1}, S_k, C_{k-1})$ (by definition of $\mathcal{I}_N(\mathcal{R}_1)$ and $\mathcal{R}_N$, M_{i+1}^1 and M_{i-1}^1 follow after establishing $M_i^1 = S_k$), contradicting $M_{i-1}^2 \neq S_{k+1}$ (since $M_{i-1}^1 = M_{i-1}^2$ and $M_i^2 = C_k$, $\tilde{M}_i^2 = \mathcal{E}_k$).

We therefore conclude that $\mathcal{A}_N(\mathcal{R}_1) \cap \mathcal{A}_N(\mathcal{R}_2) = \emptyset$, i.e., the sets $\mathcal{A}_N(\mathcal{R})$ for $\mathcal{R} \in \mathcal{R}_N$ are disjoint; it remains to show that any game $\mathcal{G} \in \Omega_N$ is in some set $\mathcal{A}_N(\mathcal{R})$. For $\mathcal{G} \in \Omega_N$, let $\mathcal{R}$ be the game resulting from replacing every instance of the sequence (M, S_k, C_{k-1}), $M \neq S_{k+1}$ in game $\mathcal{G}$ by the sequence (M, C_k) (for $k \geq 2$). The resulting game $\mathcal{R}$ is such that $(M_i, M_{i+1}) = (S_k, C_{k-1}) \implies M_{i-1} = S_{k+1}$ for all $i \in [\lambda]$ and $k \geq 2$: any sequence $(M_{i-1}, M_i, M_{i+1}) = (M, S_k, C_{k-1})$ with $M \neq S_{k+1}$ in $\mathcal{R}$ necessarily results from having replaced (S_{k-1}, C_{k-2}) for C_{k-1} in game $\mathcal{G}$ (as (S_k, C_{k-1}) would have been replaced by S_{k+1} otherwise), but we know this does not occur by the description above, so we indeed have $\mathcal{R} \in \mathcal{R}_N$. Also, $\mathcal{G} \in \mathcal{A}_N(\mathcal{R})$: we can reverse all the replacements $(S_k, C_{k-1}) \leftrightarrow C_k$ made in achieving $\mathcal{R}$ from $\mathcal{G}$, since the resulting C_k moves correspond to $\mathcal{E}_k$ (for some $k \geq 2$) in the base sequence $\mathcal{M}(\mathcal{R})$ as the preceding move is not S_{k+1}.

Therefore, applying Lemma 6 and the law of total probability,

$$\begin{aligned}\mu_N(\text{length is } z \bmod Z) &= \sum_{\mathcal{R} \in \mathcal{R}_N} \mu_N\left(\text{length is } z \bmod Z \mid \mathcal{A}_N(\mathcal{R})\right) \mu_N\left(\mathcal{A}_N(\mathcal{R})\right)\\ \mathbb{P}_N(\text{length is } z \bmod Z) &= \sum_{\mathcal{R} \in \mathcal{R}_N} \mathbb{P}_N\left(\text{length is } z \bmod Z \mid \mathcal{A}_N(\mathcal{R})\right) \mathbb{P}_N\left(\mathcal{A}_N(\mathcal{R})\right)\end{aligned} \tag{1}$$

so we can reduce proving Theorem 8 to establishing that the conditional probabilities for $\mathcal{R} \in \mathcal{R}_N$, with respect to both measures, overwhelmingly tend to $1/2$ in the limit.

4.3 Analysis

Define the random variable $m_N : \Omega_N \to \mathbb{N}$ first on $\mathcal{R} \in \mathcal{R}_N$ by $m_N(\mathcal{R}) = |\mathcal{I}_N(\mathcal{R})|$, denoting the number of terms $\mathcal{E}_k$ (for $k \geq 2$) in the base sequence of $\mathcal{R}$, then lift to arbitrary $\mathcal{G} \in \Omega_N$ by letting $m_N(\mathcal{G}) = m_N(\mathcal{R})$ for the unique $\mathcal{R} \in \mathcal{R}_N$ such that $\mathcal{G} \in \mathcal{A}_N(\mathcal{R})$ (see Lemma 6).[7] Fix $\mathcal{R} \in \mathcal{R}_N$, and observe that on the event $\mathcal{A}_N(\mathcal{R})$, $m_N(\mathcal{G})$ for $\mathcal{G} \in \mathcal{A}_N(\mathcal{R})$ is the fixed constant $|\mathcal{I}_N(\mathcal{R})|$. Now construct corresponding Bernoulli random variables $X_1^{\mathcal{R}}, X_2^{\mathcal{R}}, \ldots, X_{|m_N(\mathcal{R})|}^{\mathcal{R}} : \Omega_N \to \{0, 1\}$ for each of the instances of terms of form $\mathcal{E}_k$ (for $k \geq 2$) in the base sequence of $\mathcal{R}$: here, $X_i^{\mathcal{R}}(\mathcal{G}) = 1$ if and only if $\mathcal{G} \in \mathcal{A}_N(\mathcal{R})$ and $\mathcal{G}$ is achieved by the ith instance of $\mathcal{E}_k$ in the base sequence $\mathcal{R}$ being the $(1, k)$-prefix (S_k, C_{k-1}). Say we replaced this $\mathcal{E}_k$ with (S_k, C_{k-1}), and let n_i denote the number of playable moves in the game available in the turn immediately after playing S_k. We can say more about the random variables $X_i^{\mathcal{R}}$ defined here.

Lemma 7 *Fix $\mathcal{R} \in \mathcal{R}_N$. Define the random variables $X_1^{\mathcal{R}}, X_2^{\mathcal{R}}, \ldots, X_{|m_N(\mathcal{R})|}^{\mathcal{R}}$ as above. When conditioned on $\mathcal{A}_N(\mathcal{R})$, the variables $X_1^{\mathcal{R}}, X_2^{\mathcal{R}}, \ldots, X_{|m_N(\mathcal{R})|}^{\mathcal{R}}$ are independent Bernoulli random variables with each $X_i^{\mathcal{R}}$ having parameter $1/2$ under the measure μ_N, and parameter $p_i = \frac{1}{1+n_i}$ under the measure $\mathbb{P}_N$. Explicitly,*

$$\mu_N\left(X_i^{\mathcal{R}} = 1 \mid \mathcal{A}_N(\mathcal{R})\right) = \frac{1}{2}, \qquad \mathbb{P}_N\left(X_i^{\mathcal{R}} = 1 \mid \mathcal{A}_N(\mathcal{R})\right) = \frac{1}{1+n_i}.$$

Proof Fix a random variable $X_i^{\mathcal{R}}$, and observe that for every particular setting of all other $\mathcal{E}_k$ terms in $\mathcal{R}$, there exist exactly two games in $\mathcal{A}_N(\mathcal{R})$ faithful to this setting, corresponding to choosing C_k and the $(1, k)$-prefix (S_k, C_{k-1}) for the ith such $\mathcal{E}_k$. It follows immediately that under the uniform measure μ_N, we indeed have $\mu_N\left(X_i^{\mathcal{R}} = 1 \mid \mathcal{A}_N(\mathcal{R})\right) = \frac{1}{2}$, since in particular there exists a bijection between games in the subset $\mathcal{A}_N(\mathcal{R})$ with $X_i^{\mathcal{R}} = 0$ and $X_i^{\mathcal{R}} = 1$ and all games in Ω_N are given equal probability under the measure μ_N.

Under the probability measure $\mathbb{P}_N$, it is straightforward to observe that the game replacing the ith instance of $\mathcal{E}_k$ by the $(1, k)$-prefix (S_k, C_{k-1}) requires an additional decision with probability $\frac{1}{n_i}$ of yielding the desired C_{k-1}, and it thus follows that the parameter p_i of the Bernoulli random variable $X_i^{\mathcal{R}}$ is given by

$$\frac{p_i}{1-p_i} = \frac{1}{n_i} \implies p_i = \frac{1}{1+n_i}.$$

To establish independence, it suffices to show that for any subset $S \subseteq [m_N(\mathcal{R})]$, we have the identity

[7] In particular, $m_N(\mathcal{R}) = \log_2(\mathcal{A}_N(\mathcal{R}))$.

$$\begin{aligned}\mathbb{P}_N\left(X_i^{\mathcal{R}} = 1 \text{ iff } i \in S \mid \mathcal{A}_N(\mathcal{R})\right) &= \prod_{i\in S}\mathbb{P}_N\left(X_i^{\mathcal{R}} = 1 \mid \mathcal{A}_N(\mathcal{R})\right)\\ &\quad\cdot\prod_{j\notin S}\mathbb{P}_N\left(X_j^{\mathcal{R}} = 0 \mid \mathcal{A}_N(\mathcal{R})\right)\\ &= \prod_{i\in S} p_i \cdot \prod_{j\notin S}(1-p_j).\end{aligned}$$

We can relate the conditional probabilities $\{\mathbb{P}_N\left(X_i^{\mathcal{R}} = 1 \text{ iff } i \in S\ \mathcal{A}_N(\mathcal{R})\right) : S \in [m_N(\mathcal{R})]\}$ whenever $S_1 = S_2 \cup \{j\}$ (for $j \in [m_N(\mathcal{R})]$): all choices for each term $\mathcal{E}_k$ but one are consistent (namely, $X_j^{\mathcal{R}} = 1$ for the numerator in the following), and thus

$$\frac{\mathbb{P}_N\left(X_i^{\mathcal{R}} = 1 \text{ iff } i \in S_1 \mid \mathcal{A}_N(\mathcal{R})\right)}{\mathbb{P}_N\left(X_i^{\mathcal{R}} = 1 \text{ iff } i \in S_2 \mid \mathcal{A}_N(\mathcal{R})\right)} = \frac{1}{n_j} = \frac{p_j}{1-p_j} = \frac{\prod_{i\in S_1} p_i \cdot \prod_{j\notin S_1}(1-p_j)}{\prod_{i\in S_2} p_i \cdot \prod_{j\notin S_2}(1-p_j)}.$$

Now, since we have the identity

$$\begin{aligned}\sum_{S\subseteq[m_N(\mathcal{R})]}\mathbb{P}_N\left(X_i^{\mathcal{R}} = 1 \text{ iff } i \in S \mid \mathcal{A}_N(\mathcal{R})\right) = 1 &= \prod_{i=1}^{m_N(\mathcal{R})}(p_i + (1-p_i))\\ &= \sum_{S\subseteq[m_N(\mathcal{R})]}\left(\prod_{i\in S} p_i \cdot \prod_{j\notin S}(1-p_j)\right)\end{aligned}$$

and quotients between summands corresponding to two sets differing by one element are the same, the summands on both sides for any subset $S \subseteq [m_N(\mathcal{R})]$ are necessarily equal. More specifically, letting $p_1 = \mathbb{P}_N\left(X_i^{\mathcal{R}} = 1 \text{ iff } i \in \emptyset \mid \mathcal{A}_N(\mathcal{R})\right)$ and $p_2 = \prod_{i\in\emptyset} p_i \cdot \prod_{j\notin\emptyset}(1-p_j)$, by incrementally including elements to some $S \subseteq [m_N(\mathcal{R})]$ we can write the corresponding summands on the left and right hand sides as the same multiple of p_1 and p_2, respectively. This reduces to $p_1 = p_2$, and thus summands corresponding to the same S are equal. Thus, we have the desired identity for any subset $S \subseteq [m_N(\mathcal{R})]$.

Lemma 7 yields the following easy observation.

Corollary 1 *Say $F_n \le N < F_{n+1}$, fix some $\mathcal{R} \in \mathcal{R}_N$, and define the random variables $X_1^{\mathcal{R}}, X_2^{\mathcal{R}}, \ldots, X_{m_N(\mathcal{R})}^{\mathcal{R}}$ as above. Under the uniform measure μ_N, whenever $\mathcal{R} \in \mathcal{R}_N$ is such that $m_N(\mathcal{R}) > 0$, μ_N (Player 1 wins $\mid \mathcal{A}_N(\mathcal{R})) = 1/2$. Under the probability measure $\mathbb{P}_N$, when conditioned on the event $\mathcal{A}_N(\mathcal{R})$, there are at most $2n$ distinct values of the parameters $p_i = \frac{1}{1+n_i}$ amongst the random variables $X_i^{\mathcal{R}}$, and for any such i, $\frac{1}{2n-2} \le p_i \le 1/2$.*

Proof If $m_N(\mathcal{R}) > 0$, then the statement μ_N (Player 1 wins $\mid \mathcal{A}_N(\mathcal{R})) = 1/2$ follows immediately by applying the law of total probability on all settings of the Bernoulli random variables $(X_2^{\mathcal{R}}, \ldots, X_{m_N(\mathcal{R})}^{\mathcal{R}}) \in \{0,1\}^{m_N(\mathcal{R})-1}$, namely since this further conditioning always yields a conditional probability of $1/2$ (see the proof of

Lemma 7). At any point of a game $\mathcal{G} \in \Omega_N$ satisfying $F_n \leq N < F_{n+1}$, there are at most $2n - 3$ (and thus certainly at most $2n$) playable moves (and thus at most $2n - 3$ distinct values of n_i, and thus p_i): the combine moves $C_1, C_2, \ldots, C_{n-1}$ and the splitting moves $S_2, S_3, \ldots, S_{n-1}$. In particular, since $n_i \in [2n-3]$ for all $i \in [m_N(\mathcal{R})]$, it follows that $\frac{1}{1+(2n-3)} \leq p_i \leq \frac{1}{1+1}$, i.e., $\frac{1}{2n-2} \leq p_i \leq \frac{1}{2}$.

We will also make use of the following lemma.

Lemma 8 *Say $F_n \leq N < F_{n+1}$.*[8] *For any $c \in (0, \varphi)$ (with φ denoting the golden ratio),*

$$\lim_{N\to\infty} \mu_N \left(m_N(\mathcal{G}) \geq c^n\right) = 1, \qquad \lim_{N\to\infty} \mathbb{P}_N \left(m_N(\mathcal{G}) \geq c^n\right) = 1.$$

Proof Fix $c \in (0, \varphi)$, and consider a game $\mathcal{G} \in \Omega_N$: for the representative $\mathcal{R} \in \mathcal{R}_N$ such that $\mathcal{G} \in \mathcal{A}_N(\mathcal{R})$, every occurrence of the sequence (C_1, C_1, C_2) corresponds to an element of $\mathcal{I}_N(\mathcal{R})$ (specifically the move C_2, as it is not preceded by S_3, and the latter two moves are maintained in $\mathcal{R}$; see the proof of Lemma 6), i.e., $m_N(\mathcal{G}) = |\mathcal{I}_N(\mathcal{R})|$ is at least the number of occurrences of the sequence (C_1, C_1, C_2) in $\mathcal{G}$. Letting $\mathcal{N}(\mathcal{G})$ denote the number of occurrences of the sequence (C_1, C_1, C_2) in the game $\mathcal{G}$, it thus suffices to show that

$$\lim_{N\to\infty} \mu_N \left(\mathcal{N}(\mathcal{G}) \geq c^n\right) = 1, \qquad \lim_{N\to\infty} \mathbb{P}_N \left(\mathcal{N}(\mathcal{G}) \geq c^n\right) = 1.$$

Proceed studying the probability measure $\mathbb{P}_N$; the analysis carries over exactly[9] when $\mathbb{P}_N$ is replaced with μ_N. By Corollary 1, at any move where $h_1 \geq 5$, the probability of achieving the sequence (C_1, C_1, C_2) is at least $\left(\frac{1}{2n}\right)^3 = \frac{1}{8n^3}$, and any sequence of three moves can decrease the height of bin 1 by at most 6 (via three consecutive C_1 moves); observe that the occurrence of C_2 in a sequence (C_1, C_1, C_2) within a game $\mathcal{G}$ necessarily corresponds to an instance of the delimiter $\mathcal{E}_2$ in the base sequence of $\mathcal{G}$ (e.g. see the proof of Lemma 6). Thus, study the sequences of moves in game $\mathcal{G}$ (which necessarily exist by the preceding discussion) given by the triples

$$(M_1, M_2, M_3), (M_4, M_5, M_6), \ldots, (M_{3\lfloor N/6\rfloor-2}, M_{3\lfloor N/6\rfloor-1}, M_{3\lfloor N/6\rfloor})$$

each of which independently takes on the value (C_1, C_1, C_2) with probability at least $\frac{1}{8n^3}$. Thus, letting $Y_N \stackrel{d}{=} \mathrm{Bin}\left(\lfloor N/6\rfloor, p = \frac{1}{8n^3}\right)$, we have that

$$\mathbb{P}\left(Y_N \geq c^n\right) \leq \mathbb{P}_N\left(\mathcal{N}(\mathcal{G}) \geq c^n\right)$$

from which $\mathbb{E}[Y_N] = \frac{\lfloor N/6\rfloor}{8n^3}$ together with a Chernoff bound with $\delta = \frac{1}{2}$ yields for N large,

---

[8] It is perhaps more appropriate to think of n as a function $n(N)$.

[9] Certainly, the analysis is much looser than necessary when $\mathbb{P}_N$ is replaced with μ_N.

$$\mathbb{P}\left(Y_N \leq c^n\right) \leq \mathbb{P}\left(Y_N \leq \frac{\lfloor N/6 \rfloor}{16n^3}\right) \leq \exp\left(-\frac{\lfloor N/6 \rfloor}{32n^3}\right) \xrightarrow{N\to\infty} 0$$

since it is straightforward to verify that $c^n \leq \frac{\lfloor N/6 \rfloor}{16n^3}$ for large N (e.g. use Binet).

Corollary 2 *Lemma 8 immediately yields that for any* $d \in \mathbb{N}$,

$$\lim_{N\to\infty} \mu_N\left(m_N(\mathcal{G}) \geq n^d\right) = 1, \qquad \lim_{N\to\infty} \mathbb{P}_N\left(m_N(\mathcal{G}) \geq n^d\right) = 1$$

i.e., with probability approaching 1 *as* $N \to \infty$*, the number of decisions made to achieve* $\mathcal{G}$ *from the set* $\mathcal{A}_N(\mathcal{R})$ *containing it is superpolynomial in* n*. This is more convenient for later.*

The proof of Lemma 8 yields the following observations on the likeliest collection $\mathcal{A}_N(\mathcal{R})$ and the number of representative games, which will be referenced again in Sect. 5.

Corollary 3 *The probability of the likeliest collection* $\mathcal{A}_N(\mathcal{R})$ *vanishes as* $N \to \infty$*, i.e.,*

$$\lim_{N\to\infty} \max_{\mathcal{G}\in\Omega_N} \mu_N(\mathcal{A}_N(\mathcal{G})) = \lim_{N\to\infty} \max_{\mathcal{G}\in\Omega_N} \mathbb{P}_N(\mathcal{A}_N(\mathcal{G})) = 0$$

and the number of representatives satisfies $|\mathcal{R}_N| \xrightarrow{N\to\infty} \infty$.

Proof Again, proceed on the probability measure $\mathbb{P}_N$, as the analysis carries over exactly for μ_N. The proof of Lemma 2 (taking $d = 1$) determines there are at least $m \geq n$ instances of (C_1, C_1, C_2) in the first $\lfloor N/6 \rfloor$ moves of a random game $\mathcal{G} \in \Omega_N$, with probability tending to 1 in the limit $N \to \infty$: we proceed studying such a game $\mathcal{G}$. Writing $\mathcal{G} \in \mathcal{A}_N(\mathcal{R})$, every latter instance of C_1 in one of these m sequences is in[10] any game in $\mathcal{A}_N(\mathcal{R})$, and the move C_2 could have been played instead of this instance of C_1. For $i = 1, \ldots, m$, denote $\mathcal{B}_i^N(\mathcal{R}) \subseteq \Omega_N$ to be those games consistent with a game in $\mathcal{A}_N(\mathcal{R})$ up to the ith such instance of C_1, but for which the move C_2 is played instead. It is immediately observed that the sets $\mathcal{B}_i^N(\mathcal{R})$ are disjoint and that $\mathbb{P}_N(\mathcal{A}_N(\mathcal{R})) \leq \mathbb{P}_N(\mathcal{B}_i^N(\mathcal{R}))$ for each $i = 1, \ldots, m$, so we have

$$\begin{aligned} n \cdot \mathbb{P}_N(\mathcal{A}_N(\mathcal{R})) \leq m \cdot \mathbb{P}_N(\mathcal{A}_N(\mathcal{R})) &\leq \sum_{i=1}^{m} \mathbb{P}_N(\mathcal{B}_i^N(\mathcal{R})) \leq 1 \\ \implies \mathbb{P}_N(\mathcal{A}_N(\mathcal{R})) &\leq \frac{1}{n} \xrightarrow{N\to\infty} 0 \end{aligned}$$

[10] This is in the sense that this instance of C_1 is not compressed in achieving $\mathcal{R}$ from $\mathcal{G}$ (see the proof of Lemma 6), and is not replaced with a term of form $\mathcal{E}_k$ in achieving the base sequence $\mathcal{M}(\mathcal{R})$.

so we have $\lim_{N\to\infty}\max_{\mathcal{G}\in\Omega_N}\mathbb{P}_N(\mathcal{A}_N(\mathcal{G}))=0$, and $|\mathcal{R}_N|\xrightarrow{N\to\infty}\infty$ follows immediately.

Finally, we need the following simple result concerning the behavior of a binomial random variable with sufficiently large variance. Certainly, the case $Z=2$ in Lemma 9 corresponds to studying the expressions $\mathbb{P}(\mathcal{B}$ is odd) and $\mathbb{P}(\mathcal{B}$ is even).

Lemma 9 *Consider a binomial random variable $\mathcal{B}=Bin(m,p)$. For any values of $\epsilon>0$ and $Z\in\mathbb{N}$, there exists a constant $N(\epsilon,Z)$ such that if $var(\mathcal{B})=mp(1-p)\geq N(\epsilon,Z)$ (i.e., if the variance of $\mathcal{B}$ is sufficiently large), then for any $z\in\{0,1,\dots,Z-1\}$,*

$$\left|\mathbb{P}(\mathcal{B}\equiv z \bmod Z)-\frac{1}{Z}\right|\leq\epsilon.$$

Proof The value $\mathbb{P}(\mathcal{B}=k)$ increases on $k\leq\lfloor(m+1)p\rfloor$ and decreases on $k\geq\lfloor(m+1)p\rfloor$, and $\max_{0\leq k\leq m}\mathbb{P}(\mathcal{B}=k)=\mathbb{P}(\mathcal{B}=\lfloor(m+1)p\rfloor)\xrightarrow{mp(1-p)\to\infty}0$ (use Stirling); denoting $p_k=\mathbb{P}(\mathcal{B}=k)$, write $p_0\leq p_1\leq\cdots\leq p_{\lfloor(m+1)p\rfloor}$ and $p_{\lfloor(m+1)p\rfloor}\geq p_{\lfloor(m+1)p\rfloor+1}\geq\cdots\geq p_m$. Fix $\epsilon>0$, and choose $N(\epsilon,Z)$ such that $mp(1-p)\geq N(\epsilon,Z)$ implies $\mathbb{P}(\mathcal{B}=\lfloor(m+1)p\rfloor)<\frac{\epsilon}{2}$. Certainly, for any distinct values $z_1<z_2$ in $\{0,1,\dots,Z-1\}$, the number of terms $p_0,p_1,\dots,p_{\lfloor(m+1)p\rfloor}$ with indices equal to z_1 modulo Z and equal to z_2 modulo Z is either the same or there exists one more such term corresponding to z_1: we can switch the "dominant modulus" by removing the term of largest index equal to either z_1 or z_2 modulo Z, for which the corresponding term $p_k\leq\frac{\epsilon}{2}$ by choice of the constant $N(\epsilon,Z)$. The analogous statement extends to $p_{\lfloor(m+1)p\rfloor},p_{\lfloor(m+1)p\rfloor+1},\dots,p_m$, so it follows that we can write

$$|\mathbb{P}(\mathcal{B}\equiv z_1 \bmod Z)-\mathbb{P}(\mathcal{B}\equiv z_2 \bmod Z)|\leq\epsilon$$

which immediately yields the desired statement.

We are now ready to proceed with the proof of Theorem 8.

Proof *(Proof of Theorem* 8) Fix an integer $Z\geq2$ corresponding to the number of players in a Z-player Zeckendorf game, and some value $z\in\{0,1,\dots,Z-1\}$. As in the proof of Lemma 8, we strictly concern ourselves with the probability measure $\mathbb{P}_N$, as the analysis carries over exactly for the uniform measure μ_N (again, it is much looser than necessary for μ_N). For the probability measure $\mathbb{P}_N$, since $m_N(\mathcal{G})$ is fixed at $|\mathcal{I}_N(\mathcal{R})|=m_N(\mathcal{R})$ on any set $\mathcal{A}_N(\mathcal{R})$ for $\mathcal{R}\in\mathcal{R}_N$, for $N\geq N_1(\epsilon)$,

$$\sum_{\substack{\mathcal{R}\in\mathcal{R}_N\\ m_N(\mathcal{R})<2n^3}}\mathbb{P}_N(\mathcal{A}_N(\mathcal{R}))\leq\epsilon. \tag{2}$$

Consider a fixed representative $\mathcal{R}\in\mathcal{R}_N$ for which $m_N(\mathcal{R})\geq2n^3$: by Corollary 1, there are at most $2n$ distinct values of the parameters $p_i=\frac{1}{1+n_i}$ of the corresponding random variables $X_1^{\mathcal{R}},X_2^{\mathcal{R}},\dots,X_{m_N(\mathcal{R})}^{\mathcal{R}}$, and thus by the pigeonhole principle, the number of instances of the value of p_i with largest multiplicity is at least $\frac{2n^3}{2n} = n^2 \xrightarrow{N\to\infty} \infty$; henceforth call this p. Let us say there are $m \geq n^2$ instances of this value of p_i: we can study the sum $\mathcal{B}_{\mathcal{R}} = \sum_{i=1}^{m} Y_i$, where the Y_i correspond to those random variables $X_i^{\mathcal{R}}$ with this corresponding success probability p. Furthermore, by Lemma 7, the random variables Y_i are independent when conditioned on the event $\mathcal{A}_N(\mathcal{R})$, so $\mathcal{B}_{\mathcal{R}} \stackrel{d}{=} \text{Bin}(m, p)$ under this conditional distribution. From the bounds on p from Corollary 1, it follows that the variance of this binomial random variable, when conditioned on the event $\mathcal{A}_N(\mathcal{R})$, has the exploding lower bound

$$\text{var}(\mathcal{B}_{\mathcal{R}}) = mp(1-p) \geq \frac{n^2}{2(2n-2)} \xrightarrow{N\to\infty} \infty$$

i.e., the binomial random variable $\mathcal{B}_{\mathcal{R}}$ has variance exploding in the limit $N \to \infty$ (for any $\mathcal{R} \in \mathcal{R}_N$). By Lemma 9, the random variable $\mathcal{B}_{\mathcal{R}}$ will take a value equal to z modulo Z with probability approaching $\frac{1}{Z}$ as $N \to \infty$. Thus, studying the quantities given by $\mathbb{P}_N$ (Game length equals z mod $Z \mid \mathcal{A}_N(\mathcal{R})$), we can further condition (upon the conditioning $\mathcal{A}_N(\mathcal{R})$) on all random variables $X_i^{\mathcal{R}}$ not corresponding to those Y_i constituting a summand in the binomial random variable $\mathcal{B}_{\mathcal{R}}$ and appeal to the law of total probability:

$$\begin{aligned}&\mathbb{P}_N\,(\text{Game length is } z \bmod Z \mid \mathcal{A}_N(\mathcal{R}))\\ &= \sum_{\substack{\text{assignments } A\\ \text{to } X_i^{\mathcal{R}} \neq Y_j \forall j}} \mathbb{P}_N\,(\text{Game length is } z \bmod Z \mid \mathcal{A}_N(\mathcal{R}), A) \cdot \mathbb{P}_N\,(A \mid \mathcal{A}_N(\mathcal{R}))\end{aligned}$$

where each term $\mathbb{P}_N$ (Game length equals z mod $Z \mid \mathcal{A}_N(\mathcal{R}), A$) is understood as the probability that the length of the game is equal to z mod Z when replacing those $\mathcal{E}_k$ which correspond to the random variables Y_i, while leaving all other moves of the game fixed: this is precisely given by the binomial random variable $\mathcal{B}_{\mathcal{R}}$ added to some fixed length determined by the setting of the $X_i^{\mathcal{R}}$ which are not of the form Y_j for some j. Letting $\ell(A)$ denote this length for the assignment A, we can thus write

$$\mathbb{P}_N\,(\text{Game length is } z \bmod Z \mid \mathcal{A}_N(\mathcal{R}), A) = \mathbb{P}_N\,(\mathcal{B}_{\mathcal{R}} + \ell(A) \text{ is odd} \mid \mathcal{A}_N(\mathcal{R}), A)$$

By Lemma 7, the random variables Y_i constituting the Bernoulli trials in $\mathcal{B}_{\mathcal{R}}$ are independent from the random variables $X_i^{\mathcal{R}}$ that were fixed when conditioned on $\mathcal{A}_N(\mathcal{R})$ and the assignment A, and thus we can apply Lemma 9 to deduce that these conditional probabilities are arbitrarily close to $\frac{1}{Z}$ for $N \geq N_2(\epsilon, Z)$ by the preceding discussion (for sufficiently large $N_2(\epsilon, Z) \in \mathbb{N}$). Importantly, this is uniform over all such terms in the sum, in the sense that we can choose $N_2(\epsilon, Z)$ such that we achieve the same guarantee for any such assignment A of binary values to the random variables $X_i^{\mathcal{R}}$ not constituting $\mathcal{B}_{\mathcal{R}}$. Thus, for any $\mathcal{R} \in \mathcal{R}_N$ satisfying $m_N(\mathcal{R}) \geq 2n^3$ and $N \geq N_2(\epsilon, Z)$, we have the bound

$$\left| \mathbb{P}_N\,(\text{Game length equals } z \bmod Z \mid \mathcal{A}_N(\mathcal{R})) - \frac{1}{Z} \right| \leq \epsilon.$$

Therefore, for $N \geq \max\{N_1(\epsilon), N_2(\epsilon, Z)\}$, we can take Eqs. (1) and (2) to achieve the bound on the probability $\mathbb{P}_N$(Game length equals z mod Z) given by

$$\begin{aligned}
&\mathbb{P}_N(\text{Game length is } z \bmod Z) \\
&= \sum_{\mathcal{R}\in\mathcal{R}_N} \mathbb{P}_N\left(\text{Game length is } z \bmod Z \mid \mathcal{A}_N(\mathcal{R})\right) \cdot \mathbb{P}_N\left(\mathcal{A}_N(\mathcal{R})\right) \\
&\leq \epsilon + \sum_{\substack{\mathcal{R}\in\mathcal{R}_N \\ m_N(\mathcal{R})\geq 2n^3}} \mathbb{P}_N\left(\text{Game length is } z \bmod Z \mid \mathcal{A}_N(\mathcal{R})\right) \cdot \mathbb{P}_N\left(\mathcal{A}_N(\mathcal{R})\right) \\
&\leq \epsilon + \sum_{\substack{\mathcal{R}\in\mathcal{R}_N \\ m_N(\mathcal{R})\geq 2n^3}} \left(\frac{1}{Z} + \epsilon\right) \cdot \mathbb{P}_N\left(\mathcal{A}_N(\mathcal{R})\right) + \epsilon \leq \frac{1}{Z} + 2\epsilon
\end{aligned}$$

and similarly, we have

$$\begin{aligned}
&\mathbb{P}_N(\text{Game length is } z \bmod Z) \\
&= \sum_{\mathcal{R}\in\mathcal{R}_N} \mathbb{P}_N\left(\text{Game length is } z \bmod Z \mid \mathcal{A}_N(\mathcal{R})\right) \cdot \mathbb{P}_N\left(\mathcal{A}_N(\mathcal{R})\right) \\
&\geq \sum_{\substack{\mathcal{R}\in\mathcal{R}_N \\ m_N(\mathcal{R})\geq 2n^3}} \mathbb{P}_N\left(\text{Game length is } z \bmod Z \mid \mathcal{A}_N(\mathcal{R})\right) \cdot \mathbb{P}_N\left(\mathcal{A}_N(\mathcal{R})\right) \\
&\geq \sum_{\substack{\mathcal{R}\in\mathcal{R}_N \\ m_N(\mathcal{R})\geq 2n^3}} \left(\frac{1}{Z} - \epsilon\right) \cdot \mathbb{P}_N\left(\mathcal{A}_N(\mathcal{R})\right) \geq \left(\frac{1}{Z} - \epsilon\right)(1-\epsilon)
\end{aligned}$$

so we conclude that for $N \geq \max\{N_1(\epsilon), N_2(\epsilon, Z)\}$,

$$\epsilon^2 - \frac{Z+1}{Z}\epsilon \leq \mathbb{P}_N\left(\text{Game length equals } z \bmod Z\right) - \frac{1}{Z} \leq 2\epsilon$$

which yields the desired limit by sending $\epsilon \downarrow 0$.

As previously mentioned, the $Z = 2$ case of Theorem 8 immediately yields Theorem 9.

4.4 Extending the Partition on Ω_N

Theorem 9 establishes that if both players advance randomly, the two-player Zeckendorf game is fair in the limit $N \to \infty$: a key stage of the proof involved partitioning Ω_N into collections of games where the games in any given collection only differ via certain interchanges of $(1, k)$-prefixes with the corresponding C_k. We can naturally extend this partition of Ω_N (i.e., strictly larger classes in the partition) to encompass arbitrary (ℓ, k)-prefixes or arbitrary (ℓ, k)-suffixes. Section 5 will establish that for these enlarged partitions of Ω_N, we can achieve an analogue of Conjecture 1 for the resulting sets in the partition, in the sense that with high probability in the limit, the corresponding distribution is "nearly Gaussian," in the sense of vanishing Kolmogorov-Smirnov distance (when mean and variance are normalized to be 0 and 1, respectively) with the standard normal.[11]

We first study the generalization for arbitrary (ℓ, k)-prefixes. Define $\mathcal{R}_N^{\mathcal{P}} \in \mathcal{F}_N$ to be the collection of all Zeckendorf games on input N such that any combine move C_k, for $k \geq 2$, cannot be compressed by a $(1, k)$-prefix (and thus by any prefix of C_k). Explicitly,

$$\mathcal{R}_N^{\mathcal{P}} = \{(M_1, \ldots, M_\lambda) \in \Omega_N : M_i = C_k \implies M_{i-1} \neq S_{k+1}, \ \forall i \in [\lambda], k \geq 2\}.$$

For a game $\mathcal{R} = (M_1, M_2, \ldots, M_\lambda) \in \mathcal{R}_N^{\mathcal{P}}$ with number of moves λ, construct the subset of indices $\mathcal{I}_N(\mathcal{R})$ to denote all combine moves in $\mathcal{R}$:

$$\mathcal{I}_N(\mathcal{R}) = \{i \in [\lambda] : M_i = C_k \text{ for some } k \geq 2\}.$$

Now, construct the *formal* sequence of moves $\mathcal{M}(\mathcal{R}) = (\tilde{M}_1, \tilde{M}_2, \ldots, \tilde{M}_\lambda)$ by replacing M_i by a symbol $\mathcal{E}_k^\ell$ for all $i \in \mathcal{I}_N(\mathcal{R})$: the subscript is the corresponding $k \geq 2$, while $\ell \geq 0$ denotes the longest (ℓ, k)-prefix that $M_i = C_k$ can be expanded into; note in particular that $\mathcal{M}(\mathcal{R})$ contains no combine moves. Call $\mathcal{M}(\mathcal{R})$ the **base sequence** of $\mathcal{R} \in \mathcal{R}_N$, and let $\mathcal{A}_N(\mathcal{R}) \in \mathcal{F}_N$ denote the collection of all Zeckendorf games resulting from replacing each instance of $\mathcal{E}_k^\ell$ in $\mathcal{M}$ by an (l, k)-prefix for some $l \leq \ell$. We establish the following analogue of Lemma 6, which yields a strictly broader partition of Ω_N (in the sense that each set in the partition given by Lemma 5 is a union of sets in the partition given by Lemma 6).

Proposition 5 *The sets* $\{\mathcal{A}_N(\mathcal{R}) : \mathcal{R} \in \mathcal{R}_N^{\mathcal{P}}\}$ *partition* Ω_N.

Proof Observe that a base sequence $\mathcal{M}(\mathcal{R}) = (\tilde{M}_1, \tilde{M}_2, \ldots, \tilde{M}_\lambda)$ uniquely determines $\mathcal{R} \in \mathcal{R}_N^{\mathcal{P}}$ by its explicit split moves moves and the subscripts of each of its symbols $\mathcal{E}_k^\ell$. This is true up to any initial subsequence, in the sense that there exists at most one $\mathcal{R} \in \mathcal{R}_N^{\mathcal{P}}$ which agrees with the moves and symbols $\mathcal{E}_k^\ell$ up to subscript.

We show that the sets $\mathcal{A}_N(\mathcal{R})$ are disjoint. Take distinct $\mathcal{R}_1 = (M_1^1, \ldots, M_\lambda^1)$, $\mathcal{R}_2 = (M_1^2, \ldots, M_{\lambda'}^2) \in \mathcal{R}_N^{\mathcal{P}}$ with base sequences $\mathcal{M}_1(\mathcal{R}_1) = (\tilde{M}_1^1, \tilde{M}_2^1, \ldots, \tilde{M}_\lambda^1)$ and $\mathcal{M}_2(\mathcal{R}_2) = (\tilde{M}_1^2, \tilde{M}_2^2, \ldots, \tilde{M}_{\lambda'}^2)$, respectively. Since $\mathcal{R}_1 \neq \mathcal{R}_2$, take smallest i for which $M_i^1 \neq M_i^2$, and study the construction of any two games $\mathcal{G}_1 \in \mathcal{A}_N(\mathcal{R}_1)$ and $\mathcal{G}_2 \in \mathcal{A}_N(\mathcal{R}_2)$ as described above. If $\mathcal{G}_1$ and $\mathcal{G}_2$ are consistent prior to $\tilde{M}_i^1$ and $\tilde{M}_i^2$ (i.e., for $k \geq 2$, all $\mathcal{E}_k^\ell$ are replaced by the same prefix of C_k), which is the only way the two games remain equal up to this point, then we produce a difference in the two games on $\tilde{M}_i^1$ and $\tilde{M}_i^2$ if $M_i^1 = \tilde{M}_i^1$ and $M_i^2 = \tilde{M}_i^2$ ($M_i^1 \neq M_i^2$) or $M_i^1 \neq \tilde{M}_i^1$

[11] In this subsection and Sect. 5, we borrow much of the same notation that was used in establishing Theorem 9. It will be clear from context exactly what objects we are referring to.

and $M_i^2 \neq \tilde{M}_i^2$ ($\tilde{M}_i^1 = \mathcal{E}_{k_1}^{\ell_1}$, $\tilde{M}_i^2 = \mathcal{E}_{k_2}^{\ell_2}$, $k_1 \neq k_2$). If $M_i^1 = \tilde{M}_i^1$ and $M_i^2 \neq \tilde{M}_i^2$ (say $\tilde{M}_i^2 = \mathcal{E}_k^\ell$, so $M_i^2 = C_k$), then $\mathcal{G}_2$ is next filled with either $C_k \neq M_i^1$ ($M_i^1 = M_i^2$ if $C_k = M_i^1$) or an (l, k)-prefix for some $l \leq \ell$. In the latter case, if the two games were equal until move $i + l - 1$, necessarily $\mathcal{R}_1$ contains a prefix of length at least 1, contradicting $\mathcal{R}_1 \in \mathcal{R}_N^{\mathcal{P}}$.

We therefore conclude that $\mathcal{A}_N(\mathcal{R}_1) \cap \mathcal{A}_N(\mathcal{R}_2) = \emptyset$, i.e., the sets $\mathcal{A}_N(\mathcal{R})$ for $\mathcal{R} \in \mathcal{R}_N$ are disjoint; it remains to show that any game $\mathcal{G} \in \Omega_N$ is in some set $\mathcal{A}_N(\mathcal{R})$. For $\mathcal{G} \in \Omega_N$, let $\mathcal{R}$ be the game resulting from replacing every combine move C_k with the longest playable (ℓ, k)-prefix. It follows immediately from construction that $\mathcal{R} \in \mathcal{R}_N^{\mathcal{P}}$, and that $\mathcal{G} \in \mathcal{A}_N(\mathcal{R})$.

Similarly, we define the generalization for arbitrary (ℓ, k)-suffixes. Define $\mathcal{R}_N^{\mathcal{S}} \in \mathcal{F}_N$ to be the collection of all Zeckendorf games on input N such that any combine move C_k, for $k \geq 2$, cannot be compressed by a $(1, k)$-suffix (and thus by any suffix of C_k). Explicitly,

$$\mathcal{R}_N^{\mathcal{S}} = \{(M_1, \ldots, M_\lambda) \in \Omega_N : M_i = C_k \implies M_{i+1} \neq S_{k+1} \; \forall i \in [\lambda], k \geq 2\}$$

For a game $\mathcal{R} = (M_1, M_2, \ldots, M_\lambda) \in \mathcal{R}_N^{\mathcal{S}}$ with number of moves λ, construct the subset of indices $\mathcal{I}_N(\mathcal{R})$ to denote all combine moves in $\mathcal{R}$:

$$\mathcal{I}_N(\mathcal{R}) = \{i \in [\lambda] : M_i = C_k \text{ for some } k \geq 2\}.$$

Now, construct the *formal* sequence of moves $\mathcal{M}(\mathcal{R}) = (\tilde{M}_1, \tilde{M}_2, \ldots, \tilde{M}_\lambda)$ by replacing M_i by a symbol $\mathcal{E}_k^\ell$ for all $i \in \mathcal{I}_N(\mathcal{R})$: the subscript is the corresponding $k \geq 2$, while ℓ denotes the longest (ℓ, k)-suffix that $M_i = C_k$ can be expanded into. Call $\mathcal{M}(\mathcal{R})$ the **base sequence** of $\mathcal{R} \in \mathcal{R}_N$, and let $\mathcal{A}_N(\mathcal{R}) \in \mathcal{F}_N$ denote the collection of all Zeckendorf games resulting from replacing each instance of $\mathcal{E}_k^\ell$ in $\mathcal{M}$ by an (l, k)-suffix for some $l \leq \ell$. We establish the following analogue of Lemma 6.

Proposition 6 *The sets* $\{\mathcal{A}_N(\mathcal{R}) : \mathcal{R} \in \mathcal{R}_N^{\mathcal{S}}\}$ *partition* Ω_N.

We do not provide the proof of Proposition 6, as it is pursued analogously to the proof of Proposition 5. The easy observation that each set $\mathcal{A}_N(\mathcal{R})$, for $\mathcal{R} \in \mathcal{R}_N^{\mathcal{P}}$ or $\mathcal{R} \in \mathcal{R}_N^{\mathcal{S}}$, is a union of equivalence classes of the corresponding sets studied in the proof of Theorem 9, yields the following trivial extension of Lemma 2, where we promote the notation $\mathcal{A}_N(\mathcal{G}) = \mathcal{A}_N(\mathcal{R})$ for the unique $\mathcal{R}$ satisfying $\mathcal{G} \in \mathcal{A}_N(\mathcal{R})$.

Proposition 7 *Say* $F_n \leq N < F_{n+1}$. *For any* $c \in (0, \varphi)$ *(with* φ *the golden ratio),*

$$\lim_{N\to\infty} \mu_N\left(\log_2 |\mathcal{A}_N(\mathcal{G})| \geq c^n\right) = 1, \qquad \lim_{N\to\infty} \mathbb{P}_N\left(\log_2 |\mathcal{A}_N(\mathcal{G})| \geq c^n\right) = 1,$$

which holds for either of the understandings $\mathcal{R} \in \mathcal{R}_N^{\mathcal{P}}$ *and* $\mathcal{R} \in \mathcal{R}_N^{\mathcal{S}}$.

Finally, we have the following extension of Corollary 1. We also omit the proof of this result, as it is a straightforward generalization of the proof of the aforementioned result.

Proposition 8 *Say* $F_n \leq N < F_{n+1}$, *fix some* $\mathcal{R} \in \mathcal{R}_N^{\mathcal{P}}$, *and define random variables* $X_1^{\mathcal{R}}, X_2^{\mathcal{R}}, \ldots, X_{N-1}^{\mathcal{R}}$ *corresponding to the length of the expansion corresponding to each combine move for a game in* $\mathcal{A}_N(\mathcal{R})$. *Under the measures* μ_N *and* $\mathbb{P}_N$ *conditioned on* $\mathcal{A}_N(\mathcal{R})$, *the random variables* $X_i^{\mathcal{R}}$ *are independent.*

5 Weak Convergence of Random Game Lengths as $N \to \infty$

In pursuit of the resolution of Conjecture 1, it may be productive to ask whether there exist natural subsets of Ω_N on which the distribution of random game lengths converge weakly to a Gaussian in the limit $N \to \infty$ of infinite input. As discussed in Sect. 4.4, the partitions defined in Propositions 5 and 6 enjoy this property in the sense described in Theorem 10, restated below.

Theorem 10 *For* $\mathcal{R} \in \mathcal{R}_N^{\mathcal{P}}$, *let* $F_N^{\mathcal{R}}(x) : \mathbb{R} \to [0, 1]$ *denote the distribution function corresponding to game lengths in* $\mathcal{A}_N(\mathcal{R})$ *over the conditional distribution induced by* $\mathbb{P}_N$, *normalized to have expectation* 0 *and variance* 1. *Let* $\Phi : \mathbb{R} \to [0, 1]$ *denote the distribution function of the standard normal. Then for any* $\epsilon > 0$,

$$\lim_{N\to\infty} \mathbb{P}_N \left(\sup_{x\in\mathbb{R}} \left| F_N^{\mathcal{R}}(x) - \Phi(x) \right| \geq \epsilon \right) = 0.$$

Similarly, for $\mathcal{R} \in \mathcal{R}_N^{\mathcal{S}}$, *let* $F_N^{\mathcal{R}}(x) : \mathbb{R} \to [0, 1]$ *denote the distribution function corresponding to game lengths in* $\mathcal{A}_N(\mathcal{R})$ *over the conditional distribution induced by* $\mathbb{P}_N$, *normalized to have expectation* 0 *and variance* 1. *Then for any* $\epsilon > 0$,

$$\lim_{N\to\infty} \mathbb{P}_N \left(\sup_{x\in\mathbb{R}} \left| F_N^{\mathcal{R}}(x) - \Phi(x) \right| \geq \epsilon \right) = 0.$$

The analogous results hold for the uniform measure μ_N.

To explicitly define the distributions referenced in Theorem 10, define a probability distribution $\mathbb{P}_N^{\mathcal{R}} : 2^{\mathcal{A}_N(\mathcal{R})} \to [0, 1]$ via, for any $\mathcal{S} \in 2^{\mathcal{A}_N(\mathcal{R})}$,

$$\mathbb{P}_N^{\mathcal{R}}(\mathcal{S}) = \frac{\mathbb{P}_N\left(\mathcal{S} \cap \mathcal{A}_N(\mathcal{R})\right)}{\mathbb{P}_N\left(\mathcal{A}_N(\mathcal{R})\right)}$$

Define the random variable $L_N^{\mathcal{R}} : \mathcal{A}_N(\mathcal{R}) \to \mathbb{R}$ on the space $(\mathcal{A}_N(\mathcal{R}), 2^{\mathcal{A}_N(\mathcal{R})}, \mathbb{P}_N^{\mathcal{R}})$ by

$$L_N^{\mathcal{R}}\left((M_1, \ldots, M_\lambda)\right) = \lambda$$

for any game $(M_1, \ldots, M_\lambda) \in \mathcal{A}_N(\mathcal{R})$, i.e., $L_N^{\mathcal{R}}$ studies game lengths in $\mathcal{A}_N(\mathcal{R})$. Then

$$F_{\mathcal{R}}(x) = \mathbb{P}_N^{\mathcal{R}}\left(\frac{L_N^{\mathcal{R}} - \mathbb{E}\left[L_N^{\mathcal{R}}\right]}{\sqrt{\operatorname{var}\left(L_N^{\mathcal{R}}\right)}} \leq x\right). \tag{3}$$

In other words, when we restrict $\mathbb{P}_N$ to the sets $\mathcal{A}_N(\mathcal{R})$ in the natural sense, Theorem 10 states that the distribution of random game lengths enjoys weak convergence to a Gaussian with high probability. We write

$$\mathcal{L}_N^{\mathcal{R}} = \frac{L_N^{\mathcal{R}} - \mathbb{E}\left[L_N^{\mathcal{R}}\right]}{\sqrt{\operatorname{var}\left(L_N^{\mathcal{R}}\right)}} \tag{4}$$

to be the random variable $L_N^{\mathcal{R}}$ normalized to have mean 0 and variance 1. Throughout this section, we proceed on the measure $\mathbb{P}_N$ and the set $\mathcal{R}_N^{\mathcal{P}}$, but the analysis carries over to the uniform measure μ_N and the suffix partition $\mathcal{R}_N^{\mathcal{S}}$.

Proof *(Proof of Theorem* 10*)*
By Proposition 7, $\lim_{N\to\infty} \mathbb{P}_N\left(\log_2 |\mathcal{A}_N(\mathcal{G})| \geq 1.6^n\right) = 1$ so fix $\mathcal{R} \in \mathcal{R}_N$ such that $|\mathcal{A}_N(\mathcal{R})| > 2^{1.6^n}$, and consider the random variable $\mathcal{L}_N^{\mathcal{R}} : \mathcal{A}_N(\mathcal{R}) \to \mathbb{R}$. Define random variables[12] $X_1^{\mathcal{R}}, \ldots, X_{N-1}^{\mathcal{R}}$ to correspond to each of the combine moves in $\mathcal{R}$, with $X_i^{\mathcal{R}}$ denoting the length of the corresponding prefix for a game in $\mathcal{A}_N(\mathcal{R})$: by Proposition 8, the random variables $X_i^{\mathcal{R}}$ are independent. Expanding Eq. (4),

$$\mathcal{L}_N^{\mathcal{R}} = \frac{L_N^{\mathcal{R}} - \mathbb{E}\left[L_N^{\mathcal{R}}\right]}{\sqrt{\operatorname{var}\left(L_N^{\mathcal{R}}\right)}} = \frac{\sum_{i=1}^{N-1}\left(X_i^{\mathcal{R}} - \mathbb{E}\left[X_i^{\mathcal{R}}\right]\right)}{\sqrt{\operatorname{var}\left(L_N^{\mathcal{R}}\right)}} \tag{5}$$

where the independence (conditioned on $\mathcal{A}_N(\mathcal{R})$) of the random variables $X_i^{\mathcal{R}}$ yields that

$$\operatorname{var}\left(L_N^{\mathcal{R}}\right) = \sum_{i=1}^{N-1} \operatorname{var}\left(X_i^{\mathcal{R}} - \mathbb{E}\left[X_i^{\mathcal{R}}\right]\right)$$

and furthermore, taking the maximum over those summands $X_i^{\mathcal{R}}$ which are nonconstant,[13]

[12] Define these to be strictly positive: if the ith delimiter is replaced with a combine move, say $X_i^{\mathcal{R}} = 1$.

[13] Indeed, we understand $\mathcal{L}_N^{\mathcal{R}}$ as the sum over the nonconstant random variables amongst those included by the sum in Eq. (5).

$$\max_{1\le i\le D}\left(\frac{\mathbb{E}\left[\left|X_i^{\mathcal{R}}-\mathbb{E}\left[X_i^{\mathcal{R}}\right]\right|^3\right]}{\mathbb{E}\left[\left(X_i^{\mathcal{R}}-\mathbb{E}\left[X_i^{\mathcal{R}}\right]\right)^2\right]\cdot \operatorname{var}\left(L_N^{\mathcal{R}}\right)}\right)\le \max_{1\le i\le D}\frac{2n^3}{\operatorname{var}\left(L_N^{\mathcal{R}}\right)}\le\frac{n^3}{\operatorname{var}\left(2^{1.6^n}\right)} \quad (6)$$

and the final expression in Eq. (6) certainly vanishes as $N\to\infty$. Thus, the desired result follows immediately from the case of the Berry-Esseen theorem for independent non-identically distributed summands, which namely yields

$$\sup_{x\in\mathbb{R}}\left|F_{\mathcal{R}}(x)-\Phi(x)\right|\le C\cdot\frac{n^3}{\operatorname{var}\left(2^{1.6^n}\right)}\xrightarrow{N\to\infty}0$$

for some universal constant $C>0$.

Theorem 10 yields that, when restricted to particular natural subsets of games, we have Gaussianity. The scope of this result is admittedly restricted, especially given the result of Corollary 3.

6 Open Problems

We conclude the work with several potential directions for further inquiry.

6.1 Other Two-Player Games Based on Recurrences

Theorem 9 can be interpreted as saying that if two players proceed mindlessly, the Zeckendorf game is fair in the limit of infinite input; Theorem 8 gives that the analogous statement would be true if we were to extend to a Z-player Zeckendorf game. Many papers (such as [1, 2, 4]) have extended the paradigm of the two-player Zeckendorf game to other recurrences: we might ask which of these also enjoy this property. In particular, we pose the following conjecture, concerning the two-player Bergman game [1].

Conjecture 2 In the limit $N\to\infty$ of infinite input, the probabilities of Player 1 and Player 2 winning, under both analogous definitions of random Bergman games, is $1/2$.

The core challenges of proving Conjecture 2 are as follows. The principal difference between the two-player Bergman game and the two-player Zeckendorf game is that the move C_1 is now a split move which consumes no tokens. Thus, there are structural differences between the two games which affects the range of achievable game lengths. Lemma 2 also depends on the number of a specific move sequence which may not have an analogue in the Bergman game.

6.2 Towards Gaussianity: Other Approaches

Theorem 10 establishes that certain natural partitions of Ω_N are such that the components become arbitrarily close to being Gaussian (in the sense of vanishing Kolmogorov-Smirnov distance) with arbitrarily high probability: it is unclear how to extend this to the entirety of Ω_N. Proposition 2 remarks that all games in Ω_N can be achieved by permutations of suffixes from a shortest game, which suggests the following question.

Question 1 Can we extend the techniques in Sect. 5 to a partition founded on the greedy embedding of Proposition 2? In particular, could studying the moments of the corresponding components lead to a proof of Conjecture 1?

Another possible direction is to restrict our attention to certain subsets of moves across all of Ω_N, rather than certain subsets of Ω_N. In particular, a sufficiently strong affirmative answer to Question 2 would resolve Conjecture 1.

Question 2 Is the distribution of the number of occurrences of a particular combine or split move asymptotically Gaussian, under either the measure μ_N or $\mathbb{P}_N$?

6.3 Towards Gaussianity: Mixing

We outline one more possible approach towards proving Conjecture 1, which relies on the literature surrounding *mixing central limit theorems*: these are analogues of the central limit theorem concerning sums of dependent random variables, applicable if the dependencies amongst the summands are sufficiently well-behaved. In particular, the main result of [6] states the following.

Theorem 11 ([6]) *Let* $\{X_{N,i} : 1 \leq i \leq d_N\}$ *be a triangular array of random variables defined on the probability space* $(\Omega, \mathcal{F}, \mathbb{P})$, $\bar{X}_{N,d_N} = \frac{1}{d_N}\sum_{i=1}^{d_N} X_{N,i}$, *and* $\alpha_N : \mathbb{N} \to \mathbb{R}$ *by*

$$\alpha_N(k) = \sup_{m} \sup_{A \in \mathcal{F}_0^m(N), B \in \mathcal{F}_{m+k}^{\infty}(N)} |\mathbb{P}(A \cap B) - \mathbb{P}(A)\mathbb{P}(B)| \tag{7}$$

where $\mathcal{F}_{m_1}^{m_2}(N) = \sigma\left(X_{N,m_1}, X_{N,m_1+1}, \ldots, X_{N,m_2}\right)$ *denotes the* σ*-algebra generated by random variables* $X_{N,m_1}, X_{N,m_1+1}, \ldots, X_{N,m_2}$*. If there exists constants* $C_1, C_2 > 0$, $\delta > 0$ *such that*

$$\mathbb{E}\left[\left|X_{N,i} - \mathbb{E}[X_{N,i}]\right|^{2+\delta}\right] < C_1 \tag{8}$$

and

$$\sum_{k=0}^{\infty}(k+1)^2\alpha_N^{\frac{\delta}{4+\delta}}(k) < C_2, \tag{9}$$

then

$$\sqrt{d_N}\left(\bar{X}_{N,d_N} - \mathbb{E}\bar{X}_{N,d_N}\right) \xrightarrow[N\to\infty]{\mathcal{D}} N\left(0, var\left(\sqrt{d_N}\bar{X}_{N,d_N}\right)\right). \tag{10}$$

We can define a triangular array of Bernoulli random variables by letting $X_{N,i} = 1$ if and only if the ith move of a Zeckendorf game with input N is a splitting move (in particular, $X_{N,i}$ maps to 0 on the event that the Zeckendorf game on input N terminates prior to move i), with d_N (the number of random variables in the Nth row) the length of the longest game on input N; Conjecture 1 is thus equivalent to establishing that the row sums $\sum_{i=1}^{d_N} X_{N,i}$ weakly converge to a Gaussian, since the number of combine moves for a given input N is constant. This is precisely the statement of Eq. 10. Also, the events A and B in Eq. 7 can be understood as fixed $\{0, 1\}$-realizations of subsets of random variables in $\{X_{N,1}, \ldots, X_{N,m}\}$ and $\{X_{N,m+k}, \ldots, X_{N,d_N}\}$, respectively.

Since Eq. 8 certainly holds under this setup (the constituent random variables are Bernoulli), resolving Conjecture 1 can be reduced to the following statement.

Question 3 For the triangular array $\{X_{N,i} : 1 \leq i \leq d_N\}$ defined above, does there exist a constant $\delta > 0$ and a constant $C > 0$ such that

$$\sum_{k=0}^{\infty}(k+1)^2\alpha_N^{\frac{\delta}{4+\delta}}(k) < C \tag{11}$$

for all natural numbers $N \in \mathbb{N}$?

By the preceding discussion, answering Question 3 in the affirmative by establishing 11 immediately yields Conjecture 1. In particular, it would suffice to establish the statement of Question 4 in the affirmative, although it is not immediately clear if this should be true.

Question 4 Does there exist a constant $\epsilon > 0$ and a constant $C > 0$ such that

$$|\mathbb{P}_N(A \cap B) - \mathbb{P}_N(A)\mathbb{P}_N(B)| \leq \frac{C}{k^{2+\epsilon}} \tag{12}$$

whenever events A and B correspond to fixed $\{0, 1\}$-realizations of subsets of the Bernoulli random variables in $\{X_{N,1}, \ldots, X_{N,m}\}$ and $\{X_{N,m+k}, \ldots, X_{N,d_N}\}$, respectively?

Of course, the techniques introduced in the main body of this paper are quite elementary, and there may be other promising approaches not included in this section that would contribute towards resolving Conjecture 1.

Acknowledgements This research was conducted as part of the 2022 SMALL REU program at Williams College, and was funded by NSF Grant DMS1947438, Harvey Mudd College, and Williams College funds. We thank Professor Steven J. Miller and our colleagues from the 2022 SMALL REU program, and in particular those pursuing the Zeck Shifts project, for many helpful insights.

References

1. Benjamin Baily et al. "The Generalized Bergman Game". In: (Sept. 19, 2021). https://doi.org/10.48550/arXiv.2109.00117. arXiv: 2109.00117 [math]. URL: http://arxiv.org/abs/2109.00117
2. Paul Baird-Smith et al. "The generalized Zeckendorf game". In: *Fibonacci Quart.* 57.5 (2019), pp. 1–14. ISSN: 0015-0517. arXiv: 1809.04883 [math.NT]. URL: https://web.williams.edu/Mathematics/sjmiller/publichtml/math/papers/ZeckGameGeneralFibQ10.pdf.
3. Paul Baird-Smith et al. "The Zeckendorf Game". In: *Combinatorial and Additive Number Theory, New York Number Theory Seminar*. Springer. 2018, pp. 25–38.
4. Elżbieta Bołdyriew et al. "Extending Zeckendorf's Theorem to a Non-constant Recurrence and the Zeckendorf Game on this Non-constant Recurrence Relation'. In: arXiv:2009.12475 (2020).
5. Anna Cusenza et al. "Bounds on Zeckendorf games". In: *Fibonacci Quart.* 60.1 (2022), pp. 57–71. ISSN: 0015-0517.
6. Magnus Ekström. "A general central limit theorem for strong mixing sequences". In: *Statistics & Probability Letters* 94 (2014), pp. 236–238.
7. Ruoci Li et al. "Deterministic Zeckendorf Games". In: arXiv:2006.16457 (2020).
8. Steven J. Miller. "Private communication". In: (2022).
9. Edouard Zeckendorf. "Representations des nombres naturels par une somme de nombres de fibonacci on de nombres de lucas". In: *Bulletin de La Society Royale des Sciences de Liege* (1972), pp. 179–182.

Linear Recurrences of Order at Most Two in Nontrivial Small Divisors and Large Divisors

Hùng Việt Chu, Kevin Huu Le, Steven J. Miller, Yuan Qiu, and Liyang Shen

Abstract For each positive integer N, define

$$S'_N = \{1 < d < \sqrt{N} : d|N\} \text{ and } L'_N = \{\sqrt{N} < d < N : d|N\}.$$

Recently, Chentouf characterized all positive integers N such that the set of small divisors $\{d \leq \sqrt{N} : d|N\}$ satisfies a linear recurrence of order at most two. We nontrivially extend the result by excluding the trivial divisor 1 from consideration, which dramatically increases the analysis complexity. Our first result characterizes all positive integers N such that S'_N satisfies a linear recurrence of order at most two. Moreover, our second result characterizes all positive N such that L'_N satisfies a linear recurrence of order at most two, thus extending considerably a recent result that characterizes N with L'_N being in an arithmetic progression.

H. V. Chu
Department of Mathematics, Texas A&M University, College Station, TX 77843, USA
e-mail: hungchu1@tamu.edu

K. H. Le
Mathematics, Texas A&M University, College Station, TX 77843, USA
e-mail: kevinhle@tamu.edu

S. J. Miller (✉)
Department of Mathematics and Statistics, Williams College, Williamstown, MA 01267, USA
e-mail: sjm1@williams.edu; Steven.Miller.MC.96@aya.yale.edu

Y. Qiu
Williams College, Williamstown, MA 01267, USA
e-mail: yq1@williams.edu

L. Shen
Courant Institute of Mathematical Sciences, New York University, New York, NY 10012, USA
e-mail: Liyang.Shen@nyu.edu

M. B. Nathanson (ed.), *Combinatorial and Additive Number Theory VI*,
Springer Proceedings in Mathematics & Statistics 464,
https://doi.org/10.1007/978-3-031-65064-2_5

1 Introduction

For a positive integer N, the set of small divisors of N is

$$S_N := \{d : 1 \leq d \leq \sqrt{N}, d \text{ divides } N\}.$$

Since the case $N = 1$ is trivial, we assume throughout the paper that $N > 1$. In 2018, Iannucci characterized all positive integers N whose S_N forms an arithmetic progression (or AP, for short). Iannucci's key idea was to show that if S_N forms an AP, then the size $|S_N|$ cannot exceed 6. Observing that the trivial divisor 1 plays an important role in Iannucci's proofs (see [4, Lemma 3 and Theorem 4]), Chu [3] excluded both 1 and $\sqrt{N}$ from the definition of S_N to obtain a more general theorem that characterizes all N whose

$$S'_N := \{d : 1 < d < \sqrt{N}, d \text{ divides } N\}$$

is in an AP. Interestingly, with the trivial divisor 1 excluded, [2, Theorem 1.1] still gives that $|S'_N| \leq 5$. Recently, Chentouf generalized Iannucci's result from a different perspective by characterizing all N whose S_N satisfies a linear recurrence of order at most two. In particular, for each tuple $(u, v, a, b) \in \mathbb{Z}^4$, there is an integral linear recurrence, denoted by $U(u, v, a, b)$, of order at most two, given by

$$n_i = \begin{cases} u & \text{if } i = 1, \\ v & \text{if } i = 2, \\ an_{i-1} + bn_{i-2} & \text{if } i \geq 3. \end{cases}$$

Noting that the appearance of the trivial divisor 1 contributes nontrivially to the proof of [1, Theorem 3, Lemma 8, Theorem 10], we generalize Chentouf's result in the same manner as [2, Theorem 1.1] generalizes [4, Theorem 4]: we characterize all positive integers N whose S'_N satisfies a linear recurrence of order at most two.

Definition 1 A positive integer N is said to be small recurrent if S'_N satisfies a linear recurrence of order at most two. When $|S'_N| \leq 2$, N is vacuously small recurrent.

Theorem 1 *Let p, q, r denote prime numbers such that $p < q < r$ and k be some positive integer. A positive integer $N > 1$ is small recurrent if and only if N belongs to one of the following forms.*

1. *$N = p^k$ for some $k \geq 1$. In this case, $S'_N = \{p, p^2, \ldots, p^{\lfloor (k-1)/2 \rfloor}\}$ satisfies $U(p, p^2, p, 0)$.*
2. *$N = p^k q$ or $N = pq^k$ for some $1 \leq k \leq 3$. A restriction for $N = p^3 q$ is that either $p < q < p^2$ or $p^3 < q$.*
3. *$N = p^k q$ for some $k \geq 4$ and $q > p^k$. In this case, $S'_N = \{p, p^2, p^3, \ldots, p^k\}$ satisfies $U(p, p^2, p, 0)$.*

4. $N = p^k q$ *for some* $k \geq 4$ *and* $\sqrt{q} < p < q$. *In this case,* $S'_N = \{p, q, p^2, pq, \ldots\}$ *satisfies* $U(p, q, 0, p)$.
5. $N = pq^k$ *for some* $k \geq 4$ *and* $p < q$. *In this case,* $S'_N = \{p, q, pq, q^2, \ldots\}$ *satisfies* $U(p, q, 0, q)$.
6. $N = pq^k r$ *for some* $k \geq 2$, $p < q$, *and* $r > pq^k$. *In this case,* $S'_N=\{p, q, pq, q^2, \ldots, pq^{k-1}, q^k, pq^k\}$ *satisfies* $U(p, q, 0, q)$.
7. $N = p^2q^2$ *for some* $p < q < p^2$. *In this case,* $S'_N = \{p, q, p^2\}$.
8. $N = pqr$ *for some* $p < q < r$. *If* $r < pq$, *then* $S'_N = \{p, q, r\}$. *If* $r > pq$, *then* $S'_N = \{p, q, pq\}$.
9. $N = p^3q^2$ *for some* $p^{3/2} < q < p^2$. *In this case,* $S'_N = \{p, q, p^2, pq, p^3\}$ *satisfies* $U(p, q, 0, p)$.
10. $N = p^2qr$, *where* $p < q < p^2 < r < pq$, $(q^2 - p^3)|(pq - r)$, $(q^2 - p^3)|(rq - p^4)$, *and* $r = pq - \sqrt{(q^2 - p^3)(p^2 - q)}$. *In this case,* $S'_N = \{p, q, p^2, r, pq\}$ *satisfies* $U\left(p, q, \frac{p(pq-r)}{q^2-p^3}, \frac{rq-p^4}{q^2-p^3}\right)$.

Next, consider the set of large divisors of N

$$L_N := \{d : d \geq \sqrt{N}, d \text{ divides } N\},$$
$$L'_N := \{d : \sqrt{N} < d < N, d \text{ divides } N\}.$$

The second result of this paper is the characterization of all positive integers N whose L'_N satisfies a linear recurrence of order at most two. This considerably extends [2, Theorem 1].

Definition 2 A positive integer N is said to be large recurrent if L'_N satisfies a linear recurrence of order at most two. When $|L'_N| \leq 2$, N is vacuously large recurrent.

Theorem 2 *Let* p, q, r *denote prime numbers such that* $p < q < r$ *and* k *be some positive integer. A positive integer* $N > 1$ *is large recurrent if and only if* N *belongs to one of the following forms.*

1. $N = p^k$ *for some* $k \geq 1$. *Then* $L'_N = \{p^{\lceil (k-1)/2\rceil+1}, p^{\lceil (k-1)/2\rceil+2}, \ldots, p^{k-1}\}$ *satisfies* $U(p^{\lceil (k-1)/2\rceil+1}, p^{\lceil (k-1)/2\rceil+2}, p, 0)$.
2. $N = p^k q$ *for some* $k \geq 1$ *and* $q > p^k$. *Then* $L'_N = \{q, pq, p^2q, \ldots, p^{k-1}q\}$ *satisfies* $U(q, pq, p, 0)$.
3. $N = p^k q$ *for some* $k \geq 2$ *and* $p^{k-1} < q < p^k$. *Then*

$$L'_N = \{p^k, pq, p^2q, \ldots, p^{k-1}q\}$$

satisfies $U(p^k, pq, p, 0)$.

4. $N = p^k q$ *some for* $k \geq 3$ *and* $p < q < p^2$. *In this case,*

$$L'_N = \begin{cases} \{p^{k/2+1}, p^{k/2}q, p^{k/2+2}, \ldots, p^{k-1}q\} & \text{if } 2|k, \\ \{p^{(k-1)/2}q, p^{(k+3)/2}, p^{(k+1)/2}q, \ldots, p^{k-1}q\} & \text{if } 2 \nmid k. \end{cases}$$

Note that L'_N *satisfies* $U(p^{k/2+1}, p^{k/2}q, 0, p)$ *and* $U(p^{(k-1)/2}q, p^{(k+3)/2}, 0, p)$ *for even and odd k, respectively.*

5. $N = p^4q$ *with* $p^2 < q < p^3$, $(p^5 - q^2)|(p^2 - q)$, *and* $(p^5 - q^2)|(p^3 - q)$. *In this case,* $L'_N = \{pq, p^4, p^2q, p^3q\}$.
6. $N = p^3q^2$ *for* $p < q < p^2$. *In this case,* $L'_N = \{q^2, p^2q, pq^2, p^3q, p^2q^2\}$ *satisfies* $U(q^2, p^2q, 0, p)$.
7. $N = p^2q^2$ *for some* $p < q$. *If* $p < q < p^2$, *then* $L'_N = \{q^2, p^2q, pq^2\}$.
8. $N = pq^k$ *for some* $k \geq 2$ *and* $p < q$. *In this case,*

$$L'_N = \begin{cases} \{pq^{k/2}, q^{k/2+1}, \ldots, q^k\} & \text{if } 2|k, \\ \{q^{(k+1)/2}, pq^{(k+1)/2}, \ldots, q^k\} & \text{if } 2 \nmid k. \end{cases}$$

Observe that L'_N *satisfies* $U(pq^{k/2}, q^{k/2+1}, 0, q)$ *and* $U(q^{(k+1)/2}, pq^{(k+1)/2}, 0, q)$ *for even and odd k, respectively.*

9. $N = pq^kr$ *for some* $k \geq 1$ *and* $p < q < pq^k < r$. *In this case,* $L'_N = \{r, pr, qr, pqr, q^2r, \ldots, q^kr\}$ *satisfies* $U(r, pr, 0, q)$.

The paper is structured as follows: Sect. 2 studies the case when N has a small number of divisors and establish some preliminary results; Sect. 3 characterizes small recurrent numbers, while Sect. 4 characterizes large recurrent numbers.

2 Preliminaries

For each $N \in \mathbb{N}$ with the prime factorization $\prod_{i=1}^{\ell} p_i^{a_i}$, the divisor-counting function is

$$\tau(N) := \sum_{d|N} 1 = \prod_{i=1}^{\ell}(a_i + 1). \tag{1}$$

It is easy to verify that for $N > 1$,

$$\tau(N) := \begin{cases} 2|S'_N| + 3 = 2|L'_N| + 3 & \text{if } N \text{ is a square,} \\ 2|S'_N| + 2 = 2|L'_N| + 2 & \text{otherwise.} \end{cases} \tag{2}$$

Using (1), we can characterize all N with $\tau(N) \leq 9$; equivalently, $|S'_N|, |L'_N| \leq 3$.

1. If $\tau(N) = 2$ or 3, (1) gives that $N = p$ or p^2 for some prime p,
2. If $\tau(N) = 4$ or 5, $N = pq, p^3, p^4$ for some primes $p < q$,
3. If $\tau(N) = 6$ or 7, $N = p^5, pq^2, p^2q, p^6$ for some primes $p < q$,
4. If $\tau(N) = 8$ or 9, $N = pqr, pq^3, p^3q, p^7, p^2q^2, p^8$ for some primes $p < q < r$.

2.1 Regarding S'_N

Proposition 1 *If $|S'_N| \geq 2$, then N cannot have two (not necessarily distinct) prime factors p_1 and p_2 at least $\sqrt{N}$*

Proof Suppose that N has two prime factors p_1 and p_2 at least $\sqrt{N}$. Then $p_1 p_2 \geq N$ and $p_1 p_2$ divides N; hence, $N = p_1 p_2$, which contradicts that $|S'_N| \geq 2$.

Proposition 2 *If all elements of S'_N are divisible by some prime p and $|S'_N| \geq 4$, then either $N = p^k$ or $N = p^k q$ for some $k \geq 1$ and some prime $q > p^k$.*

Proof If all divisors (except 1) of N are divisible by p, then $n = p^k$ for some $k \geq 1$. Assume that N has a prime factor $q \neq p$. Then $q \geq \sqrt{N}$. Proposition 1 implies that N cannot have another prime factor at least $\sqrt{N}$. Hence, $N = p^k q$ for some $k \geq 1$ and $q > p^k$.

Let $p < q < r < s$ be distinct prime numbers. Write $S'_N = \{d_2, d_3, d_4, d_5, \ldots\}$. (We start with d_2 since the smallest divisor of N is usually denoted by $d_1 = 1$, which is excluded from S'_N.)

Lemma 1 *Suppose that the first 4 numbers in S'_N are $p < q < p^2 < r$. If N is small recurrent with $U(p, q, a, b)$, the following hold:*

(i) $\gcd(a, b) = 1$.
(ii) if $d_{2i} \in S'_N$, then $p | d_{2i}$; however, if $d_{2i-1} \in S'_N$, then $p \nmid d_{2i-1}$.
(iii) for $d_i, d_{2i-1} \in S'_N$, we have $\gcd(b, d_i) = \gcd(a, d_{2i-1}) = 1$.
(iv) for $d_i, d_{i+1} \in S'_N$, we have $\gcd(d_i, d_{i+1}) = 1$.
(v) for $d_{2i-1}, d_{2i+1} \in S'_N$, we have $\gcd(d_{2i-1}, d_{2i+1}) = 1$.

Proof (i) Since $r = ap^2 + bq$, we have $\gcd(a, b) | r$. Observe that $\gcd(a, b) = r$ contradicts $p^2 = aq + bp$. Hence, $\gcd(a, b) = 1$.

(ii) We prove by induction. The claim is true for $i \leq 2$. Assume that the claim holds for $i = j \geq 2$. We show that it holds for $i = j + 1$. Since $p^2 = aq + bp$, we know that $p | a$. By item (i), $p \nmid b$. Write

$$\begin{aligned} d_{2(j+1)} &= ad_{2(j+1)-1} + bd_{2(j+1)-2} = ad_{2j+1} + bd_{2j} \\ &= (a^2 + b)d_{2j} + abd_{2j-1}. \end{aligned}$$

By the inductive hypothesis, $p | d_{2j}$. Since $p | a$, we obtain $p | d_{2(j+1)}$. Furthermore, write

$$d_{2(j+1)-1} = ad_{2(j+1)-2} + bd_{2(j+1)-3} = ad_{2j} + bd_{2j-1}.$$

Since $p \nmid bd_{2j-1}$ by the inductive hypothesis and $p | ad_{2j}$, we obtain $p \nmid d_{2(j+1)-1}$. This completes our proof.

(iii) Suppose that $k = \gcd(b, d_i) > 1$ for some i. If $i = 2$, then $p | b$, which contradicts that $p | a$ and $\gcd(a, b) = 1$. If $i = 3$, then $q | b$, which contradicts $p^2 = aq + bp$.

If $i \geq 4$, write $d_i = ad_{i-1} + bd_{i-2}$. Since $k|bd_{i-2}$ and $\gcd(a, b) = 1$, we get $k|d_{i-1}$ and so, $k|\gcd(b, d_{i-1})$. By induction, we obtain $k|\gcd(b, d_3)$, which has been shown to be impossible.

Next, suppose that $k = \gcd(a, d_{2i-1}) > 1$ for some i. If $i = 2$, then $q|a$, contradicting $r = ap^2 + bq$. If $i \geq 3$,

$$k \mid d_{2i-1},\ d_{2i-1} = ad_{2i-2} + bd_{2i-3}, \text{ and } \gcd(a, b) = 1 \implies k \mid d_{2i-3}.$$

By induction, $k|d_3$; that is, $\gcd(a, d_3) > 1$, which has been shown to be impossible.

(iv) The claim holds for $i \leq 4$. For $i \geq 5$, using the recurrence $d_{i+1} = ad_i + bd_{i-1}$, we have

$$\gcd(d_i, d_{i+1}) = \gcd(d_i, bd_{i-1}) \overset{\text{iii)}}{=} \gcd(d_i, d_{i-1}).$$

By induction, we obtain $\gcd(d_i, d_{i+1}) = 1$.

(v) The claim holds for $i \leq 2$. For $i \geq 3$, we have

$$\gcd(d_{2i-1}, d_{2i+1}) = \gcd(d_{2i-1}, ad_{2i}) \overset{\text{iii)}}{=} \gcd(d_{2i-1}, d_{2i}) \overset{\text{iv)}}{=} 1.$$

This completes our proof.

Lemma 2 *Suppose that the first 4 numbers in S'_N are $p < q < r < p^2$. If N is small recurrent with $U(p, q, a, b)$, the following hold:*

(i) $\gcd(a, b) = 1$ *and* $p \nmid a$.
(ii) *For all* $d_i \in S'_N$, $\gcd(b, d_i) = 1$.
(iii) *For all* $d_i, d_{i+1} \in S'_N$, $\gcd(d_i, d_{i+1}) = 1$.
(iv) *Let* $d_i \in S'_N$. *Then* $p|d_i$ *if and only if* $i \equiv 2 \bmod 3$.

Proof (i) We have $\gcd(a, b)$ divides r because $r = aq + bp$. Furthermore, $\gcd(a, b)$ divides p because $p^2 = ar + bq$. Hence, $\gcd(a, b) = 1$.

Since $r = aq + bp$ and r is a prime, $p \nmid a$.

(ii) Suppose that $k = \gcd(b, d_i) > 1$ for some i. If $i = 2$, then $k = p$ and $p|b$. It follows from $p^2 = ar + bq$ that $p|a$, which contradicts that $\gcd(a, b) = 1$. If $i = 3$, then $k = q$ and $q|b$. It follows from $r = aq + bp$ that $q|r$, a contradiction. Hence, $i \geq 4$. Write

$$1 < \gcd(b, d_i) = \gcd(b, ad_{i-1} + bd_{i-2}) = \gcd(b, d_{i-1}).$$

By induction, we obtain $\gcd(b, d_3) > 1$, which has been shown to be impossible.

(iii) The claim holds for $i \leq 4$. Let $d_i, d_{i+1} \in S'_N$ and $i \geq 5$. Then

$$\gcd(d_i, d_{i+1}) = \gcd(d_i, ad_i + bd_{i-1}) = \gcd(d_i, bd_{i-1}) \overset{\text{ii)}}{=} \gcd(d_i, d_{i-1}).$$

By induction, $\gcd(d_i, d_{i+1}) = 1$.

(iv) The claim holds for $i \leq 4$. Suppose that the claim holds for all $i \leq j$ for some $j \geq 5$. We show that it also holds for $i = j + 1$ under the assumption that $d_{j+1} \in S'_N$. We have

$$p^2 = ar + bq = a(aq + bp) + bq = (a^2 + b)q + abp.$$

Hence, $p|(a^2 + b)$. Write

$$d_{j+1} = ad_j + bd_{j-1} = a(ad_{j-1} + bd_{j-2}) + bd_{j-1} = (a^2 + b)d_{j-1} + abd_{j-2}.$$

Therefore, $p|d_{j+1}$ if and only if $p|d_{j-2}$. By the inductive hypothesis, we know that $p|d_{j+1}$ if and only if $j + 1 \equiv 2 \bmod 3$.

The proofs of the next two lemmas are similar to those of Lemmas 1 and 2. Thus, we move their proofs to the Appendix.

Lemma 3 *Suppose that the first 4 numbers in S'_N are $p < q < r < pq$. If N is small recurrent with $U(p, q, a, b)$, the following hold:*

(i) $\gcd(a, b) = 1$ *and* $p \nmid a$.
(ii) For all $d_i \in S'_N$, $\gcd(b, d_i) = 1$.
(iii) For all $d_i, d_{i+1} \in S'_N$, $\gcd(d_i, d_{i+1}) = 1$.
(iv) Let $d_i \in S'_N$. *Then* $p|d_i$ *if and only if* $i \equiv 2 \bmod 3$.
(v) Let $d_i \in S'_N$. *Then* $q|d_i$ *if and only if* i *is odd.*

Lemma 4 *Suppose that the first 4 numbers in S'_N are $p < q < r < s$. If N is small recurrent with $U(p, q, a, b)$, the following hold:*

(i) $\gcd(a, b) = 1$.
(ii) For all $d_i \in S'_N$ *with* $i \geq 3$, $\gcd(b, d_i) = 1$.
(iii) For $d_i, d_{i+1} \in S'_N$, $\gcd(d_i, d_{i+1}) = 1$.
(iv) For $d_i \in S'_N$, $\gcd(a, d_i) = 1$.
(v) For $d_i, d_{i+2} \in S'_N$, $\gcd(d_i, d_{i+2}) = 1$.

3 Small Recurrent Numbers

We first find all small recurrent numbers with $|S'_N| \geq 4$. Then we check which N is small recurrent out of all N with $|S'_N| \leq 3$ at the end of this section. As we rely heavily on case analysis, we underline possible forms of N throughout our analysis for the ease of later summary.

3.1 The Case $|S'_N| \geq 4$

Let $d_2 = p$ for some prime p. Then d_3 is either p^2 or q for some prime $q > p$. If $d_3 = p^2$, according to Proposition 2, we know that

$$\underline{N = p^k \text{ or } N = p^k q \text{ for some } k \geq 1 \text{ and } q > p^k.}$$

Assume, for the rest of this subsection, that $d_3 = q$ for some prime $q > p$. Then $d_4 = pq, p^2, r$ for some prime $r > q$.

3.1.1 When $d_4 = pq$

Since S'_N satisfies $U(p, q, a, b)$, we get $pq = aq + bp$. So, $p|a$ and $q|b$. Write $a = pm$ for some $m \in \mathbb{Z}$ and get $b = (1 - m)q$. Since $p^2 \nmid N$ and q divides $d_5 = apq + bq$, we can write

$$d_5 = p^s q^t r_1^{\ell_1} \cdots r_k^{\ell_k},$$

where $s \leq 1$, $t \geq 1$, and r_i's are primes strictly greater than pq. If some $\ell_i \geq 1$, then $pq < r_i < d_5$ and $r_i \in S'_N$, a contradiction. Hence, $\ell_i = 0$ for all $i \leq k$ and $d_5 = p^s q^t$. Since $d_5 > pq$ and $s \leq 1$, we know that $t \geq 2$.

(a) If $s = 1$, $d_5 = pq^2 > q^2 > d_4$ and $q^2 | d_5$, so $q^2 \in S'_N$, a contradiction.
(b) If $s = 0$, $d_5 = q^t$ for some $t \geq 2$. Since d_5 is the next number after d_4 in increasing order, $d_5 = q^2$. Using the linear recurrence, we obtain $q^2 = apq + bq$, so $q = ap + b = p^2 m + (1 - m)q$. It follows that $p^2 m = mq$. We arrive at $m = 0$, $a = 0$, and $b = q$. Hence, all elements of S'_N are divisible by either p or q. If N has a prime factor $r \geq \sqrt{N}$, by Proposition 1, r is unique. We conclude that $\underline{N = pq^k \text{ or } pq^k r \text{ for some } k \geq 2,\ p < q, \text{ and } pq^k < r.}$

3.1.2 When $d_4 = p^2$

The first few divisors of N are $1 < p < q < p^2$. We have $p^2 = aq + bp$, so $p|a$. Write $a = pm$ and get $b = p - mq$. We argue for possible forms of d_5. Let r be the largest prime factor of d_5. If $r > q$, then $r > p^2$ and $d_5 = r$. Otherwise, if $r \leq q$, then $d_5 = p^\ell q^k$ for some $\ell, k \geq 0$. Suppose that $k \geq 2$. We get

$$d_5 \geq q^2 > pq > d_4 \text{ and } pq \mid N,$$

a contradiction. Hence, $k \leq 1$. If $k = 0$, then

$$d_5 = p^3 > pq > d_4 \text{ and } pq \mid N,$$

another contradiction. Therefore, $k = 1$ and $d_5 = pq$. We conclude that either $d_5 = r$ for some $r > q$ or $d_5 = pq$.

(a) If $d_5 = pq$, then $bq + ap^2 = pq$. So, $(p - mq)q + mp^3 = pq$, which gives $mq^2 = mp^3$. Hence, $m = 0$, $a = 0$, and $b = p$. We know that elements of S'_N are divisible by either p or q. If N has a prime factor r' at least $\sqrt{N}$, by Proposition 1, r' is unique. Hence, either $N = p^\ell q^k$ or $p^\ell q^k r'$ for some prime $r' > p^\ell q^k$, $\ell \geq 2$, and $k \geq 1$.

Case a.i) $N = p^\ell q^k r'$. We claim that $k = 1$. Indeed, if $k \geq 2$, then

$$q^4 < q^2 r' < N \implies q^2 < \sqrt{N} \implies q^2 \in S'_N.$$

Since $b = p$, we know that $p|d$ for all $d \geq d_4$ and $d \in S'_N$, which contradicts $q^2 \in S'_N$. Hence,

$$\underline{N = p^\ell q r' \text{ for some } \ell \geq 2, \text{ some prime } r' > p^\ell q, \text{ and } \sqrt{q} < p < q.}$$

Case a.ii) $N = p^\ell q^k$. As above, $q^2 \notin S'_N$. If $k \geq 2$, then

$$q^2 \geq \sqrt{N} \implies q^4 \geq N = p^\ell q^k = p^{\ell-2} p^2 q^k > q^{k+1}.$$

Hence, $k < 3$, which implies that $k = 2$. In this case, $N = p^\ell q^2$ and

$$q^2 > p^\ell > q^{\ell/2} \implies \ell \leq 3.$$

We conclude that one of the following holds:

- $\underline{N = p^\ell q \text{ for some } \ell \geq 2 \text{ and } \sqrt{q} < p < q,}$
- $\underline{N = p^2 q^2 \text{ for some } p < q < p^2,}$
- $\underline{N = p^3 q^2 \text{ for some } p^{3/2} < q < p^2.}$

(b) $d_5 = r$.

Proposition 3 *Suppose that the first* 4 *numbers in* S'_N *are* $p < q < p^2 < r$. *Then* $|S_N| \leq 7$. *As a result,* $|S'_N| \leq 6$.

Proof Assume that $|S_N| \geq 2i$ for some $i \geq 4$. We obtain a contradiction by showing that $|S_N| \geq 2i + 2$. By Lemma 1 item (ii), $p \nmid d_{2i-1}$, $p|d_{2i-2}$, and $p \nmid d_{2i-3}$. By Lemma 1 item (v), $\gcd(d_{2i-1}, d_{2i-3}) = 1$, so $p^2 d_{2i-1} d_{2i-3}$ divides N. Hence, $pd_{2i-3} \in S'_N$.

If $pd_{2i-3} = d_{2i-2}$, then

$$pd_{2i-3} = ad_{2i-3} + bd_{2i-4} \implies d_{2i-3} \mid bd_{2i-4},$$

which contradicts Lemma 1 items (iii) and (iv).

If $pd_{2i-3} = d_{2i}$, then

$$pd_{2i-3} = ad_{2i-1} + bd_{2i-2} = a(ad_{2i-2} + bd_{2i-3}) + bd_{2i-2}$$
$$= (a^2 + b)d_{2i-2} + abd_{2i-3}.$$

Therefore, d_{2i-3} divides $a^2 + b$. It is easy to check that for $d_j \in S'_N$, the sequence $d_j \bmod a^2 + b$ is congruent to

$$1, p, q, p^2, abp, abq, abp^2, (ab)^2 p, (ab)^2 q, (ab)^2 p^2, \ldots.$$

Hence, we can write

$$d_{2i-3} = (a^2 + b)\ell + a^k b^k s,$$

for some $\ell \in \mathbb{Z}$, some $k \geq 1$, and some $s \in \{p, q, p^2\}$. Since $d_{2i-3} | (a^2 + b)$, $d_{2i-3} | a^k b^k s$. By Lemma 1 item (iii), $d_{2i-3} | s$; that is, $d_{2i-3} \leq p^2$. However, $d_{2i-3} \geq d_5 > d_4 = p^2$, a contradiction.

We conclude that $pd_{2i-3} \geq d_{2i+2}$. Since $pd_{2i-3} \in S'_N$, we know that $d_{2i+2} \in S'_N$ and $|S_N| \geq 2i + 2$.

Proposition 4 *Suppose that the first* 4 *numbers in* S'_N *are* $p < q < p^2 < r$. *If* N *is small recurrent, then* $|S'_N| \neq 4, 6$.

Proof If $|S'_N| = 4$, then (2) gives $\tau(N) = 10$ or 11. Note that N has three distinct prime factors p, q, r and the power of p is at least 2. Since $2^3 \cdot 3 > 11$, N cannot have another prime factor besides p, q, r. Write $N = p^a q^b r^c$, for some $a \geq 2, b \geq 1, c \geq 1$. However, neither $(a + 1)(b + 1)(c + 1) = 10$ nor $(a + 1)(b + 1)(c + 1) = 11$ has a solution. Therefore, $|S'_N| \neq 4$. A similar argument gives $|S'_N| \neq 6$.

By Propositions 3 and 4, we know that $|S'_N| = 5$; that is, $\tau(N) = 12$ or 13. Using the same reasoning as in the proof of Proposition 4, we know that $\tau(N) = 12$ and $N = p^2 qr$, where $p < q < p^2 < r$.

3.1.3 When $d_4 = r$ for Some $r > q$

The possible values for d_5 are p^2, pq, s for some prime $s > r$.

(a) If $d_5 = p^2$, we can generalize the method by Chentouf.

Proposition 5 *If* N *is small recurrent and the first four numbers of* S'_N *are* $p < q < r < p^2$, *then* $|S_N| \leq 7$. *As a result,* $|S'_N| \leq 6$.

Proof Suppose that $|S_N| \geq 8$. We show that $|S_N| \geq 3i + 2$ for all $i \in \mathbb{N}$, which is a contradiction. The claim holds for $i = 2$. Assume that $|S_N| \geq 3j + 2$ for some $j \geq 2$. By Lemma 2, $p \nmid d_{3j} d_{3j+1}$ and $\gcd(d_{3j}, d_{3j+1}) = 1$. Hence, $p^2 d_{3j} d_{3j+1}$ divides N, which implies that $pd_{3j} \in S'_N$.

If $pd_{3j} = d_{3j+2} = ad_{3j+1} + bd_{3j}$, then d_{3j} divides ad_{3j+1}. By Lemma 2, $d_{3j}|a$. Observe that for $d_i \in S'_N$, the sequence d_i mod a is

$$1, p, q, bp, bq, b^2p, b^2q, \ldots.$$

Write $d_{3j} = a\ell + b^k s$, for some $\ell \in \mathbb{Z}$, some $k \geq 0$, and some $s \in \{p, q\}$. We see that $d_{3j}|b^k s$ for some $k \geq 0$ and $s \in \{p, q\}$. By Lemma 2, $d_{3j} \leq q$. However,

$$d_{3j} \geq d_6 > d_3 = q,$$

a contradiction.

If $pd_{3j} > d_{3j+2}$, then $pd_{3j} \geq d_{3(j+1)+2}$ by Lemma 2. Therefore, $|S_N| \geq 3(j+1)+2$.

Proposition 6 *There is no small recurrent N whose the first four numbers of S'_N are $p < q < r < p^2$.*

Proof By Proposition 5, $|S'_N| \in \{4, 5, 6\}$. If $|S'_N| = 4$, then $\tau(N) = 10$ or 11, none of which can be written as a product of at least three integers, each of which is at least 2. This contradicts (2) and the fact that N has three distinct prime factors. We arrive at the same conclusion when $|S'_N| = 6$. For $|S'_N| = 5$, we obtain $N = p^2qr$ for some primes $p < q < r < p^2$. However, this poses another contradiction. Observe that $(pq)^2 < p^2qr$, so the divisors in S'_N are $p < q < r < p^2 < pq$. Since $pq = ap^2 + br$, we get $p|b$, which contradicts Lemma 2 item (ii).

(b) Suppose that $d_5 = pq$.

Proposition 7 *There is no small recurrent number N such that the first four numbers of S'_N are $p < q < r < pq$.*

Proof Assume that $|S'_N| \geq 8$. Since $p^2 \notin S'_N$, p divides N exactly. By Lemma 3, d_6 is divisible neither by p nor q. Hence, $d_6 = s$ for some prime $s > pq$. The divisor d_7 is divisible by q; hence, $d_7 = qr$ or q^2. The divisor d_8 is divisible by p but not by q. So, $d_8 = pr$, which gives that d_7 must be q^2 because $d_7 < d_8$. Now $q|d_9$ and $p \nmid d_9 \implies d_9 = qr$. However, that $\gcd(d_8, d_9) = r$ contradicts Lemma 3. Therefore, $|S'_N| \in \{4, 5, 6, 7\}$. Using the same argument as in the proof of Proposition 6, we know that $|S'_N| \neq 4, 6$ and so, $|S'_N| \in \{5, 7\}$. By the above argument, if $|S'_N| \geq 5$, then d_6 is a prime greater than pq. Hence, N has at least 4 distinct prime factors, so $\tau(N)$ can be written as a product of at least 4 integers greater than 1. Clearly, (1) rules out the case $|S'_N| = 5$. If $|S'_N| = 7$, the above argument shows that $q^2|N$; hence, $\tau(N)$ can be written as a product of at least 4 integers greater than 1, one of which is greater than 2. This cannot happen as $\tau(N) \in \{16, 17\}$.

(c) Suppose that $d_5 = s$.

Proposition 8 *There is no small recurrent number N such that the first four numbers of S'_N are $p < q < r < s$.*

Proof Observe that pq and pr are in S'_N. Let $d_j = pv$ be the largest element of S'_N that is divisible by p. Clearly, $v > p$ and $j \geq 7$. By Lemma 4, d_j, d_{j-1}, and d_{j-2} are pairwise coprime. Hence, $pvd_{j-1}d_{j-2}$ divides N, so $pd_{j-2} \in S'_N$.

If $pd_{j-2} = d_{j-1}$, then $p|d_{j-1}$ and so, $p|\gcd(d_{j-1}, d_j)$, which contradicts Lemma 4 item (iii).

If $pd_{j-2} = d_j$, then $d_{j-2} = v$ and $\gcd(d_{j-2}, d_j) = v > 1$, which contradicts Lemma 4 item (v).

Therefore, we have $pd_{j-2} > d_j$, which, however, contradicts that d_j is the largest element of S'_N that is divisible by p. We conclude that there is no small recurrent number N such that the first four numbers of S'_N are $p < q < r < s$.

From the above analysis, we arrive at the following proposition.

Proposition 9 *If N is small recurrent and $|S'_N| \geq 4$, then N belongs to one of the following forms.*

(S1) $N = p^k$ *or* $N = p^k q$ *for some* $k \geq 1$ *and* $q > p^k$.
(S2) $N = pq^k$ *or* $pq^k r$ *for some* $k \geq 2$, $p < q$, *and* $pq^k < r$.
(S3) $N = p^k qr$ *for some* $k \geq 2$, *some prime* $r > p^k q$, *and* $\sqrt{q} < p < q$.
(S4) $N = p^k q$ *for some* $k \geq 2$ *and* $\sqrt{q} < p < q$.
(S5) $N = p^2 q^2$ *for some* $p < q < p^2$.
(S6) $N = p^3 q^2$ *for some* $p^{3/2} < q < p^2$.
(S7) $N = p^2 qr$, *where the first four numbers in* S'_N *are* $p < q < p^2 < r$.

These forms together establish the necessary condition for a small recurrent N to have $|S'_N| \geq 4$. We now refine each form (if necessary) to obtain a necessary and sufficient condition.

(S1)
- If $N = p^k$, then N is small recurrent with $|S'_N| \geq 4$ if $k \geq 9$. In this case, $S'_N = \{p, p^2, p^3, \ldots, p^{\lfloor (k-1)/2 \rfloor}\}$ satisfies $U(p, p^2, p, 0)$.
- If $N = p^k q$ for some $k \geq 1$ and $q > p^k$, then N is small recurrent with $|S'_N| \geq 4$ if $k \geq 4$. In this case, $S'_N = \{p, p^2, p^3, \ldots, p^k\}$ satisfies $U(p, p^2, p, 0)$.

(S2)
- If $N = pq^k$ for some $k \geq 2$ and $p < q$, then N is small recurrent with $|S'_N| \geq 4$ if $\sqrt{N} = \sqrt{pq^k} > q^2$. Hence, $N = pq^k$ for some $k \geq 4$ and $p < q$. In this case, $S'_N = \{p, q, pq, q^2, \ldots\}$ satisfies $U(p, q, 0, q)$.
- If $N = pq^k r$ for some $k \geq 2$, $p < q$, and $r > pq^k$, then N is small recurrent with $|S'_N| \geq 4$. In this case, $S'_N = \{p, q, pq, q^2, \ldots, pq^{k-1}, q^k, pq^k\}$ satisfies $U(p, q, 0, q)$.

(S3) If N belongs to (S3), then $S'_N = \{p, q, p^2, pq, \ldots, p^k, p^{k-1}q, p^k q\}$. Since $p^2 = aq + bp$, we know that $p|a$. Write $a = pm$ for some $m \in \mathbb{Z}$ and get $b = p - mq$. Hence,

$$pq = ap^2 + bq = p^3 m + (p - mq)q \implies mq^2 = p^3 m.$$

Therefore, $(m, a, b) = (0, 0, p)$. However, the largest element in S'_N, $p^k q$, is not equal to $p \cdot p^k$. We conclude that form (S3) does not give a small recurrent number.

(S4) If $N = p^k q$ for some $k \geq 2$ and $\sqrt{q} < p < q$, then the nontrivial divisors of N in increasing order is $p < q < p^2 < pq < \cdots$. In order that $|S'_N| \geq 4$, we need $(pq)^2 < p^k q$, so $q < p^{k-2}$. Hence, $k \geq 4$. In this case, $S'_N = \{p, q, p^2, pq, \ldots\}$ satisfies $U(p, q, 0, p)$.

(S5) If $N = p^2 q^2$ for some $p < q < p^2$, then $\tau(N) = 9$. However, if $|S'_N| \geq 4$, then $\tau(N) \geq 10$ by (2). We conclude that form (S5) does not give a small recurrent number.

(S6) If $N = p^3 q^2$ for some $p^{3/2} < q < p^2$, then $S'_N = \{p, q, p^2, pq, p^3\}$ satisfies $U(p, q, 0, p)$.

(S7) Let N have form (S7). Since the first four numbers of S'_N are $p < q < p^2 < r$ and $\tau(N) = 12$, we know that the fifth number in S'_N must be pq. That $p < q < p^2 < r < pq$ satisfies some $U(p, q, a, b)$ gives $a = \frac{p(pq-r)}{q^2-p^3}$, $b = \frac{rq-p^4}{q^2-p^3}$, and $r = pq - \sqrt{(q^2-p^3)(p^2-q)}$. We conclude that a number of form (S7) is small recurrent if and only if $p < q < p^2 < r < pq$, $(q^2 - p^3) | (pq - r)$, $(q^2 - p^3) | (rq - p^4)$, and $r = pq - \sqrt{(q^2-p^3)(p^2-q)}$. An example is $(p, q, r) = (2, 3, 5)$. We do not know if (2, 3, 5) is the only set of primes that satisfy all these conditions or not.

From the above analysis, we obtain the proposition, which is a refinement of Proposition 9.

Proposition 10 *Let p, q, r denote prime numbers and k be some positive integer. A positive integer N is small recurrent with $|S'_N| \geq 4$ if and only if N belongs to one of the following forms.*

1. *$N = p^k$ for some $k \geq 9$. In this case, $S'_N = \{p, p^2, p^3, \ldots, p^{\lfloor (k-1)/2 \rfloor}\}$ satisfies $U(p, p^2, p, 0)$.*
2. *$N = p^k q$ for some $k \geq 4$ and $q > p^k$. In this case, $S'_N = \{p, p^2, p^3, \ldots, p^k\}$ satisfies $U(p, p^2, p, 0)$.*
3. *$N = pq^k$ for some $k \geq 4$ and $p < q$. In this case, $S'_N = \{p, q, pq, q^2, \ldots\}$ satisfies $U(p, q, 0, q)$.*
4. *$N = pq^k r$ for some $k \geq 2$, $p < q$, and $r > pq^k$. In this case, $S'_N = \{p, q, pq, q^2, \ldots, pq^{k-1}, q^k, pq^k\}$ satisfies $U(p, q, 0, q)$.*
5. *$N = p^k q$ for some $k \geq 4$ and $\sqrt{q} < p < q$. In this case, $S'_N = \{p, q, p^2, pq, \ldots\}$ satisfies $U(p, q, 0, p)$.*
6. *$N = p^3 q^2$ for some $p^{3/2} < q < p^2$. In this case, $S'_N = \{p, q, p^2, pq, p^3\}$ satisfies $U(p, q, 0, p)$.*
7. *$N = p^2 qr$, where $p < q < p^2 < r < pq$, $(q^2 - p^3) | (pq - r)$, $(q^2 - p^3) | (rq - p^4)$, and $r = pq - \sqrt{(q^2-p^3)(p^2-q)}$. In this case, $S'_N = \{p, q, p^2, r, pq\}$ satisfies $U\left(p, q, \frac{p(pq-r)}{q^2-p^3}, \frac{rq-p^4}{q^2-p^3}\right)$.*

3.2 The Case $|S'_N| \leq 3$

If $|S'_N| \leq 3$, then $\tau(N) \leq 9$. We use the classifications of those N from the introduction to obtain the following proposition.

Proposition 11 *Let p, q, r denote prime numbers and k be some positive integer. A positive integer $N > 1$ is small recurrent with $|S'_N| \leq 3$ if and only if N belongs to one of the following forms.*

1. *$N = p^k$ for some $k \leq 8$. In this case, $S'_N = \{p, p^2, \ldots, p^{\lfloor (k-1)/2 \rfloor}\}$ satisfies $U(p, p^2, p, 0)$.*
2. *$N = pq$ for some $p < q$. In this case, $S'_N = \{p\}$.*
3. *$N = pq^2$ for some $p < q$. In this case, $S'_N = \{p, q\}$.*
4. *$N = p^2q$ for some $p < q$. If $q < p^2$, then $S'_N = \{p, q\}$. If $q > p^2$, then $S'_N = \{p, p^2\}$.*
5. *$N = pq^3$ for some $p < q$. In this case, $S'_N = \{p, q, pq\}$.*
6. *$N = p^3q$ for some $p < q$. If $p < q < p^2$, then $S'_N = \{p, q, p^2\}$. If $p^3 < q$, then $S'_N = \{p, p^2, p^3\}$. (The case $p^2 < q < p^3$ is eliminated because the three elements in S'_N would be $p < p^2 < q$. However, there is no integral solution (a, b) to $q = ap^2 + bp$.)*
7. *$N = p^2q^2$ for some $p < q$. If $p < q < p^2$, then $S'_N = \{p, q, p^2\}$. (The case $p^2 < q$ is eliminated due to the same reason as in item 6.)*
8. *$N = pqr$ for some $p < q < r$. If $r < pq$ and there is an integral solution (a, b) to $r = aq + bp$, then $S'_N = \{p, q, r\}$. If $r > pq$, then $S'_N = \{p, q, pq\}$.*

Combining Propositions 10 and 11, we obtain Theorem 1.

4 Large Recurrent Numbers

Now we characterize all positive integers N whose L'_N satisfies a linear recurrence of order at most two. By a simple observation, instead of working directly with divisors in L'_N, we work with divisors in S'_N. Again, the set of divisors of a positive integer N is $1 = d_1 < d_2 < \cdots < d_{\tau(N)}$ and the set $S'_N = \{d_2, d_3, \ldots\}$.

4.1 The Case $|L'_N| \geq 4$

Note that $|L'_N| \geq 4$ is equivalent to $|S'_N| \geq 4$.

Lemma 5 *For any $d \in L'_N$, we have $N/d \in S'_N$. If N is large recurrent with $|L'_N| \geq 4$, then*

$$ad_{i+2} + bd_{i+1} = \frac{d_{i+1}d_{i+2}}{d_i}, \forall d_i, d_{i+1}, d_{i+2} \in S'_N. \quad (3)$$

In particular, we have

$$ad_4 + bd_3 = \frac{d_3 d_4}{d_2}. \tag{4}$$

Proof If $d \in L'_N$, then $\sqrt{N} < d < N$. Then $1 < N/d < \sqrt{N}$, so $N/d \in S'_N$. Let

$$d'_i := d_{\tau(n)+1-i} = \frac{N}{d_i} \in L'_N, \forall d_i \in S'_N.$$

If N is large recurrent, then we have

$$d'_i = ad'_{i+1} + bd'_{i+2}, \forall d'_i, d'_{i+1}, d'_{i+2} \in L'_N.$$

Therefore,

$$\frac{N}{d_i} = a\frac{N}{d_{i+1}} + b\frac{N}{d_{i+2}}, \forall d_i, d_{i+1}, d_{i+2} \in S'_N,$$

which gives

$$ad_{i+2} + bd_{i+1} = \frac{d_{i+1}d_{i+2}}{d_i}, \forall d_i, d_{i+1}, d_{i+2} \in S'_N.$$

This completes our proof.

Since d_2 is a prime number p and d_3 is either p^2 or a prime number $q > p$, we consider two cases.

4.1.1 When $d_3 = p^2$

Then d_4 is either p^3 or a prime number $q > p^2$.

(a) If $d_4 = p^3$, then (4) implies that $p^2 = ap + b$.

Claim If $p \neq a$, then $S'_N = \{p, p^2, \ldots, p^k\}$ for some $k \geq 4$.

Proof We need to show that if $d_i \in S'_N$, then $d_i = p^{i-1}$. Base case: the claim holds for $i \leq 4$. Suppose that there exists a $j \geq 4$ such that $d_i = p^{i-1}$ for all $i \leq j$. Using (3), we have

$$ad_{j+1} + bp^{j-1} = ad_{j+1} + bd_j = \frac{d_{j+1}d_j}{d_{j-1}} = \frac{d_{j+1}p^{j-1}}{p^{j-2}} = pd_{j+1},$$

which, combined with $p^2 = ap + b$, gives

$$(p-a)(d_{j+1} - p^j) = 0.$$

Since $p \neq a$, we obtain $d_{j+1} = p^j$, as desired.

By Proposition 2, we know that when $p \neq a$, either $N = p^k$ or $N = p^k q$ for some $k \geq 1$ and some prime $q > p^k$.

Now suppose that $p = a$. Then $b = 0$. We can write elements in L'_N as $\{g_1, g_2, pg_2, p^2 g_2, \ldots, p^k g_2\}$ for some $k \geq 2$. Correspondingly, the set S'_N is $\{p, p^2, \ldots, p^k, p^{k+1}, p^{k+1} g_2/g_1\}$. If $p^{k+1} g_2/g_1$ is a power of p, then we have the same conclusion about N as when $p \neq a$. If $p^{k+1} g_2/g_1$ is not a power of p, then

$$\frac{p^{k+1} g_2}{g_1} = q, \text{ for some prime } q > p \implies g_1 = p^{k+1} \frac{g_2}{q}.$$

Note that $g_2/q \in \mathbb{N}$. Furthermore, we claim that $g_2/q = p$. Indeed, since $1 < g_2/q < g_1$, we know that $g_2/q \in S'_N$. If $g_2/q = q$, then

$$pq < p^{k+1} \frac{g_2}{q} = g_1,$$

which implies that $pq \in S'_N$, a contradiction. If $g_2/q = p^j$ for some $j > 1$, then

$$p^{k+2} < p^{k+1+j} = p^{k+1} \frac{g_2}{q} = g_1,$$

which implies that $p^{k+2} \in S'_N$, another contradiction. Therefore, $g_2/q = p$, and we obtain $g_1 = p^{k+2}$ and $g_2 = pq$. Hence, $N = p^{k+2} q$ for some $k \geq 2$ and $p^{k+1} < q < p^{k+2}$.

(b) If $d_4 = q$, we claim that $a \neq p$. Suppose otherwise. Applying (4) to d_2, d_3, and d_4 gives $aq + bp^2 = pq$. Hence, $a = p$ implies that $b = 0$. However, applying (3) to d_3, d_4, and d_5 gives $(p^3 - q)d_5 = 0$, a contradiction. Therefore, $a \neq p$. By (4), we have

$$d_4 = \frac{bp^2}{p - a} = q \implies q | b \implies b = kq \text{ for some } k \in \mathbb{Z} \backslash \{0\}.$$

Hence, $a = p - kp^2$. By (3) applied to d_3, d_4, and d_5,

$$d_5 = \frac{kp^2 q^2}{q - p^3 + kp^4}, \tag{5}$$

which implies that $p^2 | d_5$ since $\gcd(p^2, q - p^3 + kp^4) = 1$. Hence, $d_5 = p^3$, and (5) gives $k = p(p^3 - q)/(p^5 - q^2)$.

Case b.i) If S'_N has exactly four elements, which are p, p^2, q, p^3, then $\tau(N) = 10$, which implies that $N = p^4 q$. Hence, $L'_N = \{pq, p^4, p^2 q, p^3 q\}$ with $a = pq(p^2 - q)/(p^5 - q^2)$ and $b = pq(p^3 - q)/(p^5 - q^2)$. We conclude that $N = p^4 q (p^2 < q < p^3)$, $(p^5 - q^2) | (p^2 - q)$, and $(p^5 - q^2) | (p^3 - q)$.
Case b.ii) If $|S'_N| > 4$, then (3) gives

$$d_6 = \frac{bp^3q}{p^3 - aq} \implies q|d_6 \implies d_6 = pq.$$

However, since $(a, b) = (p(1 - kp), kq)$, we have

$$pq = d_6 = \frac{bp^3q}{p^3 - aq} = \frac{kp^2q^2}{p^2 - q(1 - kp)},$$

which gives $p^2 = q$, a contradiction.

We summarize our result when $d_3 = p^2$.

Proposition 12 *A number N is large recurrent with $|L'_N| \geq 4$ and $(d_2, d_3) = (p, p^2)$ for some prime p if and only if N belongs to one of the following forms.*

1. *$N = p^k$ for some $k \geq 9$. Then $L'_N = \{p^{\lceil (k-1)/2 \rceil + 1}, p^{\lceil (k-1)/2 \rceil + 2}, \ldots, p^{k-1}\}$ satisfies $U(p^{\lceil (k-1)/2 \rceil + 1}, p^{\lceil (k-1)/2 \rceil + 2}, p, 0)$.*
2. *$N = p^k q$ for some $k \geq 4$ and $q > p^k$. In this case, $L'_N = \{q, pq, p^2q, \ldots, p^{k-1}q\}$ satisfies $U(q, pq, p, 0)$.*
3. *$N = p^k q$ for some $k \geq 4$ and $p^{k-1} < q < p^k$. Then*

$$L'_N = \{p^k, pq, p^2q, \ldots, p^{k-1}q\}$$

   satisfies $U(p^k, pq, p, 0)$.
4. *$N = p^4 q$ with $p^2 < q < p^3$, $(p^5 - q^2)|(p^2 - q)$, and $(p^5 - q^2)|(p^3 - q)$. In this case, $L'_N = \{pq, p^4, p^2q, p^3q\}$.*

4.1.2 When $d_3 = q$

By (3),

$$p(ad_4 + bq) = d_4q \implies p|d_4.$$

Write $d_4 = kp$ for some integer k. Since $d_2 = p$ and $d_3 = q$, d_4 must be either p^2 or pq.

(a) If $d_4 = p^2$, (3) gives $ap^2 + bq = pq$. Hence, $q|a$ and $p|b$. Write $a = mq$ and $b = np$ for some integers m, n to get $mp + n = 1$. By (3), we see that

$$d_5 = \frac{bp^2q}{p^2 - aq} \implies q|d_5 \implies d_5 = pq.$$

Therefore,

$$\frac{bp^2q}{p^2 - aq} = \frac{np^3q}{p^2 - mq^2} = \frac{(1 - mp)p^3q}{p^2 - mq^2} = pq \implies m(p^3 - q^2) = 0,$$

which gives $m = 0$ and so, $(a, b) = (0, p)$. By (3), $d_{i+2} = pd_i$ for all $d_i, d_{i+2} \in S'_N$ and

$$S'_N = \{p, q, p^2, pq, \ldots\}.$$

If $|S'_N| = 4$, then $\tau(N) = 10$ and $N = p^4q$ for $p < q < p^2$. Suppose that $|S'_N| \geq 5$, then $p^3 \in S'_N$.

Case a.i) If $q^2|N$, let $k \geq 2$ and $\ell \geq 3$ be the largest power such that $q^k|N$ and $p^k|N$, respectively. Since $q^2 \notin S'_N$, we know that

$$q^4 \geq N \geq p^3q^k > q^{k+3/2} \Longrightarrow k < 5/2.$$

It follows that $k = 2$. That $q^2 < p^2q$ implies that

$$(p^2q)^2 > N \geq p^\ell q^2 \Longrightarrow 3 \geq \ell \geq 3.$$

Hence, $\ell = 3$. If N does not have any other prime divisors besides p and q, then $N = p^3q^2$ for $p < q < p^2$. If N has a prime divisor $r \neq p, q$, then $r > \sqrt{N}$. So, r must be the unique prime divisor different from p and q. We have $N = p^3q^2r$ for $p < q < p^2$ and $r > p^3q^2$. Then $q^2 \in S'_N$, a contradiction.
Case a.ii) If $q^2 \nmid N$ and N has no prime divisors other than p and q, then $N = p^kq$ some for $k \geq 2$ and $p < q < p^2$.
Case a.iii) If $q^2 \nmid N$ and there exists a prime divisor r other than p or q, then $r > \sqrt{N}$ and r is the unique prime different from p and q. Therefore, $N = p^kqr$ for some $k \geq 2$ and $p < q < p^2 < p^kq < r$. Note that the two largest elements in S'_N are $p^{k-1}q$ and p^kq. Let d be the third largest divisor in S'_N. The relation $d_{i+2} = pd_i$ for all $d_i, d_{i+2} \in S'_N$ gives that $dp = p^kq$ and so, $d = p^{k-1}q$, which contradicts that $p^{k-1}q$ is the second largest in S'_N.

(b) If $d_4 = pq$, then $p^2 \nmid N$ since $p^2 < pq$. By (4),

$$ap = q - b. \tag{6}$$

We see that d_5 is equal to q^2 or r, for some prime $r > pq$.

Case b.i) If $d_5 = q^2$, then (3) gives

$$bp = (p - a)q. \tag{7}$$

From (6) and (7), we obtain $a(p^2 - q) = 0$, so $(a, b) = (0, q)$. By (3), $d_{i+2} = qd_i$ for all $d_i, d_{i+2} \in S'_N$. Using [1, Proposition 5], we conclude that $N = pq^k$ or $N = pq^kr$ for some $k \geq 2$ and $p < q < pq^k < r$.
Case b.ii) If $d_5 = r$, then we claim that $|S'_N| > 4$. If not, $|S'_N| = 4$ implies that $\tau(N) = 10$, which contradicts that N has three distinct prime divisors. By (3), we see that

$$pq(ad_6 + br) = d_6 r,$$

so $pq|d_6$. So, $d_6 \in \{pq^2, pqr\}$. If $d_6 = pq^2$, then $q^2 < d_6$, but q^2 does not appear before d_6 in S'_N, a contradiction. If $d_6 = pqr$, then $pr < d_6$, but pr does not appear before d_6 in S'_N, again a contradiction.

Proposition 13 *A number N is large recurrent with $|L'_N| \geq 4$ and $(d_2, d_3) = (p, q)$ for some primes $p < q$ if and only if N belongs to one of the following forms.*

1. *$N = p^3 q^2$ for $p < q < p^2$. In this case, $L'_N = \{q^2, p^2 q, pq^2, p^3 q, p^2 q^2\}$ satisfies $U(q^2, p^2 q, 0, p)$.*
2. *$N = p^k q$ some for $k \geq 4$ and $p < q < p^2$. In this case,*

$$L'_N = \begin{cases} \{p^{k/2+1}, p^{k/2} q, p^{k/2+2}, \dots, p^{k-1} q\} & \text{if } 2|k, \\ \{p^{(k-1)/2} q, p^{(k+3)/2}, p^{(k+1)/2} q, \dots, p^{k-1} q\} & \text{if } 2 \nmid k. \end{cases}$$

Observe that L'_N satisfies $U(p^{k/2+1}, p^{k/2} q, 0, p)$ and $U(p^{(k-1)/2} q, p^{(k+3)/2}, 0, p)$ for even and odd k, respectively.

3. *$N = pq^k$ for some $k \geq 4$ and $p < q$. In this case,*

$$L'_N = \begin{cases} \{pq^{k/2}, q^{k/2+1}, \dots, q^k\} & \text{if } 2|k, \\ \{q^{(k+1)/2}, pq^{(k+1)/2}, \dots, q^k\} & \text{if } 2 \nmid k. \end{cases}$$

Observe that L'_N satisfies $U(pq^{k/2}, q^{k/2+1}, 0, q)$ and $U(q^{(k+1)/2}, pq^{(k+1)/2}, 0, q)$ for even and odd k, respectively.

4. *$N = pq^k r$ for some $k \geq 2$ and $p < q < pq^k < r$. In this case, $L'_N = \{r, pr, qr, pqr, q^2 r, \dots, q^k r\}$ satisfies $U(r, pr, 0, q)$.*

Combining Propositions 12 and 13, we obtain the following.

Proposition 14 *A number N is large recurrent with $|L'_N| \geq 4$ if and only if N belongs to one of the following forms.*

1. *$N = p^k$ for some $k \geq 9$. Then $L'_N = \{p^{\lceil (k-1)/2 \rceil + 1}, p^{\lceil (k-1)/2 \rceil + 2}, \dots, p^{k-1}\}$ satisfies $U(p^{\lceil (k-1)/2 \rceil + 1}, p^{\lceil (k-1)/2 \rceil + 2}, p, 0)$.*
2. *$N = p^k q$ for some $k \geq 4$ and $q > p^k$. In this case, $L'_N = \{q, pq, p^2 q, \dots, p^{k-1} q\}$ satisfies $U(q, pq, p, 0)$.*
3. *$N = p^k q$ for some $k \geq 4$ and $p^{k-1} < q < p^k$. Then*

$$L'_N = \{p^k, pq, p^2 q, \dots, p^{k-1} q\}$$

satisfies $U(p^k, pq, p, 0)$.

4. $N = p^k q$ *some for* $k \geq 4$ *and* $p < q < p^2$. *In this case,*

$$L'_N = \begin{cases} \{p^{k/2+1}, p^{k/2}q, p^{k/2+2}, \ldots, p^{k-1}q\} & \text{if } 2|k, \\ \{p^{(k-1)/2}q, p^{(k+3)/2}, p^{(k+1)/2}q, \ldots, p^{k-1}q\} & \text{if } 2 \nmid k. \end{cases}$$

Observe that L'_N *satisfies* $U(p^{k/2+1}, p^{k/2}q, 0, p)$ *and* $U(p^{(k-1)/2}q, p^{(k+3)/2}, 0, p)$ *for even and odd* k, *respectively.*

5. $N = p^4 q$ *with* $p^2 < q < p^3$, $(p^5 - q^2)|(p^2 - q)$, *and* $(p^5 - q^2)|(p^3 - q)$. *In this case,* $L'_N = \{pq, p^4, p^2q, p^3q\}$.
6. $N = p^3q^2$ *for* $p < q < p^2$. *In this case,* $L'_N = \{q^2, p^2q, pq^2, p^3q, p^2q^2\}$ *satisfies* $U(q^2, p^2q, 0, p)$.
7. $N = pq^k$ *for some* $k \geq 4$ *and* $p < q$. *In this case,*

$$L'_N = \begin{cases} \{pq^{k/2}, q^{k/2+1}, \ldots, q^k\} & \text{if } 2|k, \\ \{q^{(k+1)/2}, pq^{(k+1)/2}, \ldots, q^k\} & \text{if } 2 \nmid k. \end{cases}$$

Observe that L'_N *satisfies* $U(pq^{k/2}, q^{k/2+1}, 0, q)$ *and* $U(q^{(k+1)/2}, pq^{(k+1)/2}, 0, q)$ *for even and odd* k, *respectively.*

8. $N = pq^k r$ *for some* $k \geq 2$ *and* $p < q < pq^k < r$. *In this case,* $L'_N = \{r, pr, qr, pqr, q^2r, \ldots, q^k r\}$ *satisfies* $U(r, pr, 0, q)$.

4.2 The Case $|L'_N| \leq 3$

If $|L'_N| \leq 3$, then $\tau(N) \leq 9$. We use the classifications of those N from the introduction to obtain the following proposition

Proposition 15 *Let* p, q, r *denote prime numbers and* k *be some positive integer. A positive integer* $N > 1$ *is small recurrent with* $|L'_N| \leq 3$ *if and only if* N *belongs to one of the following forms.*

1. $N = p^k$ *for some* $k \leq 8$. *In this case,* $L'_N = \{p^{\lceil (k-1)/2 \rceil + 1}, \ldots, p^{k-1}\}$ *satisfies* $U(p^{\lceil (k-1)/2 \rceil + 1}, p^{\lceil (k-1)/2 \rceil + 2}, p, 0)$.
2. $N = pq$ *for some* $p < q$. *In this case,* $L'_N = \{q\}$.
3. $N = pq^2$ *for some* $p < q$. *In this case,* $L'_N = \{pq, q^2\}$.
4. $N = p^2q$ *for some* $p < q$. *If* $q < p^2$, *then* $L'_N = \{p^2, pq\}$. *If* $q > p^2$, *then* $L'_N = \{q, pq\}$.
5. $N = pq^3$ *for some* $p < q$. *In this case,* $L'_N = \{q^2, pq^2, q^3\}$.
6. $N = p^3q$ *for some* $p < q$. *If* $p < q < p^2$, *then* $L'_N = \{pq, p^3, p^2q\}$. *If* $p^3 < q$, *then* $L'_N = \{q, pq, p^2q\}$. *If* $p^2 < q < p^3$, *then* $L'_N = \{p^3, pq, p^2q\}$.
7. $N = p^2q^2$ *for some* $p < q$. *If* $p < q < p^2$, *then* $L'_N = \{q^2, p^2q, pq^2\}$. *The case* $p^2 < q$ *is impossible as it gives* $L'_N = \{p^2q, pq^2, q^2\}$ *and there is no integral solution* (a, b) *to* $apq^2 + bp^2q = q^2$.

8. *$N = pqr$ for some $p < q < r$. If $r > pq$, then $L'_N = \{r, pr, qr\}$. The case $r < pq$ is impossible as it gives $L'_N = \{pq, pr, qr\}$ and there is no integral solution (a, b) to $apr + bpq = qr$.*

Combining Propositions 14 and 15, we obtain Theorem 2.

5 Appendix

Proof *(Proof of Lemma 3)* (i) Since $r = aq + bp$ and $pq = ar + bq$, we know that $\gcd(a, b)|r$ and $\gcd(a, b)|pq$, respectively. Hence, $\gcd(a, b) = 1$.

Since $r = aq + bp$ and r is a prime, $p \nmid a$.

(ii) Suppose that $k = \gcd(b, d_i) > 1$ for some $d_i \in S'_N$. If $d_i = d_2$, then $p|b$. Since $pq = ar + bq$, we get $p|a$, which contradicts $\gcd(a, b) = 1$. If $d_i = d_3$, then $q|b$. Since $r = aq + bp$, we get $q|r$, a contradiction. If $d_i > d_3$, then write

$$\gcd(b, d_i) = \gcd(b, ad_{i-1} + bd_{i-2}) \overset{\text{i)}}{=} \gcd(b, d_{i-1}),$$

which, by induction, gives $1 < \gcd(b, d_i) = \gcd(b, d_3)$, which has been shown to be impossible.

(iii) The claim holds for $i \leq 4$. Let $d_i, d_{i+1} \in S'_N$ for some $i \geq 5$. We have

$$\gcd(d_i, d_{i+1}) = \gcd(d_i, ad_i + bd_{i-1}) \overset{\text{ii)}}{=} \gcd(d_i, d_{i-1}).$$

By induction, we obtain $\gcd(d_i, d_{i+1}) = 1$.

(iv) The claim holds for $i \leq 5$. Assume that it holds for all $i \leq j$ for some $j \geq 5$. We show that it holds for $i = j + 1$. We have

$$pq = ar + bq = a(aq + bp) + bq = (a^2 + b)q + abp.$$

Hence, $p|(a^2 + b)$. Write

$$d_{j+1} = ad_j + bd_{j-1} = a(ad_{j-1} + bd_{j-2}) + bd_{j-1} = (a^2 + b)d_{j-1} + abd_{j-2}.$$

Since $p|(a^2 + b)$ and $\gcd(p, ab) = 1$, we know that $p|d_{j+1}$ if and only if $p|d_{j-2}$. By the inductive hypothesis, $p|d_{j-2}$ if and only if $j - 2 \equiv 2 \bmod 3$, or equivalently, $j + 1 \equiv 2 \bmod 3$. By induction, we have the desired conclusion.

(v) The claim holds for $i \leq 5$. Assume that it holds for all $i \leq j$ for some $j \geq 5$. We show that it holds for $i = j + 1$. That $pq = ar + bq$ implies that $q|a$. Write

$$d_{j+1} = ad_j + bd_{j-1}.$$

By (ii), $q|d_{j+1}$ if and only if $q|d_{j-1}$. By the inductive hypothesis, $q|d_{j-1}$ if and only if $j+1 \equiv 1 \bmod 2$. This completes our proof.

Proof *(Proof of Lemma 4)* (i) Same as the proof of Lemma 3 item (i).

(ii) Suppose, for a contradiction, that $\gcd(b, d_i) > 1$ for some $i \geq 3$. If $i = 3$, then $q|b$. We have $r = aq + bp$. Since $q|b$, we get $q|r$, a contradiction. If $i \geq 4$, write

$$\gcd(b, d_i) \;=\; \gcd(b, ad_{i-1} + bd_{i-2}) \;=\; \gcd(b, d_{i-1}).$$

By induction, $1 < \gcd(b, d_i) = \gcd(b, d_3)$, which has been shown to be impossible.

(iii) The claim holds for $i \leq 4$. Pick $i \geq 5$. We have

$$\gcd(d_i, d_{i+1}) \;=\; \gcd(d_i, ad_i + bd_{i-1}) \;=\; \gcd(d_i, bd_{i-1}) \;\overset{\text{ii)}}{=}\; \gcd(d_i, d_{i-1}).$$

By induction, we obtain $\gcd(d_i, d_{i+1}) = \gcd(d_4, d_5) = 1$.

(iv) Assume that $\gcd(a, d_i) > 1$ for some $i \geq 2$. If $i = 2$, then $p|a$, which contradicts the primality of r and the linear recurrence $r = aq + bp$. If $i = 3$, then $q|b$, which contradicts the primality of s and the linear recurrence $s = ar + bq$. Assume that $i \geq 4$. Write

$$\gcd(a, d_i) \;=\; \gcd(a, ad_{i-1} + bd_{i-2}) \;\overset{\text{i)}}{=}\; \gcd(a, d_{i-2}).$$

By induction, either $1 < \gcd(a, d_i) = \gcd(a, d_2)$ or $1 < \gcd(a, d_i) = \gcd(a, d_3)$, neither of which is possible.

(v) The claims holds for $i \leq 3$. Pick $i \geq 4$ and suppose that $k = \gcd(d_i, d_{i+2}) > 1$. Since $d_{i+2} = ad_{i+1} + bd_i$, k divides ad_{i+1}. By (iii), $\gcd(k, d_{i+1}) = 1$, so $k|a$. However, $\gcd(a, d_i) > 1$ contradicts (iv).

References

1. A. A Chentouf, Linear recurrences of order at most two in small divisors, *J. Integer Seq.* **25** (2022).
2. H. V. Chu, When the large divisors of a natural number are in arithmetic progression, *J. Integer Seq.* **23** (2020).
3. H. V. Chu, When the nontrivial, small divisors of a natural number are in arithmetic progression, *Quaest. Math.* **45** (2022), 969–977.
4. D. E. Iannucci, When the small divisors of a natural number are in arithmetic progression, *Integers* **18** (2018).

Representing Positive Integers as a Sum of a Squarefree Number and a Small Prime

Jack Dalton and Ognian Trifonov

Abstract We prove that every positive integer n which is not equal to 1, 2, 3, 6, 11, 30, 155, or 247 can be represented as a sum of a squarefree number and a prime not exceeding $\sqrt{n}$.

1 Introduction

A positive integer is squarefree if it is not divisible by the square of a prime number. In 1931 Estermann [5] obtained an asymptotic formula for the number of representations of a positive integer as the sum of a squarefree number and a prime. As a consequence, he showed that every sufficiently large integer is the sum of a squarefree number and a prime.

In 2017 Dudek [2] showed that every integer greater than two is a sum of a squarefree number and a prime.

In 1935, Erdős [4] showed that every sufficiently large integer $\not\equiv 1 \pmod 4$ can be represented as a sum of a square of a prime and a squarefree integer. The condition $n \not\equiv 1 \pmod 4$ is needed since if p is an odd prime, then $p^2 \equiv 1 \pmod 4$. Therefore, if p is an odd prime and $n \equiv 1 \pmod 4$, then $4 \mid n - p^2$, and $n - p^2$ is not squarefree. The result of Erdős was completed by Dudek and Platt [3] who proved in 2016 that every integer $n \geq 10$ such that $n \not\equiv 1 \pmod 4$ can be represented as the sum of a square of a prime and a squarefree integer.

Thus, one can get a complete answer to the question for which positive integers n there exists a prime $p < \sqrt{n}$ such that $n - p^2$ is squarefree.

A related question is for which positive integers n, there exists a prime $p < \sqrt{n}$ such that $n - p$ is squarefree. We answer the later question completely.

J. Dalton · O. Trifonov (✉)
Department of Mathematics, University of South Carolina, 1523 Greene Street, Columbia, SC 29208, USA
e-mail: trifonov@math.sc.edu

J. Dalton
e-mail: JRDALTON@math.sc.edu

M. B. Nathanson (ed.), *Combinatorial and Additive Number Theory VI*,
Springer Proceedings in Mathematics & Statistics 464,
https://doi.org/10.1007/978-3-031-65064-2_6

Theorem 1 *Every positive integer n can be represented in the form $n = s + p$ where s is a squarefree number and p is a prime with $p \le \sqrt{n}$, except when $n \in \{1, 2, 3, 6, 11, 30, 155, 247\}$.*

Since $2^2 \mid 6-2$; $3^2 \mid 11-2$, $2^3 \mid 11-3$; $2^2 \mid 30-2$, $3^3 \mid 30-3$, $5^2 \mid 30-5$; $3^2 \mid 155-2, 2^3 \mid 155-3, 5^2 \mid 155-5, 2^2 \mid 155-7, 2^2 \mid 155-11$; and $7^2 \mid 247-2, 2^2 \mid 247-3, 11^2 \mid 247-5, 2^2 \mid 247-7, 2^2 \mid 247-11$, and $3^2 \mid 247-13$ we see that indeed the integers $n \in \{1, 2, 3, 6, 11, 30, 155, 247\}$ cannot be represented as a sum of a squarefree number and a prime not exceeding $\sqrt{n}$.

A natural question is can $\sqrt{n}$ in Theorem 1 be replaced by n^θ for $\theta < 1/2$.

In 2015, Filaseta, Graham, and the second author wrote a paper [7] which considered the distribution of various arithmetic sequences. In particular, replacing p by $-p$ in Theorem 5.1 of [7] one obtains the following theorem.

Theorem 2 (Filaseta, Graham, and T.) *There exist effectively computable constants C_0 and n_0 such that for each integer $n \ge n_0$ at least one-fifth of the primes $p \le C_0 n^{1/5} \log^2 n$ are such that $n - p$ is squarefree.*

The authors of the paper [7] did not compute the values of C_0 and n_0. However, it is certain that it is not feasible to check which integers $n \le n_0$ do not satisfy Theorem 2 by direct computation.

An easy corollary of the above theorem is that for each $\theta > 1/5$ there exists an effectively computable constant n_θ such that each integer $n \ge n_\theta$ can be represented as a sum of a squarefree number and a prime not exceeding n^θ. At present it appears that to get below the exponent $1/5$ one will need to improve the gap result about squarefree numbers.

The proof of our main result follows the method of proof of Theorem 2. However, to prove the theorem we substantially sharpened the estimates in each step and supplemented the proof with nontrivial amount of computations.

Next, we outline the proof of Theorem 1. First, we verify the theorem for all $n \le 10^9$. This computation allows us to obtain the first few terms of the following sequence.

Let b_k be the least positive integer $n > p_k$ such that none of the integers $n - p_1, \ldots, n - p_k$ are squarefree. Here and throughout the paper p_k denotes the k-th prime number.

We have $b_1 = 6$, $b_2 = 11$, $b_3 = 30$, $b_4 = b_5 = 155$, $b_6 = 247$, $b_7 = 5753$, $b_8 = b_9 = b_{10} = b_{11} = 90263$, $b_{12} = 1481287$, $b_{13} = b_{14} = b_{15} = 7409327$. Also, $b_{16} > 10^9$.

Clearly, $b_n < \prod_{k=1}^{n} p_k^2$ for $n > 1$, since by the Chinese Remainder Theorem, the system of congruences $n \equiv p_k \pmod{p_k^2}$ has a positive solution not exceeding $\prod_{k=1}^{n} p_k^2$.

One can do much better. For example, if we want to find $n > 53 = p_{16}$ such that $n - p_k$ is not squarefree for $k = 1, \ldots, 16$, one can pick $n \equiv 3 \pmod 4$. Then $4 | n - p$ for $p \in \{3, 7, 11, 19, 23, 31, 43, 47\}$. Also, if $n \equiv 8 \pmod 9$, then $9 | n - p$ for $p \in$

$\{17, 53\}$. After that, we pick $n \equiv 2 \pmod{5^2}$, $n \equiv 5 \pmod{7^2}$, $n \equiv 13 \pmod{11^2}$, $n \equiv 29 \pmod{13^2}$, $n \equiv 37 \pmod{17^2}$, and $n \equiv 41 \pmod{19^2}$. Solving the later system of congruences, we obtain that its least positive solution is 23708451225527, thus $b_{16} \leq 23708451225527$.

From now on we assume $n > 10^9$. We show that for such n, there exists a prime $p \leq \sqrt{n}$, such that $n - p$ is squarefree.

There are $\pi(\sqrt{n})$ primes which do not exceed $\sqrt{n}$. We will show that there are less than $\pi(\sqrt{n})$ primes $p \leq \sqrt{n}$ such that $n - p$ is not squarefree. Establishing the last statement, will imply that for some prime $p \leq \sqrt{n}$, $n - p = s$ where s is squarefree completing the proof of the theorem.

If for some prime $p \leq \sqrt{n}$, $n - p$ is not squarefree, then $q^2 | n - p$ for some prime q, that is

$$p \equiv n \pmod{q^2}. \tag{1}$$

Note that congruence (1) can hold only for $q < \sqrt{n}$, since $0 < n - p < n$. Furthermore, if $q | n$ the congruence (1) has at most one solution, when $p = q$. Also, the number of solutions of (1) is $\pi(\sqrt{n}; q^2, n)$.

Above we have used the standard notation $\pi(a; b, c)$ to denote the number of primes not exceeding a which are in the arithmetic sequence $x \equiv c \pmod{b}$.

We estimate the number of primes p such that congruence (1) holds for some prime q in different ways depending on the size of q. For $q \in \{2, 3, 5, 7\}$ we use the paper [1] of Bennett, Martin, O'Bryant, and Rechnitzer on explicit bounds for primes in arithmetic sequences. For $q \in [11, n^{1/8}]$, we use a version of the Brun-Titchmarsh Theorem due to Montgomery and Vaughan [8]. For $q \in [n^{1/8}, \sqrt{n}/(c \log n)]$ we use elementary bounds, and for $n \in [\sqrt{n}/(c \log n), \sqrt{n})$ we use an estimate based on first differences of values of a function. Combining the above estimates, we show that the theorem holds for $n > 59^8$. Next, considering $n \in [p_k^8, p_{k+1}^8]$ for $k = 6, \ldots, 16$ we get even sharper bounds for certain sums over primes which appear in our estimates and establish the theorem in the last 11 intervals. Since $p_6 = 13$ and $13^8 < 10^9$ this completes the proof.

2 Proof of the Main Result

Proof First, by direct computation we confirmed the theorem for $n \leq 10^9$. To save memory we check the result in intervals of the form $I_k := [k \cdot 10^7, (k+1) \cdot 10^7]$. Computing the indicator function of squarefree integers in I_k is relatively fast since for each prime p the number of multiples of p^2 in I_k is at most $10^7/p^2 + 1$ and one only needs to check multiples of p^2 with $p \leq \sqrt{(k+1) \cdot 10^7}$. Checking the theorem in I_k is also fast because for most n, $n - 2$ is squarefree and for each $n \leq 10^9$ it took no more than 17 checks to find a prime p such that $n - p$ is squarefree. Using Python on a laptop with an AMD Ryzen 5 3500U processor and 8MB of memory, checking the theorem on each I_k with $k \in [1, 99]$ took less than a minute, so the whole computation was complete in under two hours.

From now on, we assume $n > 10^9$.

We will use the notation $\sum'$ to indicate a sum over prime numbers only. For example, $\sum'_{q \le x} 1$ is the number of prime numbers up to x, that is $\pi(x)$.

We estimate the number of primes $p < \sqrt{n}$ satisfying congruence (1) in different ways depending on the size of q.

Case 1. $q \in \{2, 3, 5, 7\}$.

Let Q_1 be the number of primes less that $\sqrt{n}$ which satisfy congruence (1) for some $q \in \{2, 3, 5, 7\}$.

Here we use the following theorem which is one of the statements in Corollary 1.7 from the paper [1].

Theorem 3 (M. Bennett, G. Martin, K. O'Bryant, and A. Rechnitzer) *Let a and q be integers with $1 \le q \le 10^5$ and $\gcd(a, q) = 1$. If $x \ge 10^6$, then*

$$\left| \pi(x; q, a) - \frac{\mathrm{Li}(x)}{\varphi(q)} \right| < 0.027 \frac{x}{\log^2 x}.$$

Above, by $\mathrm{Li}(x)$ the authors mean the function defined by $\mathrm{Li}(x) = \int_2^x \frac{dt}{\log t}$. Also, everywhere in this paper $\log x$ means $\log_e x = \ln x$.

To be able to use the above theorem, we assume $n \ge 10^{12}$ and we will deal with the case $10^9 < n < 10^{12}$ later.

Using inclusion-exclusion for $q = 2$ and $q = 3$, and that congruence (1) has $\pi(\sqrt{n}; q^2, n)$ solutions for fixed q, we obtain,

$$Q_1 \le \pi(\sqrt{n}; 4, n) + \pi(\sqrt{n}; 9, n) - \pi(\sqrt{n}; 36, n) + \pi(\sqrt{n}; 25, n) + \pi(\sqrt{n}; 49, n). \tag{2}$$

When $\gcd(n, 210) = 1$, Theorem 3 applies, and we obtain

$$Q_1 \le \mathrm{Li}(\sqrt{n}) \left(\frac{1}{2} + \frac{1}{6} - \frac{1}{12} + \frac{1}{20} + \frac{1}{42} \right) + 0.54 \frac{\sqrt{n}}{\log^2 n}. \tag{3}$$

The following is Lemma 5.9 of the paper [1].

Lemma 1 *For $x \ge 1865$,*

$$\mathrm{Li}(x) < \frac{x}{\log x} \left(1 + \frac{3}{2 \log x} \right) . \tag{4}$$

Using the above lemma and (3) we get

$$Q_1 < \frac{46\sqrt{n}}{35\log n}\left(1+\frac{3}{\log n}\right)+0.54\frac{\sqrt{n}}{\log^2 n} = \frac{46\sqrt{n}}{35\log n}+\frac{1569\sqrt{n}}{350\log^2 n}. \quad (5)$$

If $7|n$, $\pi(\sqrt{n}; 49, n) \leq 1$ and we need to replace to upper bound for $\pi(\sqrt{n}; 49, n) \leq 1$ used in (5) which is $\frac{\sqrt{n}}{21\log n}\left(1+\frac{3}{\log n}\right)+0.108\frac{\sqrt{n}}{\log^2 n}$ by 1.

However, the function $f(n) = \frac{\sqrt{n}}{\log n}$ is increasing for $n > e^2$ and $f(10^{12}) = 36192.2\ldots$, so

$$\frac{\sqrt{n}}{21\log n}\left(1+\frac{3}{\log n}\right)+0.108\frac{\sqrt{n}}{\log^2 n} > 36192/21 > 1,$$

for $n \geq 10^{12}$, so Eq. (5) holds when $\gcd(n, 30) = 1$ (regardless of whether $7 \mid n$ or $7 \nmid n$).

We argue similarly when $5 \mid n$ to conclude that Eq. (5) holds when $\gcd(n, 6) = 1$ and $n \geq 10^{12}$.

When $3 \mid n$ we need to replace to upper bound for $\pi(\sqrt{n}; 9, n) - \pi(\sqrt{n}; 36, n) \leq 1$ used in (5) which is $\frac{\sqrt{n}}{12\log n}\left(1+\frac{3}{\log n}\right)+0.216\frac{\sqrt{n}}{\log^2 n}$ by 1, which is not a problem for $n \geq 10^{12}$. Thus, Eq. (5) holds when $\gcd(n, 3) = 1$ and $n \geq 10^{12}$.

We argue similarly when $2|n$ to show that Eq. (5) holds for all positive integers $n \geq 10^{12}$.

Case 2. $11 \leq q \leq n^{1/8}$

Since $n \geq 10^{12}$ we have $n^{1/8} > 11$.

Let Q_2 be the number of primes $p \leq \sqrt{n}$ such that congruence (1) holds for some $q \in [11, n^{1/8}]$. Thus,

$$Q_2 \leq \sum_{11\leq q\leq n^{1/8}}{}' \pi(\sqrt{n}; q^2, n). \quad (6)$$

In Case 2 the following theorem from the paper [8] which is a version of Brun-Titchmarsh's inequality will be helpful.

Theorem 4 (H. L. Montgomery and R. C. Vaughan) *Let $m > 0$ and l be integers with $(m, l) = 1$, and let $x > m$ be a real number. Then,* $\pi(x; m, l) \leq \frac{2x}{\varphi(m)\log(x/m)}$.

We apply Theorem 4 with $x = \sqrt{n}$, $m = q^2$, and $l = n$. When $q \nmid n$ and $q^2 < \sqrt{n}$ the conditions of the theorem hold. So, suppose that $q \le n^{1/8}$ and $q \nmid n$. Then,

$$\pi(\sqrt{n}; q^2, n) \le \frac{2\sqrt{n}}{(q-1)q\log(\sqrt{n}/q^2)}. \tag{7}$$

When $q \mid n$, then $\pi(\sqrt{n}; q^2, n) \le 1$. However, for $q \ge 2$, $\log(\sqrt{n}/q^2) \le (\log n)/2$. Also, $q(q-1) < n^{1/4}$. So,

$$\frac{2\sqrt{n}}{(q-1)q\log(\sqrt{n}/q^2)} > \frac{2n^{1/4}}{\log n}.$$

The function $f_1(n) = \dfrac{n^{1/4}}{\log n}$ is increasing for $n > e^4$ and $f_1(10^9) = 8.58\ldots$, so Eq. (7) holds when $q \mid n$, as well.

Note that

$$\frac{1}{\log(\sqrt{n}/q^2)} = \frac{2}{\log n - 4\log q} = \frac{2}{\log n}\left(\frac{1}{1-\frac{4\log q}{\log n}}\right).$$

Next, use that

$$\frac{1}{1-r} \le 1 + 2r$$

for $r \le 1/2$ with $r = \dfrac{4\log q}{\log n}$ to obtain,

$$\frac{1}{\log(\sqrt{n}/q^2)} \le \frac{2}{\log n} + \frac{16\log q}{\log^2 n}.$$

Combining the above inequality with Eq. (7) we get

$$\pi(\sqrt{n}; q^2, n) \le \frac{4\sqrt{n}}{(q-1)q\log n} + \frac{32\sqrt{n}\log q}{(q-1)q\log^2 n}.$$

Adding the above inequality for primes q in $[11, n^{1/8}]$ we obtain

$$\sum_{11\le q\le n^{1/8}}{}' \pi(\sqrt{n}; q^2, n) \le c_1(11, n^{1/8})\frac{4\sqrt{n}}{\log n} + c_2(11, n^{1/8})\frac{32\sqrt{n}}{\log^2 n}, \tag{8}$$

where we define

$$c_1(A, B) = \sum_{A\le q\le B}{}' \frac{1}{q(q-1)}, \quad c_2(A, B) = \sum_{A\le q\le B}{}' \frac{\log q}{q(q-1)}.$$

Note that

$$c_1(A, B) < C_1(A) := \sum_{A \le q}' \frac{1}{q(q-1)},$$

and

$$c_2(A, B) < C_2(A) := \sum_{A \le q}' \frac{\log q}{q(q-1)}.$$

So, in the case when we have no upper bound on n we will use the estimate

$$Q_2 \le \frac{4C_1(11)\sqrt{n}}{\log n} + \frac{32C_2(11)\sqrt{n}}{\log^2 n}. \tag{9}$$

Case 3. $n^{1/8} < q \le n^{1/4}$

Let Q_3 be the number of primes $p \le \sqrt{n}$ such that congruence (1) holds for some $q \in (n^{1/8}, n^{1/4}]$. Thus,

$$Q_3 \le \sum_{n^{1/8} < q \le n^{1/4}}' \pi(\sqrt{n}; q^2, n). \tag{10}$$

Here we use the following lemma.

Lemma 2 *Let $x > 1$, let a be an integer, and let $q > 1$ be a prime p, or a power of a prime, p^k with $p > 3$. Then,*

$$\pi(x; q, a) \le \frac{x}{3q} + 2. \tag{11}$$

Proof Since $\pi(x; q, a)$ does not change if we shift a by an integer multiple of q, we can assume $0 \le a \le q - 1$. Moreover, if $\gcd(a, q) > 1$, then $\pi(x; q, a) \le 1$ and the lemma holds.

So, we can assume $\gcd(a, q) = 1$.

Let S be the set of positive integers not exceeding x which are congruent to a modulo q. Note that $\pi(x; q, a)$ is the number of elements of S which are prime.

Denote the size of S by k, that is $k := |S|$. Then, $k = \lfloor \frac{x-a}{q} \rfloor + 1$. We consider six cases depending on what is the remainder when k is divided by 6.

Case 1. $k = 6l$.

Here we divide the elements of S into l sextuples of consecutive elements, $(a, \dots, a + 5q), \dots, (a + (6l - 6)q, \dots, a + (6l - 1)q)$. Since $\gcd(6, q) = 1$, in each sextuple we have a complete set of residues modulo 6. Therefore, in each sextuple we have exactly two elements which are relatively prime to 6. So, in each sextuple (except possible the first one) there are at most two primes. The first sextuple may contain three primes (if $a = 2$ or $a = 3$) but not four. So, if $k = 6l$, $\pi(x; q, a) \le 2l + 1$.

Case 2. $k = 6l + 1$

Here we put all elements of S, except the first one into l sextuples of consecutive elements. None of the sextuples contain 2 or 3, so if $k = 6l + 1$, $\pi(x; q, a) \le 2l + 1$.

Case 3. $k = 6l + 2$

Here we put all elements of S, except the first two into l sextuples of consecutive elements. Again, none of the sextuples contain more than two primes, so if $k = 6l + 2$, $\pi(x; q, a) \le 2l + 2$.

Case 4. $k = 6l + 3$

Here we put all elements of S, except the first three into l sextuples of consecutive elements. Again, none of the sextuples contain more than two primes. Moreover, the first three elements of S which are $a, a + q, a + 2q$ are not all prime since either $a + q$ or $a + 2q$ is even and both are greater than 2. Therefore, if $k = 6l + 3$, $\pi(x; q, a) \le 2l + 2$.

Case 5. $k = 6l + 4$

Here we put all elements of S, except the first four into l sextuples of consecutive elements.Again, none of the sextuples contain more than two primes. Moreover, the first four elements of S cannot be all prime, since as in the previous case either $a + q$ or $a + 2q$ is composite. Therefore, if $k = 6l + 4$, $\pi(x; q, a) \le 2l + 3$.

Case 6. $k = 6l + 5$

Here we put all elements of S, except the first five into l sextuples of consecutive elements. Here we claim that among the first five elements of S, $a, a + q, a + 2q, a + 3q, a + 4q$ there are at most three primes. Indeed, if a is even, then $a + 2q$ and $a + 4q$ are composite (both even and greater than 2). If a is odd, then $a + q$ and $a + 3q$ are composite (even and greater than 2). So, if $k = 6l + 5$, $\pi(x; q, a) \le 2l + 3$.

Thus, in all six cases

$$\pi(x; q, a) \le \frac{k}{3} + \frac{5}{3}.$$

Recalling that $k = \lfloor \frac{x-a}{q} \rfloor + 1$ completes the proof of the lemma. □

Using the above lemma, we obtain

$$Q_3 \le \sum_{n^{1/8} < q \le n^{1/4}}{}' \pi(\sqrt{n}; q^2, n) \le \sum_{n^{1/8} < q \le n^{1/4}}{}' \left(\frac{\sqrt{n}}{3q^2} + 2 \right) \tag{12}$$

Define $c_3(A, B) := \sum_{A < q \le B}{}' \frac{1}{q^2}$.

Then,

$$Q_3 \le \frac{c_3\left(n^{1/8}, n^{1/4}\right)\sqrt{n}}{3} + 2\pi(n^{1/4}) - 2\pi(n^{1/8}) \tag{13}$$

To estimate $c_3(A, B)$ we use the following lemma.

Lemma 3 *Suppose* $A > 2$ *and*

$$\frac{c_4 t}{\log t} \le \pi(t) \le \frac{c_5 t}{\log t} \tag{14}$$

for some constants $c_5 > c_4 > 0$ *and all* $t \ge A$. *Then,*

$$\sum_{q>A}{}' \frac{1}{q^2} < \frac{2c_5 - c_4}{A \log A}. \tag{15}$$

Proof Using a Riemann-Stieltjes integral we have

$$S(A) := \sum_{A<q}{}' \frac{1}{q^2} = \int_A^\infty \frac{d(\pi(t))}{t^2}.$$

Next, integrating by parts we get

$$S(A) = \left.\frac{\pi(t)}{t^2}\right|_A^\infty + 2\int_A^\infty \frac{\pi(t)}{t^3} dt.$$

Using that $\pi(t) \le \dfrac{c_5 t}{\log t}$ for $t \ge A$ we obtain,

$$S(A) \le -\frac{\pi(A)}{A^2} + 2c_5 \int_A^\infty \frac{1}{t^2 \log t} dt.$$

Since, $\log t$ is increasing function for $t > 2$,

$$\int_A^\infty \frac{1}{t^2 \log t} dt < \frac{1}{\log A} \int_A^\infty \frac{1}{t^2} dt = \frac{1}{A \log A}.$$

Finally, using that $\dfrac{c_4 A}{\log A} \le \pi(A)$ we obtain the lemma. □

Next, we use the bounds of Rosser and Schoenfeld [9] for $\pi(t)$.

Theorem 5 (J. Rosser and L. Schoenfeld) *We have*

$$\pi(x) > \frac{x}{\log x}\left(1 + \frac{1}{2\log x}\right) \quad \text{for } x \ge 59, \tag{16}$$

and

$$\pi(x) < \frac{x}{\log x}\left(1 + \frac{3}{2\log x}\right) \quad \text{for } x > 1. \tag{17}$$

Using Theorem 5 we get that if $A \geq 59$, we can take $c_5 = 1 + \frac{3}{2\log A}$ and $c_4 = 1 + \frac{1}{2\log A}$. Therefore, for $A \geq 59$ and $B > A$,

$$c_3(A, B) < \frac{1}{A\log A} + \frac{5}{2\log^2 A}.$$

Therefore,

$$c_3\left(n^{1/8}, n^{1/4}\right) < \frac{8\log n + 160}{n^{1/8}\log^2 n}, \tag{18}$$

for $n > 59^8$.

We obtain

$$Q_3 \leq \frac{\sqrt{n}(8\log n + 160)}{3n^{1/8}\log^2 n} + 2\pi(n^{1/4}) - 2\pi(n^{1/8}) \tag{19}$$

Case 4. $n^{1/4} < q \leq \frac{\sqrt{n}}{c\log n}$, where c is a fixed constant in the interval $(0.5, 5]$.

We already noted above that $f_1(n) = \frac{n^{1/4}}{\log n}$ is increasing for $n > e^4$ and $f(10^9) = 8.55\ldots$. Thus, $\frac{\sqrt{n}}{c\log n} > n^{1/4}$ for $n \geq 10^9$.

Let Q_4 be the number of primes $p \leq \sqrt{n}$ such that congruence (1) holds for some $q \in (n^{1/4}, \frac{\sqrt{n}}{c\log n}]$. Thus,

$$Q_4 \leq \sideset{}{'}\sum_{n^{1/4} < q \leq \frac{\sqrt{n}}{c\log n}} \pi(\sqrt{n}; q^2, n). \tag{20}$$

Note that if $q \geq n^{1/4}$, then $\pi(\sqrt{n}; q^2, n) \leq 1$ since there will be at most one positive integer congruent to n modulo q^2 not exceeding $\sqrt{n}$. Thus,

$$Q_4 \leq \pi\left(\frac{\sqrt{n}}{c\log n}\right) - \pi\left(n^{1/4}\right). \tag{21}$$

Denote $m = \frac{\sqrt{n}}{c\log n}$. We proved above that $m > n^{1/4}$ for $n > 10^{12}$, so $\log m > \log n/4$ for such n. By Theorem 5 we have

$$\pi(m) < \frac{m}{\log m}\left(1 + \frac{3}{2\log m}\right) < \frac{4\sqrt{n}}{c\log^2 n}\left(1 + \frac{6}{\log n}\right),$$

and

$$\pi\left(n^{1/4}\right) < \frac{4n^{1/4}}{\log n}\left(1 + \frac{6}{\log n}\right).$$

Combining the above two inequalities with Eqs. (19) and (21) we obtain

$$Q_3 + Q_4 < \frac{\sqrt{n}(8\log n + 160)}{3n^{1/8}\log^2 n} + \frac{4\sqrt{n}}{c\log^2 n}\left(1 + \frac{6}{\log n}\right) + \frac{4n^{1/4}}{\log n}\left(1 + \frac{6}{\log n}\right). \tag{22}$$

Case 5. $\sqrt{n}/(c\log n) < q < \sqrt{n}$

Let Q_5 be the number of primes $p \le \sqrt{n}$ such that congruence (1) holds for some $q \in (\frac{\sqrt{n}}{c\log n}, \sqrt{n}]$. Thus,

$$Q_5 \le \sideset{}{'}\sum_{\frac{\sqrt{n}}{c\log n} < q < \sqrt{n}} \pi(\sqrt{n}; q^2, n). \tag{23}$$

To estimate Q_5 we use the following.

Suppose q_1 and q_2 are distinct primes in $(\sqrt{n}/(c\log n), \sqrt{n})$ such that there exist distinct primes p_1, p_2 both not exceeding $\sqrt{n}$ such that $q_1^2 \mid n - p_1$ and $q_2^2 \mid n - p_2$. Thus, there exist positive integers k_1 and k_2 such that $n - p_1 = k_1 q_1^2$ and $n - p_2 = k_2 q_2^2$.

We claim that $k_1 \ne k_2$. Assume the opposite, that $k_1 = k_2$.

Then, $(n - p_1) - (n - p_2) = k_1(q_1^2 - q_2^2)$, that is

$$p_2 - p_1 = k_1(q_1 + q_2)(q_1 - q_2). \tag{24}$$

Note that $k_1 q_1 = k_1 q_1^2/q_1 = (n - p_1)/q_1 > (n - \sqrt{n})/\sqrt{n} = \sqrt{n} - 1$.

Therefore, $k_1(q_1 + q_2)|q_1 - q_2| > k_1 q_1 > \sqrt{n} - 1 > |p_1 - p_2|$, contradicting (24). So, $k_1 \ne k_2$.

We conclude that the contribution from $q \in (\sqrt{n}/(c\log n), \sqrt{n})$ does not exceed the number of distinct integers k such that $n - p = kq^2$. Note that for each such k we have $k = (n - p)/q^2 < (c\log n)^2$, since $q > \sqrt{n}/(c\log n)$. Thus,

$$Q_5 \le (c\log n)^2. \tag{25}$$

Now, combining (5), (9), (22), and (25) we get

$$Q_1 + Q_2 + Q_3 + Q_4 + Q_5 < \frac{46\sqrt{n}}{35\log n} + \frac{1569\sqrt{n}}{350\log^2 n} + \frac{4C_1(11)\sqrt{n}}{\log n} + \frac{32C_2(11)\sqrt{n}}{\log^2 n} +$$

$$+ \frac{\sqrt{n}(8\log n + 160)}{3n^{1/8}\log^2 n} + \frac{4\sqrt{n}}{c\log^2 n}\left(1 + \frac{6}{\log n}\right) + \frac{4n^{1/4}}{\log n}\left(1 + \frac{6}{\log n}\right) + (c\log n)^2,$$

for $n > 59^8$.

To prove the theorem it is sufficient to show $Q_1 + Q_2 + Q_3 + Q_4 + Q_5 < \pi\left(\sqrt{n}\right)$. By Theorem 5 for $n > 59^2$,

$$\pi\left(\sqrt{n}\right) > \frac{2\sqrt{n}}{\log n}\left(1+\frac{1}{\log n}\right)$$

so, it is sufficient to establish

$$\frac{2\sqrt{n}}{\log n}\left(1+\frac{1}{\log n}\right) > f_2(n), \tag{26}$$

where

$$f_2(n) := \frac{46\sqrt{n}}{35\log n} + \frac{1569\sqrt{n}}{350\log^2 n} + \frac{4C_1(11)\sqrt{n}}{\log n} + \frac{32C_2(11)\sqrt{n}}{\log^2 n} +$$

$$+\frac{\sqrt{n}(8\log n+160)}{3n^{1/8}\log^2 n} + \frac{4\sqrt{n}}{c\log^2 n}\left(1+\frac{6}{\log n}\right) + \frac{4n^{1/4}}{\log n}\left(1+\frac{6}{\log n}\right) + (c\log n)^2.$$

Dividing the inequality (26) by $\frac{\sqrt{n}}{\log n}$ and simplifying we get that the inequality is equivalent to

$$\frac{24}{35} > \frac{869}{350\log n} + 4C_1(11) + \frac{32C_2(11)}{\log n} + \frac{8\log n+160}{3n^{1/8}\log n} + \tag{27}$$

$$+\frac{4}{c\log n}\left(1+\frac{6}{\log n}\right) + 4n^{-1/4}\left(1+\frac{6}{\log n}\right) + \frac{c^2\log^3 n}{\sqrt{n}}.$$

Note the functions $\frac{1}{\log n}, \frac{1}{\log^2 n}, \frac{1}{n^{1/8}}, \frac{1}{n^{1/8}}, \frac{1}{n^{1/8}\log n}$, and $\frac{1}{n^{1/4}\log n}$ are all decreasing for $n > 1$. Moreover, the function $\frac{\log^3 n}{\sqrt{n}}$ is decreasing for $n > e^6$. Therefore, if inequality (27) holds for some $n_0 > 59^8$, then it holds for all $n \geq n_0$. Since the inequalities (26) and (27) are equivalent, then if inequality (26) holds for some $n_0 > 59^8$, then it holds for all $n \geq n_0$.

Estimating the tail of the series defining $C_1(11)$ and $C_2(11)$ in the same way we estimated the series $\sum'_{A<q<B} \frac{1}{q^2}$ we get the estimates $C_1(11) < 0.033$ and $C_2(11) < 0.1$.

Taking $c = 4$ and $n = 59^8 + 1$, we get that the left-hand-side of inequality (26) exceeds the right-hand-side by more than $95945 > 0$. Thus, the theorem holds for $n > 59^8$.

From now on, we assume $n \leq 59^8$. For such n we cannot use Theorem 5 to get lower bound for $\pi\left(n^{1/8}\right)$. However, we can estimate $c_3(A, B)$ as follows. We have

$$c_3(A, B) < c_3(A) = \sum_q{}' \frac{1}{q^2} - \sum_{q\leq A}{}' \frac{1}{q^2}.$$

Denote

$$g(A) := \sum_{q \le A}' \frac{1}{q^2}.$$

The function g is piecewise continuous and nondecreasing.

Also, it has been known since Euler [6] p. 480 (or possibly earlier) that

$$\sum_{q}' \frac{1}{q^2} := c_0 = 0.452247420041065\ldots \tag{28}$$

Replacing the estimate for $c_3\left(n^{1/8}, n^{1/4}\right)$ in Eq. (22) we get

$$Q_3 + Q_4 < \frac{\sqrt{n}\left(c_0 - g\left(n^{1/8}\right)\right)}{3} + \frac{4\sqrt{n}}{c\log^2 n}\left(1 + \frac{6}{\log n}\right) + \frac{4n^{1/4}}{\log n}\left(1 + \frac{6}{\log n}\right). \tag{29}$$

To prove the theorem now we need to establish

$$\frac{2\sqrt{n}}{\log n}\left(1 + \frac{1}{\log n}\right) > f_3(n), \tag{30}$$

where

$$f_3(n) := \frac{46\sqrt{n}}{35\log n} + \frac{1569\sqrt{n}}{350\log^2 n} + \frac{4C_1(11)\sqrt{n}}{\log n} + \frac{32C_2(11)\sqrt{n}}{\log^2 n} +$$

$$+ \frac{\sqrt{n}\left(c_0 - g\left(n^{1/8}\right)\right)}{3} + \frac{4\sqrt{n}}{c\log^2 n}\left(1 + \frac{6}{\log n}\right) + \frac{4n^{1/4}}{\log n}\left(1 + \frac{6}{\log n}\right) + (c\log n)^2.$$

Unfortunately, if we divide $f_3(n)$ by $\frac{\sqrt{n}}{\log n}$ it is not clear that the resulting function is decreasing due to the term $\frac{\sqrt{n}(c_0 - g(n^{1/8}))}{3}$, so we proceed as follows.

Let p_k be the kth prime number. Next, we will be considering the cases when $n \in (p_k^8, p_{k+1}^8]$ where $31 \le p_k \le 53$ (when $p_k = 31$ we will consider $n \in [10^{12}, 37^8]$). For such n we have $g(n^{1/8}) \ge g(p_k)$. Moreover, since $\log n$ is increasing function, $\frac{8\log p_{k+1}}{\log n} \ge 1$ for $n \in (p_k^8, p_{k+1}^8]$. Also, since now we have upper bound for n, when estimating Q_2 we be using Eq. (8), that is we will use $c_1(11, p_{k+1})$ instead of $C_1(11)$, and $c_2(11, p_{k+1})$ instead of $C_2(11)$.

We replace $f_3(n)$ by the (potentially) slightly larger function

$$f_4(n,k) := \frac{46\sqrt{n}}{35\log n} + \frac{1569\sqrt{n}}{350\log^2 n} + \frac{4c_1(11, p_{k+1})\sqrt{n}}{\log n} + \frac{32c_2(11, p_{k+1})\sqrt{n}}{\log^2 n} +$$

$$+ \frac{\sqrt{n}\left(c_0 - g\left(p_k\right)\right)(8\log p_{k+1})}{3\log n} + \frac{4\sqrt{n}}{c\log^2 n}\left(1 + \frac{6}{\log n}\right) + \frac{4n^{1/4}}{\log n}\left(1 + \frac{6}{\log n}\right) + (c\log n)^2.$$

The function $f_4(n, k)$ may be larger than $f_3(n)$ but has the property that it is decreasing on the interval $(p_k^8, p_{k+1}^8]$ after division by $\frac{\sqrt{n}}{\log n}$.

Thus, if

$$\frac{2\sqrt{n_0}}{\log n_0}\left(1+\frac{1}{\log n_0}\right) > f_4(n_0, k) \tag{31}$$

for some $p_{k+1}^8 \geq n_0 \geq \max(p_k^8, 10^{12})$, then the theorem holds for all $n \in [n_0, p_{k+1}^8]$.

In the table below we record the results of our computations in several intervals. The symbol Δ will denote the difference between the left-hand-side of inequality (31) and its right-hand-side evaluated at the left end of each interval indicated in each row.

k	Interval	$c_1(11, p_{k+1})$	$c_2(11, p_{k+1})$	c	$c_0 - g(p_k)$	Δ
16	$[53^8, 59^8]$	0.02941652	0.08277361	4	0.00352137	74613.3
15	$[47^8, 53^8]$	0.02912429	0.08158205	3.3	0.00387736	46560.4
14	$[43^8, 47^8]$	0.02876145	0.08014145	2.9	0.004330053	32612.4
13	$[41^8, 43^8]$	0.02829891	0.07836062	2.7	0.00487089	26933.3
12	$[37^8, 41^8]$	0.02774520	0.07627800	2.3	0.00546577	17911.1
11	$[10^{12}, 37^8]$	0.02713545	0.07401364	2	0.00619623	9029.5

The left end of the interval in the last row in the table above is 10^{12} rather than 31^8 since $31^8 < 10^{12}$ and we cannot use Theorem 3 to estimate $\pi(\sqrt{n}; q^2, n)$ when $\sqrt{n} < 10^6$.

When $n \leq 10^{12}$ we use another result of Bennett et al., Corollary 1.6 of [1].

Theorem 6 *Let $1 \leq q \leq 1200$ be an integer, and a be an integer coprime to q. For all $x \geq 50q^2$ we have*

$$\frac{x}{\varphi(q)\log x} < \pi(x; q, a) < \frac{x}{\varphi(q)\log x}\left(1+\frac{5}{2\log x}\right).$$

We apply the above theorem with $n \geq 10^9$, and $q \in \{2^2, 3^2, 5^2, 7^2\}$. Since $\sqrt{10^9} > 31622 > 50 \cdot 7^2 = 2450$, Theorem 6 applies when $\gcd(n, 210) = 1$ and we argue similarly to Case 1 when $\gcd(n, 210) > 1$. Substituting into Eq. (2) we obtain

$$Q_1 < \frac{46\sqrt{n}}{35\log n} + \frac{32\sqrt{n}}{7\log^2 n}. \tag{32}$$

We get a somewhat worse estimate. The constant in the second term is $32/7 = 4.57\ldots$ rather than $1569/350 = 4.48\ldots$.

So, for $n \geq 10^9$ now we need to show

$$\frac{2\sqrt{n}}{\log n}\left(1+\frac{1}{\log n}\right) > f_5(n, k), \tag{33}$$

where

$$f_5(n,k) := \frac{46\sqrt{n}}{35\log n} + \frac{32\sqrt{n}}{7\log^2 n} + \frac{4c_1(11, p_{k+1})\sqrt{n}}{\log n} + \frac{32c_2(11, p_{k+1})\sqrt{n}}{\log^2 n} +$$

$$+ \frac{\sqrt{n}\,(c_0 - g\,(p_k))\,(8\log p_{k+1})}{3\log n} + \frac{4\sqrt{n}}{c\log^2 n}\left(1 + \frac{6}{\log n}\right) + \frac{4n^{1/4}}{\log n}\left(1 + \frac{6}{\log n}\right) + (c\log n)^2.$$

We proceed exactly as before, the only difference between $f_4(n,k)$ and $f_5(n,k)$ is that the constant $1569/350$ in the second term of $f_4(n,k)$ is replaced by $32/7$ to obtain $f_5(n,k)$.

The new table with f_5 rather than f_4 is:

k	Interval	$c_1(11, p_{k+1})$	$c_2(11, p_{k+1})$	c	$c_0 - g(p_k)$	Δ
11	$[31^8, 10^{12}]$	0.02713545	0.07401364	2	0.00619623	5606.9
10	$[29^8, 31^8]$	0.02638469	0.0713027	2	0.00723681	3669.5
9	$[23^8, 29^8]$	0.02530942	0.06761027	1.6	0.00842587	910.8
8	$[19^8, 23^8]$	0.02407789	0.0634633	1.3	0.01031623	179.8
7	$[17^8, 19^8]$	0.02210161	0.05726672	1.1	0.0130863	62.1
6	$[13^8, 17^8]$	0.01917763	0.04865725	.6	0.0165465	35.3

Thus, the theorem holds for $n \geq 13^8 = 815730721$. Since we established the theorem by direct computation for $n \leq 10^9$, this completes the proof.

References

1. M. Bennett, G. Martin, K. O'Bryant, and A. Rechnitzer, Explicit bounds for primes in arithmetic progressions, *Illinois J. Math.* **62**, (2018), 427–532.
2. A. Dudek, On the sum of a prime and a square-free number, *Ramanujan J.* **42**, (2017), 233–240.
3. A. Dudek and D. Platt, On the sum of the square of a prime and a square-free number, *LMS J. Comput. Math* **19**, (2016), 16–24.
4. P. Erdős, 'The representation of an integer as the sum of the square of a prime and of a square-free integer, *J. London Math. Soc.* **4**, (1935), 243–245.
5. T. Estermann, On the representations of a number as the sum of a prime and a quadratfrei number, *J. Lond. Math. Soc.* **6**, (1931), 219–221.
6. L. Euler, Introductio in analysin infinitorum, **1**, (1748).
7. M. Filaseta, S. Graham, and O. Trifonov, Starting with gaps between k-free numbers, *Int. J. Number Theory* **11**, (2015), 1411–1435.
8. H. L. Montgomery and R. C. Vaughan, The large sieve, *Mathematika* **20** (1973), 119–134.
9. J. Rosser and L. Schoenfeld, Approximate formulas for some functions of prime numbers, *Illinois J. Math.* **6**, (1962), 64–94.

Commutative Monoid of Self-Dual Symmetric Polynomials

Leonid G. Fel

Abstract We consider a set $\mathfrak{RS}(\lambda, S_n)$ of self- and scew-reciprocal polynomials of degree mn in λ, based on polynomial invariants $I_{n,r}(\mathbf{x}^n)$ of symmetric group S_n, acting on the Euclidean space $\mathbb{E}^n$ over the field of real numbers $\mathbb{R}$, where $\mathbf{x^n} = \{x_1, \ldots, x_n\} \in \mathbb{E}^n$. We prove that $\mathfrak{RS}(\lambda, S_n)$ exhibits a commutative monoid under multiplication. Real solutions $\lambda(\mathbf{x^n})$ of scew-reciprocal equations have many remarkable properties: a homogeneity of the 1st order, a self-dual symmetry under inversion of variables $x_i \to x_i^{-1}$ and function $\lambda \to \lambda^{-1}$, a monotony of $\lambda(\mathbf{x^n})$ with respect to every x_i and others.

1 Introduction

A self-dual symmetry of positive homogeneous functions $\lambda(x_1, x_2)$ of degree 1, which are invariant under exchange variables $x_1 \leftrightarrow x_2$ and inversion of λ and x_i,

$$\lambda(x_1, x_2) \cdot \lambda\left(x_1^{-1}, x_2^{-1}\right) = 1, \tag{1.1}$$

was used by Keller [11] to study the harmonic potential in an infinite piecewise continuous plane with regular tessellation. This model is known in theoretical physics as the 2D white-black checkerboard with equal fractions of two constituents, while x_1, x_2 and λ are attributed to the partial conductivities in different zones and an effective conductivity of the whole checkerboard, respectively. Later, a self-dual symmetry was extended [5] to the 2D checkerboards with equal fractions of n constituents and effective isotropic conductivity, which is invariant under action of symmetric group S_n and inversion $\widehat{\mathbf{Q}}_n : \{x_i \to x_i^{-1},\ \lambda \to \lambda^{-1}\}$,

$$\lambda(x_1, \ldots, x_n) \cdot \lambda\left(x_1^{-1}, \ldots, x_n^{-1}\right) = 1. \tag{1.2}$$

L. G. Fel (✉)
Department of Civil Engineering, Technion, 32000 Haifa, Israel
e-mail: lfel@technion.ac.il

M. B. Nathanson (ed.), *Combinatorial and Additive Number Theory VI*,
Springer Proceedings in Mathematics & Statistics 464,
https://doi.org/10.1007/978-3-031-65064-2_7

A duality relation (1.1) was used to explain a percolation phenomenon in the white-black checkerboards with random tiling [4] and universal solution $\lambda^2 = x_1 x_2$ in such a model. For n-color ($n \geq 3$) checkerboards a universal solution of Eq. (1.2) does not exist but depends essentially on the color tiling of the infinite plane [5].

Apart from applications in theoretical physics, the relation (1.2) is of interest by itself in the following context: what kind of multivariate polynomials $H(\lambda, x_1, \ldots, x_n)$, which are invariant under action of symmetric group S_n on a set $\{x_1, \ldots, x_n\}$, gives rise to the root $\lambda(x_1, \ldots, x_n)$, satisfying (1.2), and in what algebraic structures such polynomials manifest themselves. To answer these questions, we make use of *self-* and *scew-reciprocal* polynomials in n real variables, called, for short, *self-dual*. Such polynomials appear due to a generalization of the univariate self- and scew-reciprocal polynomials, which possess many remarkable properties and occur in different areas of mathematics (coding theory [9], algebraic curves over finite fields [10], number theory [14]) and physics (Lee-Yang theorem in statistical physics [13]).

The article is organized in seven sections and two Appendices. In Sect. 1.1 we recall the definition of univariate self-dual polynomials $\mathsf{R}_m(\lambda)$ and $\mathsf{S}_m(\lambda)$ in λ of degree m and their constitutive relations (1.5).

In Sect. 2 we define multivariate polynomials $\mathsf{R}\binom{\lambda,\, \mathbf{x}^\mathbf{n}}{m, S_n}$ and $\mathsf{S}\binom{\lambda,\, \mathbf{x}^\mathbf{n}}{m, S_n}$, $\mathbf{x}^\mathbf{n} = \{x_1, \ldots, x_n\}$, in λ of degree mn with coefficients built upon the basic polynomial invariants of symmetric group S_n. In Sect. 2.1 we introduce *a unimodality number* μ as a measure of freedom to choose such (*generic*) polynomials with arbitrary coefficients and calculate μ for self-dual polynomials via partition numbers with constraints.

In Sect. 3 we prove the necessary and sufficient conditions for $\mathsf{R}\binom{\lambda,\, \mathbf{x}^\mathbf{n}}{m, S_n}$ and $\mathsf{S}\binom{\lambda,\, \mathbf{x}^\mathbf{n}}{m, S_n}$ to satisfy duality relations (3.2), which generalize relations (1.5) for univariate self-dual polynomials.

In Sect. 4 we define two sets $\mathfrak{R}(\lambda, S_n)$ and $\mathfrak{S}(\lambda, S_n)$ of generic polynomials $\mathsf{R}\binom{\lambda,\, \mathbf{x}^\mathbf{n}}{m, S_n}$ and $\mathsf{S}\binom{\lambda,\, \mathbf{x}^\mathbf{n}}{m, S_n}$ with arbitrary $m \in \mathbb{Z}_{\geq}$, respectively, and define one more set $\mathfrak{RS}(\lambda, S_n)$ of their union. We prove that both $\mathfrak{R}(\lambda, S_n)$ and $\mathfrak{RS}(\lambda, S_n)$ are multiplicative commutative monoids such that $\mathfrak{R}(\lambda, S_n) = \mathfrak{RS}(\lambda, S_n)/\mathbb{Z}_2$. In Sect. 4.1 we consider a factorization of self-dual polynomials in $\mathfrak{RS}(\lambda, S_n)$ and prove that polynomials $\mathsf{S}\binom{\lambda,\, \mathbf{x}^2}{m, S_2}$ are reducible in $\mathfrak{RS}(\lambda, n)$.

In Sect. 5 we consider a unique positive solution $\lambda(\mathbf{x}^\mathbf{n})$ of equation $\mathsf{S}\binom{\lambda,\, \mathbf{x}^\mathbf{n}}{m, S_n} = 0$, $\mathcal{S}^{s,l}_{m,n} \geq 0$, and prove that $\lambda(\mathbf{x}^\mathbf{n})$ is a homogeneous, self dual, monotonic algebraic function, which is invariant under action of symmetric group S_n and compatible with a trivial solution $\lambda(x, \ldots, x) = x$.

In Sect. 6 we define a set of *proper self-dual* polynomials $\mathsf{S}_{\odot}\binom{\lambda,\, \mathbf{x}^\mathbf{n}}{m, S_n}$, $\mathsf{R}_{\odot}\binom{\lambda,\, \mathbf{x}^\mathbf{n}}{m, S_n}$ as a subset of $\mathfrak{RS}(\lambda, S_n)$ and show that a unique positive solution $\lambda_{\odot}(\mathbf{x}^\mathbf{n})$ of equation $\mathsf{S}_{\odot}\binom{\lambda,\, \mathbf{x}^\mathbf{n}}{m, S_n} = 0$ with arbitrary n, m is bounded by arithmetic and harmonic means of the set $\{x_1, \ldots, x_n\}$.

In Sect. 7 we improve the bounds of $\lambda_{\odot}(\mathbf{x^n})$ for small $n = 3, 4$ and arbitrary m. In Sect. 8 we pose open problems.

1.1 Self-Dual Univariate Polynomials

Start with self-dual polynomials in one variable. Following [1], call a polynomial $\mathsf{R}_m(\lambda)$ *self-reciprocal*, if it may be represented as follows,

$$\mathsf{R}_m(\lambda) = \sum_{k=0}^{\lfloor m/2 \rfloor} \mathcal{R}_k^m \left(\lambda^{m-k} + \lambda^k\right), \tag{1.3}$$
$$\lambda \in \mathbb{C}, \quad \mathsf{R}_m(\lambda) \in \mathbb{R}[\lambda], \quad \mathcal{R}_k^m \in \mathbb{R}, \quad \mathcal{R}_0^m \neq 0,$$

where $\lfloor x \rfloor$ denotes a floor function. The *scew-reciprocal* polynomial $\mathsf{S}_m(\lambda)$ is defined similarly,

$$\mathsf{S}_m(\lambda) = \sum_{k=0}^{\lfloor m/2 \rfloor} \mathcal{S}_k^m \left(\lambda^{m-k} - \lambda^k\right), \tag{1.4}$$
$$\lambda \in \mathbb{C}, \quad \mathsf{S}_m(\lambda) \in \mathbb{R}[\lambda], \quad \mathcal{S}_k^m \in \mathbb{R}, \quad \mathcal{S}_0^m \neq 0,$$

where $\mathbb{R}[\lambda]$ stands for a polynomial ring. Polynomials $\mathsf{R}_m(\lambda)$, $\mathsf{S}_m(\lambda)$ satisfy the duality relations,

$$\lambda^m \mathsf{R}_m\left(\lambda^{-1}\right) - \mathsf{R}_m(\lambda) = 0, \qquad \lambda^m \mathsf{S}_m\left(\lambda^{-1}\right) + \mathsf{S}_m(\lambda) = 0. \tag{1.5}$$

The polynomial $\mathsf{S}_m(\lambda)$ is scew-reciprocal if and only if there exists a self-reciprocal polynomial $\mathsf{R}_{m-1}(\lambda)$, such that [15]

$$\mathsf{S}_m(\lambda) = (\lambda - 1)\mathsf{R}_{m-1}(\lambda). \tag{1.6}$$

For both self- and scew-reciprocal polynomials a root λ_* has a corresponding root λ_*^{-1} of the same multiplicity. The property that distinguishes $\mathsf{R}_m(\lambda)$ and $\mathsf{S}_m(\lambda)$ is the multiplicity of the root at $\lambda = 1$: in the case of simple roots, $\mathsf{R}_m(\lambda)$ has no root at $\lambda = 1$, while $\mathsf{S}_m(\lambda)$ has such root. Roots of self-reciprocal polynomials either lie on the unit circle $U = \{\lambda \in \mathbb{C} \mid |\lambda| = 1\}$ or occur in pairs conjugate to U [12]. A polynomial with real coefficients, all of whose complex roots lie on U in the complex plane (all roots are unimodular), is either self- or scew-reciprocal [15]. An inverse problem, i.e., to find a set of self-dual polynomials with roots, located within a prescribed domain, e.g., on the unit circle U, is an active research area.

2 Self-Dual Multivariate Symmetric Polynomials

Self-dual multivariate symmetric polynomials in λ with coefficients built upon the basic polynomial invariants $I_{n,r}(\mathbf{x}^n) = I_{n,r}$ of symmetric group S_n, acting on the Euclidean space $\mathbb{E}^n$

$$I_{n,r} = \sum_{i_1<i_2<\ldots<i_r}^{n} x_{i_1}x_{i_2}\cdot\ldots\cdot x_{i_r}, \quad \mathbf{x^n} = \{x_1,\ldots,x_n\} \in \mathbb{E}^n, \quad \text{i.e.} \tag{2.1}$$

$$I_{n,0} = 1, \quad I_{n,1} = \sum_{i}^{n} x_i, \quad I_{n,2} = \sum_{i<j}^{n} x_i x_j, \quad \ldots \quad I_{n,n} = \prod_{i}^{n} x_i.$$

over the field of real numbers $\mathbb{R}$, exhibit a generalization of univariate polynomials $\mathsf{R}_m(\lambda)$ and $\mathsf{S}_m(\lambda)$ in (1.3) and coincide with them if $n = 1$.

Define two polynomials in λ of degree mn, where $m \in \mathbb{Z}_{\geq}$ and $n \geq 2$,

$$\mathsf{R}\binom{\lambda,\,\mathbf{x^n}}{m,S_n} = \sum_{s=0}^{mn} A_s\binom{\mathbf{x^n}}{m,S_n}\lambda^{mn-s}, \qquad A_s\binom{\mathbf{x^n}}{m,S_n} \in \mathbb{R}[\mathbf{x^n}], \tag{2.2}$$

$$\mathsf{S}\binom{\lambda,\,\mathbf{x^n}}{m,S_n} = \sum_{s=0}^{mn} B_s\binom{\mathbf{x^n}}{m,S_n}\lambda^{mn-s}, \qquad B_s\binom{\mathbf{x^n}}{m,S_n} \in \mathbb{R}[\mathbf{x^n}], \tag{2.3}$$

where $\mathbb{R}[\mathbf{x^n}] = \mathbb{R}[x_1,\ldots,x_n]$ stands for a polynomial ring over $\mathbb{R}$, while $A_s\binom{\mathbf{x^n}}{m,S_n}$ and $B_s\binom{\mathbf{x^n}}{m,S_n}$ denote homogeneous polynomials in $\mathbf{x^n}$ of degree s, possessed special properties.

In order to specify these properties, we choose more detailed representation (Rep) for $\mathsf{R}\binom{\lambda,\,\mathbf{x^n}}{m,S_n}$ and $\mathsf{S}\binom{\lambda,\,\mathbf{x^n}}{m,S_n}$ as a combinations of monomial terms, where a symbol $\binom{\lambda,\,\mathbf{x^n}}{m,S_n}$ denotes a dependence of polynomials on its entries,

$$\mathsf{R}\binom{\lambda,\,\mathbf{x^n}}{m,S_n} = \sum_{s=0}^{\lfloor s_*\rfloor}\left[R_s^+\binom{\mathbf{x^n}}{m,S_n}\lambda^{mn-s} + R_s^-\binom{\mathbf{x^n}}{m,S_n}\lambda^s\right], \tag{2.4}$$

$$R_0^-\binom{\mathbf{x^n}}{m,S_n} = R_0^+\binom{\mathbf{x^n}}{m,S_n} I_{n,n}^m,$$

$$\mathsf{S}\binom{\lambda,\,\mathbf{x^n}}{m,S_n} = \sum_{s=0}^{\lfloor s_*\rfloor}\left[S_s^+\binom{\mathbf{x^n}}{m,S_n}\lambda^{mn-s} - S_s^-\binom{\mathbf{x^n}}{m,S_n}\lambda^s\right], \tag{2.5}$$

$$S_0^-\binom{\mathbf{x^n}}{m,S_n} = S_0^+\binom{\mathbf{x^n}}{m,S_n} I_{n,n}^m,$$

where $s_* = mn/2$ and summands $R_s^\pm\binom{\mathbf{x^n}}{m,S_n}$, $S_s^\pm\binom{\mathbf{x^n}}{m,S_n}$ in (2.4, 2.5) read

$$R_s^+\binom{\mathbf{x^n}}{m,S_n}=\sum_{l=1}^{P_n^m(s)}\mathcal{R}_{m,n}^{s,\,l}\prod_{r=0}^{n}I_{n,r}^{\alpha_{n-r}\left(\substack{mn-s,\,l\\ m,\quad S_n}\right)},$$

$$R_s^-\binom{\mathbf{x^n}}{m,S_n}=\sum_{l=1}^{P_n^m(s)}\mathcal{R}_{m,n}^{s,\,l}\prod_{r=0}^{n}I_{n,n-r}^{\alpha_r\left(\substack{s,\,l\\ m,S_n}\right)},\tag{2.6}$$

$$S_s^+\binom{\mathbf{x^n}}{m,S_n}=\sum_{l=1}^{P_n^m(s)}\mathcal{S}_{m,n}^{s,\,l}\prod_{r=0}^{n}I_{n,r}^{\alpha_{n-r}\left(\substack{mn-s,\,l\\ m,\quad S_n}\right)},$$

$$S_s^-\binom{\mathbf{x^n}}{m,S_n}=\sum_{l=1}^{P_n^m(s)}\mathcal{S}_{m,n}^{s,\,l}\prod_{r=0}^{n}I_{n,n-r}^{\alpha_r\left(\substack{s,\,l\\ m,S_n}\right)},\tag{2.7}$$

where

$$\alpha_r\binom{s,\,l}{m,S_n}=\alpha_{n-r}\binom{mn-s,\,l}{m,\quad S_n},\qquad \alpha_r\binom{s,\,l}{m,S_n}\in\mathbb{Z}_{\geq},\tag{2.8}$$
$$\mathcal{R}_{m,n}^{s,\,l},\ \mathcal{S}_{m,n}^{s,\,l}\in\mathbb{R},\qquad P_n^m(0)=1.$$

In formulas (2.6, 2.7), the function $P_n^m(s)$ denotes a partition number with constraints. Namely, $P_n^m(s)$ gives a number of non-negative integer solutions $\{\alpha^l\}$ of two Diophantine equations,

$$\sum_{r=0}^{n}r\,\alpha_r\binom{s,\,l}{m,S_n}=s,\qquad \sum_{r=0}^{n}\alpha_r\binom{s,\,l}{m,S_n}=m,\quad \alpha_r\binom{0,\,l}{m,S_n}=m\delta_{r,0},\tag{2.9}$$

which are solvable if $0\leq s\leq mn$.

In Appendix A we present examples of generic self-dual polynomials with small m,n. A definition and main properties of $P_n^m(s)$ are given in Appendix B.1.

Definition 1 Polynomials $\mathsf{R}\binom{\lambda,\,\mathbf{x^n}}{m,S_n}$ and $\mathsf{S}\binom{\lambda,\,\mathbf{x^n}}{m,S_n}$ are called generic, if all coefficients $\mathcal{R}_{m,n}^{s,\,l}$ and $\mathcal{S}_{m,n}^{s,\,l}$ in Reps (2.4, 2.5) are independent.

Two Reps (2.2, 2.3) and (2.4, 2.5) for polynomials $\mathsf{R}\binom{\lambda,\,\mathbf{x^n}}{m,S_n}$ and $\mathsf{S}\binom{\lambda,\,\mathbf{x^n}}{m,S_n}$ are related as follows,

$$A_0\binom{\mathbf{x^n}}{m,S_n}=R_0^+\binom{\mathbf{x^n}}{m,S_n},\quad\ldots,\quad A_{mn-1}\binom{\mathbf{x^n}}{m,S_n}=R_1^-\binom{\mathbf{x^n}}{m,S_n},$$
$$A_{mn}\binom{\mathbf{x^n}}{m,S_n}=R_0^-\binom{\mathbf{x^n}}{m,S_n},$$
$$B_0\binom{\mathbf{x^n}}{m,S_n}=S_0^+\binom{\mathbf{x^n}}{m,S_n},\quad\ldots,\quad B_{mn-1}\binom{\mathbf{x^n}}{m,S_n}=-S_1^-\binom{\mathbf{x^n}}{m,S_n},$$
$$B_{mn}\binom{\mathbf{x^n}}{m,S_n}=-S_0^-\binom{\mathbf{x^n}}{m,S_n}.$$

In a special case, when $mn=0 \pmod 2$, we arrive at the middle terms in (2.2, 2.3),

$$A_{s_*}\binom{\mathbf{x^n}}{m,S_n}=R_{s_*}^+\binom{\mathbf{x^n}}{m,S_n}+R_{s_*}^-\binom{\mathbf{x^n}}{m,S_n},$$

$$B_{s_*}\binom{\mathbf{x^n}}{m,S_n} = S^+_{s_*}\binom{\mathbf{x^n}}{m,S_n} - S^-_{s_*}\binom{\mathbf{x^n}}{m,S_n}. \tag{2.10}$$

Substituting (2.7) into (2.10) and changing an enumeration order of a running index r in one of two products, we obtain expression for $B_{s_*}\binom{\mathbf{x^n}}{m,S_n}$,

$$B_{s_*}\binom{\mathbf{x^n}}{m,S_n} = \sum_{l=1}^{P_n^m(s_*)} \mathcal{S}_{m,n}^{s_*,\,l}\left(\prod_{r=0}^{n} I_{n,r}^{\alpha_r\binom{s_*,\,l}{m,S_n}} - \prod_{r=0}^{n} I_{n,r}^{\alpha_{n-r}\binom{s_*,\,l}{m,S_n}}\right), \tag{2.11}$$

where $\{x_1,\ldots,x_n\} \in \mathbb{E}^n$. Among $P_n^m(s_*)$ terms, contributing to $A_{s_*}\binom{\mathbf{x^n}}{m,S_n}$, not all of them survive in $B_{s_*}\binom{\mathbf{x^n}}{m,S_n}$.

Lemma 1

$$B_{s_*}\binom{\mathbf{x^n}}{m,S_n} = 0 \quad \text{if and only if} \quad \alpha_r\binom{s_*,\,l}{m,S_n} = \alpha_{n-r}\binom{s_*,\,l}{m,S_n}, \quad 0 \le r \le n. \tag{2.12}$$

Proof First, prove the implication "$\rightarrow$". Let $B_{s_*}\binom{\mathbf{x^n}}{m,S_n} = 0$, then, by (2.11), the last equality holds for the finite number $P_n^m(s_*)$ of the fixed real coefficients $\mathcal{S}_{m,n}^{s_*,\,l}$, that may be provided only if all differences in brackets (2.11) are vanished. Equating the products, we obtain

$$\prod_{r=0}^{n} I_{n,r}^{\left[\alpha_r\binom{s_*,\,l}{m,S_n} - \alpha_{n-r}\binom{s_*,\,l}{m,S_n}\right]} = 1 \quad \text{or}$$

$$\sum_{r=0}^{n}\left[\alpha_r\binom{s_*,\,l}{m,S_n} - \alpha_{n-r}\binom{s_*,\,l}{m,S_n}\right]\log I_{n,r} = 0, \tag{2.13}$$

that holds only if all differences in brackets (2.13) are vanished. Thus, the implication "$\rightarrow$" is proven. The opposite implication "$\leftarrow$" follows if we substitute the 2nd equality (2.12) into expression (2.11) for $B_{s_*}\binom{\mathbf{x^n}}{m,S_n}$. □

Keeping in mind a partition number $U_n^m(s)$, which gives a number of solutions of Eq. (2.9) supplemented by constraints (2.12) and introduced in Appendix B.2, the expression for $B_{s_*}\binom{\mathbf{x^n}}{m,S_n}$ is given in (2.11) by replacing in the upper limit

$$P_n^m(s_*) \rightarrow P_n^m(s_*) - U_n^m(s_*).$$

2.1 *Unimodality Numbers*

Definition 2 The numbers of coefficients $\mathcal{R}_{m,n}^{s,\,l}$ and $\mathcal{S}_{m,n}^{s,\,l}$ (not necessarily independent), which appeared in self-dual polynomials $\mathsf{R}\binom{\lambda,\,\mathbf{x^n}}{m,S_n}$ and $\mathsf{S}\binom{\lambda,\,\mathbf{x^n}}{m,S_n}$, are called unimodality numbers and denoted as $\mu\left[\mathsf{R}\binom{\lambda,\,\mathbf{x^n}}{m,\,S_n}\right]$ and $\mu\left[\mathsf{S}\binom{\lambda,\,\mathbf{x^n}}{m,\,S_n}\right]$, respectively.

Next Corollary follows by Definitions 1 and 2.

Corollary 1 *The unimodality numbers* μ *of generic polynomials* $\mathsf{R}\binom{\lambda,\,\mathbf{x}^n}{m,S_n}$ *and* $\mathsf{S}\binom{\lambda,\,\mathbf{x}^n}{m,S_n}$ *with fixed* m, n *exceed such numbers of non-generic polynomials* $\mathsf{R}\binom{\lambda,\,\mathbf{x}^n}{m,S_n}$ *and* $\mathsf{S}\binom{\lambda,\,\mathbf{x}^n}{m,S_n}$, *respectively.*

We distinguish between μ for generic self- and scew-reciprocal polynomials, which are different. Their straightforward calculation from polynomial expressions in Appendix A gives,

$$\mu\left[\mathsf{R}\binom{\lambda,\,\mathbf{x}^3}{1,\,S_3}\right]=2,\quad \mu\left[\mathsf{S}\binom{\lambda,\,\mathbf{x}^3}{1,\,S_3}\right]=2,\quad \mu\left[\mathsf{R}\binom{\lambda,\,\mathbf{x}^4}{1,\,S_4}\right]=3,\quad \mu\left[\mathsf{S}\binom{\lambda,\,\mathbf{x}^4}{1,\,S_4}\right]=2,$$
$$\mu\left[\mathsf{R}\binom{\lambda,\,\mathbf{x}^3}{2,\,S_3}\right]=6,\quad \mu\left[\mathsf{S}\binom{\lambda,\,\mathbf{x}^3}{2,\,S_3}\right]=4,\quad \mu\left[\mathsf{R}\binom{\lambda,\,\mathbf{x}^4}{2,\,S_4}\right]=9,\quad \mu\left[\mathsf{S}\binom{\lambda,\,\mathbf{x}^4}{2,\,S_4}\right]=6,$$
$$\mu\left[\mathsf{R}\binom{\lambda,\,\mathbf{x}^3}{3,\,S_3}\right]=10,\quad \mu\left[\mathsf{S}\binom{\lambda,\,\mathbf{x}^3}{3,\,S_3}\right]=10,\quad \mu\left[\mathsf{R}\binom{\lambda,\,\mathbf{x}^4}{3,\,S_4}\right]=20,\quad \mu\left[\mathsf{S}\binom{\lambda,\,\mathbf{x}^4}{3,\,S_4}\right]=16,$$
$$\mu\left[\mathsf{R}\binom{\lambda,\,\mathbf{x}^3}{4,\,S_3}\right]=20,\quad \mu\left[\mathsf{S}\binom{\lambda,\,\mathbf{x}^3}{4,\,S_3}\right]=16,\quad \mu\left[\mathsf{R}\binom{\lambda,\,\mathbf{x}^4}{4,\,S_4}\right]=39,\quad \mu\left[\mathsf{S}\binom{\lambda,\,\mathbf{x}^4}{4,\,S_4}\right]=32.$$

Lemma 2 *Let the generic self-dual polynomials be given, then*

$$\mu\left[\mathsf{R}\binom{\lambda,\,\mathbf{x}^n}{m,S_n}\right]=\mu\left[\mathsf{S}\binom{\lambda,\,\mathbf{x}^n}{m,S_n}\right]=\frac{1}{2}\binom{m+n}{n},\quad mn=1 \pmod 2, \tag{2.14}$$
$$\mu\left[\mathsf{R}\binom{\lambda,\,\mathbf{x}^n}{m,S_n}\right]=\frac{1}{2}\left[\binom{m+n}{n}+P_n^m(s_*)\right],\quad mn=0 \pmod 2, \tag{2.15}$$
$$\mu\left[\mathsf{S}\binom{\lambda,\,\mathbf{x}^n}{m,S_n}\right]=\frac{1}{2}\left[\binom{m+n}{n}-U_n^m(s_*)\right],\quad mn=0 \pmod 2. \tag{2.16}$$

Proof Making use of definition (B.2) of the Gaussian polynomials, symmetry consideration (B.3) and properties (2.4, 2.5) of self- and scew-reciprocal polynomials, we arrive at (2.14–2.16) with two congruences, where a partition number $U_n^m(s_*)$ is defined in Appendix B.2. □

By Lemma 1 and (B.2, B.12), unimodality numbers are invariant under exchange indices $n \leftrightarrow m$,

$$\mu\left[\mathsf{R}\binom{\lambda,\,\mathbf{x}^n}{m,S_n}\right]=\mu\left[\mathsf{R}\binom{\lambda,\,\mathbf{x}^m}{n,S_m}\right],\qquad \mu\left[\mathsf{S}\binom{\lambda,\,\mathbf{x}^n}{m,S_n}\right]=\mu\left[\mathsf{S}\binom{\lambda,\,\mathbf{x}^m}{n,S_m}\right].$$

Keeping in mind relation (B.13) and combining it with Lemma 2, we get for $m=1$,

$$\mu\left[\mathsf{R}\binom{\lambda,\,\mathbf{x}^n}{1,\,S_n}\right]=\mu\left[\mathsf{S}\binom{\lambda,\,\mathbf{x}^n}{1,\,S_n}\right]=\frac{n+1}{2},\qquad n=1 \pmod 2,$$
$$\mu\left[\mathsf{R}\binom{\lambda,\,\mathbf{x}^n}{1,\,S_n}\right]=\frac{n}{2}+1,\quad \mu\left[\mathsf{S}\binom{\lambda,\,\mathbf{x}^n}{1,\,S_n}\right]=\frac{n}{2},\quad n=0 \pmod 2. \tag{2.17}$$

The case $n = 2$ also allows to calculate unimodality numbers explicitly. Substituting (B.2, B.14) into (2.15, 2.16), we obtain formulas,

$$\mu\left[\mathsf{S}\binom{\lambda,\ \mathbf{x^2}}{m, S_2}\right] = \frac{1}{4}\left[(m+1)^2 - \cos^2\frac{\pi m}{2}\right],$$
$$\mu\left[\mathsf{R}\binom{\lambda,\ \mathbf{x^2}}{m, S_2}\right] = \frac{1}{4}\left[(m+2)^2 - \sin^2\frac{\pi m}{2}\right], \tag{2.18}$$

which satisfy the relation,

$$\mu\left[\mathsf{S}\binom{\lambda,\ \mathbf{x^2}}{m, S_2}\right] = \mu\left[\mathsf{R}\binom{\lambda,\ \ \mathbf{x^2}}{m-1,\ S_2}\right].$$

If $n \geq 3$, formulas for $P_n^m(s_*)$ become sophisticated [2, 6] and difficult to apply.

3 Duality Relations for Polynomials $\mathsf{R}\binom{\lambda,\ \mathbf{x^n}}{m, S_n}$ and $\mathsf{S}\binom{\lambda,\ \mathbf{x^n}}{m, S_n}$

Extend relations (1.5) on multivariate self-dual polynomials. Let $\mathsf{H}\binom{\lambda,\ \mathbf{x^n}}{m, S_n}$ be a polynomial in λ with homogeneous polynomials coefficients in $\mathbf{x^n}$, invariant under action of symmetric group S_n on the Euclidean space $\mathbb{E}^n$.

Definition 3 A transformation $\widehat{\mathbf{Q}}_n$ acts on $\mathsf{H}\binom{\lambda,\ \mathbf{x^n}}{m,\ S_n}$ by inversion of variables λ and x_i,

$$\widehat{\mathbf{Q}}_n\,\mathsf{H}\binom{\lambda,\ \mathbf{x^n}}{m, S_n} = \mathsf{H}\binom{\lambda^{-1},\ \mathbf{x^{-n}}}{m,\ \ S_n}, \tag{3.1}$$
$$\widehat{\mathbf{Q}}_n I_{n,r}\left(\mathbf{x^n}\right) = I_{n,r}\left(\mathbf{x^{-n}}\right), \qquad \mathbf{x^{-n}} = \left\{\frac{1}{x_1}, \ldots, \frac{1}{x_n}\right\}.$$

The following Lemma generalizes duality relation (1.5) and gives necessary conditions for multivariate polynomials to be self-dual, i.e., to have Rep (2.4, 2.5).

Theorem 1 *Let generic polynomials* $\mathsf{R}\binom{\lambda,\ \mathbf{x^n}}{m, S_n}$, $\mathsf{S}\binom{\lambda,\ \mathbf{x^n}}{m, S_n}$ *be given according to definition (2.4, 2.5), then*

$$\lambda^{mn} I_{n,n}^m \widehat{\mathbf{Q}}_n\,\mathsf{R}\binom{\lambda,\ \mathbf{x^n}}{m, S_n} - \mathsf{R}\binom{\lambda,\ \mathbf{x^n}}{m, S_n} = 0, \tag{3.2}$$
$$\lambda^{mn} I_{n,n}^m \widehat{\mathbf{Q}}_n\,\mathsf{S}\binom{\lambda,\ \mathbf{x^n}}{m, S_n} + \mathsf{S}\binom{\lambda,\ \mathbf{x^n}}{m, S_n} = 0.$$

Proof Apply the transformation $\widehat{\mathbf{Q}}_n$ to the polynomials $\mathsf{R}\binom{\lambda,\ \mathbf{x^n}}{m, S_n}$ and $\mathsf{S}\binom{\lambda,\ \mathbf{x^n}}{m, S_n}$, defined in (2.4, 2.5), and multiply the results by $\lambda^{mn} I_{n,n}^m$,

$$\lambda^{mn} I_{n,n}^{m}\ \widehat{\mathbf{Q}}_n\ \mathsf{R}\binom{\lambda,\ \mathbf{x^n}}{m, S_n} =$$

$$I_{n,n}^{m}\sum_{s=0}^{\lfloor s_* \rfloor}\left[\lambda^s\ \widehat{\mathbf{Q}}_n\ R_s^{+}\binom{\mathbf{x^n}}{m, S_n} + \lambda^{mn-s}\ \widehat{\mathbf{Q}}_n\ R_s^{-}\binom{\mathbf{x^n}}{m, S_n}\right], \tag{3.3}$$

$$\lambda^{mn} I_{n,n}^{m}\ \widehat{\mathbf{Q}}_n\ \mathsf{S}\binom{\lambda,\ \mathbf{x^n}}{m, S_n} =$$

$$I_{n,n}^{m}\sum_{s=0}^{\lfloor s_* \rfloor}\left[\lambda^s\ \widehat{\mathbf{Q}}_n\ S_s^{+}\binom{\mathbf{x^n}}{m, S_n} - \lambda^{mn-s}\ \widehat{\mathbf{Q}}_n\ S_s^{-}\binom{\mathbf{x^n}}{m, S_n}\right]. \tag{3.4}$$

In accordance with Definition 3, calculate the invariants $I_{n,r}\left(\mathbf{x^{-n}}\right)$ and $I_{n,n-r}\left(\mathbf{x^{-n}}\right)$,

$$I_{n,r}\left(\mathbf{x^{-n}}\right) = \frac{I_{n,n-r}\left(\mathbf{x^{n}}\right)}{I_{n,n}\left(\mathbf{x^{n}}\right)}, \qquad I_{n,n-r}\left(\mathbf{x^{-n}}\right) = \frac{I_{n,r}\left(\mathbf{x^{n}}\right)}{I_{n,n}\left(\mathbf{x^{n}}\right)},$$

and, keeping in mind the relation (2.8), apply the transformation $\widehat{\mathbf{Q}}_n$ to the products

$$\widehat{\mathbf{Q}}_n\left\{\prod_{r=0}^{n} I_{n,r}^{\alpha_{n-r}\binom{mn-s,\ l}{m,\quad S_n}}\right\} = I_{n,n}^{-m}\prod_{r=0}^{n} I_{n,n-r}^{\alpha_{n-r}\binom{mn-s,\ l}{m,\quad S_n}} = I_{n,n}^{-m}\prod_{r=0}^{n} I_{n,n-r}^{\alpha_r\binom{s,\ l}{m, S_n}},$$

$$\widehat{\mathbf{Q}}_n\left\{\prod_{r=0}^{n} I_{n,n-r}^{\alpha_r\binom{s,\ l}{m, S_n}}\right\} = I_{n,n}^{-m}\prod_{r=0}^{n} I_{n,r}^{\alpha_r\binom{s,\ l}{m, S_n}} = I_{n,n}^{-m}\prod_{r=0}^{n} I_{n,r}^{\alpha_{n-r}\binom{mn-s,\ l}{m,\quad S_n}}. \tag{3.5}$$

Making use of (2.6, 2.7) and (3.5), calculate four summands in (3.3, 3.4),

$$I_{n,n}^{m}\widehat{\mathbf{Q}}_n R_s^{+}\binom{\mathbf{x^n}}{m, S_n} = R_s^{-}\binom{\mathbf{x^n}}{m, S_n}, \quad I_{n,n}^{m}\widehat{\mathbf{Q}}_n R_s^{-}\binom{\mathbf{x^n}}{m, S_n} = R_s^{+}\binom{\mathbf{x^n}}{m, S_n}$$
$$I_{n,n}^{m}\widehat{\mathbf{Q}}_n S_s^{+}\binom{\mathbf{x^n}}{m, S_n} = S_s^{-}\binom{\mathbf{x^n}}{m, S_n}, \quad I_{n,n}^{m}\widehat{\mathbf{Q}}_n S_s^{-}\binom{\mathbf{x^n}}{m, S_n} = S_s^{+}\binom{\mathbf{x^n}}{m, S_n}.$$

Substituting the last into (3.3, 3.4), we arrive at (3.2). □

Denote by $\mathbb{R}\binom{\lambda,}{m, S_n}$ and $\mathbb{S}\binom{\lambda,}{m, S_n}$ two sets of self-dual polynomials with fixed n and m and different sets of parameters $\mathcal{R}_{m,n}^{s,\ l}$, $\mathcal{S}_{m,n}^{s,\ l}$

$$\mathbb{R}\binom{\lambda,}{m, S_n} = \left\{\mathsf{R}\binom{\lambda,\ \mathbf{x^n}}{m, S_n}\right\}, \quad \mathbb{S}\binom{\lambda,}{m, S_n} = \left\{\mathsf{S}\binom{\lambda,\ \mathbf{x^n}}{m, S_n}\right\}, \tag{3.6}$$

defined in (2.6, 2.7).

Consider a set $\mathbb{H}\binom{\lambda,}{m, S_n}$ of polynomials $\mathsf{H}\binom{\lambda,\ \mathbf{x^n}}{m, S_n}$ in λ with homogeneous polynomial coefficients $H_s\binom{\mathbf{x^n}}{m, S_n}$ in $\mathbf{x^n} \in \mathbb{E}^n$ of degree s and require that $H_s\binom{\mathbf{x^n}}{m, S_n}$ be invariant under action of symmetric group S_n on $\mathbb{E}^n$ over $\mathbb{R}$, but not necessarily satisfy relations (3.2),

$$\mathbb{H}\binom{\lambda,}{m,S_n} = \left\{ \mathsf{H}\binom{\lambda,\ \mathbf{x^n}}{m,S_n} \mid \mathsf{H}\binom{\lambda,\ \mathbf{x^n}}{m,S_n} = \right.$$
$$\left. \sum_{s=0}^{\lfloor s_* \rfloor} \left[H_s^+\binom{\mathbf{x^n}}{m,S_n} \lambda^{mn-s} + H_s^-\binom{\mathbf{x^n}}{m,S_n} \lambda^s \right],\ H_s^{\pm}\binom{\mathbf{x^n}}{m,S_n} \in \mathbb{R}[\mathbf{x^n}] \right\},$$

where

$$H_s^+\binom{\mathbf{x^n}}{m,S_n} = \sum_{l=1}^{W_n(s)} \mathcal{U}_{m,n}^{s,\ l} \prod_{r=0}^{n} I_{n,r}^{\alpha_{n-r}\binom{mn-s,\ l}{m,\quad S_n}},$$
$$H_s^-\binom{\mathbf{x^n}}{m,S_n} = \sum_{l=1}^{W_n(s)} \mathcal{V}_{m,n}^{s,\ l} \prod_{r=0}^{n} I_{n,n-r}^{\alpha_r\binom{s,\ l}{m,S_n}}. \tag{3.7}$$

Exponents $\alpha_r\binom{s,\ l}{m,\ S_n}$ in (3.7) satisfy the 1st Diophantine equation in (2.9) and $W_n(s)$ is a partition function [1, 17], which accounts for a number of non-negative solutions of that equation. Two independent sets of real numbers $\{\mathcal{U}_{m,n}^{s,\ l}\}$ and $\{\mathcal{V}_{m,n}^{s,\ l}\}$ are not necessarily coincided.

By comparison (3.7) with (2.4, 2.5), we conclude that a set $\mathbb{H}\binom{\lambda,}{m,S_n}$ contains two other sets of self-dual polynomials,

$$\mathbb{R}\binom{\lambda,}{m,S_n},\ \mathbb{S}\binom{\lambda,}{m,S_n} \subset \mathbb{H}\binom{\lambda,}{m,S_n}. \tag{3.8}$$

In what follows, we give sufficient conditions for multivariate polynomials $\mathsf{H}\binom{\lambda,\ \mathbf{x^n}}{m,S_n}$ to be self-dual, i.e., to have the Rep (2.2, 2.4).

Theorem 2 *Let a polynomial* $\mathsf{H}\binom{\lambda,\ \mathbf{x^n}}{m,S_n}$ *be given in accordance with (3.7). Then*

$$\mathsf{H}\binom{\lambda,\ \mathbf{x^n}}{m,S_n} \in \mathbb{R}\binom{\lambda,}{m,S_n} \quad \text{if} \quad \lambda^{mn} I_{n,n}^m \widehat{\mathbf{Q}}_n\, \mathsf{H}\binom{\lambda,\ \mathbf{x^n}}{m,S_n} = \mathsf{H}\binom{\lambda,\ \mathbf{x^n}}{m,S_n}. \tag{3.9}$$

Proof Apply the transformation $\widehat{\mathbf{Q}}_n$ to the polynomial $\mathsf{H}\binom{\lambda,\ \mathbf{x^n}}{m,S_n}$ in (3.7) and calculate the left hand side (l.h.s.) in equality (3.9),

$$\lambda^{mn} I_{n,n}^m \widehat{\mathbf{Q}}_n\, \mathsf{H}\binom{\lambda,\ \mathbf{x^n}}{m,S_n} =$$
$$I_{n,n}^m \sum_{s=0}^{\lfloor s_* \rfloor} \left[\lambda^s\, \widehat{\mathbf{Q}}_n\, H_s^+\binom{\mathbf{x^n}}{m,S_n} + \lambda^{mn-s}\, \widehat{\mathbf{Q}}_n\, H_s^-\binom{\mathbf{x^n}}{m,S_n} \right]. \tag{3.10}$$

According to (3.7) and (3.5) we have,

$$I_{n,n}^m\, \widehat{\mathbf{Q}}_n\, H_s^+\binom{\mathbf{x^n}}{m,S_n} = I_{n,n}^{C^+} \sum_{l=1}^{W_n(s)} \mathcal{U}_{m,n}^{s,\ l} \prod_{r=0}^{n} I_{n,n-r}^{\alpha_{n-r}\binom{mn-s,\ l}{m,\quad S_n}}, \tag{3.11}$$

$$C^{+}=m-\sum_{r=0}^{n}\alpha_{n-r}\binom{mn-s,\ l}{m,\quad S_n},$$

$$I_{n,n}^{m}\,\widehat{\mathbf{Q}}_n\,H_s^{-}\binom{\mathbf{x^n}}{m,S_n}=I_{n,n}^{C^-}\sum_{l=1}^{W_n(s)}\mathcal{V}_{m,n}^{s,\ l}\prod_{r=0}^{n}I_{n,r}^{\alpha_r\binom{s,\ l}{m,S_n}},\tag{3.12}$$

$$C^{-}=m-\sum_{r=0}^{n}\alpha_r\binom{s,\ l}{m,S_n},$$

Substituting (3.7) and (3.10) into equality (3.9) and making use of (3.11, 3.12), we get

$$\sum_{l=1}^{W_n(s)}\left[I_{n,n}^{C^+}\,\mathcal{U}_{m,n}^{s,\ l}\prod_{r=0}^{n}I_{n,n-r}^{\alpha_{n-r}\binom{mn-s,\ l}{m,\quad S_n}}-\mathcal{V}_{m,n}^{s,\ l}\prod_{r=0}^{n}I_{n,n-r}^{\alpha_r\binom{s,\ l}{m,S_n}}\right]=0,$$

$$\sum_{l=1}^{W_n(s)}\left[I_{n,n}^{C^-}\,\mathcal{V}_{m,n}^{s,\ l}\prod_{r=0}^{n}I_{n,r}^{\alpha_r\binom{s,\ l}{m,S_n}}-\mathcal{U}_{m,n}^{s,\ l}\prod_{r=0}^{n}I_{n,r}^{\alpha_{n-r}\binom{mn-s,\ l}{m,\quad S_n}}\right]=0.$$

The last equalities hold for the finite number $W_n(s)$ of the fixed real coefficients $\mathcal{U}_{m,n}^{s,\ l}$ and $\mathcal{V}_{m,n}^{s,\ l}$, that may be provided only if all differences in brackets are vanished,

$$I_{n,n}^{C^+}\,\mathcal{U}_{m,n}^{s,\ l}\prod_{r-0}^{n}I_{n,n-r}^{\alpha_{n-r}\binom{mn-s,\ l}{m,\quad S_n}}=\mathcal{V}_{m,n}^{s,\ l}\prod_{r=0}^{n}I_{n,n-r}^{\alpha_r\binom{s,\ l}{m,S_n}},\tag{3.13}$$

$$I_{n,n}^{C^-}\,\mathcal{V}_{m,n}^{s,\ l}\prod_{r=0}^{n}I_{n,r}^{\alpha_r\binom{s,\ l}{m,S_n}}=\mathcal{U}_{m,n}^{s,\ l}\prod_{r=0}^{n}I_{n,r}^{\alpha_{n-r}\binom{mn-s,\ l}{m,\quad S_n}}.$$

Combining two equalities in (3.13), we obtain

$$\sum_{r=0}^{n}\left[\alpha_{n-r}\binom{mn-s,\ l}{m,\quad S_n}-\alpha_r\binom{s,\ l}{m,S_n}\right]\log\left(\frac{I_{n,n-r}I_{n,r}}{I_{n,n}}\right)=\tag{3.14}$$

$$2\log\left(\frac{\mathcal{V}_{m,n}^{s,\ l}}{\mathcal{U}_{m,n}^{s,\ l}}\right).$$

Since invariants $I_{n,r}$ are continuously dependent on $\mathbf{x^n}$, then equality (3.14) holds only if all differences in brackets and the right hand side (r.h.s.) in (3.14) are vanished,

$$\alpha_{n-r}\binom{mn-s,\ l}{m,\quad S_n}=\alpha_r\binom{s,\ l}{m,S_n},\qquad \mathcal{U}_{m,n}^{s,\ l}=\mathcal{V}_{m,n}^{s,\ l}.\tag{3.15}$$

Substituting the last equalities into (3.13), we arrive at the 2nd Diophantine equation in (2.9). However, a number of non-negative solutions of two Diophantine equations

(2.9) is $P_n^m(s)$, defined in (B.2), i.e., not all of $W_n(s)$ terms are survived in sums over l in (3.7).

Thus, substituting (3.15) into (3.7) and replacing there the upper limit $W_n(s) \to P_n^m(s)$, we arrive at definition of self-reciprocal polynomials $\mathsf{R}\binom{\lambda,\ \mathbf{x^n}}{m, S_n}$ in (2.4) and Theorem is proven. □

Theorem 3 *Let a polynomial* $\mathsf{H}\binom{\lambda,\ \mathbf{x^n}}{m, S_n}$ *be given in accordance with (3.7). Then*

$$\mathsf{H}\binom{\lambda,\ \mathbf{x^n}}{m, S_n} \in \mathbb{S}\binom{\lambda,}{m, S_n} \quad \text{if} \quad \lambda^{mn}\ I_{n,n}^m\ \widehat{\mathbf{Q}}_n\ \mathsf{H}\binom{\lambda,\ \mathbf{x^n}}{m, S_n} = -\mathsf{H}\binom{\lambda,\ \mathbf{x^n}}{m, S_n}. \tag{3.16}$$

Proof of Theorem 3 may be given by considerations, similar to proof of Theorem 2, where only equality for amplitudes in (3.15) should be changed on $\mathcal{U}_{m,n}^{s,\,l} = -\mathcal{V}_{m,n}^{s,\,l}$.

4 Commutative Monoid of Self-Dual Symmetric Polynomials

Denote by $\mathfrak{R}(\lambda, S_n)$ and $\mathfrak{S}(\lambda, S_n)$ two unions of sets $\mathbb{R}\binom{\lambda,}{m, S_n}$ and $\mathbb{S}\binom{\lambda,}{m, S_n}$, defined in (3.6),

$$\mathfrak{R}(\lambda, S_n) = \bigcup_{m \ge 0} \mathbb{R}\binom{\lambda,}{m, S_n}, \quad \mathbb{R}\binom{\lambda,}{m_1, S_n} \cap \mathbb{R}\binom{\lambda,}{m_2, S_n} = \emptyset, \quad m_1 \neq m_2, \tag{4.1}$$

$$\mathfrak{S}(\lambda, S_n) = \bigcup_{m \ge 0} \mathbb{S}\binom{\lambda,}{m, S_n}, \quad \mathbb{S}\binom{\lambda,}{m_1, S_n} \cap \mathbb{S}\binom{\lambda,}{m_2, S_n} = \emptyset, \quad m_1 \neq m_2. \tag{4.2}$$

and by $\mathfrak{RS}(\lambda, n)$ their union,

$$\mathfrak{RS}(\lambda, S_n) = \mathfrak{R}(\lambda, S_n) \cup \mathfrak{S}(\lambda, S_n). \tag{4.3}$$

Prove that an *addition* cannot be chosen as a binary operation on $\mathfrak{R}(\lambda, S_n)$ and $\mathfrak{S}(\lambda, S_n)$.

Lemma 3 *A linear combination of self-dual polynomials with different m is not self-dual,*

$$a\,\mathsf{R}\binom{\lambda,\ \mathbf{x^n}}{m_1, S_n} + b\,\mathsf{R}\binom{\lambda,\ \mathbf{x^n}}{m_2, S_n} \notin \mathfrak{R}(\lambda, S_n), \quad m_1 \neq m_2, \quad a, b \in \mathbb{R}. \tag{4.4}$$
$$a\,\mathsf{S}\binom{\lambda,\ \mathbf{x^n}}{m_1, S_n} + b\ \mathsf{S}\binom{\lambda,\ \mathbf{x^n}}{m_2, S_n} \notin \mathfrak{S}(\lambda, S_n).$$

Proof First, prove the statement (4.4) for self-reciprocal polynomials. Let, by way of contradiction, there exists m such that

$$a\,\mathsf{R}\binom{\lambda,\ \mathbf{x^n}}{m_1, S_n} + b\,\mathsf{R}\binom{\lambda,\ \mathbf{x^n}}{m_2, S_n} = c\,\mathsf{R}\binom{\lambda,\ \mathbf{x^n}}{m, S_n}, \quad m \ge m_2 \ge m_1, \quad c \in \mathbb{R}. \tag{4.5}$$

Then, applying the transformation $\lambda^{mn} I_{n,n}^m \widehat{\mathbf{Q}}_n$ to (4.5) and making use of Theorem 1, we obtain

$$a\ \left(\lambda^n I_{n,n}\right)^{m-m_1} \mathsf{R}\binom{\lambda,\ \mathbf{x}^n}{m_1, S_n} + b\ \left(\lambda^n I_{n,n}\right)^{m-m_2} \mathsf{R}\binom{\lambda,\ \mathbf{x}^n}{m_2, S_n} = c\,\mathsf{R}\binom{\lambda,\ \mathbf{x}^n}{m, S_n}. \tag{4.6}$$

By comparison the l.h.s. of equalities (4.5) and (4.6), we arrive at equality

$$a\ \left[\left(\lambda^n I_{n,n}\right)^{m-m_1} - 1\right] \mathsf{R}\binom{\lambda,\ \mathbf{x}^n}{m_1, S_n} + b\ \left[\left(\lambda^n I_{n,n}\right)^{m-m_2} - 1\right] \mathsf{R}\binom{\lambda,\ \mathbf{x}^n}{m_2, S_n} = 0,$$

which may be provided only if all differences in brackets are vanished, i.e., $m_1 = m_2 = m$. However, the last case is trivial and occur when $a + b = c$. By similar considerations one can prove a statement (4.4) for scew-reciprocal polynomials. □

Consider now a multiplication of two self-dual polynomials and extend the known statements (Corollary 6, in [15]) for univariate polynomials $\mathsf{R}_m(\lambda)$ and $\mathsf{S}_m(\lambda)$ on multivariate polynomials.

Lemma 4 *The following containments hold*

$$\mathsf{R}\binom{\lambda,\ \mathbf{x}^n}{m_1, S_n} \mathsf{R}\binom{\lambda,\ \mathbf{x}^n}{m_2, S_n},\ \mathsf{S}\binom{\lambda,\ \mathbf{x}^n}{m_1, S_n} \mathsf{S}\binom{\lambda,\ \mathbf{x}^n}{m_2, S_n} \in \mathbb{R}\binom{\lambda,}{m, S_n}, \tag{4.7}$$

$$\mathsf{R}\binom{\lambda,\ \mathbf{x}^n}{m_1, S_n} \mathsf{S}\binom{\lambda,\ \mathbf{x}^n}{m_2, S_n},\ \mathsf{S}\binom{\lambda,\ \mathbf{x}^n}{m_1, S_n} \mathsf{R}\binom{\lambda,\ \mathbf{x}^n}{m_2, S_n} \in \mathbb{S}\binom{\lambda,}{m, S_n}, \tag{4.8}$$

where $m = m_1 + m_2$.

Proof Applying the transformation $\left(\lambda^n I_{n,n}\right)^{m_1+m_2} \widehat{\mathbf{Q}}_n$ to the product of two self-reciprocal polynomials $\mathsf{R}\binom{\lambda,\ \mathbf{x}^n}{m_i, S_n}$, $i = 1, 2$, we make use of Theorem 1,

$$\begin{aligned}\left(\lambda^n I_{n,n}\right)^{m_1+m_2} \widehat{\mathbf{Q}}_n \left[\mathsf{R}\binom{\lambda,\ \mathbf{x}^n}{m_1, S_n} \mathsf{R}\binom{\lambda,\ \mathbf{x}^n}{m_2, S_n}\right] &= \\ \left(\lambda^n I_{n,n}\right)^{m_1} \widehat{\mathbf{Q}}_n \mathsf{R}\binom{\lambda,\ \mathbf{x}^n}{m_1, S_n} \cdot \left(\lambda^n I_{n,n}\right)^{m_2} \widehat{\mathbf{Q}}_n \mathsf{R}\binom{\lambda,\ \mathbf{x}^n}{m_2, S_n} &= \\ \mathsf{R}\binom{\lambda,\ \mathbf{x}^n}{m_1, S_n} \mathsf{R}\binom{\lambda,\ \mathbf{x}^n}{m_2, S_n}&.\end{aligned}$$

By Theorem 2, the containment (4.7) follows. By similar considerations and making use of Theorems 2 and 3 one can prove the other containments in (4.7, 4.8). □

Corollary 2 *A set* $\mathfrak{R}(\lambda,\ S_n)$ *is a multiplicative commutative monoid.*

Proof Consider $\mathfrak{R}(\lambda,\ S_n)$ as a set of self-reciprocal polynomials $\mathsf{R}\binom{\lambda,\ \mathbf{x}^n}{m, S_n}$, $m \in \mathbb{Z}_{\geq}$, with multiplication as an associative binary operation. According to Lemma 4 and definition (4.1), on the one hand, and to the last equality in (2.4), on the other hand, the following containments hold,

$$\mathsf{R}\binom{\lambda,\ \mathbf{x}^n}{m_1, S_n} \mathsf{R}\binom{\lambda,\ \mathbf{x}^n}{m_2, S_n} \in \mathfrak{R}(\lambda,\ S_n),\quad \mathsf{R}\binom{\lambda,\ \mathbf{x}^n}{0,\ S_n} = 1 \in \mathfrak{R}(\lambda,\ S_n). \tag{4.9}$$

Then, according to definition of algebraic structures [19], a set $\mathfrak{R}(\lambda, S_n)$ exhibits a commutative monoid under multiplication. □

In contrast to $\mathsf{R}\binom{\lambda,\, \mathbf{x^n}}{m, S_n}$, the polynomials $\mathsf{S}\binom{\lambda,\, \mathbf{x^n}}{m, S_n}$ do not generate a multiplicative monoid.

Theorem 4 *A set $\mathfrak{RS}(\lambda, S_n)$ is a multiplicative commutative monoid such that*

$$\mathfrak{R}(\lambda, S_n) = \mathfrak{RS}(\lambda, S_n)/\mathbb{Z}_2. \tag{4.10}$$

Proof By Lemma 4, a product of any two self-dual polynomials is always contained either in $\mathfrak{R}(\lambda, n)$ or $\mathfrak{S}(\lambda, n)$ sets, and, by definition (4.3), also in $\mathfrak{RS}(\lambda, n)$. This fact and the existence (4.9) of unity $\mathsf{R}\binom{\lambda,\, \mathbf{x^n}}{0, S_n}$ in $\mathfrak{R}(\lambda, S_n)$ makes $\mathfrak{RS}(\lambda, S_n)$ a multiplicative monoid. On the other hand, a bijection $\mathfrak{R}(\lambda, S_n) \Longleftrightarrow \mathfrak{S}(\lambda, S_n)$ between two sets follows by the one-to-one correspondence between polynomials $\mathsf{R}\binom{\lambda,\, \mathbf{x^n}}{m, S_n}$ and $\mathsf{S}\binom{\lambda,\, \mathbf{x^n}}{m, S_n}$. This manifests a relationship (4.10) between two multiplicative commutative monoids of self-dual polynomials. □

4.1 Factorization of Self-Dual Symmetric Polynomials

In view of Lemma 4, the existence of the commutative monoid makes reasonable a question about factorization of $\mathsf{R}\binom{\lambda,\, \mathbf{x^n}}{m, S_n}$ and $\mathsf{S}\binom{\lambda,\, \mathbf{x^n}}{m, S_n}$, $m = m_1 + m_2$, in $\mathfrak{RS}(\lambda, n)$,

$$\mathsf{R}\binom{\lambda,\, \mathbf{x^n}}{m, S_n} = \mathsf{R}\binom{\lambda,\ \ \mathbf{x^n}}{m_1,\ S_n}\mathsf{R}\binom{\lambda,\ \ \mathbf{x^n}}{m_2,\ S_n} = \mathsf{S}\binom{\lambda,\ \ \mathbf{x^n}}{m_1,\ S_n}\mathsf{S}\binom{\lambda,\ \ \mathbf{x^n}}{m_2,\ S_n}, \tag{4.11}$$

$$\mathsf{S}\binom{\lambda,\, \mathbf{x^n}}{m, S_n} = \mathsf{R}\binom{\lambda,\ \ \mathbf{x^n}}{m_1,\ S_n}\mathsf{S}\binom{\lambda,\ \ \mathbf{x^n}}{m_2,\ S_n} = \mathsf{S}\binom{\lambda,\ \ \mathbf{x^n}}{m_1,\ S_n}\mathsf{R}\binom{\lambda,\ \ \mathbf{x^n}}{m_2,\ S_n}. \tag{4.12}$$

A *division* as a binary operation cannot be introduced in $\mathfrak{RS}(\lambda, n)$ since that would lead to non-polynomial (rational) functions. However, if a ratio of two self-dual polynomials, in a special case, is a polynomial, then the last is self-dual. A monoid $\mathfrak{RS}(\lambda, n)$ comprises two kind of self-dual polynomials: those, which can be represented as (4.11, 4.12), and those, which cannot.

Definition 4 Self-dual polynomials $\mathsf{R}\binom{\lambda,\, \mathbf{x^n}}{m, S_n}$ and $\mathsf{S}\binom{\lambda,\, \mathbf{x^n}}{m, S_n}$ are called reducible in $\mathfrak{RS}(\lambda, n)$ if they can be represented as (4.11, 4.12), otherwise they are called irreducible.

By Definitions 2 and 4, the following Lemma 5 holds.

Lemma 5 *Let a self-dual polynomial* $\mathsf{H}\binom{\lambda,\, \mathbf{x^n}}{m, S_n}$ *be reducible in $\mathfrak{RS}(\lambda, n)$ such that the equality*

$$\mathsf{H}\binom{\lambda,\ \mathbf{x^n}}{m, S_n} = \mathsf{H}_1\binom{\lambda,\ \mathbf{x^n}}{m_1,\ S_n}\mathsf{H}_2\binom{\lambda,\ \mathbf{x^n}}{m_2,\ S_n}, \quad m = m_1 + m_2, \tag{4.13}$$

coincides with one of equalities (4.11, 4.12) for self-dual polynomials $\mathsf{H}_1, \mathsf{H}_2$. *Then the following inequality holds*

$$\mu\left[\mathsf{H}\binom{\lambda,\ \mathbf{x^n}}{m, S_n}\right] \le \mu\left[\mathsf{H}_1\binom{\lambda,\ \mathbf{x^n}}{m_1, S_n}\right] + \mu\left[\mathsf{H}_2\binom{\lambda,\ \mathbf{x^n}}{m_2, S_n}\right]. \tag{4.14}$$

Illustrate an appearance of inequality in (4.14). Let a polynomial $\mathsf{R}\binom{\lambda,\ \mathbf{x^2}}{2,\ S_2}$ be reducible, then

$$\mathsf{R}\binom{\lambda,\ \mathbf{x^2}}{2,\ S_2} = \mathsf{R}\binom{\lambda,\ \mathbf{x^2}}{1,\ S_2}\mathsf{R}\binom{\lambda,\ \mathbf{x^2}}{1,\ S_2}, \quad \mu\left[\mathsf{R}\binom{\lambda,\ \mathbf{x^2}}{1,\ S_2}\right] = 2, \quad \mu\left[\mathsf{R}\binom{\lambda,\ \mathbf{x^2}}{2,\ S_2}\right] = 3,$$

where $\mathsf{R}\binom{\lambda,\ \mathbf{x^2}}{1,\ S_2}$ is presented in Appendix A, while a non generic polynomial $\mathsf{R}\binom{\lambda,\ \mathbf{x^2}}{2,\ S_2}$ is given below,

$$\mathsf{R}\binom{\lambda,\ \mathbf{x^2}}{2,\ S_2} = M_1\lambda^4 + M_2 I_{2,1}\lambda^3 + (M_3 I_{2,1}^2 + 2M_1 I_{2,2})\lambda^2 + M_2 I_{2,1} I_{2,2}\lambda + M_1 I_{2,2}^2.$$

Self-dual polynomials in two variables take a particular place in the reducibility problem in $\mathfrak{RS}(\lambda, n)$. Next Lemma generalizes relation (1.6) for two-variable polynomials.

Lemma 6 *Generic scew-reciprocal polynomials* $\mathsf{S}\binom{\lambda,\ \mathbf{x^2}}{m, S_2}$, $m \ge 2$, *are reducible in* $\mathfrak{RS}(\lambda, n)$.

$$\mathsf{S}\binom{\lambda,\ \mathbf{x^2}}{m, S_2} = \mathsf{R}\binom{\lambda,\ \ \mathbf{x^2}}{m-1,\ S_2}\mathsf{S}\binom{\lambda,\ \mathbf{x^2}}{1,\ S_2}. \tag{4.15}$$

Proof Consider a generic polynomial $\mathsf{S}\binom{\lambda,\ \mathbf{x^2}}{m, S_2}$, $m \ge 2$, according to (2.5–2.8),

$$\mathsf{S}\binom{\lambda,\ \mathbf{x^2}}{m, S_2} = \tag{4.16}$$
$$\sum_{s=0}^{m}\sum_{l}^{P_2^m(s)} S_{m,2}^{s,\ l}\left[I_{2,1}^{\alpha_1\binom{s,\ l}{m,S_2}} I_{2,2}^{\alpha_2\binom{s,\ l}{m,S_2}}\lambda^{2m-s} - I_{2,1}^{\alpha_1\binom{s,\ l}{m,S_2}} I_{2,2}^{\alpha_0\binom{s,\ l}{m,S_2}}\lambda^s\right],$$

where

$$m = \alpha_0\binom{s,\ l}{m, S_2} + \alpha_1\binom{s,\ l}{m, S_2} + \alpha_2\binom{s,\ l}{m, S_2}, \tag{4.17}$$
$$s = \alpha_1\binom{s,\ l}{m, S_2} + 2\alpha_2\binom{s,\ l}{m, S_2}.$$

Diophantine equations (4.17) give $m - s = \alpha_0\binom{s,\ l}{m, S_2} - \alpha_2\binom{s,\ l}{m, S_2}$, so we represent the polynomial in (4.16) as,

$$\mathsf{S}\binom{\lambda,\,\mathbf{x^2}}{m,S_2} = \sum_{s=0}^{m-1} K_s\lambda^s\left(\lambda^{2(m-s)} - I_{2,2}^{m-s}\right),$$

$$K_s = \sum_{l}^{P_2^m(s)} \mathcal{S}_{m,2}^{s,\,l} I_{2,1}^{\alpha_1\binom{s,\,l}{m,S_2}} I_{2,2}^{\alpha_2\binom{s,\,l}{m,S_2}}.$$

A difference $\lambda^{2(m-s)} - I_{2,2}^{m-s}$ is always divisible by $\lambda^2 - I_{2,2}$, which up to a real non-zero coefficient represents, in fact, a scew-reciprocal polynomial $\mathsf{S}\binom{\lambda,\,\mathbf{x^2}}{1,\,S_2}$. Thus, the polynomial $\mathsf{S}\binom{\lambda,\,\mathbf{x^2}}{m,S_2}$ is divisible by $\mathsf{S}\binom{\lambda,\,\mathbf{x^2}}{1,\,S_2}$ in $\mathfrak{RS}(\lambda, n)$ and (4.15) holds. □

By formulas (2.18), it follows $\mu\left[\mathsf{S}\binom{\lambda,\,\mathbf{x^2}}{1,\,S_2}\right] = 1$, that, by Lemma 5, distinguishes $\mathsf{S}\binom{\lambda,\,\mathbf{x^2}}{1,\,S_2}$ from others self-dual polynomials in $\mathfrak{RS}(\lambda, n)$,

$$\mu\left[\mathsf{S}\binom{\lambda,\,\mathbf{x^n}}{m,\,S_n}\right] > 1,\quad m \geq 2, \qquad \mu\left[\mathsf{R}\binom{\lambda,\,\mathbf{x^n}}{m,\,S_n}\right] > 1,\quad m \geq 1.$$

Lemma 7 *Generic self-reciprocal polynomials* $\mathsf{R}\binom{\lambda,\,\mathbf{x^2}}{m,S_2}$, $m \geq 4$, *are irreducible in* $\mathfrak{RS}(\lambda, n)$.

Proof Let, by way of contradiction, the generic polynomial $\mathsf{R}\binom{\lambda,\,\mathbf{x^2}}{m,S_2}$, $m \geq 4$, be reducible. Keeping in mind the first possible factorization in (4.11), define a difference

$$\mathcal{M}_{RR}^{R}\binom{\lambda,\,\mathbf{x^n}}{m,S_n} = \mu\left[\mathsf{R}\binom{\lambda,\,\mathbf{x^n}}{m,S_n}\right] - \mu\left[\mathsf{R}\binom{\lambda,\,\mathbf{x^n}}{m_1,S_n}\right] - \mu\left[\mathsf{R}\binom{\lambda,\,\mathbf{x^n}}{m_2,S_n}\right], \tag{4.18}$$

where $m = m_1 + m_2$. Calculate it for $n = 2$, substituting (2.18) into (4.18),

$$\mathcal{M}_{RR}^{R}\binom{\lambda,\,\mathbf{x^2}}{m,S_2} = \frac{1}{2}\left(m_1m_2 + \sin^2\frac{\pi m_1}{2}\,\sin^2\frac{\pi m_2}{2}\right) - 1,$$
$$m_1 + m_2 \geq 4,\quad m_1, m_2 \geq 1,$$

which never vanishes, that contradicts to Lemma 5. Define another difference

$$\mathcal{M}_{SS}^{R}\binom{\lambda,\,\mathbf{x^n}}{m,S_n} = \mu\left[\mathsf{R}\binom{\lambda,\,\mathbf{x^n}}{m,S_n}\right] - \mu\left[\mathsf{S}\binom{\lambda,\,\mathbf{x^n}}{m_1,S_n}\right] - \mu\left[\mathsf{S}\binom{\lambda,\,\mathbf{x^n}}{m_2,S_n}\right], \tag{4.19}$$

where $m = m_1 + m_2$. It is related to the second possible factorization in (4.11). Calculating it for $n = 2$ by substituting (2.18) into (4.19), we obtain

$$\mathcal{M}_{SS}^{R}\binom{\lambda,\,\mathbf{x^2}}{m,S_2} = \frac{1}{2}\left(m_1m_2 + m_1 + m_2 + 1 + \cos^2\frac{\pi m_1}{2}\,\cos^2\frac{\pi m_2}{2}\right),$$
$$m_1, m_2 \geq 1,$$

which is positive, that again contradicts to Lemma 5. Thus, Lemma 7 is proven. □

Beyond the scope of the present paper, there is left a study of reducibility of self-dual polynomials in n variables that requires sophisticated analysis of $\mathcal{M}^R_{RR}\binom{\lambda,\,\mathbf{x^n}}{m,S_n}$ and $\mathcal{M}^R_{SS}\binom{\lambda,\,\mathbf{x^n}}{m,S_n}$ as well as the difference

$$\mathcal{M}^S_{RS}\binom{\lambda,\,\mathbf{x^n}}{m,S_n}=\mu\left[\mathsf{S}\binom{\lambda,\,\mathbf{x^n}}{m,S_n}\right]-\mu\left[\mathsf{R}\binom{\lambda,\,\mathbf{x^n}}{m_1,S_n}\right]-\mu\left[\mathsf{S}\binom{\lambda,\,\mathbf{x^n}}{m_2,S_n}\right], \tag{4.20}$$

where $m = m_1 + m_2$, for arbitrary $n \geq 3$. We will address this problem in a separate paper.

5 Scew-Reciprocal Polynomial Equations

In Lemma 6 we have shown that the generic polynomial equation $\mathsf{S}\binom{\lambda,\,\mathbf{x^2}}{m,\,S_2} = 0$ has always a universal (independent of m) solution $\lambda\left(\mathbf{x^2}\right) = \sqrt{x_1x_2}$, which possesses many remarkable algebraic properties (homogeneity of the 1st order, duality, monotony *etc*), which will be listed below in details. But if $n \geq 3$, all roots $\lambda\left(\mathbf{x^n}\right)$ of generic scew-reciprocal polynomial $\mathsf{S}\binom{\lambda,\,\mathbf{x^n}}{m,S_n}$ lose their universality, substantially depending on m, more precisely, on the set of coefficients $\mathcal{S}^{s,\,l}_{m,n}$. Even so, a question of validity of the rest algebraic properties, mentioned above, for $\lambda\left(\mathbf{x^n}\right)$ is left open. In this section we study these properties.

5.1 Symmetry, Homogeneity, Duality and Compatibility

Prove several statements for a real root $\lambda\left(\mathbf{x^n}\right)$ of generic scew-reciprocal polynomial $\mathsf{S}\binom{\lambda,\,\mathbf{x^n}}{m,S_n}$.

Lemma 8 *Let* $\lambda\left(\mathbf{x^n}\right)$ *be a real solution of equation* $\mathsf{S}\binom{\lambda,\,\mathbf{x^n}}{m,S_n}=0$. *Then* $\lambda\left(\mathbf{x^n}\right)$ *is invariant under action of symmetric group* S_n

$$\lambda\left(\left\{S_n\,\mathbf{x^n}\right\}\right) = \lambda\left(\mathbf{x^n}\right), \tag{5.1}$$

and is a homogeneous function of the 1st degree

$$\lambda\left(k\,\mathbf{x^n}\right) = k\,\lambda\left(\mathbf{x^n}\right), \qquad k\,\mathbf{x^n} = \{kx_1,\ldots,kx_n\}, \tag{5.2}$$

Proof The invariance of $\lambda(\mathbf{x^n})$ under action of symmetric group S_n follows due to invariance of equation $\mathsf{S}\binom{\lambda,\,\mathbf{x^n}}{m,S_n}=0$ under permutations of variables x_j since the coefficients of this equation are built up of polynomial invariants of S_n.

Rewrite equation $\mathsf{S}\binom{\lambda,\,k\mathbf{x^n}}{m,\,S_n}=0$ for the real function $\lambda(k\,\mathbf{x^n})$, making use of (2.5, 2.7),

$$\sum_{s=0}^{\lfloor s_*\rfloor}\left[S_s^+\binom{k\mathbf{x^n}}{m,\,S_n}\lambda^{mn-s}(k\mathbf{x^n})-S_s^-\binom{k\mathbf{x^n}}{m,\,S_n}\lambda^s(k\mathbf{x^n})\right]=0, \tag{5.3}$$

where expressions for $S_s^{\pm}\binom{k\mathbf{x^n}}{m,\,S_n}$ read

$$S_s^+\binom{k\mathbf{x^n}}{m,\,S_n}=k^{a^+}\sum_{l=1}^{P_n^m(s)}S_{m,n}^{s,\,l}\prod_{r=0}^{n}I_{n,r}^{\alpha_{n-r}\binom{mn-s,\ l}{m,\ \ S_n}}, \tag{5.4}$$

$$a^+=\sum_{r=0}^{n}r\,\alpha_{n-r}\binom{mn-s,\ l}{m,\ \ S_n},$$

$$S_s^-\binom{k\mathbf{x^n}}{m,\,S_n}=k^{a^-}\sum_{l=1}^{P_n^m(s)}S_{m,n}^{s,\,l}\prod_{r=0}^{n}I_{n,n-r}^{\alpha_r\binom{s,\ l}{m,S_n}}, \tag{5.5}$$

$$a^-=\sum_{r=0}^{n}(n-r)\,\alpha_r\binom{s,\ l}{m,S_n}.$$

However, according to (2.8, 2.9), we may replace $a^+\to s,\ a^-\to mn-s$. Substituting these values into formulas (5.4, 5.5) and into Eq. (5.3), we arrive at equation

$$k^{mn}\sum_{s=0}^{\lfloor s_*\rfloor}\left[S_s^+\binom{\mathbf{x^n}}{m,S_n}\left(\frac{\lambda(k\mathbf{x^n})}{k}\right)^{mn-s}-S_s^-\binom{\mathbf{x^n}}{m,S_n}\left(\frac{\lambda(k\mathbf{x^n})}{k}\right)^{s}\right]=0. \tag{5.6}$$

By comparison of Eq. (5.6) with expression (2.5) for scew-reciprocal polynomial $\mathsf{S}\binom{\lambda,\,\mathbf{x^n}}{m,S_n}$, we conclude that relation (5.2) holds and Lemma is proven. □

Lemma 9 *Let* $\lambda(\mathbf{x^n})$ *be a real solution of equation* $\mathsf{S}\binom{\lambda,\,\mathbf{x^n}}{m,S_n}=0$*, then*

$$\lambda(x_1,\ldots,x_n)=\frac{1}{\lambda\left(x_1^{-1},\ldots,x_n^{-1}\right)}. \tag{5.7}$$

Proof Apply a transformation $\widehat{\mathbf{Q}}_n$ to equation $\mathsf{S}\binom{\lambda,\,\mathbf{x^n}}{m,S_n}=0$ in accordance with (2.5, 3.1),

$$\sum_{s=0}^{\lfloor s_* \rfloor} \left[S_s^+ \binom{\mathbf{x}^{-\mathbf{n}}}{m,\ S_n} \left(\frac{1}{\lambda}\right)^{mn-s} - S_s^- \binom{\mathbf{x}^{-\mathbf{n}}}{m,\ S_n} \left(\frac{1}{\lambda}\right)^{s} \right] = 0. \tag{5.8}$$

By comparison of Eq. (5.8) with expression (2.5) for scew-reciprocal polynomial $\mathsf{S}\binom{\lambda,\ \mathbf{x}^{\mathbf{n}}}{m, S_n}$, we conclude that relation (5.7) holds and Lemma is proven. □

Lemma 10 *A real solution $\lambda(\mathbf{x}^{\mathbf{n}})$ of equation $\mathsf{S}\binom{\lambda,\ \mathbf{x}^{\mathbf{n}}}{m, S_n}=0$ satisfies the following equality*

$$\lambda(x, \ldots, x) = x. \tag{5.9}$$

Proof Substitute $x_i = x$ into equality (5.7) and multiply the latter by x^2. Then, by Lemma 9, we arrive at (5.9). □

We call a solution $\lambda(\mathbf{x}^{\mathbf{n}})$ *compatible* if equality (5.9) holds.

Making use of formulas (2.5, 2.7, 2.8), write a scew-reciprocal polynomial as follows,

$$\mathsf{S}\binom{\lambda,\ \mathbf{x}^{\mathbf{n}}}{m, S_n} = \sum_{s=0}^{\lfloor s_* \rfloor} \sum_{l=1}^{P_n^m(s)} S_{m,n}^{s,\ l} \left(\lambda^{mn-s} \prod_{r=0}^{n} I_{n,r}^{\alpha_r\binom{s,\ l}{m, S_n}} - \lambda^s \prod_{r=0}^{n} I_{n,n-r}^{\alpha_r\binom{s,\ l}{m, S_n}} \right). \tag{5.10}$$

Lemma 11 *Let $\lambda(\mathbf{x}^{\mathbf{n}})$ be a real solution of equation $\mathsf{S}\binom{\lambda,\ \mathbf{x}^{\mathbf{n}}}{m, S_n}=0$ and polynomial invariants $I_{n,1}, \ldots, I_{n,n}$ satisfy identities*

$$I_{n,r}^{n}\ I_{n,n}^{n-2r} = I_{n,n-r}^{n}, \quad 2r \le n, \tag{5.11}$$

simultaneously. Then

$$\lambda\left(\mathbf{x}^{\mathbf{n}}\right) = \sqrt[n]{I_{n,n}}. \tag{5.12}$$

Proof Substitute relations (2.8, 2.9) and (5.11) into (5.10)

$$\mathsf{S}\binom{\lambda,\ \mathbf{x}^{\mathbf{n}}}{m, S_n} = \sum_{s=0}^{\lfloor s_* \rfloor} \sum_{l=1}^{P_n^m(s)} S_{m,n}^{s,\ l} \lambda^s \prod_{r=0}^{n} I_{n,r}^{\alpha_r\binom{s,\ l}{m, S_n}} \left(\lambda^{mn-2s} - \left(\sqrt[n]{I_{n,n}}\right)^{mn-2s} \right).$$

Note, that $\mathsf{S}\binom{\lambda,\ \mathbf{x}^{\mathbf{n}}}{m, S_n}$ vanishes if $\lambda = \sqrt[n]{I_{n,n}}$, $s \neq s_*$, and vanishes identically if $s = s_*$ irrespectively to the parity of mn. Thus, $\lambda(\mathbf{x}^{\mathbf{n}}) = \sqrt[n]{I_{n,n}}$ is a solution of equation $\mathsf{S}\binom{\lambda,\ \mathbf{x}^{\mathbf{n}}}{m, S_n} = 0$. □

Lemma 12 *If there exists such λ_*, that for every x_i one can find such x_j, which satisfy*

$$x_i \cdot x_j = \lambda_*^2, \qquad 1 \le i, j \le n, \tag{5.13}$$

then

$$I_{n,r}^n \, I_{n,n}^{n-2r} = I_{n,n-r}^n, \quad 2r \le n, \quad \lambda\left(\mathbf{x^n}\right) = \lambda_*.$$

Proof Let (5.13) be true, then, multiplying such products over all i, j, we have $I_{n,n} = \lambda_*^n$. Calculate now the product $I_{n,n}^{(n-2r)/n} I_{n,r}$ according to definition (2.1),

$$I_{n,n}^{(n-2r)/n} \sum_{i_1<\ldots<i_r}^{n} x_{i_1}\ldots x_{i_r} = \sum_{j_1<\ldots<j_r}^{n} \frac{I_{n,n}^{1-2r/n}\lambda_*^{2r}}{x_{j_1}\ldots x_{j_r}} =$$

$$\sum_{j_1<\ldots<j_r}^{n} \frac{I_{n,n}}{x_{j_1}\ldots x_{j_r}} = \sum_{j_1<\ldots<j_{n-r}}^{n} x_{j_1}\ldots x_{j_{n-r}} = I_{n,n-r}.$$

Thus, Lemma is proven. □

5.2 Positiveness and Monotony

In this section we study a special type of scew-reciprocal equations which provide a unique solution $\lambda(\mathbf{x^n}) > 0$ in the positive hyperoctant $\mathbb{E}_>^n = \{\mathbf{x^n} \mid x_i > 0,\ 1 \le i \le n\}$ of the Euclidean space. Make use of the Rep (5.10) for the polynomial $\mathsf{S}\binom{\lambda,\,\mathbf{x^n}}{m,S_n}$ and require

$$\mathcal{S}_{m,n}^{s,\,l} \ge 0. \tag{5.14}$$

By Descartes' rule of signs [3], the polynomial $\mathsf{S}\binom{\lambda,\mathbf{x^n}}{m,S_n}$ has a unique positive solution

$$\lambda\left(\mathbf{x^n}\right) \ge 0, \qquad \mathbf{x^n} \in \mathbb{E}_>^n. \tag{5.15}$$

This conclusion does not depend on the existence in the polynomial $\mathsf{S}\binom{\lambda,\mathbf{x^n}}{m,S_n}$ of the non-zero term $B_{s_*}\binom{\mathbf{x^n}}{m,S_n}$ given in (2.11), when $mn = 0 \pmod 2$.

Lemma 13 *Let $\lambda(\mathbf{x^n})$ be a positive homogeneous function of degree 1 and satisfy the duality relation (5.7). Then*

$$\partial\lambda/\partial x_i > 0, \quad 1 \le i \le n, \qquad \mathbf{x^n} \in \mathbb{E}_>^n. \tag{5.16}$$

Proof Consider a positive homogeneous function $\lambda(\mathbf{x^n})$ of degree 1 and apply Euler's homogeneous function theorem,

$$\widehat{\mathbf{L}}_n\lambda = \lambda, \qquad \widehat{\mathbf{L}}_n = \sum_{j=1}^{n} x_j \frac{\partial}{\partial x_j}, \tag{5.17}$$

and differentiate Eq. (5.17) w.r.t. variable x_i,

$$\widehat{\mathbf{L}}_n \frac{\partial \lambda}{\partial x_i} = 0, \qquad i = 1, \ldots, n. \tag{5.18}$$

Equations (5.18) manifest the derivatives $\partial\lambda/\partial x_i$ as homogeneous functions of degree 0. Choose the following Rep with differentiable function ϕ satisfying Eq. (5.18),

$$\frac{\partial \lambda}{\partial x_j} = \phi\left(\frac{x_1}{x_j}, \ldots, \frac{x_{j-1}}{x_j}, \frac{x_{j+1}}{x_j}, \ldots, \frac{x_n}{x_j}\right), \quad j = 1, \ldots, n, \tag{5.19}$$

$$\lambda\left(x_1, \ldots, x_n\right) = \sum_{j=1}^{n} x_j \phi\left(\frac{x_1}{x_j}, \ldots, \frac{x_{j-1}}{x_j}, \frac{x_{j+1}}{x_j}, \ldots, \frac{x_n}{x_j}\right). \tag{5.20}$$

To study the function ϕ, put $x_j = x$ into Eq. (5.20) and compare it with equality (5.9),

$$\phi(1, \ldots, 1) = \frac{1}{n}. \tag{5.21}$$

Function $\phi\left(\frac{x_1}{x_j}, \ldots, \frac{x_{j-1}}{x_j}, \frac{x_{j+1}}{x_j}, \ldots, \frac{x_n}{x_j}\right)$ is invariant upon the action of symmetric group S_{n-1} on the Euclidean space $\mathbb{E}_j^{n-1} = \{x_1, \ldots, x_{j-1}, x_{j+1}, \ldots, x_n\}$ by permutation of $(n-1)$ variables.

Represent Eq. (5.7) as follows

$$\lambda(x_1, \ldots, x_n) \cdot \lambda\left(u_1, \ldots, u_n\right) = 1, \qquad u_j = x_j^{-1},$$

and differentiate it w.r.t. the variable x_j,

$$\frac{1}{\lambda}\frac{\partial \lambda}{\partial x_j} - \frac{\lambda}{x_j^2}\frac{\partial \lambda}{\partial u_i} = 0, \qquad \text{or}$$

$$x_j^2\phi\left(\frac{x_1}{x_j}, \ldots, \frac{x_{j-1}}{x_j}, \frac{x_{j+1}}{x_j}, \ldots, \frac{x_n}{x_j}\right) = \lambda^2\phi\left(\frac{u_1}{u_j}, \ldots, \frac{u_{j-1}}{u_j}, \frac{u_{j+1}}{u_j}, \ldots, \frac{u_n}{u_j}\right) \tag{5.22}$$

Substitute $u_j = 1/x_j$ into (5.22),

$$x_j^2\phi\left(\frac{x_1}{x_j}, \ldots, \frac{x_{j-1}}{x_j}, \frac{x_{j+1}}{x_j}, \ldots, \frac{x_n}{x_j}\right) = \lambda^2\phi\left(\frac{x_j}{x_1}, \ldots, \frac{x_j}{x_{j-1}}, \frac{x_j}{x_{j+1}}, \ldots, \frac{x_j}{x_n}\right) \tag{5.23}$$

Let, by the way of contradiction, there exists a point $P_1 = \{x_1, \ldots, x_n\}$, where $x_j = a_j > 0$, such that $\partial\lambda/\partial x_j = 0$. Then, according to definition (5.19) we get

$$\phi\left(\frac{a_1}{a_j}, \ldots, \frac{a_{j-1}}{a_j}, \frac{a_{j+1}}{a_j}, \ldots, \frac{a_n}{a_j}\right) = 0. \tag{5.24}$$

On the other hand, according to (5.23), the coordinates a_j satisfy another equation,

$$\phi\left(\frac{a_j}{a_1}, \ldots, \frac{a_j}{a_{j-1}}, \frac{a_j}{a_{j+1}}, \ldots, \frac{a_j}{a_n}\right) = 0. \tag{5.25}$$

Combining (5.24) and (5.25), we arrive at equalities $a_1 = \ldots, a_n = a$ and $\phi(1, \ldots, 1) = 0$, where the last equality contradicts (5.21). By other words, no one derivative $\partial\lambda/\partial x_j$ does vanish in the whole hyperoctant $\mathbb{E}^n_>$.

In order to fix a sign of these derivatives, it is necessary to recall (5.21) again, i.e., $\partial\lambda/\partial x_j > 0$ if $x_i = x_j$, $i, j = 1, \ldots, n$. Thus, all derivatives $\partial\lambda/\partial x_j$ are positive in $\mathbb{E}^n_>$. □

6 Proper Self-Dual Symmetric Polynomials

Making use of formulas (2.4–2.8), consider self-dual polynomials without middle terms $A_{s_*}\binom{\mathbf{x^n}}{m, S_n}$ and $B_{s_*}\binom{\mathbf{x^n}}{m, S_n}$ given in (2.10) for $mn = 0 \pmod 2$, if they do exist,

$$\mathsf{S}\binom{\lambda,\ \mathbf{x^n}}{m, S_n} = \sum_{s=0}^{\lfloor s_* \rfloor} \sum_{l=1}^{P_n^m(s)} \mathcal{S}_{m,n}^{s,\,l}\, \Theta_-^{s,\,l}\binom{\lambda,\ \mathbf{x^n}}{m, S_n}, \quad \mathcal{S}_{m,n}^{s,\,l},\ \mathcal{R}_{m,n}^{s,\,l} \ge 0, \tag{6.1}$$

$$\mathsf{R}\binom{\lambda,\ \mathbf{x^n}}{m, S_n} = \sum_{s=0}^{\lfloor s_* \rfloor} \sum_{l=1}^{P_n^m(s)} \mathcal{R}_{m,n}^{s,\,l}\, \Theta_+^{s,\,l}\binom{\lambda,\ \mathbf{x^n}}{m, S_n}, \quad \mathcal{S}_{m,n}^{s_*,\,l} = \mathcal{R}_{m,n}^{s_*,\,l} = 0,\ mn = 0 \pmod 2,$$

$$\Theta_\pm^{s,\,l}\binom{\lambda,\ \mathbf{x^n}}{m, S_n} = \lambda^s \prod_{r=0}^{n} I_{n,r}^{\alpha_r\binom{s,\ l}{m, S_n}} \left[\lambda^{mn-2s} \pm \Gamma\binom{s,\ l}{m, S_n}\right], \tag{6.2}$$

where

$$\Gamma\binom{s,\ l}{m, S_n} = \prod_{r=0}^{n} \left(\frac{I_{n,n-r}}{I_{n,r}}\right)^{\alpha_r\binom{s,\ l}{m, S_n}} = \frac{I_{n,n}^{\alpha_0}}{I_{n,1}^{\alpha_1}}\, \frac{I_{n,n-1}^{\alpha_1}}{I_{n,2}^{\alpha_2}} \cdot \ldots \cdot \frac{I_{n,2}^{\alpha_{n-2}}}{I_{n,n-1}^{\alpha_{n-1}}}\, \frac{I_{n,1}^{\alpha_{n-1}}}{I_{n,n}^{\alpha_n}}. \tag{6.3}$$

For short, in (6.3) we denote $\alpha_r = \alpha_r\binom{s,\ l}{m, S_n}$. Polynomials $\Theta_\pm^{s,\,l}\binom{\lambda,\ \mathbf{x^n}}{m, S_n}$ satisfy duality relations,

$$\lambda^{mn} I_{n,n}^{m}\, \widehat{\mathbf{Q}}_n\, \Theta_{\pm}^{s,\,l}\binom{\lambda,\,\mathbf{x^n}}{m,S_n} = \pm\, \Theta_{\pm}^{s,\,l}\binom{\lambda,\,\mathbf{x^n}}{m,S_n}.$$

A product $\Gamma\binom{s,\,l}{m,S_n}$ may be written in a more short way,

$$\Gamma\binom{s,\,l}{m,S_n} = \prod_{r=0}^{\lfloor n/2\rfloor}\left(\frac{I_{n,n-r}}{I_{n,r}}\right)^{\theta_r\binom{s,\,l}{m,S_n}}, \tag{6.4}$$

$$\theta_r\binom{s,\,l}{m,S_n} = \alpha_r - \alpha_{n-r}, \qquad \theta_r + \theta_{n-r} = 0,$$

where we denote $\theta_r = \theta_r\binom{s,\,l}{m,S_n}$. Due to the absence of middle terms (6.2), the only terms contributing to $\mathsf{S}\binom{\lambda,\,\mathbf{x^n}}{m,S_n}$ and $\mathsf{R}\binom{\lambda,\,\mathbf{x^n}}{m,S_n}$ are those with positive difference $mn - 2s$, which read,

$$\sum_{r=0}^{\lfloor n/2\rfloor}(n-2r)\theta_r = mn - 2s > 0, \qquad \sum_{r=0}^{n}\theta_r = 0. \tag{6.5}$$

Recall the Hardy-Littlewood-Pólya inequality ([8], Thm77) for basic polynomial invariants $I_{n,r}$.

Theorem 5 ([8], p. 64) *Let $a_1, \ldots, a_n$ and $b_1, \ldots, b_n$ be non-negative numbers. Then inequality*

$$\prod_{r=1}^{n}\left(\frac{I_{n,r}}{\binom{n}{r}}\right)^{a_r} \le \prod_{r=1}^{n}\left(\frac{I_{n,r}}{\binom{n}{r}}\right)^{b_r}, \tag{6.6}$$

holds if and only if the degrees a_r, b_r are satisfied $n-1$ inequalities

$$a_r + 2a_{r+1} + \ldots + (n-r+1)a_n \ge b_r + 2b_{r+1} + \ldots + (n-r+1)b_n, \tag{6.7}$$

where $1 < r \le n$, and one equality

$$a_1 + 2a_2 + \ldots + na_n = b_1 + 2b_2 + \ldots + nb_n. \tag{6.8}$$

Apply Theorem 5 to find the upper bound for $\Gamma\binom{s,\,l}{m,S_n}$.

Theorem 6 *The following inequality*

$$\Gamma\binom{s,\,l}{m,S_n} \le \left(\frac{I_{n,1}}{n}\right)^{mn-2s}, \tag{6.9}$$

holds if and only if θ_r in (6.4) satisfy a set of $n-1$ inequalities,

$$\text{if } 0 \le r < \lfloor n/2 \rfloor, \text{ then} \quad (6.10)$$
$$(r+1)\theta_0 + r\theta_1 + (r-1)\theta_2 + \ldots + 3\theta_{r-2} + 2\theta_{r-1} + \theta_r \ge 0,$$

$$\text{if } r = \lfloor n/2 \rfloor, \text{ then} \quad (6.11)$$
$$n\theta_0 + (n-2)\theta_1 + (n-4)\theta_2 + \ldots + (n - 2\lfloor n/2 \rfloor)\,\theta_{\lfloor n/2 \rfloor} = mn - 2s,$$

$$\text{if } \lfloor n/2 \rfloor < r < n-1, \quad n = 2p+1, \quad p < r < 2p, \quad p \ge 2, \text{ then} \quad (6.12)$$
$$(r+1)\theta_0 + r\theta_1 + \ldots + 2(r-p+1)\theta_{2p-r-1} + (2r-2p+1)\theta_{2p-r} +$$
$$\ldots + 3\theta_{p-1} + \theta_p \ge 0,$$

$$\text{if } \lfloor n/2 \rfloor < r < n-1, \quad n = 2q, \quad q < r < 2q-1, \quad q \ge 3, \text{ then} \quad (6.13)$$
$$(r+1)\theta_0 + r\theta_1 + \ldots + (2r-2q+3)\theta_{2q-r-2} +$$
$$2(r-q+1)\theta_{2q-r-1} + \ldots + 4\theta_{q-2} + 2\theta_{q-1} \ge 0.$$

Proof Bearing in mind a combinatorial identity $\binom{n}{r} = \binom{n}{n-r}$, in order to prove (6.9) we have to prove, according to Theorem 5 and Rep (6.3), the following inequality,

$$\left(\frac{I_{n,n}}{\binom{n}{n}}\right)^{\alpha_0} \left(\frac{I_{n,n-1}}{\binom{n}{n-1}}\right)^{\alpha_1} \cdot \ldots \cdot \left(\frac{I_{n,2}}{\binom{n}{2}}\right)^{\alpha_{n-2}} \left(\frac{I_{n,1}}{\binom{n}{1}}\right)^{\alpha_{n-1}} \le \quad (6.14)$$
$$\left(\frac{I_{n,1}}{\binom{n}{1}}\right)^{\alpha_1+c} \left(\frac{I_{n,2}}{\binom{n}{2}}\right)^{\alpha_2} \cdot \ldots \cdot \left(\frac{I_{n,n-1}}{\binom{n}{n-1}}\right)^{\alpha_{n-1}} \left(\frac{I_{n,n}}{\binom{n}{n}}\right)^{\alpha_n}, \quad c = mn - 2s.$$

Comparing two inequalities (6.14) and (6.6), we arrive at the set of relations,

$$a_1 = \alpha_{n-1}, \quad a_2 = \alpha_{n-2}, \quad \ldots, \quad a_{n-1} = \alpha_1, \quad a_n = \alpha_0, \quad (6.15)$$
$$b_1 = \alpha_1 + c, \quad b_2 = \alpha_2, \quad \ldots, \quad b_{n-1} = \alpha_{n-1}, \quad b_n = \alpha_n.$$

Substituting (6.15) into (6.7) we obtain

$$\alpha_{n-r} + 2\alpha_{n-r-1} + \ldots + (n-r+1)\alpha_0 \ge \alpha_r + 2\alpha_{r+1} + \ldots + (n-r+1)\alpha_n, \quad (6.16)$$

where $1 < r \le n$, and could be presented as a set of $n-1$ inequalities (6.10, 6.12) if $n = 1 \pmod 2$ and a set of $n-1$ inequalities (6.10, 6.13) if $n = 0 \pmod 2$. Substituting (6.15) into (6.8) we obtain

$$c = \sum_{r=0}^{n/2} (n-2r)(\alpha_r - \alpha_{n-r}). \quad (6.17)$$

A straightforward calculation of a sum in (6.17) gives

$$c=\sum_{r=0}^{n/2}(n-2r)\alpha_r-\sum_{r=n/2}^{n}(2r-n)\alpha_r=\sum_{r=0}^{n}(n-2r)\alpha_r=mn-2s,$$

that proves (6.14). A direct comparison of relations (6.10) and (6.16) shows their complete coincidence that finishes the proof of theorem. □

Example 1 The products $\Gamma\binom{s,\ l}{m,S_n}$ in the Rep (6.4): the first three $\Gamma\binom{28,\ l}{7,\ S_9}$, $\Gamma\binom{30,\ l}{8,\ S_9}$ and $\Gamma\binom{26,\ l}{9,\ S_9}$ do satisfy Theorem 6, while the last one $\Gamma\binom{19,\ l}{5,\ S_9}$ does not.

$$\Gamma\binom{28,\ l}{7,\ S_9}=I_{9,9}\frac{I_{9,8}}{I_{9,1}}\left(\frac{I_{9,7}}{I_{9,2}}\right)^{-2}\frac{I_{9,6}}{I_{9,3}}\left(\frac{I_{9,5}}{I_{9,4}}\right)^{-2},$$
$$\Gamma\binom{30,\ l}{8,\ S_9}=\left(\frac{I_{9,8}}{I_{9,1}}\right)^{2}\left(\frac{I_{9,7}}{I_{9,2}}\right)^{-1}\left(\frac{I_{9,6}}{I_{9,3}}\right)^{2}\left(\frac{I_{9,5}}{I_{9,4}}\right)^{-3},$$
$$\Gamma\binom{26,\ l}{9,\ S_9}=I_{9,9}^{3}\left(\frac{I_{9,8}}{I_{9,1}}\right)^{-1}\left(\frac{I_{9,7}}{I_{9,2}}\right)^{2}\left(\frac{I_{9,6}}{I_{9,3}}\right)^{-1}\left(\frac{I_{9,5}}{I_{9,4}}\right)^{2},$$
$$\Gamma\binom{19,\ l}{5,\ S_9}=\frac{1}{I_{9,9}}\frac{I_{9,8}}{I_{9,1}}\frac{I_{9,7}}{I_{9,2}}\frac{I_{9,6}}{I_{9,3}}\frac{I_{9,5}}{I_{9,4}}.$$

Definition 5 Self-dual polynomials, defined in (6.1, 6.2, 6.3), are called proper and denoted by $\mathsf{S}_{\odot}\binom{\lambda,\ \mathbf{x^n}}{m,S_n}$ and $\mathsf{R}_{\odot}\binom{\lambda,\ \mathbf{x^n}}{m,S_n}$ if they satisfy inequality (6.9),

$$\mathsf{S}_{\odot}\binom{\lambda,\ \mathbf{x^n}}{m,S_n}=\left\{\mathsf{S}\binom{\lambda,\ \mathbf{x^n}}{m,S_n}\ \middle|\ \mathcal{S}_{m,n}^{s,\ l}\geq 0;\ \mathcal{S}_{m,n}^{s_*,\ l}=0,\ mn=0\ (\mathrm{mod}\ 2);\ \ \text{condition (6.9)}\right\},\tag{6.18}$$
$$\mathsf{R}_{\odot}\binom{\lambda,\ \mathbf{x^n}}{m,S_n}=\left\{\mathsf{R}\binom{\lambda,\ \mathbf{x^n}}{m,S_n}\ \middle|\ \mathcal{R}_{m,n}^{s,\ l}\geq 0;\ \mathcal{R}_{m,n}^{s_*,\ l}=0,\ mn=0\ (\mathrm{mod}\ 2);\ \ \text{condition (6.9)}\right\}.$$

The proper self-dual polynomials (6.18) are no more generic, due to vanishing of the middle terms (6.2) and coefficients $\mathcal{S}_{m,n}^{s,\ l_\ominus}$, $\mathcal{R}_{m,n}^{s,\ l_\ominus}$ for those $l_\ominus$, where inequality (6.9) is broken,

$$\mathcal{S}_{m,n}^{s,\ l_\ominus}=\mathcal{R}_{m,n}^{s,\ l_\ominus}=0,\quad \text{if}\quad \Gamma\binom{s,\ l_\ominus}{m,S_n}>\left(I_{n,1}/n\right)^{mn-2s}.\tag{6.19}$$

E.g., according to Example 1, the generic scew-reciprocal polynomial $\mathsf{S}\binom{\lambda,\ \mathbf{x^9}}{5,\ S_9}$ is not a proper.

Theorem 7 *Let $\lambda_{\odot}(\mathbf{x^n})$ be a real solution of equation $\mathsf{S}_{\odot}\binom{\lambda,\ \mathbf{x^n}}{m,S_n}=0$. Then $\lambda_{\odot}(\mathbf{x^n})$ is bounded by arithmetic and harmonic means of the set $\{x_1,\ldots,x_n\}$,*

$$\frac{n I_{n,n}}{I_{n,n-1}} \le \lambda_{\odot}\left(\mathbf{x^n}\right) \le \frac{I_{n,1}}{n}, \tag{6.20}$$

Proof Write a proper scew-reciprocal equation according to (6.1),

$$\sum_{s=0}^{\lfloor s_* \rfloor} \sum_{l=1}^{P_n^m(s)} \mathcal{S}_{m,n}^{s,\,l} \lambda_{\odot}^s \prod_{r=0}^{n} I_{n,r}^{\alpha_r\binom{s,\,l}{m,S_n}} \left[\lambda_{\odot}^{mn-2s} - \Gamma\binom{s,\,l}{m,S_n}\right] = 0, \tag{6.21}$$

and apply inequality (6.9),

$$\sum_{s=0}^{\lfloor s_* \rfloor} \sum_{l=1}^{P_n^m(s)} \mathcal{S}_{m,n}^{s,\,l} \lambda_{\odot}^s \prod_{r=0}^{n} I_{n,r}^{\alpha_r\binom{s,\,l}{m,S_n}} \left[\lambda_{\odot}^{mn-2s} - \left(\frac{I_{n,1}}{n}\right)^{mn-2s}\right] \le 0. \tag{6.22}$$

A difference in the brackets (6.22) has an exact divisor $\lambda_{\odot} - I_{n,1}/n$, while a quotient of such division is always positive. Thus, we arrive at the r.h.s. of inequality (6.20),

$$\lambda_{\odot}\left(\mathbf{x^n}\right) \le \frac{I_{n,1}}{n}. \tag{6.23}$$

The l.h.s. of inequality (6.20) follows if we apply to (6.23) the duality transformation $\widehat{\mathbf{Q}}_n$, inverting all x_i and $\lambda_{\odot}$, i.e., $\lambda_{\odot}^{-1} \le I_{n,n-1}/(n I_{n,n})$. □

We call the bounds in (6.20) *the weak universal bounds* for $\lambda_{\odot}\left(\mathbf{x^n}\right)$.

Example 2 illustrates an improper polynomial $\mathsf{S}\binom{\lambda,\,\mathbf{x^3}}{6,\,S_3}$ where inequality (6.9) is broken.

Example 2 Choose the polynomial in a non-generic form, which does not satisfy Theorem 6,

$$\mathsf{S}\binom{\lambda,\,\mathbf{x^3}}{6,\,S_3} = \lambda^{18} + K\lambda^8\left(\lambda^2 I_{3,1}^5 I_{3,3} - I_{3,2}^5\right) - I_{3,3}^6,$$
$$\mathcal{S}_{6,3}^{s,\,l} = 0,\ s \ne 8, \quad \mathcal{S}_{6,3}^{8,\,l} = K > 0.$$

Its unique positive root $\lambda(x_1, x_2, 0)$ may be calculated explicitly, namely, $\lambda(x_1, x_2, 0) = \sqrt[10]{K}\sqrt{x_1 x_2}$. It does not satisfy inequality (6.20) if $K > (2/3)^{10} \simeq 0.017$.

6.1 Multiplication of Proper Self-Dual Polynomials

Denote by $\mathbb{R}_{\odot}\binom{\lambda,}{m,S_n}$ and $\mathbb{S}_{\odot}\binom{\lambda,}{m,S_n}$ two sets of proper self-dual polynomials, defined in (6.18), with fixed n and m and different sets of parameters $\mathcal{R}_{m,n}^{s,\,l}$, $\mathcal{S}_{m,n}^{s,\,l}$

$$\mathbb{R}_{\odot}\binom{\lambda,}{m,S_n} = \left\{\mathsf{R}_{\odot}\binom{\lambda,\,\mathbf{x^n}}{m,S_n}\right\}, \quad \mathbb{S}_{\odot}\binom{\lambda,}{m,S_n} = \left\{\mathsf{S}_{\odot}\binom{\lambda,\,\mathbf{x^n}}{m,S_n}\right\}, \tag{6.24}$$

and by $\mathfrak{R}_{\odot}(\lambda, S_n)$, $\mathfrak{S}_{\odot}(\lambda, S_n)$ and $\mathfrak{R}_{\odot}\mathfrak{S}_{\odot}(\lambda, n)$ three unions of sets $\mathbb{R}_{\odot}\binom{\lambda,}{m,S_n}$ and $\mathbb{S}_{\odot}\binom{\lambda,}{m,S_n}$,

$$\mathfrak{R}_{\odot}(\lambda, S_n) = \bigcup_{m\geq 0} \mathbb{R}_{\odot}\binom{\lambda,}{m,S_n}, \qquad \mathfrak{S}_{\odot}(\lambda, S_n) = \bigcup_{m\geq 0} \mathbb{S}_{\odot}\binom{\lambda,}{m,S_n},$$
$$\mathfrak{R}_{\odot}\mathfrak{S}_{\odot}(\lambda, S_n) = \mathfrak{R}_{\odot}(\lambda, S_n) \cup \mathfrak{S}_{\odot}(\lambda, S_n). \tag{6.25}$$

By definitions (6.18, 6.24) and (6.25), the following containments hold

$$\mathfrak{R}_{\odot}(\lambda, S_n) \subset \mathfrak{R}(\lambda, S_n), \quad \mathfrak{S}_{\odot}(\lambda, S_n) \subset \mathfrak{S}(\lambda, S_n),$$
$$\mathfrak{R}_{\odot}\mathfrak{S}_{\odot}(\lambda, S_n) \subset \mathfrak{R}\mathfrak{S}(\lambda, S_n). \tag{6.26}$$

Keeping in mind Theorem 4 and Corollary 2, consider how do the requirements (6.18) affect the multiplication of proper self-dual polynomials.

Theorem 8 *Neither* $\mathfrak{R}_{\odot}(\lambda, S_n)$ *nor* $\mathfrak{R}_{\odot}\mathfrak{S}_{\odot}(\lambda, S_n)$ *exhibit a commutative monoid.*

Proof In view of Theorem 4 and containments (6.26), it is enough to prove the first part of the statement in Theorem 8.

Start with multiplication of two proper self-reciprocal polynomials $\mathsf{R}_{1\odot}\binom{\lambda,\,\mathbf{x^n}}{m,S_n}$ and $\mathsf{R}_{2\odot}\binom{\lambda,\,\mathbf{x^n}}{m,S_n}$ with different sets of non-negative coefficients ${}_1\mathcal{R}^{s,\,l}_{m,n}$, ${}_2\mathcal{R}^{s,\,l}_{m,n} \geq 0$, which vanished if $s = s_*$. According to (5.1, 5.2, 5.3), we get

$$\mathsf{R}_{1\odot}\binom{\lambda,\,\mathbf{x^n}}{m,S_n}\mathsf{R}_{2\odot}\binom{\lambda,\,\mathbf{x^n}}{m,S_n} =$$
$$\sum_{s_1,s_2=0}^{\lfloor s_*\rfloor}\sum_{l_1=1}^{P_n^m(s_1)}\sum_{l_2=1}^{P_n^m(s_2)} {}_1\mathcal{R}^{s_1,l_1}_{m,n}\,{}_2\mathcal{R}^{s_2,l_2}_{m,n}\,\lambda^{s_1+s_2}\,\Pi_{\oplus}\binom{s_1,m,\,l_1}{s_2,m,\,l_2}, \tag{6.27}$$

where

$$\Pi_{\oplus}\binom{s_1,m,\,l_1}{s_2,m,\,l_2} = \prod_{r=0}^{n} I_{n,r}^{(\alpha_r\binom{s_1,\,l_1}{m,S_n}+\alpha_r\binom{s_2,\,l_2}{m,S_n})}\left[\lambda^{2(mn-s_1-s_2)} + \Gamma\binom{s_1,\,l_1}{m,S_n}\Gamma\binom{s_2,\,l_2}{m,S_n}\right] +$$
$$\prod_{r=0}^{n} I_{n,r}^{\alpha_r\binom{s_1,\,l_1}{m,S_n}+\alpha_r\binom{s_2,\,l_2}{m,S_n}}\left[\Gamma\binom{s_1,\,l_1}{m,S_n}\lambda^{mn-2s_2} + \Gamma\binom{s_2,\,l_2}{m,S_n}\lambda^{mn-2s_1}\right]. \tag{6.28}$$

A brief analysis of formulas (6.27, 6.28) shows that $\mathsf{R}_{1\odot}\binom{\lambda,\,\mathbf{x^n}}{m,S_n}\mathsf{R}_{2\odot}\binom{\lambda,\,\mathbf{x^n}}{m,S_n}$ is a non-proper self-reciprocal polynomial in λ of degree $2mn$. Indeed, when $s = s_1 = s_2$ in (6.28), the r.h.s. in formula (6.27) contains a non-vanishing term of degree mn,

$$\lambda^{mn} \sum_{s=0}^{\lfloor s_* \rfloor} \sum_{l_1,l_2=1}^{P_n^m(s)} {}_1\mathcal{R}_{m,n}^{s,l_1}\, {}_2\mathcal{R}_{m,n}^{s,l_2} \times$$
$$\prod_{r=0}^{n} I_{n,r}^{\alpha_r\binom{s,\,l_1}{m,S_n}+\alpha_r\binom{s,\,l_2}{m,S_n}} \left[\Gamma\binom{s,\,l_1}{m,S_n} + \Gamma\binom{s,\,l_2}{m,S_n}\right]. \tag{6.29}$$

Then by (6.24) and Definition 5, the following containment is forbidden

$$\mathsf{R}_{1\odot}\binom{\lambda,\,\mathbf{x^n}}{m,S_n}\mathsf{R}_{2\odot}\binom{\lambda,\,\mathbf{x^n}}{m,S_n} \not\subset \mathbb{R}_\odot\binom{\lambda,}{2m,S_n}. \tag{6.30}$$

Thus, by (6.25), neither $\mathfrak{R}_\odot(\lambda, S_n)$ nor $\mathfrak{R}_\odot\mathfrak{S}_\odot(\lambda, S_n)$ exibit a commutative monoid. □

7 Strong Universal Bounds for $\lambda_\odot(\mathbf{x^n})$ with Small n

In this Section we show that the upper and lower bounds in Theorem 7 may be improved essentially for small $n = 3, 4$.

7.1 $n = 3$

Consider a proper scew-reciprocal polynomial $\mathsf{S}_\odot\binom{\lambda,\,\mathbf{x^3}}{m,S_3}$ and let $\lambda_\odot(\mathbf{x^3})$ be its unique positive root. According to (6.1, 6.4), the polynomial $\Theta_-^{s,\,l}\binom{\lambda,\,\mathbf{x^3}}{m,S_3}$ reads,

$$\Theta_-^{s,\,l}\binom{\lambda,\,\mathbf{x^3}}{m,S_3} = \lambda_\odot^s \prod_{r=0}^{3} I_{3,r}^{\alpha_r\binom{s,\,l}{m,S_3}} \left[\lambda_\odot^{3m-2s} - \Gamma\binom{s,\,l}{m,S_3}\right], \tag{7.1}$$
$$\Gamma\binom{s,\,l}{m,S_3} = I_{3,3}^{\theta_0}\left(\frac{I_{3,2}}{I_{3,1}}\right)^{\theta_1},$$

where $3m - 2s = 3\theta_0 + \theta_1 > 0$ according to (6.5) and $\theta_0 \geq 0$ according to (6.10).

Introduce a function $\Upsilon(\rho_3, \mathbf{x^3}) = \Gamma^{\beta_3}\binom{s,\,l}{m,S_3}$, where $\beta_3 = (3\theta_0 + \theta_1)^{-1}$, i.e.,

$$\Upsilon(\rho_3, \mathbf{x^3}) = \left[I_{3,3}\left(\frac{I_{3,2}}{I_{3,1}}\right)^{\rho_3}\right]^{1/(3+\rho_3)}, \quad \rho_3 = \frac{\theta_1}{\theta_0}, \quad -3 < \rho_3 \leq \infty, \tag{7.2}$$

and define its upper $\Omega\left(\mathbf{x}^3\right)$ and lower $\omega\left(\mathbf{x}^3\right)$ bounds,

$$\begin{aligned}\Omega\left(\mathbf{x}^3\right) &= \max\{\Upsilon\left(\rho_3, \mathbf{x}^3\right), \ \rho_\mu \le \rho_3 \le \infty\},\\ \omega\left(\mathbf{x}^3\right) &= \min\{\Upsilon\left(\rho_3, \mathbf{x}^3\right), \ \rho_\mu \le \rho_3 \le \infty\},\end{aligned} \tag{7.3}$$

where $[\rho_\mu, \infty]$ is a range of a positive monotony of $\Upsilon\left(\rho_3, \mathbf{x}^3\right)$ in any variable x_i, and $-3 < \rho_\mu$.

Prove that $\rho_\mu = -1$. For this purpose calculate $\partial\Upsilon\left(\rho_3, \mathbf{x}^3\right)/\partial x_1$,

$$\begin{aligned}\frac{\partial\Upsilon\left(\rho_3, \mathbf{x}^3\right)}{\partial x_1} &= \frac{\Upsilon\left(\rho_3, \mathbf{x}^3\right)}{3+\rho_3}\frac{V\left(\rho_3, \mathbf{x}^3\right)}{x_1 I_{3,2} I_{3,1}},\\ V\left(\rho_3, \mathbf{x}^3\right) &= I_{3,1}I_{3,2} + \rho_3 x_1(x_2^2 + x_3^2 + x_2x_3),\end{aligned} \tag{7.4}$$

and require its positiveness. The derivative in (7.4) is positive if $0 \le \rho_3 \le \infty$. However, this range might be enlarged up to $-1 \le \rho_3 \le \infty$. Indeed, assume $\rho_3 = -1 + \varepsilon_3$, then

$$V\left(\rho_3, \mathbf{x}^3\right) = (1-\varepsilon_3)(x_1 I_{3,2} + x_2x_3 I_{3,1}) + \varepsilon_3 I_{3,1}I_{3,2} \ge 0, \qquad \text{if} \quad 0 \le \varepsilon_3 \le 1.$$

On the other hand, if $\varepsilon_3 < 0$ then we can choose such values $x_3 \to 0$ and $x_1/x_2 + \varepsilon_3 < 0$ that

$$V\left(\rho_3, \mathbf{x}^3\right) = (1-\varepsilon_3)x_1^2x_2 + \varepsilon_3 x_1x_2(x_1+x_2) = x_1x_2^2\left(\varepsilon_3 + \frac{x_1}{x_2}\right) < 0.$$

Thus, a range of a positive monotony of $\Upsilon\left(\rho_3, \mathbf{x}^3\right)$ in any variable is $-1 \le \rho_3 \le \infty$. To find the extremal values $\rho_{3*}^{\pm}$, corresponded to $\Omega\left(\mathbf{x}^3\right)$ and $\omega\left(\mathbf{x}^3\right)$, we check out the derivative w.r.t. ρ_3,

$$\frac{(3+\rho_3)^2}{\Upsilon\left(\rho_3, \mathbf{x}^3\right)} \cdot \frac{\partial\Upsilon\left(\rho_3, \mathbf{x}^3\right)}{\partial\rho_3} = \ln\left(\left(\frac{I_{3,2}}{I_{3,1}}\right)^3 \frac{1}{I_{3,3}}\right) = \begin{cases} > 0, \ I_{3,2}^3 > I_{3,3}I_{3,1}^3, \\ = 0, \ I_{3,2}^3 = I_{3,3}I_{3,1}^3, \\ < 0, \ I_{3,2}^3 < I_{3,3}I_{3,1}^3, \end{cases} \tag{7.5}$$

where, by Lemma 12, an equality $I_{3,3}I_{3,1}^3 = I_{3,2}^3$ in (7.5) is equivalent to another equality

$$(x_1^2 - x_2x_3)(x_2^2 - x_3x_1)(x_3^2 - x_1x_2) = 0. \tag{7.6}$$

Relations (7.5) manifests a decomposition of the Euclidean space $\mathbb{E}_>^3$

$$\mathbb{E}_>^3 = \mathbb{P}_+^3 \cup \mathbb{P}_0^2 \cup \mathbb{P}_-^3$$

in three different subspaces

$$\begin{aligned} \mathbb{P}_{+}^{3} &= \left\{\mathbb{E}_{>}^{3} \mid I_{3,2}^{3} > I_{3,3} I_{3,1}^{3}\right\}, \quad \dim \mathbb{P}_{+}^{3} = 3, \\ \mathbb{P}_{0}^{2} &= \left\{\mathbb{E}_{>}^{3} \mid I_{3,2}^{3} = I_{3,3} I_{3,1}^{3}\right\}, \quad \dim \mathbb{P}_{0}^{2} = 2, \\ \mathbb{P}_{-}^{3} &= \left\{\mathbb{E}_{>}^{3} \mid I_{3,2}^{3} < I_{3,3} I_{3,1}^{3}\right\}, \quad \dim \mathbb{P}_{-}^{3} = 3, \end{aligned} \tag{7.7}$$

where $\mathbb{P}_0^2$ is a surface which separates $\mathbb{P}_+^3$ and $\mathbb{P}_-^3$ and is given by Eq. (7.6). In every subspace the function $\Upsilon\left(\rho_3, \mathbf{x}^3\right)$ is monotone: increasing with ρ_3 in $\mathbb{P}_+^3$, decreasing with ρ_3 in $\mathbb{P}_-^3$ and independent of ρ_3 in $\mathbb{P}_0^2$.

This allows to assign the extremal values of $\Upsilon\left(\rho_3, \mathbf{x}^3\right)$ to the endpoints of the monotony range, $[-1, \infty]$. Substituting the latter into (7.2), we obtain the bounds

$$\begin{aligned} \Omega\left(\mathbf{x}^3\right) &= \max\left\{\frac{I_{3,2}}{I_{3,1}}, \left(\frac{I_{3,1} I_{3,3}}{I_{3,2}}\right)^{1/2}\right\}, \\ \omega\left(\mathbf{x}^3\right) &= \min\left\{\frac{I_{3,2}}{I_{3,1}}, \left(\frac{I_{3,1} I_{3,3}}{I_{3,2}}\right)^{1/2}\right\}, \end{aligned} \tag{7.8}$$

which are stronger than those in Theorem 7.

$$a) \quad \frac{3 I_{3,3}}{I_{3,2}} \le \omega\left(\mathbf{x}^3\right), \qquad b) \quad \Omega\left(\mathbf{x}^3\right) \le \frac{I_{3,1}}{3}. \tag{7.9}$$

Indeed, making use of the Newton-Maclaurin inequalites [8] for symmetric polynomials $I_{3,r}$,

$$\frac{I_{3,1}}{3} \ge \sqrt{\frac{I_{3,2}}{3}} \ge \sqrt[3]{I_{3,3}}, \qquad \left(\frac{I_{3,2}}{3}\right)^2 \ge \frac{I_{3,1} I_{3,3}}{3},$$

we obtain

$$\frac{3 I_{3,3}}{I_{3,2}} \le \frac{I_{3,2}}{I_{3,1}}, \qquad \left(\frac{I_{3,1} I_{3,3}}{I_{3,2}}\right)^{1/2} \le \frac{I_{3,1}}{3},$$

that proves (7.9). The next Theorem presents *the strong universal bounds* for $\lambda_{\odot}\left(\mathbf{x}^3\right)$.

Theorem 9 *A unique positive solution* $\lambda_{\odot}\left(\mathbf{x}^3\right)$ *of equation* $\mathsf{S}_{\odot}\left(\begin{smallmatrix}\lambda, & \mathbf{x}^3 \\ m, & S_3\end{smallmatrix}\right) = 0$ *is bounded*

$$\omega\left(\mathbf{x}^3\right) \le \lambda_{\odot}\left(\mathbf{x}^3\right) \le \Omega\left(\mathbf{x}^3\right). \tag{7.10}$$

Proof We follow an approach, used in proof of Theorem 7. Consider a proper scew-reciprocal equation (6.21) and apply definitions of the function $\Upsilon\left(\rho_3, \mathbf{x}^3\right)$ in (7.2) and its bounds in (7.3). Then two inequalities follow,

$$\sum_{s=0}^{\lfloor s_* \rfloor} \sum_{l=1}^{P_3^m(s)} S_{m,3}^{s,\,l} \lambda_{\odot}^{s} \prod_{r=0}^{3} I_{3,r}^{\alpha_r \binom{s,\,l}{m,S_3}} \left[\lambda_{\odot}^{3m-2s} - \Omega^{3m-2s}\left(\mathbf{x^3}\right)\right] \le 0, \tag{7.11}$$

$$\sum_{s=0}^{\lfloor s_* \rfloor} \sum_{l=1}^{P_3^m(s)} S_{m,3}^{s,\,l} \lambda_{\odot}^{s} \prod_{r=0}^{3} I_{3,r}^{\alpha_r \binom{s,\,l}{m,S_3}} \left[\lambda_{\odot}^{3m-2s} - \omega^{3m-2s}\left(\mathbf{x^3}\right)\right] \ge 0. \tag{7.12}$$

The differences in the brackets (7.11) and (7.12) have exact divisors $\lambda_{\odot} - \Omega\left(\mathbf{x^3}\right)$ and $\lambda_{\odot} - \omega\left(\mathbf{x^3}\right)$, respectively, while a quotient of such division is always positive. Thus, we arrive at inequalities (7.10) and Theorem is proven. □

Formulas (7.8) and (7.10) are dual in the following sense:

$$\text{if} \quad \left(\frac{I_{3,1} I_{3,3}}{I_{3,2}}\right)^{1/2} \le \lambda_{\odot}\left(\mathbf{x^3}\right) \le \frac{I_{3,2}}{I_{3,1}}, \quad \mathbf{x^3} \in \mathbb{P}_+^3,$$

$$\text{then} \quad \frac{I_{3,2}}{I_{3,1}} \le \lambda_{\odot}\left(\mathbf{x^3}\right) \le \left(\frac{I_{3,1} I_{3,3}}{I_{3,2}}\right)^{1/2}, \quad \mathbf{x^3} \in \mathbb{P}_-^3,$$

and vice versa, i.e., the actual expressions will be exchanged for the lower and upper bounds in subspaces $\mathbb{P}_+^3$ and $\mathbb{P}_-^3$, while if $\mathbf{x^3} \in \mathbb{P}_0^2$ then $\Omega\left(\mathbf{x^3}\right) = \omega\left(\mathbf{x^3}\right)$.

7.2 $n = 4$

The case $n = 4$ may be treated similarly to the previous case $n = 3$. Consider a proper scew-reciprocal polynomial $\mathsf{S}_{\odot}\binom{\lambda,\,\mathbf{x^4}}{m,S_4}$ and let $\lambda_{\odot}\left(\mathbf{x^4}\right)$ be its unique positive root. According to (6.1, 6.4), the polynomial $\Theta_{-}^{s,\,l}\binom{\lambda,\,\mathbf{x^4}}{m,S_4}$ reads,

$$\Theta_{-}^{s,\,l}\binom{\lambda,\,\mathbf{x^4}}{m,S_4} = \lambda_{\odot}^{s} \prod_{r=0}^{4} I_{4,r}^{\alpha_r \binom{s,\,l}{m,S_4}} \left[\lambda_{\odot}^{4m-2s} - \Gamma\binom{s,\,l}{m,S_4}\right], \tag{7.13}$$

$$\Gamma\binom{s,\,l}{m,S_4} = I_{4,4}^{\theta_0} \left(\frac{I_{4,3}}{I_{4,1}}\right)^{\theta_1},$$

where $4m - 2s = 4\theta_0 + 2\theta_1 > 0$ according to (6.5) and $\theta_0 \ge 0$ according to (6.10).

Introduce a function $\Upsilon\left(\rho_4, \mathbf{x^4}\right) = \Gamma^{\beta_4}\binom{s,\,l}{m,S_4}$, where $\beta_4 = (4\theta_0 + 2\theta_1)^{-1}$, i.e.,

$$\Upsilon\left(\rho_4, \mathbf{x^4}\right) = \left[I_{4,4}\left(\frac{I_{4,3}}{I_{4,1}}\right)^{\rho_4}\right]^{1/(4+2\rho_4)}, \quad \rho_4 = \frac{\theta_1}{\theta_0}, \quad -2 < \rho_4 \le \infty, \tag{7.14}$$

and define its upper $\Omega\left(\mathbf{x}^4\right)$ and lower $\omega\left(\mathbf{x}^4\right)$ bounds,

$$\begin{aligned}\Omega\left(\mathbf{x}^4\right) &= \max\{\Upsilon\left(\rho_4, \mathbf{x}^4\right),\ \rho_\mu \le \rho_4 \le \infty\},\\ \omega\left(\mathbf{x}^3\right) &= \min\{\Upsilon\left(\rho_4, \mathbf{x}^4\right),\ \rho_\mu \le \rho_4 \le \infty\},\end{aligned} \tag{7.15}$$

where $[\rho_\mu, \infty]$ is a range of a positive monotony of $\Upsilon\left(\rho_4, \mathbf{x}^4\right)$ in any variable x_i, and $-2 < \rho_\mu$.

Let us find the first derivative $\partial\Upsilon/\partial x_1$ and require its positiveness,

$$\begin{aligned}&\frac{4+2\rho_4}{\Upsilon\left(\rho_4, \mathbf{x}^4\right)} \cdot \frac{\partial\Upsilon\left(\rho_4, \mathbf{x}^4\right)}{\partial x_1} = \\ &\frac{1}{x_1} + \rho_4\left(\frac{x_2x_3+x_3x_4+x_4x_2}{I_{4,3}} - \frac{1}{I_{4,1}}\right) = \frac{V\left(\rho_4, \mathbf{x}^4\right)}{x_1 I_{4,3} I_{4,1}},\end{aligned} \tag{7.16}$$

where

$$V\left(\rho_4, \mathbf{x}^4\right) = I_{4,1}I_{4,3} + \rho_4 x_1\left[x_2^2(x_3+x_4) + x_3^2(x_4+x_2+x_4^2(x_2+x_3) + 2x_2x_3x_4\right]$$

that gives monotony for $0 \le \rho_4 \le \infty$. This range can also be enlarged up to $-1 \le \rho \le \infty$. Indeed, define $\rho = -1 + \varepsilon_4$ and prove that if $0 \le \varepsilon_4 \le 1$, then $V\left(\rho_4, \mathbf{x}^4\right) \ge 0$,

$$\begin{aligned}V\left(\rho_4, \mathbf{x}^4\right) = (1-\varepsilon_4)\left[x_1^2(x_2x_3 + x_3x_4 + x_4x_2 + x_2x_3x_4(x_2+x_3+x_4) +\right.\\ \left. 2x_2x_3x_4\right] + \varepsilon_4 I_{4,1}I_{4,3} \ge 0,\end{aligned}$$

On the other hand, if $\varepsilon_4 < 0$ we can choose such finite values x_1, x_2, x_3 and $x_4 \to 0$ that

$$V_4\left(\rho, \mathbf{x}^4\right) = x_1^2x_2x_3\left[1 + \varepsilon_4\left(\frac{x_2}{x_1} + \frac{x_3}{x_1}\right)\right] \le 0, \quad \text{if} \quad \frac{1}{\varepsilon_4} + \frac{x_2}{x_1} + \frac{x_3}{x_1} \le 0.$$

Thus, a range of a positive monotony of $\Upsilon\left(\rho, \mathbf{x}^4\right)$ in any variable is $-1 \le \rho \le \infty$. To find the extremal values $\rho_{4*}^{\pm}$ corresponding to $\Omega\left(\mathbf{x}^4\right)$ and $\omega\left(\mathbf{x}^4\right)$, we should check the derivative w.r.t ρ

$$\frac{(4+2\rho)^2}{\Upsilon\left(\rho, \mathbf{x}^4\right)} \cdot \frac{\partial\Upsilon\left(\rho, \mathbf{x}^4\right)}{\partial\rho} = \ln\left[\left(\frac{I_{4,3}}{I_{4,1}}\right)^2 \frac{1}{I_{4,4}}\right] = \begin{cases} > 0, & I_{4,3}^2 > I_{4,4}I_{4,1}^2,\\ = 0, & I_{4,3}^2 = I_{4,4}I_{4,1}^2,\\ < 0, & I_{4,3}^2 < I_{4,4}I_{4,1}^2,\end{cases} \tag{7.17}$$

where, by Lemma 12, an equality $I_{4,4}I_{4,1}^2 = I_{4,3}^2$ in (7.5) is equivalent to another equality

$$(x_1x_2 - x_3x_4)(x_1x_3 - x_2x_4)(x_1x_4 - x_2x_3) = 0. \tag{7.18}$$

Relation (7.17) manifests a decomposition of the Euclidean space $\mathbb{E}^4_>$

$$\mathbb{E}^4_> = \mathbb{P}^4_+ \cup \mathbb{P}^3_0 \cup \mathbb{P}^4_-,$$

in three different subspaces

$$\begin{aligned} \mathbb{P}^4_+ &= \left\{\mathbb{E}^4_> \mid I^2_{4,3} > I_{4,4} I^2_{4,1}\right\}, \quad \dim \mathbb{P}^4_+ = 4, \\ \mathbb{P}^3_0 &= \left\{\mathbb{E}^4_> \mid I^2_{4,3} = I_{4,4} I^2_{4,1}\right\}, \quad \dim \mathbb{P}^3_0 = 3, \\ \mathbb{P}^4_- &= \left\{\mathbb{E}^4_> \mid I^2_{4,3} < I_{4,4} I^2_{4,1}\right\}, \quad \dim \mathbb{P}^4_- = 4, \end{aligned} \tag{7.19}$$

where $\mathbb{P}^3_0$ is a hypersurface, which separates $\mathbb{P}^4_+$ and $\mathbb{P}^4_-$ and is given by Eq. (7.18). In every subspace the function $\Upsilon\left(\rho, \mathbf{x^4}\right)$ is monotone: increasing with ρ_4 in $\mathbb{P}^4_+$, decreasing with ρ_4 in $\mathbb{P}^4_-$ and independent of ρ_4 in $\mathbb{P}^3_0$. This allows to assign the extremal values $\rho^{\pm}_{4*}$ to the endpoints of the monotony range, $\rho^{\pm}_{4*} = -1, \infty$. Substituting the latter into (7.15), we obtain the bounds

$$\begin{aligned} \Omega\left(\mathbf{x^4}\right) &= \max\left\{\left(\frac{I_{4,3}}{I_{4,1}}\right)^{1/2}, \left(\frac{I_{4,1} I_{4,4}}{I_{4,3}}\right)^{1/2}\right\}, \\ \omega\left(\mathbf{x^4}\right) &= \min\left\{\left(\frac{I_{4,3}}{I_{4,1}}\right)^{1/2}, \left(\frac{I_{4,1} I_{4,4}}{I_{4,3}}\right)^{1/2}\right\}, \end{aligned} \tag{7.20}$$

which are stronger than those in Theorem 7, i.e.,

$$a) \quad \frac{4 I_{4,4}}{I_{4,3}} \le \omega\left(\mathbf{x^4}\right), \qquad b) \quad \Omega\left(\mathbf{x^4}\right) \le \frac{I_{4,1}}{4}. \tag{7.21}$$

Indeed, making use of the Newton-Maclaurin inequalites [8] for symmetric polynomials $I_{4,r}$,

$$\frac{I_{4,1}}{4} \ge \sqrt{\frac{I_{4,2}}{6}} \ge \sqrt[3]{\frac{I_{4,3}}{4}} \ge \sqrt[4]{I_{4,4}}, \qquad \left(\frac{I_{4,2}}{6}\right)^2 \ge \frac{I_{4,1} I_{4,3}}{16}, \quad \left(\frac{I_{4,3}}{4}\right)^2 \ge \frac{I_{4,2} I_{4,4}}{6},$$

we obtain,

$$\frac{4 I_{4,4}}{I_{4,3}} \le \left(\frac{I_{4,3}}{I_{4,1}}\right)^{1/2}, \qquad \left(\frac{I_{4,1} I_{4,4}}{I_{4,3}}\right)^{1/2} \le \frac{I_{4,1}}{4},$$

that proves (7.21). The next Theorem presents *the strong universal bounds* for $\lambda_\odot\left(\mathbf{x^4}\right)$.

Theorem 10 *A unique positive solution* $\lambda_\odot\left(\mathbf{x^4}\right)$ *of equation* $\mathsf{S}_\odot\binom{\lambda,\, \mathbf{x^4}}{m,\, S_4} = 0$ *is bounded*

$$\omega\left(\mathbf{x^4}\right) \le \lambda_{\odot}\left(\mathbf{x^4}\right) \le \Omega\left(\mathbf{x^4}\right). \tag{7.22}$$

Proof Following an approach, used in proof of Theorem 9, consider a proper scew-reciprocal equation (6.21) and apply definitions of the function $\Upsilon\left(\rho_4, \mathbf{x^4}\right)$ in (7.14) and its bounds in (7.15).

Then two inequalities follow,

$$\sum_{s=0}^{\lfloor s_* \rfloor}\sum_{l=1}^{P_4^m(s)} S_{m,4}^{s,\,l}\lambda_{\odot}^s\prod_{r=0}^{4} I_{4,r}^{\alpha_r\binom{s,\,l}{m,S_4}}\left[\lambda_{\odot}^{4m-2s} - \Omega^{4m-2s}\left(\mathbf{x^4}\right)\right] \le 0, \tag{7.23}$$

$$\sum_{s=0}^{\lfloor s_* \rfloor}\sum_{l=1}^{P_4^m(s)} S_{m,4}^{s,\,l}\lambda_{\odot}^s\prod_{r=0}^{4} I_{4,r}^{\alpha_r\binom{s,\,l}{m,S_4}}\left[\lambda_{\odot}^{4m-2s} - \omega^{4m-2s}\left(\mathbf{x^4}\right)\right] \ge 0. \tag{7.24}$$

The differences in the brackets (7.23) and (7.24) have exact divisors $\lambda_{\odot} - \Omega\left(\mathbf{x^4}\right)$ and $\lambda_{\odot} - \omega\left(\mathbf{x^4}\right)$, respectively, while a quotient of such division is always positive. Thus, we arrive at inequalities (7.22) and Theorem is proven. □

Formulas (7.20) and (7.22) are dual in the following sense:

$$\text{if} \quad \left(\frac{I_{4,1}I_{4,4}}{I_{4,3}}\right)^{1/2} \le \lambda_{\odot}\left(\mathbf{x^4}\right) \le \left(\frac{I_{4,3}}{I_{4,1}}\right)^{1/2}, \quad \mathbf{x^4} \in \mathbb{P}_+^4,$$

$$\text{then} \quad \left(\frac{I_{4,3}}{I_{4,1}}\right)^{1/2} \le \lambda_{\odot}\left(\mathbf{x^4}\right) \le \left(\frac{I_{4,1}I_{4,4}}{I_{4,3}}\right)^{1/2}, \quad \mathbf{x^4} \in \mathbb{P}_-^4,$$

and vice versa, i.e., the actual expressions will be exchanged for the lower and upper bounds in subspaces $\mathbb{P}_+^4$ and $\mathbb{P}_-^4$, while if $\mathbf{x^4} \in \mathbb{P}_0^3$ then $\Omega\left(\mathbf{x^4}\right) = \omega\left(\mathbf{x^4}\right)$.

8 Open Problems and Questions

1. Extend the polynomials $\mathsf{R}\binom{\lambda,\, \mathbf{x^n}}{m, S_n}$ and $\mathsf{S}\binom{\lambda,\, \mathbf{x^n}}{m, S_n}$ to self-dual multivariate polynomials $\mathsf{R}\binom{\lambda,\, \mathbf{x^n}}{m,\, G}$ and $\mathsf{S}\binom{\lambda,\, \mathbf{x^n}}{m,\, G}$, which are invariant under action of the finite group $G \subset S_n$.
2. Are all self-dual functions $\lambda\left(\mathbf{x^n}\right)$ algebraic ?
3. Find strong universal bounds for $\lambda_{\odot}\left(\mathbf{x^n}\right)$, $n \ge 5$.
4. Prove Conjecture 1 in Appendix B.1 for partition numbers $P_n^m(s_*)$.

A Appendix: Generic Self-Dual Polynomials with $m, n \leq 4$.

$$\mathsf{S}\binom{\lambda,\ \mathbf{x}^2}{1,\ S_2} = K_1\lambda^2 - K_1 I_{2,2}, \quad \mathsf{R}\binom{\lambda,\ \mathbf{x}^2}{1,\ S_2} = L_1\lambda^2 + L_2 I_{2,1}\lambda + L_1 I_{2,2},$$

$$\mathsf{S}\binom{\lambda,\ \mathbf{x}^3}{1,\ S_3} = K_1\lambda^3 + K_2 I_{3,1}\lambda^2 - K_2 I_{3,2}\lambda - K_1 I_{3,3}, \quad \mathsf{R}\binom{\lambda,\ \mathbf{x}^3}{1,\ S_3} = L_1\lambda^3 + L_2 I_{3,1}\lambda^2 + L_2 I_{3,2}\lambda + L_1 I_{3,3},$$

$$\mathsf{S}\binom{\lambda,\ \mathbf{x}^4}{1,\ S_4} = K_1\lambda^4 + K_2 I_{4,1}\lambda^3 - K_2 I_{4,3}\lambda - K_1 I_{4,4},$$

$$\mathsf{R}\binom{\lambda,\ \mathbf{x}^4}{1,\ S_4} = L_1\lambda^4 + L_2 I_{4,1}\lambda^3 + L_3\lambda^2 + L_2 I_{4,3}\lambda + L_1 I_{4,4},$$

$$\mathsf{S}\binom{\lambda,\ \mathbf{x}^3}{2,\ S_3} = K_1\lambda^6 + K_2 I_{3,1}\lambda^5 + \left(K_3 I_{3,1}^2 + K_4 I_{3,2}\right)\lambda^4 - \left(K_3 I_{3,2}^2 + K_4 I_{3,1} I_{3,3}\right)\lambda^2 - K_2 I_{3,2} I_{3,3}\lambda - K_1 I_{3,3}^2,$$

$$\mathsf{R}\binom{\lambda,\ \mathbf{x}^3}{2,\ S_3} = L_1\lambda^6 + L_2 I_{3,1}\lambda^5 + \left(L_3 I_{3,1}^2 + L_4 I_{3,2}\right)\lambda^4 + (L_5 I_{3,1} I_{3,2} + L_6 I_{3,3})\lambda^3 + \left(L_3 I_{3,2}^2 + L_4 I_{3,1} I_{3,3}\right)\lambda^2 + L_2 I_{3,2} I_{3,3}\lambda + L_1 I_{3,3}^2,$$

$$\mathsf{S}\binom{\lambda,\ \mathbf{x}^4}{2,\ S_4} = K_1\lambda^8 + K_2 I_{4,1}\lambda^7 + \left(K_3 I_{4,1}^2 + K_4 I_{4,2}\right)\lambda^6 + \left(K_5 I_{4,1} I_{4,2} + K_6 I_{4,3}\right)\lambda^5 - \left(K_5 I_{4,3} I_{4,2} + K_6 I_{4,1} I_{4,4}\right)\lambda^3 - \left(K_3 I_{4,3}^2 + K_4 I_{4,2} I_{4,4}\right)\lambda^2 - K_2 I_{4,3} I_{4,4}\lambda - K_1 I_{4,4}^2,$$

$$\mathsf{R}\binom{\lambda,\ \mathbf{x}^4}{2,\ S_4} = L_1\lambda^8 + L_2 I_{4,1}\lambda^7 + \left(L_3 I_{4,1}^2 + L_4 I_{4,2}\right)\lambda^6 + \left(L_5 I_{4,1} I_{4,2} + L_6 I_{4,3}\right)\lambda^5 + \left(L_7 I_{4,2}^2 + L_8 I_{4,1} I_{4,3} + L_9 I_{4,4}\right)\lambda^4 + \left(L_5 I_{4,3} I_{4,2} + L_6 I_{4,1} I_{4,4}\right)\lambda^3 + \left(L_3 I_{4,3}^2 + L_4 I_{4,2} I_{4,4}\right)\lambda^2 + L_2 I_{4,3} I_{4,4}\lambda + L_1 I_{4,4}^2,$$

$$\mathsf{S}\binom{\lambda,\ \mathbf{x}^3}{3,\ S_3} = K_1\lambda^9 + K_2 I_{3,1}\lambda^8 + \left(K_3 I_{3,1}^2 + K_4 I_{3,2}\right)\lambda^7 + \left(K_5 I_{3,1} I_{3,2} + K_6 I_{3,3} + K_7 I_{3,1}^3\right)\lambda^6 + \left(K_8 I_{3,1}^2 I_{3,2} + K_9 I_{3,2}^2 + K_{10} I_{3,1} I_{3,3}\right)\lambda^5 - \left(K_8 I_{3,1} I_{3,2}^2 + K_9 I_{3,1}^2 I_{3,3} + K_{10} I_{3,2} I_{3,3}\right)\lambda^4 - \left(K_5 I_{3,1} I_{3,2} I_{3,3} + K_6 I_{3,3}^2 + K_7 I_{3,2}^3\right)\lambda^3 - \left(K_3 I_{3,2}^2 I_{3,3} + K_4 I_{3,1} I_{3,3}^2\right)\lambda^2 - K_2 I_{3,2} I_{3,3}^2\lambda - K_1 I_{3,3}^3,$$

$$\mathsf{R}\binom{\lambda,\ \mathbf{x}^3}{3,\ S_3} = L_1\lambda^9 + L_2 I_{3,1}\lambda^8 + \left(L_3 I_{3,1}^2 + L_4 I_{3,2}\right)\lambda^7 + \left(L_5 I_{3,1} I_{3,2} + L_6 I_{3,3} + L_7 I_{3,1}^3\right)\lambda^6 + \left(L_8 I_{3,1}^2 I_{3,2} + L_9 I_{3,2}^2 + L_{10} I_{3,1} I_{3,3}\right)\lambda^5 + \left(L_8 I_{3,1} I_{3,2}^2 + L_9 I_{3,1}^2 I_{3,3} + L_{10} I_{3,2} I_{3,3}\right)\lambda^4 + \left(L_5 I_{3,1} I_{3,2} I_{3,3} + L_6 I_{3,3}^2 + L_7 I_{3,2}^3\right)\lambda^3 + \left(L_3 I_{3,2}^2 I_{3,3} + L_4 I_{3,1} I_{3,3}^2\right)\lambda^2 + L_2 I_{3,2} I_{3,3}^2\lambda + L_1 I_{3,3}^3.$$

$$\mathsf{S}\binom{\lambda,\ \mathbf{x}^3}{4,\ S_3} = K_1\lambda^{12} + K_2 1 I_{3,1}\lambda^{11} + \left(K_3 I_{3,1}^2 + K_4 I_{3,2}\right)\lambda^{10} + \left(K_5 I_{3,1}^3 + K_6 I_{3,1} I_{3,2} + K_7 I_{3,3}\right)\lambda^9 + \left(K_8 I_{3,1} I_{3,3} + K_9 I_{3,1}^2 I_{3,2} + K_{10} I_{3,1}^4 + K_{11} I_{3,2}^2\right)\lambda^8 + \left(K_{12} I_{3,1} I_{3,2}^2 + K_{13} I_{3,1}^2 I_{3,3} + K_{14} I_{3,1}^3 I_{3,2} + K_{15} I_{3,2} I_{3,3}\right)\lambda^7 + K_{16}\left(I_{3,1}^3 I_{3,3} - I_{3,2}^3\right)\lambda^6 - \left(K_{12} I_{3,1}^2 I_{3,2} I_{3,3} + K_{13} I_{3,2}^2 I_{3,3} + K_{14} I_{3,1} I_{3,2}^3 + K_{15} I_{3,1} I_{3,3}^2\right)\lambda^5 - \left(K_8 I_{3,2} I_{3,3}^2 + K_9 I_{3,1} I_{3,2}^2 I_{3,3} + K_{10} I_{3,2}^4 + K_{11} I_{3,1}^2 I_{3,3}^2\right)\lambda^4 - \left(K_5 I_{3,2}^3 I_{3,3} + K_6 I_{3,1} I_{3,2} I_{3,3}^2 + K_7 I_{3,3}^3\right)\lambda^3 - \left(K_3 I_{3,2}^2 I_{3,3}^2 + K_4 I_{3,3}^3 I_{3,1}\right)\lambda^2 - K_2 I_{3,2} I_{3,3}^3\lambda - K_1 I_{3,3}^4.$$

$$
\begin{aligned}
\mathsf{R}\binom{\lambda,\ \mathbf{x}^3}{4,\ S_3} = {} & L_1\lambda^{12}+L_2 1 I_{3,1}\lambda^{11}+\left(L_3 I_{3,1}^2+L_4 I_{3,2}\right)\lambda^{10}+\left(L_5 I_{3,1}^3+L_6 I_{3,1}I_{3,2}+L_7 I_{3,3}\right)\lambda^9+\\
& \left(L_8 I_{3,1}I_{3,3}+L_9 I_{3,1}^2 I_{3,2}+L_{10}I_{3,1}^4+L_{11}I_{3,2}^2\right)\lambda^8+\\
& \left(L_{12}I_{3,1}I_{3,2}^2+L_{13}I_{3,1}^2 I_{3,3}+L_{14}I_{3,1}^3 I_{3,2}+L_{15}I_{3,2}I_{3,3}\right)\lambda^7+\\
& \left(L_{16}I_{3,1}^3 I_{3,3}+L_{17}I_{3,1}^2 I_{3,2}^2+L_{18}I_{3,1}I_{3,2}I_{3,3}+L_{19}I_{3,2}^3+L_{20}I_{3,3}^2\right)\lambda^6+\\
& \left(L_{12}I_{3,1}^2 L_{3,2}I_{3,3}+L_{13}I_{3,2}^2 I_{3,3}+L_{14}I_{3,1}I_{3,2}^3+L_{15}I_{3,1}I_{3,3}^2\right)\lambda^5+\\
& \left(L_8 I_{3,2}I_{3,3}^2+L_9 I_{3,1}I_{3,2}^2 I_{3,3}+L_{10}I_{3,2}^4+L_{11}I_{3,1}^2 I_{3,3}^2\right)\lambda^4+\\
& \left(L_5 I_{3,2}^3 I_{3,3}+L_6 I_{3,1}I_{3,2}I_{3,3}^2+L_7 I_{3,3}^3\right)\lambda^3+\left(L_3 I_{3,2}^2 I_{3,3}^2+L_4 I_{3,3}^3 I_{3,1}\right)\lambda^2+\\
& L_2 I_{3,2}I_{3,3}^3\lambda+L_1 I_{3,3}^4.
\end{aligned}
$$

B Appendix: Restricted Partitions with Constraints

We recall known facts from the theory of restricted partitions related to Gaussian's polynomials following [1]. We present also new results, which are relevant in calculations of unimodality numbers μ of self- dual polynomials in Sect. 2.

B.1 Partition Number $P_n^m(s)$

Consider two Diophantine equations (2.9) and denote by $P_n^m(s)$ a cardinality of the set $\Delta_n^m(s)$,

$$
\Delta_n^m(s) = \bigcup_{l=1}^{P_n^m(s)} \left\{\alpha^l\right\},\quad \left\{\alpha^l\right\} = \left\{\alpha_0\binom{s,\ l}{m, S_n},\ \alpha_1\binom{s,\ l}{m, S_n},\ \ldots,\ \alpha_n\binom{s,\ l}{m, S_n}\right\} \tag{B.1}
$$

of non-negative integer solutions $\left\{\alpha^l\right\}$ of equations (2.9). There exists a number-theoretical problem, related to equations (2.9), namely, to solve the Diophantine equation and inequality,

$$
\sum_{r=1}^{n} r\ \alpha_r\binom{s,\ l}{m, S_n} = s,\qquad \sum_{r=1}^{n} \alpha_r\binom{s,\ l}{m, S_n} \le m,
$$

which differ from (2.9) by the lower summation limit. Solutions of these Diophantine equation and inequality $\left\{\alpha_1\binom{s,\ l}{m, S_n},\ \ldots,\alpha_n\binom{s,\ l}{m, S_n}\right\}$ are similar to (B.1) and a number of solutions coincides with $P_n^m(s)$. In other words, $P_n^m(s)$ gives a number of partitions of s into *at most* n parts, each $\le m$. Its generating function $G(n, m; z)$ is known due to Schur [20] and is referred to as *the Gaussian polynomial* [1],

$$G(n,m;z)=\frac{\prod_{j=1}^{n+m}(1-z^j)}{\prod_{j=1}^{n}(1-z^j)\prod_{j=1}^{m}(1-z^j)}=\sum_{s\geq 0}^{mn} P_n^m(s)z^s, \tag{B.2}$$

$$P_n^m(s)=P_m^n(s)>0,$$

with combinatorial [6] and integral [2] formulas for $P_n^m(s)$. The sequence of coefficients $P_n^m(s)$, in the range $0\leq s\leq mn$, is symmetric with respect to (w.r.t.) its middle $s_*=mn/2$,

$$P_n^m(s)=P_n^m(mn-s),\qquad P_n^m(s)=0,\quad s>mn, \tag{B.3}$$

and *unimodal* (see analytic [20] and combinatorial [16] proofs) if $m>1$

$$1\leq P_n^m(1)\leq\ldots\leq P_n^m(\lfloor s_*\rfloor)\geq\ldots\geq P_n^m(mn-1)\geq 1, \tag{B.4}$$

$$P_n^m(0)=P_n^m(mn)=1,$$

but not always log-concave [21]. A special case occurs when $m=1$ and $P_n^1(s)=1$, $0\leq s\leq n$. Making use of definition (B.2), we obtain

$$\sum_{s=0}^{mn} P_n^m(s)=\lim_{z\to 1} G(n,m;z)=\binom{m+n}{n}, \tag{B.5}$$

that results in the upper bound for $P_n^m(s_*)$

$$P_n^m(s_*)<\binom{m+n}{n}-mn. \tag{B.6}$$

In what follows, we study Diophantine equations (2.9) at $s=s_*$, $mn=0 \pmod 2$.

If $n=2$, then two Eq. (2.9) are equivalent to another pair of decoupled equations,

$$\alpha_1\binom{s_*,\ l}{m,S_n}+2\alpha_2\binom{s_*,\ l}{m,S_n}=m,\qquad \alpha_0\binom{s_*,\ l}{m,S_n}=\alpha_2\binom{s_*,\ l}{m,S_n}. \tag{B.7}$$

The 1st equation in (B.7) is not supplemented by any constraint and $P_2^m(m)$ coincides with a number of partitions of m into 1 and 2. Explicit formulas for such partitions were found in [7],

$$P_2^m(m)=\frac{1}{2}\left(m+1+\cos^2\frac{\pi m}{2}\right). \tag{B.8}$$

If $n\geq 3$, formulas for $P_n^m(s_*)$ become sophisticated [2, 6] and difficult to apply.

Conjecture 1 Let $m_1 n = m_2 n = 0 \pmod 2$ and $n \geq 3$. Then for $m_1, m_2 \geq 1$ it holds

$$P_n^{m_1}(s_{1*}) + P_n^{m_2}(s_{2*}) < P_n^m(s_*), \quad m = m_1 + m_2, \quad s_* = s_{1*} + s_{2*}. \tag{B.9}$$

B.2 Partition Number $U_n^m(s)$

Consider another partition number $U_n^m(s)$, which gives a number of solutions of two Diophantine equations (2.9) supplemented by additional constraints,

$$\alpha_r\binom{s,\ l}{m, S_n} = \alpha_{n-r}\binom{s,\ l}{m, S_n}, \quad 0 \leq r \leq n. \tag{B.10}$$

It obviously satisfies inequality $U_n^m(s) \leq P_n^m(s)$. Relations (B.10) imply two equalities

$$r\alpha_r\binom{s,\ l}{m, S_n} + (n-r)\alpha_{n-r}\binom{s,\ l}{m, S_n} = n\alpha_r\binom{s,\ l}{m, S_n},$$

$$\alpha_r\binom{s,\ l}{m, S_n} + \alpha_{n-r}\binom{s,\ l}{m, S_n} = 2\alpha_r\binom{s,\ l}{m, S_n}.$$

Substituting the last equalities into (2.9) for $s = s_*$, we obtain two coinciding Diophantine equations which looks differently dependent on the parity of m and n, separately,

a) $n = 1, m = 0 \pmod 2$,

$$\alpha_0\binom{s_*,\ l}{m, S_n} + \alpha_1\binom{s_*,\ l}{m, S_n} + \ldots + \alpha_{\frac{n-1}{2}}\binom{s_*,\ l}{m, S_n} = \frac{m}{2}, \tag{B.11}$$

b) $n = 0, m = 1 \pmod 2$,

$$\alpha_0\binom{s_*,\ l}{m, S_n} + \alpha_1\binom{s_*,\ l}{m, S_n} + \ldots + \alpha_{\frac{n}{2}-1}\binom{s_*,\ l}{m, S_n} + \frac{1}{2}\alpha_{\frac{n}{2}}\binom{s_*,\ l}{m, S_n} = \frac{m}{2},$$

c) $n = 0, m = 0 \pmod 2$,

$$\alpha_0\binom{s_*,\ l}{m, S_n} + \alpha_1\binom{s_*,\ l}{m, S_n} + \ldots + \alpha_{\frac{n}{2}-1}\binom{s_*,\ l}{m, S_n} + \frac{1}{2}\alpha_{\frac{n}{2}}\binom{s_*,\ l}{m, S_n} = \frac{m}{2}.$$

Thus, $U_n^m(s_*)$ is given by the number of solutions of a single Diophantine equation (B.11) in three different cases. To find the number $U_n^m(s_*)$, recall the classical result [18]: the Diophantine equation $\xi_1 + \ldots + \xi_p = q, \xi_j \in \mathbb{Z}_{\geq}$, has exactly $\binom{p+q-1}{q}$ solutions. By comparison of the last equation with (B.11), we obtain formulas for $U_n^m(s_*)$ through the binomial coefficients,

$$
\begin{aligned}
&a)\ \ n = 1, m = 0 \pmod 2, && U_n^m(s_*) = \binom{(m+n-1)/2}{(n-1)/2}, \qquad \text{(B.12)}\\
&b)\ \ n = 0, m = 1 \pmod 2, && U_n^m(s_*) = \binom{(m+n-1)/2}{(m-1)/2},\\
&c)\ \ n = 0, m = 0 \pmod 2, && U_n^m(s_*) = \binom{(m+n)/2}{n/2},
\end{aligned}
$$

where the last two cases differ by the parity of the last term in Diophantine's equations (B.11), $\alpha_{\frac{n}{2}}\binom{s_*,\ l}{m, S_n} = 1 \pmod 2$ in (B.11b) and $\alpha_{\frac{n}{2}}\binom{s_*,\ l}{m, S_n} = 0 \pmod 2$ in (B.11c).

If $m = 1$ and $n = 0 \pmod 2$, then by (B.12b) we have $U_n^1(s_*) = 1$ and therefore

$$
U_n^1(s_*) = P_n^1(s_*),
$$

or by exchange symmetry (B.2)

$$
U_1^m(s_*) = P_1^m(s_*). \qquad \text{(B.13)}
$$

If $n = 2$, the last two formulas in (B.12) give together with (B.8)

$$
\begin{aligned}
U_2^m(m) &= \frac{m+1}{2},\ \ m = 1 \pmod 2,\\
U_2^m(m) &= \frac{m}{2} + 1,\ \ m = 0 \pmod 2\ \rightarrow\ U_2^m(s_*) = P_2^m(s_*). \qquad \text{(B.14)}
\end{aligned}
$$

For the first time, inequality $U_n^m(s) \le P_n^m(s)$ appears for $s = s_*$ when $m, n \ge 3$ and the least value of product $mn = 12$, e.g., the middle term $K_{16}\left(I_{3,1}^3 I_{3,3} - I_{3,2}^3\right)\lambda^6$ in expression for $\mathsf{S}\binom{\lambda,\ \mathbf{x}^3}{4,\ S_3}$ in Appendix A does not vanish,

References

1. G.E. Andrews, *The theory of partitions*, Addison-Wesley Publ., Reading, 1976
2. D. Andrica & O. Bagdasar, *A new formula for the coefficients of Gaussian polynomials*, An. St. Univ. Ovidius Constanta, **27**, 125-36 (2019)
3. L.E. Dickson, First Course in the Theory of Equations, Wiley, NY, 1922
4. A.M. Dykhne, *Conductivity of a two-dimensional two-phase systems*, Sov. Phys. JETP **32**, 63-65 (1970)
5. L.G. Fel, V.Sh. Machavariani & D.J. Bergman, *Isotropic conductivity of two-dimensional three-component symmetric composites*, J. Phys. **A**, **33**, 6669-6681 (2000)
6. L.G. Fel, *Gaussian Polynomials and Restricted Partition Functions with Constraints*, arXiv:1611.09931v2
7. L.G. Fel & B.M. Rubinstein, *Sylvester waves in the Coxeter groups*, Ramanujan J., **6**, 307-329 (2002)
8. G.H. Hardy, J.E. Littlewood and G. Polya, Inequalities, Univ. Press, Cambridge, 1959

9. D. Joyner & J.-L. Kim, Selected unsolved problems in coding theory, Birkhäuser, NY (2011).
10. D. Joyner, *Zeros of some self-reciprocal polynomials*, in Excursions in Harmonic Analysis, **1**, Birkhaüser, 2013, pp. 329-351
11. J.B. Keller, *A theorem on the conductivity of a composite*, J. Math. Phys. **5**, 548-549 (1964)
12. S.-H. Kim & C.W. Park, *On the zeros of certain self-reciprocal polynomials*, J. Math. Anal. Appl., **339**, 240-247 (2008)
13. T.-D. Lee & C.-N. Yang, *Statistical theory of equations of state and phase transitions II. Lattice gas and Ising model*, Phys.Rev., **87**, 404, 410 (1952).
14. J. McKee & C. Smyth, Number theory and polynomials, Univ. Press, Cambridge, (2008)
15. I. Markovsky & Sh. Rao, *Palindromic polynomials, time-reversible systems and conserved quantities*, in Proceed. 16th Conf. on Control and Automation, Ajaccio, France, 125 (2008)
16. K. O'Hara, *Unimodality of Gaussian coefficients*, J. Comb, Theory, Ser. A, **53**, 29-52 (1990)
17. B.Y. Rubinstein & L.G. Fel, *Restricted partition functions as Bernoulli and Euler polynomials of higher order*, Ramanujan J., **11**, 331-347 (2006)
18. J. Riordan, Introduction to Combinatorial Analysis, Princeton Univ. Press, Princeton New Jersey, 1980, p. 104.
19. A. Rosenfeld, An introduction to algebraic structures, Holden-Day, San Francisco, 1968
20. I.J. Schur, *Vorlesungen über InvariantenTheorie*, in Grundlehren der mathematischen Wissenschaften, **B. 143**, ed. by B. Eckmann and B. van der Waerden, Springer, Berlin, 1968
21. R.P. Stanley, *Log-concave and unimodal sequences in algebra, combinatorics, and geometry*, Ann. NY Acad. Sci. **576** (1989), #1 Graph Theory, 500-535.

Symmetric (Not Complete Intersection) Numerical Semigroups and Syzygy Identities

Leonid G. Fel

Abstract We consider symmetric (not complete intersection) numerical semigroups $S_m = \langle d_1, \ldots, d_m \rangle$ of arbitrary *edim*, minimally generated by a set of m positive integers, such that $\gcd(d_1, \ldots, d_m) = 1$. We derive identities for degrees of syzygies of such semigroups and find the lower bound for their Frobenius numbers that generalizes recent results for $m = 4, 5, 6$.

1 Introduction

Studying the rational representation of the Hilbert series of numerical semigroups $S_m = \langle d_1, \ldots, d_m \rangle$ and the quasi-polynomial representation of the restricted partition function [3], we derived two sets of polynomial and quasi-polynomial identities for degrees of syzygies in S_m. A part of polynomial identities of degrees $1 \le n \le m-2$ was known earlier [8], but a new additional identity of degree $m-1$ turned out to be an important tool in study of symmetric not complete intersection (CI) semigroups. Such semigroups with small embedding dimension (*edim*) $m = 4, 5, 6$ were studied in [2, 4, 5], respectively. In the present paper we study the general case of symmetric (not CI) semigroups with arbitrary *edim*.

The paper is organized in two parts, studying symmetric (not CI) semigroups with even and odd *edim*, separately. These parts are preceded by an introductory section on polynomial identities for degrees of syzygies in numerical semigroups [3, 6]. For short, we call them *syzygy identities*.

Let a numerical semigroup S_m be minimally generated by a set of natural numbers $\{d_1, \ldots, d_m\}$, where $\gcd(d_1, \ldots, d_m) = 1$. Its generating function $H(S_m; z)$,

$$H(S_m; z) = \sum_{s \in S_m} z^s, \qquad z < 1, \qquad 0 \in S_m, \tag{1.1}$$

L. G. Fel (✉)
Department of Civil Engineering, Technion, 32000 Haifa, Israel
e-mail: lfel@technion.ac.il

M. B. Nathanson (ed.), *Combinatorial and Additive Number Theory VI*,
Springer Proceedings in Mathematics & Statistics 464,
https://doi.org/10.1007/978-3-031-65064-2_8

is referred to as *the Hilbert series* of S_m and has a rational representation,

$$H(S_m; z) = \frac{Q(S_m; z)}{\prod_{i=1}^{m}\left(1 - z^{d_i}\right)}, \quad C_{k,j} \in \mathbb{Z}_>, \quad 1 \le k \le m-1, \quad 1 \le j \le \beta_k, \tag{1.2}$$

where

$$Q(S_m; z) = 1 - \sum_{j=1}^{\beta_1} z^{C_{1,j}} + \sum_{j=1}^{\beta_2} z^{C_{2,j}} - \ldots \pm \sum_{j=1}^{\beta_{m-1}} z^{C_{m-1,j}}, \tag{1.3}$$
$$\sum_{k=0}^{m-1} (-1)^k \beta_k = 0, \quad \beta_0 = 1,$$

and β_k and $C_{k,j}$ denote partial Betti's numbers and degrees of the k syzygies, respectively. The largest degree g_m of the $(m-1)$-th syzygy is related to the Frobenius number F_m of the semigroup,

$$g_m = F_m + \sigma_1, \qquad g_m = \max \mathsf{PF}(S_m), \tag{1.4}$$
$$\mathsf{PF}(S_m) = \left\{C_{m-1,1}, \ldots, C_{m-1,\beta_{m-1}}\right\}, \quad \sigma_1 = \sum_{j=1}^{m} d_j,$$

where $\mathsf{PF}(S_m)$ stands for a set of pseudo-Frobenius numbers and cardinality $\#\mathsf{PF}(S_m)$, called a type of semigroup S_m. Denote by $\mathbf{C}_k(S_m)$ the alternating power sum of syzygies degrees,

$$\mathbf{C}_k(S_m) = \sum_{j=1}^{\beta_1} C_{1,j}^k - \sum_{j=1}^{\beta_2} C_{2,j}^k + \ldots - (-1)^{m-1} \sum_{j=1}^{\beta_{m-1}} C_{m-1,j}^k, \tag{1.5}$$

and write the polynomial identities (Theorem 1 in [3]) for a semigroup S_m,

$$\mathbf{C}_0(S_m) = 1, \qquad \mathbf{C}_r(S_m) = 0, \quad 1 \le r \le m-2, \tag{1.6}$$
$$\mathbf{C}_{m-1}(S_m) = (-1)^m (m-1)! \pi_m,$$

where $\pi_m = \prod_{i=1}^{m} d_i$. For $r \ge m$ the alternating power sums $\mathbf{C}_r(S_m)$ were calculated in [6],[1]

$$\mathbf{C}_n(S_m) = \frac{(-1)^m n!}{(n-m)!} \pi_m K_{n-m}(S_m), \qquad K_p(S_m) > 0, \quad p \ge 0, \tag{1.7}$$

[1] In formulas (23) of paper [6], there exist two misprints: in formula for $K_5(S_m)$, instead of $35\delta_2\delta_4 - 42\delta_2^3$ it must be written $35\delta_2^3 - 42\delta_2\delta_4$, and in formula for $K_6(S_m)$, instead of $35\sigma_2\sigma_4 - 42\sigma_2^3$ it must be written $35\sigma_2^3 - 42\sigma_2\sigma_4$.

where $K_p(S_m)$ is a linear combination of genera $G_0(S_m), \ldots, G_p(S_m)$ of the semigroup, $G_p(S_m)=\sum_{s\in\Delta_m} s^p$. Here $\Delta_m = \mathbb{Z}_> \setminus S_m$ and $G_0(S_m) = \#\Delta_m$ denote a set of gaps and a genus of S_m, respectively. E.g., according to [6], we have $K_0(S_m) = G_0(S_m) + (\sigma_1 - 1)/2$.

The lower bound of largest syzygies degree g_m in semigroups with arbitrary *edim* is given by [9]

$$g_m \geq \sqrt[m-1]{(m-1)!\pi_m}.$$

If a numerical semigroup S_m is symmetric, then degrees $C_{k,j}$ of syzygies and Betti's numbers β_k in (1.2, 1.3) are related as follows,

$$\beta_k = \beta_{m-k-1}, \quad \beta_{m-1} = 1, \quad C_{k,j} + C_{m-k-1,j} = g_m, \qquad (1.8)$$
$$\#\mathsf{PF}(S_m) = 1, \quad G_0 = \frac{1}{2}(F_m + 1).$$

A special kind of symmetric semigroups with the first Betti number

$$\beta_1 = m - 1$$

is referred to as complete intersection (CI); its Betti's numbers are well known, $\beta_k = \binom{m-1}{k}$, and exhibits a unimodal sequence, satisfying the Stanley conjecture 4B [11]. The lower bound of largest syzygies degree g_m in CI with arbitrary *edim* is also well known [3] and is given by

$$g_m \geq (m-1)\sqrt[m-1]{\pi_m}.$$

Keeping in mind the known formulas in [2–5] for the lower bounds of g_m, $m \leq 6$, in the present paper we establish the similar expressions in symmetric (not CI) semigroups for arbitrary m. To unify the previously developed approaches for $m \leq 6$ in a general one, we start in every section with semigroups S_m of small m and continue for arbitrary *edim*.

2 Symmetric (Not CI) Semigroups S_m, $m = 0 \pmod 2$

2.1 Symmetric (CI) Semigroups S_2

The Hilbert series $H(\mathsf{S}_2; z)$ is given by

$$H(\mathsf{S}_2; z) = \frac{Q_2(z)}{\prod_{i=1}^{2}(1 - z^{d_i})}, \qquad Q_4(z) = 1 - z^{g_2}. \qquad (2.1)$$

Syzygy identity (1.6) for S_2 reads $g_2 = \pi_2$.

2.2 Symmetric (Not CI) Semigroups S_4

Due to Bresinsky [1], the Hilbert series $H(\mathsf{S}_4; z)$ is given by

$$H(\mathsf{S}_4; z) = \frac{Q_4(z)}{\prod_{i=1}^{4}(1 - z^{d_i})},$$
$$Q_4(z) = 1 - \sum_{j=1}^{\beta_1} z^{x_j} + \sum_{j=1}^{\beta_1} z^{g_4 - x_j} - z^{g_4}, \quad \beta_1 = 5. \tag{2.2}$$

Syzygy identities (1.6) for S_4 read for $r \le 3$

$$\sum_{j=1}^{\beta_1} x_j^r - \sum_{j=1}^{\beta_1} (g_4 - x_j)^r + g_4^r = 0, \quad r \le 2,$$
$$\sum_{j=1}^{\beta_1} x_j^3 - \sum_{j=1}^{\beta_1} (g_4 - x_j)^3 + g_4^3 = 3!\ \pi_4. \tag{2.3}$$

Only two of three identities in (2.3) are not trivial,

$$\sum_{j=1}^{\beta_1} x_j = \mathcal{B}_4 g_4, \qquad \sum_{j=1}^{\beta_1} x_j^2(3g_4 - 2x_j) = \mathcal{B}_4 g_4^3 - 3!\ \pi_4, \tag{2.4}$$
$$\mathcal{B}_4 = \frac{\beta_1 - 1}{2} = 2.$$

Rescaling all variables, $x_j/g_4 = y_j$, where $0 < y_j < 1$, we obtain

$$\sum_{j=1}^{\beta_1} f_1(y_j) = \mathcal{B}_4, \qquad \sum_{j=1}^{\beta_1} f_3(y_j) = \mathcal{B}_4 - 3!\frac{\pi_4}{g_4^3}, \tag{2.5}$$
$$f_1(t) = t, \qquad f_3(t) = t^2(3 - 2t), \quad 0 < t < 1.$$

Since $f_3(t) > 0$, already the 2nd equation in (2.5) gives roughly a lower bound for $g_4 > \sqrt[3]{3\pi_4}$. In fact, much more delicate approach (see [4]), keeping in mind the Newton-Maclaurin inequalities for symmetric invariants [7], gives much stronger bound

$$g_4 > \lambda_4 \sqrt[3]{\pi_4}, \qquad \lambda_4 = \sqrt[3]{25}. \tag{2.6}$$

2.3 Symmetric (Not CI) Semigroups S_6

Consider a symmetric (not CI) numerical semigroup S_6, minimally generated by six positive integers. Its Hilbert series $H(\mathsf{S}_6; z)$ with independent Betti's numbers β_1, β_2 reads:

$$H(\mathsf{S}_6; z) = \frac{Q_6(z)}{\prod_{i=1}^{6}\left(1 - z^{d_i}\right)},$$
$$Q_4(z) = 1 - \sum_{j=1}^{\beta_1} z^{x_{1,j}} + \sum_{j=1}^{\beta_2} z^{x_{2,j}} - \sum_{j=1}^{\beta_2} z^{g_6 - x_{2,j}} + \sum_{j=1}^{\beta_1} z^{g_6 - x_{1,j}} - z^{g_6}.$$

Syzygy identities (1.6) for S_6 read for $r \le 5$

$$\sum_{j=1}^{\beta_1} x_{1,j}^r - \sum_{j=1}^{\beta_2} x_{2,j}^r + \sum_{j=1}^{\beta_2} (g_6 - x_{2,j})^r - \sum_{j=1}^{\beta_1} (g_6 - x_{1,j})^r + g_6^r = 0, \quad r \le 4, \quad (2.7)$$
$$\sum_{j=1}^{\beta_1} x_{1,j}^5 - \sum_{j=1}^{\beta_2} x_{2,j}^5 + \sum_{j=1}^{\beta_2} (g_6 - x_{2,j})^5 - \sum_{j=1}^{\beta_1} (g_6 x_{1,j})^5 + g_6^5 = 5!\, \pi_6. \quad (2.8)$$

Only three of five identities in (2.7, 2.8) are not trivial, these are for $r = 1, 3, 5$. Rescaling the variables, $x_{1,j}/g_6 = y_{1,j}$ and $x_{2,j}/g_6 = y_{2,j}$, we arrive at

$$\sum_{j=1}^{\beta_2} f_r(y_{2,j}) - \sum_{j=1}^{\beta_1} f_r(y_{1,j}) = \mathcal{B}_6, \quad r = 1, 3, \quad (2.9)$$
$$\sum_{j=1}^{\beta_2} f_5(y_{2,j}) - \sum_{j=1}^{\beta_1} f_5(y_{1,j}) = \mathcal{B}_6 - 5!\, \frac{3\pi_6}{g_6^5},$$
$$\mathcal{B}_6 = \frac{\beta_2 - \beta_1 + 1}{2}, \qquad f_5(t) = t^3(10 - 15t + 6t^2).$$

As it was shown in [5], the lower bound for g_6 is given by

$$g_6 \ge \lambda_6 \sqrt[5]{\pi_6}, \qquad \lambda_6 = \sqrt[5]{\frac{5!\,3}{\alpha_6 \mathcal{B}_6}}, \qquad \alpha_6 \simeq 0.06259, \quad (2.10)$$

and even may be slightly improved, if we keep in mind the Newton-Maclaurin inequalities for symmetric invariants [7].

2.4 Symmetric (Not CI) Semigroups S_{2n} with Arbitrary n

Consider a symmetric (not CI) numerical semigroup S_{2n}, minimally generated by $2n$ positive integers. Its Hilbert series $H(\mathsf{S}_{2n}; z)$ with independent Betti's numbers $\beta_1, \ldots, \beta_{n-1}$ is given:

$$H(\mathsf{S}_{2n}; z) = Q_{2n}(z) \prod_{i=1}^{2n} \left(1 - z^{d_i}\right)^{-1}, \qquad \text{where} \tag{2.11}$$

$$\begin{aligned} Q_{2n}(z) = 1 - \sum_{j=1}^{\beta_1} z^{x_{1,j}} + \ldots - (-1)^n \sum_{j=1}^{\beta_{n-1}} z^{x_{n-1,j}} + \\ (-1)^n \sum_{j=1}^{\beta_{n-1}} z^{g_{2n} - x_{n-1,j}} \pm \ldots + \sum_{j=1}^{\beta_1} z^{g_{2n} - x_{1,j}} - z^{g_{2n}}, \end{aligned} \tag{2.12}$$

Syzygy identities (1.6) for S_{2n} read

$$\begin{aligned} &\sum_{j=1}^{\beta_1} x_{1,j}^r - \sum_{j=1}^{\beta_2} x_{2,j}^r + \ldots + (-1)^n \sum_{j=1}^{\beta_{n-1}} x_{n-1,j}^r - (-1)^n \sum_{j=1}^{\beta_{n-1}} (g_{2n} - x_{n-1,j})^r \mp \ldots \\ &+ \sum_{j=1}^{\beta_2} (g_{2n} - x_{2,j})^r - \sum_{j=1}^{\beta_1} (g_{2n} - x_{1,j})^r + g_{2n}^r = 0, \quad 1 \le r < 2n - 1, \end{aligned} \tag{2.13}$$

and

$$\begin{aligned} &\sum_{j=1}^{\beta_1} x_{1,j}^{2n-1} - \sum_{j=1}^{\beta_2} x_{2,j}^{2n-1} + \ldots + \\ &(-1)^n \sum_{j=1}^{\beta_{n-1}} x_{n-1,j}^{2n-1} - (-1)^n \sum_{j=1}^{\beta_{n-1}} (g_{2n} - x_{n-1,j})^{2n-1} \mp \ldots \\ &+ \sum_{j=1}^{\beta_2} (g_{2n} - x_{2,j})^{2n-1} - \sum_{j=1}^{\beta_1} (g_{2n} - x_{1,j})^{2n-1} + g_{2n}^{2n-1} = (2n-1)!\pi_{2n}. \end{aligned} \tag{2.14}$$

To simplify identities (2.13, 2.14), denote by Y_{2n} a set of rescaled syzygics degrees of S_{2n},

$$\begin{aligned} Y_{2n} &= \left\{y_{1,1}, \ldots, y_{1,\beta_1},\ y_{2,1}, \ldots, y_{2,\beta_2},\ \ldots, y_{n-1,1}, \ldots, y_{n-1,\beta_{n-1}}\right\}, \\ y_{i,j} &= \frac{x_{i,j}}{g_{2n}}, \quad 1 \le i \le n-1, \quad 0 \le y_{i,j} \le 1. \end{aligned} \tag{2.15}$$

Making the partial summation in the l.h.s. of syzygy identities (2.13, 2.14) and collecting together the power terms for every variable $y_{i,j}$ separately, we arrive at following alternating sums

$$\sum_{j=1}^{\beta_{n-1}} f_{2l-1}(y_{n-1,j}) - \sum_{j=1}^{\beta_{n-2}} f_{2l-1}(y_{n-2,j}) + \ldots + (-1)^n \sum_{j=1}^{\beta_1} f_{2l-1}(y_{1,j}), \qquad (2.16)$$

where $1 \le l \le n$ and a polynomial $f_{2l-1}(t)$ denotes another alternating power sum (see [10]),

$$f_{2l-1}(t) = t^l \sum_{k=1}^{l} (-1)^{k-1} \binom{l+k-2}{k-1} \binom{2l-1}{l-k} t^{k-1}, \quad 0 \le t \le 1, \quad \text{e.g.,} \qquad (2.17)$$

$$\begin{aligned}
f_1(t) &= t, \qquad (2.18)\\
f_3(t) &= t^2(3-2t),\\
f_5(t) &= t^3(10-15t+6t^2),\\
f_7(t) &= t^4(35-84t+70t^2-20t^3),\\
f_9(t) &= t^5(126-420t+540t^2-315t^3+70t^4),\\
f_{11}(t) &= t^6(462-1980t+3465t^2-3080t^3+1386t^4-252t^5),\\
f_{13}(t) &= t^7(1716-9009t+20020t^2-24024t^3+16380t^4-6006t^5+924t^6).
\end{aligned}$$

Making use of Mathematica Software, we establish another representation for the polynomial $f_{2l-1}(t)$,

$$f_{2n-1}(t) = n\binom{2n-1}{n-1} \mathsf{B}[t,n,n], \quad \mathsf{B}[t,a,b] = \int_0^t u^{a-1}(1-u)^{b-1} du, \qquad (2.19)$$

where $\mathsf{B}[t,a,b]$ denotes an incomplete beta function. According to (2.19), polynomials $f_{2l-1}(t)$ are non-negative in interval [0, 1] (see Fig. 1),

$$0 \le f_{2l-1}(t) \le 1, \quad 0 \le t \le 1, \quad l \ge 1, \qquad (2.20)$$

and possess many remarkable properties (see Appendix A). For our purpose, Lemma 3 in Appendix A manifests,

$$\frac{\sqrt{t f_{2n-1}(t)}}{f_{2n-3}(t)} \ge A_{2n}, \qquad 0 \le t \le 1, \quad n \ge 3, \qquad 0 < A_{2n} < 1. \qquad (2.21)$$

In Appendix A, Table 3, we present also a few first values of A_{2n}.

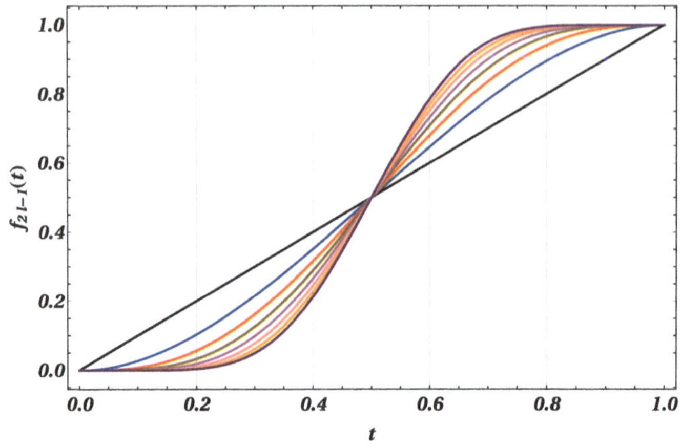

Fig. 1 Plot of the polynomials $f_{2l-1}(t)$, $t \in [0, 1]$, with different l: (*black*) $l = 1$, (*blue*) $l = 2$, (*red*) $l = 3$, (*brown*) $l = 4$, (*magenta*) $l = 5$, (*pink*) $l = 6$, (*orange*) $l = 7$, (*purple*) $l = 8$

Consider the set Y_{2n}, defined in (2.15), and disjoin it in two subsets Y_{2n}^u and Y_{2n}^v of odd and even syzygy classes, respectively,

$$Y_{2n} = Y_{2n}^u \bigcup Y_{2n}^v, \qquad \text{where} \tag{2.22}$$

$$Y_{2n}^u = \bigcup_{j=1}^{2j\le n} \left\{y_{2j-1,1}, \ldots, y_{2j-1,\beta_{2j-1}}\right\}, \qquad Y_{2n}^v = \bigcup_{j=1}^{2j\le n-1} \left\{y_{2j,1}, \ldots, y_{2j,\beta_{2j}}\right\}.$$

Their cardinalities $B_{2n}^u = \#Y_{2n}^u$ and $B_{2n}^v = \#Y_{2n}^v$ read

$$B_{2n}^u = \sum_{j=1}^{2j\le n} \beta_{2j-1}, \qquad B_{2n}^v = \sum_{j=1}^{2j\le n-1} \beta_{2j}. \tag{2.23}$$

In Table 1 we present a few first numbers B_{2n}^u and B_{2n}^v for semigroups S_{2n}.

For convenience, rename the variables $y_{i,j}$ by u_j and v_j and enumerate them as follows,

Table 1 Numbers B_{2n}^u and B_{2n}^v for semigroups S_{2n}

S_{2n}	S_4	S_6	S_8	S_{10}	S_{12}	S_{14}	S_{16}
B_{2n}^u	5	β_1	$\beta_1 + \beta_3$	$\beta_1 + \beta_3$	$\beta_1 + \beta_3 + \beta_5$	$\beta_1 + \beta_3 + \beta_5$	$\beta_1 + \beta_3 + \beta_5 + \beta_7$
B_{2n}^v	0	β_2	β_2	$\beta_2 + \beta_4$	$\beta_2 + \beta_4$	$\beta_2 + \beta_4 + \beta_6$	$\beta_2 + \beta_4 + \beta_6$

$$Y_{2n}^u = \left\{u_1, \ldots, u_{\beta_1}, \quad u_{\beta_1+1}, \ldots, u_{\beta_1+\beta_3}, \quad u_{\beta_1+\beta_3+1}, \quad \ldots, \quad u_{B_{2n}^u-1}, u_{B_{2n}^u}\right\},$$
$$Y_{2n}^v = \left\{v_1, \ldots, v_{\beta_2}, \quad v_{\beta_2+1}, \ldots, v_{\beta_2+\beta_4}, \quad v_{\beta_2+\beta_4+1}, \quad \ldots, \quad v_{B_{2n}^v-1}, v_{B_{2n}^v}\right\}. \quad (2.24)$$

Collecting together in the l.h.s. of (2.13, 2.14) the terms, which are free of power terms $y_{i,j}^k$, and renaming the latter in accordance with (2.24), we arrive at the syzygy identities as follows,

$$\sum_{j=1}^{B_{2n}^u} f_{2l-1}(u_j) - \sum_{j=1}^{B_{2n}^v} f_{2l-1}(v_j) = (-1)^n \mathcal{B}_{2n}, \qquad 1 \le l < n-1, \tag{2.25}$$

$$\sum_{j=1}^{B_{2n}^u} f_{2n-1}(u_j) - \sum_{j=1}^{B_{2n}^v} f_{2n-1}(v_j) = (-1)^n \mathcal{C}_{2n}, \tag{2.26}$$

$$\mathcal{C}_{2n} = \mathcal{B}_{2n} - (2n-1)! \binom{2n-3}{n-2} \frac{\pi_{2n}}{g_{2n}^{2n-1}},$$

where $\mathcal{B}_{2n}$ reads

$$\mathcal{B}_{2n} = \frac{(-1)^n}{2}\left(B_{2n}^u - B_{2n}^v - 1\right), \qquad \text{or} \tag{2.27}$$

$$\mathcal{B}_{2n} = \frac{\beta_{n-1} - \beta_{n-2} + \ldots + (-1)^n \beta_1 - (-1)^n}{2}.$$

In what follows, we consider symmetric (not CI) numerical semigroups S_{4n} and S_{4n+2} separately.

2.4.1 Lower Bound of g_{4n}, $n \ge 2$

Syzygy identities (2.25, 2.26) for symmetric (not CI) semigroup S_{4n} read

$$\sum_{j=1}^{B_{4n}^u} f_{2l-1}(u_j) - \sum_{j=1}^{B_{4n}^v} f_{2l-1}(v_j) = \mathcal{B}_{4n}, \quad 1 \le l < 2n-1, \tag{2.28}$$

$$\sum_{j=1}^{B_{4n}^u} f_{4n-1}(u_j) - \sum_{j=1}^{B_{4n}^v} f_{4n-1}(v_j) = \mathcal{C}_{4n}, \tag{2.29}$$

$$\mathcal{B}_{4n} = \frac{\beta_{2n-1} - \ldots + \beta_1 - 1}{2}, \qquad \mathcal{C}_{4n} = \mathcal{B}_{4n} - (4n-1)! \binom{4n-3}{2n-2} \frac{\pi_{4n}}{g_{4n}^{4n-1}}.$$

Combining inequality (2.21) and Cauchy-Schwarz's inequality, summate $\sqrt{t_j f_{2l-1}(t_j)}$ over $j \le B_{4n}^u$,

$$\left(A_{2l}\sum_{j=1}^{B_{4n}^u} f_{2l-3}(t_j)\right)^2 \leq \left(\sum_{j=1}^{B_{4n}^u}\sqrt{t_j f_{2l-1}(t_j)}\right)^2 \leq \sum_{j=1}^{B_{4n}^u} f_1(t_j)\sum_{j=1}^{B_{4n}^u} f_{2l-1}(t_j), \quad (2.30)$$

and substitute $l = 2n$ into (2.30). Making use of (2.28, 2.29) and replacing $t_j \to v_j$, we obtain,

$$A_{4n}^2\left(\mathcal{B}_{4n}+\sum_{j=1}^{B_{4n}^v} f_{4n-3}(v_j)\right)^2 \leq \left(\mathcal{B}_{4n}+\sum_{j=1}^{B_{4n}^v} f_1(v_j)\right)\left(\mathcal{C}_{4n}+\sum_{j=1}^{B_{4n}^v} f_{4n-1}(v_j)\right).$$

Denote by V_k the power sum $V_k = \sum_{j=1}^{B_{4n}^v} v_j^k$ of degree k and substitute it into the last inequality,

$$A_{4n}^2\left[\mathcal{B}_{4n}+F_{4n-3}(V)\right]^2 \leq \left[\mathcal{B}_{4n}+F_1(V)\right]\left[C_{4n}^u+F_{4n-1}(V)\right], \quad (2.31)$$

where, in accordance with (2.17),

$$F_{2n-1}(V)=\sum_{k=1}^{n}(-1)^{k-1}\binom{n+k-2}{k-1}\binom{2n-1}{n-k}V_{n+k-1}, \quad \text{e.g.,} \quad (2.32)$$
$$F_1(V)=V_1, \qquad F_3(V)=3V_2-2V_3, \quad \text{etc.}$$

On the other side, summating $\sqrt{t_j f_{2l-1}(t_j)}$ over $j \leq B_{4n}^v$, and combining inequality (2.21) and Cauchy-Schwarz's inequality, we obtain

$$A_{4n}^2\left(\sum_{j=1}^{B_{4n}^v} f_{4n-3}(v_j)\right)^2 \leq \sum_{j=1}^{B_{4n}^v} f_1(v_j)\sum_{j=1}^{B_{4n}^v} f_{4n-1}(v_j). \quad (2.33)$$

In notations V_k, the last inequality reads,

$$A_{4n}^2 F_{4n-3}^2(V) \leq F_1(V)\cdot F_{4n-1}(V). \quad (2.34)$$

Represent the both inequalities (2.31) and (2.34) as follows:

$$-\frac{\mathcal{C}_{4n}}{A_{4n}^2}+\frac{(\mathcal{B}_{4n}+F_{4n-3}(V))^2}{\mathcal{B}_{4n}+F_1(V)} \leq \frac{F_{4n-1}(V)}{A_{4n}^2}, \quad (2.35)$$

$$\frac{F_{4n-3}^2(V)}{F_1(V)} \leq \frac{F_{4n-1}(V)}{A_{4n}^2}. \quad (2.36)$$

Inequality (2.36) holds always, while inequality (2.35) is not valid for every set Y_{2n}^v. In order to make both inequalities consistent, we have to find a range of g_{4n} where both inequalities are satisfied. To satisfy them, it is enough to require that inequality

(2.36) implies inequality (2.35), i.e.,

$$-\frac{\mathcal{C}_{4n}}{A_{4n}^2}+\frac{(\mathcal{B}_{4n}+F_{4n-3}(V))^2}{\mathcal{B}_{4n}+F_1(V)}\le\frac{F_{4n-3}^2(V)}{F_1(V)}. \tag{2.37}$$

Simplifying (2.37), we get

$$\frac{(F_{4n-3}(V)-F_1(V))^2}{(\mathcal{B}_{4n}+F_1(V))\,F_1(V)}\ge 1-\frac{\mathcal{C}_{4n}}{A_{4n}^2\mathcal{B}_{4n}}. \tag{2.38}$$

The last inequality holds always if the r.h.s of (2.38) is negative. Substituting expression for $\mathcal{C}_{4n}$ from (2.29) into (2.38) and applying the above claim, we arrive at the lower bound,

$$g_{4n}\ge\lambda_{4n}\sqrt[4n-1]{\pi_{4n}}, \tag{2.39}$$

where

$$\lambda_{4n}=\left(\frac{(4n-1)!}{\alpha_{4n}\mathcal{B}_{4n}}\binom{4n-3}{2n-2}\right)^{1/(4n-1)},\qquad \alpha_{4n}=1-A_{4n}^2.$$

2.4.2 Lower Bound of g_{4n+2}, $n\ge 1$

Syzygy identities (2.25, 2.26) for symmetric (not CI) semigroup S_{4n+2} read

$$\sum_{j=1}^{B_{4n+2}^{v}}f_{2l-1}(v_j)-\sum_{j=1}^{B_{4n+2}^{u}}f_{2l-1}(u_j)=\mathcal{B}_{4n+2},\quad 1\le l<2n, \tag{2.40}$$

$$\sum_{j=1}^{B_{4n+2}^{v}}f_{4n+1}(v_j)-\sum_{j=1}^{B_{4n+2}^{u}}f_{4n+1}(u_j)=\mathcal{C}_{4n+2}, \tag{2.41}$$

$$\mathcal{B}_{4n+2}=\frac{\beta_{2n}-\ldots-\beta_1+1}{2},$$

$$\mathcal{C}_{4n+2}=\mathcal{B}_{4n+2}-(4n+1)!\binom{4n-1}{2n-1}\frac{\pi_{4n+2}}{g_{4n+2}^{4n+1}},$$

which is similar to (2.28, 2.29). Moreover, the further reasoning in the present section completely follow the approach developed in the previous Sect. 2.4.1 with necessary replacements

$$F_{4n-3}(V) \to F_{4n-1}(U), \qquad F_{4n-1}(V) \to F_{4n+1}(U),$$

$$V_k \to U_k = \sum_{j=1}^{B^v_{4n}} u_j^k, \qquad (A, \mathcal{B}, \mathcal{C})_{4n} \to (A, \mathcal{B}, \mathcal{C})_{4n+2}.$$

Skipping the intermediate calculations, which follow formulas (2.30–2.38), we present the final expression for the lower bound g_{4n+2},

$$g_{4n+2} \geq \lambda_{4n+2} \sqrt[4n+1]{\pi_{4n+2}}, \tag{2.42}$$

where

$$\lambda_{4n+2} = \left(\frac{(4n+1)!}{\alpha_{4n+2}\mathcal{B}_{4n+2}} \binom{4n-1}{2n-1} \right)^{1/(4n+1)}, \qquad \alpha_{4n+2} = 1 - A^2_{4n+2}.$$

Note that a substitution $n = 1$ into (2.42) yields formula (2.10).

We finish this section by expression of the lower bound g_{2n} for symmetric (not CI) semigrops S_{2n},

$$g_{2n} \geq \lambda_{2n} \sqrt[2n-1]{\pi_{2n}}, \qquad \lambda_{2n} = \left(\frac{(2n-1)!}{\alpha_{2n}\mathcal{B}_{2n}} \binom{2n-3}{n-2} \right)^{1/(2n-1)}, \tag{2.43}$$

which unifies both formulas (2.39) and (2.42). In Appendix A, we present a few first values of α_{2n}.

3 Symmetric (Not CI) Semigroups S_m, $m = 1 (\mathrm{mod}\ 2)$

3.1 Symmetric (CI) Semigroups S_3

Consider a symmetric numerical semigroup S_3, which is always CI and minimally generated by three positive integers. Its Hilbert series $H(\mathsf{S}_3; z)$ reads:

$$H(\mathsf{S}_3; z) = \frac{Q_3(z)}{\prod_{i=1}^3 \left(1 - z^{d_i}\right)}, \qquad Q_3(z) = 1 - z^x - z^{g_3 - x} + z^{g_3}.$$

Rescaling $y = x/g_3$, present a non-trivial syzygy identity for S_3

$$y^2 + (1-y)^2 - 1 = -2!\frac{\pi_3}{g_3^2}, \qquad \text{or} \qquad y(1-y) = \frac{2!}{2}\frac{\pi_3}{g_3^2}. \tag{3.1}$$

Quadratic equation (3.1) has a positive solution if $g_3 \geq \lambda_3 \sqrt{\pi_3}$, $\lambda_3 = 2$, that gives a lower bound for g_3.

3.2 Symmetric (Not CI) Semigroups S_5

Consider a symmetric (not CI) numerical semigroup S_5, minimally generated by five positive integers. Its Hilbert series $H(\mathsf{S}_5; z)$ reads:

$$H(\mathsf{S}_5; z) = \frac{Q_5(z)}{\prod_{i=1}^{5}\left(1 - z^{d_i}\right)}, \tag{3.2}$$

$$Q_5(z) = 1 - \sum_{j=1}^{\beta_1} z^{x_{1,j}} + \sum_{j=1}^{\beta_1-1}\left(z^{x_{2,j}} + z^{g_5 - x_{2,j}}\right) - \sum_{j=1}^{\beta_1} z^{g_5 - x_{1,j}} + z^{g_5},$$

Polynomial identities for degrees of syzygies

$$\sum_{j=1}^{\beta_1} x_{1,j}^r - \sum_{j=1}^{\beta_1-1}\left[x_{2,j}^r + (g_5 - x_{2,j})^r\right] + \sum_{j=1}^{\beta_1}(g_5 - x_{1,j})^r - g_5^r = 0, \quad 1 \le r \le 3, \tag{3.3}$$

$$\sum_{j=1}^{\beta_1} x_{1,j}^4 - \sum_{j=1}^{\beta_1-1}\left[x_{2,j}^4 + (g_5 - x_{2,j})^4\right] + \sum_{j=1}^{\beta_1}(g_5 - x_{1,j})^4 - g_5^4 = -4!\pi_5. \tag{3.4}$$

Rescale $y_{i,j} = x_{i,j}/g_5$, $i = 1, 2$, and present two (among ... (3.3, 3.4)) non trivial identities with $r = 2, 4$,

$$\sum_{j=1}^{\beta_1-1} y_{2,j}(1 - y_{2,j}) - \sum_{j=1}^{\beta_1} y_{1,j}(1 - y_{1,j}) = 0,$$

$$\sum_{j=1}^{\beta_1-1} y_{2,j}^2(1 - y_{2,j})^2 - \sum_{j=1}^{\beta_1} y_{1,j}^2(1 - y_{1,j})^2 = \frac{4!}{2}\frac{\pi_5}{g_5^4}.$$

In [4], it was shown that the lower bound for g_5 is given by

$$g_5 \ge \lambda_5 \sqrt[4]{\pi_5}, \qquad \lambda_5 = 4\sqrt[4]{\frac{3(\beta_1 - 1)}{4\beta_1}}. \tag{3.5}$$

3.3 Symmetric (Not CI) Semigroups S_{2n+1} with Arbitrary n

Consider a symmetric numerical semigroup S_{2n+1}, which is not CI and minimally generated by $2n + 1$ positive integers. Its Hilbert series $H(\mathsf{S}_{2n+1}; z)$ is given:

$$H\left(\mathsf{S}_{2n+1};z\right)=Q_{2n+1}(z)\prod_{i=1}^{2n+1}\left(1-z^{d_i}\right)^{-1},\qquad \text{where} \tag{3.6}$$

$$Q_{2n+1}(z)=1-\sum_{j=1}^{\beta_1}z^{x_{1,j}}+\ldots+(-1)^n\sum_{j=1}^{2\mathcal{B}_{2n+1}}\left(z^{x_{n-1,j}}+z^{g_{2n+1}-x_{n-1,j}}\right)\pm\ldots-$$
$$-\sum_{j=1}^{\beta_1}z^{g_{2n+1}-x_{1,j}}+t^{g_{2n+1}}.$$

Syzygy identities for S_{2n+1} read, $1\le r<2n$,

$$\sum_{j=1}^{\beta_1}x_{1,j}^r-\ldots+(-1)^n\sum_{j=1}^{\beta_{n-1}}x_{n-1,j}^r-(-1)^n\sum_{j=1}^{2\mathcal{B}_{2n+1}}\left(x_{n,j}^r+(g_{2n+1}-x_{n,j})^r\right)+$$
$$(-1)^n\sum_{j=1}^{\beta_{n-1}}(g_{2n+1}-x_{n-1,j})^r\pm\ldots+\sum_{j=1}^{\beta_1}(g_{2n+1}-x_{1,j})^r-g_{2n+1}^r=0, \tag{3.7}$$

$$\sum_{j=1}^{\beta_1}x_{1,j}^{2n}-\ldots+(-1)^n\sum_{j=1}^{\beta_{n-1}}x_{n-1,j}^{2n}-(-1)^n\sum_{j=1}^{2\mathcal{B}_{2n+1}}\left(x_{n,j}^{2n}+(g_{2n+1}-x_{n,j})^{2n}\right)+$$
$$(-1)^n\sum_{j=1}^{\beta_{n-1}}(g_{2n+1}-x_{n-1,j})^{2n}\pm\ldots+\sum_{j=1}^{\beta_1}(g_{2n+1}-x_{1,j})^{2n}-g_{2n+1}^{2n}=$$
$$-(2n)!\pi_{2n+1}, \tag{3.8}$$

where

$$\mathcal{B}_{2n+1}=\frac{1}{2}\left(\beta_{n-1}-\beta_{n-2}+\ldots+(-1)^n\beta_1-(-1)^n\right).$$

Only n of $2n$ identities in (3.7, 3.8) are not trivial, these are for $r=2l$, $1\le l\le n$. Rescale variables $y_{i,j}=x_{i,j}/g_{2n+1}$, $i=1,\ldots,n$ and denote by Y_{2n+1} a set of syzygies degrees of S_{2n+1},

$$Y_{2n+1}= \tag{3.9}$$
$$\left\{y_{1,1},\ldots,y_{1,\beta_1},\ y_{2,1},\ldots,y_{2,\beta_2},\ y_{n-1,1},\ldots,y_{n-1,\beta_{n-1}},\ y_{n,1},\ldots,y_{n,2\mathcal{B}_{2n+1}}\right\}.$$

Making the partial summation in the l.h.s. of syzygy identities (3.7, 3.8) and collecting together the power terms for every variable $y_{i,j}$ separately, we arrive at following syzygy identities, $1\le l<n$,

Table 2 Numbers B^u_{2n+1} and B^v_{2n+1} for semigroup S_{2n+1}

S_{2n+1}	S_3	S_5	S_7	S_9	S_{11}	S_{13}	S_{15}
B^v_{2n+1}	0	$\beta_1 - 1$	β_2	$\beta_1 + \beta_3 - 1$	$\beta_2 + \beta_4$	$\beta_1 + \beta_3 + \beta_5 - 1$	$\beta_2 + \beta_4 + \beta_6$
B^u_{2n+1}	1	β_1	$\beta_2 + 1$	$\beta_1 + \beta_3$	$\beta_2 + \beta_4 + 1$	$\beta_1 + \beta_3 + \beta_5$	$\beta_2 + \beta_4 + \beta_6 + 1$

$$\sum_{j=1}^{2\mathcal{B}_{2n+1}} \phi_{2l}(y_{n,j}) - \sum_{j=1}^{\beta_{n-1}} \phi_{2l}(y_{n-1,j}) + \ldots - (-1)^n \sum_{j=1}^{\beta_1} \phi_{2l}(y_{1,j}) = 0, \tag{3.10}$$

$$\sum_{j=1}^{2\mathcal{B}_{2n+1}} \phi_{2n}(y_{n,j}) - \sum_{j=1}^{\beta_{n-1}} \phi_{2n}(y_{n-1,j}) + \ldots - (-1)^n \sum_{j=1}^{\beta_1} \phi_{2n}(y_{1,j}) = \frac{(2n)!}{2}\frac{\pi_{2n+1}}{g_{2n+1}^{2n}}, \tag{3.11}$$

where a polynomial $\phi_{2l}(t)$ reads

$$\phi_{2l}(t) = \phi_2^l(t), \qquad \phi_2(t) = t(1-t), \qquad \phi_{2l}(t) \le 2^{-2l}. \tag{3.12}$$

Disjoin Y_{2n+1} in two subsets Y^u_{2n+1} and Y^v_{2n+1} of odd and even syzygy classes, respectively,

$$Y_{2n+1} = Y^u_{2n+1} \bigcup Y^v_{2n+1}, \qquad \text{where} \tag{3.13}$$

$$Y^u_{2n+1} = \bigcup_{j=1}^{2j\le n+1} \left\{y_{2j-1,1}, \ldots, y_{2j-1,\beta_{2j-1}}\right\}, \qquad Y^v_{2n+1} = \bigcup_{j=1}^{2j\le n} \left\{y_{2j,1}, \ldots, y_{2j,\beta_{2j}}\right\}.$$

Their cardinalities $B^u_{2n+1} = \#Y^u_{2n+1}$ and $B^v_{2n+1} = \#Y^v_{2n+1}$ are given in two congruence classes,

$$B^u_{4n+1} = \sum_{j=1}^{n} \beta_{2j-1}, \qquad B^v_{4n+1} = \sum_{j=1}^{n} \beta_{2j-1} - 1,$$

$$B^u_{4n+3} = \sum_{j=1}^{n} \beta_{2j} + 1, \qquad B^v_{4n+3} = \sum_{j=1}^{n} \beta_{2j}, \tag{3.14}$$

such that $B^u_{4n+1} - B^v_{4n+1} = B^u_{4n+3} - B^v_{4n+3} = 1$.

In Table 2 we present a few first numbers B^u_{2n+1} and B^v_{2n+1} for semigroup S_{2n+1}.

Rename the variables $y_{i,j}$ by u_j and v_j and enumerate them as follows,

$$Y^u_{2n+1} = \left\{u_1, \ldots, u_{\beta_1},\ \ u_{\beta_1+1}, \ldots, u_{\beta_1+\beta_3},\ \ u_{\beta_1+\beta_3+1},\ \ \ldots,\ \ u_{2B^u_{2n+1}-1}, u_{2B^u_{2n+1}}\right\},$$

$$Y^v_{2n+1} = \left\{v_1, \ldots, v_{\beta_2},\ \ v_{\beta_2+1} \ldots, v_{\beta_2+\beta_4},\ \ v_{\beta_2+\beta_4+1},\ \ \ldots,\ \ v_{2B^v_{2n+1}-1}, v_{2B^v_{2n+1}}\right\}. \quad (3.15)$$

Then we obtain syzygy identities

$$\sum_{j=1}^{B^v_{2n+1}} \phi_{2l}(v_j) - \sum_{j=1}^{B^u_{2n+1}} \phi_{2l}(u_j) = 0, \quad 1 \le l < n, \quad 0 < v_j, u_j < 1, \quad (3.16)$$

$$\sum_{j=1}^{B^v_{2n+1}} \phi_{2n}(v_j) - \sum_{j=1}^{B^u_{2n+1}} \phi_{2n}(u_j) = (-1)^n \frac{(2n)!}{2} \frac{\pi_{2n+1}}{g^{2n}_{2n+1}}, \quad (3.17)$$

To make worth of inequalities in Appendix B, we introduce new variables according to definition (3.12) of functions $\phi_2(t) = t(1-t)$,

$$L_j = u_j(1-u_j), \qquad M_j = v_j(1-v_j), \quad (3.18)$$

and substitute them into (3.16, 3.17),

$$\sum_{j=1}^{B^v_{2n+1}} M^l_j - \sum_{j=1}^{B^u_{2n+1}} L^l_j = 0, \quad 1 \le l < n,$$
$$\sum_{j=1}^{B^v_{2n+1}} M^n_j - \sum_{j=1}^{B^u_{2n+1}} L^n_j = (-1)^n \frac{(2n)!}{2} \frac{\pi_{2n+1}}{g^{2n}_{2n+1}}. \quad (3.19)$$

By comparison of Eqs. (3.19) and (B.2) in Appendix B, we have a correspondence

$$N_x = B^v_{2n+1}, \qquad N_y = B^u_{2n+1}, \qquad K_n = \frac{(2n)!}{2} \frac{\pi_{2n+1}}{g^{2n}_{2n+1}}, \quad n \ge 2, \quad (3.20)$$
$$\Lambda_n\left(B^v_{2n+1}\right) = 1 - \prod_{k=1}^{n-1}\left(1 - \frac{k}{B^v_{2n+1}}\right), \quad \Lambda_n\left(B^u_{2n+1}\right) = 1 - \prod_{k=1}^{n-1}\left(1 - \frac{k}{B^u_{2n+1}}\right).$$

We give a few expressions of $\Lambda_n\left(B^v_{2n+1}\right)$ and $\Lambda_n\left(B^u_{2n+1}\right)$ for small $n = 2, 3, 4$,

$$\Lambda_2\left(B_5^v\right)=\frac{1}{\beta_1-1},\quad \Lambda_2\left(B_5^u\right)=\frac{1}{\beta_1},$$
$$\Lambda_3\left(B_7^v\right)=\frac{3\beta_2-2}{\beta_2^2},\quad \Lambda_3\left(B_7^u\right)=\frac{3\beta_2+1}{(\beta_2+1)^2},$$
$$\Lambda_4\left(B_9^v\right)=\frac{6(\beta_1+\beta_3)^2-23(\beta_1+\beta_3)+23}{(\beta_1+\beta_3-1)^3},$$
$$\Lambda_4\left(B_9^u\right)=\frac{6(\beta_1+\beta_3)^2-11(\beta_1+\beta_3)+6}{(\beta_1+\beta_3)^3}. \tag{3.21}$$

Combining (3.20) with the upper bound for K_n, found in Appendix B, formula (B.14), we arrive at the lower bound for g_{2n+1}

$$g_{2n+1}\geq\lambda_{2n+1}\sqrt[2n]{\pi_{2n+1}}, \tag{3.22}$$

where

$$\lambda_{2n+1}=\frac{2}{\sqrt{B_{2n+1}^u}}\left(\frac{(n-1)!(2n)!}{2\Delta_n(B_{2n+1}^v,B_{2n+1}^u)}\right)^{1/2n},$$

and

$$\Delta_n(B_{2n+1}^v,B_{2n+1}^u)=\Lambda_n\left(B_{2n+1}^v\right)-\Lambda_n\left(B_{2n+1}^u\right). \tag{3.23}$$

Substituting (3.21, 3.23) into (3.22), we give a few expressions of λ_{2n+1} for small $n=2,3,4$,

$$\lambda_5=2\sqrt[4]{\frac{12\,(\beta_1-1)}{\beta_1}},\qquad \lambda_7=2\sqrt[6]{\frac{6!\,\beta_2^2}{(\beta_2^2-1)(3\beta_2+2)}}, \tag{3.24}$$
$$\lambda_9=2\sqrt[8]{\frac{3\cdot 8!\,(\beta_{1,3}-1)^3}{\beta_{1,3}(\beta_{1,3}-2)(\beta_{1,3}-3)(6\beta_{1,3}^2-4\beta_{1,3}+1)}},\qquad \beta_{1,3}=\beta_1+\beta_3.$$

Symmetric CI semigroups S_3 are also described by the last equation in (3.19); according to Table 2, here we have to put $n=1$ and $B_3^v=0$, $B_3^u=1$. Combining (3.19) and (3.16) we obtain quadratic equation

$$g_3^2u_1(1-u_1)=\pi_3,\qquad 0<u_1<1, \tag{3.25}$$

which necessary leads to the lower bound $g_3\geq\lambda_3\sqrt{\pi_3}$, where $\lambda_3=2$, in accordance with Sect. 3.1. Another expression for λ_5 in (3.24) is also coincided with that, presented in (3.5).

A Appendix: Polynomials $f_{2l-1}(t)$

In this Appendix we discuss the basic properties of polynomials $f_{2l-1}(t)$, defined in (2.19), which are necessary in the present paper.

Lemma 1 *Polynomials $f_{2l-1}(t)$, being shifted by $1/2$ in t and f directions, become antisymmetric,*

$$f_{2l-1}\left(t+\frac{1}{2}\right)-\frac{1}{2}=-f_{2l-1}\left(-t+\frac{1}{2}\right)+\frac{1}{2}, \tag{A.1}$$

Proof Consider a sum of two beta functions,

$$\mathsf{B}\left[x+\frac{1}{2},n,n\right]+\mathsf{B}\left[-x+\frac{1}{2},n,n\right]=$$
$$\frac{1}{2^{2n-1}}\int_{-\pi/2}^{\pi/2}(\cos\omega)^{2n-1}\,d\omega=\frac{1}{n}\binom{2n-1}{n-1}^{-1}. \tag{A.2}$$

Then, according to definition (2.19) of polynomials $f_{2l-1}(t)$, identity (A.2) results in (A.1). □

Antisymmetricity relation (A.1) has several other spinoffs,

$$f_{2l-1}(0)=0,\quad f_{2l-1}\left(\frac{1}{2}\right)=\frac{1}{2},\quad f_{2l-1}(1)=1,$$
$$\left.\frac{df_{2l-1}(t)}{dt}\right|_{l>1}=0,\quad t=0,1. \tag{A.3}$$

The last equality follows also by straightforward calculation of the first derivative,

$$\frac{df_{2l-1}(t)}{dt}=l\binom{2l-1}{l-1}\phi^{l-1}(t),\qquad \phi(t)=t(1-t). \tag{A.4}$$

By formula (A.4), the first derivative arrives its maximal value at $t=1/2$ with asymptotic behavior

$$\left.\frac{df_{2l-1}(t)}{dt}\right|_{t=1/2}=\frac{l}{4^{l-1}}\binom{2l-1}{l-1},\qquad \left.\frac{df_{2l-1}(t)}{dt}\right|_{t=1/2}\stackrel{l\to\infty}{\longrightarrow}2\sqrt{\frac{l}{\pi}}, \tag{A.5}$$

and satisfies the following inequality,

$$\left.\frac{df_{2l_2-1}(t)}{dt}\right|_{t=1/2}>\left.\frac{df_{2l_1-1}(t)}{dt}\right|_{t=1/2},\quad \text{iff}\quad l_2\geq l_1\geq 1. \tag{A.6}$$

A straightforward calculation of difference $f_{2l+1}(t) - f_{2l-1}(t)$ gives

$$f_{2l+1}(t) - f_{2l-1}(t) = -\binom{2l-1}{l}(1-2t)\phi^l(t), \tag{A.7}$$

that leads to important inequalities: for every $l_2 > l_1$ it holds

$$\begin{aligned} &a) \quad f_{2l_2-1}(t) < f_{2l_1-1}(t), \quad 0 < t < 1/2, \\ &b) \quad f_{2l_2-1}(t) > f_{2l_1-1}(t), \quad 1/2 < t < 1. \end{aligned} \tag{A.8}$$

Consider in interval [0, 1] a non-negative differentiable function $\rho_n(t)$,

$$\rho_n(t) = \frac{\sqrt{t f_{2n-1}(t)}}{f_{2n-3}(t)}, \quad n \geq 3, \qquad \rho_n\left(\frac{1}{2}\right) = \rho_n(1) = 1, \tag{A.9}$$

which behaves in vicinity of $t = 0$ as follows,

$$\rho_3(t) \simeq \frac{\sqrt{10}}{3}\left(1 - \frac{t}{12} + \mathcal{O}\left(t^2\right)\right), \qquad \rho_n(t) \simeq t^{-(n-3)/2}, \quad n \geq 4. \tag{A.10}$$

Its derivative

$$\begin{aligned} \frac{d\rho_n(t)}{dt} = \frac{1}{f_{2n-3}^2(t)} \Bigg[& \frac{f_{2n-3}(t)}{2\sqrt{t f_{2n-1}(t)}} \left(f_{2n-1}(t) + t\frac{df_{2n-1}(t)}{dt} \right) - \\ & \sqrt{t f_{2n-1}(t)} \frac{df_{2n-3}(t)}{dt} \Bigg], \end{aligned} \tag{A.11}$$

according to (A.3, A.4, A.5), attains at $t = 0, 1/2, 1$ the following values

$$\begin{aligned} &\left.\frac{d\rho_n(t)}{dt}\right|_{t=0} < 0, \quad \left.\frac{d\rho_n(t)}{dt}\right|_{t=1} \frac{1}{2}, \quad \left.\frac{d\rho_n(t)}{dt}\right|_{t=1/2} = 1 - R_n, \\ &R_n = \frac{2n-3}{2^{2n-3}}\binom{2n-3}{n-2}. \end{aligned} \tag{A.12}$$

Lemma 2 *The following inequality holds,*

$$\left.\frac{d\rho_n(t)}{dt}\right|_{t=1/2} < 0, \qquad n \geq 3. \tag{A.13}$$

Proof Consider derivative $d\rho_n(t)/dt$ at $t = 1/2$ and estimate R_n, defined in (A.12), if $n \geq 3$,

$$R_n = \frac{2n-3}{2^{2n-3}} \frac{(2n-3)!!(2n-4)!!}{(n-1)!(n-2)!} = \frac{2n-3}{2^{n-1}} \frac{(2n-3)!!}{(n-1)!} = \tag{A.14}$$
$$\frac{2n-3}{2^{n-1}} \prod_{k=1}^{n-1} \frac{2(n-k)-1}{n-k} = \frac{2n-3}{2^{n-1}} \prod_{p=1}^{n-1} \frac{2p-1}{p}, \quad \rightarrow$$
$$R_3 = \frac{9}{8} \quad \text{and} \quad R_n > \frac{2n-3}{2^{n-1}} \prod_{p=1}^{n-1} \frac{2p}{p+1} = 2 - \frac{3}{n} > 1, \quad n \geq 4,$$

Combining (A.14) and equality $d\rho_n(t)/dt = 1 - R_n$ at $t = 1/2$ from (A.12), we arrive at (A.13). □

By Rolle's theorem, Lemma 2 together with equality in (A.12) results in existence of at least one point $t_* \in (1/2, 1)$ where $\frac{d\rho_n(t)}{dt}\Big|_{t=t_*} = 0$ (see Figure 2). At this point $\rho_n(t)$ arrives its minimal value and is bounded from below,

$$A_{2n} \leq \rho_n(t) \leq 1, \qquad \frac{1}{2} \leq t \leq 1, \quad n \geq 3. \tag{A.15}$$

In Table 3 we present ten first values of A_{2n} and $\alpha_{2n} = 1 - A_{2n}^2$, which were calculated numerically.

In contrast to the interval [1/2, 1], the behavior of the derivative $d\rho_n(t)/dt$ at the interval [0, 1/2] is difficult to study analytically because of the function $\mathsf{B}[t, n, n]$ arisen in formula (A.11). However, based on observations of numerical calculations (see Fig. 3), we put forward conjecture on behavior of functions $f_{2n-1}(t)$ at the interval [0, 1/2], that allows to conclude about the behavior of $\rho_n(t)$ at the entire interval [0, 1].

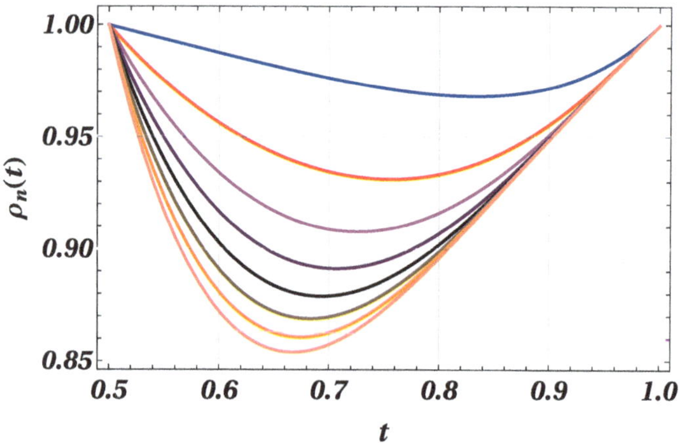

Fig. 2 Plot of the functions $\rho_l(t)$, $t \in [1/2, 1]$, with different l: (*in blue*) $l = 3$, (*in red*) $l = 4$, (*in magenta*) $l = 5$, (*in purple*) $l = 6$, (*in black*) $l = 7$, (*in brown*) $l = 8$, (*in orange*) $l = 9$, (*in pink*) $l = 10$

Table 3 Numbers A_{2n} and α_{2n} for semigroups S_{2n}

S_{2n}	S_6	S_8	S_{10}	S_{12}	S_{14}	S_{16}	S_{18}	S_{20}	S_{22}	S_{24}
A_{2n}	0.9682	0.931	0.908	0.891	0.879	0.869	0.861	0.857	0.848	0.844
α_{2n}	0.0626	0.1332	0.1755	0.2061	0.2273	0.2448	0.2587	0.2655	0.2809	0.2876

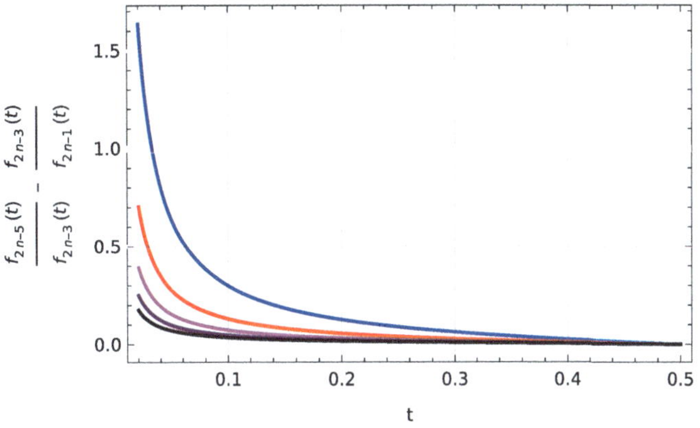

Fig. 3 Plot of the function $\frac{f_{2n-5}(t)}{f_{2n-3}(t)} - \frac{f_{2n-3}(t)}{f_{2n-1}(t)}$, $t \in [0, 1/2]$, with different n: *(in blue)* $n = 3$, *(in red)* $n = 4$, *(in magenta)* $n = 5$, *(in purple)* $n = 6$, *(in black)* $n = 7$

Conjecture 1 *Functions $f_{2n-1}(t)$ satisfy the following inequality,*

$$\frac{f_{2n-5}(t)}{f_{2n-3}(t)} > \frac{f_{2n-3}(t)}{f_{2n-1}(t)}, \qquad 0 \le t < \frac{1}{2}, \quad n \ge 3. \tag{A.16}$$

Since, according to (A.8), there holds inequality $f_{2n-5}(t) > f_{2n-3}(t)$ at interval $[0, 1/2]$, we multiply the l.h.s. of inequality (A.16) by the ratio $f_{2n-5}(t)/f_{2n-3}(t)$ and get after simple manipulations,

$$\frac{t\ f_{2n-1}(t)}{f_{2n-3}^2(t)} > \frac{t\ f_{2n-3}(t)}{f_{2n-5}^2(t)}. \tag{A.17}$$

Continuing (A.17) rightward and making worth of notation in (A.9), we arrive at sequence of inequalities

$$\rho_n(t) > \rho_{n-1}(t) > \ldots > \rho_3(t), \qquad 0 \le t < \frac{1}{2}, \quad n \ge 3. \tag{A.18}$$

Calculating $\rho_3(t)$ with help of $f_3(t)$, $f_5(t)$ from (2.18), we get $\rho_3(t) = (1-t)(1-2t)$, that together with (A.18) yields $\rho_n(t) > 1$, if $0 \le t < 1/2$. Combining the latter with (A.15), we arrive at Lemma.

Lemma 3 *The following inequality holds,*

$$\rho_n(t) \geq A_{2n}, \qquad 0 \leq t \leq 1, \quad n \geq 3.$$

B Appendix: Newton-Maclaurin's Inequalities for Symmetric Power Sums

Denote by X_k and Y_k two power sums,

$$X_k = \sum_{j=1}^{N_x} x_j^k, \qquad Y_k = \sum_{j=1}^{N_y} y_j^k, \qquad x_j, y_j \geq 0, \tag{B.1}$$

and consider n equations for $N_x + N_y$ real variables x_j, y_j,

$$X_k - Y_k = 0, \quad 1 \leq l < n, \qquad X_n - Y_n = (-1)^n K_n, \quad K_n > 0. \tag{B.2}$$

Introduce symmetric polynomials $\mathcal{Q}_k = \sum_{i_1<i_2<\ldots<i_k}^r q_{i_1} q_{i_2} \ldots q_{i_k}$, where $q_j \geq 0$,

$$\mathcal{Q}_0 = 1, \quad \mathcal{Q}_1 = \sum_{i=1}^r q_i, \quad \mathcal{Q}_2 = \sum_{i<j}^r q_i q_j, \quad \ldots, \quad \mathcal{Q}_r = \prod_{i=1}^r q_i, \tag{B.3}$$

which are related by Newton's recursion identities to power sums $Q_k = \sum_{j=1}^r q_j^k$,

$$\begin{aligned} p\,\mathcal{Q}_p &= \sum_{k=1}^{p} (-1)^{k-1} Q_k \mathcal{Q}_{p-k}, \qquad \mathcal{Q}_1 = Q_1, \qquad \text{e.g.,} \\ 2\mathcal{Q}_2 &= Q_1^2 - Q_2, \\ 6\mathcal{Q}_3 &= Q_1^3 - 3Q_1Q_2 + 2Q_3, \\ 24\mathcal{Q}_4 &= Q_1^4 - 6Q_1^2Q_2 + 8Q_1Q_3 + 3Q_2^2 - 6Q_4, \\ 120\mathcal{Q}_5 &= Q_1^5 - 10Q_1^3Q_2 + 20Q_1^2Q_3 - 30Q_1Q_4 + 15Q_1Q_2^2 - 20Q_2Q_3 + 24Q_5, \end{aligned} \tag{B.4}$$

A simple analysis shows that for any $p < r$ the following formula holds,

$$\begin{aligned} p!\,\mathcal{Q}_p &= Q_1^p - \Phi_p(Q_1, \ldots, Q_{p-1}) - (-1)^p (p-1)!\, Q_p, \quad p \geq 2, \\ &\Phi_2(Q_1) = 0, \end{aligned} \tag{B.5}$$

where the multivariate polynomial $\Phi_p(Q_1, \ldots, Q_{p-1})$ comprises monomials $\prod_{k=1}^{p-1} Q_k^{a_k}$ with integer degrees $a_k \geq 0$, satisfying Diophantine's equation $\sum_{k=1}^{p-1} k a_k = p$ and excluding the solution $a_1 = p$.

Recall the Newton-Maclaurin inequalities [7] for polynomials $\mathcal{Q}_k$,

$$\frac{Q_1}{r} \geq \left(\frac{Q_2}{\binom{r}{2}}\right)^{1/2} \geq \left(\frac{Q_3}{\binom{r}{3}}\right)^{1/3} \geq \ldots \geq (Q_r)^{1/r}. \tag{B.6}$$

Choose $p < r$ and present one of inequalities in (B.6),

$$\left(\frac{Q_1}{r}\right)^p > \frac{Q_p}{\binom{r}{p}}, \qquad \text{or} \qquad \frac{r!\, Q_1^p}{(r-p)!\, r^p} > p!\, Q_p, \tag{B.7}$$

in terms of power sums Q_p. Substituting (B.5) into inequality (B.7) and simplifying the last, we obtain

$$\Lambda_p(r)\, Q_1^p < \Phi_p(Q_1, \ldots, Q_{p-1}) + (-1)^p (p-1)!\, Q_p, \quad p \geq 2, \qquad \text{where} \tag{B.8}$$

$$\Lambda_p(r) = 1 - \prod_{k=1}^{p-1} \left(1 - \frac{k}{r}\right), \qquad \Lambda_p(r) > 0.$$

To resolve Eq. (B.2), apply inequalities (B.8) to the power sums X_k,

$$\Lambda_n(N_x)\, X_1^n < \Phi_n(X_1, \ldots, X_{n-1}) + (-1)^n (n-1)!\, X_n. \tag{B.9}$$

Substituting (B.2) into (B.9), we obtain

$$\Lambda_n(N_x)\, Y_1^n - (n-1)!\, K_n < \Phi_n(Y_1, \ldots, Y_{n-1}) + (-1)^n (n-1)!\, Y_n. \tag{B.10}$$

On the other hand, applying inequalities (B.8) to the power sums Y_k, we get

$$\Lambda_n(N_y)\, Y_1^n < \Phi_n(Y_1, \ldots, Y_{n-1}) + (-1)^n (n-1)!\, Y_n. \tag{B.11}$$

In order to make both inequalities (B.10) and (B.11) consistent, we have to find a range for Y_1 where they are satisfied. To provide both inequalities to be correct, it is sufficient (but not necessary) to require,

$$\Lambda_n(N_x)\, Y_1^n - (n-1)!\, K_n \geq \Lambda_n(N_y)\, Y_1^n.$$

Thus, we arrive at lower bound for Y_1,

$$Y_1 \geq D_n, \quad D_n = \left(\frac{(n-1)!}{\Delta_n(N_x, N_y)} K_n\right)^{1/n}, \tag{B.12}$$
$$\Delta_n(N_x, N_y) = \Lambda_n(N_x) - \Lambda_n(N_y).$$

To make the next step toward estimates, discussed in Sect. 3.3, represent y_j, introduced in the power sum (B.1), via new variable u_j,

$$y_j = u_j(1 - u_j),$$

and substitute it into inequality (B.12)

$$\mathcal{U}_1 - \mathcal{U}_2 \geq D_n, \qquad \mathcal{U}_k = \sum_{j=1}^{N_y} u_j^k, \qquad 0 \leq u_j \leq 1. \tag{B.13}$$

Making use of Newton-Maclaurin's inequality for the power sums, $\mathcal{U}_1^2 \leq N_y\, \mathcal{U}_2$, we rewrite inequality in (B.13) as follows,

$$\mathcal{U}_1^2 - N_y\, \mathcal{U}_1 + N_y D_n \leq 0.$$

The last inequality holds if $N_y \geq 4D_n$, or combining the latter with definition of D_n in (B.12), we arrive at the upper bound for K_n,

$$K_n \leq \left(\frac{N_y}{4}\right)^n \frac{\Delta_n(N_x, N_y)}{(n-1)!}. \tag{B.14}$$

C Appendix: Examples of Symmetric (Not CI) Semigroups

$$\begin{aligned}
&S_4 = \langle 8, 13, 15, 17\rangle,\\
&\qquad \beta_1 = 5, \qquad 35 > 34.2\\
&S_5 = \langle 6, 10, 14, 15, 19\rangle,\\
&\qquad \beta_1 = 7, \quad \beta_2 = 12, \qquad 87 > 79.2\\
&S_6 = \langle 7, 9, 11, 12, 13, 15\rangle,\\
&\qquad \beta_1 = 13, \quad \beta_2 = 31, \qquad 84 > 62.86\\
&S_7 = \langle 10, 11, 14, 15, 18, 19, 23\rangle,\\
&\qquad \beta_1 = 19, \quad \beta_2 = 57, \quad \beta_3 = 78, \qquad 137 > 60.4\\
&S_8 = \langle 10, 11, 14, 15, 18, 19, 23, 25\rangle,\\
&\qquad \beta_1 = 20, \quad \beta_2 = 65, \quad \beta_3 = 106, \qquad 162 > 92.35\\
&S_9 = \langle 10, 11, 14, 15, 18, 19, 23, 25, 28\rangle,\\
&\qquad \beta_1 = 21, \quad \beta_2 = 76, \quad \beta_3 = 146, \quad \beta_4 = 180, \qquad 190 > 53.95,\\
&S_{10} = \langle 10, 11, 14, 15, 18, 19, 23, 25, 28, 32\rangle,\\
&\qquad \beta_1 = 22, \quad \beta_2 = 85, \quad \beta_3 = 181, \quad \beta_4 = 258, \qquad 222 > 117.22.
\end{aligned}$$

References

1. H. Bresinsky, *Symmetric semigroups of integers generated by four elements*, Manuscripta Math., **17** (1975), 205-219.
2. L.G. Fel, *On Frobenius numbers for symmetric (not complete intersection) semigroups generated by four elements*, Semigroup Forum, **93** (2016), 423-426.
3. L.G. Fel, *Restricted partition functions and identities for degrees of syzygies in numerical semigroups*, Ramanujan J., **43** (2017), 465-491.
4. L.G. Fel, *Symmetric (not complete intersection) semigroups generated by five elements*, Integers: The Electronic J. of Comb. Number Theory, **18** (2018), # A44.
5. L.G. Fel, *Symmetric (not complete intersection) semigroups generated by six elements*, In *Numerical Semigroups*, Springer INdAM Series, **40** (2020) 93-109.
6. L.G. Fel, *Genera of Numerical Semigroups and Polynomial Identities for Degrees of Syzygies*, In *Combinatorial and Additive Number Theory 5*, Springer Proceedings in Mathematics & Statistics, **395** (2022), 153-178.
7. G.H. Hardy, J.E. Littlewood and G. Polya, *Inequalities*, Univ. Press, Cambridge, 1959.
8. J. Herzog and M. Kühl, *On the Betti numbers of finite pure and linear resolutions*, Comm. Algebra, **12** (1984), 1627-1646.
9. H.G. Killingbergtrø, *Betjening av figur i Frobenius' problem*, (In Norwegian), Normat (Nordisk Matematisk Tidskrift), **2** (2000), 75-82.
10. *On-Line Encyclopedia of Integer Sequences*, Sequence A091811, https:oeis.org A091811
11. R. P. Stanley, *Log-concave and unimodal sequences in algebra, combinatorics, and geometry*, in: Graph Theory and its Applications: East and West, Ann. New York Acad. Sci. **576**, New York Acad. Sci., New York (1989), 500-535.

Geometric Progressions in the Sets of Values of Reducible Cubic Forms

Bartosz Głowacki and Maciej Ulas

Abstract Let $a, Q \in \mathbf{Q}$ be given and consider the set $\mathcal{G}(a, Q) = \{aQ^i : i \in \mathbf{N}\}$ of terms of geometric progression with 0th term equal to a and the quotient Q. Let $f \in \mathbf{Q}(x, y)$ and $\mathcal{V}_f$ be the set of finite values of f. It is known that for any quadratic form f, there are infinitely many pairs a, Q such that $\mathcal{G}(a, Q) \subset \mathcal{V}_f$. In this note we prove that the same statement is true for reducible cubic forms. The main tool in the proof is the density of rational points on the quartic surface $z^3 = (x^2 - u)(y^2 - u)$, where $u \in \mathbf{Q} \setminus \{0\}$.

1 Introduction

For $a \in \mathbf{Q} \setminus \{0\}$ and $Q \in \mathbf{Q} \setminus \{-1, 0, 1\}$ let us consider the set

$$\mathcal{G}(a, Q) = \{aQ^i : i \in \mathbf{N}\},$$

i.e., an infinite set of values of geometric progression with initial term equal to a and the quotient equal to Q. Next, for $f \in \mathbf{Q}(x, y)$ define the set

$$\mathcal{V}_f = \{f(u, v) \in \mathbf{Q} : u, v \in \mathbf{Q} \text{ for which } f(u, v) \text{ determined}\}$$

In a recent paper [8] the second author studied the following problem.

Problem 1 Let $f \in \mathbf{Q}(x, y)$ be given. Do there exist rational numbers a, Q such that $\mathcal{G}(a, Q) \subset \mathcal{V}_f$?

B. Głowacki · M. Ulas (✉)
Faculty of Mathematics and Computer Science, Institute of Mathematics, Jagiellonian University, Łojasiewicza 6, 30 - 348 Kraków, Poland
e-mail: maciej.ulas@uj.edu.pl

B. Głowacki
e-mail: bartosz16.glowacki@student.uj.edu.pl

M. B. Nathanson (ed.), *Combinatorial and Additive Number Theory VI*,
Springer Proceedings in Mathematics & Statistics 464,
https://doi.org/10.1007/978-3-031-65064-2_9

This problem was investigated for the particular rational function $f(x, y) = (y^2 - x^3)/x$. Main motivation for these investigations was a connection of Problem 1 with an equivalent problem related to elliptic curves. Some theoretical as well as experimental results were obtained in this case. Moreover, in the cited paper, some general results were presented. More precisely, it is proved that the Problem 1 has a solution for rational functions of the form

$$f(x, y) = \frac{xg_1(y) + g_2(y)}{xh_1(y) + h_2(y)},$$

where $g_1, g_2, h_1, h_2 \in \mathbf{Q}(y)$ and $g_1h_2 \neq h_1g_2$ in $\mathbf{Q}(y)$, as well for a general rational function which is homogenous of degree 1. Problem 1 has also a positive solution for a general homogenous quadratic form. Thus, it is natural to ask what can be proved in the case of cubic forms. It is clear that in the case of f which is homogenous of degree d, to get $\mathcal{G}(a, Q) \subset \mathcal{V}_f$, it is enough to consider only the existence of x_i, y_i such that $f(x_i, y_i) = aQ^i$ for $i = 0, 1, \ldots, d-1$. Indeed, this is consequence of the equalities $f(Q^n x_i, Q^n y_i) = aQ^{dn+i}$ for $i = 0, 1, \ldots, d-1$ and $n \in \mathbf{N}$. In particular, if $f \in \mathbf{Z}[x, y]$ is a cubic form then it is enough to work with the system of equations

$$f(x_0, y_0) = a, \quad f(x_1, y_1) = aQ, \quad f(x_2, y_2) = aQ^2.$$

Let us note that in the case of a homogenous cubic form, if a, Q are fixed, then the equation $f(x, y) = aQ^i$ has only finitely many solutions (x, y, i) in co-prime integers x, y and $i \in \mathbf{N}$. This was proved by Mahler in [3] (generalizing an earlier result of Thue [7] concerning the equation $f(x, y) = A$, where $A \in \mathbf{Z} \setminus \{0\}$ is fixed). It is a difficult question whether for any choice of the cubic form this system, or equivalently Problem 1, has a solution in rational numbers $x_i, y_i, i = 0, 1, 2, a \in \mathbf{Q} \setminus \{0\}$ and $Q \in \mathbf{Q} \setminus \{-1, 0, 1\}$. The aim of the paper is to prove that this is the case for reducible (over $\mathbf{Q}$) cubic forms.

Let us describe the content of the paper in some details. In Sect. 2 we present basic reductions and observations concerning Problem 1 in case when f is a reducible homogeneous cubic form. In Sect. 3 we investigate the existence of rational points on the quartic surface $S_u : z^3 = (x^2 - u)(y^2 - u)$, where $u \in \mathbf{Q} \setminus \{0\}$ is fixed. We prove that for each $u \in \mathbf{Q} \setminus \{0\}$, the set of rational curves on S_u is infinite, and the set of rational points on S_u is dense in the set of real points lying on S_u (the density is in the Euclidean topology). The existence of a rational curve on S_u is used in the proof of main result of the paper, which says that Problem 1 has a solution for reducible cubic forms with three distinct roots. Finally, in the last section we formulate several question and present results of our computational experiments.

2 Basic Results and Observations

Let $f \in \mathbf{Z}[x, y]$ be a homogeneous form of degree d. First of all let us note that if $L_i = a_i x + b_i y \in \mathbf{Q}[x, y]$ is a linear form for $i = 1, 2$ and L_1, L_2 are independent, then $\mathcal{G}(a, Q) \subset \mathcal{V}_f$ if and only if $\mathcal{G}(a, Q) \subset \mathcal{V}_g$, where $g(x, y) = f(L_1(x, y), L_2(x, y))$. Next, it is also clear that $\mathcal{G}(a, Q) \subset \mathcal{V}_f$ if and only if $\mathcal{G}_{a/C,Q} \subset \mathcal{V}_{f/C}$, where C is any non-vanishing coefficient of f. As a consequence, if f is a reducible (over $\mathbf{Q}$) cubic form, we have three possibilities:

1. The polynomial $f(x, 1)$ has a triple root and after linear change of variables we can assume that $f(x, y) = x^3$. Thus, in this case, we have that $\mathcal{G}(a, Q) \subset \mathcal{V}_f$ if and only if a and Q are non-zero cubes of rational numbers.
2. The polynomial $f(x, 1)$ has two rational roots from which one is a double root. Thus, after a linear change of variables, we can assume that $f(x, y) = x^2 y$. As a consequence we get that for any non-zero $a, Q \in \mathbf{Q}$ we have that $\mathcal{G}(a, Q) \subset \mathcal{V}_f$. Indeed, for any sequence $(x_n)_{n\in\mathbf{N}}$ of non-zero rational numbers we can take $y_n = aQ^n/x_n^2$ and observe that $f(x_n, y_n) = aQ^n$.
3. The polynomial $f(x, 1)$ has a rational root of multiplicity 1. Thus, after a linear change of variables, we can assume that $f(x, y) = x(x^2 + Ay^2)$ for some square-free non-zero integer A.

Let $f(x, y) = x(x^2 + Ay^2)$. In the light of the remarks above, for a given $A \in \mathbf{Z} \setminus \{0\}$, we are interested in finding cube-free rational numbers a, Q such that

$$f(x_0, y_0) = a, \quad f(x_1, y_1) = aQ, \quad f(x_2, y_2) = aQ^2 \tag{1}$$

has a solution in x_i, y_i with $x_i y_i \neq 0$ for $i = 0, 1, 2$. In other words, for $i = 0, 1, 2$ we are interested in finding rational points on genus 1 curve $C_i(a, Q) : x(x^2 + Ay^2) = aQ^i$. However, for our purposes it is easier to work with equivalent elliptic curves. To formulate an appropriate equivalence, we first note the following.

Lemma 1 *Let $A, U \in \mathbf{Q} \setminus \{0\}$. The curve $C : x(x^2 + Ay^2) = U$ is birationally equivalent with the elliptic curve $E : y^2 = x^3 - A^3U^2$.*

Proof The appropriate mappings between C and E are as follows:

$$\psi : C \ni (x, y) \mapsto \left(\tfrac{AU}{x}, \tfrac{y}{Ax}\right) \in E,$$
$$\psi^{-1} : E \ni (x, y) \mapsto \left(\frac{AU}{x}, \frac{A^2Uy}{x}\right) \in C,$$

and our result follows.

Applying the above lemma to each member of the system (1), instead of working with system (1), we will work with the equivalent system of equations

$$Y_0^2 - X_0^3 = -A^3a^2, \quad Y_1^2 - X_1^3 = -A^3a^2Q^2, \quad Y_2^2 - X_2^3 = -A^3a^2Q^4 \tag{2}$$

Table 1 Ranks and the generators for the elliptic curves $E_i(1, 9, 2): y^2 = x^3 - 9^2 \cdot 2^{2i} x, i = 0, 1, 2$

i	$rank E_i(1, 9, 2)(\mathbf{Q})$	generators of $E_i(1, 9, 2)(\mathbf{Q})$
0	1	(13, 46)
1	1	(10, 26)
2	1	(193, 2681)

and look for solutions $a, Q \in \mathbf{Q}$ and $X_i, Y_i \in \mathbf{Q}$ satisfying $X_i Y_i \neq 0$ for $i = 0, 1, 2$. In other words we have the following equivalence

$$\mathcal{G}(a, Q) \subset \mathcal{V}_f \iff \mathcal{G}(-A^3a^2, Q^2) \subset \mathcal{V}_g,$$

where $g(x, y) = y^2 - x^3$. Thus, the condition $\mathcal{G}(-A^3a^2, Q^2) \subset \mathcal{V}_g$ is equivalent with the property that for each $i \in \mathbf{N}$, the elliptic curve

$$E_i(A, a, Q): \; y^2 = x^3 - A^3a^2Q^{2i}$$

contains a rational point with $xy \neq 0$. Actually, in the sequel we require that the curve has a positive rank (and thus infinitely many rational points).

Let us fix $A = 1$. It is not difficult to find an example of rational numbers a, Q such that $\mathcal{G}(-a^2, Q^2)$ is infinite and $\mathcal{G}(-a^2, Q^2) \subset \mathcal{V}_g$. Indeed, let us take $Q = 2$ and $a = 9$. We thus consider elliptic curves $E_i(1, 9, 2): \; y^2 = x^3 - 9^2 \cdot 2^{2i}$. One can compute the rank of $E_i(1, 9, 2)$ and the set of generators of infinite order. We collect the computed data in Table 1.

As a consequence we get that $\mathcal{G}(-9^2, 2^2) \subset \mathcal{V}_g$ and equivalently $\mathcal{G}(9, 2) \subset \mathcal{V}_f$.

3 Rational Points on the Surface S: $z^3 = (x^2 - u)(y^2 - u)$ and Problem 1 for Reducible Cubic Forms

In this section we prove that for each reducible cubic form with three different roots there are infinitely many $a, Q \in \mathbf{Q} \setminus \{0\}$ such that $\mathcal{G}(a, Q) \subset \mathcal{V}_f$. However, before we do that we will prove a result, of independent interest, concerning the existence of rational points on the quartic surface $S_u: \; z^3 = (x^2 - u)(y^2 - u)$.

Theorem 1 *Let $u \in \mathbf{Q} \setminus \{0\}$ and consider the surface $S_u: \; z^3 = (x^2 - u)(y^2 - u)$. For any $u \in \mathbf{Q} \setminus \{0\}$, there are infinitely many rational curves lying on S_u. In particular, the set of rational points on the surface S_u is dense in the set of all real points of S_u.*

Proof Let $F_u(x, y, z) = z^3 - (x^2 - u)(y^2 - u)$. First of all we show that S_u contains a rational curve. To see this, let us consider the intersection of S_u with the surface $L: \; z^3 = 2x(x - y)^3$. From the factorization

$$(x^2 - u)(y^2 - u) - 2x(x - y)^3 = (x^2 - 2xy + u)(2x^2 - 2xy + y^2 - u)$$

it follows that $S_u \cap L = C_{1,u} \cup C_{2,u}$, where

$$C_{1,u} \quad : \; x^2 - 2xy + u = 0, \; z^3 = 2x(x-y)^3,$$
$$C_{2,u} : \; 2x^2 - 2xy + y^2 - u = 0, \; z^3 = 2x(x-y)^3.$$

We thus see that $S_u \cap L$ defines a reducible curve and one can check that $C_{1,u}$ is a rational curve. To see this, we note that the surface L is rational with the rational parametrization given by

$$\phi : \; \mathbf{Q}^2 \ni (t_1, t_2) \mapsto (x, y, z) = (4t_1^3, t_2, 2t_1(4t_1^3 - t_2)) \in L(\mathbf{Q}).$$

The inverse mapping is given by

$$\phi^{-1} : \; L(\mathbf{Q}) \ni (x, y, z) \mapsto \left(\frac{z}{x-y}, y\right) \in \mathbf{Q}^2.$$

On the Fig. 1 one can see shapes of the surfaces S_1 and L and on Fig. 2 the curve defined by the intersection of S_1 and L.

Using the parametrization of L we get that the equation for $C_{1,u}$ reduces to the equation $16t_1^6 - 8t_2t_1^3 + u = 0$, which can be easily solved with respect to the variable t_2. We have $t_2 = \frac{16t_1^6 + u}{8t_1^3}$. We thus get that the parametrization of $C_{1,u}$ is given by

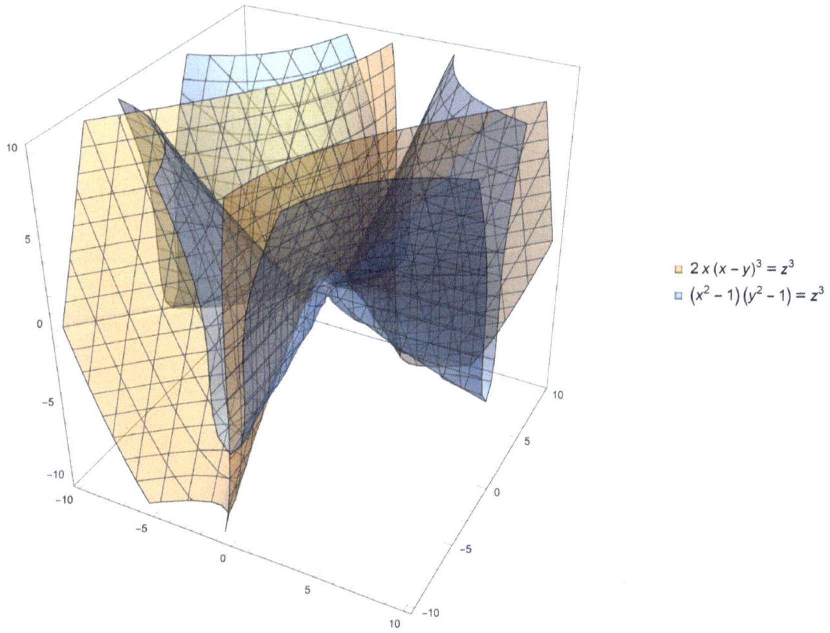

Fig. 1 Intersection of surfaces S_1 and L

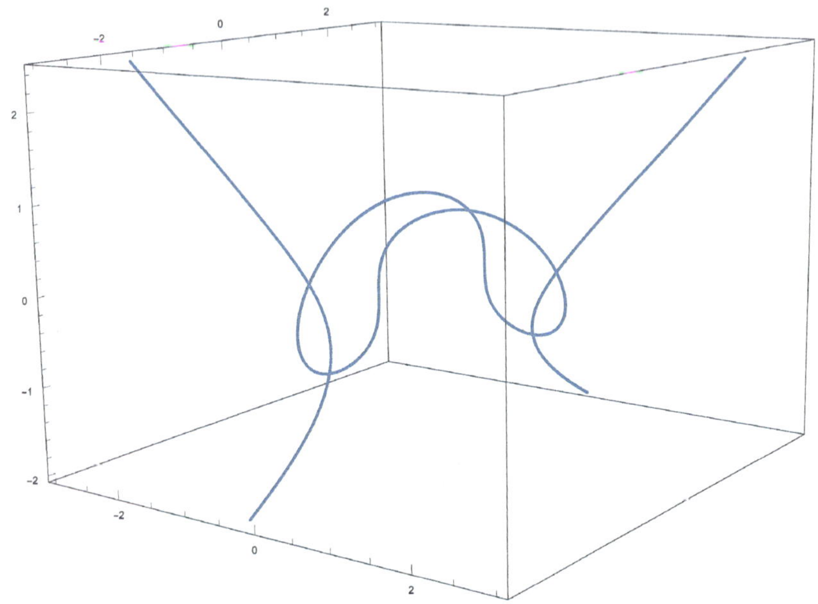

Fig. 2 The curves $C_{1,1} \cup C_{2,1}$

$$x = 4t^3, \quad y = \frac{16t^6 + u}{8t^3}, \quad z = \frac{16t^6 - u}{4t^2},$$

where, to simplify the notation, we put $t_1 = t$. We thus proved that the rational curve $C_{1,u}$ lies on the surface S_u. In particular, for each $u \in \mathbf{Q} \setminus \{0\}$ the surface S_u contains infinitely many rational points.

We are ready to prove that there are infinitely many rational curves on S_u. To see this, we look on S_u as an elliptic curve defined over function field $\mathbf{Q}(x)$. Next, we consider the curve $S'_u : z^3 = (16t^6 - u)(y^2 - u)$ coming from the S_u after the base change $x = \chi(t) = 4t^3$. Note that the set $\chi(\mathbf{Q})$ is dense in $\mathbf{Q}$—a property which will be used later. We note that S'_u is birational with an elliptic curve given by the Weierstrass equation

$$\mathcal{E}_u : Y^2 = X^3 + u(16t^6 - u)^4,$$

which is defined over rational function field $\mathbf{Q}(t)$ (remember that u is fixed). The curve $\mathcal{E}_u$ contains $\mathbf{Q}(t)$ rational point

$$P = \left(\frac{(16t^6 - u)^2}{4t^2}, \frac{(16t^6 - u)^2 (16t^6 + u)}{8t^3} \right).$$

The point P is of infinite order on $\mathcal{E}_u$. Indeed, this follows from function field variant of Nagell-Lutz theorem, which says that if $\mathcal{E} : y^2 = x^3 + A(t)x + B(t)$ is an elliptic curve over function field $\mathbf{Q}(t)$ and $A, B \in \mathbf{Z}[t]$, then the torsion points on $\mathcal{E}$ have polynomial coordinates. As a consequence, for $n \in \mathbf{N}$ we can compute $nP = (X_n(t), Y_n(t))$ and get infinitely many $\mathbf{Q}(t)$-rational points on $\mathcal{E}_u$ (and thus on S_u). Now, regarding $\mathcal{E}_u$ as an elliptic surface in the space with coordinates (X, Y, t) we see that each rational curve (X_n, Y_n, t) is included in the Zariski closure, say $\mathcal{R}_u$, of the set of rational points on $\mathcal{E}_u$. Because this closure consists of only finitely many components, it has dimension two, and as the surface $\mathcal{E}_u$ is irreducible, $\mathcal{R}_u$ is the whole surface. Thus the set of rational points on $\mathcal{E}_u$ is dense in the Zariski topology and the same is true for S'_u and also S_u. This follows from the fact that S_u comes from $\mathcal{E}_u \simeq S'_u$ after a non-invertible change of variables.

To obtain the density of the set $S_u(\mathbf{Q})$ in the Euclidean topology, we use two results: a theorem of Hurwitz [1] (see also [6, p. 78] and a theorem of Silverman [4, p. 368]. The theorem of Hurwitz states that if an elliptic curve E defined over $\mathbf{Q}$ has positive rank and at most one torsion point of order two (defined over $\mathbf{Q}$) then the set $E(\mathbf{Q})$ is dense in $E(\mathbf{R})$. The same result holds if E has three torsion points of order two under the assumption that we have a rational point of infinite order on the bounded branch of the set $E(\mathbf{R})$.

Silverman's theorem states that if $\mathcal{E}$ is an elliptic curve defined over $\mathbf{Q}(t)$ with positive rank, then for all but finitely many $t_0 \in \mathbf{Q}$, the curve $\mathcal{E}_{t_0}$ obtained from the curve $\mathcal{E}$ by specialization at $t = t_0$ has positive rank. From this result we see that for all but finitely many $t_0 \in \mathbf{Q}$ the elliptic curve $\mathcal{E}_{u,t_0}$ is of positive rank. Let us denote by G the set of $t_0 \in \mathbf{Q}$ such that the specialization P_{t_0} of the point P at $t = t_0$ is of finite order on the curve $\mathcal{E}_{u,t_0}$.

From the characterization of the possible torsion subgroups of elliptic curve given by the equation $E : y^2 = x^3 + D$, we know that the possible order of a torsion point on E is bounded by 6 [4, p. 323]. Moreover, there is no $D \in \mathbf{Q}$ such that E has three two-torsion points, i.e. the graph of the real part of E is connected. In essence one can compute explicitly the set G as the set of rational roots of denominators of X-coordinates of the points nP for $n = 1, 2, 3, 4, 6$. In any way, we know that the set G is finite. As a consequence of our reasoning, we deduce that for all $t_0 \in \mathbf{Q} \setminus G$, the curve $\mathcal{E}_{u,t_0}$ is of positive rank. The Hurwitz theorem now implies that the set $\mathcal{E}_{u,t_0}(\mathbf{Q})$ is dense in the set $\mathcal{E}_{u,t_0}(\mathbf{R})$. It follows that $S'_{u,t_0}(\mathbf{Q})$ is dense in $S'_{u,t_0}(\mathbf{R})$ because the image of $\mathbf{Q}$ by the function $\chi(t) = 4t^3$ is dense in the Euclidean topology of $\mathbf{R}$ (since $\overline{\chi(\mathbf{Q})} = \mathbf{R}$). If we put now $H = \chi(\mathbf{Q}) \subset \mathbf{Q}$ we get that for all but finitely many $t_0 \in H$, the set $S_{u,t_0}(\mathbf{Q})$ is dense in $S_{u,t_0}(\mathbf{R})$. Because H is dense in $\mathbf{R}$, it follows that the set $S_u(\mathbf{Q})$ of rational points on S_u is dense in the Euclidean topology in $S_u(\mathbf{R})$. Our theorem is proved.

Remark 1 The crucial moment in the proof of Theorem 1 is the existence of a surface L which intersect S_u along a reducible curve with a rational branch. An inquisitive reader one can ask: how the surface L was found? To find L, we used an experimental approach. More precisely, for $u \in \{1, \ldots, 50\}$ we looked for rational points $(x, y, z) = (p/s, q/s, r/s)$ lying on the surface S_u. We solved the resulting equation defining the surface S_u, with respect to p and got an algebraic function

$$p = p_u(q, r, s) = \sqrt{\frac{s\left(q^2su + r^3 - s^3u^2\right)}{q^2 - s^2u}}.$$

Next, we looked for integer values of $p_u(q, r, s)$ in the range $q, r \in \{1, \ldots, 200\}$ and $s \in \{1, \ldots, q-1\}$ (under the assumption $\gcd(q, r, s) = 1$). We found that only for $u \in \{23, 42, 47, 77, 83, 86\}$ there are no solutions in the considered range. Moreover, we observed that for $u \in \mathcal{U}$, where

$$U = \{8, 10, 12, 13, 14, 15, 17, 24, 28, 40, 44, 45, 48, 63, 65, 73, 76, 88, 89, 97\}$$

there is a solution satisfying the condition $p_u(q, r, s) = q + 1$. For example, if $u = 10$ there is a rational point $(x, y, z) = (39/12, 38/12, 3/12)$ lying on S_{10}, i.e., in this case we have $p_{10}(38, 3, 12) = 39 = 38 + 1$. We collected the solutions satisfying the condition $p_u(q, r, s) = q + 1$ in Table 2.

We expected that such a strong condition (which holds for 20 values of $u \in \{1, \ldots, 100\}$) can not be a coincidence, and concentrated on finding solutions of the equation $q + 1 = p_u(q, r, s)$. This equation is equivalent to

$$\left(\frac{(2q+1)^2 - 4s^2u - 1}{4}\right)^2 = s(r^3 + su),$$

i.e., $s(r^3 + su)$ need to be square. To make $s(r^3 + su)$ a square one can take $s = \frac{r^3}{t^2 - u}$, where t is a rational parameter. Then, the equation under consideration reduces to $F_u(t, q, r)F_u(t, q, -r)/(t^2 - u) = 0$, where

$$F_u(t, q, r) = ur^6 + t(t^2 - u)r^3 - q(1 + q)(t^2 - u)^2.$$

A quick computation revels that if we take $t = r^3/2$, then the equation $F_u(t, q, r) = 0$ has two roots

Table 2 Rational points $(x, y, z) = (p/s, q/s, r/s)$ lying on S_u and satisfying the conditions $p_u(q, r, s) \in \mathbf{Z}$ and $p_u(q, r, s) = q + 1$

u	x	y	z	u	x	y	z	u	x	y	z
8	4	3	2	17	$\frac{25}{6}$	$\frac{149}{36}$	$\frac{13}{36}$	65	4	3	14
10	$\frac{13}{4}$	$\frac{19}{6}$	$\frac{1}{4}$	24	8	7	10	73	8	7	6
12	4	$\frac{7}{2}$	1	28	$\frac{23}{4}$	$\frac{11}{2}$	$\frac{9}{4}$	73	10	9	6
13	3	$\frac{5}{2}$	3	40	8	7	6	76	$\frac{37}{4}$	$\frac{55}{6}$	$\frac{17}{4}$
14	4	$\frac{15}{4}$	$\frac{1}{2}$	44	$\frac{43}{9}$	$\frac{14}{3}$	$\frac{70}{9}$	88	10	$\frac{19}{2}$	3
14	$\frac{23}{6}$	$\frac{137}{36}$	$\frac{25}{36}$	45	3	$\frac{17}{6}$	11	89	8	7	10
15	4	$\frac{31}{8}$	$\frac{1}{4}$	48	10	$\frac{19}{2}$	13	97	5	4	18
17	4	3	2	63	11	$\frac{43}{4}$	$\frac{29}{2}$	97	10	$\frac{89}{9}$	$\frac{4}{3}$

$$q = \frac{r^6 + 4u}{r^6 - 4u}, \quad q = -\frac{2r^6}{r^6 - 4u}$$

Taking the first one, we get that the rational expressions

$$q = \frac{r^6 + 4u}{r^6 - 4u}, \quad s = \frac{4r^3}{r^6 - 4u}$$

satisfy the equation $p_u(q, r, s) = q + 1$. Moreover, we get that

$$x = \frac{p}{s} = \frac{p_u(q, r, s)}{s} = \frac{1}{2}r^3, \quad y = \frac{q}{s} = \frac{r^6 + 4u}{4r^3}, \quad z = \frac{r}{s} = \frac{r^6 - 4u}{4r^2}$$

satisfy the equation defining the surface S_u, i.e., there is a rational curve, say C, lying on S_u. Eliminating r and u form the equations defining C we found that the variables x, y, z satisfy the equation $2x(x - y)^3 = z^3$, which is exactly the equation defining the surface L.

After this digression, we are ready to prove the main result of the paper.

Theorem 2 *Let $f(x, y) = x(x^2 + Ay^2)$, where $A \in \mathbf{Z} \setminus \{0\}$. For any A there are infinitely many pairs a, Q such that $\mathcal{G}(a, Q) \subset \mathcal{V}_f$.*

Proof We are interested in finding rational numbers x_i, y_i, a, Q which solve the system (1). To find solutions, we apply Lemma 1 to each equation in (1) and we get the equivalent system

$$Y_0^2 - X_0^3 = -A^3a^2, \quad Y_1^2 - X_1^3 = -A^3a^2Q^2, \quad Y_2^2 - X_2^3 = -A^3a^2Q^4. \tag{3}$$

To solve (3) we make the following substitutions

$$X_0 = p_0a, \quad Y_0 = q_0a, \quad X_1 = p_1Q, \quad Y_1 = q_1Q, \quad X_2 = p_2aQ, \quad Y_2 = q_2aQ^2$$

and after necessary simplifications we get that

$$a = \frac{q_0^2 + A^3}{p_0^3}, \quad Q = \frac{q_1^2 + A^3a^2}{p_1^3} = \frac{ap_2^3}{q_2^2 + A^3}.$$

Thus, for any $p_0 \in \mathbf{Q} \setminus \{0\}$ and $q_0 \in \mathbf{Q}$, we can define $a = \frac{q_0^2+A^3}{p_0^3}$. To get the value of Q we need to consider the equation $\frac{q_1^2+A^3a^2}{p_1^3} = \frac{ap_2^3}{q_2^2+A^3}$ which is equivalent to

$$(q_1^2 + A^3a^2)((aq_2)^2 + A^3a^2) = (ap_1p_2)^3.$$

We thus arrived to the equation defining the surface S_u considered in Theorem 1, with $u = -A^3a^2$ and $x = q_1, y = aq_2, z = ap_1p_2$. Using the parametrization of the rational curve C_1 constructed in the proof of Theorem 1, we get that

$$p_2 = \frac{a^2A^3 + 16t^6}{4ap_1t^2}, \quad q_1 = 4t^3, \quad q_2 = \frac{16t^6 - a^2A^3}{8at^3}.$$

Tracing back all our substitutions we get that the values of a, Q we are looking for are

$$a = \frac{q_0^2 + A^3}{p_0^3}, \qquad Q = \frac{A^3(A^3 + q_0^2)^2 + 16p_0^6t^6}{p_0^6p_1^3}$$

and the corresponding solutions of the system (1) are

$$\begin{aligned} x_0 &= \tfrac{A}{p_0}, \quad x_1 = \tfrac{A(A^3+q_0^2)}{p_0^3p_1}, \quad x_2 = \tfrac{4At^2(A^3+q_0^2)}{p_0^3p_1^2}, \\ y_0 &= \tfrac{q_0}{Ap_0}, \quad y_1 = \tfrac{4t^3}{Ap_1}, \quad y_2 = -\tfrac{A^9+2A^6q_0^2+A^3q_0^4-16p_0^6t^6}{2Ap_0^6p_1^2t}. \end{aligned}$$

As an immediate consequence of Theorem 2 is the following.

Corollary 1 *Let $f(x, y) = x(x^2 + Ay^2)$ and write*

$$\mathcal{B}_A := \{a \in \mathbf{N} : \text{ there are infinitely many } Q \in \mathbf{Q} \setminus \{-1, 0, 1\} \text{ such that } \mathcal{G}(a, Q) \subset \mathcal{V}_f\}$$

and put $\mathcal{A}(x) = \mathcal{A} \cap [1, x]$. Then $\mathcal{B}_A(x) \geq \sqrt{x} - A^3$.

Proof From the proof of Theorem 2 we know that if $a = (q_0^2 + A^3)/p_0^3$, then there is a rational Q such that $\mathcal{G}(a, Q) \subset \mathcal{V}_f$. If we take $p_0 = p_1 = 1$, then for given A one can take

$$a = q_0^2 + A^3, \quad Q = A^3(A^3 + q_0^2)^2 + 16p_0^6t^6,$$

where $t \in \mathbf{Q}$. Thus, for $q_0 \in (0, \sqrt{x - A^3})$ we get that $0 < a \leq x$ and get the result.

4 Questions

The main tool in the proof of existence of $a, Q \in \mathbf{Q}$ such that $\mathcal{G}(a, Q) \subset \mathcal{V}_f$, where $f(x, y) = x(x^2 + Ay^2)$ was the existence of a rational curve lying on the surface $S_u : z^3 = (x^2 - u)(y^2 - u)$. It is an interesting question whether for any $u \in \mathbf{Q} \setminus \{0\}$ and a given $n \in \mathbf{N}_+$, there is a rational curve lying on the surface

$$S_u(n) : z^n = (x^2 - u)(y^2 - u).$$

The case $n = 2$ is easy because the surface $S_u(2)$ contains trivial rational curve $x = y = t, z = t^2 - u$, which can be used to get a rational parametrization of the form

$$x = t_1, \quad y = \frac{t_1u - 2t_2u - t_1(t_1 - t_2)^2}{u - t_1^2 + t_2^2}, \quad z = \frac{(t_1^2 - u)(u - (t_1 - t_2)^2)}{u - t_1^2 + t_2^2}.$$

Let us also note that each rational point lying on $S_u(2)$ leads to the triple of rational numbers $x^2 - u, z, y^2 - u$, which lie in geometric progression.

The case $n = 3$ was investigated in the present paper. The next interesting case is $n = 4$. We were unable to show that for each $u \in \mathbf{Q}$ there is a rational curve lying on $S_u(4)$. Let us also note that the existence of a rational point (x, y, z) on $S(4)$ leads to the triple of rational numbers $x^2 - u, z^2, y^2 - u$, which lie in geometric progression. We formulate the following.

Question 1 Let $u \in \mathbf{Q} \setminus \{0\}$ be fixed. Is the set of rational points (x, y, z) with $z \neq 0$ and lying on $S_u(4)$, infinite?

In this context we also formulate the following.

Question 2 For which $u \in \mathbf{Z} \setminus \{0\}$ there are infinitely many integer points on the surface S_u?

We do not know any value of u such that $S_u(\mathbf{Z})$ is infinite. However, we expect that at least $S_1(\mathbf{Z})$ is infinite.

The following questions were motivated by computational experiments. In the computations we used standard procedures related to elliptic curves, which are implemented in PARI/GP [5]. As is described by the creators, PARI/GP is a cross platform and open-source computer algebra system designed for fast computations in number theory. It is especially useful with computations related to elliptic curves. The function `ellrank` attempts to compute the rank of a given elliptic curve E. The function returns the vector $[r, R, s, L]$. For the rank of E we have the inequalities $r \leq rank E(\mathbf{Q}) \leq R$, the number s is related to the Tate-Shafarevitch group. Moreover, L is a list of independent, non-torsion rational points on the curve E. To initialize the curve we use the procedure `ellinit`. Here is a sample session:

```
gp> E0=ellinit([0,−2^3·11^2·3^0])
gp> E1=ellinit([0,−2^3·11^2·3^2])
gp> E2=ellinit([0,−2^3·11^2·3^4])
gp> ellrank(E0)
[1, 1, 0, [[33, 187]]]
gp> ellrank(E1)
[2, 2, 0, [[22, 44], [66, 528]]]
gp> ellrank(E2)
[1, 1, 0, [[198, 2772]]]
```

In particular, we have $rank E_1(\mathbf{Q}) = 2$ and an infinite part of $E_1(\mathbf{Q})$ is generated by the points $(22, 44)$, $(66, 528)$. Note that for $i = 0, 1, 2$ the curve $E_i(2, 11, 3)$ has positive rank and thus $\mathcal{G}(11, 3) \subset \mathcal{V}_f$ for $f(x, y) = x(x^2 + 3y^2)$.

It should be noted, that in some cases PARI/GP is not able to compute the rank exactly, like in case of the elliptic curve $H : \ y^2 = x^3 - 34^3 \cdot 2^2 \cdot 13^2$:

```
gp> E=ellinit([0,−34³·2²·13²];
gp> ellrank(E)
[0, 2, 0, []]
```

Thus, $0 \leq rankE(\mathbf{Q}) \leq 2$. In such case we invoke the procedure `ellanalyticrank` which computes the pair $[r, d]$, where r is an analytic rank of an elliptic curve and d the value of d-th derivative at 1 of a L-function associated to E. We have

```
gp> ellanalyticrank(E)
gp> [0, 8.9980371211404273322149099456397283257]
```

Thus, invoking Kolyvagin's [2] together with Wiles' [9] result on modularity of elliptic curves over **Q**, we know that the rank of E is equal to 0. In consequence, we have $\mathcal{G}(2, 13) \not\subset \mathcal{V}_f$ for $f(x, y) = x(x^2 + 34y^2)$.

We proved that for each $A \subset \mathbf{Z} \setminus \{0\}$, there are a, Q such that the elliptic curve $E_i(A, a, Q)$ has positive rank for $i \in \mathbf{N}$. In this direction one can formulate the following, more difficult, questions.

Question 3 Let $A \in \mathbf{Z} \setminus \{0\}$ and $Q \in \mathbf{Z} \setminus \{-1, 0, 1\}$ be given. Does there exist $a \in \mathbf{Q} \setminus \mathbf{Q}^3$ such that the elliptic curve $E_i(A, a, Q)$ has a positive rank for each $i \in \mathbf{N}$?

We performed a small numerical search and found that for each $A \in \{1, \ldots, 1000\}$ and $Q \in \{2, \ldots, 26\}$ there is an $a \in \mathbf{Z}$ such that $E_i(A, a, Q)$ has a positive rank for each $i \in \mathbf{N}$. Although modest, our computations suggest that the answer to Question 3 is positive. In fact, we expect that for each pair A, Q there are infinitely many a's which are square-free and $E_i(A, a, Q)$ has a positive rank for each $i \in \mathbf{N}$. In Table 3, for a square-free $A \in \{1, \ldots, 15\}$ and a cube-free $Q \in \{2, \ldots, 12\}$, we present the smallest a such that $E_i(A, a, Q)$ has a positive rank for each $i \in \mathbf{N}$.

Table 3 The smallest value of a such that $E_i(A, a, Q)$ has a positive rank for a square-free $A \in \{1, \ldots, 15\}$ and $Q \in \{2, \ldots, 11\}$ and each $i \in \mathbf{N}$

	Q									
A	2	3	4	5	6	7	9	10	11	12
1	9	22	9	2	23	9	22	11	2	11
2	3	11	3	4	11	3	11	9	3	4
3	17	35	17	11	10	3	35	21	13	11
5	3	14	3	3	3	3	14	3	4	7
6	3	28	3	11	7	1	28	7	20	14
7	11	22	11	11	1	3	22	34	1	11
10	9	7	9	9	11	9	7	2	2	13
11	1	5	1	4	2	2	5	4	1	1
13	9	10	9	2	17	9	10	14	2	5
14	3	4	3	3	15	3	4	22	4	10

Table 4 The smallest value of Q such that $E_i(A, a, Q)$ has a positive rank for a square-free $A \in \{1, \ldots, 15\}$ and $a \in \{1, \ldots, 10\}$ and each $i \in \mathbf{N}$

	a									
A	1	2	3	4	5	6	7	8	9	10
1	x	5	x	x	13	x	5	x	2	5
2	x	x	2	5	2	2	x	x	5	2
3	x	x	7	x	x	7	x	x	x	6
5	x	x	2	6	10	2	11	x	x	x
6	7	x	2	14	x	2	2	7	x	x
7	6	x	7	x	x	6	x	6	x	x
10	x	10	x	x	x	x	3	x	2	x
11	2	2	x	2	3	x	12	2	6	x
13	x	5	x	x	12	x	5	x	2	3
14	x	x	2	3	x	2	x	x	x	12

Question 4 Let $A, a \in \mathbf{Z} \setminus \{0\}$ be given. Let us suppose that the elliptic curve $E_0(A, a, Q) : y^2 = x^3 - A^3a^2$ has positive rank. Does there exist $Q \in \mathbf{Q} \setminus \mathbf{Q}^3$ such that the elliptic curve $E_i(A, a, Q)$ has a positive rank for each $i \in \mathbf{N}$?

In this case our expectations are exactly the same as in case of Question 3, i.e., for all admissible pairs A, a one can find an appropriate value of Q. We collect some data in Table 4. We denote the pairs (A, a) for which the curve $E_0(A, a, Q) : y^2 = x^3 - A^3a^2$ has rank 0 by "x".

Finally, we can formulate the following.

Question 5 Let $a \in \mathbf{Z} \setminus \{0\}$ and $Q \in \mathbf{Z} \setminus \{-1, 0, 1\}$ be given. Does there exist $A \in \mathbf{Q} \setminus \mathbf{Q}^2$ such that the elliptic curve $E_i(A, a, Q)$ has a positive rank for each $i \in \mathbf{N}$?

Again, a numerical search confirms that for each cube-free $a \in \{1, \ldots, 1000\}$ and $Q \in \{2, \ldots, 26\}$ one can find an appropriate value of A (Table 5).

Finally, we formulate the following.

Problem 2 Investigate Problem 1 for irreducible cubic forms.

To see an example, let us take $f(x, y) = x^3 + 2y^3$ and note that $\mathcal{G}(1, 17) \subset \mathcal{V}_f$. Indeed, we have

$$f(-1, 1) = 1, \quad f(1, 2) = 17, \quad f(7, -3) = 17^2.$$

However, the curve $C_0 : f(x, y) = 1$ has only finitely many rational points. Indeed, C_0 is birationally equivalent with the elliptic curve $y^2 = x^3 - 27$, which has rank equal to 0. On the other hand, we have that $\mathcal{G}(1333, 17) \subset \mathcal{V}_f$. Indeed, we have

$$f(-19, 16) = 1, \quad f(-35, 32) = 1333 \cdot 17, \quad f(-67, 70) = 1333 \cdot 17^2.$$

Table 5 The smallest square-free value of A such that $E_i(A, a, Q)$ has a positive rank for a cube-free $a \in \{1, \ldots, 11\}$ and $Q \in \{2, \ldots, 12\}$ and each $i \in \mathbf{N}$

	Q									
a	2	3	4	5	6	7	9	10	11	12
1	11	23	11	42	7	6	23	26	7	11
2	11	21	11	1	11	11	21	10	1	30
3	2	23	2	5	5	2	23	5	2	30
4	11	14	11	2	5	33	14	5	2	2
5	2	11	2	42	11	2	11	5	5	13
6	2	21	2	5	7	2	21	5	21	2
7	6	10	6	1	6	6	10	6	1	5
9	1	23	1	2	11	1	23	2	2	2
10	2	13	2	1	3	13	13	26	33	14
11	1	2	1	3	2	1	2	1	7	1

This time, for each $i \in \{0, 1, 2\}$ the curve $C_i : x^3 + 3y^3 = 1333 \cdot 17^i$ has infinitely many rational points. Indeed, we have

$$\begin{aligned} C_0 \simeq E_0 &: y^2 = x^3 - 47976003, \\ C_1 \simeq E_1 &: y^2 = x^3 - 13865064867, \\ C_2 \simeq E_2 &: y^2 = x^3 - 4007003746563. \end{aligned}$$

We have that $rank E_0(\mathbf{Q}) = 2$ with the generators of infinite part $G_{0,1} = (9331/25, 250604/125)$, $G_{0,2} = (147963/49, 56865780/343)$. Moreover, $rank E_1(\mathbf{Q}) = rank E_2(\mathbf{Q}) = 1$ and the corresponding generators of infinite parts of $E_1(\mathbf{Q})$ and $E_2(\mathbf{Q})$ are

$$\begin{aligned} G_{1,1} &= (3323376921/1254400, 96638332841469/1404928000), \\ G_{2,1} &= ((354730964169/21996100, 44646918022393947/103161709000), \end{aligned}$$

respectively.

Acknowledgements The research of the second author is supported by the grant of the National Science Centre (NCN), Poland, no. UMO-2019/34/E/ST1/00094.

References

1. Hurwitz, A.: Über ternäre diophantische Gleichungen dritten Grades, Vierteljahrschrift d. Naturforsch. Ges. Zürich 62, 207–229 (1917).
2. Kolyvagin, V.: Finiteness of $E(\mathbf{Q})$ and $X(E, \mathbf{Q})$ for a class of Weil curves, Math. USSR Izv. 32 (3), 523–541 (1989).

3. Mahler, K.: Zur Approximation algebraischer Zahlen. I, Math. Ann. 107(1), 691–730 (1933).
4. Silverman, J.: The Arithmetic of Elliptic Curves, Springer-Verlag, New York, 1986.
5. The PARI Group, PARI/GP version `2.15.2`, Univ. Bordeaux, 2022, http://pari.math.u-bordeaux.fr
6. Skolem, Th.: Diophantische Gleichungen, New York : Chelsea Publishing Company, 1950.
7. Thue, A.: Über Annäherungswerte algebraischer Zahlen, J. Reine Angew. Math., 135, 284–305 (1909)
8. Ulas, M.: Geometric progressions in the sets of values of rational functions, Indag. Math. 35 (4), 622–637 (2024).
9. Wiles, A.: Modular elliptic curves and Fermat's last theorem, Ann. Math. 141 (3), 443–551 (1995)
10. Wolfram Research, Inc., Mathematica, Version 13.2. (Champaign, IL, 2022).

Uniform Approximation by Polynomials with Integer Coefficients via the Bernstein Lattice

C. Sinan Güntürk and Weilin Li

Abstract Let $\mathscr{C}_{\mathbb{Z}}([0, 1])$ be the metric space of real-valued continuous functions on $[0, 1]$ with integer values at 0 and 1, equipped with the uniform (supremum) metric d_∞. It is a classical theorem in approximation theory that the ring $\mathbb{Z}[X]$ of polynomials with integer coefficients, when considered as a set of functions on $[0, 1]$, is dense in $\mathscr{C}_{\mathbb{Z}}([0, 1])$. In this paper, we offer a strengthening of this result by identifying a substantially small subset $\bigcup_n \mathscr{B}_n$ of $\mathbb{Z}[X]$ which is still dense in $\mathscr{C}_{\mathbb{Z}}([0, 1])$. Here $\mathscr{B}_n$, which we call the "Bernstein lattice," is the lattice generated by the polynomials

$$p_{n,k}(x) := \binom{n}{k} x^k (1-x)^{n-k}, \quad k = 0, \ldots, n.$$

Quantitatively, we show that for any $f \in \mathscr{C}_{\mathbb{Z}}([0, 1])$,

$$d_\infty(f, \mathscr{B}_n) \le \frac{9}{4}\omega_f(n^{-1/3}) + 2n^{-1/3}, \quad n \ge 1,$$

where ω_f stands for the modulus of continuity of f. We also offer a more general bound which can be optimized to yield better decay of approximation error for specific classes of continuous functions.

1 Introduction and the Statement of the Main Theorem

The topic of approximation using polynomials with integer coefficients has a mature history going back to the early 20th century. The earliest known result appears to be by Pál [11] who showed in 1914 that for any $0 < \alpha < 1$, any $f \in \mathscr{C}([-\alpha, \alpha])$

C. S. Güntürk (✉)
NYU Courant Institute, New York, NY, USA
e-mail: gunturk@courant.nyu.edu

W. Li
CUNY City College, New York, NY, USA
e-mail: wli6@ccny.cuny.edu

M. B. Nathanson (ed.), *Combinatorial and Additive Number Theory VI*,
Springer Proceedings in Mathematics & Statistics 464,
https://doi.org/10.1007/978-3-031-65064-2_10

can be approximated uniformly by elements of $\mathbb{Z}[X]$, provided that $f(0) \in \mathbb{Z}$. Here $\mathscr{C}([a,b])$ is the metric space of continuous real-valued functions on the interval $[a,b]$, equipped with the uniform (supremum) distance

$$d_\infty(f,g) := \|f-g\|_{[a,b]} := \sup_{[a,b]} |f-g|,$$

and $\mathbb{Z}[X]$ stands for the commutative ring of polynomials with integer coefficients which we consider as a subset of $\mathscr{C}([a,b])$.

The sufficient condition of Pál is clearly also necessary since any polynomial with integer coefficients is integer-valued at 0, which is a property preserved in the limit. Hence, for $\alpha < 1$, the conditions $f \in \mathscr{C}([-\alpha,\alpha])$ and $f(0) \in \mathbb{Z}$ together characterize uniform approximability on $[-\alpha,\alpha]$ by elements of $\mathbb{Z}[X]$.

Kakeya [6] extended Pál's result to the space $\mathscr{C}([-1,1])$ via the necessary and sufficient condition

$$f(-1),\, f(0),\, f(1),\, \frac{f(-1)+f(1)}{2} \in \mathbb{Z}.$$

The fourth of these arithmetic conditions is easily checked to be satisfied by elements of $\mathbb{Z}[X]$ and their limits.

Other early important results of the theory developed in the 1920s and 30s include those by Fekete, Chlodovsky, Bernstein and Kantorovich (see [1, 7]). For an extensive treatment of the subject, we refer to Ferguson's classical text [8]. One of the important characteristics of the theory is that uniform approximation is only possible on intervals of length less than 4, and then only with certain arithmetic conditions on f the number of which increases without bound as the length of the interval increases towards 4. Kakeya showed (also in [6]) that approximating $f \in \mathscr{C}([-2,2])$ arbitrarily well by elements of $\mathbb{Z}[X]$ is impossible unless $f \in \mathbb{Z}[X]$ because metrically, $\mathbb{Z}[X]$ is a discrete subset of $\mathscr{C}([-2,2])$. Indeed, any non-zero element of $\mathbb{Z}[X]$ has its supremum norm on $[-2,2]$ equal to at least 1 (and at least 2 if it is not a constant polynomial). In the other extreme, there are no arithmetic conditions to be met when approximation is sought on closed intervals containing no integers, as shown by Chlodovsky [1].

An important case between these two extremes is the interval $[0,1]$ where the characterizing condition reduces to integer valuations at the boundary points 0 and 1, as for example, demonstrated by Chlodovsky [1] and Kantorovich [7]. The corresponding class is

$$\mathscr{C}_\mathbb{Z}([0,1]) := \Big\{ f \in \mathscr{C}([0,1]) : f(0),\, f(1) \in \mathbb{Z} \Big\}.$$

Note that $\mathscr{C}_\mathbb{Z}([0,1])$ is the disjoint union of countably many translates of

$$\mathscr{C}_0([0,1]) := \Big\{ f \in \mathscr{C}([0,1]) : f(0) = f(1) = 0 \Big\}.$$

Indeed, we have

$$\mathscr{C}_{\mathbb{Z}}([0,1]) = \bigcup_{(m,n)\in\mathbb{Z}^2} \mathscr{C}_0([0,1]) + \ell_{m,n},$$

where $\ell_{m,n}$ stands for the linear polynomial satisfying $\ell_{m,n}(0) = m$, $\ell_{m,n}(1) = n$.

Let us define $\mathscr{P}_n$ to be the space of polynomials (more precisely, polynomial functions) of degree at most n where, from now on, the domain is understood to be $[0, 1]$. If we pick the standard power basis $(\pi_k)_0^n$ for $\mathscr{P}_n$, where $\pi_k(x) := x^k$, then the set of polynomials in $\mathscr{P}_n$ with coefficients in $\mathbb{Z}$ corresponds to the lattice generated by $(\pi_k)_0^n$. We will denote this lattice by Π_n. Chlodovsky's density result can now be stated as

$$\bigcup_n \Pi_n, \text{ i.e. } \mathbb{Z}[X] \text{ is dense in } \mathscr{C}_{\mathbb{Z}}([0,1]). \tag{1}$$

It is natural to ask the rate at which $d_\infty(f, \Pi_n) \to 0$ given any $f \in \mathscr{C}_{\mathbb{Z}}([0,1])$. Kantorovich in [7] gave an elementary algorithmic proof of Chlodovsky's theorem that came with a quantitative approximation error bound (see also [8] and [10]). In our notation, Kantorovich's elementary bound states that for all $f \in \mathscr{C}_{\mathbb{Z}}([0,1])$,

$$d_\infty(f, \Pi_n) := \inf_{P\in\Pi_n} \|f - P\|_{[0,1]} \le \|f - B_n\|_{[0,1]} + \frac{1}{n},$$

where $B_n := B_n(f)$ is the Bernstein polynomial of f defined by

$$B_n(x) := \sum_{k=0}^{n} f\left(\frac{k}{n}\right)\binom{n}{k} x^k (1-x)^{n-k}. \tag{2}$$

It is well known (see, e.g. [9]) that

$$\|f - B_n\|_{[0,1]} \le \frac{5}{4}\omega_f(n^{-1/2}), \tag{3}$$

where ω_f stands for the modulus of continuity of f on $[0, 1]$ given by

$$\omega_f(\delta) := \sup\{|f(x) - f(y)| : x, y \in [0,1],\ |x - y| \le \delta\},$$

therefore it follows that $d_\infty(f, \Pi_n) \le \frac{5}{4}\omega_f(n^{-1/2}) + n^{-1}$. Kantorovich actually proved a stronger result in [7], showing that

$$d_\infty(f, \Pi_n) \le 2d_\infty(f, \mathscr{P}_n) + \frac{1}{n}.$$

In other words, up to the additive term $1/n$ and the constant 2, polynomials with integer coefficients are nearly as expressive as arbitrary polynomials (of degree at most

n). This was shown by employing the unconstrained polynomial of best approximation of f as a surrogate instead of the Bernstein polynomial (see [8, 9]). Since $d_\infty(f, \mathscr{P}_n) \le 3\,\omega_f(n^{-1})$ and $\omega_f(\delta) \ge \frac{1}{2}\omega_f(1)\delta$ for all non-constant continuous functions (see e.g. [2, 9]) it follows that

$$d_\infty(f, \Pi_n) \le C_f \omega_f(n^{-1}),$$

where $C_f \le 6 + 2/\omega_f(1)$ is a constant that may possibly depend on f. (See also [8, Ch. 12] as well as [12].)

Finally, as is the case for approximation by polynomials with arbitrary coefficients, the rate of approximation by elements of Π_n can be improved under various regularity (e.g. differentiability) assumptions (again, see [8]).

In this paper, we are interested in a different form of strengthening the original density theorem of Chlodovsky (and its proof by Kantorovich) by seeking constructive approximations from much smaller lattices. The method of Kantorovich is based on rounding the coefficients of an approximating polynomial in the basis $x^k(1-x)^{n-k}$, $k = 0, \dots, n$, of $\mathscr{P}_n$. Calling the lattice generated by this basis Π_n°, clearly we have $\Pi_n^\circ \subset \Pi_n$, but the simple identity

$$x^k = x^k(x + (1-x))^{n-k} = \sum_{l=0}^{n-k} \binom{n-k}{l} x^{k+l}(1-x)^{n-(k+l)}$$

also shows that $\pi_k \in \Pi_n^\circ$ for all $k = 0, \dots, n$, implying that in fact $\Pi_n = \Pi_n^\circ$. Meanwhile, the lattice $\mathscr{B}_n$ generated by the polynomials

$$p_{n,k}(x) := \binom{n}{k} x^k (1-x)^{n-k}, \quad k = 0, \dots, n, \tag{4}$$

is a much smaller (coarser) sublattice of Π_n since it stretches the kth basis polynomial $x^k(1-x)^{n-k}$ by the integer factor $\binom{n}{k}$ which can get exponentially large in n. In particular, the fundamental cell of $\mathscr{B}_n$ consists of

$$\prod_{k=0}^{n} \binom{n}{k} = e^{\frac{1}{2}n^2(1+o(1))}$$

distinct elements of Π_n.

We call $\mathscr{B}_n$ the *Bernstein lattice* (of order n). In this paper, we strengthen (1) by showing that

$$\bigcup_n \mathscr{B}_n \text{ is dense in } \mathscr{C}_{\mathbb{Z}}([0,1]). \tag{5}$$

In fact, we prove this result quantitatively by means of our main theorem stated below:

Main Theorem. *For any $f \in \mathscr{C}_{\mathbb{Z}}([0,1])$ and positive integer n, we have*

$$d_\infty(f, \mathscr{B}_n) \leq \frac{5}{4}\omega_f(n^{-1/2}) + \rho(f, n)$$

where

$$\begin{aligned}\rho(f,n) &:= \min_{t\in[0,n/2]\cap\mathbb{Z}} \left\{\max\left(\omega_f\left(\frac{t}{n}\right), \frac{1}{2(n+1-2t)}\right) + \frac{1}{\sqrt{2(t+1)}}\right\} \\ &\leq \omega_f\left(\frac{1}{2}n^{-1/3}\right) + 2n^{-1/3}.\end{aligned}$$

In particular, we have

$$d_\infty(f, \mathscr{B}_n) \leq \frac{9}{4}\omega_f(n^{-1/3}) + 2n^{-1/3}.$$

Note. Our motivation and methods were partly developed in the papers [3] and [4], where we were able to establish pointwise approximation on (0, 1) for all $f \in \mathscr{C}([0, 1])$, however, our method could not take advantage of the smaller class $\mathscr{C}_\mathbb{Z}([0, 1])$ to make the error bound go to 0 uniformly on [0, 1]. The present paper achieves the missing uniform approximation on [0, 1] by means of a more powerful approximation idea designed for $\mathscr{C}_\mathbb{Z}([0, 1])$ with which we are able to control the error bound effectively all the way to the boundary points. However, the price paid for this uniformity is a lower rate of approximation.

2 Proof of the Main Theorem

Before we start the proof, let us recall (see, e.g. [2, 9]) the following basic polynomial identities which correspond to the central moments, up to degree two, of the Bernstein basis $(p_{n,k})_{k=0}^n$:

$$\sum_{k=0}^{n} p_{n,k}(x) = 1, \tag{6}$$

$$\sum_{k=0}^{n}(k-nx)p_{n,k}(x) = 0, \tag{7}$$

$$\sum_{k=0}^{n}(k-nx)^2 p_{n,k}(x) = nx(1-x). \tag{8}$$

We also note that $p_{n,k} \geq 0$ on [0, 1]. When interpreting the proof, it is helpful to keep in mind that $p_{n,k}$ peaks at k/n, and that the identity (6) corresponds to a smooth partition of unity.

To begin the proof, let $f \in \mathscr{C}_\mathbb{Z}([0, 1])$ and n be a positive integer. For $n = 1$, we can approximate f by the linear interpolant $\ell := f(0)p_{1,0} + f(1)p_{1,1} \in \mathscr{B}_1$ which clearly remains within $\omega_f(1)$ of f, so that the stated approximation error bound holds trivially. Therefore, without loss of generality, let $n \geq 2$. We will approximate f by a polynomial

$$Q_n := \sum_{k=0}^{n} q_k p_{n,k} \in \mathscr{B}_n$$

which will be constructed with the help of two auxiliary polynomials B_n and P_n in $\mathscr{P}_n$ approximating f. Note that B_n and P_n will in general not be in $\mathscr{B}_n$, but they will facilitate the approximation and the rounding processes. The first polynomial B_n is the standard Bernstein polynomial of f of order n defined by (2), i.e.

$$B_n := \sum_{k=0}^{n} f\left(\frac{k}{n}\right) p_{n,k}.$$

Once P_n and Q_n are defined, we will simply utilize the triangle inequality

$$\|f - Q_n\|_{[0,1]} \leq \|f - B_n\|_{[0,1]} + \|B_n - P_n\|_{[0,1]} + \|P_n - Q_n\|_{[0,1]},$$

hence the proof will consist of individual steps bounding each error term.
1. For $\|f - B_n\|_{[0,1]}$, it will suffice to use the generic bound (3) stated earlier.
2. The second approximating polynomial P_n will be a perturbation of B_n given by

$$P_n := \sum_{k=0}^{n} y_k p_{n,k}$$

where

$$y_k := \begin{cases} f(0), & 0 \leq k < t_n, \\ f\left(\frac{k}{n}\right) + \varepsilon_n, & t_n \leq k \leq n - t_n, \\ f(1), & n - t_n < k \leq n, \end{cases} \tag{9}$$

under the assumption that t_n is a nonnegative integer not exceeding $n/2$ and

$$\varepsilon_n := \frac{1}{n + 1 - 2t_n} \left(\left[\sum_{k=t_n}^{n-t_n} f\left(\frac{k}{n}\right) \right]_{\mathbb{Z}} - \sum_{k=t_n}^{n-t_n} f\left(\frac{k}{n}\right) \right). \tag{10}$$

Here $[w]_{\mathbb{Z}}$ stands for any integer such that $|w - [w]_{\mathbb{Z}}| \leq 1/2$, though choosing the closest integer is not critical for our argument.

The significance of the definition (9) will become clear when we use the y_k to construct the integer coefficients q_k of the final polynomial $Q_n \in \mathscr{B}_n$ by means of a recurrence relation. For now, we only note that y_k is a uniformly small perturbation of $f(k/n)$. Indeed,

$$\left| f\left(\frac{k}{n}\right) - y_k \right| \le \begin{cases} \omega_f(\frac{k}{n}), & 0 \le k < t_n, \\ |\varepsilon_n|, & t_n \le k \le n - t_n, \\ \omega_f(\frac{n-k}{n}), & n - t_n < k \le n, \end{cases}$$

and $|(n+1-2t_n)\varepsilon_n| \le 1/2$, so that we have

$$\max_{0 \le k \le n} \left| f\left(\frac{k}{n}\right) - y_k \right| \le \max\left(\omega_f\left(\frac{t_n}{n}\right), \frac{1}{2(n+1-2t_n)} \right).$$

Therefore, using (6), it follows that

$$\|B_n - P_n\|_{[0,1]} \le \max\left(\omega_f\left(\frac{t_n}{n}\right), \frac{1}{2(n+1-2t_n)} \right). \tag{11}$$

3. We now turn to the construction of $Q_n := \sum_{k=0}^{n} q_k p_{n,k} \in \mathscr{B}_n$ and the accompanying bound on $\|P_n - Q_n\|_{[0,1]}$. We set $u_{-1} := 0$ and define the $q_k \in \mathbb{Z}$, $k = 0, \dots, n$, according to the assignment rule

$$q_k := \left[u_{k-1} + y_k\right]_{\mathbb{Z}}, \tag{12}$$

$$u_k := u_{k-1} + y_k - q_k. \tag{13}$$

It is apparent from this recurrence that $|u_k| \le 1/2$ for all k. Furthermore, noting that $y_k = f(0) \in \mathbb{Z}$ for $k = 0, \dots, t_n - 1$, it follows recursively that $q_k = y_k = f(0)$ and $u_k = 0$ for $k = 0, \dots, t_n - 1$. Summing (13) for $k = t_n, \dots, n - t_n$ and using that $u_{t_n - 1} = 0$, we obtain

$$u_{n-t_n} = \sum_{k=t_n}^{n-t_n} y_k - \sum_{k=t_n}^{n-t_n} q_k.$$

The definition of y_k given in (9) and the definition of ε_n given in (10) together yield

$$\sum_{k=t_n}^{n-t_n} y_k = (n+1-2t_n)\varepsilon_n + \sum_{k=t_n}^{n-t_n} f\left(\frac{k}{n}\right) = \left[\sum_{k=t_n}^{n-t_n} f\left(\frac{k}{n}\right) \right]_{\mathbb{Z}}$$

which shows that $u_{n-t_n} \in \mathbb{Z}$. Given that all u_k are bounded by $1/2$ in absolute value, it follows that $u_{n-t_n} = 0$. At this point, noting that $y_k = f(1) \in \mathbb{Z}$ for $k = n - t_n + 1, \dots, n$, it follows recursively that $q_k = y_k = f(1)$ and $u_k = 0$ for $k = n - t_n + 1, \dots, n$. Hence,

$$
\begin{aligned}
P_n - Q_n &= \sum_{k=t_n}^{n-t_n} (y_k - q_k) p_{n,k} \\
&= \sum_{k=t_n}^{n-t_n} (u_k - u_{k-1}) p_{n,k} \\
&= \sum_{k=t_n}^{n-t_n-1} u_k (p_{n,k} - p_{n,k+1}), \qquad (14)
\end{aligned}
$$

where in the last equality we have used the fact that $u_{t_n-1} = u_{n-t_n} = 0$. Note that if $t_n = n/2$, then the sum in (14) contains no terms (and $P_n = Q_n$), so from now on we will assume $t_n < n/2$, which guarantees that there is at least one term in this sum.

In order to analyze this representation, we next note the identity (see e.g. [2, 9])

$$
x(1-x)\Big(p_{n,k}(x) - p_{n,k+1}(x)\Big) = \left(\frac{k+1}{n+1} - x\right) p_{n+1,k+1}(x), \qquad (15)
$$

which holds for all $k \in \mathbb{Z}$, with the natural convention that $p_{n,k} := 0$ whenever $k < 0$ or $k > n$ (following the corresponding convention for $\binom{n}{k}$). Plugging this formula in (14) and noting that $|u_k| \leq 1/2$, we obtain

$$
\begin{aligned}
x(1-x)|P_n(x) - Q_n(x)| &\leq \frac{1}{2} \sum_{k=t_n}^{n-t_n-1} \left| \frac{k+1}{n+1} - x \right| p_{n+1,k+1}(x) \\
&= \frac{1}{2} S_{n+1,t_n+1}(x), \qquad (16)
\end{aligned}
$$

where

$$
S_{n,t}(x) := \sum_{k=t}^{n-t} \left| \frac{k}{n} - x \right| p_{n,k}(x). \qquad (17)
$$

(Here, note that $t_n < n/2$ implies $t_n + 1 \leq (n+1)/2$.)

We claim that for all $x \in [0, 1]$ and $0 < t \leq n/2$, we have the bound

$$
S_{n,t}(x) \leq \frac{x(1-x)}{\sqrt{t/2}}. \qquad (18)
$$

With this bound, (16) implies (after canceling the term $x(1-x)$ for $0 < x < 1$ and using continuity at $x = 0$ and $x = 1$) that

$$
\|P_n - Q_n\|_{[0,1]} \leq \frac{1}{\sqrt{2(t_n+1)}}. \qquad (19)
$$

It remains to prove our claim (18). Applying Cauchy-Schwarz to (17) we have

$$S_{n,t}(x) \le \left(\sum_{k=t}^{n-t}\left(\frac{k}{n}-x\right)^2 p_{n,k}(x)\right)^{1/2}\left(\sum_{k=t}^{n-t} p_{n,k}(x)\right)^{1/2}. \tag{20}$$

For the first term in this product, (8) implies

$$\sum_{k=t}^{n-t}\left(\frac{k}{n}-x\right)^2 p_{n,k}(x) \le \frac{x(1-x)}{n}. \tag{21}$$

For the second term, we start by noting that

$$k(n-k) \ge t(n-t), \quad k = t, \ldots, n-t,$$

which is a consequence of the fact that the function $g(y) := y(1-y)$ reaches its minimum on the interval $[\tau, 1-\tau]$ at $y = \tau$ and $y = 1-\tau$, $0 \le \tau \le 1/2$. Hence

$$\begin{aligned}\sum_{k=t}^{n-t} p_{n,k}(x) &\le \frac{1}{t(n-t)}\sum_{k=t}^{n-t} k(n-k)p_{n,k}(x)\\ &\le \frac{1}{t(n-t)}\sum_{k=0}^{n} k(n-k)p_{n,k}(x)\\ &= \frac{n(n-1)x(1-x)}{t(n-t)}\\ &< \frac{2(n-1)x(1-x)}{t},\end{aligned} \tag{22}$$

where in the second last step we used the identity

$$\sum_{k=0}^{n} k(n-k)p_{n,k}(x) = n(n-1)x(1-x)$$

which can be easily derived from the moment relations (6)–(8) by means of the identity

$$k(n-k) = -(k-nx)^2 + n(1-2x)(k-nx) + n^2x(1-x).$$

Injecting the bounds (21) and (22) in (20) yields the claimed bound in (18).
4. To complete the proof, it remains to analyze the sum of these three bounds. With the results of the steps **1–3**, we have

$$\|f - Q_n\|_{[0,1]} \le \frac{5}{4}\omega_f(n^{-1/2}) + \max\left(\omega_f\left(\frac{t_n}{n}\right), \frac{1}{2(n+1-2t_n)}\right) + \frac{1}{\sqrt{2(t_n+1)}}.$$

So far, the value of t_n, $0 \le t_n \le n/2$, is arbitrary, noting that Q_n depends only on f and t_n according to the scheme presented in steps **2** and **3**. Hence it follows that $d_\infty(f, \mathscr{B}_n)$ is bounded above by

$$\frac{5}{4}\omega_f(n^{-1/2}) + \rho(f,n)$$

where

$$\rho(f,n) := \min_{t\in[0,n/2]\cap\mathbb{Z}} \left\{ \max\left(\omega_f\left(\frac{t}{n}\right), \frac{1}{2(n+1-2t)}\right) + \frac{1}{\sqrt{2(t+1)}} \right\}.$$

We can choose the convenient value $t = t_n := \lfloor \frac{1}{2} n^{2/3} \rfloor$. With this choice, it is straightforward to check that for all $n \ge 2$, we have

$$\frac{1}{2(n+1-2t_n)} + \frac{1}{\sqrt{2(t_n+1)}} \le \frac{2}{n^{1/3}}.$$

Hence it follows that for all $n \ge 2$, we have the bound

$$d_\infty(f, \mathscr{B}_n) \le \frac{5}{4}\omega_f(n^{-1/2}) + \omega_f(\frac{1}{2}n^{-1/3}) + 2n^{-1/3}$$

which, in particular, is bounded above by the simpler expression

$$\frac{9}{4}\omega_f(n^{-1/3}) + 2n^{-1/3}.$$

This concludes our proof. □

Notes and Comments.

1. Evaluating $\rho(f,n)$ via the optimal choice of $t = t_n$ requires more precise information on ω_f. It can be seen that for Lipschitz functions, the scaling relation $t_n \sim n^{2/3}$ is essentially optimal, hence our choice. However, the situation is different for functions of lower Hölder regularity. If $\omega_f(\delta) \le C\delta^{1/2}$, then the choice $t_n \sim n^{1/2}$ yields the upper bound $O(n^{-1/4})$ instead of the inferior $O(n^{-1/6})$ that follows from the choice $t_n \sim n^{2/3}$. It can be checked that for Hölder continuous functions with exponent $0 < \alpha \le 1$ (i.e. $\omega_f(\delta) \le C\delta^\alpha$), the optimal exponent θ_α where $t_n \sim n^{\theta_\alpha}$ is given by

$$\theta_\alpha = \begin{cases} \frac{1}{2}, & 0 < \alpha \le \frac{1}{2}, \\ \frac{2\alpha}{2\alpha+1}, & \frac{1}{2} \le \alpha \le 1, \end{cases}$$

yielding the bound

$$d_\infty(f, \mathscr{B}_n) = O\left(n^{-\min(\frac{\alpha}{2}, \frac{\alpha}{2\alpha+1})}\right).$$

2. It is natural to ask if the rate of approximation can be improved for differentiable functions, as was the case for the pointwise results in [3]. This appears to be nontrivial, if at all possible, the investigation of which we leave for future work.
3. Currently we do not have any expectations on the optimality of our upper bound since we do not know any class of functions with a matching lower bound. As a particular case, it would be interesting to derive a lower bound on the supremum of $d_\infty(f, \mathscr{B}_n)$ over all Lipschitz functions $f \in \mathscr{C}_\mathbb{Z}([0, 1])$ with Lipschitz norm at most 1. A lower bound in this spirit (with an additional constraint that limits the size of the coefficients of the approximation in $\mathscr{B}_n$) was established in [5], but without the integer boundary constraint.

References

1. I Chlodovsky. Une rèmarque sur la représentation des fonctions continues par des polynomes à coefficients entiers. *Math. Sb.*, 32(3):472–475, 1925.
2. Ronald A DeVore and George G Lorentz. *Constructive Approximation*, volume 303. Springer Science & Business Media, 1993.
3. C. Sinan Güntürk and Weilin Li. Approximation with one-bit polynomials in Bernstein form. *Constructive Approximation*, 57(2):601-30, 2023.
4. C. Sinan Güntürk and Weilin Li. Approximation of functions with one-bit neural networks. *arXiv preprint* arXiv:2112.09181, 2021.
5. C. Sinan Güntürk and Weilin Li. Near-optimality of $\Sigma\Delta$ quantization for L^2-approximation with polynomials in Bernstein form. *Fourteenth International Conference on Sampling Theory and Applications*, 2023.
6. Soichi Kakeya. On approximate polynomials. *Tohoku Mathematical Journal, First Series*, 6:182–186, 1914.
7. L. V. Kantorovich. Some remarks on the approximation of functions by means of polynomials with integral coefficients. *Izv. Akad. Nauk SSSR*, 7:1163–1168, 1931.
8. Ferguson Le Baron O. *Approximation by polynomials with integral coefficients*, volume 17. American Mathematical Soc., 1980.
9. George G Lorentz. *Bernstein polynomials*. Chelsea, 2nd edition., 1986.
10. George G Lorentz, Manfred v Golitschek, and Yuly Makovoz. *Constructive approximation: advanced problems*, volume 304. Springer, 1996.
11. Julius Pál. Zwei kleine bemerkungen. *Tohoku Mathematical Journal, First Series*, 6:42–43, 1914.
12. Roald Mikhailovich Trigub. Approximation of functions by polynomials with integer coefficients. *Izvestiya Rossiiskoi Akademii Nauk. Seriya Matematicheskaya*, 26(2):261–280, 1962.

On the Distribution of Subset Sums of Certain Sets in $\mathbb{Z}_p^2$ and in $\mathbb{N}^2$

Norbert Hegyvári

Abstract A given subset A of natural numbers is said to be complete if every element of N is the sum of distinct terms taken from A. This topic is strongly connected to the knapsack problem which is known to be NP complete. Interestingly if A and B are complete sequences then $A \times B$ is not necessarily complete in $\mathbb{N}^2$. In this paper we consider a modular version of this problem, motivated by the communication complexity problem of [2].

2020 Mathematics Subject Classification: Primary 11B30 · 11B39 · Secondary 11B75

1 Introduction

Throughout the paper $\mathbb{N}$ denotes the natural numbers, $\mathbb{Z}$ denotes the set of integers, and let $\mathbb{Z}_p := \mathbb{Z}/p\mathbb{Z}$ represented by the elements $\{1, \ldots, p-1\}$. We write $[N] = \{0, 1, \ldots, N\}$.

Let S be any additive semigroup. For sets $A, B \subset S$ the sum is defined by $A \pm B := \{a \pm b : a \in A;\ b \in B\}$, For any $X \subseteq S$ let

$$FS(X) := \{\sum_{i=1}^{\infty} \varepsilon_i x_i \ : \ x_i \in X,\ \varepsilon_i \in \{0, 1\},\ \sum_{i=1}^{\infty} \varepsilon_i < \infty\},$$

and one can extend the notion of subset sums $FS(X)$ for higher dimension, i.e. when $X \subseteq S^k$; $k \geq 1$. A set $X \subseteq S$ is said to be complete respect to S if every element of S is the sum of distinct terms taken from X. If $S = \mathbb{N}$, there is a wide literature on the subject (A good survey can be found e.g. in [6, 10, 12]).

N. Hegyvári (✉)
Institute of Mathematics, Eötvös University, Pázmány st. 1/c, Budapest H-1117, Hungary
e-mail: hegyvari@renyi.hu

Alfréd Rényi Institute of Mathematics, P.O.Box 127, Budapest H-1364, Hungary

M. B. Nathanson (ed.), *Combinatorial and Additive Number Theory VI*,
Springer Proceedings in Mathematics & Statistics 464,
https://doi.org/10.1007/978-3-031-65064-2_11

This type of problems are highly connected to the knapsack or subset-sum problem. The problem is to determine, given positive integers $a_1, a_2, \ldots, a_n$ and an integer m, whether there is a subset of the set $\{a_j\}$ that sums up to m. It is well known as an NP-complete problem. This topic is connected to some communication complexity problem too (see e.g. [2]) which motivates the present work.

Completeness in higher dimensions is very different from the one-dimensional case. For example, a classic result is that if a series contains more than $c\sqrt{n}$ elements up to $n > n_0$, then it is complete (see [12]). While in the two-dimensional case, one can give a dense set A for which $FS(A)$ does not even contain an arithmetic progression (see [8, 9]). Further results can be found in [1, 4].

Let F_n be the nth Fibonacci number, i.e. let $F_0 = 0$; $F_1 = 1$ and for $n \geq 0$ $F_{n+2} = F_{n+1} + F_n$. For negative integers let $F_{-n} = (-1)^{n-1} F_n$ when the recursive formula also holds. In section 2 we will investigate the distribution of the subset sums of three sets in $\mathbb{Z}_p^2$.

More precisely let $A_1 = \{2^n\}_{n=0}^{\infty} \times \{F_n\}_{n=0}^{\infty}$, $A_2 = \{2^n\}_{n=0}^{\infty} \times \{2^n\}_{n=0}^{\infty}$ and $A_3 = \{F_n\}_{n=0}^{\infty} \times \{F_n\}_{n=0}^{\infty}$. We will show that for almost all prim p, $FS(A_i) \pmod p = Z_p^2$ $(i = 1, 2, 3)$. For $X \subseteq \mathbb{N}^2$, $X \pmod p = \{(v_1, v_2) \in Z_p^2 : \exists (x_1, x_2) \in X;\ v_i \equiv x_i \pmod p;\ i = 1, 2\}$.

Furthermore we extend the analysis to sets of types $A_4 = \{b^n\}_{n=0}^{\infty} \times \{F_n\}_{n=0}^{\infty}$ and $A_5 = \{b^n\}_{n=0}^{\infty} \times \{2^n\}_{n=0}^{\infty}$, where $b > 2$ integer. Although the set $\{b^n\}_{n=0}^{\infty}$ is incomplete for $b > 2$, note that there can be an unbounded number of times b^n as one of the coordinates in the sets A_4 and A_5. Thus, since all positive integers can be represented in the b-ary expansion, this question fits the previous ones.

In [2] we also consider the above questions in $\mathbb{N}^2$. Interestingly, the distribution of the above sets is different than in $\mathbb{Z}_p^2$. For example, for $i = 1, 2$ and 3, we describe sets E_i for which $\mathbb{N}^2 \setminus E_i \subseteq FS(A_i)$. Furthermore, we showed that E_i contains arbitrary "large" squares which contain no elements from $FS(A_i)$, and squares of size $k \times k$ which contain $k \log k$ many elements from $FS(A_i)$.

2 Modular Case

In this section we will prove the following three theorems:

Theorem 1 *There exists a $\delta > 0$ such that for all but $O(x/(\log x)^{1+\delta})$ primes $p \leq x$, $|FS(A_4) \pmod p| \geq \frac{p^2}{b}$.* □

When $b = 2$, the previous theorem says that $|FS(A_1) \pmod p| \geq \frac{p^2}{2}$. But in this case we get more:

Theorem 2 *There exists a $\delta > 0$ such that for all but $O(x/(\log x)^{1+\delta})$ primes $p \leq x$, $FS(A_1) \pmod p = \mathbb{Z}_p^2$.* □

The following theorem can be proved in a similar way as Theorem 2

Theorem 3 *There exists a* $\delta > 0$ *such that for all but* $O(x/(\log x)^{1+\delta})$ *primes* $p \le x$, $FS(A_i) \pmod{p} = \mathbb{Z}_p^2$ *(*$i = 2, 3$*).* □

Proof (*Proof of Theorem* 2) Let $\mathcal{P}_1$ be the sets of all primes up to x for which $|\{2^n\}_{n=0}^{\infty} \pmod{p}| > 2\sqrt{p}$ and let $\mathcal{P}_2$ be the sets of all primes up to x for which $|\{F_n\}_{n=0}^{\infty} \pmod{p}| > 2\sqrt{p}$.

Denote by $\mathcal{P} = \mathcal{P}_1 \cap \mathcal{P}_2$. We will check that for all but $O(x/(\log x)^{1+\delta})$ primes $p \le x$ belong to $\mathcal{P}$.

We need the following lemmas

Lemma 1 *There exists a* $\delta > 0$ *such that for all but* $O(\frac{x}{(\log x)^{1+\delta}})$ *primes* $p \le x$ *we have*

$$|\{F_n\}_{n=0}^{\infty} \pmod{p}| > 3\sqrt{p}, \tag{1}$$

where $\delta > 0.086072$ *is admissible if* x *is large enough.* □

It is a slight modification of [7, Lemma 3.1].

Lemma 2 *There exist a* $\delta > 0$ *such that*

$$|\{b^n\}_{n=0}^{\infty} \pmod{p}| > 2\sqrt{p} \tag{2}$$

for all but $O(x/(\log x)^{1+\delta})$ *primes* $p \le x$*. One can choose* δ *as in the previous lemma.*

This lemma is a simple consequence of Theorem 3 in [5].

Using these lemmas we get that for all but $O(x/(\log x)^{1+\delta})$ primes $p \le x$ both (2) and (1) hold. Denote the set of these primes by $\mathcal{P}$ and we get that for every $p \in \mathcal{P}$ both $T := |\{b^n\}_{n=0}^{\infty} \pmod{p}| > 2\sqrt{p}$ and $R := |\{F_n\}_{n=0}^{\infty} \pmod{p}| > 2\sqrt{p}$ hold. (We only need the factor 3 of the Lemma 1 when we assume that $b > 2$, see Sect. 2.1). The third ingredient of our proof is the following

Lemma 3 (Olson) *Let* $A = \{a_1, \dots, a_k\}$ *be distinct non-zero residue classes modulo the prime* p*. If* $k > 2\sqrt{p}$ *then* $FS(A) = \mathbb{Z}_p$*.* □

See the proof in [11].

From now on in this section we assume that $b = 2$. Write $\{F_n\}_{n=0}^{\infty} \pmod{p} = \{F_{j_1}, F_{j_2}, \dots, F_{j_R}\}$ and $\{2^n\}_{n=0}^{\infty} \pmod{p} = \{2^{i_1}, 2^{i_2}, \dots, 2^{i_T}\}$. Let (v_1, v_2) any point in $\mathbb{Z}_p^2$. By Lemma 3 we can write v_1 in the form $v_1 = \sum_{m=1}^{T} \varepsilon_m 2^{i_m}$; $\varepsilon_m \in \{0, 1\}$, and v_2 in the form $v_2 = \sum_{k=1}^{R} \eta_k F_{j_k}$; $\eta_k \in \{0, 1\}$.

Assume first that $\sum_{m=1}^{T} \varepsilon_m > \sum_{k=1}^{R} \eta_k$. We are going to find a subset $\{F'_{j_n}\}$ for which $\sum_{n=1}^{V} \zeta_n F'_{j_n} = v_2$, $\zeta_n > 0$ and $\sum_n \zeta_n = T$. Note that we allow ζ_n to be bigger than one, i.e. we allow some F'_j to be repeated. Then the elements $(2^{i_1}, F'_{j_1}), (2^{i_2}, F'_{j_2}), \dots, (2^{i_T}, F'_{j_T})$ are pairwise different (because their first coordinates are pairwise different) and $\sum_{t=1}^{T} (2^{i_t}, F'_{j_t}) = (v_1, v_2)$, as we wanted.

So take F_{i_1} and find an element F_r, where $r > \sum_{m=1}^{T} \varepsilon_m$ and $F_r \equiv F_{i_1} \pmod{p}$. Since the sequence F_n is periodic $\pmod{p}$, there exists such an r. Now change F_r to $F_{r-1} + F_{r-2}$. Repeat this changing process altogether $\sum_{m=1}^{T} \varepsilon_m - \sum_{k=1}^{R} \eta_k$ many times (i.e. change F_{r-1} to $F_{r-2} + F_{r-3}$, etc.). Since $r > \sum_{m=1}^{T} \varepsilon_m$ this process is terminated. Then we can matching the terms of powers of two to the (possible repeated) elements of the appropriate Fibonacci elements.

When $\sum_{m=1}^{T} \varepsilon_m = \sum_{k=1}^{R} \eta_k$, we are done. If $\sum_{m=1}^{T} \varepsilon_m < \sum_{k=1}^{R} \eta_k$ then the process is the same; just note that we replace a term 2^m by $2^{m-1} + 2^{m-1}$.

2.1 The Case When $b > 2$

In the general case we give a bound for the cardinality of $FS(A_4) \pmod{p}$, where $A_4 = \{b^n\}_{n=0}^{\infty} \times \{F_n\}_{n=0}^{\infty}$ □

Proof (*Proof of Theorem* 1) Let $p \in \mathcal{P}$. Split $\{F_n\}_{n=0}^{\infty} \pmod{p} = \{F_{j_1}, F_{j_2}, \dots,\}$ into two disjoint parts $G_1 \cup G_2$ where $|G_2| = b - 1$. Write $\{b^n\}_{n=0}^{\infty} \pmod{p} = \{b^{i_1}, b^{i_2}, \dots, b^{i_T}\}$, $G_1 = \{Fi_1, F_{i_2}, \dots, F_{i_V}\}$ and $G_2 = \{F_{s_1}, F_{s_2}, \dots, F_{s_{b-1}}\}$. By the assumption and Lemma 2 we get that $V = |G_1| > 2\sqrt{p}$. By Lemma 3 for any $(v_1, v_2) \in \mathbb{Z}_p^2$, v_1 can be written as $\sum_{k=1}^{T} \varepsilon_k b^{i_k}$; $\varepsilon_k \in \{0, 1\}$, and v_2 in the form $v_2 = \sum_{k=1}^{V} \eta_k F_{j_k}$; $\eta_k \in \{0, 1\}$.

Let us assume that $\sum_{m=1}^{T} \varepsilon_m \geq \sum_{k=1}^{V} \eta_k$. Then follow the previous process.

So let us assume that $\sum_{m=1}^{T} \varepsilon_m < \sum_{k=1}^{V} \eta_k$. Then take an m large enough for which $b^m \equiv b^{i_1} \pmod{p}$, and change $b^m = b^{m-1} + b^{m-1} + \dots + b^{m-1}(b \text{ times})$. Repeat this process increasing the number of terms in each step by b elements. If one step the number of terms is exactly $\sum_{k=1}^{V} \eta_k$, then we are done. Otherwise one repetition we overstep $\sum_{k=1}^{V} \eta_k$ by r many elements and clearly $\min r \leq b - 1$. Then completing the sum $\sum_{k=1}^{V} \eta_k F_{j_k}$ by the terms $F_{s_1}, F_{s_2}, \dots, F_{s_r}$ we can represent $(v_1, v_2 + \sum_{t=1}^{r} F_{s_t})$ in $FS(A_4) \pmod{p}$.

Consider a coloring of $\mathbb{Z}_p^2$ according the values of $\sum_{t=1}^{r} F_{s_t}$; $(r = 0, 1, \dots b - 1)$, i.e. let (v_1, v_2) be color r, if $(v_1, v_2 + \sum_{t=1}^{r} F_{s_t})$ in $FS(A_4) \pmod{p}$. Note that the coloring is not unique; (v_1, v_2) and (v_1', v_2') can have the same color. For clarity, choose the color r to be minimal. Denote by L_r, the set of points of color r. Now choose the $L_{r_0} \subseteq \mathbb{Z}_p^2$ for which $|L_{r_0}|$ is maximal. We have $|L_{r_0}| \geq \frac{p^2}{b}$. Then $L_{r_0} + \sum_{t=1}^{r_0} F_{s_t} \subseteq FS(A_4) \pmod{p}$, and $|L_{r_0} + \sum_{t=1}^{r_0} F_{s_t}| = |L_{r_0}|| \geq \frac{p^2}{b}$. □

3 Integer Lattice Case

3.1 Large Grid in $P(A_4) \cap \mathbb{N}^2$

Recall that $A_4 = \{b^n\}_{n=0}^{\infty} \times \{F_n\}_{n=0}^{\infty}$. Since $A_4(N) := |\{b^n\}_{n=0}^{\infty} \cap [N]| \leq \frac{\log_2 N}{\log_2 b}$ hence $|P(A_4(N))| \leq 2^{\frac{\log_2 N}{\log_2 b}} = N^{1/\log_2 b}$. Furthermore since $P(A_4(N)) \subset P(\{b^n\}_{n=0}^{\infty}) \times P(\{F_n\}_{n=0}^{\infty}) = P(\{b^n\}_{n=0}^{\infty}) \times N$ we obtain that the cardinality of $P(A_4) \cap [N] \times [N]$ is at most $N^{1/\log_2 b} \cdot N = o(N^2)$ for $b > 2$. So we can not use a Szemerédi-type argument to detect a structure of $P(A_4)$ in the square $[N] \times [N]$.

Nevertheless we will show a "large" grid of it:

Theorem 4 *Let $b \geq 31$. There is a grid G in $P(A_4) \cap ([N] \times [N])$ with cardinality a least $|G| \geq c(1 + o(1))N^{1/2(1+1/\log_2 \varrho+1/\log_2 b)}$.* □

Proof Let $X := \{F_2 < \cdots < F_{2t} < F_{2t+2} < \cdots < F_{2m} \leq N\}$ be the every second Fibonacci numbers up to N, and let $Y := \{1 < b < \cdots < b^k < \cdots < b^n\}$ be the sequence of the powers of b up to N.

Let $Z := \left\{\sum_{1 \leq i_1 < \cdots < i_t \leq m} F_{2i_j} : m/2 < t \leq m\right\}$. We need the following well-known lemma which sometimes called Zeckendorf theorem:

Lemma 4 *Each positive integer n has a unique representation as the sum of distinct Fibonacci numbers if no two consecutive Fibonacci numbers are used in the representation.* □

It implies that every integer $z \in Z$ has a unique representation as the sum of distinct Fibonacci numbers.

Let $W := \{\sum_{r=1}^{s} c_r b^{i_r} : c_r \geq b/2;\ n/2 \leq s \leq n\}$. We will show that

$$Z \times W \subseteq P(A_4) \cap [N] \times [N],$$

i.e. $G := Z \times W$ is a grid in the sumset. Let $Q = (z, w) \in Z \times W$. Clearly $Q \in [N] \times [N]$. We have to show that $Q \in P(A_4)$ also holds. Write $z = \sum_{1 \leq i_1 < \cdots < i_t \leq m} F_{2i_j} : m/2 < t \leq m$ and $w = \sum_{r=1}^{s} c_r b^{i_r} : c_r \geq b/2;\ n/2 \leq s \leq n$.

For any $z \in Z$ denote by $T_F(z)$ the number of terms in the representation of z, i.e. let $T_F(z) = t$ and denote by by $T_F(w)$; $w \in W$ the number of terms in the representation of w, i.e. let $T_F(w) = \sum_{r=1}^{s} c_r$. By the definition of T_F and T_b, an easy calculation shows

Fact (i) $\max_{z \in Z} T_F(z) \leq \frac{\log_2 N}{\log_2 \varrho}$, (ii) $\min_{w \in W} T_b(w) \geq \frac{b \log_2 N}{4 \log_2 b}$.

Furthermore for every $z \in Z$, $w \in W$

$$1 \leq \frac{T_b(w)}{T_F(z)} \leq 4b \frac{\log_2 \varrho}{\log_2 b},$$

provided $b \geq 31$.

Now the process is similar as in the proof of Theorem 2; let $R := \lceil \frac{T_b(w)}{T_F(z)} \rceil$ and let F_{2i_j} be any term at the representation of z. We split each F_{2i_j} in the form $F_{2i_j-1} + F_{2i_j-2} = F_{2i_j-1} + F_{2i_j-3} + F_{2i_j-4} = F_{2i_j-1} + F_{i_j-3} \cdots + + F_{i_j-(2k-1)}$, where $k \leq R$ to be the number of terms in the (multi) representation of z will be exactly $T_b(w)$. We can do it if N is big enough. It easy to see that the multiplicity $mult(F_{2i_j})$ of each term after the splitting process is at most $mult(F_{2i_j}) \leq R - 2 < R$. Now we use an important tool which gives a sufficiently condition for the pairwise different matching of the co-ordinates:

Lemma 5 ([4, Proposition 1]) *Let $X_1, \ldots, X_s$ be disjoint finite sets and $Y_1, \ldots, Y_t$ be disjoint finite sets too. Let*

$$U = \bigcup_{i=1}^{s} X_i, \quad V = \bigcup_{j=1}^{t} Y_j,$$

with $|U| = |V|$ and suppose that $1 \leq |X_i| \leq \sqrt{|U|}$ for $i = 1, 2, \ldots s$ and $1 \leq |Y_j| \leq \sqrt{|V|}$ for $j = 1, 2, \ldots t$. Then there exists a bipartite graph $G(X, Y)$ fulfilling the following conditions:

(a) there are no two edges (x_1, y_1); (x_2, y_2) for which $x_1, x_2 \in X_i$; $y_1, y_2 \in Y_j$ for some i and j;
(b) $G(X, Y)$ is a matching. □

We identify the set X_r to b^{i_r} such a way that let $|X_r| = c_r$ the number of the multiplicity of b^{i_r} and we associate Y_j to $mult(F_{2i_j})$, i.e. let $|Y_j| = mult(F_{2i_j})$. Since $c_r \leq b$ and $mult(F_{2i_j}) \leq R \leq \frac{T_b(w)}{T_F(z)} \leq 4b\frac{\log_2 \varrho}{\log_2 b}$ we obtain that if N is big enough then the assumptions of the lemma hold. So by adding the paired b powers and F Fibonacci numbers as coordinates we get the point (z, w).

Now we turn to the estimation of the size of the grid $G = Z \times W$.

To simplify the calculation, we can assume that b and n are even. We have that the representation of integers in base b is unique. Furthermore by the symmetry of the binomial coefficient

$$|W| \geq \left(\frac{b}{2}\right)^{n/2} \cdot \sum_{k=n/2}^{n} \binom{n}{k} = \frac{1}{2} 2^{n/2} b^{n/2}.$$

Recall that $n = \lfloor \frac{\log_2 N}{\log_2 b} \rfloor$, we have

$$|W| \geq c_1(1 + o(1)) N^{1/2(1+1/\log_2 b)},$$

where the constant can be chosen as $c_1 = \frac{1}{4\log_2 b}$.

Finally we bound the cardinality of Z. Recall now that $m = \frac{1}{2}\lfloor \frac{\log_2 N}{\log_2 \varrho} \rfloor$. So we have

$$|Z| \geq \sum_{m/2 \leq k \leq m} \binom{m}{k} \geq \frac{1}{2} N^{\frac{1}{2\log_2 \varrho}}.$$

4 Concluding Remarks

In [7] the authors derived from their lemma that for all but $o(\pi(x))$ every residue class modulo $p \leq x$ can be represented as a sum of 32 Fibonacci numbers, i.e. the sequence $\{\overline{F_n}\}_p$ is an additive bases of order 32 modulo p, where $\overline{F_n} = F_n \pmod{p}$. Let G be any additive semigroup. We say that a sequence $\{a_n\} \subseteq G$ is a *restricted* bases of order s if every element of G can be written as a sum of s many pairwise different element of $\{a_n\}$. The following problem seems to be hard: is it true, that there exist a $K \in N$ and an infinite sequence of primes $\{p_i\}$ such that the sequence $\{\overline{F_n}\}_{p_i}$ is a restricted bases of order K modulo p_i?

Acknowledgements I express my thanks to Mel B. Nathanson for his help with the presentation. This work was supported by the National Research, Development and Innovation Fund of Hungary through project no. grant K-146387.

References

1. B. Bakos, M. Pálfy, Some results on an encryption method using subset-sums of pseudo-recursive sequences Discrete Mathematics Letters 5 : 1 pp. 63-67. , 5 p. (2021)
2. B. Bakos, N. Hegyvári and M. Pálfy, On a communication complexity problem in combinatorial number theory, Moscow Journal of Combinatorics, Vol. 10 (2021), No. 4, 297-302 https://doi.org/10.2140/moscow.2021.10.297
3. B. Bakos, N. Hegyvári and M. Pálfy On the distribution of subset sums of certain sets in $\mathbb{N}^2$, in progress
4. Chen, YG ; Fang, JH ; Hegyvári, N, Erdős-Birch type question in $\mathbb{N}^r$ J. of Number Theory 187 pp. 233-249. , 17 p. (2018)
5. P. Erdős and M. R. Murty, On the order of a mod p, in: Number Theory (Ottawa, ON, 1996), CRM Proceedings and Lecture Notes, 19 (American Mathematical Society, Providence, RI, 1999), pp. 87-97.
6. P. Erdős, R.L. Graham, Old and new problems and results in combinatorial number theory, Monographies de L'Enseignement Mathématique
7. Garcia V, Luca F and Mejia J., On sums of Fibonacci numbers modulo p, Bulletin of the Australian Mathematical Society 83(03):413-419, https://doi.org/10.1017/S0004972710001693
8. Hegyvári, N. Complete sequences in $\mathbb{N}^2$ European Journal of Combinatorics (17) (1996) 741-749.
9. Hegyvári, N Subset sums in $\mathbb{N}^2$ Combinatorics Probability and Computing (5) (1996) 393-402.
10. N.B. Nathanson, Additive Number Theory Springer-Verlag New York 1996
11. J. E. Olson, An addition theorem modulo p, J. Combinatorial Theory 5(1968), 45-52
12. T. Tao, V. Vu, *Additive Combinatorics*. Cambridge University Press, Cambridge 2006

On a Polynomial Reciprocity Theorem of Carlitz

Brad Isaacson

Abstract Carlitz proved a powerful reciprocity theorem for generalized Dedekind-Rademacher sums. Among its many consequences was an interesting polynomial reciprocity theorem which holds under a certain restriction of its parameters. Carlitz remarked that it was unclear how this restriction could be removed. In this paper, we remove this restriction and obtain a generalization of Carlitz's polynomial reciprocity theorem. As a corollary, we get the polynomial reciprocity theorem of Beck and Kohl.

1 Introduction

Let $[x]$ denote the greatest integer $\leq x$ and $\{x\}$ denote the fractional part of x. In [2], Carlitz proved the following polynomial reciprocity theorem in the indeterminates u, v.

Theorem 1 (Carlitz's polynomial reciprocity theorem) *([2, Eq. (1.16)]). Let h, k be positive integers and x, y be real numbers such that*

$$0 \leq x < 1, \quad 0 \leq y < 1. \tag{1}$$

Setting $z = kx + hy$, we have

$$(1 - v) \sum_{a=0}^{k-1} u^{h - [\frac{ha+z}{k}]} v^a - (1 - u) \sum_{b=0}^{h-1} v^{k - [\frac{kb+z}{h}]} u^b = u^h - v^k.$$

Remark The reciprocity formula given by [2, Eq. (1.16)] includes the condition that $(h, k) = 1$, but this is not used in the proof and therefore can be dropped.

On page 355 of [2], Carlitz made the following remark regarding this polynomial reciprocity theorem: "It is not evident how the restriction (1) can be removed." In this

B. Isaacson (✉)
Mathematics Department, New York City College of Technology, 300 Jay St., Brooklyn, NY 11201, USA
e-mail: bisaacson@citytech.cuny.edu

M. B. Nathanson (ed.), *Combinatorial and Additive Number Theory VI*,
Springer Proceedings in Mathematics & Statistics 464,
https://doi.org/10.1007/978-3-031-65064-2_12

paper, we remove this restriction and obtain a generalization of Carlitz's polynomial reciprocity theorem.

Theorem 2 *Let h, k be positive integers and x, y be any real numbers. We have*

$$(1-v)\sum_{a\,(\mathrm{mod}\,k)} u^{h(1+[\frac{a+y}{k}])-\left[\frac{h(a+y)}{k}+x\right]} v^{a-k[\frac{a+y}{k}]}$$
$$-(1-u)\sum_{b\,(\mathrm{mod}\,h)} v^{k(1+[\frac{b+x}{h}])-\left[\frac{k(b+x)}{h}+y\right]} u^{b-h[\frac{b+x}{h}]} = \frac{u^h-v^k}{u^{[x]}v^{[y]}},$$

where a and b run over complete residue systems modulo k and h, respectively.

Observe that Theorem 2 includes Theorem 1 as a special case. As evidenced by the proof of Theorem 2, it is easy to remove the restriction (1) from Theorem 1. We speculate how Carlitz might not have seen this. Consider the following corollary of Theorem 1.

Corollary 3 *Let h, k be positive integers and z be any real number such that*

$$0 \le z < h+k. \tag{2}$$

We have

$$(1-v)\sum_{a=0}^{k-1} u^{h-[\frac{ha+z}{k}]} v^a - (1-u)\sum_{b=0}^{h-1} v^{k-[\frac{kb+z}{h}]} u^b = u^h - v^k.$$

The proof follows immediately from Theorem 1 as the restriction (1) implies the restriction (2). However, it is unclear how to remove the restriction (2) from Corollary 5 due to the loss of generality from depending on z rather than x and y. If Carlitz had this polynomial reciprocity formula in mind rather than his Theorem 1, then his statement about the non-obvious nature of removing restrictions makes perfect sense.

As a corollary of Theorem 2, when $(h, k) = 1$, we get the following polynomial reciprocity formula of Beck and Kohl. Let $\lceil x \rceil$ denote the smallest integer $\ge x$.

Corollary 4 ([1, Corollary 2]) *Let h, k be positive integers such that $(h, k) = 1$, and x, y be any real numbers. We have*

$$u(1-v)\sum_{a=\lceil y\rceil}^{\lceil y\rceil+k-1} u^{\left[\frac{ha+kx-hy}{k}\right]} v^a + v(1-u)\sum_{b=\lceil x\rceil}^{\lceil x\rceil+h-1} v^{\left[\frac{kb-kx+hy}{h}\right]} u^b$$
$$= u^{\lceil x\rceil} v^{\lceil y\rceil}(1-u^h v^k) - u^c v^d (1-u)(1-v),$$

where $c \in \mathbb{Z}$ is (uniquely) determined by the conditions

$$kc \equiv kx - hy \ (\mathrm{mod}\, h) \quad and \quad x \leq c < x + h,$$

and $d := \frac{kc+hy-kx}{h}$. *If* $kx - hy \notin \mathbb{Z}$, *then the last term on the right-hand side needs to be omitted.*

This result follows from Theorem 2 by choosing appropriate residue systems modulo k and h, replacing y by $-y$ and u by $1/u$, and using the fact that $-[-x] = [x] + 1 - \delta(x)$ where $\delta(x) = 1$ if $x \in \mathbb{Z}$, and 0 otherwise, from where the correction term $u^c v^d (1-u)(1-v)$ arises. A complete proof is provided in Sect. 3.

From Theorem 1, Carlitz established many interesting results. In Sect. 4, we provide generalizations of these results with the restriction (1) removed.

2 Proof of Theorem 2

Proof of Theorem 2: Let h, k be positive integers and x, y be any real numbers. From Theorem 1, we have

$$(1-v)\sum_{a=0}^{k-1} u^{h-\left[\frac{h(a+\{y\})}{k}+\{x\}\right]} v^a - (1-u)\sum_{b=0}^{h-1} v^{k-\left[\frac{k(b+\{x\})}{h}+\{y\}\right]} u^b = u^h - v^k.$$

Since $\left[\frac{a+\{y\}}{k}\right], \left[\frac{b+\{x\}}{h}\right] = 0$ for $0 \leq a \leq k-1, 0 \leq b \leq h-1$, we have

$$(1-v)\sum_{a=0}^{k-1} u^{h\left(1+\left[\frac{a+\{y\}}{k}\right]\right)-\left[\frac{h(a+\{y\})}{k}+\{x\}\right]} v^{a-k\left[\frac{a+\{y\}}{k}\right]}$$
$$-(1-u)\sum_{b=0}^{h-1} v^{k\left(1+\left[\frac{b+\{x\}}{h}\right]\right)-\left[\frac{k(b+\{x\})}{h}+\{y\}\right]} u^{b-h\left[\frac{b+\{x\}}{h}\right]} = u^h - v^k.$$

Now observe that the summations only depend on the residue classes of $a \ (\mathrm{mod}\, k)$ and $b \ (\mathrm{mod}\, h)$, respectively. Hence, we have

$$(1-v)\sum_{a \ (\mathrm{mod}\, k)} u^{h\left(1+\left[\frac{a+\{y\}}{k}\right]\right)-\left[\frac{h(a+\{y\})}{k}+\{x\}\right]} v^{a-k\left[\frac{a+\{y\}}{k}\right]}$$
$$-(1-u)\sum_{b \ (\mathrm{mod}\, h)} v^{k\left(1+\left[\frac{b+\{x\}}{h}\right]\right)-\left[\frac{k(b+\{x\})}{h}+\{y\}\right]} u^{b-h\left[\frac{b+\{x\}}{h}\right]} = u^h - v^k.$$

Replacing a and b by $a + [y]$ and $b + [x]$, respectively, we get

$$u^{[x]}v^{[y]}(1-v)\sum_{a\,(\mathrm{mod}\,k)} u^{h(1+[\frac{a+y}{k}])-\left[\frac{h(a+y)}{k}+x\right]}v^{a-k[\frac{a+y}{k}]}$$
$$-u^{[x]}v^{[y]}(1-u)\sum_{b\,(\mathrm{mod}\,h)} v^{k(1+[\frac{b+x}{h}])-\left[\frac{k(b+x)}{h}+y\right]}u^{b-h[\frac{b+x}{h}]}=u^h-v^k,$$

from which we get the assertion of the theorem. □

3 Proof of Corollary 6

Proof of Corollary 6: Let h,k be positive integers such that $(h,k)=1$, and x,y be any real numbers. Applying Theorem 2 with residue systems $a=-[y],-[y]+1,\cdots,-[y]+k-1$ and $b=-[x],-[x]+1,\cdots,-[x]+h-1$, we get

$$(1-v)\sum_{a=-[y]}^{-[y]+k-1} u^{h-\left[\frac{ha+kx+hy}{k}\right]}v^a-(1-u)\sum_{b=-[x]}^{-[x]+h-1} v^{k-\left[\frac{kb+kx+hy}{h}\right]}u^b=\frac{u^h-v^k}{u^{[x]}v^{[y]}}.$$

Replacing y with $-y$, noting that $-[-y]=\lceil y\rceil$ and $[x]=\lceil x\rceil-1+\delta(x)$ where $\delta(x)=1$ if $x\in\mathbb{Z}$, and 0 otherwise, we get

$$(1-v)\sum_{a=\lceil y\rceil}^{\lceil y\rceil+k-1} u^{h-\left[\frac{ha+kx-hy}{k}\right]}v^a-(1-u)\sum_{b=-\lceil x\rceil+1-\delta(x)}^{-\lceil x\rceil+h-\delta(x)} v^{k-\left[\frac{kb+kx-hy}{h}\right]}u^b$$
$$=u^{-\lceil x\rceil+1-\delta(x)}v^{\lceil y\rceil}(u^h-v^k).$$

Replacing u by $1/u$ and multiplying both sides by u^{h+1}, we get

$$u(1-v)\sum_{a=\lceil y\rceil}^{\lceil y\rceil+k-1} u^{\left[\frac{ha+kx-hy}{k}\right]}v^a+(1-u)\sum_{b=-\lceil x\rceil+1-\delta(x)}^{-\lceil x\rceil+h-\delta(x)} v^{k-\left[\frac{kb+kx-hy}{h}\right]}u^{h-b}$$
$$=u^{\lceil x\rceil+\delta(x)}v^{\lceil y\rceil}(1-u^hv^k).$$

Replacing b by $h-b$ in the second summation, noting that $-[-x]=[x]+1-\delta(x)$, we obtain

$$u(1-v)\sum_{a=\lceil y\rceil}^{\lceil y\rceil+k-1} u^{\left[\frac{ha+kx-hy}{k}\right]}v^a+v(1-u)\sum_{b=\lceil x\rceil+\delta(x)}^{\lceil x\rceil+h-1+\delta(x)} v^{\left[\frac{kb-kx+hy}{h}\right]-\delta\left(\frac{kb-kx+hy}{h}\right)}u^b$$
$$=u^{\lceil x\rceil+\delta(x)}v^{\lceil y\rceil}(1-u^hv^k).$$

Upon simplifying, we get

$$u(1-v)\sum_{a=\lceil y\rceil}^{\lceil y\rceil+k-1} u^{\left[\frac{ha+kx-hy}{k}\right]} v^a + v(1-u)\sum_{b=\lceil x\rceil}^{\lceil x\rceil+h-1} v^{\left[\frac{kb-kx+hy}{h}\right]} u^b$$
$$= u^{\lceil x\rceil} v^{\lceil y\rceil}(1-u^h v^k) - (1-u)(1-v)\sum_{b=\lceil x\rceil}^{\lceil x\rceil+h-1} \delta\left(\frac{kb-kx+hy}{h}\right) v^{\frac{kb-kx+hy}{h}} u^b,$$

from which we get the assertion of the corollary. □

4 Additional Results

From Theorem 1, Carlitz established many interesting results which are given below. In this section, we generalize these results by removing the restriction (1). We omit the proofs since they are similar to that of Theorem 2; namely, we apply Carlitz's results to the fractional part of x and y, add 0 in the form of $\left[\frac{a+\{y\}}{k}\right]$ and $\left[\frac{b+\{x\}}{h}\right]$ to obtain summations over complete residue systems modulo k and h, then replace a and b by $a+[y]$ and $b+[x]$, respectively.

We define the Bernoulli polynomials $B_n(x)$ by the generating function

$$\frac{ze^{xz}}{e^z-1} = \sum_{n=0}^{\infty} B_n(x)\frac{z^n}{n!},$$

where $B_n = B_n(0)$ are the Bernoulli numbers.

In the following, let r, s be nonnegative integers. Carlitz proved the following two corollaries of Theorem 1.

Corollary 5 ([2]). *With the same assumptions as in Theorem 1, we have*

$$\sum_{a=0}^{k-1}\left(h-\left[\frac{ha+z}{k}\right]\right)^r\left((a+1)^s-a^s\right) - \sum_{b=0}^{h-1}\left(k-\left[\frac{kb+z}{h}\right]\right)^s\left((b+1)^r-b^r\right)$$
$$= -h^r\delta_{s,0} + k^s\delta_{r,0}.$$

In particular, for $r, s > 0$, Corollary 7 implies

$$\sum_{a=0}^{k-1}\left(h-\left[\frac{ha+z}{k}\right]\right)^r\left((a+1)^s-a^s\right) = \sum_{b=0}^{h-1}\left(k-\left[\frac{kb+z}{h}\right]\right)^s\left((b+1)^r-b^r\right).$$

Corollary 6 ([2]). *With the same assumptions as in Theorem 1, for any real numbers ξ and η, we have*

$$s\sum_{a=0}^{k-1}(a+\eta)^{s-1}B_r\left(h+\xi-\left[\frac{ha+z}{k}\right]\right)-r\sum_{b=0}^{h-1}(b+\xi)^{r-1}B_s\left(k+\eta-\left[\frac{kb+z}{h}\right]\right)$$
$$=B_s(k+\eta)B_r(\xi)-B_r(h+\xi)B_s(\eta).$$

Setting $\xi=1-h$ and $\eta=1-k$, Corollary 8 reduces to

$$s\sum_{a=0}^{k-1}(a+1-k)^{s-1}B_r\left(1-\left[\frac{ha+z}{k}\right]\right)-r\sum_{b=0}^{h-1}(b+1-h)^{r-1}B_s\left(1-\left[\frac{kb+z}{h}\right]\right)$$
$$=B_s(1)B_r(1-h)-B_r(1)B_s(1-k).$$

Since $B_n(1-x)=(-1)^nB_n(x)$, we get

$$s\sum_{a=0}^{k-1}(k-a-1)^{s-1}B_r\left(\left[\frac{ha+z}{k}\right]\right)-r\sum_{b=0}^{h-1}(h-b-1)^{r-1}B_s\left(\left[\frac{kb+z}{h}\right]\right)$$
$$=-B_sB_r(h)+B_rB_s(k).$$

Here, we present the analogous results with the restriction (1) removed.

Corollary 7 *With the same assumptions as in Theorem 2, we have*

$$\sum_{a\,(\mathrm{mod}\,k)}\left(h\left(1+\left[\frac{a+y}{k}\right]\right)-\left[\frac{h(a+y)}{k}+\{x\}\right]\right)^r$$
$$\times\left(\left(a+[y]-k\left[\frac{a+y}{k}\right]+1\right)^s-\left(a+[y]-k\left[\frac{a+y}{k}\right]\right)^s\right)$$
$$-\sum_{b\,(\mathrm{mod}\,h)}\left(k\left(1+\left[\frac{b+x}{h}\right]\right)-\left[\frac{k(b+x)}{h}+\{y\}\right]\right)^s$$
$$\times\left(\left(b+[x]-h\left[\frac{b+x}{h}\right]+1\right)^r-\left(b+[x]-h\left[\frac{b+x}{h}\right]\right)^r\right)$$
$$=-h^r\delta_{s,0}+k^s\delta_{r,0}.$$

Hence, for $r,s>0$, Corollary 7 implies

$$\sum_{a \,(\mathrm{mod}\, k)} \left(h\left(1+\left[\frac{a+y}{k}\right]\right) - \left[\frac{h(a+y)}{k} + \{x\}\right]\right)^{r}$$
$$\times \left(\left(a+[y]-k\left[\frac{a+y}{k}\right]+1\right)^{s} - \left(a+[y]-k\left[\frac{a+y}{k}\right]\right)^{s}\right)$$
$$= \sum_{b \,(\mathrm{mod}\, h)} \left(k\left(1+\left[\frac{b+x}{h}\right]\right) - \left[\frac{k(b+x)}{h} + \{y\}\right]\right)^{s}$$
$$\times \left(\left(b+[x]-h\left[\frac{b+x}{h}\right]+1\right)^{r} - \left(b+[x]-h\left[\frac{b+x}{h}\right]\right)^{r}\right).$$

Corollary 8 *With the same assumptions as in Theorem 2, for any real numbers ξ and η, we have*

$$s \sum_{a \,(\mathrm{mod}\, k)} \left(a+[y]+\eta-k\left[\frac{a+y}{k}\right]\right)^{s-1} B_r\left(h\left(1+\left[\frac{a+y}{k}\right]\right)+\xi-\left[\frac{h(a+y)}{k}+\{x\}\right]\right)$$
$$-r \sum_{b \,(\mathrm{mod}\, h)} \left(b+[x]+\xi-h\left[\frac{b+x}{h}\right]\right)^{r-1} B_s\left(k\left(1+\left[\frac{b+x}{h}\right]\right)+\eta-\left[\frac{k(b+x)}{h}+\{y\}\right]\right)$$
$$= B_s(k+\eta)B_r(\xi) - B_r(h+\xi)B_s(\eta).$$

Setting $\xi = 1-h-[x]$ and $\eta = 1-k-[y]$, Corollary 8 reduces to

$$s \sum_{a \,(\mathrm{mod}\, k)} \left(a+1-k\left(1+\left[\frac{a+y}{k}\right]\right)\right)^{s-1} B_r\left(1+h\left[\frac{a+y}{k}\right]-\left[\frac{h(a+y)}{k}+x\right]\right)$$
$$-r \sum_{b \,(\mathrm{mod}\, h)} \left(b+1-h\left(1+\left[\frac{b+x}{h}\right]\right)\right)^{r-1} B_s\left(1+k\left[\frac{b+x}{h}\right]-\left[\frac{k(b+x)}{h}+y\right]\right)$$
$$= B_s(1-[y])B_r(1-h-[x]) - B_r(1-[x])B_s(1-k-[y]),$$

and since $B_n(1-x) = (-1)^n B_n(x)$, we obtain

$$s \sum_{a \,(\mathrm{mod}\, k)} \left(k\left(1+\left[\frac{a+y}{k}\right]\right)-a-1\right)^{s-1} B_r\left(\left[\frac{h(a+y)}{k}+x\right]-h\left[\frac{a+y}{k}\right]\right)$$
$$-r \sum_{b \,(\mathrm{mod}\, h)} \left(h\left(1+\left[\frac{b+x}{h}\right]\right)-b-1\right)^{r-1} B_s\left(\left[\frac{k(b+x)}{h}+y\right]-k\left[\frac{b+x}{h}\right]\right)$$
$$= -B_s([y])B_r(h+[x]) + B_r([x])B_s(k+[y]).$$

Acknowledgements The author is grateful to Max Planck Institute for Mathematics in Bonn for its hospitality and financial support.

References

1. M. Beck and F. Kohl, Rademacher-Carlitz polynomials, *Acta Arith.* **163**(4) (2014), 379–393.
2. L. Carlitz, Some theorems on generalized Dedekind-Rademacher sums, *Pacific J. Math.* **75** (1978), 347–358.

Petersson-Knopp Type Identities for Generalized Dedekind-Rademacher Sums Attached to Three Dirichlet Characters

Brad Isaacson

Abstract In a previous paper, we defined the generalized Dedekind-Rademacher sum attached to three Dirichlet characters

$$\Phi_{r,s}\begin{pmatrix} a & b & c \\ x & y & z \\ \psi & \delta & \chi \end{pmatrix} = \sum_{h \,(\mathrm{mod}\, cm_\chi)} \chi(h)\mathscr{B}_{r,\psi}\left(am_\psi \frac{h+z}{cm_\chi} + x\right)\mathscr{B}_{s,\delta}\left(bm_\delta \frac{h+z}{cm_\chi} + y\right)$$

$$(a, b, c \in \mathbb{N} \text{ and } x, y, z \in \mathbb{R}),$$

where ψ, δ, χ are Dirichlet characters modulo m_ψ, m_δ, m_χ, respectively, and $\mathscr{B}_{r,\psi}(x), \mathscr{B}_{s,\delta}(x)$ are generalized Bernoulli functions. In this paper, we prove Petersson-Knopp type identities for the sum $\Phi_{r,s}\begin{pmatrix} a & b & c \\ 0 & 0 & 0 \\ \psi & \delta & \chi \end{pmatrix}$ which generalize the results of Petersson-Knopp, Can and others.

1 Introduction

Let $\overline{B}_k(x)$, $\widetilde{B}_k(x)$ denote the kth Bernoulli functions defined by

$$\overline{B}_k(x) = \begin{cases} 0 & \text{if } k = 1, x \in \mathbb{Z}, \\ B_k(x - [x]) & \text{otherwise,} \end{cases}$$

$$\widetilde{B}_k(x) = B_k(x - [x]),$$

B. Isaacson (✉)
Mathematics Department, New York City College of Technology, 300 Jay St., Brooklyn, NY 11201, USA
e-mail: bisaacson@citytech.cuny.edu

M. B. Nathanson (ed.), *Combinatorial and Additive Number Theory VI*,
Springer Proceedings in Mathematics & Statistics 464,
https://doi.org/10.1007/978-3-031-65064-2_13

where $[x]$ is the greatest integer $\leq x$ and $B_k(x)$ is the kth Bernoulli polynomial defined by the generating function

$$\frac{te^{tx}}{e^t - 1} = \sum_{n=0}^{\infty} B_n(x) \frac{t^n}{n!}.$$

For positive integers h and k, the classical Dedekind sum $s(h, k)$ is defined by

$$s(h, k) = \sum_{\mu \pmod{k}} \overline{B}_1\left(\frac{\mu}{k}\right)\overline{B}_1\left(\frac{h\mu}{k}\right).$$

This sum plays an important role in the transformation formula of the Dedekind η-function. For an excellent introduction and many references to Dedekind sums, see [15].

Appearing in Dedekind's work [5] was the following fundamental identity.

$$s(ph, k) + \sum_{\mu \pmod{d}} s\,(h + \mu k, pk) = (p + 1)s(h, k) \tag{1}$$

for any prime number p. In [8], Knopp generalized (1) in the following way:

$$\sum_{d \mid n} \sum_{\mu \pmod{d}} s\left(\frac{n}{d}h + \mu k, dk\right) = \sigma(n)s(h, k) \quad (n \in \mathbb{N}), \tag{2}$$

where $\sigma(n)$ is the sum of the positive divisors of n. For elementary proofs of (2), see [6, 12]. Observe that (2) reduces to (1) when n is a prime p.

Remark 1 In a letter to Emil Grosswald, Hans Petersson stated without proof the identity given by Equation (2) under the conditions of $(h, k) = 1$, $k \equiv 0 \pmod{n}$ and $h \equiv 1 \pmod{n}$. Since Knopp removed these restrictions in [8], the Equation (2) is commonly referred to as the Petersson-Knopp identity.

The sum $s(h, k)$ was later generalized by various mathematicians and the corresponding Petersson-Knopp type identities analagous to (2) were obtained, for example, see [2, 4, 9–11, 13, 16]. In this paper, we prove Petersson-Knopp type identities for one such generalization of $s(h, k)$ that was introduced by the present author in [7]. Before stating our results, we introduce some notation.

Let χ be a Dirichlet character modulo m. Let $\overline{B}_{k,\chi}(x)$, $\widetilde{B}_{k,\chi}(x)$ denote the kth generalized Bernoulli functions defined by

$$\begin{aligned} \overline{B}_{k,\chi}(x) &= m^{k-1} \sum_{a \pmod{m}} \chi(a)\overline{B}_k\left(\frac{a + x}{m}\right), \\ \widetilde{B}_{k,\chi}(x) &= m^{k-1} \sum_{a \pmod{m}} \chi(a)\widetilde{B}_k\left(\frac{a + x}{m}\right). \end{aligned} \tag{3}$$

Note that if χ is the trivial character, denoted $\mathbb{1}$, the generalized Bernoulli functions $\overline{B}_{k,\chi}(x)$, $\widetilde{B}_{k,\chi}(x)$ reduce to the ordinary Bernoulli functions $\overline{B}_k(x)$, $\widetilde{B}_k(x)$, respectively.

Remark 2 Our $\overline{B}_{k,\chi}(x)$ corresponds to $\overline{B}_{k,\overline{\chi}}(x)$ in the notation of Berndt [3] and others.

Let χ, ψ, δ be Dirichlet characters modulo m_χ, m_ψ, m_δ, respectively. For $\mathscr{B} = \overline{B}$ or $\widetilde{B}$, we define the *generalized Dedekind-Rademacher sum attached to three Dirichlet characters* by

$$\Phi_{r,s}\begin{pmatrix} a & b & c \\ x & y & z \\ \psi & \delta & \chi \end{pmatrix} = \sum_{h \,(\mathrm{mod}\, cm_\chi)} \chi(h)\mathscr{B}_{r,\psi}\left(am_\psi \frac{h+z}{cm_\chi} + x\right)\mathscr{B}_{s,\delta}\left(bm_\delta \frac{h+z}{cm_\chi} + y\right)$$

$$(a, b, c \in \mathbb{N},\ x, y, z \in \mathbb{R}).$$

This sum was introduced in [7] where its reciprocity and related polynomial reciprocity theorems were derived. In this paper, we prove the following Petersson-Knopp type identities for the sum $\Phi_{r,s}\begin{pmatrix} a & b & c \\ 0 & 0 & 0 \\ \psi & \delta & \chi \end{pmatrix}$.

Theorem 1 *Let a, b, c be three positive integers, ψ, δ, χ Dirichlet characters modulo m_ψ, m_δ, m_χ (respectively), and r, s nonnegative integers. Let n be a positive integer prime to m_ψ, m_δ, and m_χ. We have*

$$(i)\ \sum_{d \mid n} \psi(d)\delta(d)\overline{\chi}(d)d^{r+s-2} \sum_{\alpha \,(mod\, d)} \sum_{\beta \,(mod\, d)} \Phi_{r,s}\begin{pmatrix} \frac{n}{d}a + \alpha cm_\chi & \frac{n}{d}b + \beta cm_\chi & dc \\ 0 & 0 & 0 \\ \psi & \delta & \chi \end{pmatrix}$$

$$= n\overline{\chi}(n)\sigma_{r+s-1}(n; \psi, \delta, \chi)\Phi_{r,s}\begin{pmatrix} a & b & c \\ 0 & 0 & 0 \\ \psi & \delta & \chi \end{pmatrix},$$

where $\sigma_m(n; \psi, \delta, \chi) = \sum_{d \mid n} d^m \psi(d)\delta(d)\chi(d)$.

(ii) Additionally, if $(n, b) = 1$, we have

$$\sum_{d \mid n} \psi(d)\overline{\chi}(d)d^{r-1} \sum_{\alpha \,(mod\, d)} \Phi_{r,s}\begin{pmatrix} \frac{n}{d}a + \alpha cm_\chi & b & dc \\ 0 & 0 & 0 \\ \psi & \delta & \chi \end{pmatrix}$$

$$= n^{1-s}\overline{\chi}(n)\overline{\delta}(n)\sigma_{r+s-1}(n; \psi, \delta, \chi)\Phi_{r,s}\begin{pmatrix} a & b & c \\ 0 & 0 & 0 \\ \psi & \delta & \chi \end{pmatrix}.$$

Remark 3 The Petersson-Knopp type identities given by Can [4, Theorem 3.1] is the special case of Theorem 1 for $\delta = \mathbb{1}$.

The layout of this paper is as follows. In Sect. 2, we prove preliminary results such as multiplication formulas for generalized Bernoulli functions and properties of the Möbius μ function. In Sect. 3, we prove Theorem 1 by following the method of Can in [4].

2 Preliminaries

In this section, we discuss all relevant background information needed to prove Theorem 1 given in the Introduction. We fix the notation. Let χ be a Dirichlet character modulo m_χ. We remind the reader that $\mathscr{B} = \overline{B}$ or $\widetilde{B}$ given by (3).

We state the multiplication formula for ordinary Bernoulli functions which follows from Raabe's multiplication formula [14] for ordinary Bernoulli polynomials.

$$\sum_{\lambda \,(\mathrm{mod}\, b)} \mathscr{B}_n\left(x + \frac{a\lambda}{b}\right) = \frac{(a,b)^n}{b^{n-1}}\mathscr{B}_n\left(\frac{bx}{(a,b)}\right) \quad (a, b \in \mathbb{N}) \tag{4}$$

We next prove the multiplication formula for generalized Bernoulli functions.

Lemma 1 *Let χ be a Dirichlet character modulo m_χ, n be a nonnegative integer, and a, b positive integers. If $(b/(a,b), m_\chi) = 1$, we have*

$$\sum_{\lambda \,(\mathrm{mod}\, b)} \mathscr{B}_{n,\chi}\left(x + \frac{am_\chi\lambda}{b}\right) = \frac{(a,b)^n}{b^{n-1}}\overline{\chi}\left(\frac{b}{(a,b)}\right)\mathscr{B}_{n,\chi}\left(\frac{bx}{(a,b)}\right).$$

Proof Let $\alpha = a/(a,b)$ and $\beta = b/(a,b)$. By (3) and (4), we have

$$\begin{aligned}\sum_{\lambda \,(\mathrm{mod}\, b)} \mathscr{B}_{n,\chi}\left(x + \frac{am_\chi\lambda}{b}\right) &= m_\chi^{n-1} \sum_{\lambda \,(\mathrm{mod}\, b)} \sum_{j \,(\mathrm{mod}\, m_\chi)} \chi(j)\mathscr{B}_n\left(\frac{j+x}{m_\chi} + \frac{a\lambda}{b}\right) \\ &= \frac{(a,b)^n m_\chi^{n-1}}{b^{n-1}} \sum_{j \,(\mathrm{mod}\, m_\chi)} \chi(j)\mathscr{B}_n\left(\frac{b(j+x)}{(a,b)m_\chi}\right).\end{aligned}$$

Replacing j by $(b/(a,b))^{-1}j$ where $(b/(a,b))^{-1}$ denotes the inverse of $(b/(a,b))$ modulo m_χ, which is permissible since $(b/(a,b), m_\chi) = 1$, we get

$$\sum_{\lambda \,(\mathrm{mod}\, b)} \mathscr{B}_{n,\chi}\left(x + \frac{am_\chi \lambda}{b}\right) = \frac{(a,b)^n m_\chi^{n-1}}{b^{n-1}} \sum_{j \,(\mathrm{mod}\, m_\chi)} \chi(j) \mathscr{B}_n\left(\frac{j + bx/(a,b)}{m_\chi}\right)$$
$$= \frac{(a,b)^n}{b^{n-1}} \overline{\chi}\left(\frac{b}{(a,b)}\right) \mathscr{B}_{n,\chi}\left(\frac{bx}{(a,b)}\right).$$

□

We now need two properties of the Möbius μ function defined by

$$\mu(n) = \begin{cases} (-1)^{\nu(n)} & \text{if } n \text{ is square-free,} \\ 0 & \text{otherwise,} \end{cases}$$

where $\nu(n)$ denotes the number of distinct prime divisors of n. The first property is

$$\sum_{d \mid n} \mu(d) = \begin{cases} 1 & \text{if } n = 1, \\ 0, & \text{otherwise,} \end{cases} \tag{5}$$

which can be found in any analytic number theory book, such as [1, Theorem 2.1]. The next property we need is the following.

Lemma 2 *For $a, d \in \mathbb{N}$, we have*

$$\sum_{\substack{h=1 \\ (h,d)=1}}^{ad} f(h) = \sum_{t \mid d} \mu(t) \sum_{j=1}^{ad/t} f(tj).$$

Proof Let $e(n)$ denote the unit function defined by $e(1) = 1$ and $e(n) = 0$ for $n \geq 2$. By (5), we get

$$\sum_{\substack{h=1 \\ (h,d)=1}}^{ad} f(h) = \sum_{h=1}^{ad} f(h) e((h,d)) = \sum_{h=1}^{ad} f(h) \sum_{t \mid (h,d)} \mu(t) = \sum_{t \mid d} \mu(t) \sum_{j=1}^{ad/t} f(tj).$$

□

The last property we need is

$$\Phi_{r,s}\begin{pmatrix} ka & kb & kc \\ x & y & z \\ \psi & \delta & \chi \end{pmatrix} = k\Phi_{r,s}\begin{pmatrix} a & b & c \\ x & y & z \\ \psi & \delta & \chi \end{pmatrix} \quad (k \in \mathbb{N}) \tag{6}$$

which follows immediately from the definition of $\Phi_{r,s}$.

3 Proof of Theorem 1

In this section, we prove Theorem 1 given in the Introduction. We remind the reader of the notation. Let χ, ψ, δ be Dirichlet characters modulo m_χ, m_ψ, m_δ, respectively. Let φ denote the Euler phi function. For $\mathscr{B} = \overline{B}$ or $\widetilde{B}$ given by (3), we have

$$\Phi_{r,s}\begin{pmatrix} a & b & c \\ x & y & z \\ \psi & \delta & \chi \end{pmatrix} = \sum_{h \,(\mathrm{mod}\, cm_\chi)} \chi(h)\mathscr{B}_{r,\psi}\left(am_\psi \frac{h+z}{cm_\chi} + x\right)\mathscr{B}_{s,\delta}\left(bm_\delta \frac{h+z}{cm_\chi} + y\right).$$

Theorem 2 *Let a, b, c be three positive integers, ψ, δ, χ Dirichlet characters modulo m_ψ, m_δ, m_χ (respectively), and r, s nonnegative integers. Let n be a positive integer prime to m_ψ, m_δ, and m_χ. We have*

$$(i) \ \sum_{d\mid n}\psi(d)\delta(d)\overline{\chi}(d)d^{r+s-2}\sum_{\alpha\,(mod\, d)}\sum_{\beta\,(mod\, d)}\Phi_{r,s}\begin{pmatrix} \frac{n}{d}a+\alpha cm_\chi & \frac{n}{d}b+\beta cm_\chi & dc \\ 0 & 0 & 0 \\ \psi & \delta & \chi \end{pmatrix}$$
$$= n\overline{\chi}(n)\sigma_{r+s-1}(n;\psi,\delta,\chi)\Phi_{r,s}\begin{pmatrix} a & b & c \\ 0 & 0 & 0 \\ \psi & \delta & \chi \end{pmatrix},$$

where $\sigma_m(n;\psi,\delta,\chi) = \sum_{d\mid n} d^m\psi(d)\delta(d)\chi(d)$.
(ii) Additionally, if $(n, b) = 1$, we have

$$\sum_{d\mid n}\psi(d)\overline{\chi}(d)d^{r-1}\sum_{\alpha\,(mod\, d)}\Phi_{r,s}\begin{pmatrix} \frac{n}{d}a+\alpha cm_\chi & b & dc \\ 0 & 0 & 0 \\ \psi & \delta & \chi \end{pmatrix}$$
$$= n^{1-s}\overline{\chi}(n)\overline{\delta}(n)\sigma_{r+s-1}(n;\psi,\delta,\chi)\Phi_{r,s}\begin{pmatrix} a & b & c \\ 0 & 0 & 0 \\ \psi & \delta & \chi \end{pmatrix}.$$

Proof From Lemma 1, we have

$$
\begin{aligned}
&\sum_{d\mid n}\psi(d)\delta(d)\overline{\chi}(d)d^{r+s-2}\sum_{\alpha\,(\mathrm{mod}\,d)}\sum_{\beta\,(\mathrm{mod}\,d)}\Phi_{r,s}\begin{pmatrix}\frac{n}{d}a+\alpha cm_\chi & \frac{n}{d}b+\beta cm_\chi & dc\\ 0 & 0 & 0\\ \psi & \delta & \chi\end{pmatrix}\\
&=\sum_{d\mid n}\psi(d)\delta(d)\overline{\chi}(d)d^{r+s-2}\\
&\quad\times\sum_{\alpha\,(\mathrm{mod}\,d)}\sum_{\beta\,(\mathrm{mod}\,d)}\sum_{h\,(\mathrm{mod}\,dcm_\chi)}\chi(h)\mathscr{B}_{r,\psi}\left(\frac{nam_\psi h}{d^2cm_\chi}+\frac{\alpha m_\psi h}{d}\right)\mathscr{B}_{s,\delta}\left(\frac{nbm_\delta h}{d^2cm_\chi}+\frac{\beta m_\delta h}{d}\right)\\
&=\sum_{d\mid n}\psi(d)\delta(d)\overline{\chi}(d)\\
&\quad\times\sum_{h\,(\mathrm{mod}\,dcm_\chi)}\chi(h)(h,d)^{r+s}\overline{\psi}\left(\frac{d}{(h,d)}\right)\overline{\delta}\left(\frac{d}{(h,d)}\right)\mathscr{B}_{r,\psi}\left(\frac{nam_\psi h}{(h,d)dcm_\chi}\right)\mathscr{B}_{s,\delta}\left(\frac{nbm_\delta h}{(h,d)dcm_\chi}\right)\\
&=\sum_{d\mid n}\overline{\chi}(d)\sum_{l\mid d}\psi(l)\delta(l)l^{r+s}\sum_{\substack{h\,(\mathrm{mod}\,dcm_\chi)\\(h,d)=l}}\chi(h)\mathscr{B}_{r,\psi}\left(\frac{nam_\psi h}{ldcm_\chi}\right)\mathscr{B}_{s,\delta}\left(\frac{nbm_\delta h}{ldcm_\chi}\right)\\
&=\sum_{d\mid n}\overline{\chi}(d)\sum_{l\mid d}\psi(l)\delta(l)\chi(l)l^{r+s}\sum_{\substack{h\,\left(\mathrm{mod}\,\frac{dcm_\chi}{l}\right)\\ \left(h,\frac{d}{l}\right)=1}}\chi(h)\mathscr{B}_{r,\psi}\left(\frac{nam_\psi h}{dcm_\chi}\right)\mathscr{B}_{s,\delta}\left(\frac{nbm_\delta h}{dcm_\chi}\right).
\end{aligned}
\tag{7}
$$

By Lemma 2, (5) and (6), the right-hand side of (7) becomes

$$
\begin{aligned}
&\sum_{d\mid n}\overline{\chi}(d)\sum_{l\mid d}\psi(l)\delta(l)\chi(l)l^{r+s}\sum_{t\mid\frac{d}{l}}\mu(t)\sum_{j\,\left(\mathrm{mod}\,\frac{dcm_\chi}{lt}\right)}\chi(tj)\mathscr{B}_{r,\psi}\left(\frac{nam_\psi tj}{dcm_\chi}\right)\mathscr{B}_{s,\delta}\left(\frac{nbm_\delta tj}{dcm_\chi}\right)\\
&\quad=\sum_{d\mid n}\sum_{l\mid d}\sum_{t\mid\frac{d}{l}}\overline{\chi}(d/lt)\psi(l)\delta(l)l^{r+s}\mu(t)\Phi_{r,s}\begin{pmatrix}\frac{na}{l} & \frac{nb}{l} & \frac{dc}{lt}\\ 0 & 0 & 0\\ \psi & \delta & \chi\end{pmatrix}\\
&\quad=\sum_{elt\mid n}\overline{\chi}(e)\psi(l)\delta(l)l^{r+s}\mu(t)\Phi_{r,s}\begin{pmatrix}\frac{na}{l} & \frac{nb}{l} & ec\\ 0 & 0 & 0\\ \psi & \delta & \chi\end{pmatrix}\\
&\quad=\sum_{el\mid n}\overline{\chi}(e)\psi(l)\delta(l)l^{r+s}\Phi_{r,s}\begin{pmatrix}\frac{na}{l} & \frac{nb}{l} & ec\\ 0 & 0 & 0\\ \psi & \delta & \chi\end{pmatrix}\sum_{t\mid\frac{n}{el}}\mu(t)\\
&\quad=\sum_{el=n}\overline{\chi}(e)\psi(l)\delta(l)l^{r+s}\Phi_{r,s}\begin{pmatrix}\frac{na}{l} & \frac{nb}{l} & ec\\ 0 & 0 & 0\\ \psi & \delta & \chi\end{pmatrix}\\
&\quad=n\overline{\chi}(n)\sum_{l\mid n}\psi(l)\delta(l)\chi(l)l^{r+s-1}\Phi_{r,s}\begin{pmatrix}a & b & c\\ 0 & 0 & 0\\ \psi & \delta & \chi\end{pmatrix}.
\end{aligned}
$$

Thus we get the assertion (i). We next prove the assertion (ii). Similar reasoning to the above yields

$$\begin{aligned}&\sum_{d\mid n}\psi(d)\overline{\chi}(d)d^{r-1}\sum_{\alpha\,(\mathrm{mod}\,d)}\Phi_{r,s}\begin{pmatrix}\frac{n}{d}a+\alpha c m_\chi & b & dc\\ 0 & 0 & 0\\ \psi & \delta & \chi\end{pmatrix}\\ &=\overline{\chi}(n)\sum_{l\mid n}\psi(l)\chi(l)l^r\Phi_{r,s}\begin{pmatrix}\frac{na}{l} & b & \frac{nc}{l}\\ 0 & 0 & 0\\ \psi & \delta & \chi\end{pmatrix}.\end{aligned} \tag{8}$$

By the multiplication formula Lemma 1, observe that

$$\Phi_{r,s}\begin{pmatrix}\frac{na}{l} & b & \frac{nc}{l}\\ 0 & 0 & 0\\ \psi & \delta & \chi\end{pmatrix}=\frac{(b,n/l)^s}{(n/l)^{s-1}}\overline{\delta}\left(\frac{n/l}{(b,n/l)}\right)\Phi_{r,s}\begin{pmatrix}a & \frac{b}{(b,n/l)} & c\\ 0 & 0 & 0\\ \psi & \delta & \chi\end{pmatrix}.$$

Hence, if $(n,b)=1$, we get

$$\Phi_{r,s}\begin{pmatrix}\frac{na}{l} & b & \frac{nc}{l}\\ 0 & 0 & 0\\ \psi & \delta & \chi\end{pmatrix}=(n/l)^{1-s}\overline{\delta}\,(n/l)\,\Phi_{r,s}\begin{pmatrix}a & b & c\\ 0 & 0 & 0\\ \psi & \delta & \chi\end{pmatrix}$$

for any divisor l of n. Plugging this into the right-hand side of (8), we get

$$\begin{aligned}&\sum_{d\mid n}\psi(d)\overline{\chi}(d)d^{r-1}\sum_{\alpha\,(\mathrm{mod}\,d)}\Phi_{r,s}\begin{pmatrix}\frac{n}{d}a+\alpha c m_\chi & b & dc\\ 0 & 0 & 0\\ \psi & \delta & \chi\end{pmatrix}\\ &=n^{1-s}\overline{\chi}(n)\overline{\delta}(n)\sum_{l\mid n}\psi(l)\delta(l)\chi(l)l^{r+s-1}\Phi_{r,s}\begin{pmatrix}a & b & c\\ 0 & 0 & 0\\ \psi & \delta & \chi\end{pmatrix}.\end{aligned}$$

□

Acknowledgements The author is grateful to Max Planck Institute for Mathematics in Bonn for its hospitality and financial support.

References

1. T. M. Apostol, *Introduction to Analytic Number Theory*, Springer Verlag, New York-Heidelberg, 1979.
2. T. M. Apostol and T. H. Vu, Identities for sums of Dedekind type, *J. Number Theory*, **14**(3) (1982), 391–396.
3. B. C. Berndt, Character analogues of the Poisson and Euler-MacLaurin summation formulas with applications, *J. Number Theory*, **7**(4) (1975), 413–445.
4. M. Can, Higher-order character Dedekind sum, *Turkish J. Math.*, **44**(3) (2020), 998–1015.

5. R. Dedekind, *Erläuterungen zu der Riemannschen Fragmenten über die Grenzfälle der elliptischen Funktionen*, Gesammelte Math. Werke 1, Braunschweig, 1930, pp. 159–173.
6. L. A. Goldberg, An elementary proof of the Petersson-Knopp theorem on Dedekind sums, *J. Number Theory*, **12**(4) (1980), 541–542.
7. B. Isaacson, Reciprocity formulae for generalized Dedekind-Rademacher sums attached to three Dirichlet characters and related polynomial reciprocity formulae, *Int. J. Number Theory*, **20**(6) (2024), 1463–1492.
8. M. I. Knopp, Hecke operators and an identity for the Dedekind sums, *J. Number Theory*, **12**(1) (1980), 2–9.
9. K. Kozuka, Dedekind type sums attached to Dirichlet characters, *Kyushu J. Math.*, **58** (2004), 1–24.
10. K. Kozuka, Knopp-type identities for generalized multiple Dedekind-type sums, *Kyushu J. Math.*, **68** (2014), 405–419.
11. H. Liu, Some generalizations of Knopp's identity, *J. Fac. Sci. Univ. Tokyo, Sect. IA Math.*, **38**(2) (2007), 179–188.
12. L. Parson, Dedekind sums and Hecke operators, *Math. Proc. Camb. Philos. Soc.*, **88**(1) (1980), 11–14.
13. L. Parson and K. H. Rosen, Hecke operators and Lambert series, *Math. Scand.*, **49** (1981), 5–14.
14. J. L. Raabe, Zurückführung einiger Summen und bestimmten Integrale auf die Jacob-Bernoullische Funktion, *J. Reine Angew. Math.*, **42** (1851), 348–367.
15. H. Rademacher and E. Grosswald, *Dedekind sums*, vol. 16, American Mathematical Soc., 1972.
16. Z. Zheng, The Petersson-Knopp Identity for the Homogeneous Dedekind Sums, *J. Number Theory*, **57**(2) (1996), 223–230.

Explicit Bounds for Large Gaps Between Cubefree Integers

Angel Kumchev, Wade McCormick, Nathan McNew, Ariana Park, Russell Scherr, and Willow Ziehr

Abstract We obtain explicit forms of the current best known asymptotic upper bounds for gaps between cubefree integers. In particular, we show that the interval $(x, x + 5x^{1/7} \log x]$ contains a cubefree integer for any $x \geq 2$. The constant 5 can be improved further, if x is assumed to be larger than a very large constant.

1 Introduction

If $k \geq 2$ is a fixed integer, an integer n is called *k-free* if it is not divisible by p^k for any prime p. In the special cases $k = 2$ and $k = 3$, the k-free integers are also known, respectively, as *squarefree* and *cubefree*. The distribution of k-free integers, especially in the squarefree case, has been studied for a long time and by many mathematicians.

An important part of that research concerns the study of the size of gaps between consecutive k-free numbers. Its history originates with Fogels' result [5] that if $\theta >$

A. Kumchev (✉) · N. McNew · R. Scherr
Department of Mathematics, Towson University, Towson 21252, MD, USA
e-mail: akumchev@towson.edu

N. McNew
e-mail: nmcnew@towson.edu

R. Scherr
e-mail: rscher4@students.towson.edu

W. McCormick
Department of Mathematics, Northwestern University, Evanston 60208, IL, USA
e-mail: s1r1a6@u.northwestern.edu

A. Park
School of Mathematics, University of Minnesota Twin Cities, Minneapolis 55455, MN, USA
e-mail: park2968@umn.edu

W. Ziehr
Department of Mathematics, University of California Berkeley, Berkeley 94720, CA, USA
e-mail: simon_ziehr@berkeley.edu

M. B. Nathanson (ed.), *Combinatorial and Additive Number Theory VI*,
Springer Proceedings in Mathematics & Statistics 464,
https://doi.org/10.1007/978-3-031-65064-2_14

$2/5$ the interval $(x, x + x^{\theta}]$ contains a squarefree integer for all sufficiently large x. In 1951, Halberstam and Roth [6] proved that if $k \geq 2$, the interval $(x, x + x^{\theta}]$ contains a k-free integer for any $\theta > 1/(2k)$ and for all sufficiently large x. Around the same time, Erdős [1] proved that there exist infinitely many intervals $(x, x + h]$, with

$$h \gg \frac{\log x}{\log \log x},$$

which contain no squarefree integers. Together, these results inspired the conjecture that for any fixed $\varepsilon > 0$, the interval $(x, x + x^{\varepsilon}]$ contains a squarefree integer for sufficiently large x. This conjecture appears to be far beyond the reach of present methods, with the sharpest asymptotic bounds known to date going back to the early 1990s. In 1991, Filaseta and Trifonov [4] proved that there exists a constant $c > 0$ such that the interval $(x, x + cx^{1/5} \log x]$ contains a squarefree integer for all sufficiently large x. Shortly after, Trifonov [10] generalized this result and proved that, for each $k \geq 3$, there exists a constant $c = c(k) > 0$ such that the interval $(x, x + cx^{1/(2k+1)} \log x]$ contains a k-free integer for all sufficiently large x. The reader interested in the detailed history of the problem can consult the survey [3] by Filaseta, Graham, and Trifonov or the introduction of our earlier work [7].

In recent years, the number-theoretic community has shown an increased interest in numerically explicit results, so the present authors set on a quest to obtain fully explicit versions of the results of Filaseta and Trifonov [4, 10]. In [7], we proved the following explicit theorem about the gaps between squarefree integers.

Theorem 1 *For any $x \geq 2$, the interval $(x, x + 11x^{1/5} \log x]$ contains a squarefree integer.*

In the present paper, we continue our investigations and obtain a similar theorem on the gaps between cubefree integers.

Theorem 2 *For any $x \geq 2$, the interval $(x, x + 5x^{1/7} \log x]$ contains a cubefree integer.*

The focus of the above theorems is on providing explicit intervals that work for all x. The price we pay for this universality are the somewhat elevated values of the constants 11 and 5 in the theorems. The next result is a version of [7, Theorem 2]. It provides a variant of our main result above, which reduces the constant 5 in Theorem 2, assuming that one is willing to accept a result that holds only for sufficiently large x.

Theorem 3 *Every interval of the form*

- $(x, x + 2x^{1/7} \log x]$ *contains a cubefree number for* $x \geq e^{550}$;
- $(x, x + x^{1/7} \log x]$ *contains a cubefree number for* $x \geq e^{2300}$;
- $(x, x + \frac{1}{2}x^{1/7} \log x]$ *contains a cubefree number for* $x \geq e^{75\,000}$.

Mossinghoff, Oliveira e Silva and Trudgian [8] investigated long gaps between squarefree numbers numerically. Their computational work establishes the size of the longest gaps up to 10^{18}, which are all dramatically smaller than the bounds that we get in this paper. The largest gap that they find is a string of 18 consecutive non-squarefree numbers, the first of which is 125 781 000 834 058 568. As a result of their work, we can assume $x \geq 10^{18} > e^{41}$ throughout the rest of this paper.

Our proof of Theorem 2 relies also on several propositions giving results with larger exponents, which are however superior to the theorem for small x. Those results may also be of independent interest. For example, we show that the interval $(x, x + 2x^{1/5}]$ always contains a cubefree integer (Proposition 2) and the intervals $(x, x + 10x^{1/6}]$ and $(x, x + 8.5x^{1/6}]$ contain a cubefree integer for $x \geq e^{95}$ and $x \geq e^{191}$ respectively (Proposition 3).

We conclude this introduction by remarking that the interested reader can find the `SageMath` code for the computational part of our work at https://urldefense.com/v3/__https://github.com/agreatnate/explicit-k-free-integer-bounds__;!!M2cvx14AM25G!y3OW0BO-pjCnI10PBhKGUfeqSwun8ujt4nbF6md TwOTWILVWBUoh3hg649Z-DgMpea2wFeF_lwlcSqt2wXaQ3B2hi8rVmQ4I$. This should allow a motivated reader not only to check the numerical values that appear below, but also the explore the possibilities for future improvements.

Notation

Throughout the paper, for a real number θ, we use $\lfloor \theta \rfloor$ to denote the greatest integer less than or equal to θ; also, $\{\theta\} = \theta - \lfloor \theta \rfloor$. We write $|A|$ for the size of the set A, and $\pi(x)$ for the prime counting function. Finally, we use $c_1, c_2, \ldots$ to denote constants that appear in the proofs. Those constants tend to depend on various parameters introduced throughout our arguments (such as λ, δ and m); we may indicate such dependencies by labeling our constants as functions of said parameters—see the constant $c_2(m)$ in (24), for example.

2 Preliminaries

Outline of the Method

Let $N(x, h)$ be the number of integers in $(x, x + h]$ that are not cubefree. To prove any of our theorems, it suffices to show that $N(x, h) < h - 1$ for the respective choices of x and h. We first sieve this interval of the cubes of very small primes. Let J to be a parameter to be chosen later. The number of integers in $(x, x + h]$ divisible by the cube of a prime up to J is at most

$$h\left(1 - \prod_{p \leq J}\left(1 - \frac{1}{p^3}\right)\right) + 2^{\pi(J)}$$

$$= h\left(1 - \prod_{p \le J}\left(1 - \frac{1}{p^3}\right) + \frac{2^{\pi(J)}}{h}\right) =: h\sigma_0'(h, J).$$

We then count separately the integers divisible by p^3 for each prime $p > J$. We find that

$$N(x, h) \le h\sigma_0'(h, J) + \sum_{p > J}\left(\left\lfloor \frac{x+h}{p^3} \right\rfloor - \left\lfloor \frac{x}{p^3} \right\rfloor\right), \tag{1}$$

where the sum on the right is over all primes greater than J. To bound the latter sum, we introduce a parameter H, which we will later choose as $H = mh$, with $m \ge 1$ of moderate size, and we use this parameter to split the sum in (1) as follows:

$$\left(\sum_{J < p \le H} + \sum_{p > H}\right)\left(\left\lfloor \frac{x+h}{p^3} \right\rfloor - \left\lfloor \frac{x}{p^3} \right\rfloor\right) =: \Sigma_1 + \Sigma_2. \tag{2}$$

Beginning with the contribution of the small primes, we find that

$$\begin{aligned} \Sigma_1 &\le \sum_{J < p \le H}\left(\frac{h}{p^3} + 1\right) \le h \sum_{p > J} \frac{1}{p^3} + \pi(H) \\ &< h\left(\sigma_1 - \sum_{p \le J} \frac{1}{p^3}\right) + \pi(H), \end{aligned} \tag{3}$$

where σ_1, the sum of the reciprocals of the cubes of all primes, satisfies

$$\sigma_1 < 0.1748. \tag{4}$$

Hence,

$$N(h, x) \le h\big(\sigma_0(h, J) + \sigma_1\big) + \pi(H) + \Sigma_2, \tag{5}$$

where

$$\sigma_0(h, J) = 1 - \prod_{p \le J}\left(1 - \frac{1}{p^3}\right) - \sum_{p \le J} \frac{1}{p^3} + \frac{2^{\pi(J)}}{h}. \tag{6}$$

The term $\pi(H)$ in (5) can be bounded with the help of the following lemma due to Rosser and Schoenfeld [9, (3.2)].

Lemma 1 *For any $x > 1$, one has*

$$\pi(x) < \frac{x}{\log x}\left(1 + \frac{1.5}{\log x}\right). \tag{7}$$

Applying this lemma, we see that

$$\pi(H) < \sigma_2(h, m)h, \quad \sigma_2(h, m) := \frac{m}{\log(mh)}\left(1 + \frac{1.5}{\log(mh)}\right). \tag{8}$$

The estimation of the sum Σ_2 occupies the remainder of the paper. We remark that primes $p > \sqrt[3]{2x}$ do not contribute to that sum. Moreover, if $p > h^{1/3}$, we get

$$0 \le \left\lfloor \frac{x+h}{p^3} \right\rfloor - \left\lfloor \frac{x}{p^3} \right\rfloor \le \frac{h}{p^3} + 1 < 2.$$

Thus, the finite sum Σ_2 counts the primes $p \in \left(H, \sqrt[3]{2x}\right]$ for which there exists an integer m with

$$\frac{x}{p^3} < m \le \frac{x+h}{p^3},$$

or equivalently, such that $\{xp^{-3}\} > 1 - hp^{-3}$ (where $\{\cdot\}$ is the fractional part). Thus,

$$\Sigma_2 \le \left|S(H, \sqrt[3]{2x})\right|, \tag{9}$$

where

$$S(M, N) := \left\{u \in \mathbb{Z} : M < u \le N,\ \gcd(u, 2) = 1,\ 1 - \frac{h}{u^3} \le \left\{\frac{x}{u^3}\right\} < 1\right\}. \tag{10}$$

Putting together (5), (8) and (9), we reduce the problem of bounding $N(x, h)$ to choosing H so that

$$\left|S(H, \sqrt[3]{2x})\right| \le h\sigma_3(h, m), \tag{11}$$

for some bounded function $\sigma_3(h, m)$ such that

$$\sigma_0(h, J) + \sigma_1 + \sigma_2(h, m) + \sigma_3(h, m) < 1 - \frac{1}{h}. \tag{12}$$

3 Some Lemmas

Our bounds on $|S(M, N)|$ use the simple idea that if the minimum distance between distinct elements of a set of integers $\mathcal{A}$ is at least d, then

$$|\mathcal{A} \cap (M, N]| \leq d^{-1}(N - M) + 1. \tag{13}$$

In particular, we will be interested in the special case of (13) when $\mathcal{A}$ is a set of the form

$$S(M) := S(M, \lambda M),$$

where $\lambda > 1$ is a constant and M is a large parameter, with $H \leq M \leq \sqrt[3]{2x}$. As we pointed out in the introduction, the computational work in [8] allows us to assume that x is large. Thus, we assume in the remainder of the paper that

$$x \geq e^{41}, \quad 1000 \leq h \leq H \leq 2x^{1/3}. \tag{14}$$

We will the next lemma to sum the ensuing estimates for $|S(M)|$ to obtain bounds of the form (11). For the proof of this result see [2, Lemma 1] and the comments in [7, Lemma 2].

Lemma 2 *Suppose that A_1, A_2, A_3, b_1, b_2 are positive reals and u, v, λ are real numbers with $0 < u < v < 1 < \lambda$. Assume that for all $M \in [x^u, x^v]$ the estimate*

$$|S(M)| \leq A_1 M^{b_1} + A_2 M^{-b_2} + A_3$$

holds. Then

$$|S(x^u, x^v)| \leq A_1' x^{b_1 v} + A_2' x^{-b_2 u} + A_3' \log x + A_3,$$

where

$$A_1' = \frac{A_1}{1 - \lambda^{-b_1}}, \quad A_2' = \frac{A_2}{1 - \lambda^{-b_2}}, \quad A_3' = A_3 \cdot \frac{v - u}{\log \lambda}.$$

Lemma 3 *Suppose that $H \leq M$. If u and $u + a$ are distinct elements of $S(M)$, then*

$$a > 0.3333x^{-1} M^4. \tag{15}$$

Proof This is a variant of [7, Lemma 4], and the proof is similar to the proof of that result, so we will be brief. Let $f(u) = xu^{-3}$. If $u, u + a \in S(M)$, we have

$$f(u) = n_1 - \theta_1, \quad f(u + a) = n_2 - \theta_2, \tag{16}$$

with $n_1, n_2 \in \mathbb{Z}$, $0 < \theta_1, \theta_2 < hM^{-3} < 10^{-5}$. Thus, the mean-value theorem yields a number $\xi \in (u, u + a)$ such that

$$|f(u + a) - f(u)| = a|f'(\xi)| = \frac{3ax}{\xi^4} > \frac{3x}{(\lambda M)^4}.$$

If $n_1 = n_2$, we deduce that

$$\frac{3x}{(\lambda M)^4} < |\theta_2 - \theta_1| < \frac{h}{M^3} <\leq \frac{1}{M^2},$$

which contradicts (14); thus, $|n_1 - n_2| \geq 1$. So, we have

$$0.9999 \leq 1 - |\theta_2 - \theta_1| \leq |f(u+a) - f(u)| = 3ax\xi^{-4} < 3axM^{-4},$$

and the lemma follows. □

Applying (13) to the result of the last lemma, we obtain the following bound on the size of $S(M)$.

Corollary 1 *Under the hypotheses of Lemma 3, we have*

$$|S(M)| \leq 0.3333^{-1}(\lambda - 1)xM^{-3} + 1.$$

We now use the above lemma to prove the following alternative bound, which is stronger than (15) for $M \leq 1.5x^{2/7}$. This is the first of several results that make use of polynomial identities, similar to (18) below, which are inspired by the theory of Padé approximations. Such identities play a central role in our forthcoming work on gaps between k-free integers for general k.

Lemma 4 *Let $\lambda \leq 1.2$, and suppose that $3\lambda^5 h \leq H \leq M$. If u and $u + a$ are distinct elements of $S(M)$, then*

$$a > 0.7934x^{-1/3}M^{5/3}. \tag{17}$$

Proof We recall the algebraic identity

$$\frac{a^3(2u+a)}{u^3(u+a)^3} = \frac{u+2a}{(u+a)^3} - \frac{u-a}{u^3}. \tag{18}$$

From this (and defining $n_1, n_2, \theta_1, \theta_2$ as in (16)), we deduce that

$$\begin{aligned}\frac{a^3(2u+a)x}{u^3(u+a)^3} &= f(u+a)(u+2a) - f(u)(u-a)\\ &= (n_2 - \theta_2)(u+2a) - (n_1 - \theta_1)(u-a) = n + \theta, \end{aligned}\tag{19}$$

where $n \in \mathbb{Z}$ and $\theta = (u-a)\theta_1 - (u+2a)\theta_2$. In particular, using that $0 < \theta_i < hM^{-3}$ and $u, u+a \in (M, 1.2M]$, we get

$$|\theta| \leq a(2\theta_1 + \theta_2) + u|\theta_1 - \theta_2| < (u+3a)hM^{-3} \leq 1.6hM^{-2}. \tag{20}$$

Next, we will prove that $n \neq 0$. We have

$$\frac{a^3(2u+a)x}{u^3(u+a)^3} > \frac{2a^3ux}{u^3(u+a)^3} \geq \frac{2a^3x}{(\lambda M)^5} > 0.6666\lambda^{-5}M^{-1}, \tag{21}$$

on recalling (15) and the trivial bound $a^2 \geq 1$. On the other hand, if $n = 0$, the right side of (21) equals θ (which must be positive), and (20) and (21) together yield

$$0.6666\lambda^{-5} < 1.6hM^{-1} \leq 1.6hH^{-1},$$

which contradicts the hypothesis on H. Thus, we must have $|n| \geq 1$.

Similarly to (21), we find that

$$\frac{a^3(2u+a)x}{u^3(u+a)^3} \leq \frac{2a^3(u+a)x}{u^3(u+a)^3} = \frac{2a^3x}{u^3(u+a)^2} \leq \frac{2a^3x}{M^5}. \tag{22}$$

On the other hand, the hypotheses of the lemma and (14) yield

$$|\theta| < 1.6hH^{-2} < 10^{-3},$$

so

$$\frac{a^3(2u+a)x}{u^3(u+a)^3} \geq |n| - |\theta| > 0.999.$$

Combining the last inequality with (22), we deduce

$$a^3 > 0.4995x^{-1}M^5, \tag{23}$$

and the conclusion of the lemma follows. □

We can use this to obtain the following.

Corollary 2 *Under the hypotheses of Lemma 4, we have*

$$|S(M)| < 1.2604(\lambda - 1)x^{1/3}M^{-2/3} + 1.$$

Next, we consider any three distinct elements $u, u+a, u+b$ of $S(M)$, with $0 < a < b$, and obtain lower bounds on b.

Lemma 5 *Let $\lambda \leq 1.04$, $m \geq 3$, and suppose that $mh \leq H \leq M$. If $0 < a < b$ and $u, u+a, u+b$ are elements of $S(M)$, then*

$$ab^4 \geq c_2(m)x^{-1}M^6, \tag{24}$$

where $c_2(m) = 0.4 - 0.05m^{-1}$.

Proof We begin with the identity

$$\frac{a^5}{u^3(u+a)^3} = \frac{6u^2 - 3au + a^2}{u^3} - \frac{6u^2 + 15ua + 10a^2}{(u+a)^3}. \tag{25}$$

By substitution, this yields also the two companion identities

$$\frac{b^5}{u^3(u+b)^3} = \frac{6u^2 - 3bu + b^2}{u^3} - \frac{6u^2 + 15ub + 10b^2}{(u+b)^3}; \tag{26}$$

$$\frac{(b-a)^5}{(u+a)^3(u+b)^3} = \frac{6(u+a)^2 - 3(b-a)(u+a) + (b-a)^2}{(u+a)^3} - \frac{6(u+a)^2 + 15(u+a)(b-a) + 10(b-a)^2}{(u+b)^3}. \tag{27}$$

In order to cancel out the higher order terms of u, we subtract (25) and (27) from (26). This gives

$$\begin{aligned} &\frac{(b-a)(a+b-3u)}{u^3} + \frac{b(5a-b+3u)}{(u+a)^3} + \frac{a(a-5b-3u)}{(u+b)^3} \\ &= \frac{b^5}{u^3(u+b)^3} - \frac{a^5}{u^3(u+a)^3} - \frac{(b-a)^5}{(u+a)^3(u+b)^3}. \end{aligned} \tag{28}$$

For $u_1 = u$, $u_2 = u + a$, and $u_3 = u + b$, let

$$f(u_i) = n_i - \theta_i, \quad 0 < \theta_i < hM^{-3} \qquad (i = 1, 2, 3).$$

Multiplying both sides of (28) by x, we have that

$$(b-a)(a+b-3u)f(u_1) + b(5a-b+3u)f(u_2) + a(a-5b-3u)f(u_3) =: n - \theta,$$

where

$$n = (b-a)(a+b-3u)n_1 + b(5a-b+3u)n_2 + a(a-5b-3u)n_3,$$

(note that n must be even, as a and b are even) and

$$\begin{aligned} |\theta| &= |b(3u-b)(\theta_2-\theta_1) + 5ab(\theta_2-\theta_3) + a(3u-a)(\theta_1-\theta_3)| \\ &< hM^{-3}(3bu + 3au + 3ab - (a-b)^2) < 3bhM^{-3}(2u+a) \\ &< 6bhM^{-3}(u+b) < 6\lambda bhM^{-2}. \end{aligned} \tag{29}$$

Next, we show that $n \neq 0$. Suppose not. A direct check reveals that

$$\frac{b^5}{u^3(u+b)^3} - \frac{a^5}{u^3(u+a)^3} - \frac{(b-a)^5}{(u+a)^3(u+b)^3} = \frac{P(u;a,b)}{u^3(u+a)^3(u+b)^3},$$

where

$$\begin{aligned} P(u) &= 5ab(b-a)(b^2-ab+a^2)u^3 + 3ab(b^4-a^4)u^2 \\ &\quad + 3a^2b^2(b^3-a^3)u + a^3b^3(b^2-a^2) \end{aligned}$$

$$> 5ab(b-a)(b^2-ab+a^2)u^3 \geq 5a^2b^2(b-a)u^3 > 10a^3bu^3. \tag{30}$$

In particular, the two sides of (28) are positive. When $n = 0$, this entails that $\theta < 0$, and

$$-\theta = \frac{P(u;a,b)x}{u^3(u+a)^3(u+b)^3} > \frac{10a^3bu^3x}{u^3(u+a)^3(u+b)^3} \geq \frac{10b(0.4995M^5)}{(\lambda M)^6},$$

after appeals to (30) and (23). Hence, comparing the bound above to (29),

$$4.995\lambda^{-6}bM^{-1} < 6\lambda bhM^{-2} \leq 6\lambda bh(HM)^{-1},$$

or

$$m = \frac{H}{h} < \frac{6\lambda^5}{4.995}$$

which contradicts the hypotheses of the lemma. As such, we must have that $|n| \geq 2$.

On the other hand, bounding the numerator $P(u;a,b)$ from above, we find that

$$\begin{aligned} P(u) &\leq 5ab(b-a)(b^2-ab+a^2)u^3 + 3ab^5u^2 + 3a^2b^5u + a^3b^5 \\ &\leq 5ab^3(b-a)u^3 + 3ab^5u^2 + 3ab^6u + ab^7 \\ &< ab^4(5u^3+3bu^2+3b^2u+b^3) < 5ab^4(u+b)^3. \end{aligned}$$

Thus,

$$n - \theta = \frac{P(u;a,b)x}{u^3(u+a)^3(u+b)^3} \leq \frac{5ab^4x}{u^3(u+a)^3} < 5ab^4xM^{-6}. \tag{31}$$

Further, recalling that $b < (\lambda - 1)M$, we deduce that

$$|\theta| \leq 6\lambda bhM^{-2} < 0.25hM^{-1} \leq 0.25hH^{-1}. \tag{32}$$

From (31) and (32), we deduce that

$$5ab^4xM^{-6} \geq \frac{P(u;a,b)x}{u^3(u+a)^3(u+b)^3} \geq 2 - |\theta| > 2 - 0.25hH^{-1},$$

which implies the conclusion of the lemma. □

Corollary 3 *Under the hypotheses of Lemma 5, we have*

$$|S(M)| \leq 2(\lambda-1)c_2(m)^{-1/5}x^{1/5}M^{-1/5} + 2.$$

4 Main Bounds on $|S(M)|$

In this section, we first study a special family of quadruples $u, u+a, u+b, u+a+b$ of elements of $S(M)$. The special form of the spacing between the four numbers allows us to use several algebraic identities to obtain a series of lower bounds on products of the form $a^i b^j$. These yield bounds on b that are stronger than those for general quadruples in $S(M)$. Later in the section, we will average these bounds over b.

Lemma 6 *Let $\lambda \le 1.04$, $m \ge 3$, and suppose that $mh \le H \le M$. If $0 < a \le b$ and $u, u+a, u+b, u+a+b$ are elements of $S(M)$, then*

$$a^3 b > 0.0999 x^{-1} M^6. \tag{33}$$

Proof Note that under the hypotheses of the lemma, we have

$$a \le \frac{1}{2}(a+b) < 0.5(\lambda - 1)M. \tag{34}$$

Recall that, by (19) and (20), we have

$$\frac{a^3(2u+a)x}{u^3(u+a)^3} = n_1 + \theta_1$$

with $n_1 \in \mathbb{Z}$ and $|\theta_1| < 1.6hM^{-2}$. Indeed, using (34) we can strengthen this to $|\theta_1| < 1.06hM^{-2}$. Similarly,

$$\frac{a^3(2u+2b+a)x}{(u+b)^3(u+a+b)^3} = n_2 + \theta_2$$

with $n_2 \in \mathbb{Z}$ and $|\theta_2| < 1.06hM^{-2}$. Combining these identities, we find that

$$\frac{a^3(2u+a)x}{u^3(u+a)^3} - \frac{a^3(2u+2b+a)x}{(u+b)^3(u+a+b)^3} = n + \theta$$

with $|\theta| \le |\theta_1| + |\theta_2| < 2.12hM^{-2}$. Now, we note that when the right side of the above identity is combined into a single fraction, the numerator can be written as

$$a^3 x[4bu^3(u+a)^3 + 3abu^2(u+a)^3 + 3b(2u+a)u^3(u+a)^2 + \text{other positive terms}].$$

Since

$$3abu^2(u+a)^3 + 3b(2u+a)u^3(u+a)^2 > 6bu^3(u+a)^3,$$

we can use (23) to obtain the lower bound

$$\begin{aligned}
n+\theta &\geq \frac{10a^3bu^3(u+a)^3x}{u^3(u+a)^3(u+b)^3(u+a+b)^3} \\
&= \frac{10a^3bx}{(u+b)^3(u+a+b)^3} \geq \frac{20a^3x}{(\lambda M)^6} \geq \frac{9.99}{\lambda^6 M}.
\end{aligned}$$

In particular, if $n = 0$, we see that

$$9.99\lambda^{-6} \leq M\theta < 2.12hM^{-1} < 2.12m^{-1},$$

which contradicts the hypotheses. Thus, we must have $|n| \geq 1$.

We now observe that (note the first equality can be checked using a computer algebra system)

$$\begin{aligned}
&(2u+a)(u+b)^3(u+a+b)^3 - (2u+2b+a)u^3(u+a)^3 \\
&\quad = 10bu^3(u+a+b)^3 + b(3(a+b)^2+7b^2+9ab)u^2(u+a+b)^2 \\
&\qquad + b^2(3(a+b)^3+ab^2+2a^2b-b^3)u(u+a+b) + ab^3(a+b)^3 \\
&\quad < 10bu^3(u+a+b)^3 + 10b(a+b)^2u^2(u+a+b)^2 \\
&\qquad + 3b^2(a+b)^3u(u+a+b) + ab^3(a+b)^3 \\
&\quad < 10bu^3(u+a+b)^3 + b^2u^2(u+a+b)^3 + b^3u(u+a+b)^3 + b^4(a+b)^3 \\
&\quad < b(u+a+b)^3(10u^3+bu^2+b^2u+b^3) < 10b(u+b)^3(u+a+b)^3.
\end{aligned}$$

Thus,

$$1-|\theta| \leq n+\theta \leq \frac{10a^3bx}{u^3(u+a)^3} \leq 10a^3bxM^{-6}.$$

Since $|\theta| \leq 2.12hH^{-2} < 0.001$, we conclude that

$$0.999 < 10a^3bxM^{-6},$$

and the lemma follows. □

Lemma 7 *Let $\lambda \leq 1.04$, $m \geq 8$, and suppose that $mh \leq H \leq M$. If $0 < a \leq b$ and $u, u+a, u+b, u+a+b$ are elements of $S(M)$, then*

$$a^3b^3 > 0.0664x^{-1}M^7. \tag{35}$$

Proof First, we observe that if $b \geq 0.01M$, Lemma 6 gives

$$a^3b^3 \geq 0.0999b^2x^{-1}M^6 \geq 0.000009x^{-1}M^8 > 0.07x^{-1}M^7,$$

since $M \geq mh \geq 8000$. Thus, we may assume for the remainder of the proof that $0 < a \leq b < 0.01M$. Let $u_1 = u$, $u_2 = u+a$, $u_3 = u+b$, and $u_4 = u+a+b$, and recall that by the definition of the set $S(M)$, there exist integers $n_1, \dots, n_4$ and reals $\theta_1, \dots, \theta_4$ such that

$$f(u_i) = n_i - \theta_i, \quad 0 < \theta_i < hM^{-3} \qquad (1 \le i \le 4). \tag{36}$$

We begin by constructing a rational function of the form

$$R(u; a, b) = \frac{P_1(u; a, b)}{u^3} + \frac{P_2(u; a, b)}{(u+a)^3} + \frac{P_3(u; a, b)}{(u+b)^3} + \frac{P_4(u; a, b)}{(u+a+b)^3}, \tag{37}$$

where $P_i(u; a, b)$ are homogeneous quadratic polynomials, which are at most linear in u. Clearly, any such rational function can be rewritten as

$$R(u; a, b) = \frac{C(u; a, b)}{u^3(u+a)^3(u+b)^3(u+a+b)^3}, \tag{38}$$

for some homogeneous polynomial $C(u; a, b)$ of total degree 11, which has at most degree 10 in u. We choose the polynomials $P_i(u; a, b)$ as to minimize the degree of C with respect to u. The identities

$$\frac{u+2a}{(u+a)^3} - \frac{u-a}{u^3} = \frac{a^3(2u+a)}{u^3(u+a)^3}, \quad \frac{1}{u^3} - \frac{1}{(u+a)^3} = \frac{a(3u^2+3ua+a^2)}{u^3(u+a)^3},$$

imply that any choice of the form

$$P_1(u, a, b) = \alpha(-u+a) - \beta, \qquad P_2(u, a, b) = \alpha(u+2a) + \beta,$$
$$P_3(u, a, b) = P_1(u+b, a, b) + 2\beta, \qquad P_4(u, a, b) = P_2(u+b, a, b) - 2\beta,$$

where α, β depend only on a, b, reduces the u-degree of C to at most 7. The choice $\alpha = 3b$ and $\beta = a^2$ then ensures that the coefficients of u^7 and u^6 in $C(u; a, b)$ also cancel out. Thus, we choose

$$P_1(u, a, b) = -3bu + 3ab - a^2, \qquad P_2(u, a, b) = 3bu + 6ab + a^2,$$
$$P_3(u, a, b) = -3bu - 3b^2 + 3ab + a^2, \qquad P_4(u, a, b) = 3bu + 3b^2 + 6ab - a^2.$$

With the above choice, a straightforward (but tedious) direct calculation reveals that

$$C(u; a, b) = 6a^3b(5b^2 - a^2)v^5 - 15a^3b(a+b)(5b^2 - a^2)v^4 + D(v; a, b), \tag{39}$$

where $v = u + a + b$ and $D(v; a, b)$ is a homogeneous polynomial of total degree 11, which is cubic in v. In particular, the coefficient of v^3 in $D(v; a, b)$ is

$$72a^3b^5 + 150a^4b^4 + 52a^5b^3 - 30a^6b^2 - 12a^7b < 74a^3b^3(a+b)^2,$$

on recalling that $0 < a \le b$. Note that we have also

$$72a^3b^5 + 150a^4b^4 + 52a^5b^3 - 30a^6b^2 - 12a^7b > 58a^3b^3(a+b)^2.$$

Similarly, the coefficient of v^2 is bounded above and below as

$$-36a^3b^3(a+b)^3 < 3a^8b + 18a^7b^2 - 3a^6b^3 - 93a^5b^4 - 108a^4b^5 - 33a^3b^6 < -27a^3b^3(a+b)^3;$$

the coefficient of v is bounded above and below as

$$6a^3b^4(a+b)^3 < -3a^8b^2 - 6a^7b^3 + 18a^6b^4 + 48a^5b^5 + 33a^4b^6 + 6a^3b^7 < 14a^3b^4(a+b)^3;$$

and the constant (in v) terms are bounded as

$$-4.25a^4b^5(a+b)^2 < a^8b^3 - 6a^6b^5 - 8a^5b^6 - 3a^4b^7 < -3a^4b^5(a+b)^2.$$

Moreover, since $M < u < v \le \lambda M$ and $0 < a, b < 0.01M$, we find that

$$a^4b^5(a+b)^2 < 0.01a^3b^4(a+b)^3v, \quad a^3b^4(a+b)^3v < 0.01a^3b^3(a+b)^3v^2,$$
$$a^3b^3(a+b)^3v^2 < 0.02a^3b^3(a+b)^2v^3.$$

From these and the earlier bounds on the coefficients of $D(v; a, b)$, we deduce that

$$0 < D(v; a, b) < 74a^3b^3(a+b)^2v^3. \tag{40}$$

Inserting this upper bound into (39) we now have

$$\begin{aligned} C(u; a, b) &= 6a^3b(5b^2 - a^2)v^5 - 15a^3b(a+b)(5b^2 - a^2)v^4 + D(v; a, b) \\ &\le a^3bv^3\big(6(5b^2 - a^2)v^2 - 15(a+b)(5b^2 - a^2)v + 74b^2(a+b)^2\big). \end{aligned}$$

We expand $v = u + a + b$ (and use $a + b < 0.02u$) to bound the term in parentheses above as

$$\begin{aligned} &6(5b^2 - a^2)v^2 - 15(a+b)(5b^2 - a^2)v + 74b^2(a+b)^2 \\ &\quad = 6(5b^2 - a^2)u^2 - 3(a+b)(5b^2 - a^2)u - 9(a+b)^2(5b^2 - a^2) + 74b^2(a+b)^2 \\ &\quad \le 6(5b^2 - a^2)u^2 - 3(a+b)(5b^2 - a^2)u + 38b^2(a+b)^2 \\ &\quad < 6(5b^2 - a^2)u^2 - 12(a+b)b^2u + (a+b)b^2u < 30b^2u^2. \end{aligned}$$

Thus $C(u; a, b) < 30a^3b^3u^2v^3$. We can also use (40) to bound C from below:

$$\begin{aligned} C(u; a, b) &= 3a^3b(5b^2 - a^2)v^4(2v - 5(a+b)) + D(v; a, b) \\ &> 12a^3b^3v^4(2u - 3(a+b)) \\ &> 12a^3b^3v^2\big(u^2 + 2(a+b)u\big)(2u - 3(a+b)) \end{aligned}$$

$$> 12a^3b^3uv^2\big(2u^2 + 0.98(a+b)u\big) > 24a^3b^3u^3v^2.$$

From these bounds and (38), we deduce that

$$18.238a^3b^3M^{-7} < \frac{24a^3b^3}{(\lambda M)^7} \le R(u;a,b) \le 30a^3b^3M^{-7}. \tag{41}$$

On the other hand, multiplying both sides of (37) by x, we get from (36) that

$$xR(u;a,b) = n - \theta$$

where $n \in \mathbb{Z}$ is even (since all of the coefficients in each of the P_i are even) and

$$\theta = \sum_{i=1}^{4} P_i(u;a,b)\theta_i.$$

We can bound $|\theta|$ as follows:

$$\begin{aligned}|\theta| &\le 3bu|\theta_1 - \theta_2| + 3b(u+b)|\theta_3 - \theta_4| \\ &\qquad + 3ab(\theta_1 + 2\theta_2 + \theta_3 + 2\theta_4) + a^2|\theta_1 - \theta_2 - \theta_3 + \theta_4| \\ &< \big(3b(u+b) + 3bu + 18ab + 2a^2\big)hM^{-3} \\ &< (6b(u+a+b) + 11ab)hM^{-3} < (6\lambda + 0.11)bhM^{-2},\end{aligned}$$

upon recalling that $u + a + b < \lambda M$ and $a < 0.01M$.

Next, we will show that $n \ne 0$. Suppose that $n = 0$. Then we have

$$18.238a^3b^3xM^{-7} \le |Cx| = |\theta| \le (6\lambda + 0.11)bhM^{-2} \le 6.35bhM^{-2}.$$

Combining this inequality with (33), we obtain

$$1.822b^2M^{-1} < 18.238a^3b^3xM^{-7} \le 6.35b(mM)^{-1},$$

which is a contradiction under the hypotheses of the lemma. Thus $n \ge 2$, and we get

$$Cx = n - \theta > 2 - 6.35bhM^{-2} > 2 - 0.0635m^{-1} > 1.992.$$

Combining this with the upper bound in (41), we get that

$$1.992 < Cx < 30a^3b^3xM^{-7},$$

and the desired conclusion follows. □

Our last lemma differs from the spacing lemmas established hitherto. In this lemma, instead of proving that the distance b between the two pairs exceeds some lower bound in terms of x, M, and possibly, a, we establish a kind of a dichotomy for b: either $b \geq B_1$ for some lower bound B_1, or $b \leq B_2$, with B_2 significantly smaller than B_1. Note, however, that this dichotomy is not complete, and conditions (42) and (43) below are not always mutually exclusive—when M is close to H and $m \leq 15$, the two inequalities may hold simultaneously.

Lemma 8 *Let $\lambda \leq 1.04$, $m \geq 8$, and suppose that $mh \leq H \leq M$. If $0 < a \leq b$ and $u, u+a, u+b, u+a+b$ are elements of $S(M)$, then at least one of the conditions*

$$a^5 b < 2\lambda^9 h x^{-1} M^6, \tag{42}$$

or

$$a^5 b > (\tfrac{1}{3} - 2\lambda^2 m^{-1}) x^{-1} M^7, \tag{43}$$

must hold.

Proof As in the proof of the last lemma, let $u_1 = u$, $u_2 = u + a$, $u_3 = u + b$, and $u_4 = u + a + b$, and recall (36). We rely on the identity

$$\begin{aligned} \frac{a^5 x}{u^3(u+a)^3} &= \frac{(6u^2 - 3au + a^2)x}{u^3} - \frac{(6u^2 + 15ua + 10a^2)x}{(u+a)^3} \\ &= (6u^2 - 3ua + a^2) f(u) - (6u^2 + 15ua + 10a^2) f(u+a) = n' - \theta', \end{aligned}$$

with $n' \in \mathbb{Z}$, even and

$$\theta' = (6u^2 - 3ua + a^2)\theta_1 - (6u^2 + 15ua + 10a^2)\theta_2.$$

Using this relation and the analogous one for the pairs $u + b$, $u + a + b$, we find that

$$\frac{xa^5}{u^3(u+a)^3} - \frac{xa^5}{(u+b)^3(u+a+b)^3} = n + \theta, \tag{44}$$

with $n \in \mathbb{Z}$ even and

$$\begin{aligned} |\theta| &< (6u^2 + 6(u+b)^2 + 12au + 15ab + 11a^2) h M^{-3} \\ &< (6(u+a+b)^2 + 6(u+a)^2) h M^{-3} < 12\lambda^2 h M^{-1}. \end{aligned}$$

By the mean-value theorem,

$$(u+b)^3(u+a+b)^3 - u^3(u+a)^3 = 3b\xi^2(\xi + a)^2(2\xi + a)$$

for some $\xi \in (u, u+b)$, so

$$6bu^3(u+a)^2 < (u+b)^3(u+a+b)^3 - u^3(u+a)^3 < 6b(u+b)^2(u+a+b)^3.$$

We use this to bound the left side of (44) and get that

$$n+\theta \le \frac{6a^5bx}{u^3(u+a)^3(u+b)} < 6a^5bxM^{-7}.$$

Similarly, we have

$$n+\theta \ge \frac{6a^5bx}{(u+a)(u+b)^3(u+a+b)^3} > 6\lambda^{-7}a^5bxM^{-7}.$$

Since $M \ge H$, we have that $|\theta| < 12\lambda^2 m^{-1}$. So, it follows from the upper bound on $n+\theta$ that if $b \le (\frac{1}{3} - 2\lambda^2 m^{-1})a^{-5}x^{-1}M^7$, we have

$$n+\theta < 6a^5bxM^{-7} < 2 - |\theta|;$$

hence, $n < 2$. On the other hand, if $b \ge 2\lambda^9 a^{-5}hx^{-1}M^6$, then using the lower bound for $n+\theta$, we have that

$$n+\theta > 6\lambda^{-7}a^5bxM^{-7} \ge 12\lambda^2 hH^{-1} > |\theta|,$$

implying that $n > 0$. Since $n > 0$ and $n < 2$ cannot occur simultaneously when n is even, the lemma follows. □

Let

$$A = (0.4 - 0.05m^{-1})^{1/5}x^{-1/5}M^{6/5}. \tag{45}$$

In Lemma 5, we proved that $b \ge A$ whenever $u, u+a, u+b$ are distinct elements of $S(M)$. Therefore, if $u_0, u_1, \ldots, u_s$ are the elements of $S(M)$, listed in increasing order, the set $S'(M) = \{u_0, u_2, u_4, \ldots\}$ has no gaps $< A$ and satisfies

$$|S(M)| \le 2|S'(M)|. \tag{46}$$

We will use (46) and Lemmas 7 and 8 to prove the following result.

Proposition 1 *Suppose* $h = 5x^{1/7}\log x$, *let* $\lambda = 1.04$ *and* $x \ge e^{200}$, *and suppose that* $11h \le M \le x^{2/7}$. *Then*

$$|S(M)| \le h(\sigma_3(M) + \sigma_4(M)), \tag{47}$$

where

$$\sigma_3(M) = \big(0.4212x^{1/7} + 0.1900x^{-1/7}M\big)h^{-1} + 0.0374x^{1/21}M^{-1/3}, \tag{48}$$

and

$$\sigma_4(M) = \begin{cases} 0.9519x^{-2/3}M^{11/3} & \text{if } M \le 18x^{1/6}, \\ 5.1698x^{-1/15}M^{1/15} & \text{if } M > 18x^{1/6}. \end{cases} \tag{49}$$

Remark 1 Notice that when x is relatively small, the condition $M \leq 18x^{1/6}$ is impossible, and so only the second condition will be used in the range of "small" values of x.

The proof of this proposition uses the set

$$T(M;a) = \{u : u, u+a \text{ are consecutive elements of } S'(M)\}$$

to bound $|S'(M)|$. The starting point is the elementary identity

$$|S'(M)| = 1 + \sum_{a=1}^{\infty} |T(M;a)| = 1 + \sum_{a \geq A} |T(M;a)|, \tag{50}$$

which is a direct consequence of the definition of $T(M;a)$. Further, for any $B \geq A$, we have

$$\sum_{a \geq B} a|T(M;a)| \leq \sum_{a \geq A} a|T(M;a)| \leq (\lambda - 1)M + 1,$$

so

$$\sum_{a \geq B} |T(M;a)| \leq (\lambda - 1)MB^{-1} + B^{-1}.$$

Applying this inequality to the right side of (50), we find that, for any parameter $B \geq 2$,

$$|S'(M)| \leq 1.5 + (\lambda - 1)MB^{-1} + \sum_{A \leq a < B} |T(M;a)|. \tag{51}$$

Proof of Proposition 1

We select

$$B = \delta x^{-1/7} M, \quad \delta = 0.19, \tag{52}$$

in the application of (51). Fix an integer a, with $A \leq a \leq B$. By Lemma 8, if we consider an interval I of length

$$|I| \leq (\tfrac{1}{3} - 2\lambda^2 m^{-1})x^{-1}M^7,$$

we must have $b < 2\lambda^9 h a^{-5} x^{-1} M^6$ for any elements $u, u+b \in T(M;a) \cap I$. Then we can use (35) to get

$$\begin{aligned} |T(M;a) \cap I| &\leq \frac{2\lambda^9 h a^{-5} x^{-1} M^6}{(0.0664)^{1/3} a^{-1} x^{-1/3} M^{7/3}} + 1 \\ &< 7.0298 a^{-4} h x^{-2/3} M^{11/3} + 1. \end{aligned}$$

Since we need at most

$$\frac{(\lambda - 1)M}{(\frac{1}{3} - 2\lambda^2 m^{-1})x^{-1}M^7 a^{-5} x^{-1} M^7} + 1 < 0.2929a^5 x M^{-6} + 1$$

intervals of length $|I|$ to cover $(M, \lambda M]$, we conclude that

$$\begin{aligned} |T(M; a)| &\leq (0.2929a^5 x M^{-6} + 1)(7.0298a^{-4} h x^{-2/3} M^{11/3} + 1) \\ &\leq 2.0591 a h x^{1/3} M^{-7/3} + 0.2929 a^5 x M^{-6} + 7.0298 a^{-4} h x^{-2/3} M^{11/3} + 1. \end{aligned} \tag{53}$$

Next, we use this to bound the right side of (51). With our choice of parameters, (51) yields

$$|S'(M)| \leq 1.5 + 0.04\delta^{-1} x^{1/7} + \sum_{\substack{A \leq a < B \\ a \text{ even}}} |T(M; a)|. \tag{54}$$

Next, we sum each of the four terms on the right side of (53) over $a \in [A, B)$. Recall the inequality

$$\sum_{k \leq K} k^s < \frac{(K+1)^{s+1}}{s+1} \qquad (s > 0),$$

and note that $B = \delta M x^{-1/7} \geq 11 h \delta x^{-1/7} > 10.45 \log x > 2090$. Thus,

$$\sum_{\substack{2 \leq a \leq B \\ a \text{ even}}} a^s < \frac{(B+2)^{s+1}}{2(s+1)} < \frac{(1.001B)^{s+1}}{2(s+1)},$$

and we deduce

$$\begin{aligned} 2.0591 h x^{1/3} M^{-7/3} \sum_{\substack{A \leq a < B \\ a \text{ even}}} a &< \frac{2.0591(1.001B)^2}{4} x^{1/3} M^{-7/3} \\ &< 0.0187 h x^{1/21} M^{-1/3}, \end{aligned} \tag{55}$$

$$0.2929 x M^{-6} \sum_{\substack{A \leq a < B \\ a \text{ even}}} a^5 < \frac{0.2929 \cdot (1.001B)^6}{12} x M^{-6} < 0.00001 x^{1/7}. \tag{56}$$

Combining (52)–(56), we conclude that

$$|S'(M)| \leq h\sigma_3'(M) + 7.0298 h x^{-2/3} M^{11/3} \sum_{\substack{A \leq a < B \\ a \text{ even}}} a^{-4}, \tag{57}$$

where

$$\sigma_3'(M) = \left(0.2106x^{1/7} + 0.095x^{-1/7}M\right)h^{-1} + 0.0187x^{1/21}M^{-1/3}. \tag{58}$$

We estimate the sum on the right side of (57) in different ways, depending on the size of M. When $M \le 18x^{1/6}$, we use that

$$\sum_{\substack{A \le a < B \\ a \text{ even}}} a^{-4} < \frac{\zeta(4)}{16} = \frac{\pi^4}{1440} < 0.0677. \tag{59}$$

On the other hand, when $M > 18x^{1/6}$, we have $A > 0.8306 \cdot 18^{6/5} > 26.6513$, so

$$\sum_{\substack{A \le a < B \\ a \text{ even}}} a^{-4} < \frac{1}{3 \cdot 2^4}\left(\frac{A}{2} - 1\right)^{-3} < 0.2107A^{\;3} < 0.3677x^{3/5}M^{-18/5}. \tag{60}$$

The proposition follows from (46) and (57)–(60). □

5 Proof of Theorem 2

As stated in the introduction, the theorem can be checked by brute force for $x \le e^{41}$. Thus, we focus on proving it for $x \ge e^{41}$.

5.1 *Large x*

Let $x \ge e^{200}$ and set $H = 11h$ in (2) and (11). We will use Proposition 1 and Lemma 2 to bound $|S(H, x^{2/7})|$.

Suppose first that $H \le 18x^{1/6}$, (in this case we can assume $x \ge e^{284}$) we again split $S(H, x^{2/7})$ according to the two cases in (49). When we apply Lemma 2 to the bound (47) for $M \in [H, 18x^{1/6}]$, we find that

$$\begin{aligned}
|S(H, 18x^{1/6})| &< h\left(\frac{0.9519 \cdot 18^{11/3}x^{-1/18}}{1 - 1.04^{-11/3}} + \frac{0.0374x^{1/21}H^{-1/3}}{1 - 1.04^{-1/3}}\right) + \frac{3.8x^{1/42}}{1 - 1.04^{-1}} \\
&\quad + 0.4212x^{1/7}\left(\frac{\log x}{42\log(1.04)} - \frac{\log(\frac{55}{18}\log x)}{\log(1.04)} + 1\right) \\
&< 0.1552h + 0.2557x^{1/7}\log x - 72.2396x^{1/7} < 0.2064h.
\end{aligned}$$

Similarly, when we apply Lemma 2 to (47) for $M \in [18x^{1/6}, x^{2/7}]$, we get

$$\begin{aligned}|S(18x^{1/6}, x^{2/7})| &< h\left(\frac{5.1698x^{-1/21}}{1-1.04^{-1/15}} + \frac{0.0374 \cdot 18^{-1/3}x^{-1/126}}{1-1.04^{-1/3}}\right) + \frac{0.19x^{1/7}}{1-1.04^{-1}}\\ &\quad + 0.4212x^{1/7}\left(\frac{5\log x}{42\log(1.04)} - \frac{\log 18}{\log(1.04)} + 1\right)\\ &< 0.1180h + 1.2785x^{1/7}\log x - 25.6791x^{1/7} < 0.3737h.\end{aligned}$$

Hence,

$$|S(H, x^{2/7})| < 0.2064h + 0.3737h = 0.5801h. \tag{61}$$

Next, we consider the case $H > 18x^{1/6}$, noting that we must have $x \le e^{285}$. We use the latter case of Proposition 1 for M in the full range $(H, x^{2/7}]$, and we find that

$$\begin{aligned}|S(H, x^{2/7})| &< h\left(\frac{5.1698x^{-1/21}}{1-1.04^{-1/15}} + \frac{0.0374x^{1/21}H^{-1/3}}{1-1.04^{-1/3}}\right) + \frac{0.19x^{1/7}}{1-1.04^{-1}}\\ &\quad + 0.4212x^{1/7}\left(\frac{\log x}{7\log(1.04)} - \frac{\log(55\log x)}{\log(1.04)} + 1\right)\\ &< 0.2742h + 1.5342x^{1/7}\log x - 94.5742x^{1/7} < 0.5148h, \end{aligned} \tag{62}$$

on noting that $94.5742x^{1/7} > 0.0663h$ when $x \le e^{285}$.

To complete the estimation of $|S(H, \sqrt[3]{2x})|$, we apply Lemma 2 to the bound in Corollary 1 for $M \in [x^{2/7}, \sqrt[3]{2x}]$. This yields

$$|S(x^{2/7}, \sqrt[3]{2x})| \le \frac{0.1202x^{1/7}}{1-1.04^{-3}} + \frac{\log x}{21\log(1.04)} + \frac{\log 2}{3\log(1.04)} + 1 < 0.0011h. \tag{63}$$

Combining (61)–(63), we obtain (11) with

$$\sigma_3 = \begin{cases} 0.5812 & \text{if } H \le 18x^{1/6},\\ 0.5159 & \text{if } H > 18x^{1/6},\end{cases}$$

for all $x \ge e^{200}$. Taking $J = 100$ in (6), we have $\sigma_0(h, 100) \le -0.0066$ in the same range. Furthermore, we have

$$\sigma_2(h, 11) < \begin{cases} 0.2256 & \text{if } x \ge e^{284},\\ 0.3020 & \text{if } x \ge e^{200}.\end{cases}$$

Thus,

$$\sigma_0(h, 100) + \sigma_1 + \sigma_2(h, 11) + \sigma_3(h, 11) < \begin{cases} 0.9750 & \text{if } H \le 18x^{1/6}, \\ 0.9861 & \text{if } H > 18x^{1/6}, \end{cases}$$

which establishes (12), and therefore the theorem, for $x \ge e^{200}$.

5.2 Intermediate x

Suppose that $x \ge e^{41}$. We take $h = 2x^{1/5}$, $H = 2.7h$, and $\lambda = 1.06$, and apply Lemma 2 to the result of Corollary 2. We obtain

$$\begin{aligned} |S(H, \sqrt[3]{2x})| &\le \frac{1.2604(0.06)(xH^{-2})^{1/3}}{1 - 1.06^{-2/3}} + \frac{\log \sqrt[3]{2x} - \log H}{\log(1.06)} + 1 \\ &< 0.3225h + \frac{2 \log x}{15 \log(1.06)} - \frac{\log(5.4) - \frac{1}{3} \log 2}{\log(1.06)} + 1 < 0.3354h. \end{aligned}$$

That is, (11) holds with $\sigma_3(h, 2.7) = 0.3354$. As $\sigma_2(h, 2.7) \le 0.3146$, and $\sigma_0(h, 6) < -0.0047$ we get

$$\sigma_0(h, 6) + \sigma_1 + \sigma_2(h, 2.7) + \sigma_3(h, 2.7) < 0.8201.$$

This establishes that the interval $(x, x + 2x^{1/5}]$ contains a cubefree integer for all $x \ge e^{41}$. Moreover, recalling the results of [8], we obtain the following proposition.

Proposition 2 *For any $x \ge 2$, the interval $(x, x + 2x^{1/5}]$ contains a cubefree integer.*

Since $2x^{1/5} \le 5x^{1/7} \log x$ for $x \le e^{95.8}$ the main result follows from Proposition 2 for $x \le e^{95}$. Next, we prove that $h = 10x^{1/6}$ is admissible when $x \ge e^{95}$. With this choice of h, we let $H = 4h$ and $\lambda = 1.03$. An application of Lemma 2 to the bound of Corollary 3 yields

$$|S(H, \sqrt[3]{2x})| \le \frac{0.0726(xH^{-1})^{1/5}}{1 - 1.03^{-1/5}} + \frac{2 \log x}{6 \log(1.03)} < 0.5891h,$$

or $\sigma_3(h, 4) = 0.5891$ in (11). Since $\sigma_0(h, 20) < -0.0066$ and $\sigma_2(h, 4) < 0.2207$ for $x \ge e^{95}$, we deduce that

$$\sigma_0(h, 20) + \sigma_1 + \sigma_2(h, 4) + \sigma_3(h, 4) < 0.9780.$$

Since $10x^{1/6} \le 5x^{1/7} \log x$ in the range $x \le e^{191.6}$, this establishes the theorem for $x \le e^{191}$.

By the same method we show that $h = 8.5x^{1/6}$ is admissible when $x \geq e^{191}$. With this choice of h, we let $H = 5h$ and $\lambda = 1.01$. In this case, applying Lemma 2 to the bound of Corollary 3 yields

$$|S(H, \sqrt[3]{2x})| \leq \frac{0.0242(xH^{-1})^{1/5}}{1 - 1.01^{-1/5}} + \frac{2\log x}{6\log(1.01)} < 0.6767h,$$

or $\sigma_3(h, 5) = 0.6767$ in (11). Since $\sigma_0(h, 20) < -0.0066$ and $\sigma_2(h, 5) < 0.1465$ for $x \geq e^{191}$, we deduce that

$$\sigma_0(h, 20) + \sigma_1 + \sigma_2(h, 5) + \sigma_3(h, 5) < 0.9914.$$

Since $8.5x^{1/6} \leq 5x^{1/7}\log x$ in the range $x \leq e^{200.3}$, this completes the proof of Theorem 2.

As we close this section, we take a moment to record the following result, which we just proved.

Proposition 3 *For any $x \geq e^{95}$, the interval $(x, x + 10x^{1/6}]$ contains a cubefree integer, and for any $x \geq e^{191}$, the interval $(x, x + 8.5x^{1/6}]$ contains a cubefree integer.*

6 Asymptotic Results

We conclude by noting a few of the explicit bounds that can be obtained by these methods if one no longer requires the bounds to be admissible for all values of $x \geq 2$, allowing instead results valid for sufficiently large values of x.

Some of the possible results that can be obtained by tweaking the parameters used in the proof of Theorem 2 are given in the statement of Theorem 3. To prove any of those results, we reset the parameters m, J, λ, δ that appear in the proofs of Proposition 1 and Theorem 2 and then update the various constants. (When x is as large as in Theorem 3, the inequality $H \leq 18x^{1/6}$ always holds, so only the first case in the proof of Theorem 2 can occur.) To establish the claims of Theorem 3, we always select $J = 20$, $\lambda = 1.002$, and $m = \sqrt{\log x_0}$, where x_0 is the lower bound on x in each result; we only vary the choice of δ. For example, when $h = 2x^{1/7}\log x$, $x \geq e^{550}$ (hence, $m = 23.4520\cdots$), and $\delta = 0.38$, we have

$$\sigma_0(h, 20) + \sigma_1 + \sigma_2(h, m) + \sigma_3(h, m) < 0.9914.$$

For $h = x^{1/7}\log x$ and $x \geq e^{2300}$, the choice $\delta = 0.66$ yields an upper bound of 0.9919; and for $h = \frac{1}{2}x^{1/7}\log x$ and $x \geq e^{75\,000}$, $\delta = 0.9$ gives a bound of 0.9977.

Acknowledgements This work is the result of an REU project that took place on the campus of Towson University during the summer of 2022, with the financial support of the National Science Foundation under grants DMS-2136890 and DMS-2149865. The authors also acknowledge financial support from the Fisher College of Science and Mathematics and TU's Mathematics Department. We also thank M. Filaseta, O. Trifonov, and T. Trudgian for commenting on early drafts of this work.

References

1. P. Erdős, *Some problems and results in elementary number theory*, Publ. Math. Debrecen **2** (1951), 103–109.
2. M. Filaseta, *Short interval results for squarefree numbers*, J. Number Theory **35** (1990), no. 2, 128–149.
3. M. Filaseta, S. Graham, and O. Trifonov, *Starting with gaps between k-free numbers*, Int. J. Number Theory **11** (2015), no. 5, 1411–1435.
4. M. Filaseta and O. Trifonov, *On gaps between squarefree numbers. II*, J. London Math. Soc. (2) **45** (1992), no. 2, 215–221.
5. E. Fogels, *On average values of arithmetic functions*, Proc. Cambridge Philos. Soc. **37** (1941), 358–372.
6. H. Halberstam and K. F. Roth, *On the gaps between consecutive k-free integers*, J. London Math. Soc. **26** (1951), 268–273.
7. A. Kumchev, W. McCormick, N. McNew, A. Park, R. Scherr, S. Ziehr, *Explicit bounds for large gaps between squarefree integers*, J. Number Theory **254** (2024), 336–357.
8. M. J. Mossinghoff, T. Oliveira e Silva, and T. S. Trudgian, *The distribution of k-free numbers*, Math. Comp. **90** (2021), no. 328, 907–929.
9. J. B. Rosser and L. Schoenfeld, *Approximate formulas for some functions of prime numbers*, Illinois J. Math. **6** (1962), 64–94.
10. O. Trifonov, *On gaps between k-free numbers*, J. Number Theory **55** (1995), no. 1, 46–59.

The Family of a-Floor Quotient Partial Orders

Jeffrey C. Lagarias and David Harry Richman

Abstract An approximate divisor order is a partial order on the positive integers $\mathbb{N}^+$ that refines the divisor order and is refined by the additive total order. A previous paper studied such a partial order on $\mathbb{N}^+$, produced using the floor function. A positive integer d is a floor quotient of n, denoted $d \preccurlyeq_1 n$, if there is a positive integer k such that $d = \lfloor n/k \rfloor$. The floor quotient relation defines a partial order on the positive integers. This paper studies a family of partial orders, the a-floor quotient relations $\preccurlyeq_a$, for $a \in \mathbb{N}^+$, which interpolate between the floor quotient order and the divisor order on $\mathbb{N}^+$. The paper studies the internal structure of these orders.

Mathematics Subject Classification. 06A06 · 11A05 (primary) · 05A16 · 15B36 · 26D07 (secondary)

1 Introduction

Denote the positive integers $\mathbb{N}^+ = \{1, 2, 3, \ldots\}$. We call a partial order $\mathcal{P} = (\mathbb{N}^+, \preccurlyeq)$ on the positive integers an *approximate divisor order* if:

1. It refines the (multiplicative) divisor partial order $\mathcal{D} = (\mathbb{N}^+, \mid)$, i.e.

$$d \mid n \quad \Rightarrow \quad d \preccurlyeq n,$$

   in which $d \mid n$ denotes d divides n.
2. It is refined by the additive total order $\mathcal{A} = (\mathbb{N}^+, \leq)$, i.e.

$$d \preccurlyeq n \quad \Rightarrow \quad d \leq n.$$

J. C. Lagarias (✉)
Department of Mathematics, University of Michigan, Ann Arbor, MI 48109-1043, USA
e-mail: lagarias@umich.edu

D. H. Richman
Department of Mathematics, University of Washington, Seattle, WA 98195-4350, USA
e-mail: hrichman@uw.edu

M. B. Nathanson (ed.), *Combinatorial and Additive Number Theory VI*,
Springer Proceedings in Mathematics & Statistics 464,
https://doi.org/10.1007/978-3-031-65064-2_15

In a previous paper [6] we studied a partial order on the positive integers $\mathbb{N}^+$ called the floor quotient partial order, defined using the floor function $\lfloor x \rfloor$, which sends x to the greatest integer no larger than x.

Definition 1.1 A positive integer d is called a *floor quotient* of n, written $d \preccurlyeq_1 n$, if $d = \left\lfloor \frac{n}{k} \right\rfloor$ for some positive integer k.

The floor quotient order is an approximate divisor order. It has many more approximate divisors of a given integer n than the usual divisor order, about $2\sqrt{n}$, as noted by Cardinal [2, Prop. 4] and Heyman [4].

1.1 The a-Floor Quotient Order

This paper studies an infinite family of approximate divisor orders, constructed from the floor quotient relation, called *a-floor quotients*, where $a \geq 1$ is an integer parameter.

Definition 1.2 A positive integer d is called an *a-floor quotient* of n, written $d \preccurlyeq_a n$, if and only if $ad \preccurlyeq_1 an$ holds.

The floor quotient relation is the case $a = 1$ of this relation. The 1-floor quotient relation is fundamental in that all the a-floor quotient relations are directly constructible from it; they are embedded inside the 1-floor quotient relation, via the dilation map $d \to ad$ used in its definition.

Theorem 1.3 *The a-floor quotient relation $\preccurlyeq_a$ defines a partial order $\mathcal{Q}_a := (\mathbb{N}^+, \preccurlyeq_a)$. For each $a \geq 1$, the partial order $\mathcal{Q}_a$ is an approximate divisor order.*

Theorem 1.3 follows as a consequence of a more detailed result (Theorem 3.1) showing the equivalence of six properties characterizing a-floor quotients. These six properties parallel the six equivalence for the 1-floor quotient proved in [6, Theorem 3.1]. The first two of these are:

(1) (Cutting property) $ad = \left\lfloor \frac{an}{k} \right\rfloor$ for some integer k.
(2) (Covering property) For the half-open unit interval $I_d = [d, d + 1/a)$, there is some integer k such that the interval kI_d covers I_n, i.e. $k[d, d + 1/a) \supseteq [n, n + 1/a)$.

The partial order axioms for $\mathcal{Q}_a = (\mathbb{N}^+, \preccurlyeq_a)$ follow easily from the covering property (2). The cutting property (1) easily shows that the floor quotient order refines the divisor partial order, and is refined by the additive total order.

If d is an a-floor quotient of n, then we will call n an *a-floor multiple* of d; we will also call any associated value k such that $an = \left\lfloor \frac{ad}{k} \right\rfloor$ an *a-cutting length* of (d, n). There may be more than one a-cutting length k yielding a given a-floor quotient pair (d, n). The a-floor quotient relation drops the extra information contained in the value of k.

The study of the a-floor quotients is motivated by the fact that the divisor order has a dilation symmetry preserving the order structure: $d \mid n$ implies $ad \mid an$ for all $a \geq 1$. The floor quotient order $\preccurlyeq_1$ does not have this dilation symmetry, and applying the a-dilation results in a new partial order, the a-floor quotient order. The next two subsections state some of the main results concerning this family of partial orders.

1.2 Hierarchy of a-Floor Quotient Orders

A main result states that the family of a-floor quotient orders for $a \in \mathbb{N}^+$ interpolates monotonically between the floor quotient order $\mathcal{Q}_1 = (\mathbb{N}^+, \preccurlyeq_1)$ and the divisor partial order $\mathcal{D} = (\mathbb{N}^+, \mid)$.

Theorem 1.4 *(a-Floor quotient order hierarchy).*

(1) If $a < b$ (in the additive order) then the a-floor quotient partial order $\mathcal{Q}_a$ strictly refines the b-floor quotient partial order $\mathcal{Q}_b$. That is, the a-floor quotient orders form a nested decreasing sequence of order relations on $\mathbb{N}^+$ as the parameter a increases.
(2) The intersection of all the a-floor quotient partial order relations, $\bigcap_{a=1}^{\infty} \mathcal{Q}_a$, is the divisor relation $\mathcal{D} = (\mathbb{N}^+, \mid)$.

Theorem 1.4 follows directly from Theorem 3.2, proved in Sect. 3.2.

Section 4 gives results on the rate of convergence of an initial interval $\mathcal{Q}_a[1, n]$ of the a-floor quotient order to the divisor order, as a increases. (Recall that an *interval* $\mathcal{P}[d, n]$ of a partial order $\mathcal{P}$ on $\mathbb{N}^+$ is $\mathcal{P}[d, n] = \{e : d \preccurlyeq_{\mathcal{P}} e \preccurlyeq_{\mathcal{P}} n\}$, and is empty if $d \not\preccurlyeq_{\mathcal{P}} n$.) The next result determines the range of a for which a given (d, n) belongs to the a-floor quotient partial order. To state it, we define the *scaling set* $\mathcal{S}(d, n)$ of a given (d, n) as

$$\mathcal{S}(d, n) := \{a \in \mathbb{N}^+ : \ d \preccurlyeq_a n\}.$$

We show the following result.

Theorem 1.5 *(Structure of scaling sets). For positive integers d and n, the set $\mathcal{S}(d, n)$ of all a where $d \preccurlyeq_a n$ is either $\mathbb{N}^+$, or an initial interval $\mathcal{A}[1, m]$ of the additive order $\mathcal{A} = (\mathbb{N}^+, \leq)$, or the empty set. In other words, the following trichotomy holds.*

(i) If $d \mid n$, then $d \preccurlyeq_a n$ for all $a \in \mathbb{N}^+$.
(ii) If $d \nmid n$ but $d \preccurlyeq_1 n$, then $d \preccurlyeq_a n$ if and only if $a < \dfrac{\lfloor n/d \rfloor}{\delta(d, n)}$, where $\delta(d, n) = n - d \lfloor n/d \rfloor$.
(iii) If $d \not\preccurlyeq_1 n$, then $d \not\preccurlyeq_a n$ for all $a \in \mathbb{N}^+$.

This result is proved in Sect. 4 as Theorem 4.4.

We use Theorem 1.5 to obtain a result on stabilization for an interval of the order.

Theorem 1.6 *(a-Floor quotient order stabilization). The a-floor quotient partial order on the interval $\mathcal{Q}_a[d, n]$ is identical to the divisor partial order on the interval $\mathcal{D}[d, n]$ when $a \geq \lfloor \frac{n}{d} \rfloor$. That is,*

$$(\mathcal{Q}_a[d, n], \preccurlyeq_a) = (\mathcal{D}[d, n], |) \quad \textit{for all} \quad a \geq \left\lfloor \frac{n}{d} \right\rfloor.$$

If in addition $d \mid n$, then the a-floor quotient order on the interval $\mathcal{Q}_a[d, n]$ is identical to the divisor partial order on $\mathcal{D}[d, n]$ when $a \geq \lfloor \frac{n}{d+1} \rfloor$.

The result is proved in Sect. 4 as Theorem 4.5.

We may reinterpret these results as making statements about the two-variable zeta functions of these partial orders as a varies. The two-variable partial order zeta function $\zeta_{\mathcal{Q}_a} : \mathbb{N}^+ \times \mathbb{N}^+ \to \mathbb{Z}$ of the a-floor quotient order $\mathcal{Q}_a$ is given by

$$\zeta_{\mathcal{Q}_a}(d, n) = \begin{cases} 1 & \text{if } d \preccurlyeq_a n, \\ 0 & \text{otherwise.} \end{cases} \tag{1.1}$$

In the sequel we abbreviate notation to rename this function $\zeta_a(d, n)$. Theorem 1.4 asserts that as $a \to \infty$, the zeta functions $\zeta_a(\cdot, \cdot)$ converge pointwise to the two-variable partial order zeta function of the divisor order $\mathcal{D}$, which is

$$\zeta_{\mathcal{D}}(d, n) = \begin{cases} 1 & \text{if } d \mid n \\ 0 & \text{otherwise.} \end{cases} \tag{1.2}$$

Since Theorem 1.4 says the divisor order $\mathcal{D}$ corresponds to the limit $a = +\infty$, we may rename it $\mathcal{Q}_\infty$, and abbreviate notation of $\zeta_{\mathcal{D}}(d, n)$ to $\zeta_\infty(d, n)$. Theorem 1.6 asserts that the convergence for ζ_a to ζ_∞ is monotonically downward pointwise as a increases, with the convergence on $\{1, \ldots, n\}$ stabilized for all $a \geq n$.

A related topic concerns the behavior of the Möbius function of the a-floor quotient partial orders. The partial order Möbius function is the inverse of the zeta function in the incidence algebra of the partial order, under convolution. The results above imply that the Möbius functions of the a-floor quotient partial orders, denoted $\mu_a(\cdot, \cdot)$, converge pointwise (as a increases) to the Möbius function $\mu_{\mathcal{D}}(\cdot, \cdot)$ of the divisor order. The function $\mu_{\mathcal{D}}$ is given by

$$\mu_{\mathcal{D}}(d, n) = \begin{cases} \mu(\frac{n}{d}) & \text{if } d \mid n, \\ 0 & \text{otherwise,} \end{cases} \tag{1.3}$$

where $\mu(n)$ is the classical Möbius function. Since the a-floor quotient relation $\preccurlyeq_a$ converges to the divisor relation as $a \to \infty$, for each (m, n) the Möbius value $\mu_a(m, n)$ converges to $\mu_{\mathcal{D}}(m, n)$ as $a \to \infty$. The convergence property as $a \to \infty$ motivates study of the a-floor quotient Möbius functions as a possible source of new insight into the behavior of the classical Möbius function.

We do not study the Möbius function μ_a in any detail in this paper, making only a few remarks in Sect. 8. The Möbius value $\mu_a(d, n)$ always stabilizes at $\mu_{\mathcal{D}}(d, n)$ for $a \geq \lfloor \frac{n}{d} \rfloor$, Examples 8.1 and 8.2 show that the convergence to $\mu_{\mathcal{D}}(d, n)$ is generally not monotonic in a.

1.3 Structure of a-Floor Quotient Initial Intervals and Almost Complementation

The n-*floor reciprocal map* $\Phi_n(k) = \lfloor \frac{n}{k} \rfloor : \mathcal{A}[1, n] \to \mathcal{A}[1, n]$ acting on the additive partial order interval $\mathcal{A}[1, n] = \{1, \ldots, n\}$, leaves both the partial order intervals $\mathcal{Q}_1[1, n]$ and $\mathcal{D}[1, n]$ invariant, and it acts as an involution on $\mathcal{Q}_1[1, n]$ and $\mathcal{D}[1, n]$. Under this map Φ_n the initial intervals $\mathcal{Q}_1[1, n]$ of the 1-floor quotient order have an *almost complementation* property, described in Sect. 2.3. Namely, we have $\mathcal{Q}_1[1, n] = \mathcal{Q}_1^-(n) \cup \mathcal{Q}_1^+(n)$ where the two sets of "small" elements $\mathcal{Q}_1^-(n)$ and "large" elements $\mathcal{Q}_1^+(n)$ have at most one element in common, and if there is a common element, it is $s = \lfloor \sqrt{n} \rfloor$. The almost complementation property states that: the two sets $\mathcal{Q}_1^+(n)$ and $\mathcal{Q}_1^-(n)$ are the same size, with Φ_n giving a bijective map between them, see Sect. 2.4.

We can define analogues of the almost complementing sets: for $a \geq 1$ we define

$$\mathcal{Q}_a^+(n) := \mathcal{Q}_1^+(n) \cap \mathcal{Q}_a[1, n] \quad \text{and} \quad \mathcal{Q}_a^-(n) := \mathcal{Q}_1^-(n) \cap \mathcal{Q}_a[1, n].$$

Using this definition, for each fixed $a \geq 2$ we find that the almost complementation property breaks down for a-floor quotients on initial intervals $\mathcal{Q}_a[1, n]$, for sufficiently large n. That is, for each $a \geq 2$, the n-floor reciprocal map Φ_n does not generally preserve the interval $\mathcal{Q}_a[1, n]$. We show that the map Φ_n always gives an injection $\Phi_n : \mathcal{Q}_a^+(n) \to \mathcal{Q}_a^-(n)$, so that in cardinality $|\mathcal{Q}_a^+(n)| \leq |\mathcal{Q}_a^-(n)|$. For $a \geq 2$, this inequality is generally strict, as shown by example for $a = 2$ in Fig. 2 in Sect. 5.2. We conclude there is an asymmetry for $a \geq 2$, that $\mathcal{Q}_a^+(n)$ is in general smaller than $\mathcal{Q}_a^-(n)$, and we give a heuristic to suggest a quantitative version of this asymmetry.

We study the sizes of $\mathcal{Q}_a^-(n)$ and $\mathcal{Q}_a^+(n)$. In Sect. 5.1 we prove that $|\mathcal{Q}_a^-(n)| \geq \lfloor \sqrt{\frac{n}{a}} \rfloor$ and that $|\mathcal{Q}_a[1, n]| > \sqrt{\frac{n}{a}}$ for all $n > 1$. In Sect. 5.2 we formulate a heuristic probabilistic argument suggesting that $\mathcal{Q}_a^-(n)$ should have expected size about $(\frac{2}{\sqrt{a}} - \frac{1}{a})\sqrt{n}$ while $\mathcal{Q}_a^+(n)$ should have expected size about $\frac{1}{a}\sqrt{n}$. The heuristic predicts $|\mathcal{Q}_a[1, n]|$ should have expected size about $2\sqrt{\frac{n}{a}}$. Numerical data for $a = 2$ for small n is given which supports these predictions, and this data exhibits significant fluctuations of $|\mathcal{Q}_a[1, n]|$ around any expected size function. The heuristic argument given in Sect. 5.2, if interpreted as an expected value estimate, suggests that

$$\liminf_{n \to \infty} \frac{|\mathcal{Q}_a^+(n)|}{|\mathcal{Q}_a^-(n)|} \leq \frac{1}{2\sqrt{a} - 1},$$

may hold for fixed $a \geq 1$.

In Sect. 7 we prove an averaged version of this heuristic.

Theorem 1.7 *(Average size of a-floor initial interval, variable n). For all integers $a \geq 1$ and all $x \geq 1$,*

$$\frac{1}{x}\sum_{n=1}^{x} |\mathcal{Q}_a[1,n]| = \frac{4}{3}\sqrt{\frac{x}{a}} + O\left(\log x + a\right), \tag{1.4}$$

in which the O-constant is independent of a and x.

Theorem 1.7 is proved, relabeled as Theorem 7.1, in Sect. 7.

The proof of Theorem 7.1 makes use of a classification of integers which are a-floor multiples of a given d, given in Sect. 6. The set of a-floor multiples $\mathcal{M}_a(d)$ of d is given by $\mathcal{M}_a(d) := \{n \in \mathbb{N}^+ : d \preccurlyeq_a n\}$.

Theorem 1.8 *(Numerical semigroup of a-floor multiples). The set $\mathcal{M}_a(d)$ of a-floor multiples of d is a numerical semigroup. It has the following properties.*

(1) The largest integer not in $\mathcal{M}_a(d)$, its Frobenius number, is $(d-1)(ad+1)$.
(2) The number of positive integers not in $\mathcal{M}_a(d)$ is $\frac{1}{2}(d-1)(ad+2)$.
(3) The minimal generating set of $\mathcal{M}_a(d)$ is $\langle \gamma_0, \gamma_1, \ldots, \gamma_{d-1}\rangle$ where

$$\gamma_j = (ja+1)d + j = j(ad+1) + d \quad \textit{for} \quad 0 \leq j \leq d-1.$$

Theorem 1.8 is proved as Theorem 6.3 in Sect. 6. Its proof generalizes the $a = 1$ case proved in [6, Theorem 4.2].

1.4 Prior Work

The study of the 1-floor quotient relation in [6] was motivated by work of J.-P. Cardinal [2]. The main topic of Cardinal's paper is the construction of a commutative quotient algebra $\mathcal{A} = \mathcal{A}_n$ of integer matrices, for each $n \geq 1$, whose rows and columns are indexed by the floor quotients of n, which form the initial interval $\mathcal{Q}_1[1,n]$ of the 1-floor quotient partial order. Cardinal [2] showed that certain matrices $\mathcal{M} = \mathcal{M}_n$ of the algebra $\mathcal{A}_n$ have explicit connections with the Mertens function $M(n) = \sum_{j=1}^{n} \mu(j)$ and he related behavior of the norm of his matrices to the Riemann hypothesis [2, Theorem 24], see also [3].

Cardinal proved the commutativity of the algebra $\mathcal{A}_n$ using a floor function identity which states the commutativity of certain dilated floor functions acting on the real line [2, Lemma 6]: for positive integers k, ℓ one has the identities

$$\left\lfloor \frac{1}{k}\left\lfloor \frac{n}{\ell}\right\rfloor\right\rfloor = \left\lfloor \frac{1}{\ell}\left\lfloor \frac{n}{k}\right\rfloor\right\rfloor = \left\lfloor \frac{n}{k\ell}\right\rfloor. \tag{1.5}$$

These identities are the basis of the existence of the floor quotient partial order; they imply the transitivity property of the floor quotient partial order. In [5], the authors with T. Murayama classified pairs of dilated floor functions $(\lfloor \alpha x \rfloor, \lfloor \beta x \rfloor)$ that commute under composition as functions on $\mathbb{R}$. They showed that the identities (1.5) are quite special: they are the only nontrivial cases where the commutativity property of dilated floor functions occurs. (The trivial cases are where $\alpha = 0$, or $\beta = 0$ or $\alpha = \beta$.)

The existence of Cardinal's algebra seems related to the existence of the complementation structure in the 1-floor quotient partial order. We do not know of an analogue of Cardinal's algebra for the a-floor quotient order for $a \geq 2$.

1.5 Contents of Paper

This paper has two main observations.

(a) As $a \to \infty$ the a-floor quotient orders converge monotonically pointwise (in a precise sense) to the divisor order on $\mathbb{N}^+$. We obtain quantitative information on the rate of convergence.

(b) The 1-floor quotient order $\mathcal{Q}_1$ and the divisor order $\mathcal{D}$ have an almost complementation symmetry on initial intervals of their respective orders, but the a-floor quotient orders $\mathcal{Q}_a$ for $a \geq 2$ break this almost complementation symmetry. We obtain quantitative information on the amount of asymmetry.

A more detailed description follows.

1. Section 2 recalls results on the 1-floor quotient order and the divisibility order on $\mathbb{N}^+$, which is a distributive lattice. The divisibility order has a dilation symmetry which is broken for the 1-floor quotient order, whose breaking leads to the existence of the a-floor quotient orders for $a \geq 2$.
2. Section 3 gives six equivalent characterizations of the a-floor quotient relation in Theorem 3.1. Section 3.2 shows that if $a < b$ in the additive order, then the order $(\mathbb{N}^+, \preccurlyeq_a)$ strictly refines $(\mathbb{N}^+, \preccurlyeq_b)$.
3. Section 4 studies the variation of the structure of a-floor quotient relations, as $a \geq 1$ varies. We describe the set of a for which a given pair (d, n) satisfies $d \preccurlyeq_a n$, showing that this set is an initial interval in the additive order. We determine a quantitative bound on the rate at which the a-floor quotient orders converge to the divisor order as $a \to \infty$ on $\{1, \ldots, n\}$, see Theorem 4.5. We introduce the *stabilization index* of an interval $\mathcal{Q}_1[d, n]$ as the least value of a where the $(\preccurlyeq_a)$-interval is order-isomorphic to $\mathcal{D}[d, n]$.
4. Section 5 studies the size and structure of the initial intervals $\mathcal{Q}_a[1, n]$ for fixed a. We give a decomposition $\mathcal{Q}_a[1, n] = \mathcal{Q}_a^-(n) \cup \mathcal{Q}_a^-(n)$. The n-floor reciprocal map $\Phi_n : \mathcal{Q}_a^+(n) \to \mathcal{Q}_a^-(n)$ is well-defined and injective. In particular, $\left|\mathcal{Q}_a^+(n)\right| \leq \left|\mathcal{Q}_a^-(n)\right|$ and we show by example that this inequality is generally strict for $a \geq 2$. We show $|\mathcal{Q}_a[1, n]| > \sqrt{\frac{n}{a}}$. We present data for $a = 2$ showing that the values $|\mathcal{Q}_a[1, n]|$ are not monotone in n and show considerable fluctuation. We formulate

a heuristic argument suggesting that (on average) $|\mathcal{Q}_a[1, n]|$ should be of size about $2\sqrt{\frac{n}{a}}$.

5. Section 6 shows in Theorem 6.3 that the set of a-floor multiples of a given integer d forms a numerical semigroup, and characterizes the resulting class of numerical semigroups.
6. Section 7 studies the average size of $|\mathcal{Q}_a[1, n]|$ for $1 \le n \le x$. It shows this average size is asymptotic to $\frac{4}{3}\sqrt{\frac{x}{a}}$ as $x \to \infty$. This answer is compatible with the heuristic for the size $|\mathcal{Q}_a[1, n]|$ of expected size $2\sqrt{\frac{n}{a}}$ given in Sect. 5.
7. Section 8 addresses areas for further work.

   (i) Study of the Möbius function of the a-floor quotient order;
   (ii) Structure of the incidence algebra of the a-floor quotient poset;
   (iii) Possible existence of a family of approximate divisor orders interpolating between the 1-floor quotient order and the additive total order on $\mathbb{N}^+$.

2 Preliminary Results on 1-floor Quotient Order

We review properties of the 1-floor quotient order and its relation to the divisibility partial order, shown in [6]. A general reference for facts on partial orders is Stanley [14, Chap. 3].

2.1 *The 1-Floor Quotient Order Compared with Divisibility Order*

The *divisibility partial order* $\mathcal{D} := (\mathbb{N}^+, \mid)$ is a distributive lattice in the sense of Birkhoff. That is, it has a well-defined join function $\vee$ (least common multiple), and a well-defined meet function $\wedge$ (greatest common divisor), which obey the distributive law $x \wedge (y \vee z) = (x \vee y) \wedge (x \vee z)$. It also has a well-defined rank function with the rank of an element d is the length of a minimal set of covering relations to the minimal element 1; the element 1 is assigned rank 0. We let $\mathcal{D}[d, n]$ denote an interval of the divisor order, i.e.

$$\mathcal{D}[d, n] := \{e \in \mathbb{N}^+ : d \mid e \text{ and } e \mid n\}.$$

Each interval of the divisibility partial order with the induced order relation is itself a ranked distributive lattice.

In contrast, it was shown by examining various intervals in [6, Sect. 2] that the 1-floor quotient poset $\mathcal{Q}_1 = (\mathbb{N}^+, \preccurlyeq_1)$ does not have well-defined join (least upper bound) or meet functions (greatest lower bound), so the 1-floor quotient partial order on $\mathbb{N}^+$ is not a lattice. In addition, it does not have a well-defined rank function.

There exist similar examples for the 2-floor quotient relation on $\mathbb{N}^+$, showing that it is not a lattice and does not have a well-defined rank function. We expect the same features hold for the a-floor quotient relation for each $a \geq 1$.

2.2 Scaling-Invariance Properties: Divisor and 1-floor Quotient Order

The finite intervals $\mathcal{D}[d, n]$ of the divisor poset $\mathcal{D} = (\mathbb{N}^+, \mid)$ have a scaling-invariance property: for each pair (d, n), and each $a \geq 1$, there is an isomorphism of partial order intervals

$$\mathcal{D}[d, n] \simeq \mathcal{D}[ad, an]. \tag{2.1}$$

For an interval $I = \mathcal{D}[d, n]$ with endpoints (d, n) we call $w(I) = w(\mathcal{D}[d, n]) = \frac{n}{d}$ the *width* of the interval I. The width is a scale-invariant quantity for the divisor poset: any two intervals having the same width are order-isomorphic. In particular, intervals with the same width have the same cardinality. (If $d \nmid n$ the cardinality is zero.)

The nonempty intervals $\mathcal{Q}_1[d, n]$ of the 1-floor quotient order usually break scale-invariance. That is, given $d \preccurlyeq_1 n$, the order interval $\mathcal{Q}_1[d, n]$ is generally not order-isomorphic to $\mathcal{Q}_1[ad, an]$; in particular, the cardinality of $\mathcal{Q}_1[ad, an]$ varies with $a \geq 1$. If $d \nmid n$ while $d \preccurlyeq_1 n$, then $\mathcal{Q}_1[d, n]$ is nonempty, while for large enough a, we have $\mathcal{Q}_1[ad, an] = \emptyset$. (See Example 4.9 for variation with a.)

2.3 Self-Duality and Almost Complementation Properties: Divisor Order

The divisor poset $\mathcal{D} = (\mathbb{N}^+, \mid)$ has two important symmetry properties, a self-duality property and an (almost) complementation property on each of its finite intervals. We discuss it here for the initial interval $\mathcal{D}[1, n]$.

(1) The divisor poset has on each of its initial divisor intervals $\mathcal{D}[1, n]$ a *self-duality property*, stating that the divisor partial order on $\mathcal{D}[1, n]$ is isomorphic to the dual divisor partial order $\mid^*$ on $\mathcal{D}[1, n]$ given by

$$d_1 \mid^* d_2 \quad \leftrightarrow \quad d_2 \mid d_1.$$

That is:

$$(\mathcal{D}[1, n], \mid) \simeq (\mathcal{D}[1, n], \mid^*).$$

This isomorphism is established using the *n-reciprocal map* $\Phi_n : \mathcal{D}[1, n] \to \mathcal{D}[1, n]$ given by $\Phi_n(d) = \frac{n}{d}$. This map is *order-reversing*: If $d \mid e$ in the divisor order on $\mathcal{D}[1, n]$ then $\frac{n}{e} \mid \frac{n}{d}$. The map Φ_n is also an *involution*: $\Phi_n^{\circ 2}(d) = d$ on the domain $\mathcal{D}(1, n)$, hence it is a bijection.

(2) The divisor partial order $\mathcal{D}$ has an *almost complementation* property of each interval $\mathcal{D}[1, n]$, with respect to the involution Φ_n. It divides the order interval $\mathcal{D}[1, n]$ in two almost complementary parts

$$\mathcal{D}[1, n] = \mathcal{D}^-(n) \bigcup \mathcal{D}^+(n),$$

in which

$$\mathcal{D}^-(n) = \{d \in \mathcal{D}[1, n] : 1 \le d \le \sqrt{n}\}$$

and

$$\mathcal{D}^+(n) := \{d \in \mathcal{D}[1, n] : 1 \le \Phi_n(d) \le \sqrt{n}\}$$

It is clear that $\Phi_n(\mathcal{D}^+(n)) = \mathcal{D}^-(n)$ and $\Phi_n(\mathcal{D}^-(n)) = \mathcal{D}^+(n)$.
We have

$$\mathcal{D}^+(n) \bigcap \mathcal{D}^-(n) = \begin{cases} \{s\} & \text{if } n = s^2 \\ \emptyset & \text{if } n \ne s^2 \quad \text{for any } s \ge 1. \end{cases}$$

If $n = s^2$ then the common element s is the unique fixed point of the involution Φ_n.

The "almost complementation" property is different from the notion of a complementation on the (Birkhoff) lattice $\mathcal{D}[1, n]$, which requires that each element a of the interval has a complement a' whose join with a is the maximal element n and whose meet is the minimal element 1. A divisor order interval $\mathcal{D}[d, n]$ has a complement in the (Birkhoff) lattice sense if and only if $\frac{n}{d}$ is squarefree.

2.4 Self-Duality and Almost Complementation Properties: 1-Floor Quotient Order

The 1-floor quotient order retains weaker forms of self-duality and almost complementation properties, with respect to the *n-floor reciprocal map* $\Phi_n : \mathcal{Q}_1[1, n] \to \mathcal{Q}_1[1, n]$, given by

$$\Phi_n(k) = \left\lfloor \frac{n}{k} \right\rfloor. \tag{2.2}$$

The map Φ_n acts as an involution on the set $\mathcal{Q}_1[1, n]$. The restriction of Φ_n to the smaller domain $\mathcal{D}[1, n]$ of divisors of n agrees with the n-reciprocal map Φ_n defined in Sect. 2.2, since $\lfloor \frac{n}{d} \rfloor = \frac{n}{d}$ for divisors d of n. Moreover, $\mathcal{Q}_1[1, n]$ is the maximal subset of $\{1, 2, \ldots, n\}$ on which Φ_n acts as an involution ([6, Lemma 4.1]).

The involution Φ_n acting on the initial interval $\mathcal{Q}_1[1, n]$ of the floor quotient order is not order-reversing, unlike its action on the divisor order interval $\mathcal{D}[1, n]$. The map Φ_n has instead the weaker property that: the image under Φ_n of two comparable elements $d \preccurlyeq_1 e$ in the floor quotient order either reverses their order $\Phi_n(e) \preccurlyeq_1 \Phi_n(d)$

or else gives two incomparable elements, which we write $\Phi_n(e) \perp_1 \Phi_n(d)$. The incomparable case can only occur when $d \le \sqrt{n}$.

The following result summarizes the almost complementation action on 1-floor quotients. This result is due to Cardinal [2, Proposition 4] (stated in a different notation). The statement below combines [6, Lemma 4.1] and [6, Proposition 4.2].

Proposition 2.1 *(Almost complementation for initial 1-floor quotient intervals) The n-floor reciprocal map Φ_n acts as an involution when restricted to the domain $\mathcal{Q}_1[1, n]$ of all floor quotients of n, i.e.*

$$\Phi_n^{\circ 2}(d) = d \quad \textit{for all} \quad d \in \mathcal{Q}_1[1, n]. \tag{2.3}$$

Given an integer $n \ge 1$, let

$$\mathcal{Q}_1^-(n) = \{d : d \preccurlyeq_1 n,\ d \le \sqrt{n}\} \quad \textit{and} \quad \mathcal{Q}_1^+(n) = \{d : d \preccurlyeq_1 n,\ \left\lfloor \frac{n}{d} \right\rfloor \le \sqrt{n}\}.$$

1. *We have $\mathcal{Q}[1, n] = \mathcal{Q}_1^-(n) \bigcup \mathcal{Q}_1^+(n)$. The map $\Phi_n : k \mapsto \lfloor n/k \rfloor$ restricts to a bijection sending $\mathcal{Q}_1^+(n)$ to $\mathcal{Q}_1^-(n)$ and to a bijection sending $\mathcal{Q}_1^-(n)$ to $\mathcal{Q}_1^+(n)$.*
2. *Moreover, letting $s = \lfloor \sqrt{n} \rfloor$, one has*

$$\mathcal{Q}_1^-(n) = \{1, 2, \ldots, s\}, \quad \textit{and} \quad \mathcal{Q}_1^+(n) = \{n, \left\lfloor \frac{n}{2} \right\rfloor, \left\lfloor \frac{n}{3} \right\rfloor, \ldots, \left\lfloor \frac{n}{s} \right\rfloor\}. \tag{2.4}$$

The intersection of $\mathcal{Q}_1^-(n)$ and $\mathcal{Q}_1^+(n)$ is

$$\mathcal{Q}_1^-(n) \bigcap \mathcal{Q}_1^+(n) = \begin{cases} \{s\} & \textit{if } n < s(s+1) \\ \emptyset & \textit{if } n \ge s(s+1). \end{cases}$$

Remark 2.2 (1) Part (2) of Proposition 2.1 determines the cardinality of $|\mathcal{Q}_1[1, n]|$ as follows. Set $s = \lfloor \sqrt{n} \rfloor$.

(i) If $s^2 \le n < s(s+1)$, then $\lfloor \frac{n}{s} \rfloor = s$ and $|\mathcal{Q}_1[1, n]| = 2s - 1$. The element $s \in \mathcal{Q}_1[1, n]$ is the unique fixed point of the map Φ_n on $\mathcal{Q}_1[1, n]$.

(ii) If $s(s+1) \le n < (s+1)^2$, then $\lfloor \frac{n}{s} \rfloor = s + 1$ and $|\mathcal{Q}_1[1, n]| = 2s$. The map Φ_n has no fixed points on $\mathcal{Q}_1[1, n]$.

(2) The almost complementation property of 1-floor quotients in Proposition 2.1 fails to hold for a-floor quotients with $a \ge 2$, see Sect. 5.

3 The Family of a-Floor Quotient Orders

In this section we prove six equivalent characterizations of the a-floor quotient relation, given in Theorem 3.1. We deduce Theorem 1.3, showing that for each a, the a-floor quotient relation is a partial order that is an approximate divisor order. We then show that the a-floor order refines the b-floor order whenever $a \le b$, and that as $b \to \infty$ the order $(\mathbb{N}^+, \preccurlyeq_b)$ approaches the divisor order on $\mathbb{N}^+$ (Theorem 3.2). It follows that the 1-floor quotient relation is maximally refined among all these partial orders. (That is, it has the largest number of incidences in its order relation).

3.1 Characterizations of a-Floor Quotients

The a-floor quotient has a set of equivalent characterizations, which parallel and generalize those for the 1-floor quotient given in [6, Theorem 3.1].

Theorem 3.1 *(Characterizations of a-floor quotients). For a fixed integer $a \ge 1$, the following conditions on positive integers d, n are equivalent to $d \preccurlyeq_a n$:*

(1) (a-Cutting property) *There exists a positive integer k such that $ad = \left\lfloor \frac{an}{k} \right\rfloor$.*

(2) (a-Covering property) *There exists a positive integer k such that*

$$k[d, d + 1/a) \supseteq [n, n + 1/a),$$

where both sides denote half-open intervals in $\mathbb{R}$.

(3) (a-Intersection property) *There exists a positive integer k such that*

$$k[d, d + 1/a) \cap [n, n + 1/a) \neq \emptyset.$$

(4) (a-Strong remainder property) *There is a positive integer k such that*

$$n = dk + r \quad \text{where the remainder } r \text{ satisfies } 0 \le r < \min\left(d, \frac{k}{a}\right).$$

(5) (a-Tipping-point property) *There holds the strict inequality $\left\lfloor \frac{n}{d} \right\rfloor > \left\lfloor \frac{n}{d + 1/a} \right\rfloor$.*

(6) (a-Reciprocal-duality property) *There holds the identity*

$$ad = \left\lfloor \frac{an}{\lfloor n/d \rfloor} \right\rfloor. \tag{3.1}$$

Equivalently

$$d \le \frac{n}{\lfloor n/d \rfloor} < d + \frac{1}{a}. \tag{3.2}$$

Proof of Theorem 3.1. (1) $\Leftrightarrow$ (2). Suppose $ad = \lfloor \frac{an}{k} \rfloor$ for some k, which is property (1). Then $an = k(ad) + r$ with $0 \le r \le k - 1$, so $[an, an + 1) \subset [kad, kad + k) = k[ad, ad + 1)$. Dividing by a, we obtain

$$[n, n + \tfrac{1}{a}) \subset k[d, d + \tfrac{1}{a}).$$

This is the a-covering property (2) for (d, n). These steps are all reversible, so the properties are equivalent.

(2) $\Leftrightarrow$ (3). The direction (2) $\Rightarrow$ (3) is immediate. For (3) $\Rightarrow$ (2) if $k[d, d + \frac{1}{a})$ contains some $x \in [n, n + \frac{1}{a})$ then, multiplying by a, we have $k[ad, ad + 1)$ contains some $y = ax \in [an, an + 1)$. But then it contains the full half-open interval $[an, an + 1)$, because the interval $k[ad, ad + 1)$ has integer endpoints.

(1) $\Leftrightarrow$ (5). For positive integers d, n, we define the *a-cutting length set* to be the set of integers

$$\mathcal{K}_a(d, n) := \{k \in \mathbb{N}^+ : ad = \left\lfloor \frac{an}{k} \right\rfloor\}. \tag{3.3}$$

Property (1) says that this set is nonempty. We have

$$\mathcal{K}_a(d, n) = \{k \in \mathbb{N}^+ : ad \leq \frac{an}{k} < ad + 1\} = \{k \in \mathbb{N}^+ : d \leq \frac{n}{k} < d + \frac{1}{a}\}.$$

The lower bound $d \leq \frac{n}{k}$ is equivalent to $k \leq \left\lfloor \frac{n}{d} \right\rfloor$. The upper bound $\frac{n}{k} < d + 1/a$ is equivalent to $\left\lfloor \frac{n}{d+1/a} \right\rfloor < k$. Therefore

$$\mathcal{K}_a(d, n) = \{k \in \mathbb{N}^+ : \left\lfloor \frac{n}{d + 1/a} \right\rfloor < k \leq \left\lfloor \frac{n}{d} \right\rfloor\}, \tag{3.4}$$

so $\mathcal{K}_a n$ is nonempty if and only if $\left\lfloor \frac{n}{d+1/a} \right\rfloor < \left\lfloor \frac{n}{d} \right\rfloor$, which is the tipping point property (5).

(1) $\Leftrightarrow$ (6). The direction (6) $\Rightarrow$ (1) is clear. Now suppose (1) holds. The proof of (1) $\Leftrightarrow$ (5) established (3.4), showing that

$$ad = \left\lfloor \frac{an}{k} \right\rfloor \quad \text{is equivalent to} \quad \left\lfloor \frac{n}{d + 1/a} \right\rfloor < k \leq \left\lfloor \frac{n}{d} \right\rfloor.$$

Property (1) says that $\mathcal{K}_a(d, n) = \{k \in \mathbb{N}^+ : ad = \left\lfloor \frac{an}{k} \right\rfloor\}$ is nonempty. Thus if $\mathcal{K}_a(d, n)$ is nonempty then it must contain $k = \left\lfloor \frac{n}{d} \right\rfloor$, which is the involution-duality property (6).

(4) $\Leftrightarrow$ (6). Suppose property (4) holds, so there is a positive integer k such that $n = dk + r$ where $0 \leq r < \min(d, k/a)$. Then $\left\lfloor \frac{n}{d} \right\rfloor = \left\lfloor k + \frac{r}{d} \right\rfloor = k$, hence

$$\left\lfloor \frac{an}{\lfloor n/d \rfloor} \right\rfloor = \left\lfloor \frac{adk + ar}{k} \right\rfloor = \left\lfloor ad + \frac{ar}{k} \right\rfloor = ad$$

which is the identity in property (6).

It remains to show that if property (4) does not hold for k, then property (6) does not hold. For a given (d, n), dividing n by d with remainder gives $n = dk + r$ for integers k, r with $0 \le r < d$ and $k = \lfloor n/d \rfloor$. Property (4) does not hold exactly when the remainder r satisfies $r \ge k/a$. In that case

$$\frac{an}{\lfloor n/d \rfloor} = \frac{adk + ar}{k} = ad + \frac{ar}{k} \ge ad + 1,$$

which implies $\left\lfloor \frac{an}{\lfloor n/d \rfloor} \right\rfloor > ad$, so that property (6) does not hold. □

We can now prove that each a-floor quotient relation is a partial order that is an approximate divisor order.

Proof of Theorem 1.3. We show all a-floor quotient orders are partial orders. This was already shown for $a = 1$ in [6]. Assuming the property for $a = 1$, the partial order property is inherited for general $a \ge 2$ by the definition that $d \preccurlyeq_a n$ if and only if $ad \preccurlyeq_1 an$. The restriction of the partial order $\preccurlyeq_1$ to the subset $a\mathbb{N}^+$ is automatically a partial order.

Alternatively, the partial order properties of reflexivity, symmetry and transitivity can be deduced directly from characterizations given in Theorem 3.1 above.

If $d \mid n$, then $ad \mid an$, so $ad \preccurlyeq_1 an$, which by definition implies $a \preccurlyeq_a n$. Thus $\preccurlyeq_a$ refines the divisor order.

If $d \preccurlyeq_a n$, then the condition in Theorem 3.1 (1) implies that $d \le n$ in the additive order. Hence $\preccurlyeq_a$ is refined by the additive total order. We conclude that $\mathcal{Q}_a = (\mathbb{N}^+, \preccurlyeq_a)$ is an approximate divisor order, for any $a \ge 1$. □

3.2 Hierarchy of a-Floor Quotient Orders: Proof of Theorem 1.4

We derive Theorem 1.4 from the following result, which reformulates its condition (2) in a more convenient form.

Theorem 3.2 *(a-floor quotient order hierarchy). The a-floor quotient partial orders $\mathcal{Q}_a$ have the following properties.*

(1) If $a < b$ (in the additive order) then the a-floor quotient partial order $\mathcal{Q}_a$ strictly refines the b-floor quotient partial order $\mathcal{Q}_b$.

(2) The intersection of the a-floor quotient partial orders $\mathcal{Q}_a$ over all $a \ge 1$ is the divisor partial order $\mathcal{D} = (\mathbb{N}^+, \mid)$. Namely,

$$\bigcap_{a \ge 1} \{(d, n) \in \mathbb{N}^+ \times \mathbb{N}^+ : d \preccurlyeq_a n\} = \{(d, n) \in \mathbb{N}^+ \times \mathbb{N}^+ : d \mid n\}.$$

Proof (1) Theorem 3.1 (6) states that

$$d \preccurlyeq_a n \quad \text{if and only if} \quad d \le \frac{n}{\lfloor n/d \rfloor} < d + \frac{1}{a}. \tag{3.5}$$

If $d \preccurlyeq_b n$ holds, it follows that $d \preccurlyeq_a n$ for all integers $a \le b$. Thus $\preccurlyeq_a$ refines $\preccurlyeq_b$.

It remains to show that for $a < b$ the refinement is strict, i.e. there exists a pair (d, n) having $d \preccurlyeq_a n$ but $d \not\preccurlyeq_b n$. A suitable choice is $(d, n) = (2, 2b + 1)$, so $\lfloor n/d \rfloor = b$. The condition (3.5) becomes

$$2 \preccurlyeq_a 2b + 1 \quad \text{if and only if} \quad 2 \le \frac{2b+1}{b} = 2 + \frac{1}{b} < 2 + \frac{1}{a}.$$

Thus $2 \preccurlyeq_a 2b + 1$ assuming $a < b$, but $2 \not\preccurlyeq_b 2b + 1$.

(2) We show that

$$\bigcap_{a \ge 1} \{(d, n) \in \mathbb{N}^+ \times \mathbb{N}^+ : d \preccurlyeq_a n\} = \{(d, n) \in \mathbb{N}^+ \times \mathbb{N}^+ : d \mid n\}. \tag{3.6}$$

Since each a-floor quotient order $\preccurlyeq_a$ refines the divisor order, we have the inclusion

$$\bigcap_{a \ge 1} \{(d, n) \in \mathbb{N}^+ \times \mathbb{N}^+ : d \preccurlyeq_a n\} \supseteq \{(d, n) \in \mathbb{N}^+ \times \mathbb{N}^+ : d \mid n\}.$$

On the other hand, if $d \preccurlyeq_a n$ for all $a \ge 1$, then condition (3.5) implies that

$$\frac{n}{\lfloor n/d \rfloor} = d \quad \Leftrightarrow \quad n = d \lfloor n/d \rfloor,$$

so d divides n. This shows the reverse inclusion. □

4 Hierarchy of a-Floor Quotient Orders: Stabilization Bounds

We study the effect of varying a on intervals of the a-floor quotient order.

4.1 a-Floor Scaling Sets

Definition 4.1 For a pair (d, n) of positive integers, the *a-floor scaling set* $\mathcal{S}(d, n)$ is the set of a for which d is an a-floor quotient of n, i.e.

$$\mathcal{S}(d, n) := \{a \in \mathbb{N}^+ : \ d \preccurlyeq_a n\}.$$

The a-floor scaling set $\mathcal{S}(d, n)$ may be infinite, finite, or empty.

Example 4.2 We have

$$\mathcal{S}(5, 15) = \mathbb{N}^+, \quad \mathcal{S}(5, 16) = \{1, 2\}, \quad \mathcal{S}(5, 17) = \{1\}, \quad \mathcal{S}(5, 18) = \emptyset.$$

Definition 4.3 Given positive integers d and n, the *quotient discrepancy* $\delta(d, n)$ is

$$\delta(d, n) = n - d \left\lfloor \frac{n}{d} \right\rfloor. \tag{4.1}$$

Note that $\delta(d, n)$ always satisfies $0 \le \delta(d, n) < d$. We have $\delta(d, n) = 0$ if and only if d divides n, and $\delta(d, n) < \lfloor n/d \rfloor$ if and only if d is a floor quotient of n. Moreover, if d is a floor quotient of n, then $\delta(d, n) = \delta(\lfloor n/d \rfloor, n)$.

Theorem 4.4 *(Structure of a-floor scaling sets). For positive integers d and n, the set $\mathcal{S}(d, n)$ of all a where $d \preccurlyeq_a n$ is either $\mathbb{N}^+$, or an initial interval $\mathcal{A}[1, m]$ of $\mathbb{N}^+$ of the additive order $\mathcal{A} = (\mathbb{N}^+, \le)$, or the empty set. The following trichotomy holds.*

(i) If $d \mid n$, then $\mathcal{S}(d, n) = \mathbb{N}^+$.
(ii) If $d \preccurlyeq_1 n$ and $d \nmid n$, then $\mathcal{S}(d, n) = \{1, 2, 3, \cdots, \tilde{a}(d, n)\}$, where $\tilde{a}(d, n)$ is the largest integer such that

$$\tilde{a}(d, n) < \frac{\lfloor n/d \rfloor}{\delta(d, n)}$$

and $\delta(d, n)$ is defined by (4.1).
(iii) If $d \not\preccurlyeq_1 n$, then $\mathcal{S}(d, n) = \emptyset$.

Proof Theorem 1.4 (1) implies that $\mathcal{S}(d, n)$ must be an initial interval in the additive order, i.e. if $b \in \mathcal{S}(d, n)$ and $a < b$ then $a \in \mathcal{S}(d, n)$.

(i) Suppose $d \mid n$. Then $d \preccurlyeq_a n$ for all $a \ge 1$, by Theorem 1.3. Therefore $\mathcal{S}(d, n) = \mathbb{N}^+$.

(ii) Suppose $d \preccurlyeq_1 n$ but $d \nmid n$. Using criterion (3.5), we have $a \in \mathcal{S}(d, n)$ if and only if

$$\frac{n}{\lfloor n/d \rfloor} < d + \frac{1}{a} \quad \Leftrightarrow \quad a < \frac{\lfloor n/d \rfloor}{n - \lfloor n/d \rfloor d} =: \frac{\lfloor n/d \rfloor}{\delta(d, n)}.$$

The quantity $\delta(d, n) = n - d \lfloor n/d \rfloor$ is positive by the assumption that d does not divide n. In this case we define $\tilde{a}(d, n)$ to the largest integer that is strictly less than $\frac{\lfloor n/d \rfloor}{\delta(d,n)}$.

(iii) If $d \not\preccurlyeq_1 n$, then Theorem 3.2 implies that $d \not\preccurlyeq_a n$ for all $a \ge 1$. Therefore in this case $\mathcal{S}(d, n) = \emptyset$. □

4.2 Hierarchy Bounds for Stabilization to Divisor Order

Theorem 4.5 *(a-floor quotient order stabilization). The a-floor quotient partial order on the interval $\mathcal{Q}_a[d, n]$, is identical to the divisor partial order $\mid$ on the interval $\mathcal{D}[d, n]$, whenever $a \ge \left\lfloor \frac{n}{d} \right\rfloor$. That is,*

$$(\mathcal{Q}_a[d,n], \preccurlyeq_a) = (\mathcal{D}[d,n], |) \quad \textit{for all} \quad a \geq \left\lfloor \frac{n}{d} \right\rfloor.$$

If in addition $d \mid n$, then the a-floor quotient partial order on $(\mathcal{Q}_a[d,n], \preccurlyeq_a)$ is identical to the divisor partial order on $(\mathcal{D}[d,n], |)$, for $a \geq \left\lfloor \frac{n}{d+1} \right\rfloor$.

Proof The result follows using Theorem 4.4 (ii). Since all the order relations in $\mathcal{D}[d,n]$ are included in $\mathcal{Q}_a[d,n]$ for each $a \geq 1$, if a is such that $\mathcal{Q}_a[d,n]$ is not order-isomorphic to $\mathcal{D}[d,n]$, then there must exist $e, m \in \mathcal{Q}_a[d,n]$ having an extra order relation

$$d \preccurlyeq_a e \preccurlyeq_a m \preccurlyeq_a n \quad \text{with} \quad e \nmid m.$$

The assumption $e \nmid m$ implies that $\delta(e,m) = m - e\lfloor m/e \rfloor \geq 1$. Since $a \in \mathcal{S}(e,m)$, Theorem 4.4 (ii) implies that necessarily

$$a < \frac{\lfloor m/e \rfloor}{\delta(e,m)} \leq \left\lfloor \frac{m}{e} \right\rfloor. \tag{4.2}$$

Since $m \leq n$ and $e \geq d$, this implies $a < \left\lfloor \frac{n}{d} \right\rfloor$ as claimed in the first part of the theorem.

For the second part of the theorem, assuming $d \mid n$, we treat two cases. Any pair (e,m) with $e \nmid m$ cannot be (d,n), so either $e \geq d+1$ and $m = n$, or else $e \geq d$ and $m \leq \frac{n}{2}$.

Case 1. In the first case $(e,m) := (e,n)$ has $e \nmid n$. Let $k = k_{min}(d,n)$ denote the least integer $k \geq d$ such that $k \nmid n$, so that $e \geq k_{min}(d,n)$. Now (4.2) gives, for $k = k_{min}(d,n)$,

$$a < \frac{\lfloor n/e \rfloor}{\delta(e,n)} \leq \left\lfloor \frac{n}{e} \right\rfloor \leq \left\lfloor \frac{n}{k} \right\rfloor.$$

Thus for all pairs (e,n) occurring this subcase stabilization to the divisor order occurs for $a \geq \left\lfloor \frac{n}{k_{min}(d,n)} \right\rfloor$ Since $k_{min}(d,n) \geq d+1$, we obtain stabilization for these (e,m) for all $a \geq \left\lfloor \frac{n}{d+1} \right\rfloor$.

Case 2. In the second case (e,m) satisfies $e \nmid m$ with $m \leq \frac{n}{2}$. Now (4.2) gives

$$a < \frac{\lfloor m/e \rfloor}{\delta(e,m)} \leq \left\lfloor \frac{m}{e} \right\rfloor \leq \left\lfloor \frac{n}{2e} \right\rfloor \leq \left\lfloor \frac{n}{2d} \right\rfloor.$$

Thus stabilization to the divisor order occurs for all (e,m) occurring in this subcase in the range $a \geq \left\lfloor \frac{n}{2d} \right\rfloor$.

We conclude that for $d \mid n$ we have stabilization for $a \geq \max(\left\lfloor \frac{n}{k_{min}(d,n)} \right\rfloor, \left\lfloor \frac{n}{2d} \right\rfloor)$. Since $k_{min}(n,d) \geq d+1$ hence $\left\lfloor \frac{n}{k_{min}} \right\rfloor \leq \left\lfloor \frac{n}{d+1} \right\rfloor$, and $\left\lfloor \frac{n}{2d} \right\rfloor \leq \left\lfloor \frac{n}{d+1} \right\rfloor$, we obtain stabilization of $(\mathcal{Q}_a[d,n], \preccurlyeq_a) = (\mathcal{D}[d,n], |)$ to the divisor order $(\mathcal{D}[d,n], |)$ for $a \geq \left\lfloor \frac{n}{d+1} \right\rfloor$. □

A weaker condition than equality-as-posets asks how large a must be to guarantee an equality of the underlying sets of $\mathcal{Q}_a[d,n]$ and $\mathcal{D}[d,n]$.

Example 4.6 The equality as sets of $\mathcal{Q}_a[1,n]$ and $\mathcal{D}[1,n]$ can occur without isomorphism of the order relations.

For example, for $a = 5$ the interval $\mathcal{Q}_5[1,26] = \{1,2,13,26\}$ is equal to $\mathcal{D}[1,26]$ as a set. However, the order relation $\preccurlyeq_5$ restricts to a total order $1 \preccurlyeq_5 2 \preccurlyeq_5 13 \preccurlyeq_5 26$, whereas the divisor order on $\{1,2,13,26\}$ is not a total order. On this example Theorem 4.5 has $k_{min} = 3$, $2d = 2$, and asserts agreement with the divisor order for $a \geq \left\lfloor \frac{26}{2} \right\rfloor = 13$. The stabilization with the divisor order in fact occurs at $a = 7$.

The following example gives intervals $\mathcal{Q}_a[1,10]$ in the a-floor quotient order for various a. Recall that $\mathcal{Q}_a[d,n] := \frac{1}{a}\left(\mathcal{Q}_1[ad,an] \bigcap a\mathbb{N}^+\right)$.

Example 4.7 The a-floor quotient interval $\mathcal{Q}_a[1,10]$ is shown in the table below, for $a = 1,2,3,\ldots$.

a	elements in $\mathcal{Q}_a[1,10]$	size
1	$\{1,2,3,5,10\}$	5
2	$\{1,2,3,5,10\}$	5
3	$\{1,2,5,10\}$	4
$a \geq 4$	$\{1,2,5,10\}$	4

Theorem 4.5 asserts that stabilization to the divisor order occurs by $a = \left\lfloor \frac{n}{2} \right\rfloor = 5$. Stabilization occurs at $a \geq 3$.

Example 4.8 The a-floor quotient interval $\mathcal{Q}_a[1,11]$ is shown in the table below, for $a = 1,2,3\ldots$

a	elements in $\mathcal{Q}_a[1,11]$	size
1	$\{1,2,3,5,11\}$	5
2	$\{1,2,11\}$	3
3	$\{1,2,11\}$	3
$a \geq 5$	$\{1,11\}$	2

Theorem 4.5 asserts that stabilization to the divisor order occurs by $a = \left\lfloor \frac{n}{k} \right\rfloor = \left\lfloor \frac{11}{2} \right\rfloor = 5$.

The rescaled interval $a\mathcal{Q}_a[1,n]$ is always contained in the set $\mathcal{Q}_1[ad,an]$; compare the Example 4.7 for $\mathcal{Q}_a[1,10]$ with the full interval $\mathcal{Q}_1[ad,an]$ (from Example 5.2 of [6]), given below.

Example 4.9 The floor quotient intervals $\mathcal{Q}[a, 10a]$ for various a are shown in the table below.

a	elements in $\mathcal{Q}_1[a, 10a]$	size
1	$\{1, 2, 3, 5, 10\}$	5
2	$\{2, 4, 5, 6, 10, 20\}$	6
3	$\{3, 6, 7, 10, 15, 30\}$	6
4	$\{4, 8, 13, 20, 40\}$	5
9	$\{9, 18, 45, 90\}$	4

4.3 *Improved Bounds on Hierarchy Stabilization*

The *set-stabilization threshold* of the integer n is the smallest value of a such that the interval $\mathcal{Q}_a[1, n]$ is equal, as a set, to the interval $\mathcal{D}[1, n]$. The *poset-stabilization threshold* of n is the smallest value of a such that the interval $\mathcal{Q}_a[1, n]$, as a poset, is equal to the poset interval $\mathcal{D}[1, n]$.

In general, the set-stabilization threshold is smaller than the poset-stabilization threshold. See Fig. 1 for a comparison of the set- and poset-stabilization threshold for $n = 1, 2, \ldots, 100$.

5 Action of Floor Reciprocal Map: Broken Reciprocal Duality

The reciprocal-duality property for the 1-floor quotient partial order on $\mathcal{Q}_1[1, n]$ is weakened to the property (6) given in Theorem 3.1. In particular, the sets $\mathcal{Q}_a^+(n)$ and

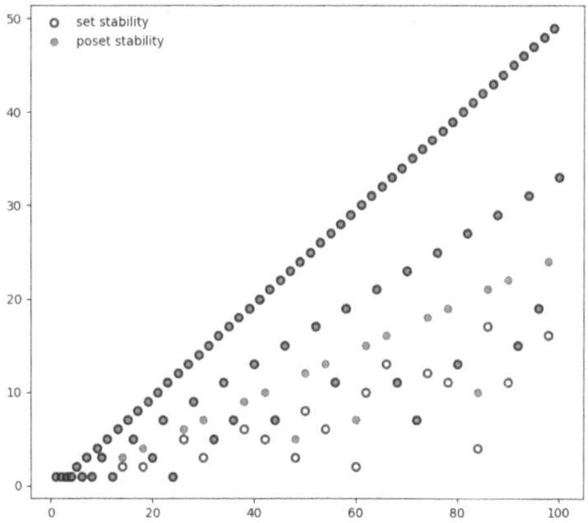

Fig. 1 Plot of set- and poset-stabilization thresholds for $\mathcal{Q}_a[1, n]$ for $1 \leq n \leq 100$. The horizontal axis is the n-variable and the vertical axis is the a-variable

$\mathcal{Q}_a^-(n)$ are not in 1-to-1 correspondence. The reciprocal duality symmetry (at the level of sets) on the initial interval $\mathcal{Q}_a[1,n]$ for $n \geq 1$ is broken for $a \geq 2$.

We show the n-floor reciprocal map $\Phi_n : \mathcal{Q}_a^+(n) \to \mathcal{Q}_a^-(n)$ is injective, but in the reverse direction the map Φ_n only sends $\mathcal{Q}_a^-(n)$ into $\mathcal{Q}_1^+(n)$. The image of $\mathcal{Q}_a^-(n)$ is always contained in $\mathcal{Q}_1^+(n)$, but for $a \geq 2$, is usually not in $\mathcal{Q}_a^+(n)$.

5.1 Asymmetry of Initial a-Floor Quotient Intervals

We have $\mathcal{Q}_a[1,n] \subseteq \mathcal{Q}_1[1,n]$, which is proved in Theorem 3.2. It guarantees that each $d \in \mathcal{Q}_a[1,n]$ can be written as $d = \lfloor \frac{n}{k} \rfloor$ for a unique $k \in \mathcal{Q}_1[1,n]$; however it need not hold that $k \in \mathcal{Q}_a[1,n]$.

Since the a-floor quotient interval $\mathcal{Q}_a[1,n]$ is contained in $\mathcal{Q}_1[1,n]$, we have a decomposition into "small" and "large" quotients

$$\mathcal{Q}_a[1,n] = \mathcal{Q}_a^-(n) \bigcup \mathcal{Q}_a^+(n),$$

with

$$\mathcal{Q}_a^-(n) := \mathcal{Q}_a[1,n] \bigcap \mathcal{Q}_1^-(n) = \{d \in \mathcal{Q}_a[1,n] : d \leq \sqrt{n}\}$$

and

$$\mathcal{Q}_a^+(n) := \mathcal{Q}_a[1,n] \bigcap \mathcal{Q}_1^+(n) = \{d \in \mathcal{Q}_a[1,n] : \left\lfloor \frac{n}{d} \right\rfloor \leq \sqrt{n}\}.$$

Lemma 5.1 *The function Φ_n maps the set $\mathcal{Q}_a^+(n)$ injectively into the set $\mathcal{Q}_a^-(n)$. Consequently, for all $n \geq 1$,*

$$\left|\mathcal{Q}_a^+(n)\right| \leq \left|\mathcal{Q}_a^-(n)\right|. \tag{5.1}$$

Proof Suppose $d \in \mathcal{Q}_a[1,n]$. Then Property (4) of Theorem 3.1 says this is equivalent to

$$n = dk + r \quad \text{where} \quad k = \Phi_n(d) = \left\lfloor \frac{n}{d} \right\rfloor \quad \text{and} \quad 0 \leq r < \min(d, k/a).$$

Note that $k = \Phi_n(d)$ and $d \preccurlyeq_1 n$ implies $d = \Phi_n(k)$, by Proposition 2.1.

If $d \in \mathcal{Q}_a^+(n)$, then $k \leq d$ so the bound $r < k/a$ also implies $r < d/a$. Moreover, since $a \geq 1$ the bound $r < k/a$ implies $r < k$. Thus

$$n = kd + r \quad \text{where} \quad d = \Phi_n(k) = \left\lfloor \frac{n}{k} \right\rfloor \quad \text{and} \quad 0 \leq r < \min(k, d/a).$$

By Property (4) of Theorem 3.1, this implies $k = \Phi_n(d) \in \mathcal{Q}_a[1,n]$. Since $k \leq \sqrt{n}$, we have $k \in \mathcal{Q}_a^-(n)$ as desired.

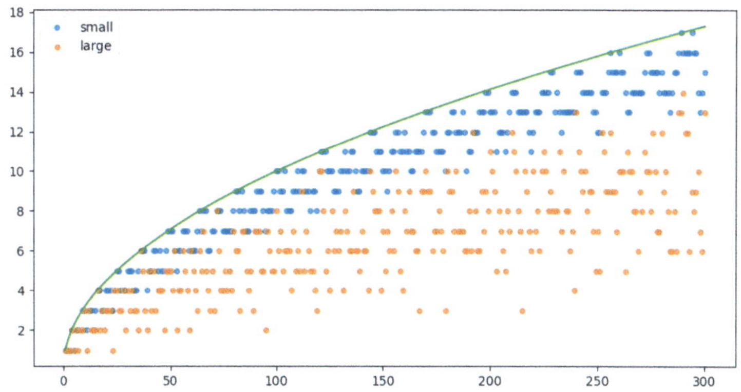

Fig. 2 Plot of the number of small 2-floor quotients $\left|\mathcal{Q}_2^-(n)\right|$ (blue) versus number of large 2-floor quotients $\left|\mathcal{Q}_2^+(n)\right|$ (orange) for n up to 300. The upper envelope line is $y = \sqrt{n}$

Thus Φ_n maps $\mathcal{Q}_a^+(n)$ into $\mathcal{Q}_a^-(n)$, and injectivity is inherited from the behavior of Φ_n on $\mathcal{Q}_1[1, n]$. □

For $a \geq 2$ the set $\mathcal{Q}_a^+(n)$ is generally significantly smaller than the set $\mathcal{Q}_a^-(n)$, see Fig. 2 for $a = 2$. Since these sets are contained in $\mathcal{Q}_1^+(n)$ and $\mathcal{Q}^-(n)$, respectively, they are each of size at most $\sqrt{n}$.

Lemma 5.2 *Let $a \geq 1$. For every $n \geq 1$, all $m \in [1, \sqrt{\frac{n}{a}}]$ belong to $\mathcal{Q}_a[1, n]$. In particular,*

$$\left|\mathcal{Q}_a^-(n)\right| \geq \left\lfloor \sqrt{\frac{n}{a}} \right\rfloor. \tag{5.2}$$

Additionally $|\mathcal{Q}_a[1, n]| > \sqrt{\frac{n}{a}}$ if $n > 1$.

Proof Let m be in the range $1 \leq m \leq \left\lfloor \sqrt{\frac{n}{a}} \right\rfloor$. We wish to show that $m \preccurlyeq_a n$, so by definition of the a-floor quotient relation, it suffices to show that $am \preccurlyeq_1 an$.

The condition $m \leq \sqrt{n/a}$ implies that $am \leq \sqrt{an}$. It follows that $am \preccurlyeq_1 an$ as desired, by applying Proposition 2.1.

For the final assertion, if $n \geq 2$ then $\mathcal{Q}_a[1, n]$ includes the element $n \in \mathcal{Q}_a^+(n) \setminus \mathcal{Q}_1^-(n)$. □

Lemma 5.3 *For each $a \geq 2$, there is some n such that $\mathcal{Q}_a^-(n)$ is strictly larger in size than $\mathcal{Q}_a^+(n)$.*

Proof Suppose $n = a^2 + a + 1$. Then $a \in \mathcal{Q}_a^-(n)$, while $a + 1 = \Phi_n(a) \notin \mathcal{Q}_a^+(n)$. □

5.2 Total Size of Initial a-Floor Quotient Intervals: Heuristic

We study the total size of the initial a-floor quotient interval $|\mathcal{Q}_a[1, n]|$, as n varies. In the case $a = 1$ these values are non-decreasing in n, and their total number is $2\sqrt{n} + O(1)$.

We present data for 2-floor quotients in Fig. 3. The interval sizes $|\mathcal{Q}_2[1, n]|$ are scattered, exhibiting non-monotonic behavior. The occurrence of some scatter of values is not surprising, in that as $a \to \infty$ the interval $\mathcal{Q}_a[1, n]$ converges to the set $\mathcal{D}[1, n]$ consisting of divisors of n, and the number of divisors of n varies non-monotonically in n.

We formulate a probabilistic heuristic that makes a prediction for the (expected) size of the sets $\mathcal{Q}_a^+(n)$, $\mathcal{Q}_a^-(n)$ and $\mathcal{Q}_a[1, n]$, viewing a as fixed and $n \gg 1$.

Recall that by the reciprocal-duality property (6) of Theorem 3.1 the value d is an a-floor quotient of n if and only if, writing $k = \lfloor n/d \rfloor$, we have $\lfloor an/k \rfloor = ad$. We have three regions to consider. (Note that the "small" and "large" terminology here differs from previous usage for $\mathcal{Q}_a^-(n)$ and $\mathcal{Q}_a^+(n)$.)

(1) (small region) $1 \le d \le \sqrt{\frac{n}{a}}$.
In this case we have $d \in \mathcal{Q}_a[1, n]$ by Lemma 5.2. Thus $\mathcal{Q}_a[1, n]$ contains exactly

$$E_{small} = \left\lfloor \sqrt{\frac{n}{a}} \right\rfloor = \sqrt{\frac{n}{a}} + O(1)$$

such elements in this region, all of them in $\mathcal{Q}_a^-(n)$.

(2) (medium region) $\sqrt{\frac{n}{a}} < d \le \sqrt{n}$.
In this region, the cutting length $k = \lfloor n/d \rfloor$ is in the range $\sqrt{n} \le k < \lfloor \sqrt{an} \rfloor$. The number of integers k in this range is $(\sqrt{a} - 1)\sqrt{n} + O(1)$.

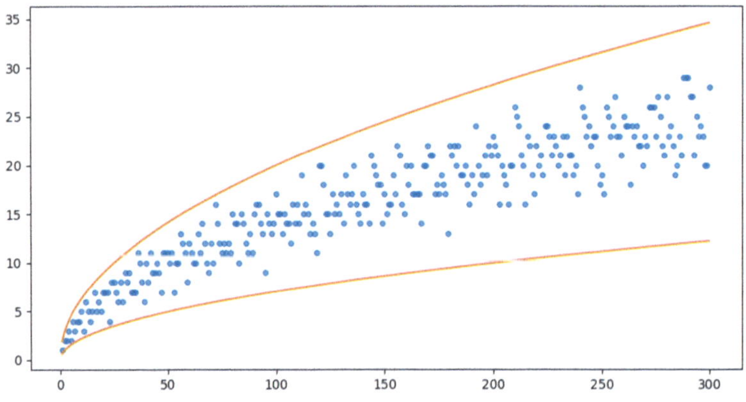

Fig. 3 Plot of the total number of 2-floor quotients $|\mathcal{Q}_2[1, n]|$ for n up to 300. The upper smooth curve is $y = 2\sqrt{n}$ and the lower curve is $y = \sqrt{n/2}$

The element d corresponding to k is an a-floor quotient if and only if $\lfloor \frac{an}{k} \rfloor$ is a multiple of a, in which case $d = \frac{1}{a} \lfloor \frac{an}{k} \rfloor$. (Note that over this range of k, all values $\lfloor \frac{an}{k} \rfloor$ are distinct.) As a heuristic, for fixed n we introduce a probabilistic model for accepting the cutting length k as generating a "random a-floor quotient" (d', n) as follows. For each k we replace the value $r := \lfloor an/k \rfloor \pmod a$ by an (independently drawn) random residue $\tilde{r} \pmod a$ uniformly distributed in $\{0, 1, \ldots, a-1\}$. If $\tilde{r} \equiv 0 \pmod a$, then we accept k as giving a "random a-floor quotient," which would be $d' = \frac{1}{a}(\lfloor \frac{an}{k} \rfloor - r)$. The model will treat all d' as distinct, although repeated values of d' could occur. That is, for a sample of the model we count the number of k in $\sqrt{n} \le k < \lfloor \sqrt{an} \rfloor$ having $\tilde{r} \equiv 0 \pmod a$. The probability that $\tilde{r}$ is a multiple of a is $\frac{1}{a}$, so the expected number of "random a-floor quotients" in this region is

$$E_{medium} = \frac{1}{a} |\{k : \sqrt{n} \le k < \lfloor \sqrt{an} \rfloor\}| = \left(\frac{1}{\sqrt{a}} - \frac{1}{a}\right)\sqrt{n} + O\left(\frac{1}{a}\right).$$

(3) (large region) $\sqrt{n} \le d \le n$.
In this region, the cutting length $k = \lfloor \frac{n}{d} \rfloor$ is in the range $1 \le k \le \lfloor \sqrt{n} \rfloor$, which consists of $\sqrt{n} + O(1)$-many integers. We apply again the heuristic probabilistic model of "random a-floor quotients" above for $1 \le k \le \lfloor \sqrt{n} \rfloor$, finding that the expected number of "random a-floor quotients" assigned from this region is

$$E_{large} = \frac{1}{a} \left|\{k : 1 \le k \le \lfloor \sqrt{n} \rfloor\}\right| = \frac{1}{a}\sqrt{n} + O\left(\frac{1}{a}\right).$$

The heuristic (2) (taken together with the rigorous bound (1)) suggests that the number of elements of $\mathcal{Q}_a^-(n)$ is approximately $(\frac{2}{\sqrt{a}} - \frac{1}{a})\sqrt{n}$. The heuristic (3) suggests that the number of elements of $\mathcal{Q}_a^+(n)$ is approximately $\frac{1}{a}\sqrt{n}$. Combining the heuristics on the two ranges predicts that the expected number of elements of $\mathcal{Q}_a[1, n]$ is approximately $2\sqrt{\frac{n}{a}}$.

In comparison to the heuristic for $\mathcal{Q}_a[1, n]$, Lemma 5.2 yields a rigorous lower bound for the number of elements in $\mathcal{Q}_a[1, n]$, that is half the value predicted combining (1), (2), (3) for the "expected" number of elements.

Remark 5.4 (1) It is an interesting question, not addressed here, to determine information on the size of the scatter of values for total a-floor quotients $|\mathcal{Q}_a[1, n]|$ visible in Fig. 3, around their "expected size" $f_a(n) := 2\sqrt{\frac{n}{a}}$.

(2) One may also ask whether, for fixed a, asymptotic formulas $|\mathcal{Q}_a^-(n)| \sim (\frac{2}{\sqrt{a}} - \frac{1}{a})\sqrt{n}$ and $|\mathcal{Q}_a^+(n)| \sim \frac{1}{a}\sqrt{n}$ might hold as $n \to \infty$.

6 Floor Multiples for a-Floor Quotient Orders

Recall that the 1-floor multiples of a fixed positive integer d are the integers n such that $d \preccurlyeq_1 n$. In [6] it was shown that the 1-floor multiples of d form a numerical semigroup. Here we extend the results to a-floor multiples of d.

Definition 6.1 We call n an *a-floor multiple of d* if d is an a-floor quotient of n. We let $\mathcal{M}_a(d)$ denote the set of floor multiples of d, i.e.

$$\mathcal{M}_a(d) := \{n \in \mathbb{N}^+ : d \preccurlyeq_a n\}.$$

Recall that a *numerical semigroup* is a subset of $\mathbb{N}^+$ that is closed under addition and contains all but finitely many elements of $\mathbb{N}^+$. The *Frobenius number* of a numerical semigroup is the largest integer not belonging to the semigroup. We determine the Frobenius numbers and minimal generating sets of these numerical semigroups. For general information on numerical semigroups and of the Frobenius number of a numerical semigroup, see Ramírez Alfonsín [8] and Assi and Garcia-Sánchez [1].

6.1 Semigroup of 1-Floor Multiples

The numerical semigroup property was proved for the case $a = 1$ in [6, Theorem 3.8].

Theorem 6.2 *(1-floor multiple numerical semigroup). The set $\mathcal{M}_1(d)$ of floor multiples of d is a numerical semigroup. It has the following properties.*

(1) The largest integer not in the numerical semigroup $\mathcal{M}_1(d)$, its Frobenius number, is $d^2 - 1$.
(2) There are exactly $\frac{1}{2}(d-1)(d+2) = \frac{1}{2}(d^2+d-2)$ positive integers not in $\mathcal{M}_1(d)$.
(3) The minimal generating set of $\mathcal{M}_1(d)$ has d generators, with $\mathcal{M}_1(d) = \langle \gamma_0, \gamma_1, \ldots, \gamma_{d-1} \rangle$ where $\gamma_j = (j+1)d + j = j(d+1) + d$ for $0 \le j \le d-1$.

6.2 Semigroup of a-Floor Multiples

We treat the case of general a.

Theorem 6.3 *(a-floor multiple numerical semigroup). The set $\mathcal{M}_a(d)$ of a-floor multiples of d is a numerical semigroup. It has the following properties.*

(1) The largest integer not in $\mathcal{M}_a(d)$, its Frobenius number, is $(d-1)(ad+1)$.
(2) The number of positive integers not in $\mathcal{M}_a(d)$ is $\frac{1}{2}(d-1)(ad+2)$.

(3) The minimal generating set of $\mathcal{M}_a(d)$ is $\langle \gamma_0, \gamma_1, \ldots, \gamma_{d-1} \rangle$ where

$$\gamma_j = (ja+1)d + j = j(ad+1) + d \quad \textit{for} \quad 0 \le j \le d-1.$$

The proof of Theorem 6.3 follows the same outline as the $a = 1$ case given in [6, Theorem 4.2].

Lemma 6.4 *The set $\mathcal{M}_a(d)$ of a-floor multiples of d is closed under addition.*

Proof By definition of the relation $\preccurlyeq_a$, the map $n \mapsto an$ gives a bijection from $\mathcal{M}_a(d)$ to the intersection $a\mathbb{N}^+ \cap \mathcal{M}_1(ad)$. This map respects addition, so it suffices to check that the image of $\mathcal{M}_a(d)$ is closed under addition. The intersection $a\mathbb{N}^+ \cap \mathcal{M}_1(ad)$ is closed under addition because $a\mathbb{N}^+$ and $\mathcal{M}_1(ad)$ are both closed under addition. □

Lemma 6.5 *Fix positive integers a and d. Let $\mathcal{M}_a(d)$ denote the set of a-floor multiples of d.*

(1) For any positive integer n, write $n = kd + j$ for unique integers j, k satisfying $0 \le j < d$. Then $n \in \mathcal{M}_a(d)$ if and only if $aj < k$.
(2) If $n \ge ad^2 - (a-1)d$, then $n \in \mathcal{M}_a(d)$.

Proof (1) The a-reciprocal duality property, Theorem 3.1 (6), states that

$$d \preccurlyeq_a n \qquad \text{if and only if} \qquad \frac{n}{\lfloor n/d \rfloor} < d + \frac{1}{a}.$$

Given that $n = kd + j$ with $j \in \{0, 1, \ldots d-1\}$, we have

$$\left\lfloor \frac{n}{d} \right\rfloor = k = \frac{n-j}{d}.$$

The expression $n/\lfloor n/d \rfloor$ can be rewritten as

$$\frac{n}{\lfloor n/d \rfloor} = \frac{n}{(n-j)/d} = d\left(1 + \frac{j}{n-j}\right) = d\left(1 + \frac{j}{kd}\right).$$

Thus the a-reciprocal duality property implies that

$$d \preccurlyeq_a n \qquad \text{if and only if} \qquad d\left(1 + \frac{j}{kd}\right) < d\left(1 + \frac{1}{ad}\right),$$

which is equivalent to the condition $aj < k$ as claimed.

(2) Suppose $n \ge ad^2 - (a-1)d$, so

$$n = kd + j \qquad \text{for } j \in \{0, 1, \ldots, d-1\} \text{ and } k \ge ad - a + 1.$$

We then have

$$aj \le a(d-1) < k.$$

Thus $d \preccurlyeq_a n$ by part (1). □

Proof of Theorem 6.3. Lemma 6.4 shows $\mathcal{M}_a(d)$ is an additive semigroup, and Lemma 6.5 (2) then shows $\mathcal{M}_a(d)$ is a numerical semigroup.

(1) The condition in Lemma 6.5 (1) implies that $n = ad^2 - (a-1)d - 1$ is not in $\mathcal{M}_a(d)$; in this case $n = kd + j$ has $k = a(d-1)$ and $j = d-1$ with $aj = k$. On the other hand, Lemma 6.5 (2) shows that all integers $n \ge ad^2 - (a-1)d$ belong to $\mathcal{M}_a(d)$.

(2) To count the size of $\mathbb{N}^+ \setminus \mathcal{M}_a(d)$, we apply the condition in Lemma 6.5 (1). That condition counts the set $n = kd + j$ with $0 \le k \le ad$ and $0 \le j \le d-1$ such that $aj \ge k \ge 0$, excluding the pair $(j,k) = (0,0)$, because $n = 0$ is excluded from $\mathbb{N}^+$. We sum over $0 \le j \le d-1$, noting for $j = 0$ there are no solutions, and there are $aj + 1$ solutions for $1 \le j \le d-1$. We obtain

$$\left|\mathbb{N}^+ \setminus \mathcal{M}_a(d)\right| = \sum_{j=1}^{d-1}(aj+1) = a\left(\frac{d(d-1)}{2}\right) + d - 1 = \frac{1}{2}(d-1)(ad+2).$$

(3) Recall that $\gamma_i = ((i-1)a+1)d + i - 1$. Each generator γ_i has $\gamma_i = kd + j$ with $j = i-1$ and $k = (i-1)a+1$, so $aj < k$, whence by Lemma 6.5 (1) γ_i is in $\mathcal{M}_a(d)$. Thus we have the containment $\langle \gamma_1, \ldots, \gamma_d \rangle \subset \mathcal{M}_a(d)$. The reverse containment $\mathcal{M}_a(d) \subset \langle \gamma_1, \ldots, \gamma_d \rangle$ holds because $d \in \mathcal{M}_a(d)$ and the generating set $\gamma_1, \gamma_2, \ldots, \gamma_d$ contains the minimal element of $\mathcal{M}_a(d)$ in each residue class modulo d, which also follows from Lemma 6.5.

It remains to prove that $\gamma_1, \ldots, \gamma_d$ is minimal as a generating set of $\mathcal{M}_a(d)$, i.e. that no strict subset of $\{\gamma_1, \ldots, \gamma_d\}$ generates $\mathcal{M}_a(d)$. Since $\gamma_1 < \gamma_2 < \cdots < \gamma_d$, it suffices to show that γ_i is not contained in the numerical semigroup $\langle \gamma_1, \ldots, \gamma_{i-1} \rangle$.

Suppose for the sake of contradiction that $\gamma_i \in \langle \gamma_1, \ldots, \gamma_{i-1} \rangle$. Then for some $\ell < i$, we have $\gamma_i - \gamma_\ell \in \langle \gamma_1, \ldots, \gamma_{i-1} \rangle$. However, we have the inclusion $\langle \gamma_1, \ldots, \gamma_{i-1} \rangle \subset \mathcal{M}_a(d)$, and Lemma 6.5(2) implies that

$$\gamma_i - \gamma_\ell = (i-\ell)ad - (i-\ell) = kd + j \notin \mathcal{M}_a(d),$$

since $aj = k$. This contradiction implies that $\gamma_i \notin \langle \gamma_1, \ldots, \gamma_{i-1} \rangle$, so $\gamma_1, \ldots, \gamma_d$ is the minimal generating set of $\mathcal{M}_a(d)$ as claimed. □

7 Average Size of Intervals $|\mathcal{Q}_a[1, n]|$

We use the structure of floor multiples to estimate the average size of a-floor quotient initial intervals for fixed a and variable n. For context, recall that when $a = 1$ the number of 1-floor quotients satisfies

$$|\mathcal{Q}_1[1,n]| = 2\sqrt{n} + O(1).$$

Hence, the average size of $|\mathcal{Q}_1[1,n]|$ in the range $1 \le n \le x$ is asymptotic to

$$\frac{1}{x}\sum_{n=1}^{x} |\mathcal{Q}_1[1,n]| = \frac{1}{x}\left(\int_{t=0}^{x} 2\sqrt{t}\,dt + O(x)\right) = \frac{4}{3}\sqrt{x} + O(1).$$

7.1 Averaged Number of a-Floor Quotients

We have the following estimate, averaging over all $n \le x$.

Theorem 7.1 *For all integers $a \ge 1$ and all $x \ge 1$,*

$$\frac{1}{x}\sum_{n=1}^{x} |\mathcal{Q}_a[1,n]| = \frac{4}{3}\sqrt{\frac{x}{a}} + O\left(\log x + a\right), \tag{7.1}$$

in which the O-constant is independent of a and x.

The proof of Theorem 7.1 is based on study of an auxiliary counting function.

Definition 7.2 The *a-floor-multiple counting function for d*, denoted $\sigma_{a,d}(x)$, is

$$\sigma_{a,d}(x) := |\{n : d \preccurlyeq_a n,\ n \le x\}| = |\mathcal{M}_a(d) \cap \{n \le x\}|\,. \tag{7.2}$$

The usefulness of this function is based on the identity

$$\sum_{n=1}^{x} |\mathcal{Q}_a[1,n]| = \sum_{d=1}^{x} \sigma_{a,d}(x), \tag{7.3}$$

which is valid for all $a \ge 1$ and $x \ge 1$.

The result (7.1) is proved in three lemmas in Sect. 7.3 which estimate the contributions of $\sigma_{a,d}(x)$ on the right side of (7.3) for three regions of d that parallel the regions of the heuristic in Sect. 5.2.

The estimate in Theorem 7.1 is compatible with the heuristic for $|\mathcal{Q}_a[1,n]|$ being of size $2\sqrt{\frac{n}{a}}$, as given in Sect. 5.2, in the sense that the integrated version of the heuristic matches the main term on the right side of (7.1). That is,

$$\frac{1}{x}\sum_{n=1}^{x} 2\sqrt{\frac{n}{a}} = \frac{4}{3}\sqrt{\frac{x}{a}} + O\left(\frac{1}{\sqrt{ax}}\right). \tag{7.4}$$

7.2 Counting a-Floor Multiples

We give a formula for the a-floor multiple counting function values $\sigma_{a,d}(x)$.

Proposition 7.3 *Let $a \geq 1$ and $d \geq 1$. Then, for all $x \geq 1$,*

$$\sigma_{a,d}(x) = \sum_{j=0}^{d-1} \left\lfloor \frac{x - j(ad+1)}{d} \right\rfloor^{+}, \tag{7.5}$$

where $\lfloor x \rfloor^{+} = \max\{\lfloor x \rfloor, 0\}$.

Proof The structure of a-floor multiples n given in Theorem 6.3 shows the minimal a-floor multiple in each residue class $j \pmod d$ (taking $0 \leq j \leq d-1$) is $\gamma_j = j(ad+1) + d$. After this value, every shift by a multiple of d gives an a-floor quotient. The a-floor multiples of d in this residue class, no larger than x, are counted exactly by $\left\lfloor \frac{x-j(ad+1)}{d} \right\rfloor^{+}$. Summing over all residue classes $(\bmod\ d)$ gives the result. □

The next result gives a size estimate for $\sigma_{a,d}(x)$, which exhibits a phase transition at $x = ad^2$. For fixed (a, d) it initially increases in x following a parabola, and then flattens out to linear growth in x, when the numerical semigroup takes all values.

Proposition 7.4 *For all $a \geq 1$ and $d \geq 1$, there holds*

$$\sigma_{a,d}(x) = \begin{cases} \frac{1}{2}a\left(\frac{x}{ad}\right)^2 + O(ad) & \text{if } 0 < x \leq ad^2, \\ x - \frac{1}{2}ad^2 + \frac{1}{2}ad + O(d) & \text{if } ad^2 \leq x. \end{cases}$$

The O-constants are independent of a and d.

Proof For the saturation region above the phase change, $x \geq ad^2$, we have

$$x \geq ad^2 \geq (d-1)(ad+1) = ad^2 - (a-1)d - 1.$$

Consequently $\left\lfloor \frac{x-j(ad+1)}{d} \right\rfloor^{+} = \left\lfloor \frac{x-j(ad+1)}{d} \right\rfloor$ holds for the full range $0 \leq j \leq d-1$. Then Proposition 7.3 gives

$$\begin{aligned} \sigma_{a,d}(x) &= \sum_{j=0}^{d-1} \left\lfloor \frac{x - j(ad+1)}{d} \right\rfloor \\ &= \sum_{j=0}^{d-1} \left(\frac{x}{d} - \frac{j}{d}(ad+1) + O(1) \right) \\ &= x - \frac{1}{2}(d-1)(ad+1) + O(d) \end{aligned}$$

$$= x - \frac{1}{2}ad^2 + \frac{1}{2}ad + O\left(d\right).$$

For the region below the phase change, where $\ell ad \leq x < (\ell + 1)ad$ for some ℓ with $0 \leq \ell < d$, we have $\left\lfloor \frac{x-j(ad+1)}{d} \right\rfloor^{+} = \left\lfloor \frac{x-j(ad+1)}{d} \right\rfloor$ for $0 \leq j \leq \ell - 1$, but equals $O\left(a\right)$ for $j = \ell$, and equals 0 for $j \geq \ell + 1$. Consequently

$$\begin{aligned} \sigma_{a,d}(x) &= \sum_{j=0}^{d-1} \left\lfloor \frac{x - j(ad+1)}{d} \right\rfloor^{+} \\ &= \left(\sum_{j=0}^{\ell-1} \left\lfloor \frac{x - j(ad+1)}{d} \right\rfloor \right) + O\left(a\right) \\ &= \sum_{j=0}^{\ell-1} \left(\left\lfloor \frac{x}{d} \right\rfloor - \frac{j}{d}(ad+1) + O(1) \right) + O\left(a\right). \end{aligned}$$

Since $\ell \left\lfloor \frac{x}{d} \right\rfloor = a\ell^2 + O\left(a\ell\right) = a(\frac{x}{ad})^2 + O\left(a\ell\right)$, we obtain

$$\begin{aligned} \sigma_{a,d}(x) &= \ell \left\lfloor \frac{x}{d} \right\rfloor - \frac{(\ell-1)\ell}{2} \left(a + \frac{1}{d} \right) + O\left(a + \ell\right) \\ &= a(\frac{x}{ad})^2 - \frac{1}{2}a \left(\frac{x}{ad} \right)^2 + O\left(ad\right), \\ &= \frac{1}{2}a(\frac{x}{ad})^2 + O\left(ad\right), \end{aligned}$$

using $\left| \frac{x}{ad} - \ell \right| \leq 1$ at the second to last line.

The sum for $\sigma_{a,d}(x)$ is vacuous for $x \leq 0$. □

7.3 Total Size of Initial a-Floor Quotient Intervals

We estimate the size of sums $\sigma_{a,d}(x)$ over the small, medium, and large regions of d, where $1 \leq d \leq x$. Some of these sums may be empty for small x.

Lemma 7.5 *(Small region). For any integer $a \geq 1$ and any real $x \geq 1$,*

$$\sum_{d=1}^{\sqrt{x/a}} \sigma_{a,d}(x) = \frac{5}{6\sqrt{a}} x^{3/2} + O\left(x\right), \tag{7.6}$$

where the constant implied by the O-symbol is independent of both a and x.

Proof For $1 \leq d \leq \sqrt{\frac{x}{a}}$ we have $x \geq ad^2$. Applying Proposition 7.4 yields

$$\begin{aligned}\sum_{d=1}^{\sqrt{x/a}}{}' \sigma_{a,d}(x) &= \sum_{d=1}^{\sqrt{x/a}} \left(x - \frac{1}{2}ad^2 + \frac{1}{2}ad + O(d)\right)\\ &= \left(\frac{x^{3/2}}{\sqrt{a}} + O(x)\right) - \left(\frac{1}{6}\frac{x^{3/2}}{\sqrt{a}} + O(x)\right) + \left(\frac{1}{4}x + O\left(\sqrt{ax}\right)\right) + O\left(\frac{x}{a}\right)\\ &= \frac{5}{6\sqrt{a}}x^{3/2} + O(x),\end{aligned}$$

which is (7.6). □

Lemma 7.6 *(Medium region). For any integer $a \geq 1$ and any real $x \geq 1$,*

$$\sum_{d>\sqrt{x/a}}^{\sqrt{x}} \sigma_{a,d}(x) = \frac{1}{2\sqrt{a}}x^{3/2} - \frac{1}{2a}x^{3/2} + O\left(x\sqrt{a} + a\sqrt{x}\right), \tag{7.7}$$

where the constant implied by the O-symbol is independent of both a and x.

Proof By hypothesis $\sqrt{x/a} < d \leq \sqrt{x}$, so that $x < ad^2 \leq ax$, whence

$$\left\lfloor \frac{x}{ad} \right\rfloor \leq \frac{x}{ad} < d \leq \frac{x}{d}. \tag{7.8}$$

Recall that for all integers $a \geq 1$ and any $x \geq 1$,

$$\begin{aligned}\sum_{d>\sqrt{x/a}}^{\sqrt{x}} \sigma_{a,d}(x) &= \sum_{d>\sqrt{x/a}}^{\sqrt{x}} \left(\sum_{j=0}^{d-1} \left\lfloor \frac{x - j(ad+1)}{d} \right\rfloor^{+}\right)\\ &= \sum_{d>\sqrt{x/a}}^{\sqrt{x}} \left((\sum_{j=0}^{\lfloor x/ad \rfloor} \left\lfloor \frac{x - j(ad+1)}{d} \right\rfloor) + O\left(\frac{1}{d}\left\lfloor \frac{x}{ad} \right\rfloor\right)\right).\end{aligned}$$

where the cutoff in the inner sum can be made since $x - j(ad+1) < 0$ holds for all $j \geq \lfloor \frac{x}{ad} \rfloor + 1$. Only the term $j = \lfloor \frac{x}{ad} \rfloor$ can contribute to the remainder term, and only when

$$0 > x - j(ad+1) = (x - \left\lfloor \frac{x}{ad} \right\rfloor ad) - \left\lfloor \frac{x}{ad} \right\rfloor \geq - \left\lfloor \frac{x}{ad} \right\rfloor$$

requiring a correction term at most $O\left(\frac{1}{d}\lfloor \frac{x}{ad} \rfloor\right)$. The implied O-constant is independent of x, a and d. Consequently,

$$\sum_{d>\sqrt{x/a}}^{\sqrt{x}} \sigma_{a,d}(x) = \sum_{d>\sqrt{x/a}}^{\sqrt{x}} \left(\sum_{j=0}^{\lfloor x/ad \rfloor} \left(\left\lfloor \frac{x}{d} \right\rfloor - j(a + \frac{1}{d}) + O(1) + O\left(\frac{1}{d}\left\lfloor \frac{x}{ad} \right\rfloor\right)\right)\right)$$

$$= \sum_{d>\sqrt{x/a}}^{\sqrt{x}} \left(\sum_{j=0}^{\lfloor x/ad \rfloor} \left(\left\lfloor \frac{x}{d} \right\rfloor - j(a + \frac{1}{d}) \right) + O(1) \right),$$

using $0 < \left\lfloor \frac{x}{ad} \right\rfloor < d$ from (7.8) in the last line. Next, using this bound again on the first line,

$$\begin{aligned}
\sum_{d>\sqrt{x/a}}^{\sqrt{x}} \sigma_{a,d}(x) &= \sum_{d>\sqrt{x/a}}^{\sqrt{x}} \left(\sum_{j=0}^{\lfloor x/ad \rfloor} \left(\left\lfloor \frac{x}{d} \right\rfloor - j(a + \frac{1}{d}) \right) \right) + O\left(\sum_{d>\sqrt{x/a}}^{\sqrt{x}} d \right) \\
&= \sum_{d>\sqrt{x/a}}^{\sqrt{x}} \left(\sum_{j=0}^{\lfloor x/ad \rfloor} \left(\left\lfloor \frac{x}{d} \right\rfloor - j(a + \frac{1}{d}) \right) \right) + O\left(x(1 - \frac{1}{a}) + \frac{x}{\sqrt{a}} \right).
\end{aligned}$$

We evaluate the inner sums to obtain

$$\begin{aligned}
\sum_{d>\sqrt{x/a}}^{\sqrt{x}} \sigma_{a,d}(x) &= \sum_{d>\sqrt{x/a}}^{\sqrt{x}} \left((\left\lfloor \frac{x}{ad} \right\rfloor + 1) \left\lfloor \frac{x}{d} \right\rfloor - (a + \frac{1}{d}) \frac{1}{2} \left\lfloor \frac{x}{ad} \right\rfloor (\left\lfloor \frac{x}{ad} \right\rfloor + 1) \right) + O\left(x(1 - \frac{1}{a}) + \frac{x}{\sqrt{a}} \right) \\
&= \sum_{d>\sqrt{x/a}}^{\sqrt{x}} \left((\left\lfloor \frac{x}{ad} \right\rfloor \left\lfloor \frac{x}{d} \right\rfloor - \frac{a}{2} \left\lfloor \frac{x}{ad} \right\rfloor^2) + O\left(a(1 + \left\lfloor \frac{x}{ad} \right\rfloor) \right) \right) + O\left(x(1 - \frac{1}{a}) + \frac{x}{\sqrt{a}} \right).
\end{aligned}$$

We remove the floor functions, adding new remainder terms, and simplify, obtaining

$$\begin{aligned}
\sum_{d>\sqrt{x/a}}^{\sqrt{x}} \sigma_{a,d}(x) &= \frac{1}{2a} \sum_{d>\sqrt{x/a}}^{\sqrt{x}} \frac{x^2}{d^2} + \sum_{d>\sqrt{x/a}}^{\sqrt{x}} O\left(a + \left\lfloor \frac{x}{d} \right\rfloor \right) + O\left(x(1 - \frac{1}{\sqrt{a}}) \right) \\
&= \frac{x^2}{2a} \sum_{d>\sqrt{x/a}}^{\sqrt{x}} \frac{1}{d^2} + O\left(\sqrt{x}a + x\log(\sqrt{a}) + x(1 - \frac{1}{\sqrt{a}}) \right) \\
&= \frac{x^2}{2a} \left(\sqrt{\frac{a}{x}} - \frac{1}{\sqrt{x}} \right) + O\left(x(1 + \log a) + a\sqrt{x} \right) \\
&= \frac{x^2}{2a} \left(\sqrt{\frac{a}{x}} - \frac{1}{\sqrt{x}} \right) + O\left(x\sqrt{a} + a\sqrt{x} \right),
\end{aligned}$$

which gives (7.7) on the domain $x \geq 1$ and integer $a \geq 1$. □

Lemma 7.7 *(Large region). For any $a \geq 1$ and any real $x \geq 1$,*

$$\sum_{d>\sqrt{x}}^{x} \sigma_{a,d}(x) = \frac{1}{2a} x^{3/2} + O(x \log x + ax), \tag{7.9}$$

where the constant implied by the O-symbol is independent of both a and x.

Proof The arguments parallel the last lemma, except the estimates of remainder terms change, since (7.8) does not hold; instead $\sqrt{x} < d \le x$. Note that $ad > x$ can occur.

As before we obtain

$$\sum_{d>\sqrt{x}}^{x} \sigma_{a,d}(x) = \sum_{d>\sqrt{x}}^{x} \left(\sum_{j=0}^{\lfloor \frac{x}{ad} \rfloor} \left\lfloor \frac{x - j(ad+1)}{d} \right\rfloor + O\left(\frac{1}{d} \left\lfloor \frac{x}{ad} \right\rfloor\right)\right)$$

$$= \sum_{d>\sqrt{x}}^{x} \left(\sum_{j=0}^{\lfloor \frac{x}{ad} \rfloor} \left\lfloor \frac{x}{d} \right\rfloor - j(a + \frac{1}{d}) + O(1) + O\left(\frac{1}{d} \left\lfloor \frac{x}{ad} \right\rfloor\right)\right),$$

But now we have

$$\sum_{d>\sqrt{x}}^{x} \sigma_{a,d}(x) = \sum_{d>\sqrt{x}}^{x} \left(\sum_{j=0}^{\lfloor \frac{x}{ad} \rfloor} \left(\left\lfloor \frac{x}{d} \right\rfloor - j(a + \frac{1}{d}) \right)\right) + O\left(\sum_{d>\sqrt{x}}^{x} (\left\lfloor \frac{x}{ad} \right\rfloor + \frac{1}{d} \left\lfloor \frac{x}{ad} \right\rfloor^2)\right)$$

$$= \sum_{d>\sqrt{x}}^{x} \left(\sum_{j=0}^{\lfloor \frac{x}{ad} \rfloor} \left(\left\lfloor \frac{x}{d} \right\rfloor - j(a + \frac{1}{d}) \right)\right) + O\left(\frac{x}{a} \log x + \frac{x}{2a^2}\right),$$

in which the implied O-constants are independent of x and a. The remainder term above is estimated using

$$\sum_{d>\sqrt{x}}^{x} \frac{x^2}{a^2 d^3} = \frac{x^2}{a^2} \int_{\sqrt{x}}^{x} \frac{1}{t^3} dt + O(\frac{\sqrt{x}}{a^2}) = \frac{x}{2a^2} + O(\frac{\sqrt{x}}{a^2}).$$

Evaluating the inner sums yields

$$\sum_{d>\sqrt{x}}^{x} \sigma_{a,d}(x) = \sum_{d>\sqrt{x}}^{x} \left((\left\lfloor \frac{x}{ad} \right\rfloor \left\lfloor \frac{x}{d} \right\rfloor - \frac{a}{2} \left\lfloor \frac{x}{ad} \right\rfloor^2) + O\left(a(1 + \left\lfloor \frac{x}{ad} \right\rfloor)\right)\right) + O\left(\frac{x}{a}(\log x + 1)\right).$$

We remove the floor functions, adding new remainder terms, and simplify, obtaining

$$\sum_{d>\sqrt{x}}^{x} \sigma_{a,d}(x) = \frac{1}{2a} \sum_{d>\sqrt{x}}^{x} \frac{x^2}{d^2} + \sum_{d>\sqrt{x}}^{x} O\left(a + \left\lfloor \frac{x}{d} \right\rfloor\right) + O\left(\frac{x}{a} \log x\right)$$

$$= \frac{x^2}{2a} \sum_{d>\sqrt{x}}^{x} \frac{1}{d^2} + O\left(x \log x + ax\right)$$

$$= \frac{x^2}{2a} \left(\frac{1}{\sqrt{x}} - \frac{1}{x} + O(\frac{1}{x^{3/2}})\right) + O\left(x \log x + ax\right),$$

which implies (7.9), with the constant implied by the O-symbol independent of both x and a, over the parameter domain $D = \{(x, a) : 1 \le x < \infty,\ 1 \le a < \infty,\ a \in \mathbb{N}^+\}$. □

7.4 Proof of Theorem 7.1

Proof of Theorem 7.1. The result

$$\sum_{n=1}^{x} |\mathcal{Q}_a[1, n]| = \frac{4}{3\sqrt{a}} x^{3/2} + O\left(x \log x + ax\right). \tag{7.10}$$

follows from the identity (7.3), combining the estimates of the three Lemmas 7.5, 7.6 and 7.7. The constant implied by the O-symbol is independent of both x and a, over the parameter domain $D = \{(x, a) : 1 \le x < \infty,\ 1 \le a < \infty,\ a \in \mathbb{N}^+\}$. Since $x \ge 1$, dividing both sides by x yields the result. □

Remark 7.8 We have the identity, valid for $1 \le y \le x$,

$$\sum_{d=1}^{y} \sigma_{a,d}(x) = \sum_{n=1}^{x} |\mathcal{Q}_a[1, n] \cap \{1, 2, \ldots, y\}|. \tag{7.11}$$

We may then rewrite Lemma 7.5 as asserting, for any $a \ge 1$ and $x \ge 2$,

$$\sum_{n=1}^{x} \left| \mathcal{Q}_a[1, n] \cap [1, \sqrt{\frac{x}{a}}] \right| = \frac{5}{6\sqrt{a}} x^{3/2} + O\left(x\right). \tag{7.12}$$

On the other hand, one can prove

$$\sum_{n=1}^{x} \left| \mathcal{Q}_a[1, n] \cap [1, \sqrt{\frac{n}{a}}] \right| = \frac{2}{3\sqrt{a}} x^{3/2} + O\left(x\right), \tag{7.13}$$

using the fact that $\left|\mathcal{Q}_a[1, n] \cap [1, \sqrt{\frac{n}{a}}]\right| = \sqrt{\frac{n}{a}} + O(1)$ holds, by Lemma 5.2. Combining (7.13) with Theorem 7.1, one then deduces

$$\sum_{n=1}^{x} \left| \mathcal{Q}_a[1, n] \cap [\sqrt{\frac{n}{a}}, n] \right| = \frac{2}{3\sqrt{a}} x^{3/2} + O\left(x \log x + ax\right). \tag{7.14}$$

It follows that the median of the elements of $Q_a[1, n]$, is, on average close to $\sqrt{\frac{n}{a}}$.

8 Concluding Remarks

We conclude with three directions for further investigation.

8.1 a-Floor Quotient Möbius Functions

The zeta function and Möbius functions of a partial order were introduced by Rota [9], though they were long studied for distributive lattices, see Weisner [15] and Ore [7, Chaps. 10, 11].

The two-variable Möbius function of the 1-floor quotient partial order was studied by [6]. The results there showed the function $\mu_1(1, n)$ has infinitely many sign changes. Numerical studies indicate it takes large values, suggesting that $|\mu_1(1, n)| \geq n^{0.6}$ infinitely often, but it has not even been proved that $|\mu_1(1, n)|$ is unbounded. It is an interesting problem to study the Möbius functions of the a-floor quotient partial orders $\mathcal{Q}_a = (\mathbb{N}^+, \preccurlyeq_a)$, particularly addressing the Möbius functions of their initial intervals $\mathcal{Q}_a[1, n]$. Here we note the following facts.

1. Theorem 1.6 implies that as a increases, the value of $\mu_a(d, n)$ starts at $\mu_1(d, n)$ and arrives at the classical Möbius function value $\mu(\frac{n}{d})$ for $a \geq \lfloor \frac{n}{d} \rfloor$, with the convention $\mu(x) = 0$ if x is not an integer.
2. One can show a polynomial upper bound on the size of the a-floor quotient Möbius function $|\mu_a(d, n)| \leq (\frac{n}{d})^{\alpha_0}$ where $\alpha_0 \approx 1.729$ is the unique solution to $\zeta(\sigma) = 2$ on the region $\sigma > 1$, using methods similar to that used in [6] for the 1-floor quotient. The upper bounds for the 1-floor quotient $\mu_1(1, n)$ used information on the number of chains connecting 1 to n. Since the 1-floor quotient order refines the a-floor quotient order, each a-floor quotient interval has fewer chains than that of the 1-floor quotient so the 1-floor quotient upper bounds apply.
3. The behavior of the Möbius function $\mu_a(d, n)$ of the a-floor quotient order may exhibit new features over that of the 1-floor quotient. As observed in Sect. 5, the duality involution for the 1-floor quotient breaks down for the a-floor quotient for $a \geq 2$.

Examples show the convergence of $\mu_a(d, n)$ to $\mu(d, n)$, as $a \to \infty$, is not monotonic on a. In fact there exist d, n and $a > 1$ such that

$$|\mu_a(d, n) - \mu(d, n)| > |\mu_1(d, n) - \mu_(d, n)|.$$

Example 8.1 For $(d, n) = (1, 13)$ we have $\mu_1(1, 13) = 0$ and $\mu_2(1, 13) = 1$, while $\mu_a(1, 13) = \mu(13) = -1$ holds for all sufficiently large a (in particular, for $a \geq \lfloor \frac{13}{2} \rfloor = 6$ by Theorem 1.6). A table listing all values of $\mu_a(1, 13)$, as a varies, is given below.

a	1 2 3 4 5 6 7 ⋯
$\mu_a(1, 13)$	0 1 1 0 0 −1 −1 ⋯

Example 8.2 For $(d, n) = (1, 21)$, we have $\mu_1(1, 21) = -1$ and $\mu_2(1, 21) = 2$, while $\mu_a(1, 21) = \mu(21) = 1$ for sufficiently large a. The table below lists all values of $\mu_a(1, 21)$, as a varies.

a	1 2 3 4 5 ⋯ 9 10 11 ⋯
$\mu_a(1, 21)$	−1 2 3 2 2 ⋯ 2 1 1 ⋯

8.2 Incidence Algebras of Floor Quotient Orders

The incidence algebra of a poset $(\mathbb{N}^+, \preccurlyeq)$ is the set of $\mathbb{Z}$-valued functions on the set of pairs (d, n) such that $d \preccurlyeq n$ under the operations of addition and convolution, see Smith [10, 11]. For general structures of incidence algebras over a ring R, which are in general non-commutative associative algebras, see Stanley [13] and Spiegel and O'Donnell [12]. In general, zeta functions and Möbius functions of (locally finite) partial orders are members of the incidence algebra.

Recall that the 1-floor quotient incidence algebra $\mathcal{I}(\mathcal{Q}_1[1, n])$ can be realized as a noncommutative ring of square integer matrices of size $|\mathcal{Q}_1[1, n]|$. It has been shown by several authors that $s_1(n) := |\mathcal{Q}_1[1, n]|$ is of size $s_1(n) = 2\sqrt{n} + O(1)$, ([2, 4, 6, Corollary 4.4]). Allowing $\mathbb{Q}$-coefficients, it was shown in [6, Theorem 4.15] that the incidence algebra of $\mathcal{Q}_1[1, n]$ has dimension $\dim_{\mathbb{Q}}(\mathcal{I}_n(\mathbb{Q})) \sim \frac{16}{3} n^{3/4}$, as $n \to \infty$.

One may study analogous questions concerning the structure of the incidence algebras $\mathcal{I}(\mathcal{Q}_a[d, n])$ of intervals of a-floor quotient partial orders. Each a-floor quotient incidence algebra $\mathcal{I}(\mathcal{Q}_a[1, n])$ can be realized as a noncommutative ring of square integer matrices of size $s_a(n) := |\mathcal{Q}_a[1, n]|$. This paper showed for $a \geq 2$ that $s_a(n)$ is smaller than $s(n) = |\mathcal{Q}_1[1, n]| \approx 2\sqrt{n}$, and that the values $s_a(n)$ do not vary smoothly as function of n. It is an interesting question to estimate the dimension of the incidence algebra $\dim_{\mathbb{Q}}(\mathcal{I}(\mathcal{Q}_a[1, n]))$ as a $\mathbb{Q}$-vector space, which counts the total number of incidences, as a function of a and n.

8.3 Approximate Divisor Orders Interpolating Towards the Additive Total Order

The paper [6, Sect. 7.4] noted that the 1-floor quotient order is in some sense "halfway between" the multiplicative divisor order and the additive total order on $\mathbb{N}^+$. For a quantitative comparison of incidence relations, see [6, Sect. 2.6].

The a-floor quotient orders provide a sequence of approximate divisor orders interpolating between the 1-floor quotient order and the divisor order. We ask: Is there an interesting family of approximate divisor orders on $\mathbb{N}^+$ that interpolate in the opposite direction between the 1-floor quotient order and the additive total order?

A positive answer to this question, if there is one, probably will require dropping the condition of monotonicity of convergence of the zeta functions of such a family to the zeta function of the additive order.

Acknowledgements We thank the four reviewers for helpful comments. The first author was partially supported by NSF grant DMS-1701576, by a Chern Professorship at MSRI in Fall 2018, and by a 2018 Simons Fellowship in Mathematics in January 2019. MSRI is supported in part by a grant from the National Science Foundation. The second author was partially supported by NSF grant DMS-1600223, a Rackham Predoctoral Fellowship, and an AMS–Simons Travel Grant.

References

1. A. Assi and P. A. García-Sánchez, *Numerical semigroups and applications*, Springer, Cham, 2016.
2. J.-P. Cardinal, *Symmetric matrices related to the Mertens function*, Linear Algebra Appl. **432** (2010), no. 1, 161–172.
3. J.-P. Cardinal and M. Overholt, *A variant of Möbius inversion*, Exp. Math. 29 (2020), no. 3, 247–252.
4. R. Heyman, *Cardinality of a floor function set*, Integers **19** (2019), A67, 7 pp.
5. J. C. Lagarias, T. Murayama, and D. H. Richman, *Dilated floor functions that commute*, Amer. Math. Monthly **123** (2016), no. 10, 1033–1038.
6. J. C. Lagarias and D. H. Richman, *The floor quotient partial order*, Adv. in Appl. Math. **153** (2024), 102615, 54. https://doi.org/10.1016/j.aam.2023.102615
7. Ø. Ore, *Theory of graphs*, American Mathematical Society (AMS), Providence, RI, 1962.
8. J. L. Ramírez Alfonsín, *The Diophantine Frobenius problem*, Oxford University Press, Oxford, 2005.
9. G.-C. Rota, *On the foundations of combinatorial theory. I: Theory of Möbius functions*, Wahrscheinlichkeitstheor. Verw. Geb. **2** (1964), 340–368.
10. D. A. Smith, *Incidence functions as generalized arithmetic functions*. I, Duke Math. J. **34** (1967), 617–633.
11. D. A. Smith, *Incidence functions as generalized arithmetic functions. II, III*, Duke Math. J. **36** (1969), 15–30, 353–367.
12. E. Spiegel and C. J. O'Donnell, *Incidence algebras*, Pure Appl. Math., vol. **206**, Marcel Dekker, New York, NY, (1997).
13. R. P. Stanley, *Structure of incidence algebras and their automorphism groups*, Bull. Amer. Math. Soc. **76** (1970), 1236–1239.
14. R. P. Stanley, *Enumerative combinatorics*. Vol. **1**, 2nd ed., Cambridge University Press, Cambridge, 2012.
15. L.Weisner, *Abstract theory of inversion of finite series*, Trans. Amer. Math. Soc. **38** (1935), 474–484.

The Muirhead-Rado Inequality, 1: Vector Majorization and the Permutohedron

Melvyn B. Nathanson

Abstract Let $\mathbf{a}$ and $\mathbf{b}$ be vectors in $\mathbf{R}^n$ with nonnegative coordinates. Permuting the coordinates, one can assume that $a_1 \geq \cdots \geq a_n$ and $b_1 \geq \cdots \geq b_n$. The vector $\mathbf{a}$ majorizes the vector $\mathbf{b}$ if $\sum_{i=1}^k b_i \leq \sum_{i=1}^k a_i$ for all $k \in \{1, \ldots, n-1\}$ and $\sum_{i=1}^n b_i = \sum_{i=1}^n a_i$. This paper exposes theorems of Hardy-Littlewood-Pólya and of Rado: The following are equivalent: (1) The vector $\mathbf{a}$ majorizes the vector $\mathbf{b}$, (2) $P\mathbf{a} = \mathbf{b}$ for some doubly stochastic matrix P, and (3) $\mathbf{b}$ is in the S_n-permutohedron generated by $\mathbf{a}$.

2020 Mathematics Subject Classification: 05E05 · 11B83 · 15B51 · 26D05 · 26D15 · 52A20 · 52A30 · 52A40

1 Permutation Matrices and the Permutohedron

This paper describes majorization results of Hardy-Littlewood-Pólya [2, 3] and Rado [8] that will be applied in [7] to prove beautiful inequalitites of Muirhead [6] and Rado [8] that generalize the arithmetic and geometric mean inequality. An encyclopedic reference is Marshall-Olkin-Arnold [4].

A matrix $P = (p_{i,j})$ with real coordinates is *nonnegative* if $p_{i,j} \geq 0$ for all i and j. A nonnegative $n \times n$ matrix $P = (p_{i,j})$ is *column stochastic* if its column sums satisfy

$$\text{colsum}_j(P) = \sum_{i=1}^{n} p_{i,j} = 1 \qquad \text{for all } j \in \{1, \ldots, n\}.$$

Supported in part by PSC-CUNY Research Award Program grant 66197-00 54.

M. B. Nathanson (✉)
Lehman College (CUNY), Bronx, New York 10468, USA
e-mail: melvyn.nathanson@lehman.cuny.edu

M. B. Nathanson (ed.), *Combinatorial and Additive Number Theory VI*,
Springer Proceedings in Mathematics & Statistics 464,
https://doi.org/10.1007/978-3-031-65064-2_16

A nonnegative $n \times n$ matrix $P = (p_{i,j})$ is *row stochastic* if its row sums satisfy

$$\text{rowsum}_i(P) = \sum_{j=1}^{n} p_{i,j} = 1 \qquad \text{for all } i \in \{1, \ldots, n\}.$$

A nonnegative $n \times n$ matrix $P = (p_{i,j})$ is *doubly stochastic* if it is both column stochastic and row stochastic.

We begin with some simple lemmas.

Lemma 1 *Let P and Q be $n \times n$ matrices. If P and Q are column stochastic, then PQ is column stochastic. If P and Q are row stochastic, then PQ is row stochastic. If P and Q are doubly stochastic, then PQ is doubly stochastic.*

Proof Let $P = (p_{i,j})$ and $Q = (q_{i,j})$. If P and Q are column stochastic $n \times n$ matrices, then for all $j \in \{1, \ldots, n\}$ we have

$$\begin{aligned}
\text{colsum}_j(PQ) &= \sum_{i=1}^{n} (PQ)_{i,j} = \sum_{i=1}^{n} \sum_{k=1}^{n} p_{i,k} q_{k,j} \\
&= \sum_{k=1}^{n} q_{k,j} \sum_{i=1}^{n} p_{i,k} = \sum_{k=1}^{n} q_{k,j} \,\text{colsum}_k(P) \\
&= \sum_{k=1}^{n} q_{k,j} = \text{colsum}_j(Q) = 1
\end{aligned}$$

and so PQ is column stochastic.

Similarly, if P and Q are row stochastic $n \times n$ matrices, then for all $i \in \{1, \ldots, n\}$ we have

$$\begin{aligned}
\text{rowsum}_i(PQ) &= \sum_{j=1}^{n} (PQ)_{i,j} = \sum_{j=1}^{n} \sum_{k=1}^{n} p_{i,k} q_{k,j} \\
&= \sum_{k=1}^{n} p_{i,k} \sum_{j=1}^{n} q_{k,j} = \sum_{k=1}^{n} p_{i,k} \,\text{rowsum}_k(Q) \\
&= \sum_{k=1}^{n} p_{i,k} = \text{rowsum}_i(P) = 1
\end{aligned}$$

and so PQ is row stochastic.

It follows that PQ is doubly stochastic if P and Q are doubly stochastic. This completes the proof. □

Let $\mathcal{E} = \{\mathbf{e}_1, \ldots, \mathbf{e}_n\}$ be the standard basis for $V = \mathbf{R}^n$. For every permutation σ in the symmetric group S_n, define the linear transformation $P_\sigma : V \to V$ by

$$P_\sigma(\mathbf{e}_j) = \mathbf{e}_{\sigma(j)}$$

for all $j \in \{1, \ldots, n\}$. If $\mathbf{x} = \sum_{j=1}^n x_j \mathbf{e}_j \in V$, then

$$\begin{aligned} P_\sigma(\mathbf{x}) &= P_\sigma\left(\sum_{j=1}^n x_j \mathbf{e}_j\right) = \sum_{j=1}^n x_j P_\sigma\left(\mathbf{e}_j\right) \\ &= \sum_{j=1}^n x_j \mathbf{e}_{\sigma(j)} = \sum_{j=1}^n x_{\sigma^{-1}(j)} \mathbf{e}_j. \end{aligned}$$

Equivalently,

$$P_\sigma \begin{pmatrix} x_1 \\ \vdots \\ x_n \end{pmatrix} = \begin{pmatrix} x_{\sigma^{-1}(1)} \\ \vdots \\ x_{\sigma^{-1}(n)} \end{pmatrix}.$$

We also denote by P_σ the matrix of this linear transformation with respect to the standard basis $\mathcal{E}$. The $n \times n$ matrix P_σ is called a *permutation matrix*. Its (i, j)th coordinate is

$$(P_\sigma)_{i,j} = \delta_{i,\sigma(j)} = \begin{cases} 1 & \text{if } i = \sigma(j) \\ 0 & \text{if } i \neq \sigma(j) \end{cases}$$

where $\delta_{i,j}$ is the Kronecker delta. For all $j \in \{1, \ldots, n\}$ we have

$$\operatorname{colsum}_j(P_\sigma) = \sum_{i=1}^n (P_\sigma)_{i,j} = \sum_{i=1}^n \delta_{i,\sigma(j)} = 1.$$

For all $i \in \{1, \ldots, n\}$ we have

$$\operatorname{rowsum}_i(P_\sigma) = \sum_{j=1}^n (P_\sigma)_{i,j} = \sum_{j=1}^n \delta_{i,\sigma(j)} = 1.$$

Thus, the permutation matrix P_σ is doubly stochastic for all $\sigma \in S_n$.

For example, if $\sigma = (1, 2, 3) \in S_3$, then

$$P_\sigma = \begin{pmatrix} 0 & 0 & 1 \\ 1 & 0 & 0 \\ 0 & 1 & 0 \end{pmatrix}.$$

Lemma 2 *For all permutations $\sigma, \tau \in S_n$,*

$$P_\sigma P_\tau = P_{\sigma\tau}.$$

Proof For all $j \in \{1, \ldots, n\}$ we have

$$\begin{aligned}(P_\sigma P_\tau)(\mathbf{e}_j) &= P_\sigma\left(P_\tau(\mathbf{e}_j)\right) = P_\sigma\left(\mathbf{e}_{\tau(j)}\right) = \mathbf{e}_{\sigma(\tau(j))} \\ &= \mathbf{e}_{(\sigma\tau)(j)} = P_{\sigma\tau}(\mathbf{e}_j).\end{aligned}$$

Equivalently, for all $i, j \in \{1, \ldots, n\}$ we have

$$(P_\sigma P_\tau)_{i,j} = \sum_{k=1}^{n} (P_\sigma)_{i,k} (P_\tau)_{k,j} = \sum_{k=1}^{n} \delta_{i,\sigma(k)}\delta_{k,\tau(j)} = \delta_{i,\sigma\tau(j)} = (P_{\sigma\tau})_{i,j}\,.$$

This completes the proof. □

Lemma 3 *Let A be a $m \times n$ matrix. Let σ be a permutation in S_n and let τ be a permutation in S_m.*

(a) The jth column of the $m \times n$ matrix AP_σ is the $\sigma(j)$th column of the matrix A.
(b) The ith row of the $m \times n$ matrix $P_\tau A$ is the $\tau^{-1}(i)$th row of the matrix A.
(c) The (i, j)th coordinate of the matrix $P_\tau A P_\sigma$ is $a_{\tau^{-1}(i),\sigma(j)}$.

Proof Matrix multiplication. □

The classical *permutohedron* is the convex hull of the set of $n!$ lattice points in $\mathbf{R}^n$ whose coordinates are the permutations of the set $\{1, 2, 3, \ldots, n\}$, that is, the convex hull of the set

$$\left\{ P_\sigma \begin{pmatrix} 1 \\ 2 \\ \vdots \\ n \end{pmatrix} : \sigma \in S_n \right\} = \left\{ \begin{pmatrix} \sigma^{-1}(1) \\ \sigma^{-1}(2) \\ \vdots \\ \sigma^{-1}(n) \end{pmatrix} : \sigma \in S_n \right\} = \left\{ \begin{pmatrix} \sigma(1) \\ \sigma(2) \\ \vdots \\ \sigma(n) \end{pmatrix} : \sigma \in S_n \right\}.$$

Every vector in this set lies on the hyperplane

$$x_1 + x_2 + \cdots + x_n = 1 + 2 + \cdots + n = \frac{n(n+1)}{2}.$$

For example, the classical permutohedron in $\mathbf{R}^2$ is the convex hull of the set

$$\left\{ P_\sigma \begin{pmatrix} 1 \\ 2 \end{pmatrix} : \sigma \in S_2 \right\}$$

that is, the line segment

$$\left\{(1-t)\begin{pmatrix}1\\2\end{pmatrix}+t\begin{pmatrix}2\\1\end{pmatrix} : 0 \leq t \leq 1\right\} = \left\{\begin{pmatrix}1+t\\2-t\end{pmatrix} : 0 \leq t \leq 1\right\}.$$

The classical permutohedron in $\mathbf{R}^3$ is the convex hull of the set

$$\left\{P_\sigma\begin{pmatrix}1\\2\\3\end{pmatrix} : \sigma \in S_3\right\} = \left\{\begin{pmatrix}1\\2\\3\end{pmatrix}, \begin{pmatrix}1\\3\\2\end{pmatrix}, \begin{pmatrix}2\\1\\3\end{pmatrix}, \begin{pmatrix}2\\3\\1\end{pmatrix}, \begin{pmatrix}3\\1\\2\end{pmatrix}, \begin{pmatrix}3\\2\\1\end{pmatrix}\right\}.$$

Let $\mathbf{a} = \begin{pmatrix}a_1\\\vdots\\a_n\end{pmatrix} \in \mathbf{R}^n$. The *permutohedron* $K(\mathbf{a})$ constructed from the vector $\mathbf{a}$ is the convex hull of the set of vectors in $\mathbf{R}^n$ whose coordinates are the permutations of the set $\{a_1, a_2, a_3, \ldots, a_n\}$, that is, the convex hull of the set

$$\left\{P_\sigma\begin{pmatrix}a_1\\a_2\\\vdots\\a_n\end{pmatrix} : \sigma \in S_n\right\} = \left\{\begin{pmatrix}a_{\sigma(1)}\\a_{\sigma(2)}\\\vdots\\a_{\sigma(n)}\end{pmatrix} : \sigma \in S_n\right\}.$$

Every vector in this set lies on the hyperplane

$$H = \left\{\begin{pmatrix}x_1\\\vdots\\x_n\end{pmatrix} \in \mathbf{R}^n : x_1 + x_2 + \cdots + x_n = a_1 + a_2 + \cdots + a_n\right\}.$$

Let G be a subgroup of S_n. The *G-permutohedron* constructed from the vector $\mathbf{a} = \begin{pmatrix}a_1\\\vdots\\a_n\end{pmatrix}$, denoted $K_G(\mathbf{a})$, is the convex hull of the set of vectors in $\mathbf{R}^n$ whose coordinates are the G-permutations of the vector $\mathbf{a}$, that is, the set

$$\left\{P_\sigma\begin{pmatrix}a_1\\a_2\\\vdots\\a_n\end{pmatrix} : \sigma \in G\right\} = \left\{\begin{pmatrix}a_{\sigma(1)}\\a_{\sigma(2)}\\\vdots\\a_{\sigma(n)}\end{pmatrix} : \sigma \in G\right\}.$$

The G-permutohedron $K_G(\mathbf{a})$ is a subset of the permutohedron $K(\mathbf{a})$.

2 Decreasing and Increasing Vectors

For every permutation $\sigma \in S_n$ we have the permutation matrix P_σ and for every vector $\mathbf{a} = \begin{pmatrix} a_1 \\ \vdots \\ a_n \end{pmatrix} \in \mathbf{R}^n$ we have the vector with reordered coordinates

$$P_\sigma(\mathbf{a}) = \begin{pmatrix} a_{\sigma^{-1}(1)} \\ \vdots \\ a_{\sigma^{-1}(n)} \end{pmatrix} \in \mathbf{R}^n.$$

The vector $P_\sigma(\mathbf{a})$ is called a *rearrangement* of the vector in $\mathbf{a}$. Of special interest are vectors with decreasing coordinates and vectors with increasing coordinates.

Let $\mathbf{x} = \begin{pmatrix} x_1 \\ \vdots \\ x_n \end{pmatrix} \in \mathbf{R}^n$. The vector $\mathbf{x}$ is *decreasing* if

$$x_1 \geq x_2 \geq \cdots \geq x_n.$$

Associated to every vector $\mathbf{a} = \begin{pmatrix} a_1 \\ \vdots \\ a_n \end{pmatrix} \in \mathbf{R}^n$ is a unique decreasing vector $\mathbf{a}^\downarrow = \begin{pmatrix} a_1^\downarrow \\ \vdots \\ a_n^\downarrow \end{pmatrix} \in \mathbf{R}^n$ obtained from $\mathbf{a}$ by a rearrangement of coordinates. For every vector $\mathbf{a} \in \mathbf{R}^n$ there is a permutation $\sigma \in S_n$ such that $P_\sigma(\mathbf{a}) = \mathbf{a}^\downarrow$. This permutation is unique if and only if the vector $\mathbf{a}$ has distinct coordinates.

The vector $\mathbf{x} = \begin{pmatrix} x_1 \\ \vdots \\ x_n \end{pmatrix} \in \mathbf{R}^n$ is *increasing* if

$$x_1 \leq x_2 \leq \cdots \leq x_n.$$

Associated to every vector $\mathbf{a} = \begin{pmatrix} a_1 \\ \vdots \\ a_n \end{pmatrix} \in \mathbf{R}^n$ is a unique increasing vector $\mathbf{a}^\uparrow = \begin{pmatrix} a_1^\uparrow \\ \vdots \\ a_n^\uparrow \end{pmatrix} \in \mathbf{R}^n$ obtained from $\mathbf{a}$ by a rearrangement of coordinates. For every vector $\mathbf{a} \in \mathbf{R}^n$ there is a permutation $\tau \in S_n$ such that $P_\tau \mathbf{a} = \mathbf{a}^\uparrow$. This permutation is unique if and only if the vector $\mathbf{a}$ has distinct coordinates.

For example, the decreasing vector obtained from the vector

$$\mathbf{a} = \begin{pmatrix} a_1 \\ a_2 \\ a_3 \\ a_4 \end{pmatrix} = \begin{pmatrix} 3 \\ 1 \\ 2 \\ 4 \end{pmatrix} \in \mathbf{R}^4.$$

is

$$\mathbf{a}^{\downarrow} = \begin{pmatrix} 4 \\ 3 \\ 2 \\ 1 \end{pmatrix} = \begin{pmatrix} a_4 \\ a_1 \\ a_3 \\ a_2 \end{pmatrix} = \begin{pmatrix} a_{\sigma^{-1}(1)} \\ a_{\sigma^{-1}(2)} \\ a_{\sigma^{-1}(3)} \\ a_{\sigma^{-1}(n)} \end{pmatrix}$$

where the rearrangement permutation $\sigma \in S_4$ is

$$\sigma^{-1} = \begin{pmatrix} 1 & 2 & 3 & 4 \\ 4 & 1 & 3 & 2 \end{pmatrix} \quad \text{and} \quad \sigma = \begin{pmatrix} 1 & 2 & 3 & 4 \\ 2 & 4 & 3 & 1 \end{pmatrix}.$$

Thus, $\mathbf{a}^{\downarrow} = P_\sigma(\mathbf{a})$ and σ is the 3-cycle $(1, 2, 4)$.

The increasing vector obtained from $\mathbf{a}$ is

$$\mathbf{a}^{\uparrow} = \begin{pmatrix} 1 \\ 2 \\ 3 \\ 4 \end{pmatrix} = \begin{pmatrix} a_2 \\ a_3 \\ a_1 \\ a_4 \end{pmatrix} = \begin{pmatrix} a_{\tau^{-1}(1)} \\ a_{\tau^{-1}(2)} \\ a_{\tau^{-1}(3)} \\ a_{\tau^{-1}(n)} \end{pmatrix}$$

where the rearrangement permutation $\tau \in S_4$ is

$$\tau^{-1} = \begin{pmatrix} 1 & 2 & 3 & 4 \\ 2 & 3 & 1 & 4 \end{pmatrix} \quad \text{and} \quad \tau = \begin{pmatrix} 1 & 2 & 3 & 4 \\ 3 & 1 & 2 & 4 \end{pmatrix}.$$

Thus, $\mathbf{a}^{\uparrow} = P_\tau(\mathbf{a})$ and τ is the 3-cycle $(1, 3, 2)$.

If $\mathbf{b} = \begin{pmatrix} 2 \\ 1 \\ 3 \\ 2 \end{pmatrix}$, then

$$\mathbf{b}^{\downarrow} = \begin{pmatrix} 3 \\ 2 \\ 2 \\ 1 \end{pmatrix} = P_{\sigma_1}(\mathbf{b}) = P_{\sigma_2}(\mathbf{b})$$

where $\sigma_1 = (1, 2, 4, 3)$ and $\sigma_2 = (1, 3)(2, 4)$, and

$$\mathbf{b}^{\uparrow} = \begin{pmatrix} 1 \\ 2 \\ 2 \\ 3 \end{pmatrix} = P_{\tau_1}(\mathbf{b}) = P_{\tau_2}(\mathbf{b})$$

where $\tau_1 = (1, 3, 4, 2)$ and $\tau_2 = (1, 2)(3, 4)$. The coordinates of $\mathbf{b}$ are not distinct and the rearrangement permutations are not unique.

Note that for all vectors $\mathbf{a}, \mathbf{b} \in \mathbf{R}^n$ and all permutations $\sigma \in S_n$ we have

$$\left(\mathbf{a}^{\downarrow} + \mathbf{b}^{\downarrow}\right)^{\downarrow} = \mathbf{a}^{\downarrow} + \mathbf{b}^{\downarrow} \quad \text{and} \quad \left(\mathbf{a}^{\uparrow} + \mathbf{b}^{\uparrow}\right)^{\uparrow} = \mathbf{a}^{\uparrow} + \mathbf{b}^{\uparrow}$$

and

$$(P_\sigma(\mathbf{a}))^{\downarrow} = \mathbf{a}^{\downarrow} \quad \text{and} \quad (P_\sigma(\mathbf{a}))^{\uparrow} = \mathbf{a}^{\uparrow}.$$

Let $\mathbf{x} = \begin{pmatrix} x_1 \\ \vdots \\ x_n \end{pmatrix} \in \mathbf{R}^n$ and let $\tau = (k, \ell) \in S_n$ be a transposition with $k < \ell$. The transposition τ *increases the vector* $\mathbf{x}$ if $x_k > x_\ell$ and $P_\tau(\mathbf{x})_k < P_\tau(\mathbf{x})_\ell$. The transposition τ *decreases the vector* $\mathbf{x}$ if $x_k < x_\ell$ and $P_\tau(\mathbf{x})_k > P_\tau(\mathbf{x})_\ell$.

Lemma 4 *For all $n \in \mathbf{N}$ and $\mathbf{b} \in \mathbf{R}^n$, there exists a finite sequence of vectors $\mathbf{b} = \mathbf{b}_0, \mathbf{b}_1, \ldots, \mathbf{b}_r = \mathbf{b}^\uparrow$ and a finite sequence of transpositions $\tau_1, \tau_2, \ldots, \tau_r \in S_n$ such that $\mathbf{b}_i = P_{\tau_i}(\mathbf{b}_{i-1})$ and τ_i increases the vector $\mathbf{b}_{i-1}$ for all $i \in \{1, \ldots, r\}$.*

Proof The proof is by induction on n. If $n = 1$ and $\mathbf{b} = (b_1)$ or if $n = 2$ and $\mathbf{b} = \binom{b_1}{b_2}$ with $b_1 \leq b_2$, then $\mathbf{b} = \mathbf{b}^\uparrow$ and there is nothing to prove. If $\mathbf{b} = \binom{b_1}{b_2}$ with $b_1 > b_2$, then the transposition $\tau_1 = (1, 2) \in S_2$ increases $\mathbf{b}$ and $P_{\tau_1}(\mathbf{b}) = \binom{b_2}{b_1} = \mathbf{b}^\uparrow$.

Let $n \geq 3$ and assume that the Lemma holds for vectors in $\mathbf{R}^{n-1}$. Let $\mathbf{b} = \begin{pmatrix} b_1 \\ \vdots \\ b_n \end{pmatrix} \in \mathbf{R}^n$. Choose $i_0 \in \{1, \ldots, n\}$ such that $b_{i_0} = \min\{b_1, b_2, \ldots, b_n\}$. If $b_1 = b_{i_0}$, then apply the Lemma to the vector $\begin{pmatrix} b_2 \\ b_3 \\ \vdots \\ b_n \end{pmatrix} \in \mathbf{R}^{n-1}$.

If $b_1 > b_{i_0}$, then the transposition $\tau_1 = (1, i_0)$ increases $\mathbf{b}$. Let $\mathbf{b}_1 = P_{\tau_1}(\mathbf{b})$. We have $b_{\tau(1)} = b_{i_0} \leq b_{\tau(i)}$ for all $i \in \{1, \ldots, n\}$. Applying the Lemma to the vector $\begin{pmatrix} b_{\tau(2)} \\ \vdots \\ b_{\tau(n)} \end{pmatrix} \in \mathbf{R}^{n-1}$ completes the proof. □

The *standard inner product* of vectors $\mathbf{x} = \begin{pmatrix} x_1 \\ \vdots \\ x_n \end{pmatrix}$ and $\mathbf{y} = \begin{pmatrix} y_1 \\ \vdots \\ y_n \end{pmatrix}$ is

$$(\mathbf{x}, \mathbf{y}) = \sum_{j=1}^{n} x_j y_j.$$

For all $\rho, \sigma \in S_n$ we have

$$\begin{aligned}
\big(P_\rho(\mathbf{x}), P_\sigma(\mathbf{y})\big) &= \sum_{j=1}^{n} x_{\rho^{-1}(j)} y_{\sigma^{-1}(j)} \\
&= \sum_{j=1}^{n} x_j y_{\sigma^{-1}\rho(j)} \\
&= \big(\mathbf{x}, P_{\rho^{-1}\sigma}(\mathbf{y})\big)
\end{aligned}$$

and so

$$(P_\sigma(\mathbf{x}), P_\sigma(\mathbf{y})) = (\mathbf{x}, \mathbf{y}).$$

Lemma 5 *Let $\mathbf{a}$ and $\mathbf{b}$ be vectors in $\mathbf{R}^n$ and let τ be a transposition in S_n. If $\mathbf{a}$ is increasing and τ increases $\mathbf{b}$, or if $\mathbf{a}$ is decreasing and τ decreases $\mathbf{b}$, then*

$$(\mathbf{a}, P_\tau(\mathbf{b})) \geq (\mathbf{a}, \mathbf{b}).$$

If **a** *is increasing and* τ *decreases* **b**, *or if* **a** *is decreasing and* τ *increases* **b**, *then*

$$(\mathbf{a}, P_\tau(\mathbf{b})) \leq (\mathbf{a}, \mathbf{b}).$$

Proof Let $\mathbf{a} = \begin{pmatrix} a_1 \\ \vdots \\ a_n \end{pmatrix}$ and $\mathbf{b} = \begin{pmatrix} b_1 \\ \vdots \\ b_n \end{pmatrix}$, and let $\tau = (k, \ell)$ with $k < \ell$. Because $\tau^{-1} = \tau$ and $b_{\tau(i)} = b_i$ for all $i \neq k, \ell$, we have

$$\begin{aligned}
(\mathbf{a}, P_\tau(\mathbf{b})) - (\mathbf{a}, \mathbf{b}) &= \sum_{i=1}^{n} a_i b_{\tau(i)} - \sum_{i=1}^{n} a_i b_i \\
&= \sum_{i=1}^{n} a_i \left(b_{\tau(i)} - b_i\right) \\
&= a_k(b_\ell - b_k) + a_\ell(b_k - b_\ell) \\
&= (a_\ell - a_k)(b_k - b_\ell).
\end{aligned}$$

If **a** is increasing and τ increases **b**, then $a_\ell \geq a_k$ and $b_k > b_\ell$, and so $(\mathbf{a}, P_\tau(\mathbf{b})) - (\mathbf{a}, \mathbf{b}) \geq 0$. If **a** is decreasing and τ decreases **b**, then $a_\ell \leq a_k$ and $b_k < b_\ell$, and so $(\mathbf{a}, P_\tau(\mathbf{b})) - (\mathbf{a}, \mathbf{b}) \leq 0$. The proof of the second inequality is similar. □

Let **a** and **b** be vectors in $\mathbf{R}^n$. Consider the set of all inner products of rearrangements of the vectors **a** and **b**, that is, the set

$$S = \{(P_\sigma(\mathbf{a}), P_\tau(\mathbf{b})) : \sigma, \tau \in S_n\}.$$

We have

$$S = \{(\mathbf{a}, P_\tau(\mathbf{b})) : \tau \in S_n\} = \left\{\left(\mathbf{a}^\uparrow, P_\tau(\mathbf{b})\right) : \tau \in S_n\right\}.$$

The following is an inequality of Hardy, Littlewood, and Pólya.

Theorem 1 *For all vectors* $\mathbf{a}, \mathbf{b} \in \mathbf{R}^n$,

$$\left(\mathbf{a}^\uparrow, \mathbf{b}^\downarrow\right) \leq (\mathbf{a}, \mathbf{b}) \leq \left(\mathbf{a}^\uparrow, \mathbf{b}^\uparrow\right) = \left(\mathbf{a}^\downarrow, \mathbf{b}^\downarrow\right).$$

Proof Let $\mathbf{a} = \begin{pmatrix} a_1 \\ \vdots \\ a_n \end{pmatrix}$ and $\mathbf{b} = \begin{pmatrix} b_1 \\ \vdots \\ b_n \end{pmatrix}$. Choose $\rho, \sigma \in S_n$ such that

$$\mathbf{a}^\uparrow = P_\rho(\mathbf{a}) = \begin{pmatrix} a_{\rho^{-1}(1)} \\ a_{\rho^{-1}(2)} \\ \vdots \\ a_{\rho^{-1}(n)} \end{pmatrix} \quad \text{and} \quad \mathbf{b}^\uparrow = P_\sigma(\mathbf{b}) = \begin{pmatrix} b_{\sigma^{-1}(1)} \\ b_{\sigma^{-1}(2)} \\ \vdots \\ b_{\sigma^{-1}(n)} \end{pmatrix}.$$

Then

$$\mathbf{a}^{\downarrow} = \begin{pmatrix} a_{\rho^{-1}(n)} \\ \vdots \\ a_{\rho^{-1}(2)} \\ a_{\rho^{-1}(1)} \end{pmatrix} \quad \text{and} \quad \mathbf{b}^{\downarrow} = \begin{pmatrix} a_{\rho^{-1}(n)} \\ \vdots \\ b_{\sigma^{-1}(2)} \\ b_{\sigma^{-1}(1)} \end{pmatrix}$$

and so

$$\left(\mathbf{a}^{\uparrow}, \mathbf{b}^{\uparrow}\right) = \sum_{j=1}^{n} a_{\rho^{-1}(j)} a_{\rho^{-1}(j)} = \left(\mathbf{a}^{\downarrow}, \mathbf{b}^{\downarrow}\right).$$

Next we prove that

$$\left(\mathbf{a}^{\uparrow}, \mathbf{b}\right) \leq \left(\mathbf{a}^{\uparrow}, \mathbf{b}^{\uparrow}\right).$$

By Lemma 4, there is a finite sequence of vectors $\mathbf{b} = \mathbf{b}_0, \mathbf{b}_1, \ldots, \mathbf{b}_r = \mathbf{b}^{\uparrow}$ and a finite sequence of transpositions $\tau_1, \tau_2, \ldots, \tau_r \in S_n$ such that $\mathbf{b}_i = P_{\tau_i}(\mathbf{b}_{i-1})$ and τ_i increases $\mathbf{b}_{i-1}$ for all $i \in \{1, \ldots, r\}$. By Lemma 5,

$$\begin{aligned}
\left(\mathbf{a}^{\uparrow}, \mathbf{b}^{\uparrow}\right) - \left(\mathbf{a}^{\uparrow}, \mathbf{b}\right) &= \left(\mathbf{a}^{\uparrow}, \mathbf{b}_r\right) - \left(\mathbf{a}^{\uparrow}, \mathbf{b}_0\right) \\
&= \sum_{i=1}^{r} \left(\left(\mathbf{a}^{\uparrow}, \mathbf{b}_i\right) - \left(\mathbf{a}^{\uparrow}, \mathbf{b}_{i-1}\right)\right) \\
&= \sum_{i=1}^{r} \left(\left(\mathbf{a}^{\uparrow}, P_{\tau_i}(\mathbf{b}_{i-1})\right) - \left(\mathbf{a}^{\uparrow}, \mathbf{b}_{i-1}\right)\right) \\
&\geq 0.
\end{aligned}$$

It follows that

$$(\mathbf{a}, \mathbf{b}) = \left(P_\rho(\mathbf{a}), P_\rho(\mathbf{b})\right) = \left(\mathbf{a}^{\uparrow}, P_\rho(\mathbf{b})\right) \leq \left(\mathbf{a}^{\uparrow}, P_\rho(\mathbf{b})^{\uparrow}\right) = \left(\mathbf{a}^{\uparrow}, \mathbf{b}^{\uparrow}\right).$$

There is a similar proof of the inequality

$$\left(\mathbf{a}^{\uparrow}, \mathbf{b}^{\downarrow}\right) \leq \left(\mathbf{a}^{\uparrow}, \mathbf{b}\right).$$

By Lemma 4, there is a finite sequence of vectors $\mathbf{b} = \mathbf{b}_0, \mathbf{b}_1, \ldots, \mathbf{b}_r = \mathbf{b}^{\uparrow}$ and a finite sequence of transpositions $\tau_1, \tau_2, \ldots, \tau_r \in S_n$ such that $\mathbf{b}_i = P_{\tau_i}(\mathbf{b}_{i-1})$ and τ_i decreases $\mathbf{b}_{i-1}$ for all $i \in \{1, \ldots, r\}$. By Lemma 5,

$$\begin{aligned}
\left(\mathbf{a}^{\uparrow}, \mathbf{b}^{\uparrow}\right) - \left(\mathbf{a}^{\uparrow}, \mathbf{b}\right) &= \left(\mathbf{a}^{\uparrow}, \mathbf{b}_r\right) - \left(\mathbf{a}^{\uparrow}, \mathbf{b}_0\right) \\
&= \sum_{i=1}^{r} \left(\left(\mathbf{a}^{\uparrow}, \mathbf{b}_i\right) - \left(\mathbf{a}^{\uparrow}, \mathbf{b}_{i-1}\right)\right)
\end{aligned}$$

$$= \sum_{i=1}^{r} \left(\left(\mathbf{a}^{\uparrow}, P_{\tau_i}(\mathbf{b}_{i-1})\right) - \left(\mathbf{a}^{\uparrow}, \mathbf{b}_{i-1}\right) \right)$$
$$\leq 0.$$

It follows that

$$\left(\mathbf{a}^{\uparrow}, \mathbf{b}^{\downarrow}\right) = \left(\mathbf{a}^{\uparrow}, P_{\rho}(\mathbf{b})^{\downarrow}\right) \leq \left(\mathbf{a}^{\uparrow}, P_{\rho}(\mathbf{b})\right) = \left(P_{\rho}(\mathbf{a}), P_{\rho}(\mathbf{b})\right) = (\mathbf{a}, \mathbf{b}).$$

This completes the proof. □

3 Vector Majorization and Doubly Stochastic Matrices

Let $\mathbf{R}_{\geq 0}$ denote the set of nonnegative real numbers. A *nonnegative vector* is a vector with nonnegative coordinates. The *nonnegative octant* in $\mathbf{R}^n$ is the set of nonnegative vectors in $\mathbf{R}^n$, that is,

$$\mathbf{R}^n_{\geq 0} = \left\{ \mathbf{a} = \begin{pmatrix} a_1 \\ \vdots \\ a_n \end{pmatrix} \in \mathbf{R}^n : a_i \geq 0 \text{ for all } i \in \{1, \ldots, n\} \right\}.$$

If $\mathbf{a} \in \mathbf{R}^n_{\geq 0}$, then $P_\sigma(\mathbf{a}) \in \mathbf{R}^n_{\geq 0}$ for all $\sigma \in S_n$. In particular, if $\mathbf{a} \in \mathbf{R}^n_{\geq 0}$, then $\mathbf{a}^{\downarrow} \in \mathbf{R}^n_{\geq 0}$ and $\mathbf{a}^{\uparrow} \in \mathbf{R}^n_{\geq 0}$.

Let $\mathbf{a}$ and $\mathbf{b}$ be vectors in $\mathbf{R}^n$ and let

$$\mathbf{a}^{\downarrow} = \begin{pmatrix} a_1^{\downarrow} \\ \vdots \\ a_n^{\downarrow} \end{pmatrix} \quad \text{and} \quad \mathbf{b}^{\downarrow} = \begin{pmatrix} b_1^{\downarrow} \\ \vdots \\ b_n^{\downarrow} \end{pmatrix}$$

be the associated decreasing vectors. Thus,

$$a_1^{\downarrow} \geq \cdots \geq a_n^{\downarrow} \quad \text{and} \quad b_1^{\downarrow} \geq \cdots \geq b_n^{\downarrow}.$$

The vector $\mathbf{a}$ *majorizes* the vector $\mathbf{b}$ if the vectors $\mathbf{a}$ and $\mathbf{b}$ have the same coordinate sum and if

$$\sum_{i=1}^{k} b_i^{\downarrow} \leq \sum_{i=1}^{k} a_i^{\downarrow} \quad \text{for all } k \in \{1, \ldots, n-1\}.$$

We write

$$\mathbf{b} \prec \mathbf{a}$$

if the vector $\mathbf{b}$ is majorized by the vector $\mathbf{a}$.

Majorization is a partial order but not a total order on the set of vectors in $\mathbf{R}^n_{\geq 0}$ with the same coordinate sum.

For example, in $\mathbf{R}^3_{\geq 0}$, we have

$$\begin{pmatrix} 5 \\ 4 \\ 3 \end{pmatrix} \prec \begin{pmatrix} 7 \\ 3 \\ 2 \end{pmatrix} \quad \text{and} \quad \begin{pmatrix} 2 \\ 6 \\ 5 \end{pmatrix} \prec \begin{pmatrix} 4 \\ 1 \\ 8 \end{pmatrix}.$$

The vectors $\mathbf{a} = \left(\begin{smallmatrix} 7 \\ 5 \\ 1 \end{smallmatrix}\right)$ and $\mathbf{b} = \left(\begin{smallmatrix} 8 \\ 3 \\ 2 \end{smallmatrix}\right)$ are decreasing with coordinate sum 13, but neither vector majorizes the other.

For every positive integer n we have the following linearly ordered majorization chain:

$$\begin{pmatrix} 1/n \\ 1/n \\ 1/n \\ \vdots \\ 1/n \\ 1/n \end{pmatrix} \prec \begin{pmatrix} 1/(n-1) \\ 1/(n-1) \\ 1/(n-1) \\ \vdots \\ 1/(n-1) \\ 0 \end{pmatrix} \prec \cdots \prec \begin{pmatrix} 1/3 \\ 1/3 \\ 1/3 \\ 0 \\ \vdots \\ 0 \end{pmatrix} \prec \begin{pmatrix} 1/2 \\ 1/2 \\ 0 \\ 0 \\ \vdots \\ 0 \end{pmatrix} \prec \begin{pmatrix} 1 \\ 0 \\ 0 \\ 0 \\ \vdots \\ 0 \end{pmatrix}.$$

We have the following majorization relations.

Lemma 6 *If $i_1, \ldots, i_k$ are integers such that $1 \leq i_1 < i_2 < \cdots < i_k \leq n$ and if $x_1, \ldots, x_n$ be real numbers such that $x_1 \geq x_2 \geq \cdots \geq x_n$, then*

$$\sum_{j=1}^{k} x_{i_j} \leq \sum_{j=1}^{k} x_j.$$

Proof We have $i_1 \geq 1$. If $j \in \{1, \ldots, k-1\}$ and $i_j \geq j$, then $i_{j+1} > i_j \geq j$ and so $i_{j+1} \geq j+1$. Therefore, $i_j \geq j$ for all $j \in \{1, \ldots, k\}$ and so $x_{i_j} \leq x_j$ for all $j \in \{1, \ldots, k\}$. The inequality follows immediately. □

The following majorization inequality is due to Chong [1].

Lemma 7 *For all integers $\ell \geq 2$ and all vectors $\mathbf{a}_1, \mathbf{a}_2, \ldots, \mathbf{a}_\ell \in \mathbf{R}^n$,*

$$(\mathbf{a}_1 + \mathbf{a}_2 + \cdots + \mathbf{a}_\ell)^{\downarrow} \prec \mathbf{a}_1^{\downarrow} + \mathbf{a}_2^{\downarrow} + \cdots \mathbf{a}_\ell^{\downarrow}.$$

Proof For all $j \in \{1, \ldots, \ell\}$, let

$$\mathbf{a}_j = \begin{pmatrix} a_{1,j} \\ \cdots \\ a_{n,j} \end{pmatrix}$$

and choose $\rho_i \in S_n$ such that

$$\mathbf{a}_j^{\downarrow} = \begin{pmatrix} a_{\rho^{-1}(1),\,j} \\ \cdots \\ a_{\rho^{-1}(n),j} \end{pmatrix}.$$

Choose $\sigma \in S_n$ such that

$$(\mathbf{a}_1 + \mathbf{a}_2 + \cdots + \mathbf{a}_\ell)^{\downarrow} = \begin{pmatrix} \sum_{j=1}^{\ell} a_{\sigma^{-1}(1),j} \\ \vdots \\ \sum_{j=1}^{\ell} a_{\sigma^{-1}(n),j} \end{pmatrix}.$$

Let

$$\mathbf{b} = (\mathbf{a}_1 + \mathbf{a}_2 + \cdots + \mathbf{a}_\ell)^{\downarrow} = \begin{pmatrix} b_1 \\ \vdots \\ b_n \end{pmatrix}$$

and

$$\mathbf{c} = \mathbf{a}_1^{\downarrow} + \mathbf{a}_2^{\downarrow} + \cdots \mathbf{a}_\ell^{\downarrow} = \begin{pmatrix} c_1 \\ \vdots \\ c_n \end{pmatrix}.$$

The coordinates in the vectors $\mathbf{a}_1^{\downarrow}, \ldots, \mathbf{a}_\ell^{\downarrow}$ are decreasing and so, by Lemma 6,

$$\sum_{i=1}^{k} a_{\sigma^{-1}(i),j} \leq \sum_{i=1}^{k} a_{\rho^{-1}(i),j}$$

for all $j \in \{1, \ldots, \ell\}$ and $k \in \{1, \ldots, n-1\}$ and so

$$\begin{aligned} \sum_{i=1}^{k} b_i &= \sum_{i=1}^{k} \sum_{j=1}^{\ell} a_{\sigma^{-1}(i),j} = \sum_{j=1}^{\ell} \sum_{i=1}^{k} a_{\sigma^{-1}(i),j} \\ &\leq \sum_{j=1}^{\ell} \sum_{i=1}^{k} a_{\rho^{-1}(i),j} = \sum_{i=1}^{k} \sum_{j=1}^{\ell} a_{\rho^{-1}(i),j} \\ &= \sum_{i=1}^{k} c_j. \end{aligned}$$

This completes the proof. □

Lemma 8 *Let* $\mathbf{a}_1, \mathbf{a}_2, \ldots, \mathbf{a}_\ell \in \mathbf{R}^n$ *and* $\mathbf{b}_1, \mathbf{b}_2, \ldots, \mathbf{b}_\ell \in \mathbf{R}^n$ *with* $\mathbf{b}_i \prec \mathbf{a}_i$ *for all* $i \in \{1, \ldots, \ell\}$. *If* $\lambda_1, \ldots, \lambda_\ell \in \mathbf{R}_{\geq 0}$, *then*

$$\lambda_1 \mathbf{b}_1 + \lambda_2 \mathbf{b}_2 + \cdots + \lambda_\ell \mathbf{b}_\ell \prec \lambda_1 \mathbf{a}_1 + \lambda_2 \mathbf{a}_2 + \cdots \lambda_\ell \mathbf{a}_\ell.$$

Let $\mathbf{a}, \mathbf{b}_1, \mathbf{b}_2, \ldots, \mathbf{b}_\ell \in \mathbf{R}^n$ *with* $\mathbf{b}_i \prec \mathbf{a}$ *for all* $i \in \{1, \ldots, \ell\}$. *If* $\lambda_1, \ldots, \lambda_\ell \in \mathbf{R}_{\geq 0}$ *and* $\sum_{i=1}^{\ell} \lambda_i = 1$, *then*

$$\lambda_1 \mathbf{b}_1 + \lambda_2 \mathbf{b}_2 + \cdots + \lambda_\ell \mathbf{b}_\ell \prec \mathbf{a}.$$

Proof This follows from Lemma 7. □

In $\mathbf{R}^n$, for all $j \in \{1, \ldots, n\}$ we have

The standard basis vectors $\{\mathbf{e}_1, \ldots, \mathbf{e}_n\}$ in $\mathbf{R}^n$ are defined by $\mathbf{e}_j = \begin{pmatrix} \delta_{1,j} \\ \vdots \\ \delta_{n,j} \end{pmatrix}$ for all $j \in \{1, \ldots, n\}$, where $\delta_{i,j}$ is the Kronecker delta. Let $\mathbf{f}_n = \begin{pmatrix} 1/n \\ \vdots \\ 1/n \end{pmatrix} \in \mathbf{R}^n$. We have $\mathbf{e}_j^{\downarrow} = \mathbf{e}_1$ for all $j \in \{1, \ldots, n\}$ and $\mathbf{f}_n^{\downarrow} = \mathbf{f}_n$.

Theorem 2 *For every vector* $\mathbf{a} = \begin{pmatrix} a_1 \\ \vdots \\ a_n \end{pmatrix}$ *in* $\mathbf{R}^n_{\geq 0}$ *with* $\sum_{i=1}^n a_i = 1$, *there is the majorization inequality*

$$\mathbf{f}_n \prec \mathbf{a} \prec \mathbf{e}_k$$

for all $k \in \{1, \ldots, n\}$. *Moreover,*

$$\mathbf{a} \prec \mathbf{f}_n \quad \textit{implies} \quad \mathbf{a} = \mathbf{f}_n$$

and

$$\mathbf{e}_1 \prec \mathbf{a} \quad \textit{implies}\ \mathbf{a} = \mathbf{e}_k \textit{ for some } k \in \{1, \ldots, n\}.$$

Proof Let $\mathbf{a}^{\downarrow} = \begin{pmatrix} a_1^{\downarrow} \\ \vdots \\ a_n^{\downarrow} \end{pmatrix}$, where $a_1^{\downarrow} \geq \cdots \geq a_n^{\downarrow} \geq 0$ and $\sum_{i=1}^n a_i^{\downarrow} = 1$. For all $k \in \{1, 2, \ldots, n\}$, we have $\mathbf{e}_k^{\downarrow} = \mathbf{e}_1 = \begin{pmatrix} \delta_{1,1} \\ \vdots \\ \delta_{n,1} \end{pmatrix}$ and

$$\sum_{i=1}^{k} a_i^{\downarrow} \leq \sum_{i=1}^{n} a_i^{\downarrow} = 1 = \delta_{1,1} = \sum_{i=1}^{k} \delta_{i,1}$$

and so $\mathbf{a} \prec \mathbf{e}_1 = \mathbf{e}_k^{\downarrow}$.

If

$$\sum_{i=1}^{k} a_i^{\downarrow} < \frac{k}{n}$$

for some $k \in \{1, \ldots, n\}$, then

$$ka_k^{\downarrow} \leq \sum_{i=1}^{k} a_i^{\downarrow} < \frac{k}{n}$$

implies

$$a_k^{\downarrow} < \frac{1}{n}$$

and so

$$1 = \sum_{i=1}^{n} a_i^{\downarrow} = \sum_{i=1}^{k} a_i^{\downarrow} + \sum_{i=k+1}^{n} a_i^{\downarrow} < \frac{k}{n} + \frac{n-k}{n} = 1$$

which is absurd. Therefore, $k/n \leq \sum_{i=1}^{k} a_i^{\downarrow}$ for all $k \in \{1, \ldots, n\}$ and $\mathbf{f}_n \prec \mathbf{a}$.

If $\mathbf{a} \prec \mathbf{f}_n$, then, for all $k \in \{1, \ldots, n\}$, we have

$$ka_k^{\downarrow} \leq \sum_{i=1}^{k} a_i^{\downarrow} \leq \frac{k}{n}$$

and so

$$a_k^{\downarrow} \leq \frac{1}{n}.$$

The coordinate sum $\sum_{k=1}^{n} a_k^{\downarrow} = 1$ implies that $a_k^{\downarrow} = 1/n$ for all k and $\mathbf{a} = \mathbf{f}_n$.

If $\mathbf{e}_1 \prec \mathbf{a}$, then $1 = \delta_{1,1} \leq a_1^{\downarrow} \leq 1$ and so $a_1^{\downarrow} = 1$. It follows from $\sum_{i=1}^{n} a_i^{\downarrow} = 1$ that $a_i^{\downarrow} = 0$ for all $i \in \{2, 3, \ldots, n\}$ and so $\mathbf{a}^{\downarrow} = \mathbf{e}_1$ and $\mathbf{a} = \mathbf{e}_k$ for some $k \in \{1, \ldots, n\}$. This completes the proof. □

Lemma 9 *Let $P = (p_{i,j})$ be a column stochastic $n \times n$ matrix. If $\mathbf{a} = \begin{pmatrix} a_1 \\ \vdots \\ a_n \end{pmatrix} \in \mathbf{R}^n_{\geq 0}$ and $P(\mathbf{a}) = \mathbf{b} = \begin{pmatrix} b_1 \\ \vdots \\ b_n \end{pmatrix}$, then $\mathbf{b} \in \mathbf{R}^n_{\geq 0}$ and*

$$\sum_{i=1}^{n} b_i = \sum_{j=1}^{n} a_j.$$

Proof For all $i \in \{1, \ldots, n\}$, we have

$$b_i = \sum_{j=1}^{n} p_{i,j} a_j \geq 0$$

and so $\mathbf{b} \in \mathbf{R}^n_{\geq 0}$. Moreover,

$$\sum_{i=1}^{n} b_i = \sum_{i=1}^{n}\sum_{j=1}^{n} p_{i,j}a_j = \sum_{j=1}^{n} a_j \sum_{i=1}^{n} p_{i,j} = \sum_{j=1}^{n} a_j \operatorname{colsum}_j(P) = \sum_{j=1}^{n} a_j.$$

This completes the proof. □

Theorem 3 *Let $P = (p_{i,j})$ be a nonnegative $n \times n$ matrix. The following are equivalent:*

(i) $P(\mathbf{a}) \prec \mathbf{a}$ *for all vectors* $\mathbf{a} \in \mathbf{R}^n_{\geq 0}$.
(ii) $P(\mathbf{e}_j) \prec \mathbf{e}_j$ *for all* $j \in \{1, \ldots, n\}$ *and* $P(\mathbf{f}_n) \prec \mathbf{f}_n$.
(iii) The matrix $P = (p_{i,j})$ *is doubly stochastic.*

Proof Clearly, (i) implies (ii).

Assume (ii). For all $j \in \{1, \ldots, n\}$, the jth column of the matrix P is

$$\operatorname{col}_j(P) = \begin{pmatrix} p_{1,j} \\ \vdots \\ p_{n,j} \end{pmatrix} = P(\mathbf{e}_j).$$

If $P(\mathbf{e}_j) \prec \mathbf{e}_j$, then

$$1 = \operatorname{colsum}(\mathbf{e}_j) = \operatorname{colsum}_j(P) = \sum_{i=1}^{n} p_{i,j}$$

and so P is column stochastic.

By Theorem 2, the majorization relation $P(\mathbf{f}_n) \prec \mathbf{f}_n$ implies $P(\mathbf{f}_n) = \mathbf{f}_n$. The ith coordinate of the vector $\mathbf{f}_n$ is $1/n$ and the ith coordinate of the vector $\mathbf{f}_n$ is $\sum_{j=1}^{n} \frac{p_{i,j}}{n}$. Therefore,

$$\frac{1}{n} = \sum_{j=1}^{n} \frac{p_{i,j}}{n} = \frac{1}{n} \operatorname{rowsum}_i(P)$$

and so $\operatorname{rowsum}_i(P) = 1$ for all $i \in \{1, \ldots, n\}$ and the matrix P is also row stochastic. Thus, (ii) implies (iii).

Let P be a doubly stochastic matrix. Let $\mathbf{a} \in \mathbf{R}^n_{\geq 0}$ and $\mathbf{b} = P(\mathbf{a})$. We shall prove that $\mathbf{b}^{\downarrow} \prec \mathbf{a}^{\downarrow}$.

There exist permutations σ and τ in S_n such that

$$P_\sigma(\mathbf{a}) = \mathbf{a}^{\downarrow} \qquad \text{and} \qquad P_\tau(\mathbf{b}) = \mathbf{b}^{\downarrow}$$

and so

$$\mathbf{b}^{\downarrow} = P_\tau(\mathbf{b}) = P_\tau P(\mathbf{a}) = P_\tau P P_{\sigma^{-1}}\left(\mathbf{a}^{\downarrow}\right) = Q\mathbf{a}^{\downarrow}.$$

The matrices P_τ and $P_{\sigma^{-1}}$ are doubly stochastic and so the matrix $Q = P_\tau P P_{\sigma^{-1}} = (q_{i,j})$ is doubly stochastic.

For $k \in \{1, \ldots, n-1\}$ and $j \in \{1, \ldots, n\}$, define the partial column sums

$$c_j^{(k)} = \sum_{i=1}^{k} q_{i,j}.$$

We have

$$0 \leq c_j^{(k)} \leq \sum_{i=1}^{n} q_{i,j} = \operatorname{colsum}_j(Q) = 1$$

and

$$\sum_{j=1}^{n} c_j^{(k)} = \sum_{j=1}^{n} \sum_{i=1}^{k} q_{i,j} = \sum_{i=1}^{k} \sum_{j=1}^{n} q_{i,j} = \sum_{i=1}^{k} \operatorname{rowsum}_i(Q) = k.$$

Let

$$\mathbf{a}^{\downarrow} = \begin{pmatrix} a_1^{\downarrow} \\ \vdots \\ a_n^{\downarrow} \end{pmatrix} \quad \text{and} \quad \mathbf{b}^{\downarrow} = \begin{pmatrix} b_1^{\downarrow} \\ \vdots \\ b_n^{\downarrow} \end{pmatrix}.$$

From Lemma 9 we obtain $\sum_{i=1}^{n} b_i^{\downarrow} = \sum_{i=1}^{n} a_i^{\downarrow}$. The inequalities $a_1^{\downarrow} \geq a_2^{\downarrow} \geq \cdots \geq a_n^{\downarrow}$ imply that

$$b_1^{\downarrow} = \sum_{j=1}^{n} q_{1,j} a_j^{\downarrow} \leq a_1^{\downarrow} \sum_{j=1}^{n} q_{1,j} = a_1^{\downarrow} \operatorname{rowsum}_1(Q) = a_1^{\downarrow}.$$

For all $k \in \{2, \ldots, n-1\}$ we have

$$\begin{aligned}
\sum_{i=1}^{k} b_i^{\downarrow} &= \sum_{i=1}^{k} \sum_{j=1}^{n} q_{i,j} a_j^{\downarrow} = \sum_{j=1}^{n} a_j^{\downarrow} \sum_{i=1}^{k} q_{i,j} = \sum_{j=1}^{n} a_j^{\downarrow} c_j^{(k)} \\
&\leq \sum_{j=1}^{k-1} c_j^{(k)} a_j^{\downarrow} + a_k^{\downarrow} \sum_{j=k}^{n} c_j^{(k)} = \sum_{j=1}^{k-1} c_j^{(k)} a_j^{\downarrow} + a_k^{\downarrow} \left(k - \sum_{j=1}^{k-1} c_j^{(k)} \right) \\
&= \sum_{j=1}^{k-1} c_j^{(k)} (a_j^{\downarrow} - a_k^{\downarrow}) + k a_k^{\downarrow} \leq \sum_{j=1}^{k-1} (a_j^{\downarrow} - a_k^{\downarrow}) + k a_k^{\downarrow} \\
&= \sum_{j=1}^{k} a_j^{\downarrow}.
\end{aligned}$$

Therefore, $\mathbf{b} \prec \mathbf{a}$ and (iii) implies (i). This completes the proof. □

4 The Hardy-Littlewood-Pólya T-transformation

Let $\mathbf{a}, \mathbf{b} \in \mathbf{R}^n$. In Theorem 3 we proved that if P is an $n \times n$ doubly stochastic matrix, then $\mathbf{a}$ majorizes $P(\mathbf{a})$, that is, if $\mathbf{b} = P(\mathbf{a})$, then $\mathbf{b} \prec \mathbf{a}$. In this section we prove the converse: If $\mathbf{b} \prec \mathbf{a}$, then there is a doubly stochastic matrix P such that $\mathbf{b} = P(\mathbf{a})$. The proof is based on the following elementary lemma.

Lemma 10 *Let a_k, a_ℓ, c_k, and c_ℓ be real numbers such that*

$$a_\ell + a_k = c_\ell + c_k \tag{1}$$

and

$$a_\ell < c_\ell \leq c_k < a_k. \tag{2}$$

There is a unique 2×2 doubly stochastic matrix $T_{k,\ell}$ such that

$$T_{k,\ell}\begin{pmatrix} a_k \\ a_\ell \end{pmatrix} = \begin{pmatrix} c_k \\ c_\ell \end{pmatrix}.$$

Moreover,

$$T_{k,\ell} = \begin{pmatrix} \lambda & 1-\lambda \\ 1-\lambda & \lambda \end{pmatrix}$$

where

$$\lambda = \frac{c_k - a_\ell}{a_k - a_\ell} = \frac{a_k - c_\ell}{a_k - a_\ell}.$$

Proof Every 2×2 doubly stochastic matrix is of the form $\begin{pmatrix} \lambda & 1-\lambda \\ 1-\lambda & \lambda \end{pmatrix}$ for some $\lambda \in [0, 1]$. We have

$$\begin{pmatrix} \lambda & 1-\lambda \\ 1-\lambda & \lambda \end{pmatrix}\begin{pmatrix} a_k \\ a_\ell \end{pmatrix} = \begin{pmatrix} c_k \\ c_\ell \end{pmatrix}$$

if and only if

$$\lambda a_k + (1-\lambda)a_\ell = c_k$$

and

$$(1-\lambda)a_k + \lambda a_\ell = c_\ell.$$

The unique solution of the first equation is

$$\lambda_1 = \frac{c_k - a_\ell}{a_k - a_\ell}$$

and the unique solution of the second equation is

$$\lambda_2 = \frac{a_k - c_\ell}{a_k - a_\ell}.$$

Equation (1) implies that $\lambda_1 = \lambda_2 = \lambda$. Inequality (2) implies that $0 < \lambda < 1$. This completes the proof. □

For example, if $\begin{pmatrix} a_2 \\ a_5 \end{pmatrix} = \begin{pmatrix} 4 \\ -10 \end{pmatrix}$ and $\begin{pmatrix} c_2 \\ c_5 \end{pmatrix} = \begin{pmatrix} 1 \\ -7 \end{pmatrix}$, then $\lambda = 11/14$ and

$$T_{2,5}\begin{pmatrix} 4 \\ -10 \end{pmatrix} = \begin{pmatrix} 11/14 & 3/14 \\ 3/14 & 11/14 \end{pmatrix}\begin{pmatrix} 4 \\ -10 \end{pmatrix} = \begin{pmatrix} 1 \\ -7 \end{pmatrix}.$$

Lemma 11 *Let $k < \ell$ and let $\mathbf{a} = \begin{pmatrix} a_1 \\ \vdots \\ a_n \end{pmatrix}$ and $\mathbf{c} = \begin{pmatrix} c_1 \\ \vdots \\ c_n \end{pmatrix}$ be vectors in $\mathbf{R}^n$ such that*

$$a_\ell + a_k = c_\ell + c_k$$

$$a_\ell < c_\ell \le c_k < a_k$$

and

$$c_i = a_i \text{ for all } i \ne k, \ell.$$

There is a doubly stochastic $n \times n$ matrix $T = (t_{i,j})$ such that $T\mathbf{a} = \mathbf{c}$.

Proof Let

$$\lambda = \frac{c_k - a_\ell}{a_k - a_\ell} = \frac{a_k - c_\ell}{a_k - a_\ell}.$$

Let $T = (t_{i,j})$ be the $n \times n$ matrix defined as follows:

$$t_{k,k} = t_{\ell,\ell} = \lambda$$

$$t_{k,\ell} = t_{\ell,k} = 1 - \lambda$$

and

$$t_{i,j} = \delta_{i,j} \qquad \text{for all } (i,j) \notin \{(k,k), (k,\ell), (\ell,k), (\ell,\ell)\}$$

where $\delta_{i,j}$ is the Kronecker delta. The matrix T looks like

$$T = \begin{pmatrix} 1 \cdots 0 & 0 & 0 \cdots 0 & 0 & 0 \cdots 0 \\ \vdots & & & & \vdots \\ 0 \cdots 1 & 0 & 0 \cdots 0 & 0 & 0 \cdots 0 \\ 0 \cdots 0 & \lambda & 0 \cdots 0 & 1-\lambda & 0 \cdots 0 \\ 0 \cdots 0 & 0 & 1 \cdots 0 & 0 & 0 \cdots 0 \\ \vdots & & & & \vdots \\ 0 \cdots 0 & 0 & 0 \cdots 1 & 0 & 0 \cdots 0 \\ 0 \cdots 0 & 1-\lambda & 0 \cdots 0 & \lambda & 0 \cdots 0 \\ 0 \cdots 0 & 0 & 0 \cdots 0 & 0 & 1 \cdots 0 \\ \vdots & & & & \vdots \\ 0 \cdots 0 & 0 & 0 \cdots 0 & 0 & 0 \cdots 1 \end{pmatrix}.$$

It follows from Lemma 10 that $T\mathbf{a} = \mathbf{c}$. One sees directly that T is doubly stochastic. This completes the proof. □

The matrix T constructed in Lemma 11 is the *Hardy-Littlewood-Pólya T-transformation.*

For example, if $n = 5$, $k = 2$, $\ell = 3$, and

$$\mathbf{a} = \begin{pmatrix} 3 \\ 4 \\ 5 \\ 8 \\ -10 \end{pmatrix} \quad \text{and} \quad \mathbf{b} = \begin{pmatrix} 3 \\ 1 \\ 5 \\ 8 \\ -7 \end{pmatrix}$$

then

$$T = \begin{pmatrix} 1 & 0 & 0 & 0 & 0 \\ 0 & 11/14 & 0 & 0 & 3/14 \\ 0 & 0 & 1 & 0 & 0 \\ 0 & 0 & 0 & 1 & 0 \\ 0 & 3/14 & 0 & 0 & 11/14 \end{pmatrix}$$

and $T\mathbf{a} = \mathbf{b}$.

The *Hamming distance* between vectors $\mathbf{a} = \begin{pmatrix} a_1 \\ \vdots \\ a_n \end{pmatrix}$ and $\mathbf{b} = \begin{pmatrix} b_1 \\ \vdots \\ b_n \end{pmatrix}$ in $\mathbf{R}^n$ is

$$d_H(\mathbf{a}, \mathbf{b}) = \text{card}\{i \in \{1, \ldots, n\} : a_i \neq b_i\}. \tag{3}$$

Lemma 12 *For all vectors* $\mathbf{a}, \mathbf{b}, \mathbf{c} \in \mathbf{R}^n$,

(i) $d_H(\mathbf{a}, \mathbf{b}) \in \{0, 1, 2, \ldots, n\}$
(ii) $d_H(\mathbf{a}, \mathbf{b}) = 0$ *if and only if* $\mathbf{a} = \mathbf{b}$
(iii) *Symmetry:* $d_H(\mathbf{a}, \mathbf{b}) = d_H(\mathbf{b}, \mathbf{a})$
(iv) *Triangle inequality:* $d_H(\mathbf{a}, \mathbf{b}) \leq d_H(\mathbf{a}, \mathbf{c}) + d_H(\mathbf{c}, \mathbf{b})$
(v) *If* $\mathbf{a} \neq \mathbf{b}$ *and* $\sum_{i=1}^n a_i = \sum_{i=1}^n b_i$, *then* $d_H(\mathbf{a}, \mathbf{b}) \geq 2$.

Proof Properties (i), (ii), and (iii) are immediate.

To prove (iv), which is the triangle inequality for vectors **a**, **b**, **c**, let

$$\begin{aligned} I &= \{i \in \{1, \ldots, n\} : a_i \neq c_i\} \\ J &= \{j \in \{1, \ldots, n\} : c_j \neq b_j\} \\ K &= \{k \in \{1, \ldots, n\} : a_k \neq b_k\} \\ L &= \{1, \ldots, n\} \setminus (I \cup J). \end{aligned}$$

If $\ell \in L$, then $\ell \notin I$ and $\ell \notin J$, and so $a_\ell = c_\ell$ and $c_\ell = b_\ell$, hence $a_\ell = b_\ell$. If $k \in K$, then $a_k \neq b_k$ and so $k \notin L$, that is, $k \in \{1, \ldots, n\} \setminus L = I \cup J$. It follows that $K \subseteq I \cup J$ and

$$d_H(\mathbf{a}, \mathbf{b}) = |K| \leq |I \cup J| \leq |I| + |J| = d_H(\mathbf{a}, \mathbf{c}) + d_H(\mathbf{c}, \mathbf{b}).$$

To prove (v), we observe that if $\mathbf{a} \neq \mathbf{b}$, then $a_{i_1} \neq b_{i_1}$ for some $i_1 \in \{1, 2, \ldots, n\}$. If $\sum_{i=1}^n a_i = \sum_{i=1}^n b_i$, then

$$\sum_{\substack{i=1 \\ i \neq i_1}}^n a_i \neq \sum_{\substack{i=1 \\ i \neq i_1}}^n b_i$$

and so $a_{i_2} \neq b_{i_2}$ for some $i_2 \in \{1, 2, \ldots, n\} \setminus \{i_1\}$. Therefore, $d_H(\mathbf{a}, \mathbf{b}) \geq 2$. This completes the proof. □

Lemma 13 *Let* **a** *and* **b** *be decreasing vectors in* $\mathbf{R}^n$ *such that* $\mathbf{b} \prec \mathbf{a}$ *and* $d_H(\mathbf{a}, \mathbf{b}) > 2$. *There exists a vector* **c** *in* $\mathbf{R}^n$ *such that*

(i) **c** *is decreasing,*
(ii) $\mathbf{b} \prec \mathbf{c} \prec \mathbf{a}$,
(iii) $d_H(\mathbf{a}, \mathbf{c}) = 2$ *and* $2 \leq d_H(\mathbf{b}, \mathbf{c}) < d_H(\mathbf{b}, \mathbf{a})$,
(iv) $T\mathbf{a} = \mathbf{c}$ *for some doubly stochastic matrix* T.

Proof Let $\mathbf{a} = \begin{pmatrix} a_1 \\ \vdots \\ a_n \end{pmatrix}$ and $\mathbf{b} = \begin{pmatrix} b_1 \\ \vdots \\ b_n \end{pmatrix}$. We have $\mathbf{a} \neq \mathbf{b}$ because $d_H(\mathbf{a}, \mathbf{b}) > 2$ and so there is a smallest integer k_1 such that $b_{k_1} \neq a_{k_1}$. Therefore, $b_i = a_i$ for all $i \in \{1, \ldots, k_1 - 1\}$ and $\sum_{i=1}^{k_1-1} b_i = \sum_{i=1}^{k_1-1} a_i$. The inequality $\sum_{i=1}^{k_1} b_i \leq \sum_{i=1}^{k_1} a_i$ implies that $b_{k_1} \leq a_{k_1}$ and so $b_{k_1} < a_{k_1}$. The equality $\sum_{i=1}^n b_i = \sum_{i=1}^n a_i$ implies that $b_{\ell_1} > a_{\ell_1}$ for some integer $\ell_1 > k_1$.

Let ℓ be the smallest integer such that $b_\ell > a_\ell$, and let k be the largest integer such that $k < \ell$ and $b_k < a_k$. Because the vectors **a** and **b** are decreasing, we have

$$a_\ell < b_\ell \leq b_k < a_k$$

$$a_i = b_i \qquad \text{for all } \; i \in \{k+1, k+2, \ldots, \ell - 1\}$$

and

$$\delta = \min(a_k - b_k, b_\ell - a_\ell) > 0.$$

Define the vector $\mathbf{c} = \begin{pmatrix} c_1 \\ \vdots \\ c_n \end{pmatrix}$ as follows:

$$\begin{aligned} c_k &= a_k - \delta \\ c_\ell &= a_\ell + \delta \\ c_i &= a_i \quad \text{for all } i \neq k, \ell. \end{aligned}$$

It follows that

$$c_1 \geq c_2 \geq \cdots \geq c_{k-1}$$

$$c_{k+1} \geq c_{k+2} \geq \cdots \geq c_{\ell-1}$$

and

$$c_{\ell+1} \geq c_{\ell+2} \geq \cdots \geq c_n.$$

We have

$$c_{k+1} = a_{k+1} = b_{k+1} \leq b_k \leq a_k - \delta = c_k < a_k \leq a_{k-1} = c_{k-1}$$

and

$$c_{\ell+1} = a_{\ell+1} \leq a_\ell < a_\ell + \delta = c_\ell \leq b_\ell \leq b_{\ell-1} = a_{\ell-1} = c_{\ell-1}$$

and so the vector $\mathbf{c}$ is decreasing. This proves (i).

If $j \in \{1, \ldots, k-1\}$, then

$$\sum_{i=1}^{j} c_i = \sum_{i=1}^{j} a_i.$$

If $j \in \{k, \ldots, \ell-1\}$, then

$$\sum_{i=1}^{j} c_i = \sum_{i=1}^{k-1} a_i + (a_k - \delta) + \sum_{i=k+1}^{j} a_i < \sum_{i=1}^{j} a_i.$$

If $j \in \{\ell, \ldots, n\}$, then

$$\sum_{i=1}^{j} c_i = \sum_{i=1}^{k-1} a_i + (a_k - \delta) + \sum_{i=k+1}^{\ell-1} a_i + (a_\ell + \delta) + \sum_{i=\ell+1}^{j} a_i = \sum_{i=1}^{j} a_i.$$

Therefore, $\mathbf{c} \prec \mathbf{a}$.

Next we prove that $\mathbf{b} \prec \mathbf{c}$. If $j \in \{1, \ldots, k-1\}$, then

$$\sum_{i=1}^{j} b_i \leq \sum_{i=1}^{j} a_i = \sum_{i=1}^{j} c_i.$$

We have

$$b_k \leq a_k - \delta = c_k$$

$$c_\ell = a_\ell + \delta \leq b_\ell$$

$$b_i = a_i = c_i \quad \text{for } i \in \{k+1, \ldots, \ell-1\}.$$

Therefore, if $j \in \{k, \ldots, \ell-1\}$, then

$$\sum_{i=1}^{j} b_i = \sum_{i=1}^{k-1} b_i + b_k + \sum_{i=k+1}^{j} b_i \leq \sum_{i=1}^{k-1} a_i + (a_k - \delta) + \sum_{i=k+1}^{j} c_i = \sum_{i=1}^{j} c_i.$$

If $j \in \{\ell, \ldots, n-1\}$, then

$$\sum_{i=1}^{j} b_i \leq \sum_{i=1}^{j} a_i = \sum_{\substack{i=1 \\ i \neq k,\ell}}^{j} a_i + (a_k - \delta) + (a_\ell + \delta) = \sum_{i=1}^{j} c_i.$$

Finally,

$$\sum_{i=1}^{n} b_i = \sum_{i=1}^{n} a_i = \sum_{\substack{i=1 \\ i \neq k,\ell}}^{n} a_i + (a_k - \delta) + (a_\ell + \delta) = \sum_{i=1}^{n} c_i.$$

Therefore, $\mathbf{b} \prec \mathbf{c}$. This proves (ii).

Because $a_\ell < c_\ell \leq c_k < a_k$ and $a_i = c_i$ for all $i \neq k, \ell$, we have

$$d_H(\mathbf{a}, \mathbf{c}) = 2.$$

The triangle inequality for Hamming distance implies

$$2 < d_H(\mathbf{a}, \mathbf{b}) \leq d_H(\mathbf{a}, \mathbf{c}) + d_H(\mathbf{b}, \mathbf{c}) = 2 + d_H(\mathbf{b}, \mathbf{c})$$

and so $d_H(\mathbf{b}, \mathbf{c}) > 0$. Lemma 12 implies $d_H(\mathbf{b}, \mathbf{c}) \geq 2$.

Because $a_k \neq b_k$ and $a_\ell \neq b_\ell$, we have

$$\begin{aligned} r = d_H(\mathbf{a}, \mathbf{b}) - 2 &= \operatorname{card}\{i : b_i \neq a_i \text{ and } i \neq k, \ell\} \\ &= \operatorname{card}\{i : b_i \neq c_i \text{ and } i \neq k, \ell\}. \end{aligned}$$

In the vector $\mathbf{c}$, we have $c_k = b_k$ or $c_\ell = b_\ell$, and so $d_H(\mathbf{b}, \mathbf{c}) = r$ or $r + 1$. Thus, $d_H(\mathbf{b}, \mathbf{c}) \leq r + 1 < d_H(\mathbf{a}, \mathbf{b})$. This proves (iii).

To prove (iv), we observe that the vectors $\mathbf{a}$ and $\mathbf{c}$ satisfy the conditions of Lemma 11 and so there is a Hardy-Littlewood-Pólya T-transformation such that $T\mathbf{a} = \mathbf{c}$. The Hardy-Littlewood-Pólya T-transformations are doubly stochastic. This completes the proof. □

For example, in $\mathbf{R}^3_{\geq 0}$ the vectors

$$\mathbf{a} = \begin{pmatrix} 8 \\ 3 \\ 1 \end{pmatrix} \quad \text{and} \quad \mathbf{b} = \begin{pmatrix} 6 \\ 4 \\ 2 \end{pmatrix}$$

satisfy $\mathbf{b} \prec \mathbf{a}$ and $d_H(\mathbf{b}, \mathbf{a}) = 3$. The construction in the proof of Lemma 13 yields the vector

$$\mathbf{c} = \begin{pmatrix} 7 \\ 4 \\ 1 \end{pmatrix}$$

which satisfies

$$\mathbf{b} \prec \mathbf{c} \prec \mathbf{a} \quad \text{and} \quad d_H(\mathbf{b}, \mathbf{c}) = d_H(\mathbf{a}, \mathbf{c}) = 2.$$

The doubly stochastic matrices

$$T_1 = \begin{pmatrix} 4/5 & 1/5 & 0 \\ 1/5 & 4/5 & 0 \\ 0 & 0 & 1 \end{pmatrix} \quad \text{and} \quad T_2 = \begin{pmatrix} 5/6 & 0 & 1/6 \\ 0 & 1 & 0 \\ 1/6 & 0 & 5/6 \end{pmatrix}$$

satisfy $T_1(\mathbf{a}) = \mathbf{c}$ and $T_2(\mathbf{c}) = \mathbf{b}$. The doubly stochastic matrix

$$T = T_2 T_1 = \begin{pmatrix} 2/3 & 1/6 & 1/6 \\ 1/5 & 4/5 & 0 \\ 2/15 & 1/30 & 5/6 \end{pmatrix}$$

satisfies $T\mathbf{a} = \mathbf{b}$.

Theorem 4 *Let* $\mathbf{a}$ *and* $\mathbf{b}$ *be distinct vectors in* $\mathbf{R}^n$ *such that* $\mathbf{b} \prec \mathbf{a}$. *For some positive integer* $r \leq d_H(\mathbf{a}, \mathbf{b})$, *there is a sequence of vectors* $\mathbf{c}_0, \mathbf{c}_1, \ldots, \mathbf{c}_r \in \mathbf{R}^n$ *such that*

$$\mathbf{b} = \mathbf{c}_r \prec \mathbf{c}_{r-1} \prec \cdots \prec \mathbf{c}_1 \prec \mathbf{c}_0 = \mathbf{a} \tag{4}$$

and

$$d_H(\mathbf{c}_{i-1}, \mathbf{c}_i) = 2 \quad \text{for all } i \in \{1, \ldots, r\}. \tag{5}$$

Proof The proof is by induction on the Hamming distance $h = d_H(\mathbf{a}, \mathbf{b})$. We have $h \geq 2$ because $\mathbf{a} \neq \mathbf{b}$.

If $h = 2$, then conditions (4) and (5) are satisfied with $r = 1$, $\mathbf{c}_0 = \mathbf{a}$, and $\mathbf{c}_1 = \mathbf{b}$.

Let $h > 2$, and assume that the Theorem holds if $d_H(\mathbf{a}, \mathbf{b}) \leq h - 1$. Let vectors $\mathbf{a}$ and $\mathbf{b}$ satisfy $\mathbf{a} \prec \mathbf{b}$ and $d_H(\mathbf{a}, \mathbf{b}) = h$. Let $\mathbf{c}_0 = \mathbf{a}$. By Lemma 13, there exists a vector $\mathbf{c}_1 \in \mathbf{R}^n$ such that

$$\mathbf{b} \prec \mathbf{c}_1 \prec \mathbf{c}_0 = \mathbf{a} \qquad \text{and} \qquad d_H(\mathbf{c}_0, \mathbf{c}_1) = 2 \tag{6}$$

and

$$2 \leq d_H(\mathbf{b}, \mathbf{c}_1) < d_H(\mathbf{a}, \mathbf{b}).$$

Thus,

$$d_H(\mathbf{b}, \mathbf{c}_1) \leq h - 1.$$

Applying the induction hypothesis to the vectors $\mathbf{b} \prec \mathbf{c}_1$, we obtain an integer $r \leq h$ and a sequence of vectors $\mathbf{c}_1, \ldots, \mathbf{c}_r$ such that

$$\mathbf{b} = \mathbf{c}_r \prec \mathbf{c}_{r-1} \prec \cdots \prec \mathbf{c}_1 \tag{7}$$

and $d_H(\mathbf{c}_{i-1}, \mathbf{c}_i) = 2$ for all $i \in \{2, \ldots, r\}$. Conjoining the majorization chains (6) and (7) completes the induction. □

Let $\mathbf{a}$ and $\mathbf{b}$ be vectors in $\mathbf{R}^n$ such that $\mathbf{b} \prec \mathbf{a}$. A sequence of vectors $\mathbf{c}_0, \mathbf{c}_1, \ldots, \mathbf{c}_r \in \mathbf{R}^n$ such that

$$\mathbf{b} = \mathbf{c}_r \prec \mathbf{c}_{r-1} \prec \cdots \prec \mathbf{c}_1 \prec \mathbf{c}_0 = \mathbf{a}$$

and

$$d_H(\mathbf{c}_{i-1}, \mathbf{c}_i) = 2 \qquad \text{for all } i \in \{1, \ldots, r\}$$

is called a *strict majorization chain of length r* from $\mathbf{a}$ to $\mathbf{b}$. For example,

$$\mathbf{b} = \begin{pmatrix} 6 \\ 4 \\ 2 \end{pmatrix} \prec \begin{pmatrix} 6 \\ 5 \\ 1 \end{pmatrix} \prec \begin{pmatrix} 8 \\ 3 \\ 1 \end{pmatrix} = \mathbf{a}$$

and

$$\mathbf{b} = \begin{pmatrix} 6 \\ 4 \\ 2 \end{pmatrix} \prec \begin{pmatrix} 6 \\ 5 \\ 1 \end{pmatrix} \prec \begin{pmatrix} 7 \\ 4 \\ 1 \end{pmatrix} \prec \begin{pmatrix} 8 \\ 3 \\ 1 \end{pmatrix} = \mathbf{a}$$

are strict majorization chains of lengths 2 and 3, respectively.

Theorem 5 *Let* $\mathbf{a}$ *and* $\mathbf{b}$ *be distinct vectors in* $\mathbf{R}^n_{\geq 0}$. *If* $\mathbf{b} \prec \mathbf{a}$, *then there is a doubly stochastic matrix* P *such that* $\mathbf{b} = P(\mathbf{a})$. *Moreover, if* $d_H(\mathbf{a}, \mathbf{b}) = h$, *then the matrix* P *is a product of at most* $h - 1$ *Hardy-Littlewood-Pólya* T*-transformations.*

Proof The proof is by induction on the Hamming distance $d_H(\mathbf{a}, \mathbf{b})$.

We have $d_H(\mathbf{a}, \mathbf{b}) \geq 2$. If $d_H(\mathbf{a}, \mathbf{b}) = 2$, then the vector $\mathbf{b}$ satisfies the conditions of the vector $\mathbf{c}$ in Lemma 11, and so $\mathbf{b} = T\mathbf{a}$ for some Hardy-Littlewood-Pólya transformation T.

Let $h = d_H(\mathbf{a}, \mathbf{b}) \geq 3$ and suppose the Theorem is true for vectors $\mathbf{a}$ and $\mathbf{b}$ with $d_H(\mathbf{a}, \mathbf{b}) \leq h - 1$. Let $\mathbf{a}, \mathbf{b} \in \mathbf{R}^n_{\geq 0}$ satisfy $\mathbf{b} \prec \mathbf{a}$ and $d_H(\mathbf{a}, \mathbf{b}) = h$. By Lemma 13, there is a vector $\mathbf{c} \in \mathbf{R}^n_{\geq 0}$ such that

$$\mathbf{b} \prec \mathbf{c} = T\mathbf{a}$$

for some Hardy-Littlewood-Pólya T-transformation, and

$$d_H(\mathbf{b}, \mathbf{c}) \leq d_H(\mathbf{b}, \mathbf{a}) - 1 \leq h - 1.$$

By the induction hypothesis, there is a doubly stochastic matrix P' such that $\mathbf{b} = P'\mathbf{c}$ and P' is the product of at most $h - 2$ Hardy-Littlewood-Pólya T-transformations. Then $P = P'T$ is a doubly stochastic matrix that is the product of at most $h - 1$ Hardy-Littlewood-Pólya T-transformations, and $\mathbf{b} = P'\mathbf{c} = P'T\mathbf{a} = P(\mathbf{a})$. This completes the proof. □

Theorem 6 *Let mba and* $\mathbf{b}$ *be vectors in* $\mathbf{R}^n_{\geq 0}$. *The vector* $\mathbf{a}$ *majorizes* $\mathbf{b}$ *if and only if there exists a doubly stochastic matrix* P *such that* $P(\mathbf{a}) = \mathbf{b}$.

Proof This follows immediately from Theorems 3 and 5. □

The following result is a fundamental theorem that relates majorization and convexity.

Theorem 7 *Let* $\mathbf{a}$ *and* $\mathbf{b}$ *be vectors in* $\mathbf{R}^n_{\geq 0}$. *Let* $K(\mathbf{a})$ *be the* S_n*-permutohedron generated by* $\mathbf{a}$. *The vector* $\mathbf{a}$ *majorizes* $\mathbf{b}$ *if and only if* $\mathbf{b} \in K(\mathbf{a})$.

Proof Let P_σ be the permutation matrix constructed from the permutation $\sigma \in S_n$. The permutohedron $K(\mathbf{a})$ is the convex hull of the set of vectors $\{P_\sigma(\mathbf{a}) : \sigma \in S_n\}$. By Theorem 7, if $\mathbf{a}$ majorizes $\mathbf{b}$, then there is a doubly stochastic matrix P such that $P(\mathbf{a}) = \mathbf{b}$. The Birkhoff-von Neumann theorem in convexity theory states that every doubly stochastic matrix is a convex combination of permutation matrices. It follows that there are permutations $\sigma_1, \ldots, \sigma_k \in S_n$ and positive numbers $\lambda_1, \ldots, \lambda_k$ such that

$$\sum_{i=1}^{k} \lambda_i = 1$$

and

$$P = \sum_{i=1}^{k} \lambda_i P_{\sigma_i}.$$

Therefore,

$$\mathbf{b} = P(\mathbf{a}) = \sum_{i=1}^{k} \lambda_i P_{\sigma_i}(\mathbf{a})$$

is a vector in $K(\mathbf{a})$.

Conversely, if $\mathbf{b} \in K(\mathbf{a})$, then there exist permutations $\sigma_1, \ldots, \sigma_k \in S_n$ and positive numbers $\lambda_1, \ldots, \lambda_k$ such that

$$\sum_{i=1}^{k} \lambda_i = 1$$

and

$$\mathbf{b} = \sum_{i=1}^{k} \lambda_i P_{\sigma_i}(\mathbf{a}) = P(\mathbf{a})$$

where the matrix $P = \sum_{i=1}^{k} \lambda_i P_{\sigma_i}$ is doubly stochastic. Theorem 3 implies that $\mathbf{a}$ majorizes $\mathbf{b}$. This completes the proof. □

We also have a majorization inequality for convex functions. Let $I \subseteq \mathbf{R}$ be a finite or infinite interval. The function $f : I \to \mathbf{R}$ is *convex* if, for all $a_1, a_2 \in I$ and $t_1, t_2 \in [0, 1]$ with $t_1 + t_2 = 1$,

$$f(t_1 a_1 + t_2 a_2) \leq t_1 f(a_1) + t_2 f(a_2). \tag{8}$$

It follows by induction on n that if $a_1, a_2, \ldots, a_n \in I$ and $t_1, t_2, \ldots, t_n \in [0, 1]$ with $t_1 + t_2 + \cdots + t_n = 1$, then

$$f(t_1 a_1 + t_2 a_2 + \cdots t_n a_n) \leq t_1 f(a_1) + t_2 f(a_2) + \cdots + t_n f(a_n). \tag{9}$$

Theorem 8 *Let f be a convex function on the interval I. Let $\mathbf{a}, \mathbf{b} \in \mathbf{R}^n_{\geq 0}$ with $\mathbf{b} \prec \mathbf{a}$. If $\mathbf{a} = \begin{pmatrix} a_1 \\ \vdots \\ a_n \end{pmatrix}$ and $\mathbf{b} = \begin{pmatrix} b_1 \\ \vdots \\ b_n \end{pmatrix}$ with $a_i, b_i \in I$ for all $i \in \{1, \ldots, n\}$, then*

$$f(b_1) + \cdots + f(b_n) \leq f(a_1) + \cdots + f(a_n).$$

Proof By Theorem 7, there is a doubly stochastic matrix $P = (p_{i,j})$ such that $\mathbf{b} = P(\mathbf{a})$. Equivalently,

$$b_i = \sum_{j=1}^{n} p_{i,j} a_j$$

for all $i \in \{1, \ldots, n\}$. Because P is doubly stochastic, we have

$$\text{rowsum}_i(P) = \sum_{j=1}^{n} p_{i,j} = 1$$

for all $i \in \{1, \ldots, n\}$ and

$$\text{colsum}_j(P) = \sum_{i=1}^{n} p_{i,j} = 1$$

for all $j \in \{1, \ldots, n\}$. The convexity of the function f implies that

$$f(b_i) = f\left(\sum_{j=1}^{n} p_{i,j} a_j\right) \leq \sum_{j=1}^{n} p_{i,j} f\left(a_j\right)$$

for all $i \in \{1, \ldots, n\}$ and so

$$\sum_{i=1}^{n} f(b_i) \leq \sum_{i=1}^{n}\sum_{j=1}^{n} p_{i,j} f\left(a_j\right) = \sum_{j=1}^{n}\left(\sum_{i=1}^{n} p_{i,j}\right) f\left(a_j\right) = \sum_{j=1}^{n} f\left(a_j\right).$$

This completes the proof. □

Acknowledgements I thank Darij Grinberg for helpful comments on this paper.

References

1. K.-M. Chong, An induction principle for spectral and rearrangement inequalities, Trans. Amer. Math. Soc. 196 (1974), 371–383.
2. G. H. Hardy, J. E. Littlewood, and G. Pólya, *Some simple inequalities satisfied by convex functions*, Messenger of Math. **58** (1929), 145–152.
3. G. H. Hardy, J. E. Littlewood, and G. Pólya, *Inequalities*, Cambridge Mathematical Library, Cambridge University Press, Cambridge, 1988, reprint of the 1952 edition.
4. A. W. Marshall, I. Olkin, and B. C, Arnold, *Inequalities: Theory of Majorization and its Applications*, Springer Series in Statistics, Springer, New York, 2011.
5. L. Mirsky, Results and problems in the theory of doubly-stochastic matrices, Z. Wahrscheinlichkeitstheorie und Verw. Gebiete 1 (1962/63), 319–334.
6. R. F. Muirhead, *Some methods applicable to identities and inequalities of symmetric algebraic functions of n letters*, Proc. Edinburgh Math. Soc. **21** (1903), 144–157.
7. M. B. Nathanson, The Muirhead-Rado inequality, 2: Symmetric means and inequalities, in: Combinatorial and Additive Number Theory VI, Springer-Nature, New York, 2024.
8. R. Rado, *An inequality*, J. London Math. Soc. **27** (1952), 1–6.

The Muirhead-Rado Inequality, 2: Symmetric Means and Inequalities

Melvyn B. Nathanson

Abstract This paper gives elementary proofs of the Muirhead and Rado inequalities.

2020 Mathematics Subject Classification: 05E05 · 11B83 · 15B51 · 26D05 · 26D15 · 52A20 · 52A30 · 52A40

1 Symmetric Means of Monomial Functions

A *nonnegative vector* is a vector with nonnegative coordinates. The *nonnegative octant* in $\mathbf{R}^n$ is

$$\mathbf{R}^n{}_{\geq 0} = \left\{ \mathbf{a} = \begin{pmatrix} a_1 \\ \vdots \\ a_n \end{pmatrix} \in \mathbf{R}^n : a_i \geq 0 \text{ for all } i \in \{1, \ldots, n\} \right\}.$$

A *positive vector* is a vector with positive coordinates. The *positive octant* in $\mathbf{R}^n$ is

$$\mathbf{R}^n{}_{>0} = \left\{ \mathbf{x} = \begin{pmatrix} x_1 \\ \vdots \\ x_n \end{pmatrix} \in \mathbf{R}^n : x_i > 0 \text{ for all } i \in \{1, \ldots, n\} \right\}.$$

Supported in part by PSC-CUNY Research Award Program grant 66197-00 54.

M. B. Nathanson (✉)
Lehman College (CUNY), Bronx, New York 10468, US
e-mail: melvyn.nathanson@lehman.cuny.edu

M. B. Nathanson (ed.), *Combinatorial and Additive Number Theory VI*,
Springer Proceedings in Mathematics & Statistics 464,
https://doi.org/10.1007/978-3-031-65064-2_17

Every nonnegative vector $\mathbf{a}$ defines the *monomial function* $\mathbf{x}^{\mathbf{a}}$ on the positive octant as follows: For $\mathbf{a} = \begin{pmatrix} a_1 \\ \vdots \\ a_n \end{pmatrix} \in \mathbf{R}^n_{\geq 0}$ and $\mathbf{x} = \begin{pmatrix} x_1 \\ \vdots \\ x_n \end{pmatrix} \in \mathbf{R}^n_{>0}$, let

$$\mathbf{x}^{\mathbf{a}} = x_1^{a_1} x_2^{a_2} \cdots x_n^{a_n}.$$

We call $\mathbf{a}$ the *exponent vector* of the monomial $\mathbf{x}^{\mathbf{a}}$. Note that $\mathbf{x}^{\mathbf{a}} > 0$ for all $\mathbf{x} \in \mathbf{R}^n_{>0}$.

Let S_n be the symmetric group. The *symmetric mean* of the monomial function $\mathbf{x}^{\mathbf{a}}$ is the function

$$[\mathbf{x}^{\mathbf{a}}]_{S_n} = \frac{1}{n!} \sum_{\sigma \in S_n} x_{\sigma(1)}^{a_1} x_{\sigma(2)}^{a_2} \cdots x_{\sigma(n)}^{a_n}.$$

For every subgroup G of S_n, the *G-symmetric mean* of the monomial $\mathbf{x}^{\mathbf{a}}$ is the function

$$[\mathbf{x}^{\mathbf{a}}]_G = \frac{1}{|G|} \sum_{\sigma \in G} x_{\sigma(1)}^{a_1} x_{\sigma(2)}^{a_2} \cdots x_{\sigma(n)}^{a_n}.$$

Let $\sigma \in S_n$. If $i \in \{1, \ldots, n\}$ and $j = \sigma(i)$, then

$$x_{\sigma(i)}^{a_i} = x_j^{a_{\sigma^{-1}(j)}}$$

and

$$\prod_{i=1}^{n} x_{\sigma(i)}^{a_i} = \prod_{j=1}^{n} x_j^{a_{\sigma^{-1}(j)}}.$$

Because $\{\sigma \in G\} = \{\sigma^{-1} \in G\}$, we have

$$\begin{aligned} [\mathbf{x}^{\mathbf{a}}]_G &= \frac{1}{|G|} \sum_{\sigma \in G} \prod_{i=1}^{n} x_{\sigma(i)}^{a_i} = \frac{1}{|G|} \sum_{\sigma \in G} \prod_{j=1}^{n} x_j^{a_{\sigma^{-1}(j)}} \\ &= \frac{1}{|G|} \sum_{\sigma \in G} \prod_{j=1}^{n} x_j^{a_{\sigma(j)}} = \frac{1}{|G|} \sum_{\sigma \in G} x_1^{a_{\sigma(1)}} x_2^{a_{\sigma(2)}} \cdots x_n^{a_{\sigma(n)}}. \end{aligned}$$

If $\mathbf{x} \in \mathbf{R}^n_{>0}$ is the constant vector with all coordinates equal to x, then

$$[\mathbf{x}^{\mathbf{a}}]_G = \frac{1}{|G|} \sum_{\sigma \in G} \prod_{j=1}^{n} x^{a_{\sigma(j)}} = \frac{1}{|G|} \sum_{\sigma \in G} x^{\sum_{i=1}^{n} a_i} = x^{\sum_{i=1}^{n} a_i}. \tag{1}$$

Let G be a subgroup of S_n and let $\mathbf{a} = \begin{pmatrix} a_1 \\ \vdots \\ a_n \end{pmatrix}$ and $\mathbf{b} = \begin{pmatrix} b_1 \\ \vdots \\ b_n \end{pmatrix}$ be nonnegative vectors such that

$$\sum_{j=1}^{n} a_j = \sum_{j=1}^{n} b_j.$$

If

$$\left[\mathbf{x}^{\mathbf{b}}\right]_G < \left[\mathbf{x}^{\mathbf{a}}\right]_G \tag{2}$$

for all vectors $\mathbf{x} \in \mathbf{R}^n_{>0}$, then (2) is called a *monomial inequality with respect to the subgroup G* determined by **a** and **b**. A *monomial inequality* is a monomial inequality with respect to the symmetric group S_n. Here are some examples. If $\mathbf{a} = \binom{7}{3} \in \mathbf{R}^2_{\geq 0}$, then $\mathbf{x}^{\mathbf{a}} = x_1^7 x_2^3$. The S_2-symmetric mean of $\mathbf{x}^{\mathbf{a}}$ is

$$[\mathbf{x}^{\mathbf{a}}]_{S_2} = \frac{1}{2}\left(x_1^7 x_2^3 + x_2^7 x_1^3\right) = \frac{1}{2}\left(x_1^7 x_2^3 + x_1^3 x_2^7\right).$$

If $\mathbf{b} = \binom{6}{4} \in \mathbf{R}^2_{\geq 0}$, then $\mathbf{x}^{\mathbf{b}} = x_1^6 x_2^4$ and the S_2-symmetric mean of $\mathbf{x}^{\mathbf{b}}$ is

$$[\mathbf{x}^{\mathbf{b}}]_{S_2} = \frac{1}{2}\left(x_1^6 x_2^4 + x_1^4 x_2^6\right).$$

It is straightforward to check that

$$x_1^6 x_2^4 + x_1^4 x_2^6 < x_1^7 x_2^3 + x_1^3 x_2^7 \tag{3}$$

for all nonconstant positive vectors $\mathbf{x} = \binom{x_1}{x_2}$ and so $[\mathbf{x}^{\mathbf{b}}]_{S_2} < [\mathbf{x}^{\mathbf{a}}]_{S_2}$ is a monomial inequality.

Another example. Associated to the vector $\mathbf{e}_1 = \begin{pmatrix} 1 \\ 0 \\ \vdots \\ 0 \end{pmatrix} \in \mathbf{R}^n{}_{\geq 0}$ is the monomial function

$$\mathbf{x}^{\mathbf{e}_1} = x_1^1 x_2^0 \cdots x_n^0 = x_1.$$

For every ordered pair of integers (i, j) with $i, j \in \{1, \ldots, n\}$, there are $(n-1)!$ permutations $\sigma \in S_n$ with $\sigma(i) = j$. In particular, for all $j \in \{1, \ldots, n\}$, there are $(n-1)!$ permutations $\sigma \in S_n$ with $\sigma(1) = j$. For all $\mathbf{x} = \begin{pmatrix} x_1 \\ \vdots \\ x_n \end{pmatrix} \in \mathbf{R}^n_{>0}$, we have

$$\begin{aligned}
[\mathbf{x}^{\mathbf{e}_1}]_{S_n} &= \frac{1}{n!} \sum_{\sigma \in S_n} x_{\sigma(1)}^1 x_{\sigma(2)}^0 \cdots x_{\sigma(n)}^0 \\
&= \frac{1}{n!} \sum_{\sigma \in S_n} x_{\sigma(1)} = \frac{1}{n!} \sum_{j=1}^{n} (n-1)! x_j \\
&= \frac{1}{n} \sum_{j=1}^{n} x_j.
\end{aligned}$$

Thus, the symmetric mean of the monomial $\mathbf{x}^{\mathbf{e}_1}$ is the *arithmetic mean* of the positive numbers $x_1, \ldots, x_n$.

Associated to the vector $\mathbf{f}_n = \begin{pmatrix} 1/n \\ 1/n \\ \vdots \\ 1/n \end{pmatrix} \in \mathbf{R}^n{}_{\geq 0}$ is the monomial function

$$\mathbf{x}^{\mathbf{f}_n} = x_1^{1/n} x_2^{1/n} \cdots x_n^{1/n} = (x_1 x_2 \cdots x_n)^{1/n}$$

which is is the *geometric mean* of the positive numbers $x_1, \ldots, x_n$. Because $x_{\sigma(1)} x_{\sigma(2)} \cdots x_{\sigma(n)} = x_1 x_2 \cdots x_n$ for all $\sigma \in S_n$, the symmetric mean

$$\begin{aligned} [\mathbf{x}^{\mathbf{f}_n}]_{S_n} &= \frac{1}{n!} \sum_{\sigma \in S_n} \left(x_{\sigma(1)} x_{\sigma(2)} \cdots x_{\sigma(n)} \right)^{1/n} \\ &= \frac{1}{n!} \sum_{\sigma \in S_n} (x_1 x_2 \cdots x_n)^{1/n} \\ &= (x_1 x_2 \cdots x_n)^{1/n} \end{aligned}$$

is also the geometric mean of the numbers $x_1, \ldots, x_n$. The arithmetic and geometric mean inequality is equivalent to the monomial inequality

$$[\mathbf{x}^{\mathbf{f}_n}]_{S_n} < [\mathbf{x}^{\mathbf{e}_1}]_{S_n}$$

for all nonconstant vectors $\begin{pmatrix} x_1 \\ \vdots \\ x_n \end{pmatrix} \in \mathbf{R}^n{}_{>0}$.

There are infinitely many ordered pairs $(\mathbf{a}, \mathbf{b})$ of nonnegative vectors such that

$$[\mathbf{x}^{\mathbf{b}}]_{S_n} < [\mathbf{x}^{\mathbf{a}}]_{S_n}$$

for all nonconstant vectors $\mathbf{x} \in \mathbf{R}^n{}_{>0}$. Muirhead's inequality (Theorem 4) and its converse (Theorem 5) classify the pairs of vectors **a** and **b** that determine monomial inequalities. Richard Rado (Theorems 9 and 10) classified monomial inequalities with respect to a subgroup G of S_n. The purpose of this paper is to prove these results.

2 Simple Inequalities

Here is a trivial fact, from which we obtain infinitely many monomial inequalities.

Theorem 1 *The product of two real numbers is positive if and only if both numbers are positive or both numbers are negative.*

Corollary 1 *Let* $\mathbf{a} = \binom{a_1}{a_2} \in \mathbf{R}^2_{>0}$ *and* $\mathbf{x} = \binom{x_1}{x_2} \in \mathbf{R}^2_{>0}$. *If* $x_1 \neq x_2$, *then*

$$x_1^{a_1}x_2^{a_2} + x_1^{a_2}x_2^{a_1} < x_1^{a_1+a_2} + x_2^{a_1+a_2}. \tag{4}$$

If $x_1 = x_2$, *then*

$$x_1^{a_1}x_2^{a_2} + x_1^{a_2}x_2^{a_1} = x_1^{a_1+a_2} + x_2^{a_1+a_2}.$$

Proof If $0 < x_2 < x_1$, then $x_1^{a_1} - x_2^{a_1} > 0$ and $x_1^{a_2} - x_2^{a_2} > 0$. If $0 < x_1 < x_2$, then $x_1^{a_1} - x_2^{a_1} < 0$ and $x_1^{a_2} - x_2^{a_2} < 0$. Thus, the numbers $x_1^{a_1} - x_2^{a_1}$ and $x_1^{a_2} - x_2^{a_2}$ are both positive or both negative. Applying Theorem 1 to these numbers gives

$$x_1^{a_1+a_2} + x_2^{a_1+a_2} - x_1^{a_1}x_2^{a_2} - x_1^{a_2}x_2^{a_1} = (x_1^{a_1} - x_2^{a_1})(x_1^{a_2} - x_2^{a_2}) > 0.$$

This proves (4). If $x_1 = x_2$, then

$$x_1^{a_1}x_2^{a_2} + x_1^{b}x_2^{a_1} = x_1^{a_1+a_2} + x_2^{a_1+a_2}.$$

This completes the proof. □

Corollary 2 *Let* ρ, δ, *and* Δ *satisfy the inequalities*

$$\rho > 0 \quad \textit{and} \quad |\delta| < \Delta.$$

Let $\mathbf{x} = \binom{x_1}{x_2} \in \mathbf{R}^2_{>0}$. *If* $x_1 \neq x_2$, *then*

$$x_1^{\rho+\delta}x_2^{\rho-\delta} + x_1^{\rho-\delta}x_2^{\rho+\delta} < x_1^{\rho+\Delta}x_2^{\rho-\Delta} + x_1^{\rho-\Delta}x_2^{\rho+\Delta}. \tag{5}$$

Proof We have $\Delta + \delta > 0$ and $\Delta - \delta > 0$. Applying Corollary 1 with $\mathbf{a} = \begin{pmatrix} \Delta + \delta \\ \Delta - \delta \end{pmatrix} \in \mathbf{R}^2_{>0}$ gives

$$x_1^{\Delta+\delta}x_2^{\Delta-\delta} + x_1^{\Delta-\delta}x_2^{\Delta+\delta} < x_1^{2\Delta} + x_2^{2\Delta}.$$

Multiplying this inequality by $(x_1x_2)^{-\Delta}$ gives

$$x_1^{\delta}x_2^{-\delta} + x_1^{-\delta}x_2^{\delta} < x_1^{\Delta}x_2^{-\Delta} + x_1^{-\Delta}x_2^{\Delta}.$$

Multiplying by $(x_1x_2)^{\rho}$ completes the proof. □

For example, letting $\rho = 5, \delta = 1$, and $\Delta = 2$ in inequality (5) gives inequality (3). Similarly, for all nonconstant vectors $\mathbf{x} \in \mathbf{R}^2_{>0}$, we have the chain of inequalities

$$\begin{aligned} 2x_1^5x_2^5 &< x_1^6x_2^4 + x_1^4x_2^6 < x_1^7x_2^3 + x_1^3x_2^7 \\ &< x_1^8x_2^2 + x_1^2x_2^8 < x_1^9x_2 + x_1x_2^9 < x_1^{10} + x_2^{10}. \end{aligned}$$

These inequalities fail if $x_1 = 1$ and $x_2 = -1$ or, more generally, if $x_1x_2 \leq 0$.

Theorem 2 *Let vectors* $\mathbf{a} = \binom{a_1}{a_2} \in \mathbf{R}^2_{\geq 0}$ *and* $\mathbf{b} = \binom{b_1}{b_2} \in \mathbf{R}^2_{\geq 0}$ *satisfy*

$$a_2 < b_2 \leq b_1 < a_1 \qquad \text{and} \qquad a_1 + a_2 = b_1 + b_2.$$

If $\mathbf{x} = \binom{x_1}{x_2} \in \mathbf{R}^2_{>0}$ *with* $x_1 \neq x_2$, *then* $[\mathbf{x}]^{\mathbf{b}}_{S_2} < [\mathbf{x}]^{\mathbf{a}}_{S_2}$ *or, equivalently,*

$$x_1^{b_1} x_2^{b_2} + x_1^{b_2} x_2^{b_1} < x_1^{a_1} x_2^{a_2} + x_1^{a_2} x_2^{a_1}. \tag{6}$$

Proof Let

$$\rho = \frac{a_1 + a_2}{2} = \frac{b_1 + b_2}{2}$$

and

$$\delta = \frac{b_1 - b_2}{2} \qquad \text{and} \qquad \Delta = \frac{a_1 - a_2}{2}.$$

We have

$$\rho > 0 \qquad \text{and} \qquad 0 \leq \delta < \Delta.$$

Applying inequality (5) with

$$\rho + \delta = b_1 \qquad \text{and} \qquad \rho - \delta = b_2$$

and

$$\rho + \Delta = a_1 \qquad \text{and} \qquad \rho - \Delta = a_2$$

gives inequality (6). This completes the proof. □

This is Muirhead's inequality for monomials in two variables. Inequality (3) is the special case $\mathbf{a} = \binom{7}{3}$ and $\mathbf{b} = \binom{6}{4}$.

3 Symmetric Means of Functions of *n* Variables

Recall the *Kronecker delta*

$$\delta_{i,j} = \begin{cases} 1 & \text{if } i = j \\ 0 & \text{if } i \neq j. \end{cases}$$

The *standard basis* for $\mathbf{R}^n$ is $\mathcal{E} = \{\mathbf{e}_1, \ldots, \mathbf{e}_n\}$, where $\mathbf{e}_j = \begin{pmatrix} \delta_{1,j} \\ \vdots \\ \delta_{n,j} \end{pmatrix}$. Every permutation $\sigma \in S_n$ defines a linear operator on $\mathbf{R}^n$ by $\sigma(\mathbf{e}_j) = \mathbf{e}_{\sigma(j)}$. If $\mathbf{x} = \sum_{j=1}^n x_j \mathbf{e}_j \in \mathbf{R}^n$, then

$$\sigma(\mathbf{x}) = \sum_{j=1}^n x_j \sigma(\mathbf{e}_j) = \sum_{j=1}^n x_j \mathbf{e}_{\sigma(j)} = \sum_{j=1}^n x_{\sigma^{-1}(j)} \mathbf{e}_j$$

and so

$$\sigma \begin{pmatrix} x_1 \\ \vdots \\ x_n \end{pmatrix} = \begin{pmatrix} x_{\sigma^{-1}(1)} \\ \vdots \\ x_{\sigma^{-1}(n)} \end{pmatrix}. \tag{7}$$

Let Ω be a subset of $\mathbf{R}^n$ that is closed under the action of S_n, that is, $\mathbf{x} \in \Omega$ implies $\sigma(\mathbf{x}) \in \Omega$ for all $\sigma \in S_n$. Let $\mathcal{F}(\Omega)$ be the set of real-valued functions defined on Ω.

For every function $f \in \mathcal{F}(\Omega)$ and every permutation σ in the symmetric group S_n, define the function $\sigma f \in \mathcal{F}(\Omega)$ by

$$(\sigma f)(\mathbf{x}) = f\left(\sigma^{-1}(\mathbf{x})\right). \tag{8}$$

For all $\sigma, \tau \in S_n$ we have

$$\begin{aligned} (\tau(\sigma f))(\mathbf{x}) &= (\sigma f)\left(\tau^{-1}(\mathbf{x})\right) = f\left(\sigma^{-1}\left(\tau^{-1}(\mathbf{x})\right)\right) \\ &= f\left(\left(\sigma^{-1}\tau^{-1}\right)(\mathbf{x})\right) = f\left((\tau\sigma)^{-1}(\mathbf{x})\right) \\ &= ((\tau\sigma) f)(\mathbf{x}). \end{aligned}$$

Thus,

$$(\tau\sigma) f = \tau(\sigma f)$$

and (8) defines an action of the group S_n on the function space $\mathcal{F}(\Omega)$.

Let G be a subgroup of S_n of order $|G|$. The *G-symmetric mean* of the function f is the function $[f]_G \in \mathcal{F}(\Omega)$ defined by

$$[f]_G = \frac{1}{|G|} \sum_{\sigma \in G} \sigma f.$$

Because $G = \{\sigma^{-1} : \sigma \in G\}$, we have

$$\begin{aligned} [f]_G(\mathbf{x}) &= \frac{1}{|G|} \sum_{\sigma \in G} (\sigma f)(\mathbf{x}) = \frac{1}{|G|} \sum_{\sigma \in G} f\left(\sigma^{-1}(\mathbf{x})\right) \\ &= \frac{1}{|G|} \sum_{\sigma \in G} f(\sigma(\mathbf{x})). \end{aligned}$$

The *symmetric mean* of the function f is the S_n-symmetric mean.

Lemma 1 *Let G be a subgroup of S_n and let τ in G. For all functions $f \in \mathcal{F}(\Omega)$,*

$$[\tau f]_G = [f]_G$$

Proof From the identity

$$G\tau = \{\sigma\tau : \sigma \in G\} = G$$

we obtain

$$[\tau f]_G = \frac{1}{|G|}\sum_{\sigma\in G}\sigma(\tau f) = \frac{1}{|G|}\sum_{\sigma\in G}(\sigma\tau) f = \frac{1}{|G|}\sum_{\sigma\in G}\sigma f = [f]_G.$$

This completes the proof. □

Lemma 2 *Let G be a subgroup of S_n. For every nonnegative vector* $\mathbf{a} = \begin{pmatrix} a_1 \\ \vdots \\ a_n \end{pmatrix} \mathbf{R}^n_{\geq 0}$ *and every permutation $\tau \in G$,*

$$[\mathbf{x}^{\mathbf{a}}]_G = [\mathbf{x}^{\tau(\mathbf{a})}]_G.$$

Proof From (7) we have

$$\tau(\mathbf{a}) = \begin{pmatrix} a_{\tau^{-1}(1)} \\ \vdots \\ a_{\tau^{-1}(n)} \end{pmatrix} \qquad \text{and} \qquad \mathbf{x}^{\tau(\mathbf{a})} = x_1^{a_{\tau^{-1}(1)}} \cdots x_n^{a_{\tau^{-1}(n)}}.$$

Because $G\tau = \{\sigma\tau : \sigma \in G\} = G$ for all $\tau \in G$, we have

$$\begin{aligned}
[\mathbf{x}^{\tau(\mathbf{a})}]_G &= \frac{1}{n!}\sum_{\sigma\in G} x_{\sigma(1)}^{a_{\tau^{-1}(1)}} \cdots x_{\sigma(n)}^{a_{\tau^{-1}(n)}} \\
&= \frac{1}{n!}\sum_{\sigma\in G} x_1^{a_{\tau^{-1}\sigma^{-1}(1)}} \cdots x_n^{a_{\tau^{-1}\sigma^{-1}(n)}} \\
&= \frac{1}{n!}\sum_{\sigma\in G} x_1^{a_{(\sigma\tau)^{-1}(1)}} \cdots x_n^{a_{(\sigma\tau)^{-1}(n)}} \\
&= \frac{1}{n!}\sum_{\sigma\in G} x_1^{a_{\sigma^{-1}(1)}} \cdots x_n^{a_{\sigma^{-1}(n)}} \\
&= [\mathbf{x}^{\mathbf{a}}]_G.
\end{aligned}$$

This completes the proof. □

Lemma 3 *Let G be a subgroup of S_n that contains the transposition $\tau = (j, k)$ with $j < k$. For every nonnegative vector* $\mathbf{a} = \begin{pmatrix} a_1 \\ \vdots \\ a_n \end{pmatrix} \in \mathbf{R}^n_{\geq 0}$ *and positive vector* $\mathbf{x} = \begin{pmatrix} x_1 \\ \vdots \\ x_n \end{pmatrix} \in \mathbf{R}^n_{>0}$,

$$[\mathbf{x}^{\mathbf{a}}]_G = \frac{1}{2|G|}\sum_{\sigma\in G}\left(x_{\sigma(j)}^{a_j}x_{\sigma(k)}^{a_k} + x_{\sigma(j)}^{a_k}x_{\sigma(k)}^{a_j}\right)\prod_{\substack{i=1\\ i\neq j,k}}^{n} x_{\sigma(i)}^{a_i}.$$

Proof The transposition $\rho = (j,k)$ acts as follows:

$$\text{if} \quad \mathbf{a} = \begin{pmatrix} a_1 \\ \vdots \\ a_{j-1} \\ a_j \\ a_{j+1} \\ \vdots \\ a_{k-1} \\ a_k \\ a_{k+1} \\ \vdots \\ a_n \end{pmatrix} \qquad \text{then} \quad \rho(\mathbf{a}) = \begin{pmatrix} a_1 \\ \vdots \\ a_{j-1} \\ a_k \\ a_{j+1} \\ \vdots \\ a_{k-1} \\ a_j \\ a_{k+1} \\ \vdots \\ a_n \end{pmatrix}.$$

By Lemma 2,

$$\begin{aligned} 2[\mathbf{x}^{\mathbf{a}}]_G &= [\mathbf{x}^{\mathbf{a}}]_G + [\mathbf{x}^{\tau(\mathbf{a})}]_G \\ &= \frac{1}{|G|}\sum_{\sigma\in G} x_{\sigma(1)}^{a_1}\cdots x_{\sigma(j)}^{a_j}\cdots x_{\sigma(k)}^{a_k}\cdots x_{\sigma(n)}^{a_n} \\ &\qquad + \frac{1}{|G|}\sum_{\sigma\in G} x_{\sigma(1)}^{a_1}\cdots x_{\sigma(j)}^{a_k}\cdots x_{\sigma(k)}^{a_j}\cdots x_{\sigma(n)}^{a_n} \\ &= \frac{1}{|G|}\sum_{\sigma\in G}\left(x_{\sigma(j)}^{a_j}x_{\sigma(k)}^{a_k} + x_{\sigma(j)}^{a_k}x_{\sigma(k)}^{a_j}\right)\prod_{\substack{i=1 \\ i\neq j,k}}^{n} x_{\sigma(i)}^{a_i}. \end{aligned}$$

Dividing by 2 completes the proof. □

Lemma 4 *Let G be a subgroup of S_n that contains the transposition $\tau = (j,k)$ with $j < k$. Let $\mathbf{u} = \begin{pmatrix} u_1 \\ \vdots \\ u_n \end{pmatrix}$ be a nonnegative vector such that $u_j > u_k$ and let*

$$\rho = \frac{u_j + u_k}{2} \qquad \textit{and} \qquad \Delta = \frac{u_j - u_k}{2}.$$

Then

$$0 < \Delta \leq \rho.$$

Let

$$0 \leq \delta < \Delta.$$

Define the nonnegative vector $\mathbf{v} = \begin{pmatrix} v_1 \\ \vdots \\ v_n \end{pmatrix}$ by

$$v_j = \rho + \delta \qquad \textit{and} \qquad v_k = \rho - \delta$$

and

$$v_i = u_i \qquad \text{if } i \neq j, k.$$

If $\mathbf{x} = \begin{pmatrix} x_1 \\ \vdots \\ x_n \end{pmatrix} \in \mathbf{R}^n{}_{>0}$ *is a nonconstant vector such that* $x_{\sigma(j)} \neq x_{\sigma(k)}$ *for some* $\sigma \in G$, *then*

$$[\mathbf{x}^{\mathbf{v}}]_G < [\mathbf{x}^{\mathbf{u}}]_G.$$

Proof We have $v_i = u_i \geq 0$ for $i \neq j, k$ and

$$0 \leq u_k = \rho - \Delta < \rho - \delta = v_k \leq v_j = \rho + \delta < \rho + \Delta = u_j$$

and so the vector is $\mathbf{v}$ is nonnegative.

Let $\mathbf{x}$ be a nonconstant vector in $\mathbf{R}^n_{>0}$. Applying Lemma 3, we have

$$\begin{aligned}
[\mathbf{x}^{\mathbf{u}}]_G &= \frac{1}{2|G|} \sum_{\sigma \in G} \left(x_{\sigma(j)}^{u_j} x_{\sigma(k)}^{u_k} + x_{\sigma(j)}^{u_k} x_{\sigma(k)}^{u_j} \right) \prod_{\substack{i=1 \\ i \neq j,k}}^{n} x_{\sigma(i)}^{u_i} \\
&= \frac{1}{2|G|} \sum_{\sigma \in G} \left(x_{\sigma(j)}^{\rho+\Delta} x_{\sigma(k)}^{\rho-\Delta} + x_{\sigma(j)}^{\rho-\Delta} x_{\sigma(k)}^{\rho+\Delta} \right) \prod_{\substack{i=1 \\ i \neq j,k}}^{n} x_{\sigma(i)}^{u_i}
\end{aligned}$$

and

$$\begin{aligned}
[\mathbf{x}^{\mathbf{v}}]_G &= \frac{1}{2|G|} \sum_{\sigma \in G} \left(x_{\sigma(j)}^{v_j} x_{\sigma(k)}^{v_k} + x_{\sigma(j)}^{v_k} x_{\sigma(k)}^{v_j} \right) \prod_{\substack{i=1 \\ i \neq j,k}}^{n} x_{\sigma(i)}^{v_i} \\
&= \frac{1}{2|G|} \sum_{\sigma \in G} \left(x_{\sigma(j)}^{\rho+\delta} x_{\sigma(k)}^{\rho-\delta} + x_{\sigma(j)}^{\rho-\delta} x_{\sigma(k)}^{\rho+\delta} \right) \prod_{\substack{i=1 \\ i \neq j,k}}^{n} x_{\sigma(i)}^{u_i}.
\end{aligned}$$

Therefore,

$$\begin{aligned}
&[\mathbf{x}^{\mathbf{u}}]_G - [\mathbf{x}^{\mathbf{v}}]_G \\
&= \frac{1}{2|G|} \sum_{\sigma \in G} \left(x_{\sigma(j)}^{\rho+\Delta} x_{\sigma(k)}^{\rho-\Delta} + x_{\sigma(j)}^{\rho-\Delta} x_{\sigma(k)}^{\rho+\Delta} - x_{\sigma(j)}^{\rho+\delta} x_{\sigma(k)}^{\rho-\delta} - x_{\sigma(j)}^{\rho-\delta} x_{\sigma(k)}^{\rho+\delta} \right) \prod_{\substack{i=1 \\ i \neq j,k}}^{n} x_{\sigma(i)}^{u_i}.
\end{aligned}$$

If $x_{\sigma(j)} = x_{\sigma(k)}$, then

$$x_{\sigma(j)}^{\rho+\Delta} x_{\sigma(k)}^{\rho-\Delta} + x_{\sigma(j)}^{\rho-\Delta} x_{\sigma(k)}^{\rho+\Delta} - x_{\sigma(j)}^{\rho+\delta} x_{\sigma(k)}^{\rho-\delta} - x_{\sigma(j)}^{\rho-\delta} x_{\sigma(k)}^{\rho+\delta} = 0.$$

By (5) of Corollary 2, if $x_{\sigma(j)} \neq x_{\sigma(k)}$, then

$$x_{\sigma(j)}^{\rho+\Delta} x_{\sigma(k)}^{\rho-\Delta} + x_{\sigma(j)}^{\rho-\Delta} x_{\sigma(k)}^{\rho+\Delta} - x_{\sigma(j)}^{\rho+\delta} x_{\sigma(k)}^{\rho-\delta} - x_{\sigma(j)}^{\rho-\delta} x_{\sigma(k)}^{\rho+\delta} > 0.$$

Because $x_{\sigma(j)} \neq x_{\sigma(k)}$ for some $\sigma \in G$, we have $[\mathbf{x}^{\mathbf{u}}]_G - [\mathbf{x}^{\mathbf{v}}]_G > 0$. This completes the proof. □

4 Muirhead's Inequality

The vector $\mathbf{a} = \begin{pmatrix} a_1 \\ \vdots \\ a_n \end{pmatrix} \in \mathbf{R}^n$ is *decreasing* if

$$a_1 \geq a_2 \geq \cdots \geq a_n.$$

Associated to every vector $\mathbf{a} = \begin{pmatrix} a_1 \\ \vdots \\ a_n \end{pmatrix} \in \mathbf{R}^n$ is a unique decreasing vector $\mathbf{a}^{\downarrow} = \begin{pmatrix} a_1^{\downarrow} \\ \vdots \\ a_n^{\downarrow} \end{pmatrix} \in \mathbf{R}^n$ obtained from $\mathbf{a}$ by a rearrangement of coordinates. Thus, $\mathbf{a}^{\downarrow} = \sigma(\mathbf{a})$ for some $\sigma \in S_n$.

For example, from the vector $\mathbf{a} = \begin{pmatrix} 3 \\ 1 \\ 2 \\ 4 \end{pmatrix} \in \mathbf{R}^4$ we obtain the decreasing vector $\mathbf{a}^{\downarrow} = \begin{pmatrix} 4 \\ 3 \\ 2 \\ 1 \end{pmatrix}$. The permutation

$$\sigma = \begin{pmatrix} 1 & 2 & 3 & 4 \\ 2 & 4 & 3 & 1 \end{pmatrix} \in S_4$$

satisfies $\mathbf{a}^{\downarrow} = \sigma(\mathbf{a})$.

Let $\mathbf{a} = \begin{pmatrix} a_1 \\ \vdots \\ a_n \end{pmatrix}$ and $\mathbf{b} = \begin{pmatrix} b_1 \\ \vdots \\ b_n \end{pmatrix}$ be vectors in $\mathbf{R}^n$, and let $\mathbf{a}^{\downarrow} = \begin{pmatrix} a_1^{\downarrow} \\ \vdots \\ a_n^{\downarrow} \end{pmatrix}$ and $\mathbf{b}^{\downarrow} = \begin{pmatrix} b_1^{\downarrow} \\ \vdots \\ b_n^{\downarrow} \end{pmatrix}$ be the corresponding decreasing vectors obtained by permutation of coordinates. The following definition is fundamental: The vector $\mathbf{a}$ *majorizes* the vector $\mathbf{b}$, denoted $\mathbf{b} \prec \mathbf{a}$, if

$$\sum_{i=1}^{k} b_i \leq \sum_{i=1}^{k} a_i \qquad \text{for all } i \in \{1, \ldots, n-1\}$$

and

$$\sum_{i=1}^{n} b_i = \sum_{i=1}^{n} a_i.$$

The vector **a** *strictly majorizes* **b** if $\mathbf{b} \prec \mathbf{a}$ and $\mathbf{b} \neq \mathbf{a}$.

The *Hamming distance* between vectors $\mathbf{a} = \begin{pmatrix} a_1 \\ \vdots \\ a_n \end{pmatrix}$ and $\mathbf{b} = \begin{pmatrix} b_1 \\ \vdots \\ b_n \end{pmatrix}$ in $\mathbf{R}^n$ is

$$d_H(\mathbf{a}, \mathbf{b}) = \text{card}\{i \in \{1, \ldots, n\} : a_i \neq b_i\}. \tag{9}$$

Lemma 5 *Let* $\mathbf{u} = \begin{pmatrix} u_1 \\ \vdots \\ u_n \end{pmatrix}$ *and* $\mathbf{v} = \begin{pmatrix} v_1 \\ \vdots \\ v_n \end{pmatrix}$ *be vectors in* $\mathbf{R}^n{}_{\geq 0}$ *such that* $\mathbf{v} \prec \mathbf{u}$ *and* $d_H(\mathbf{u}, \mathbf{v}) = 2$. *Then*

$$\left[\mathbf{x}^{\mathbf{v}}\right]_{S_n} < \left[\mathbf{x}^{\mathbf{u}}\right]_{S_n}$$

for all nonconstant vectors $\mathbf{x} \in \mathbf{R}^n_{>0}$.

Proof By Lemma 2, we can assume that the vectors **u** and **v** are decreasing.

Hamming distance $d_H(\mathbf{u}, \mathbf{v}) = 2$ implies that there are unique integers $j < k$ such that $u_i = v_i$ for all $i \neq j, k$ and

$$0 \leq u_k < v_k \leq v_j < u_j \qquad \text{and} \qquad v_j + v_k = u_j + u_k.$$

The numbers

$$\rho = \frac{u_j + u_k}{2} = \frac{v_j + v_k}{2}, \qquad \Delta = \frac{u_j - u_k}{2}$$

and

$$\delta = v_j - \frac{u_j + u_k}{2} = \frac{v_j - v_k}{2}$$

satisfy the inequality

$$0 \leq \delta < \Delta \leq \rho.$$

We have

$$\rho + \delta = \frac{u_j + u_k}{2} + \left(v_j - \frac{u_j + u_k}{2}\right) = v_j$$

and

$$\rho - \delta = \frac{v_j + v_k}{2} - \frac{v_j - v_k}{2} = v_k.$$

Applying Lemma 4 completes the proof. □

Theorem 3 *Let* **a** *and* **b** *be distinct vectors in* $\mathbf{R}^n$ *such that* $\mathbf{b} \prec \mathbf{a}$. *For some positive integer* $r \leq d_H(\mathbf{a}, \mathbf{b})$, *there is a sequence of decreasing vectors* $\mathbf{c}_0, \mathbf{c}_1, \ldots, \mathbf{c}_r \in \mathbf{R}^n_{\geq 0}$ *such that*

$$\mathbf{b} = \mathbf{c}_r \prec \mathbf{c}_{r-1} \prec \cdots \prec \mathbf{c}_1 \prec \mathbf{c}_0 = \mathbf{a} \tag{10}$$

and

$$d_H(\mathbf{c}_{i-1}, \mathbf{c}_i) = 2 \quad \textit{for all } i \in \{1, \ldots, r\}. \tag{11}$$

Proof This is Theorem 4 in Nathanson [5]. □

We observe that if $\mathbf{a} = \begin{pmatrix} a_1 \\ \vdots \\ a_n \end{pmatrix}$ and $\mathbf{b} = \begin{pmatrix} b_1 \\ \vdots \\ b_n \end{pmatrix}$ and if $\sum_{i=1}^n a_i = \sum_{i=1}^n b_i$, then, for every $x > 0$, the constant vector $\mathbf{x} = \begin{pmatrix} x \\ \vdots \\ x \end{pmatrix} \in \mathbf{R}^n_{>0}$ satisfies

$$[\mathbf{x}^{\mathbf{b}}]_{S_n} = [\mathbf{x}^{\mathbf{a}}]_{S_n} = x^{\sum_{i=1}^n a_i}.$$

In particular, if $\mathbf{b} \prec \mathbf{a}$, then $[\mathbf{x}^{\mathbf{b}}]_{S_n} = [\mathbf{x}^{\mathbf{a}}]_{S_n}$ for every constant vector $\mathbf{x}$. Muirhead's inequality is a strict inequality for nonconstant vectors.

Theorem 4 (Muirhead's inequality)
Let $\mathbf{a}$ *and* $\mathbf{b}$ *be distinct vectors in* $\mathbf{R}^n_{\geq 0}$. *If* $\mathbf{b} \prec \mathbf{a}$, *then*

$$[\mathbf{x}^{\mathbf{b}}]_{S_n} < [\mathbf{x}^{\mathbf{a}}]_{S_n}$$

for every nonconstant vector $\mathbf{x} \in \mathbf{R}^n_{>0}$.

Proof We give two proofs of this inequality. By Theorem 3, there are decreasing vectors $\mathbf{c}_1, \ldots, \mathbf{c}_r$ and a strict majorization chain

$$\mathbf{b} = \mathbf{c}_r \prec \mathbf{c}_{r-1} \prec \cdots \prec \mathbf{c}_1 \prec \mathbf{c}_0 = \mathbf{a}$$

such that

$$d_H(\mathbf{c}_{i-1}, \mathbf{c}_i) = 2 \qquad \text{for all } i \in \{1, \ldots, r\}.$$

Let $\mathbf{x}$ be a nonconstant vector in $\mathbf{R}^n_{>0}$. The symmetric group S_n contains all transpositions, and so, by Lemma 5 with $\mathbf{u} = \mathbf{c}_{i-1}$ and $\mathbf{v} = c_i$,

$$[\mathbf{x}^{\mathbf{c}_i}]_{S_n} < [\mathbf{x}^{\mathbf{c}_{i-1}}]_{S_n}$$

for all $i \in \{1, \ldots, r\}$. Therefore,

$$[\mathbf{x}^{\mathbf{b}}]_{S_n} = [\mathbf{x}^{\mathbf{c}_r}]_{S_n} < [\mathbf{x}^{\mathbf{c}_{r-1}}]_{S_n} < \cdots < [\mathbf{x}^{\mathbf{c}_1}]_{S_n} < [\mathbf{x}^{\mathbf{c}_0}]_{S_n} = [\mathbf{x}^{\mathbf{a}}]_{S_n}.$$

This completes the first proof. □

The second proof of Theorem 4 uses only the arithmetic and geometric mean inequality.

Proof Let $(c_i)_{i=1}^n$ be a sequence of real numbers such that $c_i \neq 0$ for some i and

$$\sum_{i=1}^{n} c_i = 0.$$

Let $\mathbf{x} = \begin{pmatrix} x_1 \\ \vdots \\ x_n \end{pmatrix}$ be a nonconstant positive vector. For every pair of integers (i, j) with $i, j \in \{1, \ldots, n\}$, there are $(n-1)!$ permutations $\sigma \in S_n$ with $\sigma(i) = j$ and so there are $(n-1)!$ permutations σ such that $x_{\sigma(i)}^{c_i} = x_j^{c_i}$. Therefore,

$$\begin{aligned}
\prod_{\sigma \in S_n} x_{\sigma(1)}^{c_1} x_{\sigma(2)}^{c_2} \cdots x_{\sigma(n)}^{c_n} &= \prod_{\sigma \in S_n} \prod_{i=1}^{n} x_{\sigma(i)}^{c_i} = \prod_{i=1}^{n} \prod_{\sigma \in S_n} x_{\sigma(i)}^{c_i} \\
&= \prod_{i=1}^{n} \prod_{j=1}^{n} x_j^{(n-1)! c_i} = \prod_{j=1}^{n} \prod_{i=1}^{n} x_j^{(n-1)! c_i} \\
&= \prod_{j=1}^{n} x_j^{(n-1)! \sum_{i=1}^{n} c_i} = 1.
\end{aligned}$$

The arithmetic and geometric mean inequality gives

$$1 = \left(\prod_{\sigma \in S_n} x_{\sigma(1)}^{c_1} x_{\sigma(2)}^{c_2} \cdots x_{\sigma(n)}^{c_n} \right)^{1/n!} < \frac{1}{n!} \sum_{\sigma \in S_n} x_{\sigma(1)}^{c_1} x_{\sigma(2)}^{c_2} \cdots x_{\sigma(n)}^{c_n}$$

and so

$$n! < \sum_{\sigma \in S_n} x_{\sigma(1)}^{c_1} x_{\sigma(2)}^{c_2} \cdots x_{\sigma(n)}^{c_n}.$$

Equivalently,

$$0 < \sum_{\sigma \in S_n} \left(x_{\sigma(1)}^{c_1} x_{\sigma(2)}^{c_2} \cdots x_{\sigma(n)}^{c_n} - 1 \right). \tag{12}$$

The vector $\mathbf{a} = \begin{pmatrix} a_1 \\ \vdots \\ a_n \end{pmatrix}$ strictly majorizes the vector $\mathbf{b} = \begin{pmatrix} b_1 \\ \vdots \\ b_n \end{pmatrix}$ and so $\mathbf{a} \neq \mathbf{b}$. Let

$$c_i = a_i - b_i$$

for $i \in \{1, \ldots, n\}$. Then

$$\sum_{i=1}^{n} c_i = \sum_{i=1}^{n}(a_i - b_i) = \sum_{i=1}^{n} a_i - \sum_{i=1}^{n} b_i = 0$$

and $c_i \neq 0$ for some $i \in \{1, \ldots, n\}$. We have

$$\sum_{\sigma \in S_n} \prod_{i=1}^{n} x_{\sigma(i)}^{a_i} = \sum_{\sigma \in S_n} \prod_{i=1}^{n} x_{\sigma(i)}^{b_i + c_i} = \sum_{\sigma \in S_n} \prod_{i=1}^{n} x_{\sigma(i)}^{b_i} \prod_{i=1}^{n} x_{\sigma(i)}^{c_i}.$$

Inequality (12) implies

$$\begin{aligned}
[\mathbf{x}^{\mathbf{a}}]_{S_n} - [\mathbf{x}^{\mathbf{b}}]_{S_n} &= \frac{1}{n!} \sum_{\sigma \in S_n} \prod_{i=1}^{n} x_{\sigma(i)}^{a_i} - \frac{1}{n!} \sum_{\sigma \in S_n} \prod_{i=1}^{n} x_{\sigma(i)}^{b_i} \\
&= \frac{1}{n!} \sum_{\sigma \in S_n} \prod_{i=1}^{n} x_{\sigma(i)}^{b_i} \prod_{i=1}^{n} x_{\sigma(i)}^{c_i} - \frac{1}{n!} \sum_{\sigma \in S_n} \prod_{i=1}^{n} x_{\sigma(i)}^{b_i} \\
&= \frac{1}{n!} \sum_{\sigma \in S_n} \prod_{i=1}^{n} x_{\sigma(i)}^{b_i} \left(\prod_{i=1}^{n} x_{\sigma(i)}^{c_i} - 1 \right) \\
&> 0.
\end{aligned}$$

This completes the proof. □

Next we prove the converse of Muirhead's inequality.

Theorem 5 *Let* $\mathbf{a} = \begin{pmatrix} a_1 \\ \vdots \\ a_n \end{pmatrix}$ *and* $\mathbf{b} = \begin{pmatrix} b_1 \\ \vdots \\ b_n \end{pmatrix}$ *be vectors in* $\mathbf{R}^n_{\geq 0}$. *If*

$$[\mathbf{x}^{\mathbf{b}}]_{S_n} \leq [\mathbf{x}^{\mathbf{a}}]_{S_n}$$

for every vector $\mathbf{x} \in \mathbf{R}^n_{>0}$, *then* $\mathbf{b} \prec \mathbf{a}$.

Proof For all $x > 0$, the constant vector $\mathbf{x} = \begin{pmatrix} x \\ \vdots \\ x \end{pmatrix}$ is positive and satisfies

$$\left[\mathbf{x}^{\mathbf{a}}\right]_{S_n} = x^{\sum_{i=1}^n a_i} \quad \text{and} \quad \left[\mathbf{x}^{\mathbf{b}}\right]_{S_n} = x^{\sum_{i=1}^n b_i}.$$

Suppose that $[\mathbf{x}^{\mathbf{b}}]_{S_n} \leq [\mathbf{x}^{\mathbf{a}}]_{S_n}$ for every positive vector $\mathbf{x}$. For all $x > 1$ we have

$$x^{\sum_{i=1}^n b_i} \leq x^{\sum_{i=1}^n a_i}$$

and so

$$\sum_{i=1}^{n} b_i \leq \sum_{i=1}^{n} a_i.$$

Similarly, $1/x > 0$ and

$$\frac{1}{x^{\sum_{i=1}^n b_i}} = \left(\frac{1}{x}\right)^{\sum_{i=1}^n b_i} \leq \left(\frac{1}{x}\right)^{\sum_{i=1}^n a_i} = \frac{1}{x^{\sum_{i=1}^n a_i}}$$

and so

$$\sum_{i=1}^n b_i \geq \sum_{i=1}^n a_i.$$

Therefore,

$$\sum_{i=1}^n b_i = \sum_{i=1}^n a_i.$$

By Lemma 2, we can assume that the vectors **a** and **b** are decreasing. For $k \in \{1, \ldots, n-1\}$, we define the variables

$$x_i = \begin{cases} x & \text{if } i \in \{1, \ldots, k\} \\ 1 & \text{if } i \in \{k+1, \ldots, n\} \end{cases}$$

the vector

$$\mathbf{x} = \begin{pmatrix} x_1 \\ \vdots \\ x_k \\ x_{k+1} \\ \vdots \\ x_n \end{pmatrix} = \begin{pmatrix} x \\ \vdots \\ x \\ 1 \\ \vdots \\ 1 \end{pmatrix}$$

and the polynomial

$$\left[\mathbf{x}^{\mathbf{a}}\right]_{S_n} = \frac{1}{n!} \sum_{\sigma \in S_n} x_{\sigma(1)}^{a_1} \cdots x_{\sigma(n)}^{a_n}.$$

The leading coefficient of the polynomial $[\mathbf{x}^{\mathbf{a}}]_{S_n}$ is positive and, because the vector **a** is decreasing, the degree of this polynomial is $\sum_{i=1}^k a_i$. Similarly, the polynomial

$$\left[\mathbf{x}^{\mathbf{b}}\right]_{S_n} = \frac{1}{n!} \sum_{\sigma \in S_n} x_{\sigma(1)}^{b_1} \cdots x_{\sigma(n)}^{b_n}$$

has a positive leading coefficient and degree $\sum_{i=1}^k b_i$. The inequality $\left[\mathbf{x}^{\mathbf{b}}\right]_{S_n} \leq [\mathbf{x}^{\mathbf{a}}]_{S_n}$ for all $x > 0$ implies that $\sum_{i=1}^k b_i \leq \sum_{i=1}^k a_i$ for all $k \in \{1, 2, \ldots, n\}$ and so $\mathbf{b} \prec \mathbf{a}$. This completes the proof. □

We use Muirhead's inequality to prove the following result.

Theorem 6 *Let* $(u_i)_{i=1}^n$ *and* $(v_i)_{i=1}^n$ *be decreasing sequences of positive numbers. If* $(u_i)_{i=1}^n \neq (v_i)_{i=1}^n$ *and*

$$\prod_{i=1}^{j} v_i \leq \prod_{i=1}^{j} u_i \quad \textit{for all } j \in \{1, \ldots, n\} \tag{13}$$

then

$$\sum_{i=1}^{n} v_i < \sum_{i=1}^{n} u_i. \tag{14}$$

Proof Let

$$v_{n+1} = \min(v_n, u_n) \qquad \text{and} \qquad u_{n+1} = v_{n+1}\left(\frac{\prod_{i=1}^{n} v_i}{\prod_{i=1}^{n} u_i}\right).$$

We have $v_n \geq v_{n+1} > 0$ and so the sequence $(v_i)_{i=1}^{n+1}$ is positive and decreasing. Inequality (13) with $j = n$ implies that $\prod_{i=1}^{n} u_i \geq \prod_{i=1}^{n} v_i > 0$ and so

$$0 < u_{n+1} = v_{n+1}\left(\frac{\prod_{i=1}^{n} v_i}{\prod_{i=1}^{n} u_i}\right) \leq v_{n+1} = \min(v_n, u_n) \leq u_n.$$

Thus, the sequence $(u_i)_{i=1}^{n+1}$ is also positive and decreasing. Moreover,

$$\prod_{i=1}^{n+1} u_i = u_{n+1} \prod_{i=1}^{n} u_i = v_{n+1}\left(\frac{\prod_{i=1}^{n} v_i}{\prod_{i=1}^{n} u_i}\right) \prod_{i=1}^{n} u_i = \prod_{i=1}^{n+1} v_i.$$

Let $\lambda > 0$. For all $j \in \{1, \ldots, n\}$, we have

$$\prod_{i=1}^{j} \lambda v_i \leq \prod_{i=1}^{j} \lambda u_i \quad \text{if and only if} \quad \prod_{i=1}^{j} v_i \leq \prod_{i=1}^{j} u_i$$

and

$$\sum_{i=1}^{j} \lambda v_i \leq \sum_{i=1}^{j} \lambda u_i \quad \text{if and only if} \quad \sum_{i=1}^{j} v_i \leq \sum_{i=1}^{j} u_i.$$

Choosing λ sufficiently large, we can assume that $v_i > 1$ and $u_i > 1$ for all $i \in \{1, \ldots, n, n+1\}$, and so

$$a_i = \log u_i > 0 \qquad \text{and} \qquad b_i = \log v_i > 0$$

for all $i \in \{1, \ldots, n, n+1\}$. The sequences $(b_i)_{i=1}^{n+1}$ and $(a_i)_{i=1}^{n+1}$ are decreasing sequences of positive numbers. Moreover, $(u_i)_{i=1}^{n+1} \neq (v_i)_{i=1}^{n+1}$ implies $(a_i)_{i=1}^{n+1} \neq (b_i)_{i=1}^{n+1}$.

Let

$$\mathbf{a} = \begin{pmatrix} a_1 \\ \vdots \\ a_n \\ a_{n+1} \end{pmatrix} \quad \text{and} \quad \mathbf{b} = \begin{pmatrix} b_1 \\ \vdots \\ b_n \\ b_{n+1} \end{pmatrix}.$$

For all $j \in \{1, \ldots, n\}$ we have

$$\sum_{i=1}^{j} b_i = \sum_{i=1}^{j} \log v_i = \log \prod_{i=1}^{j} v_i \leq \log \prod_{i=1}^{j} u_i = \sum_{i=1}^{j} \log u_i = \sum_{i=1}^{j} a_i$$

and

$$\sum_{i=1}^{n+1} b_i = \sum_{i=1}^{n+1} \log v_i = \log \prod_{i=1}^{n+1} v_i = \log \prod_{i=1}^{n+1} u_i = \sum_{i=1}^{n+1} \log u_i = \sum_{i=1}^{n+1} a_i.$$

Thus, the vector **a** strictly majorizes the vector **b**.

By Muirhead's inequality (Theorem 4), for every positive vector $\mathbf{x} = \begin{pmatrix} x_1 \\ \vdots \\ x_n \end{pmatrix}$, we have

$$\begin{aligned} \frac{1}{n!} \sum_{\sigma \in S_{n+1}} \prod_{i=1}^{n+1} x_{\sigma(i)}^{\log v_i} &= \frac{1}{n!} \sum_{\sigma \in S_{n+1}} \prod_{i=1}^{n+1} x_{\sigma(i)}^{b_i} = [\mathbf{x}^{\mathbf{b}}]_{S_n} \\ &< [\mathbf{x}^{\mathbf{a}}]_{S_n} = \frac{1}{n!} \sum_{\sigma \in S_{n+1}} \prod_{i=1}^{n+1} x_{\sigma(i)}^{a_i} \\ &= \frac{1}{n!} \sum_{\sigma \in S_{n+1}} \prod_{i=1}^{n+1} x_{\sigma(i)}^{\log u_i}. \end{aligned}$$

Let $x_1 = e$ and $x_i = 1$ for $i \in \{2, \ldots, n+1\}$. If $\sigma \in S_{n+1}$ and $\sigma(j) = 1$, then

$$\prod_{i=1}^{n+1} x_{\sigma(i)}^{b_i} = \prod_{i=1}^{n+1} x_{\sigma(i)}^{\log v_i} = x_1^{\log v_j} = e^{\log v_j} = v_j.$$

There are $n!$ permutations $\sigma \in S_{n+1}$ such that $\sigma(j) = 1$, and so

$$n! v_j = \sum_{\substack{\sigma \in S_{n+1} \\ \sigma(j)=1}} v_j.$$

Similarly,

$$n!u_j = \sum_{\substack{\sigma \in S_{n+1} \\ \sigma(j)=1}} u_j.$$

Therefore,

$$\begin{aligned}
n! \sum_{j=1}^{n+1} v_j &= \sum_{j=1}^{n+1} \sum_{\substack{\sigma \in S_{n+1} \\ \sigma(j)=1}} v_j = \sum_{j=1}^{n+1} \sum_{\substack{\sigma \in S_{n+1} \\ \sigma(j)=1}} x_1^{b_j} \\
&= \sum_{j=1}^{n+1} \sum_{\substack{\sigma \in S_{n+1} \\ \sigma(j)=1}} \prod_{i=1}^{n+1} x_{\sigma(i)}^{b_i} = \sum_{\sigma \in S_{n+1}} \prod_{i=1}^{n+1} x_{\sigma(i)}^{b_i} \\
&< \sum_{\sigma \in S_{n+1}} \prod_{i=1}^{n+1} x_{\sigma(i)}^{a_i} = \sum_{i=1}^{n+1} \sum_{\substack{\sigma \in S_{n+1} \\ \sigma(j)=1}} \prod_{i=1}^{n+1} x_{\sigma(i)}^{a_i} \\
&= \sum_{j=1}^{n+1} \sum_{\substack{\sigma \in S_{n+1} \\ \sigma(j)=1}} x_1^{a_j} = \sum_{j=1}^{n+1} \sum_{\substack{\sigma \in S_{n+1} \\ \sigma(j)=1}} u_j \\
&= n! \sum_{j=1}^{n+1} u_j.
\end{aligned}$$

The inequality $u_{n+1} \leq v_{n+1}$ implies

$$\sum_{i=1}^{n} v_i = \sum_{i=1}^{n+1} v_i - v_{n+1} < \sum_{i=1}^{n+1} u_i - u_{n+1} = \sum_{i=1}^{n} u_i.$$

This completes the proof. □

5 Rado's Inequality

Let G be a subgroup of S_n and let $\mathbf{a}$ be a vector in $\mathbf{R}^n$. The G-*permutohedron* generated by $\mathbf{a}$, denoted $K_G(\mathbf{a})$, is the convex hull of the finite set of vectors $\{\gamma(\mathbf{a}) : \gamma \in G\}$. The G-permutohedron $K_G(\mathbf{a})$ is a compact convex subset of $\mathbf{R}^n$. If the vector $\mathbf{a}$ is nonnegative, then $\gamma(\mathbf{a})$ is nonnegative for all $\gamma \in G$ and so every convex combination of the vectors $\gamma(\mathbf{a})$, that is, every vector in $K_G(\mathbf{a})$, is nonnegative.

We recall the following separation theorem from convexity theory.

Theorem 7 *Let K be a compact convex set in $\mathbf{R}^n$. If the vector $\mathbf{b} \in \mathbf{R}^n$ is not in K, then the set K and the vector $\mathbf{b}$ are strictly separated by a hyperplane.*

This means that there is a nonzero linear functional

$$H(\mathbf{x}) = \sum_{i=1}^{n} u_i x_i$$

and there are scalars c and δ with $\delta > 0$ such that

$$H(\mathbf{x}) \leq c \qquad \text{for all } \in K$$

and

$$H(\mathbf{b}) \geq c + \delta.$$

We also use the general form of the *arithmetic-geometric mean inequality*.

Theorem 8 *Let G be a finite set. If $\{w_\gamma : \gamma \in G\}$ and $\{t_\gamma : \gamma \in G\}$ are sets of nonnegative numbers such that $\sum_{\gamma \in G} t_\gamma = 1$, then*

$$\prod_{\gamma \in G} w_\gamma^{t_\gamma} \leq \sum_{\gamma \in G} t_\gamma w_\gamma.$$

Moreover,

$$\prod_{\gamma \in G} w_\gamma^{t_\gamma} = \sum_{\gamma \in G} t_\gamma w_\gamma.$$

if and only if $w_\gamma = w_{\gamma'}$ for all $\gamma, \gamma' \in G$.

The following result of Richard Rado extends Muirhead's inequality to subgroups of S_n.

Theorem 9 (Rado)
Let G be a subgroup of the symmetric group S_n. Let $\mathbf{a}$ be a nonnegative vector in $\mathbf{R}^n$ and let $K_G(\mathbf{a})$ be the G-permutohedron, that is, the convex hull of the set of vectors $\{\gamma(\mathbf{a}) : \gamma \in G\}$. If

$$\mathbf{b} \in K_G(\mathbf{a})$$

then $\mathbf{b}$ is nonnegative and

$$[\mathbf{x}^{\mathbf{b}}]_G \leq [\mathbf{x}^{\mathbf{a}}]_G$$

for all positive vectors $\mathbf{x} \in \mathbf{R}^n$.

Proof Let $\mathbf{a} = \begin{pmatrix} a_1 \\ \vdots \\ a_n \end{pmatrix} \in \mathbf{R}^n_{\geq 0}$ and $\mathbf{b} = \begin{pmatrix} b_1 \\ \vdots \\ b_n \end{pmatrix} \in K_G(\mathbf{a})$. Then

$$\mathbf{b} = \sum_{\gamma \in G} t_\gamma \, \gamma(\mathbf{a})$$

where $t_\gamma \in [0, 1]$ for all $\gamma \in G$ and $\sum_{\gamma \in G} t_\gamma = 1$. By (7), the ith component of the vector $\mathbf{b}$ is

$$b_i = \sum_{\gamma \in G} t_\gamma (\gamma \mathbf{a})_i = \sum_{\gamma \in G} t_\gamma \, a_{\gamma^{-1}(i)}.$$

Let $\mathbf{x} = \begin{pmatrix} x_1 \\ \vdots \\ x_n \end{pmatrix}$ be a positive vector. For every permutation $\sigma \in G$, the arithmetic and geometric mean inequality (Theorem 8) gives

$$\prod_{\gamma \in G} \left(\prod_{i=1}^{n} x_{\sigma(i)}^{a_{\gamma^{-1}(i)}} \right)^{t_\gamma} \leq \sum_{\gamma \in G} t_\gamma \left(\prod_{i=1}^{n} x_{\sigma(i)}^{a_{\gamma^{-1}(i)}} \right)$$

and so

$$\begin{aligned}
\left[\mathbf{x}^{\mathbf{b}}\right]_G &= \frac{1}{|G|} \sum_{\sigma \in G} \prod_{i=1}^{n} x_{\sigma(i)}^{b_i} = \frac{1}{|G|} \sum_{\sigma \in G} \prod_{i=1}^{n} x_{\sigma(i)}^{\sum_{\gamma \in G} t_\gamma a_{\gamma^{-1}(i)}} \\
&= \frac{1}{|G|} \sum_{\sigma \in G} \prod_{i=1}^{n} \prod_{\gamma \in G} x_{\sigma(i)}^{t_\gamma a_{\gamma^{-1}(i)}} = \frac{1}{|G|} \sum_{\sigma \in G} \prod_{\gamma \in G} \left(\prod_{i=1}^{n} x_{\sigma(i)}^{a_{\gamma^{-1}(i)}} \right)^{t_\gamma} \\
&\leq \frac{1}{|G|} \sum_{\sigma \in G} \sum_{\gamma \in G} t_\gamma \left(\prod_{i=1}^{n} x_{\sigma(i)}^{a_{\gamma^{-1}(i)}} \right) = \frac{1}{|G|} \sum_{\gamma \in G} t_\gamma \sum_{\sigma \in G} \left(\prod_{i=1}^{n} x_{\sigma(i)}^{a_{\gamma^{-1}(i)}} \right) \\
&= \frac{1}{|G|} \sum_{\gamma \in G} t_\gamma \sum_{\sigma \in G} \left(\prod_{j=1}^{n} x_{\sigma\gamma(j)}^{a_j} \right) = \frac{1}{|G|} \sum_{\gamma \in G} t_\gamma \sum_{\sigma \in G} \left(\prod_{j=1}^{n} x_{\sigma(j)}^{a_j} \right) \\
&= \frac{1}{|G|} \sum_{\sigma \in G} \left(\prod_{j=1}^{n} x_{\sigma(j)}^{a_j} \right) \sum_{\gamma \in G} t_\gamma = \frac{1}{|G|} \sum_{\sigma \in G} \left(\prod_{j=1}^{n} x_{\sigma(j)}^{a_j} \right) \\
&= \left[\mathbf{x}^{\mathbf{a}}\right]_G .
\end{aligned}$$

This completes the proof. □

Next we prove the converse of Theorem 9.

Theorem 10 (Rado)
Let G be a subgroup of the symmetric group S_n. Let $\mathbf{a}$ be a nonnegative vector in $\mathbf{R}^n$ and let $K_G(\mathbf{a})$ be the G-permutohedron. If $\mathbf{b}$ is a nonnegative vector in $\mathbf{R}^n$ such that

$$\left[\mathbf{x}^{\mathbf{b}}\right]_G \leq \left[\mathbf{x}^{\mathbf{a}}\right]_G$$

for all positive vectors $\mathbf{x} \in \mathbf{R}^n$, then

$$\mathbf{b} \in K_G(\mathbf{a}).$$

Proof Let $\mathbf{a} = \begin{pmatrix} a_1 \\ \vdots \\ a_n \end{pmatrix}$ and $\mathbf{b} = \begin{pmatrix} b_1 \\ \vdots \\ b_n \end{pmatrix}$. We shall prove that if $\mathbf{b} \notin K_G(\mathbf{a})$, then there is a positive vector $\mathbf{x} \in \mathbf{R}^n$ such that $\left[\mathbf{x}^{\mathbf{b}}\right]_G > [\mathbf{x}^{\mathbf{a}}]_G$.

By Theorem 7, if $\mathbf{b} \notin K_G(\mathbf{a})$, then the compact convex set $K_G(\mathbf{a})$ and the vector $\mathbf{b}$ are strictly separated by a hyperplane H and so there is a nonzero linear functional

$$H(\mathbf{x}) = \sum_{i=1}^{n} u_i x_i$$

and scalars c and δ with $\delta > 0$ such that

$$H(\mathbf{x}) \leq c \qquad \text{for all } \in K_G(\mathbf{a})$$

and

$$H(\mathbf{b}) \geq c + \delta.$$

For all $\gamma \in G$ we have

$$\gamma^{-1}(\mathbf{a}) = \begin{pmatrix} a_{\gamma(1)} \\ \vdots \\ a_{\gamma(n)} \end{pmatrix} \in K_G(\mathbf{a})$$

and so

$$\sum_{i=1}^{n} u_i a_{\gamma^{-1}(i)} = H(\gamma(\mathbf{a})) \leq c \leq H(\mathbf{b}) - \delta = \sum_{i=1}^{n} u_i b_i - \delta.$$

Let

$$M > |G|^{1/\delta} > 1 \qquad \text{and} \qquad x_i = M^{u_i}$$

for all $i \in \{1, \ldots, n\}$. The vector $\mathbf{x} = \begin{pmatrix} x_1 \\ \vdots \\ x_n \end{pmatrix}$ is positive. The subgroup G contains the identity permutation, and so

$$\begin{aligned} M^{\sum_{i=1}^{n} u_i b_i} &\leq \sum_{\gamma \in G} M^{\sum_{i=1}^{n} u_i b_{\gamma(i)}} = \sum_{\gamma \in G} \prod_{i=1}^{n} M^{u_i b_{\gamma(i)}} \\ &= \sum_{\gamma \in G} \prod_{i=1}^{n} x_i^{b_{\gamma(i)}} = |G| \left[\mathbf{x}^{\mathbf{b}}\right]_G. \end{aligned}$$

We have $\gamma(\mathbf{a}) \in K_G(\mathbf{a})$ and $H(\gamma(\mathbf{a})) \geq H(\mathbf{b}) - \delta$. It follows that

$$[\mathbf{x}^{\mathbf{a}}]_G = \frac{1}{|G|} \sum_{\gamma \in G} \prod_{i=1}^{n} x_i^{a_{\gamma^{-1}(i)}} = \frac{1}{|G|} \sum_{\gamma \in G} \prod_{i=1}^{n} M^{u_i a_{\gamma^{-1}(i)}}$$

$$
\begin{aligned}
&= \frac{1}{|G|} \sum_{\gamma \in G} M^{\sum_{i=1}^{n} u_i a_{\gamma^{-1}(i)}} = \frac{1}{|G|} \sum_{\gamma \in G} M^{H(\gamma(\mathbf{a}))} \\
&\leq \frac{1}{|G|} \sum_{\gamma \in G} M^{H(\mathbf{b})-\delta} = \frac{1}{|G|} M^{\sum_{i=1}^{n} u_i b_i - \delta} \\
&= \frac{1}{M^{\delta}} \frac{1}{|G|} M^{\sum_{i=1}^{n} u_i b_i} \leq \frac{1}{M^{\delta}} \left[\mathbf{x}^{\mathbf{b}}\right]_G \\
&< \left[\mathbf{x}^{\mathbf{b}}\right]_G .
\end{aligned}
$$

This completes the proof. □

References

1. G. H. Hardy, J. E. Littlewood, and G. Pólya, *Some simple inequalities satisfied by convex functions*, Messenger of Math. **58** (1929), 145–152.
2. G. H. Hardy, J. E. Littlewood, and G. Pólya, *Inequalities*, Cambridge Mathematical Library, Cambridge University Press, Cambridge, 1988, reprint of the 1952 edition.
3. A. W. Marshall, I. Olkin, and B. C. Arnold, *Inequalities: Theory of Majorization and its Applications*, Springer Series in Statistics, Springer, New York, 2011.
4. R. F. Muirhead, *Some methods applicable to identities and inequalities of symmetric algebraic functions of n letters*, Proc. Edinburgh Math. Soc. **21** (1903), 144–157.
5. M. B. Nathanson, *The Muirhead-Rado inequality, 1: Vector majorization and the permutohedron*, in: *Combinatorial and Additive Number Theory VI*, Springer, New York, 2024.
6. R. Rado, *An inequality*, J. London Math. Soc. **27** (1952), 1–6.

Increasing-Decreasing Patterns in the Iteration of an Arithmetic Function

Melvyn B. Nathanson

Abstract Let Ω be a set of positive integers and let $f : \Omega \to \Omega$ be an arithmetic function. Let $V = (v_i)_{i=1}^n$ be a finite sequence of positive integers. An integer $m \in \Omega$ has *increasing-decreasing pattern* V with respect to f if, for all odd integers $i \in \{1, \ldots, n\}$,

$$f^{v_1+\cdots+v_{i-1}}(m) < f^{v_1+\cdots+v_{i-1}+1}(m) < \cdots < f^{v_1+\cdots+v_{i-1}+v_i}(m)$$

and, for all even integers $i \in \{2, \ldots, n\}$,

$$f^{v_1+\cdots+v_{i-1}}(m) > f^{v_1+\cdots+v_{i-1}+1}(m) > \cdots > f^{v_1+\cdots+v_{i-1}+v_i}(m).$$

The arithmetic function f is *wildly increasing-decreasing* if, for every finite sequence V of positive integers, there exists an integer $m \in \Omega$ such that m has increasing-decreasing pattern V with respect to f. This paper gives a proof that the Syracuse function is wildly increasing-decreasing.

2020 Mathematics Subject Classification: 11A25 · 11B37 · 11B83 · 11D04 · 68Q99

1 Iterations and Patterns

An arithmetic function is a function whose domain is a subset Ω of the set $\mathbf{N} = \{1, 2, 3, \ldots\}$ of positive integers. Let $f : \Omega \to \Omega$ be an arithmetic function and, for all $j \in \mathbf{N}$, let $f^j : \Omega \to \Omega$ be the jth iterate of f. The function $f^0 : \Omega \to \Omega$ is the identity function. The *trajectory* of $m \in \Omega$ is the sequence of positive integers $\left(f^j(m)\right)_{j=0}^{\infty}$. The pair (Ω, f) is a discrete dynamical system.

Supported in part by PSC-CUNY Research Award Program grant 66197-00 54.

M. B. Nathanson (✉)
Lehman College (CUNY), Bronx 10468 New York, USA
e-mail: melvyn.nathanson@lehman.cuny.edu

M. B. Nathanson (ed.), *Combinatorial and Additive Number Theory VI*,
Springer Proceedings in Mathematics & Statistics 464,
https://doi.org/10.1007/978-3-031-65064-2_18

Let

$$\Omega^{\text{fix}} = \{m \in \Omega : f(m) = m\}$$

be the set of fixed points of f and let

$$\Omega^{\text{per}} = \{m \in \Omega : f^k(m) = m \text{ for some positive integer } k\}$$

be the set of periodic points of f. The *period* of $m \in \Omega^{\text{per}}$ is the smallest positive integer k such that $f^k(m) = m$. The fixed points are the points of period one.

The trajectory $\left(f^j(m)\right)_{j=0}^{\infty}$ is *eventually constant* if $f^k(m) \in \Omega^{\text{fix}}$ for some $k \in \mathbf{N}$. The trajectory $\left(f^j(m)\right)_{j=0}^{\infty}$ is *eventually periodic* if $f^k(m) \in \Omega^{\text{per}}$ for some $k \in \mathbf{N}$. The trajectory is *bounded* if and only if $f^k(m) \in \Omega^{\text{per}}$ for some $k \in \mathbf{N}$. The trajectory is *unbounded* if and only if $\lim_{k\to\infty} f^k(m) = \infty$. If $m \in \Omega$ and $m \notin \Omega^{\text{fix}}$, then either $f(m) > m$ (that is, f increases m) or $f(m) < m$ (that is, f decreases m). We are interested in the pattern of changes (increases and decreases) in the trajectory $\left(f^j(m)\right)_{j=0}^{\infty}$.

Here are some examples. Let Ω be the set of odd positive integers and let $\ell \in \Omega$. For all $m \in \Omega$, the integer $\ell m + 1$ is even. We define the arithmetic function $S_\ell : \Omega \to \Omega$ by

$$S_\ell(m) = \frac{\ell m + 1}{2^e}$$

where 2^e is the highest power of 2 that divides $\ell m + 1$. The function S_ℓ has a fixed point if and only if $\ell = 2^k - 1$ for some positive integer k, and in this case the unique fixed point is $m = 1$. The function S_5 has no fixed point, but it does have periodic points. For example, $S_5(1) = 3$ and $S_5(3) = 1$.

The *Syracuse function* (also called the *Collatz function*) is the arithmetic function

$$S(m) = S_3(m) = \frac{3m + 1}{2^e} \tag{1}$$

where 2^e is the highest power of 2 that divides $3m + 1$. The unique fixed point of this function is $m = 1$. There is a large literature on the Syracuse function (cf. Eliahou [3], Everett [7], Lagarias [11, 12], Tao [17], Terras [18]).

Let $\Omega = \mathbf{N}$. The *sum of the proper divisors function* is the arithmetic function $s : \mathbf{N} \to \mathbf{N}$ defined by

$$s(m) = \sum_{\substack{d \mid m \\ 1 \le d < m}} d.$$

The trajectory of the positive integer m, that is, $\left(s^j(m)\right)_{j=1}^{\infty}$, is called an *aliquot sequence*. The fixed points of s are the *perfect numbers* $6, 28, 496, \ldots$. If m is a point of period 2 and if $n = s(m)$, then $m \neq n$ and $s(n) = m$. The pair of integers (m, n) is called an *amicable pair*. For example, $(220, 284)$ and $(1184, 1210)$ are amicable pairs. Integers whose trajectories are periodic under s are called *sociable numbers*. There exist sociable numbers of period 4 (such as 1264460), but no sociable number of period 3 is known. The behavior of aliquot sequences is poorly understood (cf. Guy [8], P. Erdős, A. Granville, C. Pomerance, and C. Spiro [6], Pollack-Pomerance [15]).

Let Ω be a set of positive integers and let $f : \Omega \to \Omega$ be an arithmetic function. Let $V = (v_i)_{i=1}^n$ be a finite sequence of positive integers. We say that an integer m in Ω has *increasing-decreasing pattern* V with respect to f if

$$m < f(m) < f^2(m) < \cdots < f^{v_1}(m)$$

$$f^{v_1}(m) > f^{v_1+1}(m) > \cdots > f^{v_1+v_2}(m)$$

$$f^{v_1+v_2}(m) < f^{v_1+v_2+1}(m) < \cdots < f^{v_1+v_2+v_3}(m)$$

and, in general, if i is odd, then

$$f^{v_1+\cdots+v_{i-1}}(m) < f^{v_1+\cdots+v_{i-1}+1}(m) < \cdots < f^{v_1+\cdots+v_{i-1}+v_i}(m) \tag{2}$$

and if i is even, then

$$f^{v_1+\cdots+v_{i-1}}(m) > f^{v_1+\cdots+v_{i-1}+1}(m) > \cdots > f^{v_1+\cdots+v_{i-1}+v_i}(m). \tag{3}$$

The arithmetic function f is *wildly increasing-decreasing* if, for every finite sequence V of positive integers, there exists an integer $m \in \Omega$ such that m has increasing-decreasing pattern V with respect to f.

More generally, let $\mathcal{F} = (f_j)_{j=0}^{\infty}$ be a sequence of arithmetic functions $f_j : \Omega \to \Omega$. The $\mathcal{F}$-trajectory of $m \in \Omega$ is the sequence $\left(f_j(m)\right)_{j=1}^{\infty}$. Let $V = (v_i)_{i=1}^n$ be a finite sequence of positive integers. We say that an integer m in Ω has *increasing-decreasing pattern* V with respect to the sequence $\mathcal{F}$ if, for odd i,

$$f_{v_1+\cdots+v_{i-1}}(m) < f_{v_1+\cdots+v_{i-1}+1}(m) < \cdots < f_{v_1+\cdots+v_{i-1}+v_i}(m)$$

and, for even i,

$$f_{v_1+\cdots+v_{i-1}}(m) > f_{v_1+\cdots+v_{i-1}+1}(m) > \cdots > f_{v_1+\cdots+v_{i-1}+v_i}(m).$$

The sequence $\mathcal{F}$ is *wildly increasing-decreasing* if, for every finite sequence V of positive integers, there exists $m \in \Omega$ such that m has increasing-decreasing pattern V with respect to the sequence $\mathcal{F}$.

Fix an arithmetic function $f : \Omega \to \Omega$. Let $(k_j)_{j=0}^{\infty}$ be a strictly increasing sequence of nonnegative integers. If f is wildly increasing-decreasing, then the sequence of arithmetic functions $\mathcal{F} = \left(f^{k_j}\right)_{j=0}^{\infty}$ is also wildly increasing-decreasing.

It is not known if aliquot sequences are wildly increasing-decreasing. Lenstra [13] proved that, for every positive integer v_1, there are infinite many integers m such that $m < s(m) < s^2(m) < \cdots < s^{v_1}(m)$. Erdős [5] subsequently refined this result. Pomerance [16] proved that, for every positive integer v_2, there are infinitely many integers m such that $m > s(m) > s^2(m) > \cdots > s^{v_2}(m)$. It is an open problem to determine if, for every pair of positive integers v_1, v_2 there are integers m such that

$$m < s(m) < s^2(m) < \cdots < s^{v_1}(m)$$

and

$$s^{v_1}(m) > s^{v_1+1}(m) > s^{v_1+2}(m) > \cdots > s^{v_1+v_2}(m).$$

There are two competing conjectures about the trajectories of the sum of the proper divisors function $s(m)$. Catalan [1] and Dickson [2] conjectured that every aliquot sequence is bounded. Guy and Selfridge [9, 10] conjectured that infinitely many aliquot sequences go to infinity.

The goal of this paper is to prove that Syracuse function $S(m)$ defined by (1) is wildly increasing-decreasing. The proof uses integer matrices and systems of linear diophantine equations.

2 Integer Matrices

We consider vectors in $\mathbf{R}^n$. A vector is *positive* if all of its coordinates are positive and *negative* if all of its coordinates are negative. A vector is *integral* if all of its coordinates are integers.

Let $\mathbf{Z}^n$ denote the set of n-dimensional integral vectors. An integral vector is *odd* if its coordinates are odd integers and *even* if its coordinates are even integers. An integral vector is *primitive* if its coordinates are relatively prime (not necessarily pairwise relatively prime) positive integers.

Theorem 1 *Let $(a_i)_{i=1}^n$ and $(b_j)_{j=2}^{n+1}$ be sequences of nonzero real numbers. Consider the $n \times (n+1)$ matrix*

$$M = \begin{pmatrix} a_1 & -b_2 & 0 & 0 & 0 & 0 \cdots 0 & 0 & 0 \\ 0 & a_2 & -b_3 & 0 & 0 & 0 \cdots 0 & 0 & 0 \\ 0 & 0 & a_3 & -b_4 & 0 & 0 \cdots 0 & 0 & 0 \\ 0 & 0 & 0 & a_4 & -b_5 & 0 \cdots 0 & 0 & 0 \\ \vdots & & & & & & & \vdots \\ 0 & 0 & 0 & 0 & 0 & 0 \cdots 0 & a_n & -b_{n+1} \end{pmatrix}$$

(a) *The matrix M has rank n and the kernel of M has dimension 1.*
(b) *Let $(a_i)_{i=1}^n$ and $(b_j)_{j=2}^{n+1}$ be sequences of positive real numbers. If $\mathbf{x} \in$ kernel(M) and $\mathbf{x} \neq \mathbf{0}$, then the vector $\mathbf{x}$ is positive or negative.*

(c) *Let* $(a_i)_{i=1}^{n}$ *and* $(b_j)_{j=2}^{n+1}$ *be sequences of positive integers. There is a unique primitive vector* $\mathbf{z}$ *that generates kernel*(M).

(d) *Let* $(a_i)_{i=1}^{n}$ *be a sequence of odd positive integers and let* $(b_j)_{j=2}^{n+1}$ *be a sequence of even positive integers. If* $\mathbf{z} = \begin{pmatrix} z_1 \\ \vdots \\ z_n \\ z_{n+1} \end{pmatrix}$ *is the unique primitive vector in kernel*(M), *then the integers* z_i *are even for all* $i \in \{1, 2, \ldots, n\}$ *and the integer* z_{n+1} *is odd.*

(e) *Let* $(a_i)_{i=1}^{n}$ *be a sequence of odd positive integers and let* $(b_j)_{j=2}^{n+1}$ *be a sequence of even positive integers. Let* $\mathbf{h} \in \mathbf{Z}^n$ *be an odd vector. If* $M\mathbf{x} = \mathbf{h}$ *for some integral vector* $\mathbf{x} \in \mathbf{Z}^{n+1}$, *then there exists an odd positive vector* $\mathbf{g} \in \mathbf{Z}^{n+1}$ *such that*

$$M\mathbf{g} = \mathbf{h}.$$

Proof

(a) The n row vectors of the matrix M are linearly independent and so M has rank n.

(b) Let $\mathbf{x} = \begin{pmatrix} x_1 \\ \vdots \\ x_n \end{pmatrix} +1 \in \text{kernel}(M)$. If $M\mathbf{x} = \mathbf{0}$, then for all $i \in \{1, \ldots, n\}$ we have

$$a_i x_i = b_{i+1} x_{i+1}.$$

If $\mathbf{x} \neq \mathbf{0}$, then $x_{i_0} \neq 0$ for some $i_0 \in \{1, \ldots, n+1\}$. The positivity of the numbers a_i and b_i implies that if $x_{i_0} > 0$, then $x_i > 0$ for all $i \in \{1, \ldots, n+1\}$. Similarly, if $x_{i_0} < 0$, then $x_i < 0$ for all $i \in \{1, \ldots, n+1\}$. Therefore, $\mathbf{x}$ is either a positive vector or a negative vector.

(c) Consider M as a matrix with coordinates in the field $\mathbf{Q}$ of rational numbers. The kernel of M is one-dimensional. Let $\mathbf{x} \in \mathbf{Q}^{n+1}$ be a nonzero vector in the kernel of M. By (b), the coordinates of $\mathbf{x}$ are either all positive rational numbers or all negative rational numbers. Multiplying, if necessary, by -1, we can assume that the coordinates are positive rational numbers. Multiplying by a common denominator of the denominators of the coordinates, we obtain a vector whose coordinates are positive integers. Dividing by the greatest common divisor of these integers, we obtain a primitive vector in the kernel of M.

Let $\mathbf{z} = \begin{pmatrix} z_1 \\ \vdots \\ z_n \\ z_{n+1} \end{pmatrix}$ and $\mathbf{z}' = \begin{pmatrix} z'_1 \\ \vdots \\ z'_n \\ z'_{n+1} \end{pmatrix}$ be primitive vectors in kernel(M). Because the kernel is one-dimensional, there is a positive rational number p/q with $\gcd(p, q) = 1$ such that $(p/q)\mathbf{z} = \mathbf{z}'$. Multiplying by q we obtain

$$\begin{pmatrix} pz_1 \\ \vdots \\ pz_n \\ pz_{n+1} \end{pmatrix} = p\mathbf{z} = q\mathbf{z}' = \begin{pmatrix} qz'_1 \\ \vdots \\ qz'_n \\ qz'_{n+1} \end{pmatrix}$$

and so

$$pz_i = qz'_i$$

for all $i \in \{1, \ldots, n, n+1\}$. The divisibility condition $\gcd(p, q) = 1$ implies that p divides z'_i and q divides z_i for all $i \in \{1, \ldots, n, n+1\}$. Because the vectors $\mathbf{z}$ and $\mathbf{z}'$ are primitive, we have $p = q = 1$ and so $\mathbf{z} = \mathbf{z}'$. Thus, kernel(M) contains a unique primitive vector.

(d) Let $\mathbf{z} = \begin{pmatrix} z_1 \\ \vdots \\ z_n \\ z_{n+1} \end{pmatrix}$ be the unique primitive vector in kernel(M). For all $i \in \{1, \ldots, n\}$ we have

$$a_i z_i = b_{i+1} z_{i+1}.$$

Because b_i is even and a_i is odd, it follows that z_i is even for all $i \in \{1, \ldots, n\}$. We have $\gcd(z_1, \ldots, z_n, z_{n+1}) = 1$ because the vector $\mathbf{z}$ is primitive, and so z_{n+1} must be odd.

(e) Let $\mathbf{h} = \begin{pmatrix} h_1 \\ \vdots \\ h_n \end{pmatrix} \in \mathbf{Z}^n$ be an odd vector. If $\mathbf{x} = \begin{pmatrix} x_1 \\ \vdots \\ x_n \end{pmatrix} + 1 \in \mathbf{Z}^{n+1}$ and $M\mathbf{x} = \mathbf{h}$, then for all $i \in \{1, \ldots, n\}$ we have

$$a_i x_i - b_{i+1} x_{i+1} = h_i.$$

Because the integer b_{i+1} is even and the integers a_i and h_i are odd, it follows that the integer x_i is odd for all $i \in \{1, \ldots, n\}$. The integer x_{n+1} is not necessarily odd, nor are the integers $x_1, \ldots, x_n, x_{n+1}$ necessarily positive.

Let $\mathbf{z} = \begin{pmatrix} z_1 \\ \vdots \\ z_n \\ z_{n+1} \end{pmatrix}$ be the unique primitive vector in kernel(M). By (d), the integers z_i are even for all $i \in \{1, 2, \ldots, n\}$ and the integer z_{n+1} is odd. We have $M\mathbf{z} = \mathbf{0}$ and so

$$M(\mathbf{x} + k\mathbf{z}) = M\mathbf{x} = \mathbf{h}$$

for all integers k, where

$$\mathbf{x} + k\mathbf{z} = \begin{pmatrix} x_1 \\ \vdots \\ x_n \end{pmatrix} + 1 + k \begin{pmatrix} z_1 \\ \vdots \\ z_n \\ z_{n+1} \end{pmatrix} = \begin{pmatrix} x_1 + kz_1 \\ \vdots \\ x_n + kz_n \\ x_{n+1} + kz_{n+1} \end{pmatrix}.$$

The coordinates z_i are positive, and so the vector $\mathbf{x} + k\mathbf{z}$ is positive for all sufficiently large k.
For all $i \in \{1, \ldots, n\}$ the integer x_i is odd and the integer z_i is even, and so $x_i + kz_i$ is an odd integer. The integer z_{n+1} is odd, and so the integers $x_{n+1} + kz_{n+1}$ and $x_{n+1} + (k+1)z_{n+1}$ have opposite parity. It follows that for all sufficiently large k, either $\mathbf{g} = \mathbf{x} + k\mathbf{z}$ or $\mathbf{g} = \mathbf{x} + (k+1)\mathbf{z}$ is an odd positive vector such that $M\mathbf{g} = \mathbf{h}$.

This completes the proof. □

Theorem 2 *Let* $(a_i)_{i=1}^n$ *and* $(b_j)_{j=2}^{n+1}$ *be sequences of nonzero integers such that*

$$\gcd(a_i, b_j) = 1 \tag{4}$$

for all $i \in \{1, \ldots, n\}$ *and* $j \in \{2, \ldots, n+1\}$. *Let* M *be the* $n \times (n+1)$ *matrix defined in Theorem 1. The homomorphism* $M : \mathbf{Z}^{n+1} \to \mathbf{Z}^n$ *is surjective.*

Proof We give two proofs. The first proof uses induction on n. Because $(a_1, b_2) = 1$, for every integer h_1 there are integers x_1, x_2 such that $a_1x_1 + b_2x_2 = h_1$. Equivalently, for the 2×1 matrix $M = \begin{pmatrix} a_1 & b_2 \end{pmatrix}$ and for $\mathbf{h} = (h_1)$, we have

$$M \begin{pmatrix} x_1 \\ x_2 \end{pmatrix} = \begin{pmatrix} a_1 & b_2 \end{pmatrix} \begin{pmatrix} x_1 \\ x_2 \end{pmatrix} = \mathbf{h}.$$

This is the case $n = 1$.

For $n \geq 2$, let M be the $n \times (n+1)$ matrix defined in Theorem 1 and let $\mathbf{h} = \begin{pmatrix} h_1 \\ \vdots \\ h_n \end{pmatrix} \in \mathbf{Z}^n$. There exists an integral vector $\mathbf{x} = \begin{pmatrix} x_1 \\ \vdots \\ x_n \end{pmatrix} +1 \in \mathbf{Z}^{n+1}$ such that

$$M\mathbf{x} = \mathbf{h}$$

if and only if the diophantine system of n linear equations in $n+1$ variables

$$\begin{aligned} a_1x_1 - b_2x_2 &= h_1 \\ a_2x_2 - b_3x_3 &= h_2 \\ &\vdots \\ a_{i-1}x_{i-1} - b_ix_i &= h_{i-1} \\ a_ix_i - b_{i+1}x_{i+1} &= h_i \\ &\vdots \\ a_nx_n - b_{n+1}x_{n+1} &= h_n \end{aligned}$$

has a solution in integers $x_1, \ldots, x_n, x_{n+1}$.

For $n = 2$, we have two equations

$$\begin{aligned} a_1 x_1 - b_2 x_2 &= h_1 \\ a_2 x_2 - b_3 x_3 &= h_2. \end{aligned}$$

The divisibility condition (4) gives $\gcd(a_1, b_2) = \gcd(a_2, b_3) = 1$, and so both equations have solutions in integers. If (c_1, d_2) is a particular solution of the first equation, then the general solution of the first equation is

$$\begin{aligned} x_1 &= c_1 + b_2 y_1 \\ x_2 &= d_2 + a_1 y_1 \end{aligned}$$

for any integer y_1. If (c_2, d_3) is a particular solution of the second equation, then the general solution of the second equation is

$$\begin{aligned} x_2 &= c_2 + b_3 y_2 \\ x_3 &= d_3 + a_2 y_2 \end{aligned}$$

for any integer y_2. We have a simultaneous solution of the system of two equations if and only if there exist integers y_1 and y_2 such that

$$x_2 = d_2 + a_1 y_1 = c_2 + b_3 y_2$$

or, equivalently,

$$a_1 y_1 - b_3 y_2 = c_2 - d_2.$$

Because $\gcd(a_1, b_3) = 1$, this equation has a solution in integers. This proves the Theorem for $n = 2$.

Let $n \geq 3$ and assume that the Theorem is true for $n - 1$ equations in n variables. Consider the diophantine system of n equations in $n + 1$ variables such that, for all $i \in \{1, \ldots, n\}$, equation (i) is

$$a_i x_i - b_{i+1} x_{i+1} = h_i.$$

This equation has a solution in integers because $\gcd(a_i, b_{i+1}) = 1$. If (c_i, d_{i+1}) is a particular solution of the equation, then the general solution of equation (i) is

$$\begin{aligned} x_i &= c_i + b_{i+1} y_i \\ x_{i+1} &= d_{i+1} + a_i y_i \end{aligned}$$

for any integer y_i. Similarly, for all $i \in \{2, \ldots, n + 1\}$, equation $(i - 1)$ is

$$a_{i-1} x_{i-1} - b_i x_i = h_{i-1}$$

This equation has a solution in integers because $\gcd(a_{i-1}, b_i) = 1$. If (c_{i-1}, d_i) is a particular solution of the equation, then the general solution of equation $(i-1)$ is

$$\begin{aligned} x_{i-1} &= c_{i-1} + b_i y_{i-1} \\ x_i &= d_i + a_{i-1} y_{i-1} \end{aligned}$$

for any integer y_{i-1}. Equations $(i-1)$ and (i) have a simultaneous solution in integers if and only if there exist integers y_{i-1} and y_i such that

$$d_i + a_{i-1} y_{i-1} = c_i + b_{i+1} y_i$$

or, equivalently, if

$$a_{i-1} y_{i-1} - b_{i+1} y_i = c_i - d_{i-1}.$$

It follows that the original system of n equations in $n+1$ variables has a solution in integers if and only if the following system of $n-1$ equations in n variables has a solution in integers:

$$\begin{aligned} a_1 y_1 - b_3 y_2 &= c_2 - d_1 \\ a_2 y_2 - b_4 y_3 &= c_3 - d_2 \\ &\vdots \\ a_{n-1} y_{n-1} - b_{n+1} y_n &= c_n - d_{n-1}. \end{aligned}$$

The divisibility condition (4) and the induction hypothesis imply that this system of equations does have an integral solution. This completes the first proof.

The second proof uses the *Smith normal form* of an integral matrix (Marcus and Minc [14, pp. 40–48]). The Smith normal form of an $m \times n$ matrix M of rank k is the unique diagonal matrix

$$SNF(M) = \begin{pmatrix} s_1 & 0 & 0 & 0 & 0 & \cdots & 0 & 0 & 0 \\ 0 & s_2 & 0 & 0 & 0 & \cdots & 0 & 0 & 0 \\ 0 & 0 & s_3 & 0 & 0 & \cdots & 0 & 0 & 0 \\ \vdots & & & & & & & & \vdots \\ 0 & 0 & 0 & 0 & 0 & \cdots & s_k & 0 & 0 \\ 0 & 0 & 0 & 0 & 0 & \cdots & 0 & 0 & 0 \\ \vdots & & & & & & & & \vdots \\ 0 & 0 & 0 & 0 & 0 & \cdots & 0 & 0 & 0 \end{pmatrix}$$

constructed from M by integral elementary row and column operations and whose coordinates $s_1, \ldots, s_k$ are positive integers. The *kth determinantal divisor* of A is the greatest common divisor of all of the $k \times k$ minors of A. If the kth determinantal divisor of A is 1, then $s_i = 1$ for all $i = 1, \ldots, k$. If A has rank n and the nth determinantal divisor of A is 1, then $s_i = 1$ for all $i = 1, \ldots, n$ and

$$SNF(M) = \begin{pmatrix} 1\,0\,0\,0\,0 \cdots 0\,0\,0 \\ 0\,1\,0\,0\,0 \cdots 0\,0\,0 \\ 0\,0\,1\,0\,0 \cdots 0\,0\,0 \\ \vdots \qquad\qquad\quad \vdots \\ 0\,0\,0\,0\,0 \cdots 0\,1\,0 \end{pmatrix}. \tag{5}$$

The homomorphism $M : \mathbf{Z}^{n+1} \to \mathbf{Z}^n$ is surjective if and only if the homomorphism $SNF(M) : \mathbf{Z}^{n+1} \to \mathbf{Z}^n$ is surjective if and only if $\text{rank}(M) = n$ and $s_i = 1$ for all $i = 1, \ldots, n$.

Let M be the $n \times (n+1)$ matrix constructed in Theorem 1 from integer sequences $(a_i)_{i=1}^n$ and $(b_j)_{j=2}^{n+1}$ such that $\gcd(a_i, b_j) = 1$ for all $i \in \{1, \ldots, n\}$ and $j \in \{2, \ldots, n+1\}$. The determinant of the $n \times n$ minor obtained by deleting the first column of M is $(-1)^n \prod_{j=2}^{n+1} b_j$. The determinant of the $n \times n$ minor obtained by deleting the last column of M is $\prod_{i=1}^n a_i$. The divisibility condition $\gcd(a_i, b_j) = 1$ implies that these determinants are relatively prime integers and so the nth determinantal divisor of M is 1. It follows that the matrix M has Smith normal form (5) and the homomorphisms $SNF(M) : \mathbf{Z}^{n+1} \to \mathbf{Z}^n$ and $M : \mathbf{Z}^{n+1} \to \mathbf{Z}^n$ are surjective. This completes the second proof. □

Theorem 3 *Let $(v_i)_{i=1}^{n+1}$ be a sequence of positive integers. Define the positive integral sequences $(a_i)_{i=1}^n$ and $(b_j)_{j=2}^{n+1}$ as follows:*

$$a_i = 3^{v_i}$$

and

$$b_j = \begin{cases} 4^{v_j} & \text{if } j \text{ is even} \\ 2^{v_j} & \text{if } j \text{ is odd.} \end{cases}$$

Let

$$c_{n+1} = \begin{cases} 4 & \text{if } n+1 \text{ is even} \\ 2 & \text{if } n+1 \text{ is odd.} \end{cases}$$

The system of linear diophantine equations

$$\begin{aligned} 3^{v_1}x_1 - 4^{v_2}x_2 &= 1 \\ 3^{v_2}x_2 \quad 2^{v_3}x_3 &= -1 \\ 3^{v_3}x_3 - 4^{v_4}x_4 &= 1 \\ &\vdots \\ 3^{v_n}x_n - c_{n+1}^{v_{n+1}}x_{n+1} &= (-1)^{n+1} \end{aligned}$$

has a solution in odd positive integers $w_1, w_2, \ldots, w_n, w_{n+1}$.

Proof Let

$$M = \begin{pmatrix} 3^{v_1} & -4^{v_2} & 0 & 0 & 0 & 0 \cdots 0 & 0 & 0 \\ 0 & 3^{v_2} & -2^{v_3} & 0 & 0 & 0 \cdots 0 & 0 & 0 \\ 0 & 0 & 3^{v_3} & -4^{v_4} & 0 & 0 \cdots 0 & 0 & 0 \\ 0 & 0 & 0 & 3^{v_4} & -2^{v_5} & 0 \cdots 0 & 0 & 0 \\ \vdots & & & & & & & \vdots \\ 0 & 0 & 0 & 0 & 0 & 0 \cdots 0 & 3^{v_n} & -c_{n+1}^{v_{n+1}} \end{pmatrix}$$

be the $n \times (n+1)$ matrix constructed in Theorem 1. Let $h_i = (-1)^{i+1}$ for $i \in \{1, \ldots, n\}$ and $\mathbf{h} = \begin{pmatrix} h_1 \\ \vdots \\ h_n \end{pmatrix} \in \mathbf{Z}^n$. By Theorem 2, there exists an integralr vector $\mathbf{x} \in \mathbf{Z}^{n+1}$ such that $M\mathbf{x} = \mathbf{h}$. By Theorem 1(e), there exists an odd positive vector $\mathbf{w} = \begin{pmatrix} w_1 \\ \vdots \\ w_n \\ w_{n+1} \end{pmatrix} \in \mathbf{Z}^{n+1}$ such that $M\mathbf{w} = \mathbf{h}$. This completes the proof. □

3 Iterations of the Syracuse Function

Let Ω be the set of odd positive integers. The Syracuse function is the arithmetic function $S : \Omega \to \Omega$ defined by

$$S(m) = \frac{3m+1}{2^e}$$

where e is the largest integer such that 2^e divides $3m + 1$. Note that $S(m) = 1$ if and only if $m = (4^e - 1)/3$ for some positive integer e.

Lemma 1 *Let m be a positive integer such that*

$$m \equiv -1 \pmod 4$$

and let v and w be the unique positive integers with w odd such that

$$m = 2^{v+1}w - 1.$$

Then

$$m < S(m) < S^2(m) < \cdots < S^v(m) \tag{6}$$

and

$$S^v(m) = 2 \cdot 3^v w - 1. \tag{7}$$

Moreover,

$$S^{v+1}(m) = \frac{3^v w - 1}{2^e}$$

for some positive integer e.

Proof We use induction on j to prove that if

$$0 \leq j \leq v$$

then

$$S^j(m) = 2^{v+1-j}3^j w - 1 = \left(\frac{3}{2}\right)^j (m+1) - 1. \tag{8}$$

This is true for $j = 0$. For $j = 1$ we have

$$S(m) = \frac{3(2^{v+1}w - 1) + 1}{2} = \left(\frac{3}{2}\right) 2^{v+1} w - 1 = \left(\frac{3}{2}\right)(m+1) - 1.$$

If $j \leq v - 1$ and

$$S^j(m) = 2^{v+1-j}3^j w - 1$$

then

$$S^{j+1}(m) = 2^{v+1-(j+1)}3^{j+1} w - 1 = \left(\frac{3}{2}\right)^{j+1} (m+1) - 1.$$

This proves (8), and (8) implies (6).

Note that

$$S^v(m) = 2 \cdot 3^v w - 1$$

implies

$$S^{v+1}(m) = S\left(2 \cdot 3^v w - 1\right) = \frac{3^{v+1} w - 1}{2^e}$$

for some positive integer e. This completes the proof. □

Lemma 2 *Let m be a positive integer such that*

$$m \equiv 1 \pmod 8$$

and let v and w be the unique positive integers with $v \geq 3$ and w odd such that

$$m = 2^v w + 1.$$

If

$$v_0 = \left[\frac{v-1}{2}\right]$$

then

$$v = 2v_0 + 1 \quad or \quad v = 2v_0 + 2$$

and

$$m > S(m) > S^2(m) > \cdots > S^{v_0-1}(m) > S^{v_0}(m). \tag{9}$$

If $v = 2v_0 + 1$, *then*

$$S^{v_0}(m) = 2 \cdot 3^{v_0} w + 1 \tag{10}$$

If $v = 2v_0 + 2$, *then*

$$S^{v_0}(m) = 4 \cdot 3^{v_0} w + 1$$

Proof We have $v - 2j \geq 1$ if and only if

$$j \leq v_0 = \left[\frac{v-1}{2}\right]$$

and $v \geq 3$ implies $v_0 \geq 1$. We use induction on j to prove that if

$$0 \leq j \leq v_0$$

then

$$S^j(m) = 2^{v-2j} 3^j w + 1 = \left(\frac{3}{4}\right)^j (m-1) + 1. \tag{11}$$

This is true for $j = 0$. For $j = 1$ we have

$$\begin{aligned} S(m) = S\left(2^v w + 1\right) &= \frac{3(2^v w + 1) + 1}{4} \\ &= 2^{v-2} 3w + 1 = \frac{3}{4}(2^v w) + 1 \\ &= \frac{3}{4}(m-1) + 1. \end{aligned}$$

If $1 \leq j \leq v_0 - 1$ and

$$S^j(m) = 2^{v-2j} 3^j w + 1$$

then

$$v - 2(j+1) \geq v - 2v_0 = v - 2\left[\frac{v-1}{2}\right] \geq 1$$

and

$$S^{j+1}(m) = S\left(2^{v-2j} 3^j w + 1\right) = \frac{2^{v-2j} 3^{j+1} w + 4}{4}$$

$$= 2^{v-2(j+1)}3^{j+1}w + 1 = \left(\frac{3}{4}\right)^{j+1}(2^v w) + 1$$

$$= \left(\frac{3}{4}\right)^{j+1}(m-1) + 1.$$

This proves (11), and (11) implies (9).

We have

$$S^{v_0}(m) = 2^{v-2v_0}3^{v_0}w + 1$$

If $v = 2v_0 + 1$, then

$$S^{v_0}(m) = 2 \cdot 3^{v_0}w + 1$$

If $v = 2v_0 + 2$, then

$$S^{v_0}(m) = 4 \cdot 3^{v_0}w + 1$$

This completes the proof. □

Theorem 4 *The Syracuse function S is wildly increasing-decreasing.*

Proof Let

$$c_{n+1} = \begin{cases} 4 & \text{if } n+1 \text{ is even} \\ 2 & \text{if } n+1 \text{ is odd.} \end{cases}$$

Let $V = (v_i)_{i=1}^n$ be a finite sequence of positive integers. By Lemma 1, if

$$m = 2^{v_1+1}w_1 - 1$$

for some odd positive integer w_1, then the odd positive integer m increases for v_1 iterations of the Syracuse function S. From (7) we have the odd positive integer

$$S^{v_1}(m) = 2 \cdot 3^{v_1}w_1 - 1.$$

By Lemma 2, if

$$S^{v_1}(m) = 2 \cdot 4^{v_2}w_2 + 1$$

for some odd positive integer w_2, then the odd positive integer $S^{v_1}(m)$ decreases for v_2 iterations of the Syracuse function S. The integers w_1 and w_2 are solutions of the diophantine equation

$$2 \cdot 3^{v_1}w_1 - 1 = 2 \cdot 4^{v_2}w_2 + 1$$

or, equivalently,

$$3^{v_1}w_1 - 4^{v_2}w_2 = 1.$$

From (10) we have the odd positive integer

$$S^{v_1+v_2}(m) = 2 \cdot 3^{v_2} w_2 + 1.$$

By Lemma 1, if

$$S^{v_1+v_2}(m) = 2^{v_3+1} w_3 - 1$$

for some odd positive integer w_3, then the integer $S^{v_1+v_2}(m)$ increases for v_3 iterations of the Syracuse function S. The integers w_2 and w_3 are solutions of the diophantine equation

$$2 \cdot 3^{v_2} w_2 + 1 = 2^{v_3+1} w_3 - 1$$

or, equivalently,

$$3^{v_2} w_2 - 2^{v_3} w_3 = -1.$$

Continuing inductively, we obtain an odd positive integer m that has increasing-decreasing pattern V with respect to the Syracuse function S if the system of linear diophantine equations

$$\begin{aligned} 3^{v_1} w_1 - 4^{v_2} w_2 &= 1 \\ 3^{v_2} w_2 - 2^{v_3} w_3 &= -1 \\ 3^{v_3} w_3 - 4^{v_4} w_4 &= 1 \\ &\vdots \\ 3^{v_n} w_n - c_{n+1}^{v_{n+1}} w_{n+1} &= (-1)^{n+1} \end{aligned}$$

has a solution $w_1, \ldots, w_{n+1}$ in odd positive integers. This is precisely what Theorem 2 provides. This completes the proof. □

4 A Simpler Proof

I thank an anonymous referee for the reference to Everett [7] and the following variant of Theorem 4.

Theorem 5 *Let k be a positive integer and let $\Lambda = (\lambda_1, \lambda_2, \ldots, \lambda_k)$ be a k-tuple with $\lambda_i \in \{increase, decrease\}$ for all $i \in \{1, 2, \ldots, k\}$. Let $S(\Lambda)$ be the set of all odd positive integers m such that*

$$S^{i-1}(m) < S^i(m) \quad \text{if } \lambda_i = increase$$

and

$$S^{i-1}(m) > S^i(m) \quad \text{if } \lambda_i = decrease$$

for all $i \in \{1, 2, \ldots, k\}$. *There are positive integers* $\ell = \ell(\Lambda)$ *and* $r = r(\Lambda)$ *with* r *odd such that, for all* $m \geq 1$,

$$m \equiv r \pmod{2^{\ell}} \quad \textit{implies} \quad m \in S(\Lambda). \tag{12}$$

This result immediately implies thats that the Sylvester function is wildly increasing-decreasing.

Proof The proof is by induction on k.

Let $k = 1$. If $\Lambda = \{\text{increase}\}$, then choose $\ell = 2$ and $r = 3$. By Lemma 1, the congruence $m \equiv 3 \pmod{2^2}$ implies $m \in S(\Lambda)$. If $\Lambda = \{\text{decrease}\}$, then choose $\ell = 3$ and $r = 1$. By Lemma 2, the congruence $m \equiv 1 \pmod{2^3}$ implies $m \in S(\Lambda)$. This proves the Theorem for $k = 1$.

Let $k \geq 1$ and assume that for every k-tuple Λ there exist positive integers $\ell = \ell(\Lambda)$ and $r = r(\Lambda)$ with r odd that satisfy (12) for all $m \geq 1$. Let $\Lambda' = (\lambda_1, \lambda_2, \ldots, \lambda_k, \lambda_{k+1})$ be a $(k+1)$-tuple with $\lambda_i \in \{\text{increase, decrease}\}$ for all $i \in \{1, 2, \ldots, k, k+1\}$. Consider the k-tuple $\Lambda = (\lambda_2, \ldots, \lambda_k, \lambda_{k+1})$. By the induction hypothesis, there exist positive integers ℓ' and r' with r' odd such that, for all $m' \geq 1$,

$$m' \equiv r' \pmod{2^{\ell'}} \quad \text{implies} \quad m' \in S(\Lambda').$$

Suppose $\lambda_1 = \text{increase}$. The odd positive integer m is in $S(\Lambda)$ if

$$S(m) = \frac{3m+1}{2} \equiv r' \pmod{2^{\ell'}}$$

or, equivalently, if

$$3m \equiv 2r' - 1 \pmod{2^{\ell'+1}}.$$

Let $\ell = \ell' + 1$. Choose $t \in \mathbf{N}$ such that $3t \equiv 1 \pmod{2^{\ell}}$. Then t and $r = (2r' - 1)t$ are odd and ℓ and r satisfy (12).

Suppose $\lambda_1 = \text{decrease}$. The odd positive integer m is in $S(\Lambda)$ if

$$S(m) = \frac{3m+1}{4} \equiv r' \pmod{2^{\ell'}}$$

or, equivalently, if

$$3m \equiv 4r' - 1 \pmod{2^{\ell'+2}}.$$

Let $\ell = \ell' + 2$. Choose $t \in \mathbf{N}$ such that $3t \equiv 1 \pmod{2^{\ell}}$. Then t and $r = (4r' - 1)t$ are odd and ℓ and r satisfy (12). Thus, the Theorem holds for all $(k+1)$-tuples Λ. This completes the proof. □

5 Open Problems

5.1 The Infinite Constant Sequence $v_i = 1$

The Collatz conjecture states that for every positive integer m there is an integer k_m such that $S^{k_m}(m) = 1$ and so the Syracuse trajectory $\left(S^j(m)\right)_{j=0}^{\infty}$ is eventually constant. This would imply that there is no infinite sequence of positive integers $V = (v_i)_{i=1}^{\infty}$ and no positive integer m for which the trajectory $\left(S^j(m)\right)_{j=0}^{\infty}$ satisfies the increasing-decreasing conditions (2) and (3) for all $i \in \mathbf{N}$. Is it possible to prove that some particular infinite sequence $V = (v_i)_{i=1}^{\infty}$ is not the increasing-decreasing pattern of any positive integer m under iterations of the Syracuse function? For example, can the constant sequence $V = (v_i)_{i=1}^{\infty}$ with $v_i = 1$ for all i be proven impossible? This is equivalent to proving that if M is the infinite matrix with (i, j)th coordinate

$$M_{i,j} = \begin{cases} 3 & \text{if } j = i \\ -4 & \text{if } j = i + 1 \\ 0 & \text{otherwise} \end{cases}$$

for all $i, j \in \mathbf{N}$ and if $\mathbf{h} = (h_i)_{i=1}^{\infty}$ is the infinite vector with $h_i = (-1)^{i+1}$ for all $i \in \mathbf{N}$, then there exists no infinite odd positive vector $\mathbf{w} = (w_j)_{j=1}^{\infty}$ such that $M\mathbf{w} = \mathbf{h}$.

5.2 The Finite Constant Sequence $v_i = 1$

For every positive integer n there exists an odd positive integer m that has the increasing-decreasing pattern $V = (v_i)_{i=1}^{n}$ with $v_i = 1$ for all $i \in \{1, 2, \ldots, n\}$. Let m_n be the least such integer. The sequence $(m_n)_{n=1}^{\infty}$ is increasing. Is it strictly increasing? Can m_n be efficiently computed?

5.3 Sufficient Conditions for Wildly Increasing-Decreasing

The method of proof of Theorem 5 can be extended to prove that a large class of arithmetic functions is wildly increasing decreasing. Are there necessary and sufficient conditions for an arithmetic function f to be wildly increasing-decreasing?

Are there sufficient conditions for a sequence $\mathcal{F} = (f_j)_{j=0}^{\infty}$ of arithmetic functions to be wildly increasing-decreasing?

References

1. E. Catalan, Propositions et questions diverses, Bull. Soc. Math. France 16 (1888), 128–129.
2. L. E. Dickson, Theorems and tables on the sums of divisors of a number, Quart. J. Math. 44 (1913), 264-296.
3. S. Eliahou, The $3x + 1$ problem: new lower bounds on nontrivial cycle lengths, Discrete Math. 118 (1993), 45–56.
4. P. Erdős, Über die Zahlen der Form $\sigma(n) - n$ und $n - \varphi(n)$, Elem. Math 11 (1973), 83–86.
5. P. Erdős, On asymptotic properties of aliquot sequences, Math. Computation 30 (1976), 641–645.
6. P. Erdős, A. Granville, C. Pomerance, and C. Spiro, On the normal behavior of the iterates of some arithmetic functions, *Analytic Number Theory (Allerton Park, IL, 1989)*, Prog. Math. 85, Birkhäuser Boston, Boston, MA, 1990, pp. 165–204.
7. C. J. Everett, Iteration of the number-theoretic function $f(2n) = n$, $f(2n+1) = 3n+2$, Adv. Math. 25 (1977), 42–45.
8. R. K. Guy, *Unsolved Problems in Number Theory*, 3rd edition, Springer-Verlag, New York, 2004.
9. R. K. Guy and J. L. Selfridge, J. L., Interim report on aliquot series, in: Proceedings of the Manitoba Conference on Numerical Mathematics (Univ. Manitoba, Winnipeg, Man., 1971), 557–580.
10. R. K. Guy and J. L. Selfridge, J. L., What drives an aliquot sequence?, Math. Computation 29 (1975), 101–107.
11. J. C. Lagarias, The $3x + 1$ problem and its generalizations, Amer. Math. Monthly 92 (1985), 3–23.
12. J. C. Lagarias, *The ultimate challenge: the* $3x + 1$ *problem*, Amer. Math. Soc., Providence, RI, 2010.
13. H. W. Lenstra, Jr., Problem 6064, Amer. Math. Monthly 82 (1975), 1016. Solution by the proposer, Amer. Math. Monthly 84 (1977), 580.
14. M. Marcus and H. Minc, *A Survey of Matrix Theory and Matrix Inequalities*, Dover Publications, New York, 1992.
15. P. Pollack and C. Pomerance, Some problems of Erdős on the sum-of-divisors function, Trans. Amer. Math. Soc. Ser. B 3 (2016), 1–26.
16. C. Pomerance, Aliquot sequences, University of Calgary lecture, October 2, 2020, https://math.dartmouth.edu/~carlp/upintconf.pdf.
17. T. Tao, Almost all orbits of the Collatz map attain almost bounded values, Forum Math. Pi 10 (2022), Paper No. e12, 56.
18. R. Terras, A stopping time problem on the positive integers, Acta Arith. 30 (1976), 241–252.

The Rate of Convergence for Selberg's Central Limit Theorem Under the Riemann Hypothesis

Asher Roberts

Abstract We assume the Riemann hypothesis to improve upon the rate of convergence of $(\log\log\log T)^2/\sqrt{\log\log T}$ in Selberg's central limit theorem for $\log|\zeta(1/2+it)|$ given by the author in [8]. We achieve a rate of convergence of $\sqrt{\log\log\log\log T}/\sqrt{\log\log T}$ in the Dudley distance. The proof is an adaptation of the techniques used by the author in [8], based on the work of Radziwiłł and Soundararajan in [7] and Arguin et al. in [1], combined with a lemma of Selberg [11] that provides for a mollifier close to the critical line $\mathrm{Re}(s) = 1/2$ under the Riemann hypothesis.

1 Introduction

Let τ be a random point uniformly distributed on $[T, 2T]$. Selberg's central limit theorem states that as $T \to \infty$,

$$\mathbb{P}\left(\log|\zeta(\tfrac{1}{2}+i\tau)| > V\sqrt{\tfrac{1}{2}\log\log T}\right) \to \frac{1}{\sqrt{2\pi}}\int_V^\infty e^{-u^2/2}du, \quad \forall V \in \mathbb{R}.$$

In this paper, we determine the rate of convergence to a Gaussian random variable in Selberg's theorem using the Dudley (also called bounded Wasserstein, e.g. see Gaunt and Li [2]) distance between two random vectors in $\mathbb{R}^n$ defined on a probability space $(\Omega, \mathcal{F}, \mathbb{P})$ given by

$$d_{\mathcal{D}}(X,Y) = \sup_{f\in\mathcal{L}} |\mathbb{E}[f(X)] - \mathbb{E}[f(Y)]|$$

A. Roberts (✉)
Graduate Center, City University of New York, New York, NY 10016, USA
e-mail: aroberts@gradcenter.cuny.edu
URL: https://asherbroberts.com

M. B. Nathanson (ed.), *Combinatorial and Additive Number Theory VI*,
Springer Proceedings in Mathematics & Statistics 464,
https://doi.org/10.1007/978-3-031-65064-2_19

where $\mathcal{L} = \{f : \mathbb{R}^n \to \mathbb{R} \mid f$ is Lipschitz, $\|f\|_\infty \leq 1$, and $\|f\|_{\text{Lip}} \leq 1\}$. Here and throughout the paper $\|f\|_\infty$ denotes the supremum norm of f and $\|f\|_{\text{Lip}} = \sup_{x \neq y} \frac{|f(x)-f(y)|}{\|x-y\|_1}$ with $\|\cdot\|_1$ denoting the L^1 norm. We use the Dudley distance since the boundedness property allows us to use Lemma 4 in Proposition 5, while the Lipschitz property assists us with the remaining propositions in the paper.

1.1 Main Result

Theorem 1 *Assume the Riemann hypothesis and let τ be a random point distributed uniformly on $[T, 2T]$. Then for T large enough,*

$$d_{\mathcal{D}}\left(\frac{\log|\zeta(\frac{1}{2}+i\tau)|}{\sqrt{\frac{1}{2}\log\log T}}, \mathcal{Z}\right) \ll \frac{\sqrt{\log\log\log\log T}}{\sqrt{\log\log T}},$$

where $\mathcal{Z}$ is a standard normal random variable.

Theorem 1 is an improvement of the rate of convergence

$$d_{\mathcal{D}}\left(\frac{\log|\zeta(\frac{1}{2}+i\tau)|}{\sqrt{\frac{1}{2}\log\log T}}, \mathcal{Z}\right) \ll \frac{(\log\log\log T)^2}{\sqrt{\log\log T}}$$

given by the author in [8] using the Dudley distance. We modify the techniques of the author in [8] to prove Theorem 1 by mollifying $\log|\zeta(1/2+it)|$ closer to the critical line $\text{Re}(s) = 1/2$. This allows us to approximate $\log|\zeta(1/2+it)|$ by an "almost" Gaussian prime sum using fewer steps, yielding a smaller error.

1.2 Relations to Previous Results

In 1946 Selberg presented a proof of his central limit theorem for the Riemann zeta function in [9]. Later, Radziwiłł and Soundararajan used simpler techniques to prove Selberg's theorem for $\log|\zeta(\frac{1}{2}+it)|$ in [7]. It is challenging to apply these techniques to the imaginary case, due to the difficulty in mollifying the imaginary part of zeta. Nevertheless, the mollifier used in Proposition 2 of this paper can be applied to the imaginary part of zeta, but this would require a different way of moving the imaginary part of zeta off axis than we present in the proof of Proposition 1. However, our work be extended to obtain analogous results for Dirichlet L-functions, as Hsu and Wong did in [4] to prove the central limit theorem using the techniques of Radziwiłł and Soundararajan.

Our paper mainly uses the techniques employed by the author in [8], although some modifications are needed. Here the rate of convergence was obtained for zeta at two points, while our result is presented for zeta at one point for simplicity but can be similarly extended. The main difference lies in the mollification step Proposition 2. Without the Riemann hypothesis, the mollifier Selberg offers in [11] presents a sum over zeta zeros that is difficult to estimate. By assuming the Riemann hypothesis and moving off axis by $1/\log T$ (instead of by $(\log\log\log T)^2/\log T$ in [8]), we can estimate this sum to be as close to order 1 as we would like by further iterating the log in the exponent of the parameter X used.

In [10] Selberg claims a rate of convergence of $(\log\log\log T)^2/\sqrt{\log\log T}$ unconditionally for the real part of zeta using the Kolmogorov distance, a proof of which is given by Tsang in [12]. The author reproduces this rate using the techniques of Radziwiłł and Soundararajan as well as Arguin et al. in [8]. Selberg also claims a rate of $\log\log\log T/\sqrt{\log\log T}$ for the imaginary case. An explicit proof of this result is given by Wahl in [13] in the context of mod-gaussian convergence (more information is given in Kowalski and Nikeghbali [6]).

While the Riemann hypothesis allows us to improve upon Selberg's rate of convergence in [10], there exist barriers towards further improvement. Specifically, although we can push the parameter X further in Proposition 2 as stated, we are still limited in the choice of parameters X_1 and X_2 used in the Dirichlet polynomials we compare to Gaussian distributions in Proposition 5. That is, iterating the log further in those parameters results in Dirichlet polynomials that align less closely with Gaussian distributions, worsening the errors presented in Lemmas 5 and 6. Improving our overall stated rate of convergence therefore requires more sophisticated techniques in handling parameters with log iterations that approach order 1.

1.3 Structure of the Proofs

The proof of Theorem 1 follows by five successive propositions. All proofs are given in Sect. 2, and the distances between the random variables are explicit at each step of the proof. The trickiest step is Proposition 2, mollifying zeta. In particular, handling the sum over zeta zeros. However, the step that incurs the greatest error is Proposition 4, truncating the prime sum. Although we could improve the error obtained in mollifying by iterating the logarithm further in the definition of the parameter X used in Proposition 2, we would still be required to perform the same truncation in Proposition 4, in order to effectively compare moments with the Fourier transforms in Proposition 5. That is, an improvement to the rate of convergence we obtain requires an improvement in the way we truncate using Proposition 4.

To connect zeta with the aforementioned "almost" Gaussian prime sum, we first move off axis using Proposition 1, since the Riemann hypothesis will then ensure that we do not encounter a zero when mollifying zeta in Proposition 2. To simplify notation, we set

$$\sigma_0 = \frac{1}{2} + \frac{1}{\log T}, \qquad s_0 = \sigma_0 + i\tau, \qquad \text{and} \qquad \mathfrak{s} = \sqrt{\tfrac{1}{2}\log\log T}. \tag{1}$$

The contribution towards the rate of convergence in Theorem 1 by moving off axis is then given by

Proposition 1 (Moving off axis). *With the notation above, we have*

$$d_{\mathcal{D}}\left(\frac{\log|\zeta(\frac{1}{2}+i\tau)|}{\mathfrak{s}}, \frac{\log|\zeta(s_0)|}{\mathfrak{s}}\right) \ll \frac{1}{\sqrt{\log\log T}}.$$

Off axis, we are now able to approximate zeta with a mollifier in the form of a Dirichlet polynomial $\mathcal{P}(s)$ given in Proposition 2. Note that Proposition 2 is the only proposition conditional on the Riemann hypothesis. The contribution towards the rate of convergence in Theorem 1 by mollifying is then given by

Proposition 2 (Mollifying). *Assume the Riemann hypothesis. Let* $X = T^{1/(\log_4 T)^{1/4}}$ *where* $\log_\ell$ *stands for the logarithm iterated* ℓ *times, and*

$$\Lambda_X(n) = \begin{cases} \Lambda(n) & \text{for } 1 \le n \le X, \\ \Lambda(n)\dfrac{\log \frac{X^2}{n}}{\log X} & \text{for } X \le n \le X^2. \end{cases}$$

Set

$$\mathcal{P}(s_0) = \sum_{n<X^2} \frac{\Lambda_X(n)}{n^{s_0}\log n}.$$

Then

$$d_{\mathcal{D}}\left(\frac{\log|\zeta(s_0)|}{\mathfrak{s}}, \frac{\operatorname{Re}\mathcal{P}(s_0)}{\mathfrak{s}}\right) \ll \frac{\sqrt{\log_4 T}}{\sqrt{\log\log T}}.$$

We then discard the higher order primes in the sum $\mathcal{P}(s)$, whose contribution towards the rate of convergence in Theorem 1 is given by

Proposition 3 (Discarding higher order primes). *Set*

$$\mathcal{P}(s_0) = \sum_{n<X^2} \frac{\Lambda_X(n)}{n^{s_0}\log n} \qquad \text{and} \qquad P(s_0) - \operatorname{Re}\sum_{p\le X} \frac{1}{p^{s_0}}.$$

With the notation above, we have

$$d_{\mathcal{D}}\left(\frac{\operatorname{Re}\mathcal{P}(s_0)}{\mathfrak{s}}, \frac{P(s_0)}{\mathfrak{s}}\right) \ll \frac{1}{\sqrt{\log\log T}}.$$

In order to compare moments in Proposition 5, we truncate the prime sum. The more we truncate the sum, the faster the sum tends toward a Gaussian random variable. This truncation bears a contribution towards the rate of convergence in Theorem 1 given by

Proposition 4 (Truncating the prime sum). *Set*

$$P_1(s_0) = \mathrm{Re} \sum_{p \le X_1} \frac{1}{p^{s_0}},$$

$$P_2(s_0) = \mathrm{Re} \sum_{X_1 < p \le X_2} \frac{1}{p^{s_0}}, \text{ and}$$

$$P_3(s_0) = \mathrm{Re} \sum_{X_2 < p \le X} \frac{1}{p^{s_0}},$$

so that $P(s_0) = P_1(s_0) + P_2(s_0) + P_3(s_0)$ *where* $X_1 = T^{1/\log\log T}$ *and* $X_2 = T^{1/\log\log\log T}$. *With the notation above, we have*

$$d_{\mathcal{D}}\left(\frac{P(s_0)}{\mathfrak{s}}, \frac{P_1(s_0) + P_2(s_0)}{\mathfrak{s}}\right) \ll \frac{\sqrt{\log_4 T}}{\sqrt{\log\log T}}.$$

Finally, we compare our truncated sum to a Gaussian random variable, generating a contribution towards the rate of convergence in Theorem 1 given by

Proposition 5 *With the notation above, we have*

$$d_{\mathcal{D}}\left(\frac{P_1(s_0) + P_2(s_0)}{\mathfrak{s}}, \mathcal{Z}\right) \ll \frac{1}{\sqrt{\log\log T}},$$

where $\mathcal{Z}$ *is a standard normal random variable.*

Acknowledgements. I give thanks to Emma Bailey for spotting a crucial issue in one of my initial drafts for this paper, and to Prof. Louis-Pierre Arguin for his helpful insight that provided me with the confidence to fix it. I would also like to thank everyone who read through the paper and offered edits. Your feedback is very much appreciated.
Partial support is provided by grants NSF CAREER 1653602 and NSF DMS 2153803.

2 Proofs

2.1 Proof of Theorem 1

Proof With the notation above, we have by Lemma 1 to 5 and the triangle inequality,

$$\begin{aligned} d_{\mathcal{D}}\left(\frac{\log|\zeta(\frac{1}{2}+i\tau)|}{\sqrt{\frac{1}{2}\log\log T}}, \mathcal{Z}\right) &\ll d_{\mathcal{D}}\left(\frac{\log|\zeta(\frac{1}{2}+i\tau)|}{\mathfrak{s}}, \frac{\log|\zeta(s_0)|}{\mathfrak{s}}\right) \\ &\quad + d_{\mathcal{D}}\left(\frac{\log|\zeta(s_0)|}{\mathfrak{s}}, \frac{\operatorname{Re}\mathcal{P}(s_0)}{\mathfrak{s}}\right) \\ &\quad + d_{\mathcal{D}}\left(\frac{\operatorname{Re}\mathcal{P}(s_0)}{\mathfrak{s}}, \frac{P(s_0)}{\mathfrak{s}}\right) \\ &\quad + d_{\mathcal{D}}\left(\frac{P(s_0)}{\mathfrak{s}}, \frac{P_1(s_0)+P_2(s_0)}{\mathfrak{s}}\right) \\ &\quad + d_{\mathcal{D}}\left(\frac{P_1(s_0)+P_2(s_0)}{\mathfrak{s}}, \mathcal{Z}\right) \\ &\ll \frac{\sqrt{\log\log\log\log T}}{\sqrt{\log\log T}}. \end{aligned}$$

2.2 Proof of Proposition 1

Proof Denote $\log|\zeta(\frac{1}{2}+i\tau)|/\mathfrak{s}$ by $\mathcal{V}$ and $\log|\zeta(s_0)|/\mathfrak{s}$ by $\mathcal{W}$. Then

$$d_{\mathcal{D}}(\mathcal{V},\mathcal{W}) = \sup_{f\in\mathcal{L}} |\mathbb{E}[f(\mathcal{V})] - \mathbb{E}[f(\mathcal{W})]| \leq \sup_{f\in\mathcal{L}} \mathbb{E}[|f(\mathcal{V}) - f(\mathcal{W})|].$$

Since $f:\mathbb{R}\to\mathbb{R}$ where $f\in\mathcal{L}$ is Lipschitz with $\|f\|_{\text{Lip}}\leq 1$, we have $|f(v) - f(w)| \leq \|f\|_{\text{Lip}}\|v-w\|_1 \leq \|v-w\|_1$, and thus

$$|f(\mathcal{V}) - f(\mathcal{W})| \leq \|\mathcal{V}-\mathcal{W}\|_1. \tag{2}$$

Therefore, by definition of the Dudley distance, the above distance is less than or equal to $\mathbb{E}[|\mathcal{V}-\mathcal{W}|]$. We now introduce Proposition 1 of [7] as Lemma 1.

Lemma 1 *With the notation above, for T large enough and any $\sigma > 1/2$, we have*

$$\mathbb{E}\Big[\Big|\log|\zeta(\tfrac{1}{2}+i\tau)| - \log|\zeta(\sigma+i\tau)|\Big|\Big] \ll (\sigma-\tfrac{1}{2})\log T.$$

The proof of Lemma 1 uses Hadamard's factorization formula for the completed ζ-function and is given in [7]. By Lemma 1, $\mathbb{E}[|\mathcal{V} - \mathcal{W}|] \ll 1/\sqrt{\log\log T}$, and so our result follows.

2.3 Proof of Proposition 2

Proof Similar to the way we proved Proposition 1, we denote $\log|\zeta(s_0)|/\mathfrak{s}$ by $\mathcal{W}$, $\operatorname{Re}\mathcal{P}(s_0)/\mathfrak{s}$ by $\mathcal{X}$ and have

$$d_{\mathcal{D}}(\mathcal{W}, \mathcal{X}) = \sup_{f\in\mathcal{L}} |\mathbb{E}[f(\mathcal{W})] - \mathbb{E}[f(\mathcal{X})]| \leq \sup_{f\in\mathcal{L}} \mathbb{E}[|f(\mathcal{W}) - f(\mathcal{X})|]. \tag{3}$$

To compute this distance, we first approximate zeta using a mollifier of Selberg [11] given by the following lemma.

Lemma 2 *With the same assumptions as Proposition 2, and for $s \neq 1$, $s \neq \rho$, and $s \neq -2q$ $(q = 1, 2, 3, \dots)$, where ρ is a zero of ζ, we have*

$$\frac{\zeta'}{\zeta}(s) = -\sum_{n<X^2} \frac{\Lambda_X(n)}{n^s} + \frac{1}{\log X}\sum_{\rho} \frac{X^{\rho-s} - X^{2(\rho-s)}}{(s-\rho)^2} + O(1).$$

Proof The proof of the result

$$\begin{aligned}\frac{\zeta'}{\zeta}(s) = -\sum_{n<X^2} \frac{\Lambda_X(n)}{n^s} + \frac{X^{2(1-s)} - X^{1-s}}{\log X\cdot(1-s)^2} + \frac{1}{\log X}\sum_{q=1}^{\infty} \frac{X^{-2q-s} - X^{-2(2q+s)}}{(2q+s)^2} \\ + \frac{1}{\log X}\sum_{\rho} \frac{X^{\rho-s} - X^{2(\rho-s)}}{(s-\rho)^2}\end{aligned}$$

is given unconditionally in Lemma 2 of Selberg [11] using a contour integral. For $s = \sigma + it$ we have $|1-s|^2 \asymp t^2$, and so

$$\frac{X^{2(1-s)} - X^{1-s}}{\log X\cdot(1-s)^2} \ll \frac{X^{2(1-\sigma)} + X^{1-\sigma}}{t^2\log X} \ll \frac{X}{t^2\log X} \ll \frac{1}{T\log T},$$

while the sum over q is clearly $O(1)$.

We continue by applying the Riemann hypothesis, setting $s = \sigma + it$ and $\rho = 1/2 + i\gamma$, while integrating both sides of Lemma 2 to obtain

$$\int_{\sigma_0}^{2} \operatorname{Re}\frac{\zeta'}{\zeta}(s)\,d\sigma = -\operatorname{Re}\sum_{n<X^2} \Lambda_X(n)\int_{\sigma_0}^{2} n^{-s}d\sigma + \frac{1}{\log X}\operatorname{Re}\sum_{\rho}\int_{\sigma_0}^{2} \frac{X^{\rho-s} - X^{2(\rho-s)}}{(s-\rho)^2}\,d\sigma.$$

The first summand becomes

$$
\begin{aligned}
-\operatorname{Re}\sum_{n<X^2}\Lambda_X(n)\int_{\sigma_0}^{2} n^{-\sigma-it}d\sigma &= -\operatorname{Re}\sum_{n<X^2}\Lambda_X(n)n^{-it}\left[\frac{\exp(-\sigma\log n)}{-\log n}\right]_{\sigma=\sigma_0}^{\sigma=2}\\
&= -\operatorname{Re}\sum_{n<X^2}\Lambda_X(n)n^{-it}\left[\frac{n^{-\sigma}}{-\log n}\right]_{\sigma=\sigma_0}^{\sigma=2}\\
&= -\operatorname{Re}\sum_{n<X^2}\Lambda_X(n)\left[\frac{n^{-s_0}}{\log n}-\frac{n^{-2-it}}{\log n}\right]\\
&= -\operatorname{Re}\sum_{n<X^2}\frac{\Lambda_X(n)}{\log n}n^{-s_0},
\end{aligned}
$$

by the summability of $\sum n^{-2-it}$. Meanwhile, the second summand is

$$
\begin{aligned}
&\frac{1}{\log X}\operatorname{Re}\sum_{\rho}\int_{\sigma_0}^{2}\frac{X^{\rho-s}-X^{2(\rho-s)}}{(s-\rho)^2}\,d\sigma\\
&\ll \frac{1}{\log X}\sum_{\rho}\frac{1}{|s_0-\rho|^2}\int_{\sigma_0}^{2}(X^{1/2-\sigma}+X^{2(1/2-\sigma)})\,d\sigma\\
&\ll \frac{1}{(\log X)^2}\sum_{\rho}\frac{1}{|s_0-\rho|^2}.
\end{aligned}
$$

Therefore,

$$
\log|\zeta(2+it)|-\log|\zeta(s_0)| = -\operatorname{Re}\sum_{n<X^2}\frac{\Lambda_X(n)}{\log n}n^{-s_0}+O\left(\frac{1}{(\log X)^2}\sum_{\rho}\frac{1}{|s_0-\rho|^2}\right).
$$

As we obtained the inequality (2) by taking advantage of the Lipschitz nature of f, we similarly obtain

$$
|f(\mathcal{W})-f(\mathcal{X})|\le\|\mathcal{W}-\mathcal{X}\|_1,
$$

and so the expectation in (3) is less than or equal to

$$
\begin{aligned}
\mathbb{E}[|\mathcal{W}-\mathcal{X}|] &= \frac{1}{\mathfrak{s}}\mathbb{E}[|\log|\zeta(s_0)|-\operatorname{Re}\mathcal{P}(s_0)|]\\
&\ll \frac{1}{\mathfrak{s}}\mathbb{E}\left[\frac{1}{(\log X)^2}\sum_{\rho}\frac{1}{|s_0-\rho|^2}\right],
\end{aligned}
$$

since $\log|\zeta(2+it)| = O(1)$. We then proceed similarly as in the proof of Proposition 1 of Harper [3], computing the expectation by integrating

$$\int_T^{2T} \sum_{\rho:T\le\gamma\le 2T} \frac{1}{|s_0-\rho|^2}\,dt = \sum_{\rho:T\le\gamma\le 2T} \int_T^{2T} \frac{1}{|s_0-\rho|^2}\,dt.$$

Since we are assuming the Riemann hypothesis, and since we have moved off axis by $1/\log T$, we have $|s_0-\rho| > 1/\log T$ for all ρ, and therefore the integral above is

$$\le \sum_{\rho:T\le\gamma\le 2T} \int_{1/\log T}^{\infty} \frac{1}{t^2}\,dt = \log T \sum_{\rho:T\le\gamma\le 2T} 1 \ll (\log T)^2,$$

where in the end we used the Riemann-von Mangoldt formula (see, e.g., Ivić [5]), which provides an estimate of $\log T$ for the number of zeta zeros between T and $2T$. Thus

$$\frac{1}{\mathfrak{s}}\mathbb{E}[|\log|\zeta(s_0)| - \operatorname{Re}\mathcal{P}(s_0)|] \ll \frac{1}{\mathfrak{s}}\cdot\frac{1}{(\log X)^2}\cdot(\log T)^2 \ll \frac{\sqrt{\log_4 T}}{\sqrt{\log\log T}}$$

as desired.

2.4 *Proof of Proposition 3*

Proof By taking advantage of the Lipschitz nature of the Dudley distance, we obtain

$$\begin{aligned} &d_{\mathcal{D}}\left(\frac{\operatorname{Re}\mathcal{P}(s_0)}{\mathfrak{s}}, \frac{P(s_0)}{\mathfrak{s}}\right) \\ &\le \mathbb{E}\left[\left|\frac{\operatorname{Re}\mathcal{P}(s_0)}{\mathfrak{s}} - \frac{P(s_0)}{\mathfrak{s}}\right|\right]. \end{aligned}$$

Then $\Lambda_X(n) \le \Lambda(n)$, and so

$$\sum_{n<X^2} \frac{\Lambda_X(n)}{n^{s_0}\log n} \ll \sum_{n<X} \frac{\Lambda(n)}{n^{s_0}\log n},$$

since $T^{2/(\log_4 T)^{1/4}} = O(T^{1/(\log_4 T)^{1/4}})$. Therefore, computing the difference in the expectation we get

$$\frac{1}{\mathfrak{s}}|\operatorname{Re}\mathcal{P}(s_0) - P(s_0)| \ll \frac{1}{\mathfrak{s}}\left|\sum_{2\le p^2\le X} \frac{1}{2p^{2s_0}}\right| + \frac{1}{\mathfrak{s}}\left|\sum_{\substack{2\le p^k\le X,\\ k\ge 3}} \frac{\log p}{p^{ks_0}(k\log p)}\right|. \tag{4}$$

Then since

$$\left| \sum_{\substack{2 \le p^k \le X, \\ k \ge 3}} \frac{\log p}{p^{ks_0}(k \log p)} \right| \le \sum_{\substack{2 \le p^k \le X, \\ k \ge 3}} \frac{1}{3p^{k\sigma_0}} = O(1)$$

and

$$\begin{aligned}
\mathbb{E}\left[\left| \sum_{2 \le p^2 \le X} \frac{1}{2p^{2(s_0)}} \right| \right] &\le \mathbb{E}\left[\left| \sum_{2 \le p^2 \le X} \frac{1}{2p^{2(s_0)}} \right|^2 \right]^{1/2} \\
&= \frac{1}{T} \int_T^{2T} \left| \sum_{2 \le p^2 \le X} \frac{1}{2p^{2(s_0)}} \right|^2 dt \\
&= \frac{1}{T} \cdot \frac{1}{4} \sum_{p_1, p_2 \le \sqrt{X}} \int_T^{2T} \frac{1}{p_1^{2(\sigma_0 + it)} p_2^{2(\sigma_0 - it)}} dt \\
&\ll \sum_{p \le \sqrt{X}} \frac{1}{p^{4\sigma_0}} + \frac{1}{T} \sum_{\substack{p_1, p_2 \le \sqrt{X} \\ p_1 \ne p_2}} \frac{1}{p_1^{2\sigma_0} p_2^{2\sigma_0}} \sqrt{p_1 p_2} = O(1),
\end{aligned}$$

we see that (4) is $\ll 1/\mathfrak{s}$.

2.5 Proof of Proposition 4

Proof We have

$$d_{\mathcal{D}}\left(\frac{P(s_0)}{\mathfrak{s}}, \frac{P_1(s_0) + P_2(s_0)}{\mathfrak{s}} \right) \le \mathbb{E}\left[\left(\frac{P_3(s_0)}{\mathfrak{s}} \right)^2 \right]^{1/2}.$$

To evaluate this expectation, we use Lemma 5 of [8], which we introduce as Lemma 3

Lemma 3 *For any non-negative integer k we have*

$$\begin{aligned}
&\mathbb{E}\left[\left(\mathrm{Re}\left(\sum_{p \le X} \frac{1}{p^{s_0}} \right) \right)^{2k} \right] \\
&= 2^{-k} \frac{(2k)!}{k!} \left[\left(\frac{1}{2} \sum_{p \le X} \frac{1}{p^{2\sigma_0}} \right)^k + O_k \left(\frac{1}{2} \sum_{p \le X} \frac{1}{p^{2\sigma_0}} \right)^{k-2} \right] + O\left(\frac{X^{4k}}{T} \right),
\end{aligned}$$

where O_k indicates that the implied constant depends on k. If k is odd, we have

$$\mathbb{E}\left[\left(\mathrm{Re}\left(\sum_{p\le X}\frac{1}{p^{s_0}}\right)\right)^k\right] = O\left(\frac{X^{2k}}{T}\right).$$

Proof The statement follows from the proof of Lemma 5 of [8] by taking $u = 1$, $u' = 0$.

By Lemma 3, we have

$$\mathbb{E}[|P_3(\tfrac{1}{2}+it)|^{2k}] \ll \frac{(2k)!}{k!2^k}\left(\sqrt{\tfrac{1}{4}\log_4 T}\right)^{2k} + O\left(\frac{X^{4k}}{T}\right). \tag{5}$$

Here we used Mertens's theorem to estimate

$$\sum_{X_2<p\le X}\frac{1}{p} \ll \log\log X_2 - \log\log X = \log(\sqrt{\log\log\log T}) - \log((\log_4 T)^{1/4}) \ll \tfrac{1}{2}\log_4 T.$$

Thus the expectation above is less than or equal to $\frac{1}{\mathfrak{s}}(\sqrt{\log_4 T} + O(1))$ when $k = 1$.

2.6 *Proof of Proposition 5*

Proof Similar to Proposotion 6 of the author in [8], we obtain the error in approximating $P_1(s_0) + P_2(s_0)$ with a Gaussian random variable by comparing their Fourier transforms using Lemma 3.11 of [1], which we introduce as Lemma 4,

Lemma 4 *Let μ and ν be two probability measures on $\mathbb{R}$ with Fourier transform $\hat{\mu}$ and $\hat{\nu}$. There exists constant $c > 0$ such that for any function $f : \mathbb{R} \to \mathbb{R}$ with Lipschitz constant 1 and for any $R, F > 0$,*

$$\left|\int_{\mathbb{R}} f\,d\mu - \int_{\mathbb{R}} f\,d\nu\right| \le \frac{1}{F} + \|f\|_\infty\left\{(RF)\|(\hat{\mu}-\hat{\nu})\mathbb{1}_{(-F,F)}\|_\infty + \mu((-R,R)^c) + \nu((-R,R)^c)\right\}.$$

To use Lemma 4, we introduce a lemma that shows that the difference in expectations between the Fourier transforms of the prime sum and that of a Gaussian random variable is small.

Lemma 5

$$\left|\mathbb{E}\left[\exp\left(\frac{i\xi P_1(s_0)}{\mathfrak{s}}\right)\right] - \mathbb{E}[\exp(i\xi\mathcal{Z})]\right| \ll \frac{1}{(\log\log T)^2}.$$

Proof We first truncate the Fourier transform of $P_1(s_0)$ for $\xi \in \mathbb{R}$ by writing

$$\mathbb{E}\left[\exp\left(\frac{i\xi P_1(s_0)}{\mathfrak{s}}\right)\cdot \mathbb{1}(|P_1(s_0)| \le \log\log T)\right]$$
$$\ll \sum_{n\le N}\frac{1}{n!}i^n\mathbb{E}\left[\left(\frac{\xi P_1(s_0)}{\mathfrak{s}}\right)^n\cdot \mathbb{1}(|P_1(s_0)| \le \log\log T)\right] + (\log T)^{-C}$$

for a constant C, where the error term $(\log T)^{-C}$ is estimated in [8] using Stirling's formula. The event $\{|P_1(s_0)| \le \log\log T\}$ is introduced to assist with the truncation, and is subsequently removed by inserting the complement of this event, $\{|P_1(s_0)| > \log\log T\}$, and showing that the probability of the complement is small. After truncating, the moment estimate for P_1 in Lemma 3 is used to provide the error in comparing the truncated moments to a Gaussian. Full details of the proof are given in Lemma 9 of [8].

Next we introduce Lemma 6 as an analog of Lemma 5 for $P_2(s_0)$.

Lemma 6

$$\left|\mathbb{E}\left[\exp\left(\frac{i\xi P_2(s_0)}{\mathfrak{s}}\right)\right] - \mathbb{E}[\exp(i\xi\mathcal{Z})]\right| \ll \frac{1}{(\log\log T)^2}.$$

Proof We truncate the Fourier transform of $P_2(s_0)$ for $\xi \in \mathbb{R}$ by writing

$$\mathbb{E}\left[\exp\left(\frac{i\xi P_2(s_0)}{\mathfrak{s}}\right)\cdot \mathbb{1}(|P_2(s_0)| \le 2\log\log\log T)\right]$$
$$\ll \sum_{n\le N}\frac{1}{n!}i^n\mathbb{E}\left[\left(\frac{\xi P_2(s_0)}{\mathfrak{s}}\right)^n\cdot \mathbb{1}(|P_2(s_0)| \le 2\log\log\log T)\right] + (\log T)^{-D}$$

for a constant D, similar to the way we truncated the Fourier transform of $P_1(s_0)$. Note that we take $2\log\log\log T$ in the event in order to achieve an error of $1/(\log\log T)^2$, since the proof requires an estimate for $\mathbb{P}(|P_2(s_0)| > 2\log\log\log T)$ instead of for $\mathbb{P}(|P_1(s_0)| > \log\log T)$. As such,

$$\begin{aligned}
\mathbb{P}(|P_2(s_0)| > 2\log\log\log T) &\le \frac{1}{(2\log\log\log T)^{2k}}\mathbb{E}\left[|P_2(s_0)|^{2k}\right] \\
&\ll \frac{1}{(2\log\log\log T)^{2k}}\left[\frac{(2k)!}{k!2^k}\left(\sqrt{\tfrac{1}{2}\log\log\log T}\right)^{2k}\right] \\
&\ll \frac{1}{(2\log\log\log T)^{2k}}\left[\frac{(2k)^{2k}}{k^k 2^k}\frac{\exp(-2k)}{\exp(-k)}\left(\sqrt{\tfrac{1}{2}\log\log\log T}\right)^{2k}\right] \\
&= \frac{1}{(2\log\log\log T)^{2k}}\left[(2k)^k\exp(-k)2^{-k}(\log\log\log T)^k\right],
\end{aligned}$$

so we can set $k = \lfloor \log\log\log T \rfloor$ to get

$$\begin{aligned}
\mathbb{P}(|P_2(s_0)| > 2\log\log\log T) &\ll \frac{1}{(2k)^{2k}}[(2k)^k \exp(-k)2^{-k}k^k] \\
&= 2^{-2k}\exp(-k) \\
&\ll \exp(-2k) = \frac{1}{(\log\log T)^2}.
\end{aligned}$$

The rest of the proof is the same as for Lemma 5.

Now, using Lemmas 4 to 6 we have

$$\begin{aligned}
& d_{\mathcal{D}}\left[\left(\frac{P_1(s_0)+P_2(s_0)}{\mathfrak{s}}\right), \mathcal{Z}\right] \le d_{\mathcal{D}}\left[\left(\frac{P_1(s_0)}{\mathfrak{s}}\right), \mathcal{Z}\right] + d_{\mathcal{D}}\left[\left(\frac{P_2(s_0)}{\mathfrak{s}}\right), \mathcal{Z}\right] \\
& \le \frac{1}{F} + \left\{(R_1F)\left|\mathbb{E}\left[\exp\left(\frac{i\xi P_1(s_0)}{\mathfrak{s}}\right)\right] - \mathbb{E}[\exp(i\xi\mathcal{Z})]\right| \right. \\
& \quad \left. +\mathbb{P}\left(\left|\frac{P_1(s_0)}{\mathfrak{s}}\right| > R_1\right) + \mathbb{P}(|\mathcal{Z}| > R_1)\right\} \\
& \quad + \frac{1}{F} + \left\{(R_2F)\left|\mathbb{E}\left[\exp\left(\frac{i\xi P_2(s_0)}{\mathfrak{s}}\right)\right] - \mathbb{E}[\exp(i\xi\mathcal{Z})]\right| \right. \\
& \quad \left. +\mathbb{P}\left(\left|\frac{P_2(s_0)}{\mathfrak{s}}\right| > R_2\right) + \mathbb{P}(|\mathcal{Z}| > R_2)\right\}
\end{aligned}$$

for any $R_1, R_2, F > 0$. Since

$$\begin{aligned}
& \mathbb{P}\left(|P_2(s_0)| > \sqrt{\log\log\log T}\sqrt{\log\log T}\right) \\
& \le \frac{1}{(\log\log\log T)(\log\log T)}\mathbb{E}[|P_2(s_0)|^2] = \frac{1}{\log\log T}
\end{aligned}$$

we can take $R_1 = \sqrt{\log\log T}$, $R_2 = \sqrt{\log\log\log T}$ and $F = C\sqrt{\log\log T}$ to get

$$\begin{aligned}
& \frac{1}{\sqrt{\log\log T}} + \left[C\log\log T\left(\frac{1}{(\log\log T)^2}\right) + \frac{1}{\log T} + \frac{1}{(\log T)^D}\right] \\
& + \frac{1}{\sqrt{\log\log T}} + \left[C\sqrt{\log\log\log T}\sqrt{\log\log T}\left(\frac{1}{(\log\log T)^2}\right) + \frac{1}{\log\log T} + \frac{1}{(\log\log T)^D}\right] \\
& \ll \frac{1}{\sqrt{\log\log T}}.
\end{aligned}$$

References

1. Arguin, L.-P., Belius, D., Bourgade, P., Radziwiłł, M., Soundararajan, K.: Maximum of the Riemann zeta function on a short interval of the critical line. Preprint, arxiv: 1612.08575v2 (2017)
2. Gaunt, R.E., Li, S.: Bounding Kolmogorov distances through Wasserstein and related integral probability metrics. (2022). Preprint, arXiv:2201.12087.
3. Harper, A.J.: A note on the maximum of the Riemann zeta function, and log-correlated random variables. Preprint, arxiv: 1304.0677 (2013)
4. Hsu, P., Wong, P.: On Selberg's central limit theorem for Dirichlet L-functions. Journal de Théorie des Nombres de Bordeaux. 32, 685-710 (2020)
5. Ivić, A.: The Riemann Zeta Function. Wiley. (1985)
6. Kowalski, E., Nikeghbali, A.: Mod-Gaussian convergence and the value distribution of $\zeta(1/2 + it)$ and related quantities. J. London Math. Soc. 86(1), 291-319 (2012)
7. Radziwiłł, M., Soundararajan, K.: Selberg's central limit theorem for log $|\zeta(\frac{1}{2} + it)|$. Enseign. Math. 63(1-2), 1-19 (2017)
8. Roberts, A.: The multivariate rate of convergence for Selberg's central limit theorem. Preprint, arxiv: 2212.01411 (2022)
9. Selberg, A.: Contributions to the theory of the Riemann zeta-function. Arch. Math. Naturvid. 48(5), 89-155 (1946)
10. Selberg, A.: Old and new conjectures and results about a class of Dirichlet series. Proceedings of the Amalfi Conference on Analytic Number Theory (Maiori, 1989). Univ. Salerno, Salerno. 367-385 (1992)
11. Selberg, A.: On the normal density of primes in small intervals, and the difference between consecutive primes. Arch. Math. Naturvid. 47(6), 87-105 (1943)
12. Tsang, K.: The distribution of the values of the Riemann zeta-function. Thesis Princeton University (1984)
13. Wahl, M.: On the mod-Gaussian convergence of a sum over primes. Mathematische Zeitschrift. 276(3), 635-654 (2014)

Some Proofs About Sequences in the Spirit of Paul du Bois-Reymond

David A. Ross

Abstract I'll apply some ideas originally developed over 100 years ago by Paul du Bois-Reymond et al. to three examples: a Fekete-style theorem; a result about almost additive sequences due to Pólya and Szegő; and a fixed point theorem due to Boyd and Wong.

Across several papers in the nineteenth century, Paul du Bois-Reymond developed a 'calculus' of scales of infinity, that is, relative sizes in the limit of functions increasing to infinity. Other mathematicians—notably Hadamard, Borel, and Hardy—later developed his ideas further.

These results, which make it possible to treat such sequences as if they were elements of a number system including infinite and infinitesimal quantities, later contributed some motivation for the creation of Abraham Robinson's *nonstandard analysis*. Nonstandard analysis has proved useful over the years for discovery and proof of results in many areas of mathematics, including additive number theory. However, the learning curve has sometimes prevented mathematicians from embracing the technology.

The author has recently observed that du Bois-Reymond's machinery has properties that are quite like $\aleph_1$–saturation and permanence, two of the principles in nonstandard analysis that make it so powerful. The question arises whether this standard machinery can be used to give standard proofs that are motivated by nonstandard reasoning. One example of this appears in Ross [9]. In this paper we give three more examples of such results, two involving approximately subadditive sequences of real numbers and one a fixed-point theorem for complete metric spaces.[1]

[1] The author would like to emphasize than in none of these examples is the *idea* of the proof new; the novelty is the substitution of the sequence technology for infinitesimal and infinite quantities in nonstandard proofs.

D. A. Ross (✉)
Department of Mathematics, University of Hawaii, Hawaii, HI 96822 Honolulu, US
e-mail: ross@math.hawaii.edu

M. B. Nathanson (ed.), *Combinatorial and Additive Number Theory VI*,
Springer Proceedings in Mathematics & Statistics 464,
https://doi.org/10.1007/978-3-031-65064-2_20

1 Machinery

Let $\mathfrak{N}$ be the set of sequences of natural numbers, and $\mathfrak{N}_\infty$ be the set of *strictly increasing* sequences of natural numbers. For $\mathbf{s}, \mathbf{t} \in \mathfrak{N}$ write

$$\mathbf{s} \prec \mathbf{t}$$

if $\lim_{n\to\infty} \mathbf{s}(n)/\mathbf{t}(n) = 0$.

Let $\mathfrak{R}$ be the set of all sequences of nonnegative real numbers, and $\mathfrak{R}_\infty$ be the set of sequences of nonnegative real numbers with infinite limit. For $\mathbf{a}, \mathbf{b} \in \mathfrak{R}$ write

$$\begin{array}{ll} \mathbf{a} \lessapprox \mathbf{b} & \text{if for all } \epsilon > 0,\ \mathbf{a}(n) \le \mathbf{b}(n) + \epsilon \text{ for large enough n} \\ \mathbf{a} \ll \mathbf{b} & \text{if for some } \epsilon > 0,\ \mathbf{a}(n) < \mathbf{b}(n) - \epsilon \text{ for large enough n} \\ \mathbf{a} \approx \mathbf{b} & \text{if } |\mathbf{a} - \mathbf{b}| \lessapprox \mathbf{0} := \langle 0, 0, 0, \dots \rangle \\ \mathbf{a} \not\approx \mathbf{b} & \text{if } \mathbf{0} \ll |\mathbf{a} - \mathbf{b}| \\ \mathbf{a} < \infty & \text{if } \mathbf{a} \text{ is bounded} \end{array}$$

and note that $\approx$ is an equivalence relation respecting addition; in particular,

$$\text{if } \mathbf{a}, \mathbf{b} \approx \mathbf{0} \text{ then } \mathbf{a} + \mathbf{b} \approx \mathbf{0}\ .$$

Here we have adopted the convention that $\mathbf{a} + \mathbf{b}$ is the sequence $\mathbf{a}(n) + \mathbf{b}(n)$; more generally, if f is any real function on $\mathbb{R}^k$ then $f(\mathbf{a}, \mathbf{b}, \dots)$ is the sequence $f(\mathbf{a}(n), \mathbf{b}(n), \dots)$.

This paper was motivated by results like the following, due to Jacques Hadamard:

Theorem 1 *Given functions $\phi_n, \Phi \in \mathfrak{R}_\infty$ with*

$$\phi_1 \prec \phi_2 \prec \phi_3 \prec \cdots \Phi$$

we can find f with $\phi_n \prec f \prec \Phi$ for all values of n.

Given functions $\psi_n, \Phi \in \mathfrak{R}_\infty$ with

$$\psi_1 \succ \psi_2 \succ \psi_3 \succ \cdots \Phi$$

we can find f with $\psi_p \succ f \succ \Phi$ for all values of p.

Given functions $\phi_n, \psi_p \in \mathfrak{R}_\infty$ with

$$\phi_1 \prec \phi_2 \prec \phi_3 \prec \cdots \prec \psi_3 \prec \psi_2 \prec \psi_1$$

we can find f with $\phi_n \prec f \prec \psi_p$ for all values of n and p.

The proof is a straightforward 'diagonal' argument. Note that to any mathematical logician, these look like applications of $\aleph_1$–saturation.

The next result, which is the key component in the proof of Theorem 2, would not be out of place in Chapter II of Hardy [5].

Lemma 1 *Let* $\mathbf{s} \in \mathfrak{R}_\infty$*, and* $E \subseteq (0, \infty)$ *be an unbounded set. There is a nondecreasing sequence* $\mathbf{t} \in \mathfrak{R}_\infty$ *such that* $\mathbf{t} \prec \mathbf{s}$ *and* $\mathbf{t}(n) \in E$ *for every* n*.*

Proof Let $e_1 < e_2 < \dots$ be a sequence in E with $\sup_n e_n = \infty$. It does no harm to assume that $e_1 \leq \sqrt{\mathbf{s}(1)}$. Let $\mathbf{t}(n)$ be the largest e_k with $e_k \leq \sqrt{\mathbf{s}(n)}$. It is easy to see that this sequence satisfies the conclusion of the lemma.

2 Results of Fekete Type

The main result of this section is a short proof of the following.

Theorem 2 *Let* G *be an additive subgroup of* $\mathbb{R}$*,* $G_+ = G \cap (0, \infty)$*, and* $f : G_+ \to \mathbb{R}$ *be subadditive. Suppose further that* $f(x)/x$ *is bounded on* G_+*. Then*

$$\lim_{x \in G_+} \frac{f(x)}{x} = \inf_{x \in G_+} \frac{f(x)}{x} \qquad \text{exists in } [-\infty, \infty)\ .$$

Here the notation $\lim_{x \in G_0}$ means the limit as $x \to \infty$ through G_0; similar notation will be adopted for the lim inf and lim sup. Recall that a function $f : G_0 \to \mathbb{R}$ is *subadditive* provided $f(a + b) \leq f(a) + f(b)$ for all $a, b \in G_0$.

In the case when G is the group of integers this is Fekete's Lemma [3]; see Corollary 1 below. Of course, known proofs of Fekete's Lemma are neither especially difficult nor long. This one has the novelty of following easily from Lemma 1, so is in the spirit of Paul du Bois-Reymond et al.

Proof First, note that for m a positive integer and $x \in G_0$,

$$f(mx) = f(x + \cdots + x) \leq f(x) + \cdots + f(x) = mf(x)$$

so

$$\frac{f(mx)}{mx} \leq \frac{f(x)}{x}. \tag{1}$$

In particular, $\liminf_{x \in G_0} f(x)/x = \inf_{x \in G_0} f(x)/x$. It therefore suffices to show $\limsup_{x \in G_0} f(x)/x \leq \liminf_{x \in G_0} f(x)/x$.

Let $L > \liminf_{x \in G_+} f(x)/x$ and $\mathbf{s} \in \mathfrak{R}_\infty$ with $\mathbf{s}(n) \in G_+$. It suffices to show that $f(\mathbf{s})/\mathbf{s} \lessapprox L$. Note

$$E := \{x \in G_+ \mid f(x)/x < L\}$$

is unbounded in $\mathbb{R}$.

Let $\mathbf{t} \prec \mathbf{s}$ as in Lemma 1: $\mathbf{t}(n)/\mathbf{s}(n) \to 0$ as $n \to \infty$ and $f(\mathbf{t}(n))/\mathbf{t}(n) < L$ for all n.

Put $\mathbf{w} = \lfloor \mathbf{s}/\mathbf{t} \rfloor$ (which is a sequence of integers!), and $\mathbf{r} = \mathbf{s} - \mathbf{tw}$. Then:

$$\frac{f(\mathbf{s})}{\mathbf{s}} \leq \frac{f(\mathbf{tw}) + f(\mathbf{r})}{\mathbf{s}} = \frac{f(\mathbf{tw})}{\mathbf{tw}} \underbrace{\frac{\mathbf{tw}}{\mathbf{s}}}_{\leq 1} + \frac{f(\mathbf{r})}{\mathbf{s}} \leq \underbrace{\frac{f(\mathbf{t})}{\mathbf{t}}}_{<L} + \underbrace{\frac{f(\mathbf{r})}{\mathbf{r}} \frac{\mathbf{r}}{\mathbf{s}}}_{\approx \mathbf{0}}$$

which completes the proof.

Now, suppose G is the set of integers. Any element of G_0 is a positive integer m, and by Equation 1 the ratios $f(m)/m$ are bounded by $f(1)$. It follows:

Corollary 1 (Fekete) *Let* $\{a_n\}_{n \in \mathbb{N}}$ *be a sequence of real numbers satisfying*

$$a_{m+n} \leq a_m + a_n$$

for every $m, n \in \mathbb{N}$. *Then*

$$\lim_n \frac{a_n}{n} = \inf_n \frac{a_n}{n}$$

exists in $[-\infty, \infty)$.

The same argument applies to any group G of the form $G = \{n g_0 : n \in \mathbb{Z}\}$ for a fixed $g_0 > 0$. It follows:

Corollary 2 *Suppose* G *be a finitely–generated additive subgroup of* $\mathbb{R}$ *and* $f : G_0 \to \mathbb{R}$ *is subadditive. Then*

$$\lim_{x \in G_0} \frac{f(x)}{x} = \inf_{x \in G_0} \frac{f(x)}{x}$$

exists in $[-\infty, \infty)$.

Now, say that an additive subgroup G of $\mathbb{R}$ has the *divisibility property* if for any $y \in G_0$ and positive integer m there is an $x \in G_0$ with $mx = y$. For example, G might be $\mathbb{R}$ itself, or $\mathbb{Q}$, or the set of decimal fractions.

Corollary 3 *Suppose* G *be an additive subgroup of* $\mathbb{R}$ *with the divisibility property,* $f : G_0 \to \mathbb{R}$ *is subadditive, and* $f(x)/x$ *is bounded on a neighborhood of* 0. *Then*

$$\lim_{x \in G_0} \frac{f(x)}{x} = \inf_{x \in G_0} \frac{f(x)}{x}$$

exists in $[-\infty, \infty)$.

Proof Suppose $f(x)/x < M$ on $G_0 \cap (0, a)$. If $y \in G_0$ is arbitrary and m is a positive integer greater than y/a, there is an x with $mx = y$. Note $x < a$, so $f(y)/y \leq f(x)/x < M$ by Equation 1. The conclusion now follows from Theorem 2.

3 A Problem of Pólya and Szegö

In this section we consider the following result of Pólya and Szegö [8, Problem 99].

Theorem 3 *Let* $\{a_n\}_n$ *be a sequence of real numbers satisfying*

$$|a_{m+n} - a_m - a_n| \leq \alpha$$

for every $m, n \in \mathbb{N}$ *(where* α *is a fixed positive real number). Then for some* $\sigma \in \mathbb{R}$ *and all* $n \in \mathbb{N}$,

$$|a_n - n\sigma| \leq \alpha \, .$$

(In particular, $\lim_{n\to\infty} a_n/n = \sigma$.*)*

The proof here is an adaptation of a nonstandard proof by W.A.J. Luxemburg [7], who actually proved something stronger:

Theorem 4 *Let* $\{a_n\}_n$ *be a sequence of real numbers satisfying*

$$|a_{m+n} - a_m - a_n| \leq \frac{\alpha}{2}(m^p + n^p)$$

for every $m, n \in \mathbb{N}$ *(where* α *is a fixed positive real number and* $0 \leq p < 1$*). Then for some* $\sigma \in \mathbb{R}$ *and all* $n \in \mathbb{N}$,

$$|a_n - n\sigma| \leq \alpha \frac{n^p}{2 - 2^p} \, .$$

(In particular, $\lim_{n\to\infty} a_n/n = \sigma$.*)*

Proof (of Theorem 3)
Let $n \in \mathbb{N}$. Then:

$$|a_{2n} - 2a_n| = |a_{2n} - a_n - a_n| \leq \alpha, \quad |a_{4n} - 2a_{2n}| \leq \alpha, \quad |a_{8n} - 2a_{4n}| \leq \alpha, \ldots$$

and so:

$$\begin{aligned} |a_{4n} - 4a_n| &\leq |a_{4n} - 2a_{2n}| + 2|a_{2n} - 2a_n| \leq \alpha + 2\alpha = 3\alpha, \\ |a_{8n} - 8a_n| &\leq |a_{8n} - 2a_{4n}| + 2|a_{4n} - 4a_n| \leq \alpha + 2 \cdot 3\alpha = 7\alpha, \\ &\vdots \\ |a_{2^k n} - 2^k a_n| &\leq (2^k - 1)\alpha \, . \end{aligned} \tag{2}$$

Put

$$\phi_k(n) = \frac{a_{2^k n}}{2^k}$$

and note that $|a_n - \phi_k(n)| \le (1-2^{-k})\alpha$, which is $\le \alpha$. Now, replace k by any sequence $\mathbf{k} \in \mathfrak{N}_\infty$ (I'll specify a particular one later):

$$\phi_{\mathbf{k}}(m+n) = \frac{a_{2^{\mathbf{k}}(m+n)}}{2^{\mathbf{k}}} = \frac{a_{2^{\mathbf{k}}m} + a_{2^{\mathbf{k}}n} + (\text{something } \le \alpha)}{2^{\mathbf{k}}} \approx \phi_{\mathbf{k}}(m) + \phi_{\mathbf{k}}(n)$$

It follows:

$$\phi_{\mathbf{k}}(n) = \phi_{\mathbf{k}}(\underbrace{1 + \cdots + 1}_{n}) \approx n\phi_{\mathbf{k}}(1)$$

so

$$|a_n - n\phi_{\mathbf{k}}(1)| \lessapprox \alpha \text{ for all } n \, .$$

We'd be done if $\phi_{\mathbf{k}}(1)$ was a number, not a sequence. So let σ be any limit point of $\{\phi_k(1) | k \subset \mathbb{N}\} = \{a_{2^k}/2^k | k \in \mathbb{N}\}$, and let $\mathbf{k} \in \mathfrak{N}_\infty$ be a sequence so that $\phi_{\mathbf{k}(m)}(1) \to \sigma$ as $m \to \infty$. Then for any n,

$$|a_n - n\sigma| \approx |a_n - n\phi_{\mathbf{k}}(1)| \lessapprox \alpha$$

which means $|a_n - n\sigma| \le \alpha$ (since the two sides of the inequality are constants).

The more general result Theorem 4 can be obtained by more careful bookkeeping on the inequalities (2). One obtains

$$|a_{2^k n} - 2^k a_n| \le \alpha n^p \frac{2^k - 2^{pk}}{1 - 2^{p-1}}$$

then proceeds as above.

4 A Fixed-Point Theorem

In this section X will be any complete metric space (with metric d), and $T : X \to X$ will be a continuous function. In 1922 Banach proved that if T is a uniform contraction (that is, for some $r < 1$ and all $x, y \in X$ we have $d(Tx, Ty) \le r d(x, y)$) then T has an *iterative fixed point* x_∞ on X; that is,

(i) $T(x_\infty) = x_\infty$, and
(ii) For every $x \in X$, $\lim_{n\to\infty} T^n x = x_\infty$.

The proof is an exercise in undergraduate real analysis, and should be familiar to most readers of this paper.

In 1969, Boyd and Wong [1] considered the question of whether Banach's fixed point result still holds if the contraction is not uniform, but depends on distance, and proved:

Theorem 5 *Suppose (X, d) is a complete metric space, $T : X \to X$ is continuous, $\phi : [0, \infty) \to [0, \infty)$ is upper semicontinuous from the right with $\phi(s) < s$ for $s > 0$, and $d(Tx, Ty) \le \phi(d(x, y))$ for all $x, y \in X$ Then T has an iterative fixed point x_∞.*

Here *upper semicontinuous from the right* means that for all s, $\limsup_{t \to s^+} \phi(t) \le \phi(s)$. This translates naturally into the notation of this paper:

Proposition 1 *If $\phi : [0, \infty) \to [0, \infty)$ is upper semicontinuous from the right, $\gamma \in (0, \infty)$, $\mathbf{t} \in \mathfrak{R}$, $\mathbf{t} > \gamma$ and $\mathbf{t} \approx \gamma$, then $\phi(\mathbf{t}) \lessapprox \gamma$.*

The proof is an exercise.

Recently Lindstrøm and Ross [6] gave a new nonstandard proof of this result as well as generalizations to so-called *asymptotic fixed points*, and Ross [9] has used the machinery of the current paper to give a new standard proof of one such result. This section adapts the proof of Boyd and Wong in [6].

We begin with a useful fact about the function ϕ.

Proposition 2 *Suppose $\phi : [0, \infty) \to [0, \infty)$ is upper semicontinuous from the right with $\phi(s) < s$ for $s > 0$. Then for any $s > 0$ the sequence $s, \phi(s), \phi(\phi(s)) = \phi^{(2)}(s), \phi^{(3)}(s), \ldots, \phi^{(n)}(s), \ldots$ is monotonic with $\lim_{n\to\infty} \phi^{(n)}(s) = 0$.*

Proof The sequence is obviously monotonic, so $\inf_{n\to\infty} \phi^{(n)}(s) = \lim_{n\to\infty} \phi^{(n)}(s)$ exists; call it s_∞, and suppose $s_\infty > 0$. Then

$$s_\infty = \lim_{n\to\infty} \phi^{(n)}(s) \le \limsup_{t \to s_\infty^+} \phi(t) \le \phi(s_\infty) < s_\infty,$$

a contradiction.

We now suppose, for the remainder of this section, that X, T, and ϕ are as given in the hypothesis of Theorem 5.

Corollary 4 *If $x, y \in X$ and $\mathbf{s}, \mathbf{t} \in \mathfrak{N}_\infty$ then*

(i) $d(T^{\mathbf{s}}x, T^{\mathbf{s}}y) \approx \mathbf{0}$
(ii) $d(T^{\mathbf{s}}x, T^{\mathbf{s}+1}x) \approx \mathbf{0}$
(iii) $d(T^{\mathbf{s}+1}x, T^{\mathbf{t}+1}y) \approx d(T^{\mathbf{s}}x, T^{\mathbf{t}}y)$

Proof $d(T^{\mathbf{s}(n)}x, T^{\mathbf{s}(n)}y) \le \phi^{(\mathbf{s}(n))}(d(x, y)) \to 0$ as $n \to \infty$ (by Proposition 2), proving (i). For (ii), let $y = T(x)$. (iii) follows immediately from (i) and (ii).

Lemma 2 *Suppose for every $x, y \in X$ and $\mathbf{s} \le \mathbf{t} \in \mathfrak{N}_\infty$, $d(T^{\mathbf{s}}x, T^{\mathbf{t}}y) \approx \mathbf{0}$. Then T has an iterative fixed point x_∞ on X.*

Proof First suppose $x = y$. I claim that $\{T^n x\}_{n\in\mathbb{N}}$ is a Cauchy sequence. Suppose it isn't; then for some $\epsilon > 0$ we have:

$$(\forall n \in \mathbb{N})(\exists \mathbf{s}(n), \mathbf{t}(n) \ge n)[d(T^{\mathbf{s}(n)}x, T^{\mathbf{t}(n)}x) > \epsilon]$$

We can choose $\mathbf{s}(n) \leq \mathbf{t}(n)$, and assume that $\mathbf{s}, \mathbf{t}$ are both strictly increasing. Then $d(T^{\mathbf{s}}x, T^{\mathbf{t}}y) \not\approx \mathbf{0}$, which is a contradiction.

Since X is complete, $\lim_{n\to\infty} T^n x = x_\infty$ for some $x_\infty \in X$. Since T is continuous,

$$T(x_\infty) = T(\lim_{n\to\infty} T^n x) = \lim_{n\to\infty} T^{n+1} x = x_\infty$$

so x_∞ is a fixed point.

More generally, suppose y is another element of X, then $\lim_{n\to\infty} T^n y = y_\infty$ for some y_∞. But taking $\mathbf{s}$ any element of $\mathfrak{N}_\infty$, we see that $\mathbf{0} \approx d(T^{\mathbf{s}}x, T^{\mathbf{s}}y)$ means that $x_\infty = y_\infty$, proving that this is an iterative fixed point.

Proof (of Theorem 5)

Let x, y in X, $\mathbf{s} < \mathbf{t} \in \mathfrak{N}_\infty$; we want to show $d(T^{\mathbf{s}}x, T^{\mathbf{t}}y) \approx \mathbf{0}$

First, suppose $x = y$. If $d(T^{\mathbf{s}}x, T^{\mathbf{t}}x) \not\approx \mathbf{0}$, there is a $\gamma > 0$ and a subsequence of $(\mathbf{s}, \mathbf{t})$ (which we can again call $(\mathbf{s}, \mathbf{t})$) with $d(T^{\mathbf{s}}x, T^{\mathbf{t}}x) > \gamma$. For each n there is a $\mathbf{m}(n) \geq 0$ with $d(T^{\mathbf{s}(n)+\mathbf{m}(n)}x, T^{\mathbf{t}(n)}x) \geq \gamma > d(T^{\mathbf{s}(n)+\mathbf{m}(n)+1}x, T^{\mathbf{t}(n)}x)$. Then

$$\begin{aligned} d(T^{\mathbf{s}+\mathbf{m}+1}x, T^{\mathbf{t}}x) &< \gamma \leq d(T^{\mathbf{s}+\mathbf{m}}x, T^{\mathbf{t}}x) \\ &\leq d(T^{\mathbf{s}+\mathbf{m}+1}x, T^{\mathbf{t}}x) + \underbrace{d(T^{\mathbf{s}+\mathbf{m}+1}x, T^{\mathbf{s}+\mathbf{m}}x)}_{\approx \mathbf{0}} \end{aligned}$$

so

$$\begin{aligned} \gamma &\approx d(T^{\mathbf{s}+\mathbf{m}}x, T^{\mathbf{t}}x) \approx d(T^{\mathbf{s}+\mathbf{m}+1}x, T^{\mathbf{t}+1}x) \\ &= d(T(T^{\mathbf{s}+\mathbf{m}}x), T(T^{\mathbf{t}}x)) \leq \phi(d(T^{\mathbf{s}+\mathbf{m}}x, T^{\mathbf{t}}x)) \\ &\lessapprox \phi(\gamma) \qquad \text{(since } \phi \text{ is upper semicontinuous from the right)} \\ &< \gamma \end{aligned}$$

a contradiction.

For the more general case,

$$d(T^{\mathbf{s}}x, T^{\mathbf{t}}y) \approx d(T^{\mathbf{s}}x, T^{\mathbf{t}}x) + d(T^{\mathbf{t}}x, T^{\mathbf{t}}y) \approx \mathbf{0}$$

which completes the proof.

References

1. Boyd, D. W., Wong, J. S. W., *On nonlinear contractions*, Proc. Amer. Math. Soc. 20 (1969) 458–464.
2. du Bois-Reymond, P., *Ueber die Paradoxen des Infinitärcalcüls*, Math. Ann. 11, 149–167 (1877).
3. Fekete, M., Über die Verteilung der Wurzeln bei gewissen algebraischen Gleichungen mit ganzzahligen Koeffizienten. *Math Z* 17, 228–249 (1923).

4. Hadamard, J., *Sur les caractères de convergence des séries a termes positifs et sur les fonctions indéfiniment croissantes*, Acta Mathematica, volume 18 (1894)
5. Hardy, G. H., *Orders of infinity. The Infinitärcalcül of Paul du Bois-Reymond*, Cambridge Tracts in Mathematics and Mathematical Physics, No. 12, 1910
6. Lindstrøm, T., Ross, D.A., *A nonstandard approach to asymptotic fixed point theorems*, J. Fixed Point Theory Appl. 25, 35 (2023)
7. Luxemburg, W. A., *What Is Nonstandard Analysis?*, Amer. Math. Monthly 80, 38–67, (1973).
8. Pólya, G., Szegö, G., Problems and Theorems in Analysis I, Springer-Verlag Berlin Heidelberg (1998).
9. Ross, D. A., *Asymptotic fixed points and theorems of du Bois-Reymond and Hadamard*, Proc. Amer. Math. Soc., to appear

Series with Summands Involving Harmonic Numbers

Zhi-Wei Sun

Abstract For each positive integer m, the mth order harmonic numbers are given by

$$H_n^{(m)} = \sum_{0<k\leqslant n} \frac{1}{k^m} \quad (n = 0, 1, 2, \ldots).$$

We discover exact values of some series involving harmonic numbers of order not exceeding four. For example, we conjecture that

$$\sum_{k=0}^{\infty}(6k+1)\frac{\binom{2k}{k}^3}{256^k}\left(H_{2k}^{(3)} - \frac{7}{64}H_k^{(3)}\right) = \frac{25\zeta(3)}{8\pi} - G,$$

where G denotes the Catalan constant $\sum_{k=0}^{\infty}\frac{(-1)^k}{(2k+1)^2}$. This paper contains 70 conjectures posed by the author during 2022–2023.

2020 Mathematics Subject Classification. Primary 11B65 · 05A19 · Secondary 11A07 · 11B68

1 Introduction

The usual harmonic numbers are those rational numbers

$$H_n = \sum_{0<k\leqslant n} \frac{1}{k} \quad (n = 0, 1, 2, \ldots).$$

The initial version of this paper was posted to arXiv in Oct. 2022 as a preprint with the ID arXiv:2210.07238.

Z.-W. Sun (✉)
School of Mathematics, Nanjing University, Nanjing 210093, People's Republic of China
e-mail: zwsun@nju.edu.cn

M. B. Nathanson (ed.), *Combinatorial and Additive Number Theory VI*,
Springer Proceedings in Mathematics & Statistics 464,
https://doi.org/10.1007/978-3-031-65064-2_21

For each $m \in \mathbb{Z}^+ = \{1, 2, 3, \ldots\}$, the harmonic numbers of order m are defined by

$$H_n^{(m)} = \sum_{0<k\leqslant n} \frac{1}{k^m} \quad (n \in \mathbb{N} = \{0, 1, 2, \ldots\}).$$

For any $m, n \in \mathbb{Z}^+$, we clearly have

$$\sum_{k=1}^{n} \frac{1}{(2k-1)^m} = \sum_{k=1}^{2n} \frac{1}{k^m} - \sum_{j=1}^{n} \frac{1}{(2j)^m} = H_{2n}^{(m)} - \frac{1}{2^m} H_n^{(m)}.$$

J. Wolstenholme [48] established two fundamental congruences for harmonic numbers:

$$H_{p-1} \equiv 0 \pmod{p^2} \text{ and } H_{p-1}^{(2)} \equiv 0 \pmod{p}$$

for any prime $p > 3$. For series and congruences involving harmonic numbers, one may consult [29, 36, 42], [40, Sect. 10.5], and the recent preprint [5] solving various conjectures of the author.

In 2012, K. N. Boyadzhiev [7] proved that

$$\sum_{k=0}^{\infty} \binom{2k}{k} H_k x^k = \frac{2}{\sqrt{1-4x}} \log \frac{1+\sqrt{1-4x}}{2\sqrt{1-4x}} \quad \text{for } x \in \left(-\frac{1}{4}, \frac{1}{4}\right). \tag{1}$$

In 2016, H. Chen [8] deduced that

$$\sum_{k=0}^{\infty} \binom{2k}{k} H_{2k} x^k = \frac{1}{\sqrt{1-4x}} \log \frac{1+\sqrt{1-4x}}{2(1-4x)} \quad \text{for } x \in \left(-\frac{1}{4}, \frac{1}{4}\right). \tag{2}$$

It is well known that

$$2 \arcsin \frac{x}{2} = \sum_{k=0}^{\infty} \frac{\binom{2k}{k} x^{2k+1}}{(2k+1)16^k} \quad \text{for } |x| \leqslant 2;$$

in particular,

$$\sum_{k=0}^{\infty} \frac{\binom{2k}{k}}{(2k+1)16^k} = \frac{\pi}{3} \text{ and } \sum_{k=0}^{\infty} \frac{\binom{2k}{k}}{(2k+1)8^k} = \frac{\pi}{2\sqrt{2}}.$$

The author [27, Theorem 1.1(ii)] determined

$$\sum_{k=0}^{(p-3)/2} \frac{\binom{2k}{k}}{(2k+1)16^k} \text{ and } \sum_{k=(p+1)/2}^{p-1} \frac{\binom{2k}{k}}{(2k+1)16^k}$$

modulo p^2 for any prime $p > 3$. By [6], we have

$$\left(\arcsin\frac{x}{2}\right)^3 = 3\sum_{k=0}^{\infty}\frac{\binom{2k}{k}x^{2k+1}}{(2k+1)16^k}\sum_{0\leqslant j<k}\frac{1}{(2j+1)^2} \tag{3}$$

and

$$\left(\arcsin\frac{x}{2}\right)^4 = \frac{3}{2}\sum_{k=1}^{\infty}\frac{H_{k-1}^{(2)}x^{2k}}{k^2\binom{2k}{k}} \tag{4}$$

for $|x| \leqslant 2$. In particular,

$$\sum_{k=0}^{\infty}\frac{\binom{2k}{k}}{(2k+1)16^k}\left(H_{2k}^{(2)} - \frac{1}{4}H_k^{(2)}\right) = \frac{\pi^3}{648},$$
$$\sum_{k=0}^{\infty}\frac{\binom{2k}{k}}{(2k+1)8^k}\left(H_{2k}^{(2)} - \frac{1}{4}H_k^{(2)}\right) = \frac{\sqrt{2}\pi^3}{384}.$$

and

$$\sum_{k=1}^{\infty}\frac{H_{k-1}^{(2)}}{k^2\binom{2k}{k}} = \frac{\pi^4}{1944}.$$

In view of (3), we have the following result.

Theorem 1 *If $|x| < 2$, then*

$$\frac{(\arcsin(x/2))^2}{\sqrt{4-x^2}} = \sum_{k=1}^{\infty}\frac{\binom{2k}{k}x^{2k}}{16^k}\left(H_{2k}^{(2)} - \frac{1}{4}H_k^{(2)}\right). \tag{5}$$

Proof By taking derivatives of both sides of (3), we get

$$3\left(\arcsin\frac{x}{2}\right)^2 \times \frac{1/2}{\sqrt{1-(x/2)^2}} = 3\sum_{k=0}^{\infty}\frac{\binom{2k}{k}x^{2k}}{16^k}\sum_{0\leqslant j<k}\frac{1}{(2j+1)^2}$$

and hence

$$\frac{(\arcsin(x/2))^2}{\sqrt{4-x^2}} = \sum_{k=1}^{\infty}\frac{\binom{2k}{k}x^k}{16^k}\sum_{j=1}^{k}\frac{1}{(2j-1)^2},$$

which is equivalent to (5).

Motivated by the Ramanujan series

$$\sum_{k=0}^{\infty}(6k+1)\frac{\binom{2k}{k}^3}{(-512)^k} = \frac{2\sqrt{2}}{\pi} \tag{6}$$

(cf. [25]), L. Long [23] conjectured the congruence

$$\sum_{k=0}^{(p-1)/2}(6k+1)\frac{\binom{2k}{k}^3}{(-512)^k}\sum_{j=1}^{k}\left(\frac{1}{(2j-1)^2}-\frac{1}{16j^2}\right)\equiv 0 \pmod{p} \tag{7}$$

for any odd prime p, which was confirmed by H. Swisher [43] in 2015. Note that (7) can be rewritten as

$$\sum_{k=0}^{(p-1)/2}(6k+1)\frac{\binom{2k}{k}^3}{(-512)^k}\left(H_{2k}^{(2)}-\frac{5}{16}H_k^{(2)}\right)\equiv 0 \pmod{p}.$$

In 2022 C. Wei [46] deduced the two identities

$$\sum_{k=0}^{\infty}(6k+1)\frac{\binom{2k}{k}^3}{(-512)^k}\left(H_{2k}^{(2)}-\frac{5}{16}H_k^{(2)}\right)=-\frac{\sqrt{2}}{48}\pi$$

and

$$\sum_{k=0}^{\infty}(6k+1)\frac{\binom{2k}{k}^3}{256^k}\left(H_{2k}^{(2)}-\frac{5}{16}H_k^{(2)}\right)=\frac{\pi}{12}$$

conjectured by V.J.W. Guo and X. Lian [19], as well as their q-analogues.

Motivated by Bauer's series

$$\sum_{k=0}^{\infty}(4k+1)\frac{\binom{2k}{k}^3}{(-64)^k}=\frac{2}{\pi} \tag{8}$$

and Ramanujan's series

$$\sum_{k=0}^{\infty}(8k+1)\frac{\binom{2k}{k}^2\binom{4k}{2k}}{48^{2k}}=\frac{2\sqrt{3}}{\pi}, \tag{9}$$

Wei and G. Ruan [47] proved the two new identities

$$\sum_{k=1}^{\infty}(4k+1)\frac{\binom{2k}{k}^3}{(-64)^k}\sum_{j=1}^{2k}\frac{(-1)^j}{j^2}=\frac{\pi}{12}$$

and

$$\sum_{k=1}^{\infty}(8k+1)\frac{\binom{2k}{k}^2\binom{4k}{2k}}{48^{2k}}\sum_{j=1}^{k}\left(\frac{1}{(2j-1)^2}-\frac{1}{36j^2}\right)=\frac{\sqrt{3}\pi}{54},$$

i.e.,

$$\sum_{k=0}^{\infty}(4k+1)\frac{\binom{2k}{k}^3}{(-64)^k}\left(H_{2k}^{(2)}-\frac{1}{2}H_k^{(2)}\right)=-\frac{\pi}{12} \tag{10}$$

and

$$\sum_{k=0}^{\infty}(8k+1)\frac{\binom{2k}{k}^2\binom{4k}{2k}}{48^{2k}}\left(H_{2k}^{(2)}-\frac{5}{18}H_k^{(2)}\right)=\frac{\sqrt{3}\pi}{54}. \tag{11}$$

In 1997 van Hamme [44] thought that series for powers of $\pi=\Gamma(1/2)^2$ should have their p-adic analogues involving the p-adic Gamma function $\Gamma_p(x)$, where p is an odd prime. Note that for any odd prime p we have

$$\Gamma_p\left(\frac{1}{2}\right)^2=(-1)^{(p+1)/2}=-\left(\frac{-1}{p}\right),$$

where $(\frac{\cdot}{p})$ denotes the Legendre symbol. The author [38, 39, 41] found many new series for powers of π motivated by related congruences. However, van Hamme's philosophy fails for some Ramanujan-type series for $1/\pi$. For example, T. Huber, D. Schultz and D. Ye [22] used modular forms to obtain that

$$\sum_{k=0}^{\infty}(6k+1)\frac{a_k}{16^k}=\frac{16}{\pi},$$

where $a_0=1,\ a_1=4,\ a_2=20$ and

$$(n+1)^3a_{n+1}=4(2n+1)(2n^2+2n+1)a_n-16n(4n^2+1)a_{n-1}+8(2n-1)^3a_{n-2}$$

for all $n=2,3,\ldots$; but for a general odd prime p we even cannot find any pattern for $\sum_{k=0}^{p-1}(6k+1)a_k/16^k$ modulo p.

The Bernoulli numbers $B_0, B_1, B_2, \ldots$ are defined by

$$\frac{x}{e^x-1}=\sum_{n=0}^{\infty}B_n\frac{x^n}{n!}\quad(0<|x|<2\pi).$$

Equivalently,

$$B_0=1,\ \text{and}\ \sum_{k=0}^{n}\binom{n+1}{k}B_k=0\ \text{ for } n=1,2,3,\ldots.$$

In 1900 J.W.L. Glaiser [12] proved that

$$H_{p-1}\equiv-\frac{p^2}{3}B_{p-3}\pmod{p^3}\ \text{ and }\ H_{p-1}^{(2)}\equiv\frac{2}{3}pB_{p-3}\pmod{p^2}$$

for any prime $p > 3$. The Bernoulli polynomials are given by

$$B_n(x) = \sum_{k=0}^{n} \binom{n}{k} B_k x^{n-k} \quad (n \in \mathbb{N}).$$

The Euler numbers $E_0, E_1, E_2, \ldots$ are defined by

$$\frac{2}{e^x + e^{-x}} = \sum_{n=0}^{\infty} E_n \frac{x^n}{n!} \quad \left(|x| < \frac{\pi}{2}\right).$$

Clearly $E_{2n+1} = 0$ for all $n \in \mathbb{N}$. It is also known that

$$\sum_{k=0}^{n} \binom{2n}{2k} E_{2k} = 0 \quad \text{for each } n \in \mathbb{N}.$$

The Euler polynomials are given by

$$E_n(x) = \sum_{k=0}^{n} \binom{n}{k} \frac{E_k}{2^k} \left(x - \frac{1}{2}\right)^{n-k} \quad (n \in \mathbb{N}).$$

The author [28, 32] first observed that many Ramanujan-type series have corresponding congruences involving Bernoulli or Euler polynomials.

Now we introduce some notations throughout this paper. The Riemann zeta function is defined by

$$\zeta(s) = \sum_{n=1}^{\infty} \frac{1}{n^s} \quad \text{with } \Re(s) > 1.$$

The Dirichlet beta function is given by

$$\beta(m) = \sum_{k=0}^{\infty} \frac{(-1)^k}{(2k+1)^m} \quad (m = 1, 2, 3, \ldots).$$

Note that $G = \beta(2)$ is the Catalan constant. We also adopt the notation

$$K := L\left(2, \left(\frac{-3}{\cdot}\right)\right) = \sum_{n=1}^{\infty} \frac{\left(\frac{k}{3}\right)}{k^2}$$

with $(\frac{-3}{\cdot})$ the Kronecker symbol. For a prime p and an integer $a \not\equiv 0 \pmod{p}$, we use $q_p(a)$ to denote the Fermat quotient $(a^{p-1} - 1)/p$. Many congruences in later sections involve Fermat quotients.

In Sects. 2–4, we will propose 70 new conjectures on series and related congruences with summands involving not only harmonic numbers of order at most four, but also products of several binomial coefficients. All the conjectures have been checked via Mathematica.

2 Series with Summands Containing One or Two Binomial Coefficients

Conjecture 1 (2022-10-12)
(i) We have

$$\sum_{k=1}^{\infty}\frac{(-1)^{k-1}}{k^3\binom{2k}{k}}\left(H_{2k-1}^{(2)}-\frac{123}{16}H_{k-1}^{(2)}\right)=\frac{451}{40}\zeta(5)-\frac{14}{15}\pi^2\zeta(3). \tag{12}$$

(ii) For any prime $p > 5$, we have

$$\sum_{k=1}^{(p-1)/2}\frac{(-1)^{k}}{k^3\binom{2k}{k}}\left(16H_{2k-1}^{(2)}-123H_{k-1}^{(2)}\right)\equiv -542B_{p-5} \pmod{p} \tag{13}$$

and

$$p\sum_{k=1}^{p-1}\frac{(-1)^{k}}{k^2}\binom{2k}{k}\left(16H_{2k}^{(2)}-123H_{k}^{(2)}\right)\equiv 192\frac{H_{p-1}}{p^2} \pmod{p^2}. \tag{14}$$

Remark 1 In 1979 R. Apéry [3] proved the irrationality of $\zeta(3)=\sum_{n=1}^{\infty}1/n^3$ via the identity

$$\sum_{k=1}^{\infty}\frac{(-1)^{k-1}}{k^3\binom{2k}{k}}=\frac{2}{5}\zeta(3).$$

In 2014 the author [33] proved the congruence

$$\sum_{k=1}^{(p-1)/2}\frac{(-1)^{k-1}}{k^3\binom{2k}{k}}\equiv 2B_{p-3} \pmod{p}$$

for any prime $p > 5$. The author's conjectural identity (cf. [34])

$$\sum_{k=1}^{\infty}\frac{(-1)^{k-1}}{k^3\binom{2k}{k}}(H_{2k}+4H_k)=\frac{2\pi^4}{75}$$

was proved by W. Chu [9] as well as K. C. Au [4, Prop. 7.14]. After seeing an earlier arXiv version of this paper, Au [5, Corollary 2.9] confirmed the author's conjectural identity (12).

Conjecture 2 (2022-11-14) We have the identity

$$\sum_{k=0}^{\infty}\frac{\binom{2k}{k}}{8^k}\left(H_{2k}^{(3)}-\frac{1}{8}H_k^{(3)}\right)=\frac{35\sqrt{2}}{64}\zeta(3)-\frac{\sqrt{2}}{8}\pi G. \tag{15}$$

Remark 2 Applying (1) and (2) with $x = 1/8$, we see that

$$\sum_{k=0}^{\infty}\frac{\binom{2k}{k}}{8^k}H_k=-\sqrt{2}\log(12-8\sqrt{2})\ \text{and}\ \sum_{k=0}^{\infty}\frac{\binom{2k}{k}}{8^k}H_{2k}=\frac{\log(3/2+\sqrt{2})}{\sqrt{2}}.$$

In contrast with (15), we have

$$\sum_{k=0}^{\infty}\frac{\binom{2k}{k}}{8^k}\left(H_{2k}^{(2)}-\frac{1}{4}H_k^{(2)}\right)=\frac{\pi^2}{16\sqrt{2}}$$

by applying (5) with $x = \sqrt{2}$.

Conjecture 3 (2022-11-14) We have the identity

$$\sum_{k=0}^{\infty}\frac{\binom{2k}{k}}{16^k}\left(H_{2k}^{(3)}-\frac{1}{8}H_k^{(3)}\right)=\frac{2\zeta(3)}{3\sqrt{3}}-\frac{\pi K}{8}. \tag{16}$$

Remark 3 Applying (1) and (2) with $x = 1/8$, we see that

$$\sum_{k=0}^{\infty}\frac{\binom{2k}{k}}{16^k}H_k=-\frac{2}{\sqrt{3}}\log(84-48\sqrt{3})\ \text{and}\ \sum_{k=0}^{\infty}\frac{\binom{2k}{k}}{16^k}H_{2k}=\frac{\log((7+4\sqrt{3})/9)}{\sqrt{3}}.$$

In contrast with (16), we have

$$\sum_{k=0}^{\infty}\frac{\binom{2k}{k}}{16^k}\left(H_{2k}^{(2)}-\frac{1}{4}H_k^{(2)}\right)=\frac{\pi^2}{36\sqrt{3}}$$

by applying (5) with $x = 1$.

Conjecture 4 (2023-05-28)

(i) We have

$$\sum_{k=1}^{\infty}\frac{H_{3k}-H_k}{k2^k\binom{3k}{k}}=\frac{2}{5}(G+\log^2 2)-\frac{\pi^2}{24} \tag{17}$$

and

$$\sum_{k=1}^{\infty}\frac{H_{3k}-H_k}{k^2 2^k\binom{3k}{k}}=\frac{11}{4}\zeta(3)-\frac{\pi^2}{24}\log 2-\pi G. \tag{18}$$

(ii) For any prime $p > 5$ with $p \equiv 1 \pmod 4$, we have

$$p\sum_{k=1}^{p-1}\frac{H_{3k}-H_k}{k2^k\binom{3k}{k}} \equiv \frac{7}{10}q_p(2) \pmod p. \tag{19}$$

Also, for each odd prime p we have

$$p^2\sum_{k=1}^{p-1}\frac{H_{3k}-H_k}{k^22^k\binom{3k}{k}} \equiv -\frac{q_p(2)}{4} \pmod p. \tag{20}$$

Remark 4 The author's conjectural identities

$$\sum_{k=1}^{\infty}\frac{H_{2k}-H_k}{k2^k\binom{3k}{k}} = \frac{3}{10}\log^2 2+\frac{\pi}{20}\log 2-\frac{\pi^2}{60}$$

and

$$\sum_{k=1}^{\infty}\frac{H_{2k}-H_k}{k^22^k\binom{3k}{k}} = \frac{33}{32}\zeta(3)+\frac{\pi^2}{24}\log 2-\frac{\pi G}{2}$$

(cf. [40, Conjecture 10.61]) were confirmed by Au [4] in 2022.

Conjecture 5 (2023-05-28) (i) We have

$$\sum_{k=1}^{\infty}\frac{25k-3}{2^k\binom{3k}{k}}(H_{3k}-8H_{2k}+7H_k) = 2G-2(\pi+9)\log 2. \tag{21}$$

(ii) For any odd prime p, we have the congruence

$$p^2\sum_{k=1}^{p-1}\frac{25k-3}{2^k\binom{3k}{k}}(H_{3k}-8H_{2k}+7H_k) \equiv -\left(\frac{-1}{p}\right)\frac{9}{4} \pmod p. \tag{22}$$

Remark 5 In 1974 R. W. Gosper announced the identity

$$\sum_{k=0}^{\infty}\frac{25k-3}{2^k\binom{3k}{k}} = \frac{\pi}{2},$$

an elegant proof of which can be found in [1].

Conjecture 6 (2023-05-28)
(i) We have

$$\sum_{k=0}^{\infty}\frac{2^k\binom{3k}{k}}{27^k}H_{2k} = \frac{3}{2}(1+\sqrt{3})\log\left(1+\frac{\sqrt{3}}{3}\right)-\sqrt{3}\log 2 \tag{23}$$

and

$$\sum_{k=0}^{\infty}\frac{2^k\binom{3k}{k}}{27^k}H_{3k}=\frac{1+\sqrt{3}}{2}\left(2\log(1+\sqrt{3})-\frac{\log 3}{2}\right)-\sqrt{3}\log 2. \tag{24}$$

(ii) For any prime $p>3$, we have

$$\sum_{k=(p+1)/2}^{p-1}\binom{3k}{k}\left(\frac{2}{27}\right)^k\equiv\frac{1-(\frac{3}{p})}{3}\pmod p. \tag{25}$$

Remark 6 For any positive integer n, we clearly have

$$H_n=\sum_{k=0}^{n-1}\frac{1}{k+1}=\sum_{k=0}^{n-1}\int_0^1 t^k dt=\int_0^1\sum_{k=0}^{n-1}t^k dt=\int_0^1\frac{1-t^n}{1-t}dt.$$

Using this trick we can deduce that

$$\sum_{k=0}^{\infty}\frac{2^k\binom{3k}{k}}{27^k}H_k=\frac{3}{4}\left((1-\sqrt{3})\log 4-(1+\sqrt{3})\log 3\right)+2\sqrt{3}\log(1+\sqrt{3}).$$

Conjecture 7 (2023-05-28) We have

$$\sum_{k=0}^{\infty}\binom{3k}{k}\left(\frac{3+\sqrt{5}}{54}\right)^k(H_{3k}-H_{2k})=\phi(\log 3-2\log\phi), \tag{26}$$

where ϕ denotes the golden ratio $(1+\sqrt{5})/2\approx 1.618\ldots$.

Remark 7 `Mathematica` yields that

$$\sum_{k=0}^{\infty}\binom{3k}{k}\left(\frac{4x}{27}\right)^k=\frac{1}{\sqrt{1-x}}\cos\frac{\arcsin\sqrt{x}}{3}$$

for any $x\in(-1,1)$. Applying this with $x=((1+\sqrt{5})/4)^2$ we obtain that

$$\sum_{k=0}^{\infty}\binom{3k}{k}\left(\frac{3+\sqrt{5}}{54}\right)^k=\frac{\cos(\pi/10)}{\sqrt{(5-\sqrt{5})/8}}=\frac{\sqrt{(5+\sqrt{5})/8}}{\sqrt{(5-\sqrt{5})/8}}=\phi.$$

Conjecture 8 (2023-05-30) If $(1-\sqrt{2})/2\leqslant x<1/2$, then

$$\sum_{k=0}^{\infty}\binom{4k}{2k}\left(\frac{x(1-x)}{4}\right)^k(2H_{4k}-3H_{2k}+H_k)=\frac{\sqrt{1-x}}{2x-1}\log(1-x). \tag{27}$$

Remark 8 For any $x \in (-1, 1)$, we have

$$\sum_{k=0}^{\infty} \binom{4k}{2k} \left(\frac{x}{16}\right)^k = \sqrt{\frac{1+\sqrt{1-x}}{2(1-x)}}$$

which can be proved directly or via `Mathematica`. In particular,

$$\sum_{k=0}^{\infty} \binom{4k}{2k} \left(\frac{3}{64}\right)^k = \sqrt{3}.$$

For $x \in (-1, 1)$, we obviously have

$$\sum_{k=0}^{\infty} \binom{2k}{k} H_k \left(\left(\frac{x}{4}\right)^k + \left(-\frac{x}{4}\right)^k\right) = 2 \sum_{k=0}^{\infty} \binom{4k}{2k} H_{2k} \left(\frac{x}{4}\right)^{2k}$$

and

$$\sum_{k=0}^{\infty} \binom{2k}{k} H_{2k} \left(\left(\frac{x}{4}\right)^k + \left(-\frac{x}{4}\right)^k\right) = 2 \sum_{k=0}^{\infty} \binom{4k}{2k} H_{4k} \left(\frac{x}{4}\right)^{2k},$$

and hence we may find closed formulas for the two series

$$\sum_{k=0}^{\infty} \binom{4k}{2k} H_{2k} \left(\frac{x}{4}\right)^{2k} \quad \text{and} \quad \sum_{k=0}^{\infty} \binom{2k}{k} H_{4k} \left(\frac{x}{4}\right)^{2k}$$

by using (1) and (2). In particular,

$$\sum_{k=0}^{\infty} \binom{4k}{2k} \left(\frac{3}{64}\right)^k H_{2k} = 2(1+\sqrt{3}) \log(1+\sqrt{3}) - (1+3\sqrt{3}) \log 2 + \frac{\sqrt{3}-1}{2} \log 3$$

and

$$\sum_{k=0}^{\infty} \binom{4k}{2k} \left(\frac{3}{64}\right)^k H_{4k} = (2+\sqrt{3}) \log(1+\sqrt{3}) - (1+\sqrt{3}) \log 2 + \frac{\sqrt{3}-1}{4} \log 3.$$

With the aid of `Mathematica`, we obtain that

$$\begin{aligned}\sum_{k=0}^{\infty} \binom{4k}{2k} \left(\frac{3}{64}\right)^k H_k &= \sum_{k=1}^{\infty} \binom{4k}{2k} \left(\frac{3}{64}\right)^k \int_0^1 \frac{1-t^k}{1-t} dt \\ &= \int_0^1 \frac{1}{1-t} \sum_{k=1}^{\infty} \binom{4k}{2k} \left(\frac{3}{64}\right)^k (1-t^k) dt\end{aligned}$$

$$= \int_0^1 \frac{1}{1-t}\left(\sqrt{3} - \frac{\sqrt{2+\sqrt{4-3t}}}{\sqrt{4-3t}}\right) dt$$
$$= \log \frac{2+\sqrt{3}}{3} + \sqrt{3} \log \frac{7+4\sqrt{3}}{8}$$
$$= (2+4\sqrt{3})\log(1+\sqrt{3}) - \log 3 - (1+5\sqrt{3})\log 2.$$

Conjecture 9 (2022-11-14) We have the identity

$$\sum_{k=0}^{\infty} \frac{\binom{2k}{k}^2}{32^k}\left(H_{2k}^{(2)} - \frac{1}{4}H_k^{(2)}\right) = \Gamma\left(\frac{1}{4}\right)^2 \frac{\pi^2 - 8G}{32\pi\sqrt{\pi}}, \tag{28}$$

where $\Gamma(x)$ is the well-known Gamma function.

Remark 9 In contrast with (28), `Mathematica` yields that

$$\sum_{k=0}^{\infty} \frac{\binom{2k}{k}^2}{32^k} = \frac{\Gamma(1/4)}{\sqrt{2\pi}\,\Gamma(3/4)} = \frac{\Gamma(1/4)^2}{2\pi\sqrt{\pi}}.$$

For any odd prime p, Z.-H. Sun [26] proved that

$$\sum_{k=0}^{p-1} \frac{\binom{2k}{k}^2}{32^k} \equiv \begin{cases} 2x - p/(2x) \pmod{p^2} & \text{if } p = x^2 + y^2\ (x, y \in \mathbb{Z}) \text{ with } 4 \mid x - 1, \\ 0 \pmod{p^2} & \text{if } p \equiv 3 \pmod 4, \end{cases}$$

which was previously conjectured by the author (cf. [27, Conjecture 5.5]).

Conjecture 10 (2022-12-30) We have

$$\sum_{k=0}^{\infty} \frac{\binom{2k}{k}\binom{3k}{k}}{(-216)^k}(3H_{3k} - H_k) = \left(\log \frac{8}{9}\right) \sum_{k=0}^{\infty} \frac{\binom{2k}{k}\binom{3k}{k}}{(-216)^k}. \tag{29}$$

Remark 10 For any prime $p > 3$, we have

$$\sum_{k=0}^{p-1} \frac{\binom{2k}{k}\binom{3k}{k}}{(-216)^k} \equiv \left(\frac{p}{3}\right) \sum_{k=0}^{p-1} \frac{\binom{2k}{k}\binom{3k}{k}}{24^k} \pmod{p^2}$$

by [31, Corollary 1.4], and

$$\sum_{k=0}^{p-1} \frac{\binom{2k}{k}\binom{3k}{k}}{24^k} \equiv \begin{cases} \binom{(2p-2)/3}{(p-1)/3} \pmod{p^2} & \text{if } p \equiv 1 \pmod 3, \\ p/\binom{(2p+2)/3}{(p+1)/3} \pmod{p^2} & \text{if } p \equiv 2 \pmod 3, \end{cases}$$

as conjectured by the author [28, Conjecture 5.13] and proved by C. Wang and Sun [45, Theorem 1.2].

Conjecture 11 (2023-09-12)

(i) We have

$$\sum_{k=1}^{\infty}\frac{48^k}{k(2k-1)\binom{2k}{k}\binom{4k}{2k}}\left(6H_{4k-1}-9H_{2k-1}+2H_{k-1}+\frac{6}{2k-1}\right)=\frac{2\pi^3}{\sqrt{3}}. \quad (30)$$

(ii) Let $p>3$ be a prime. Then

$$\sum_{k=1}^{(p-1)/2}\frac{k\binom{2k}{k}\binom{4k}{2k}}{(2k+1)48^k}\left(6H_{4k}-9H_{2k}+2H_k-\frac{6}{2k+1}\right)\equiv 0 \pmod p \quad (31)$$

and

$$\sum_{k=1}^{p-1}\frac{k\binom{2k}{k}\binom{4k}{2k}}{(2k+1)48^k}\left(6H_{4k}-9H_{2k}+2H_k-\frac{6}{2k+1}\right)\equiv \frac{5p}{12}B_{p-2}\left(\frac{1}{3}\right) \pmod{p^2}. \quad (32)$$

Remark 11 The author's conjectural identity

$$\sum_{k=1}^{\infty}\frac{48^k}{k(2k-1)\binom{2k}{k}\binom{4k}{2k}}=\frac{15}{2}K$$

(cf. [34]) was confirmed by Au [5].

Conjecture 12 (2022-12-30) We have

$$\sum_{k=0}^{\infty}\frac{\binom{4k}{2k}\binom{2k}{k}}{128^k}(2H_{4k}-H_{2k})=\frac{(\log 2)\sqrt{\pi}}{2\Gamma(5/8)\Gamma(7/8)}. \quad (33)$$

Remark 12 `Mathematica` yields the identity

$$\sum_{k=0}^{\infty}\frac{\binom{4k}{2k}\binom{2k}{k}}{128^k}=\frac{\sqrt{\pi}}{\Gamma(5/8)\Gamma(7/8)}.$$

By [31, Corollary 1.3], for any prime $p\equiv 5,7 \pmod 8$ we have

$$\sum_{k=0}^{p-1}\frac{\binom{4k}{2k}\binom{2k}{k}}{128^k}\equiv 0 \pmod{p^2}.$$

Conjecture 13 (2022-12-30) We have

$$\sum_{k=0}^{\infty}\frac{\binom{4k}{2k}\binom{2k}{k}}{72^k}(2H_{4k}-H_{2k})=(\log 3)\sum_{k=0}^{\infty}\frac{\binom{4k}{2k}\binom{2k}{k}}{72^k} \quad (34)$$

and

$$\sum_{k=0}^{\infty}\frac{\binom{4k}{2k}\binom{2k}{k}}{576^k}(2H_{4k}-H_{2k})=\frac{1}{2}\left(\log\frac{9}{8}\right)\sum_{k=0}^{\infty}\frac{\binom{4k}{2k}\binom{2k}{k}}{576^k}. \tag{35}$$

Remark 13 Let $p>3$ be a prime. By [31, Corollary 1.4],

$$\sum_{k=0}^{p-1}\frac{\binom{4k}{2k}\binom{2k}{k}}{576^k}\equiv\left(\frac{-2}{p}\right)\sum_{k=0}^{p-1}\frac{\binom{4k}{2k}\binom{2k}{k}}{72^k} \pmod{p^2}.$$

We also have

$$\left(\frac{6}{p}\right)\sum_{k=0}^{p-1}\frac{\binom{4k}{2k}\binom{2k}{k}}{72^k}$$
$$\equiv\begin{cases}2x-\frac{p}{2x} \pmod{p^2} & \text{if } p=x^2+y^2\ (x,y\in\mathbb{Z}\ \&\ 4\mid x-1),\\ \frac{2p}{3\binom{(p+1)/2}{(p+1)/4}} \pmod{p^2} & \text{if } p\equiv 3 \pmod 4,\end{cases}$$

as conjectured by the author [28, Conjecture 5.14(iii)] and proved by Wang and Sun [45, Theorem 5.2 and Remark 5.2].

Conjecture 14 (2022-12-30) We have

$$\sum_{k=0}^{\infty}\frac{\binom{4k}{2k}\binom{2k}{k}}{(-192)^k}(2H_{4k}-H_{2k})=\frac{1}{2}\left(\log\frac{3}{4}\right)\sum_{k=0}^{\infty}\frac{\binom{4k}{2k}\binom{2k}{k}}{(-192)^k}. \tag{36}$$

Remark 14 Let $p>3$ be a prime. By [31, Corollary 1.4], we have

$$\sum_{k=0}^{p-1}\frac{\binom{4k}{2k}\binom{2k}{k}}{(-192)^k}\equiv\left(\frac{-2}{p}\right)\sum_{k=0}^{p-1}\frac{\binom{4k}{2k}\binom{2k}{k}}{48^k} \pmod{p^2}.$$

If $p=x^2+3y^2$ with $x,y\in\mathbb{Z}$ and $x\equiv 1 \pmod 3$, then

$$\sum_{k=0}^{p-1}\frac{\binom{4k}{2k}\binom{2k}{k}}{48^k}\equiv 2x-\frac{p}{2x} \pmod{p^2},$$

as conjectured by the author [28, Conjecture 5.14] and confirmed by G.-S. Mao and H. Pan [24]. If $p\equiv 2 \pmod 3$, then

$$\sum_{k=0}^{p-1}\frac{\binom{4k}{2k}\binom{2k}{k}}{48^k}\equiv\frac{3p}{2\binom{(p+1)/2}{(p+1)/6}} \pmod{p^2}$$

as conjectured by the author [28, Conjecture 5.14] and confirmed by Wang and Sun [45].

Conjecture 15 (2022-12-30) We have

$$\sum_{k=0}^{\infty}\frac{\binom{4k}{2k}\binom{2k}{k}}{(-4032)^k}(2H_{4k}-H_{2k})=\frac{1}{2}\left(\log\frac{63}{64}\right)\sum_{k=0}^{\infty}\frac{\binom{4k}{2k}\binom{2k}{k}}{(-4032)^k}. \tag{37}$$

Remark 15 Let $p>3$ be a prime with $p\neq 7$. By [31, Corollary 1.4],

$$\sum_{k=0}^{p-1}\frac{\binom{4k}{2k}\binom{2k}{k}}{(-4032)^k}\equiv\left(\frac{-2}{p}\right)\sum_{k=0}^{p-1}\frac{\binom{4k}{2k}\binom{2k}{k}}{63^k}\pmod{p^2}.$$

The author [28, Conjecture 5.14(ii)] conjectured that

$$\sum_{k=0}^{p-1}\frac{\binom{4k}{2k}\binom{2k}{k}}{63^k}$$
$$\equiv\begin{cases}(\frac{p}{3})(2x-\frac{p}{2x})\pmod{p^2} & \text{if } p=x^2+7y^2 \text{ with } x,y\in\mathbb{Z} \text{ and } (\frac{x}{7})=1,\\ 0\pmod{p} & \text{if } (\frac{p}{7})=-1.\end{cases}$$

Conjecture 16 (2022-12-30) We have

$$\sum_{k=0}^{\infty}\frac{\binom{6k}{3k}\binom{3k}{k}}{864^k}(6H_{6k}-3H_{3k}-2H_{2k}+H_k)=\frac{(\log 2)\sqrt{\pi}}{\Gamma(7/12)\Gamma(11/12)}. \tag{38}$$

Remark 16 `Mathematica` yields the identity

$$\sum_{k=0}^{\infty}\frac{\binom{6k}{3k}\binom{3k}{k}}{864^k}=\frac{\sqrt{\pi}}{\Gamma(7/12)\Gamma(11/12)}.$$

By [31, Corollary 1.3], for any prime $p>3$ with $p\equiv 3\pmod 4$ we have

$$\sum_{k=0}^{p-1}\frac{\binom{6k}{3k}\binom{3k}{k}}{864^k}\equiv 0\pmod{p^2}.$$

3 Series and Congruences with Summands Containing 3 or 4 Binomial Coefficients

In 1993 D. Zeilberger [49] used the WZ method to establish the identity

$$\sum_{k=1}^{\infty}\frac{21k-8}{k^3\binom{2k}{k}^3}=\frac{\pi^2}{6}.$$

The author [35] proved that

$$\sum_{k=1}^{(p-1)/2}\frac{21k-8}{k^3\binom{2k}{k}^3}\equiv-\left(\frac{-1}{p}\right)4E_{p-3}\pmod p$$

for any prime $p>3$.

Conjecture 17 (2022-10-11)
(i) We have the identity

$$\sum_{k=1}^{\infty}\frac{21k-8}{k^3\binom{2k}{k}^3}\left(H_{2k-1}^{(2)}-\frac{25}{8}H_{k-1}^{(2)}\right)=\frac{47\pi^4}{2880}. \tag{39}$$

(ii) For any prime $p>3$, we have

$$\sum_{k=1}^{(p-1)/2}(21k+8)\binom{2k}{k}^3\left(H_{2k}^{(2)}-\frac{25}{8}H_k^{(2)}\right)\equiv 32p\left(\frac{-1}{p}\right)E_{p-3}\pmod{p^2} \tag{40}$$

and

$$\sum_{k=0}^{p-1}(21k+8)\binom{2k}{k}^3\left(H_{2k}^{(2)}-\frac{25}{8}H_k^{(2)}\right)\equiv -48H_{p-1}+\frac{246}{5}p^4B_{p-5}\pmod{p^5}. \tag{41}$$

(iii) For each prime $p>5$, we have

$$\sum_{k=1}^{p-1}\binom{2k}{k}^3(2(21k+8)(H_{2k}-H_k)+7)\equiv -112pH_{p-1}\pmod{p^5}. \tag{42}$$

Remark 17 After seeing an earlier arXiv version of this paper, Au [5, Corollary 2.3] confirmed the author's conjectural identity (39), and proved an identity (after the proof of [5, Theorem 2.2]) which has the equivalent form

$$\sum_{k=1}^{\infty}\frac{(21k-8)(H_{2k-1}-H_{k-1})-7/2}{k^3\binom{2k}{k}^3}=\zeta(3).$$

Conjecture 18 (2022-10-11)
(i) We have

$$\sum_{k=1}^{\infty}\frac{21k-8}{k^3\binom{2k}{k}^3}\left(H_{2k-1}^{(3)}+\frac{43}{8}H_{k-1}^{(3)}\right)=\frac{711}{28}\zeta(5)-\frac{29}{14}\pi^2\zeta(3). \tag{43}$$

(ii) For any prime $p > 7$, we have

$$\sum_{k=0}^{(p-1)/2}(21k+8)\binom{2k}{k}^3\left(H_{2k}^{(3)}+\frac{43}{8}H_k^{(3)}\right)\equiv 32\left(\frac{-1}{p}\right)E_{p-3}\pmod{p} \tag{44}$$

and

$$\sum_{k=0}^{p-1}(21k+8)\binom{2k}{k}^3\left(H_{2k}^{(3)}+\frac{43}{8}H_k^{(3)}\right)\equiv -\frac{120}{7}pB_{p-3}\pmod{p^2}. \tag{45}$$

Remark 18 The identity (43) looks quite challenging.

Conjecture 19 (2022-10-13)
(i) We have

$$\sum_{k=1}^{\infty}\frac{(3k-1)(-8)^k}{k^3\binom{2k}{k}^3}\left(H_{2k-1}^{(2)}-\frac{5}{4}H_{k-1}^{(2)}\right)=-2\beta(4). \tag{46}$$

(ii) For any prime $p > 3$, we have

$$\sum_{k=0}^{(p-1)/2}(3k+1)\frac{\binom{2k}{k}^3}{(-8)^k}\left(H_{2k}^{(2)}-\frac{5}{4}H_k^{(2)}\right)\equiv\left(\frac{2}{p}\right)\frac{p}{4}E_{p-3}\left(\frac{1}{4}\right)\pmod{p^2} \tag{47}$$

and

$$\sum_{k=0}^{p-1}(3k+1)\frac{\binom{2k}{k}^3}{(-8)^k}\left(H_{2k}^{(2)}-\frac{5}{4}H_k^{(2)}\right)\equiv pE_{p-3}\pmod{p^2}. \tag{48}$$

(iii) Let p be an odd prime. Then

$$\sum_{k=0}^{(p-1)/2}\frac{\binom{2k}{k}^3}{(-8)^k}(2(3k+1)(H_{2k}-H_k)+1)\equiv\left(\frac{-1}{p}\right)2^{p-1}\pmod{p^2} \tag{49}$$

and

$$\sum_{k=0}^{p-1}(3k+1)\frac{\binom{2k}{k}^3}{(-8)^k}\left(H_{2k}^{(3)}+\frac{7}{8}H_k^{(3)}\right)\equiv 0\pmod{p}. \tag{50}$$

Remark 19 In 2008, J. Guillera [14] used the WZ method to find the identity

$$\sum_{k=1}^{\infty}\frac{(3k-1)(-8)^k}{k^3\binom{2k}{k}^3}=-2G.$$

The identity (46) provides a fast converging series for computing the constant $\beta(4)$. We are unable to find the exact values of the series

$$\sum_{k=1}^{\infty}\frac{(-8)^k}{k^3\binom{2k}{k}^3}(2(3k-1)(H_{2k-1}-H_{k-1})-1)$$

and

$$\sum_{k=1}^{\infty}\frac{(3k-1)(-8)^k}{k^3\binom{2k}{k}^3}\left(H_{2k-1}^{(3)}+\frac{7}{8}H_{k-1}^{(3)}\right).$$

Conjecture 20 (2022-10-11)

(i) We have the identity

$$\sum_{k=1}^{\infty}\frac{(3k-1)16^k}{k^3\binom{2k}{k}^3}\left(H_{2k-1}^{(2)}-\frac{5}{4}H_{k-1}^{(2)}\right)=\frac{\pi^4}{24}. \tag{51}$$

(ii) Let $p>3$ be a prime. Then

$$\sum_{k=0}^{(p-1)/2}(3k+1)\frac{\binom{2k}{k}^3}{16^k}\left(H_{2k}^{(2)}-\frac{5}{4}H_{k}^{(2)}\right)\equiv 2p\left(\frac{-1}{p}\right)E_{p-3} \pmod{p^2} \tag{52}$$

and

$$\sum_{k=0}^{p-1}(3k+1)\frac{\binom{2k}{k}^3}{16^k}\left(H_{2k}^{(2)}-\frac{5}{4}H_{k}^{(2)}\right)\equiv\frac{7}{6}p^2B_{p-3} \pmod{p^3}. \tag{53}$$

(iii) Let p be an odd prime. Then

$$\sum_{k=1}^{p-1}\frac{\binom{2k}{k}^3}{16^k}(2(3k+1)(H_{2k}-H_k)+1)\equiv\frac{4}{3}p\,q_p(2)-\frac{2}{3}p^2q_p(2)^2 \pmod{p^3}. \tag{54}$$

If $p\equiv 2 \pmod 3$, then

$$\sum_{k=0}^{(p-1)/2}\frac{\binom{2k}{k}^3}{16^k}\left(H_{2k}^{(2)}-\frac{5}{4}H_{k}^{(2)}\right)\equiv 0 \pmod{p}. \tag{55}$$

Remark 20 In 2008 Guillera [14, Identity 1] used the WZ method to establish the identity

$$\sum_{k=1}^{\infty}\frac{(3k-1)16^k}{k^3\binom{2k}{k}^3}=\frac{\pi^2}{2}.$$

Two q-analogues of this identity were given by Q.-H. Hou, C. Krattenthaler and Z.-W. Sun [21]. Guo and Lian [19] proved that the two sides of (53) are congruent modulo p^2 for any prime $p>3$. After seeing an earlier arXiv version of this paper, Au [5] confirmed the author's conjectural identity (51), and proved an identity (after the proof of [5, Corollary 2.3]) which has the equivalent form

$$\sum_{k=1}^{\infty}\frac{16^k}{k^3\binom{2k}{k}^3}\left((3k-1)(H_{2k-1}-H_{k-1})-\frac{1}{2}\right)=\frac{\pi^2}{3}\log 2+\frac{7}{6}\zeta(3).$$

Conjecture 21 (2022-10-11)
(i) We have

$$\sum_{k=1}^{\infty}\frac{(3k-1)16^k}{k^3\binom{2k}{k}^3}\left(H_{2k-1}^{(3)}+\frac{7}{8}H_{k-1}^{(3)}\right)=\frac{\pi^2}{2}\zeta(3). \tag{56}$$

(ii) For any odd prime p, we have

$$\sum_{k=1}^{(p-1)/2}(3k+1)\frac{\binom{2k}{k}^3}{16^k}\left(H_{2k}^{(3)}+\frac{7}{8}H_k^{(3)}\right)\equiv 2\left(\frac{-1}{p}\right)E_{p-3} \pmod{p} \tag{57}$$

and

$$\sum_{k=0}^{p-1}(3k+1)\frac{\binom{2k}{k}^3}{16^k}\left(H_{2k}^{(3)}+\frac{7}{8}H_k^{(3)}\right)\equiv \frac{3}{2}pB_{p-3} \pmod{p^2}. \tag{58}$$

Remark 21 Conjecture 21 looks more challenging than Conjecture 20.

Conjecture 22 (2022-10-16)
(i) We have

$$\sum_{k=1}^{\infty}\frac{(4k-1)(-64)^k}{k^3\binom{2k}{k}^3}\left(H_{2k-1}^{(2)}-\frac{1}{2}H_{k-1}^{(2)}\right)=-16\beta(4). \tag{59}$$

(ii) For any prime $p>3$, we have

$$\sum_{k=0}^{(p-1)/2}(4k+1)\frac{\binom{2k}{k}^3}{(-64)^k}\left(H_{2k}^{(2)}-\frac{1}{2}H_k^{(2)}\right)\equiv pE_{p-3} \pmod{p^2}, \tag{60}$$

$$\sum_{k=0}^{p-1}(4k+1)\frac{\binom{2k}{k}^3}{(-64)^k}\left(H_{2k}^{(2)}-\frac{1}{2}H_k^{(2)}\right)\equiv pE_{p-3} \pmod{p^2}. \tag{61}$$

Remark 22 Guillera [14] proved that

$$\sum_{k=1}^{\infty}\frac{(4k-1)(-64)^k}{k^3\binom{2k}{k}^3}=-16G.$$

The congruence (60) is motivated by (10). After seeing an earlier arXiv version of this paper, Au [5, Corollary 2.11] confirmed the author's conjectural identity (59). It seems that for any $m,n\in\mathbb{Z}^+$ and odd prime p, we have

$$\sum_{k=0}^{(p-1)/2}(4k+1)\frac{\binom{2k}{k}^n}{(-4)^{kn}}\left(H_{2k}^{(2m)}-\frac{H_k^{(2m)}}{2^{2m-1}}\right)\equiv 0 \pmod p. \tag{62}$$

Conjecture 23 (2022-12-05)
(i) We have

$$\sum_{k=1}^{\infty}\frac{8^k((10k-3)(H_{2k-1}-H_{k-1})-1)}{k^3\binom{2k}{k}^2\binom{3k}{k}}=\frac{7}{2}\zeta(3) \tag{63}$$

and

$$\sum_{k=1}^{\infty}\frac{8^k((10k-3)(H_{3k-1}-H_{k-1})-8/3)}{k^3\binom{2k}{k}^2\binom{3k}{k}}=\frac{2\pi^2\log 2+7\zeta(3)}{4}. \tag{64}$$

(ii) For any odd prime p, we have

$$\sum_{k=1}^{p-1}\frac{\binom{2k}{k}^2\binom{3k}{k}}{8^k}((10k+3)(H_{2k}-H_k)+1)\equiv\frac{63}{8}p^3B_{p-3} \pmod{p^4} \tag{65}$$

and

$$\sum_{k=1}^{p-1}\frac{\binom{2k}{k}^2\binom{3k}{k}}{8^k}(3(10k+3)(H_{3k}-H_k)+8)\equiv 9p\,q_p(2)-\frac{9}{2}p^2q_p(2)^2 \pmod{p^3}. \tag{66}$$

Remark 23 As conjectured by the author [28] and confirmed by Guillera and M. Rogers [18], we have

$$\sum_{k=1}^{\infty}\frac{(10k-3)8^k}{k^3\binom{2k}{k}^2\binom{3k}{k}}=\frac{\pi^2}{2}.$$

Conjecture 24 (2022-12-05)
(i) We have

$$\sum_{k=1}^{\infty}\frac{(-27)^k((15k-4)(3H_{3k-1}-H_{k-1})-9)}{k^3\binom{2k}{k}^2\binom{3k}{k}}=-\frac{4\pi^3}{\sqrt{3}}. \tag{67}$$

(ii) For any prime $p>3$, we have

$$\sum_{k=0}^{p-1}\frac{\binom{2k}{k}^2\binom{3k}{k}}{(-27)^k}((15k+4)(3H_{3k}-H_k)+9)\equiv 9\left(\frac{p}{3}\right)+6p^2B_{p-2}\left(\frac{1}{3}\right) \pmod{p^3}. \tag{68}$$

Remark 24 As conjectured by the author [28] and confirmed by Kh. Hessami Pilehrood and T. Hessami Pilehrood [20], we have

$$\sum_{k=1}^{\infty}\frac{(15k-4)(-27)^{k-1}}{k^3\binom{2k}{k}^2\binom{3k}{k}}=K.$$

Conjecture 25 (2023-08-26)
(i) We have

$$\sum_{k=1}^{\infty}\frac{(-27)^k}{k^3\binom{2k}{k}^2\binom{3k}{k}}\left((15k-4)(H_{2k-1}-H_{k-1})+\frac{1}{2k-1}\right)=-\frac{4\pi^3}{3\sqrt{3}}. \tag{69}$$

(ii) For any prime $p>3$, we have

$$\begin{aligned}&\sum_{k=0}^{p-1}\frac{\binom{2k}{k}^2\binom{3k}{k}}{(-27)^k}\left((15k+4)(H_{2k}-H_k)+\frac{1}{2k+1}\right)\\&\equiv\left(\frac{p}{3}\right)+\frac{4}{3}p^2B_{p-2}\left(\frac{1}{3}\right)\pmod{p^3}.\end{aligned} \tag{70}$$

Remark 25 This conjecture is similar to Conjecture 24.

Conjecture 26 (2022-12-05)
(i) We have

$$\sum_{k=1}^{\infty}\frac{64^{k-1}((11k-3)(2H_{2k-1}+H_{k-1})-4)}{k^3\binom{2k}{k}^2\binom{3k}{k}}=\frac{7}{2}\zeta(3) \tag{71}$$

and

$$\sum_{k=1}^{\infty}\frac{64^{k-1}((11k-3)(3H_{3k-1}-6H_{k-1})-7)}{k^3\binom{2k}{k}^2\binom{3k}{k}}=\frac{6\pi^2\log 2-21\zeta(3)}{8}. \tag{72}$$

(ii) For any odd prime p, we have

$$\sum_{k=1}^{p-1}\frac{\binom{2k}{k}^2\binom{3k}{k}}{64^k}((11k+3)(2H_{2k}+H_k)+4)\equiv 21p^3B_{p-3}\pmod{p^4} \tag{73}$$

and

$$\sum_{k=1}^{p-1}\frac{\binom{2k}{k}^2\binom{3k}{k}}{64^k}((11k+3)(3H_{3k}-6H_k)+7)\equiv 18p\,q_p(2)+9p^2q_p(2)^2\pmod{p^3}. \tag{74}$$

Remark 26 As conjectured by the author [28] and confirmed by Guillera [17], we have

$$\sum_{k=1}^{\infty}\frac{(11k-3)64^k}{k^3\binom{2k}{k}^2\binom{3k}{k}}=8\pi^2.$$

Conjecture 27 (2022-12-09)
(i) We have

$$\sum_{k=1}^{\infty}\frac{81^k((35k-8)(H_{4k-1}-H_{k-1})-35/4)}{k^3\binom{2k}{k}^2\binom{4k}{2k}}=12\pi^2\log 3+39\zeta(3). \tag{75}$$

(ii) For any prime $p>3$, we have

$$\begin{aligned}&\sum_{k=1}^{p-1}\frac{\binom{2k}{k}^2\binom{4k}{2k}}{81^k}(4(35k+8)(H_{4k}-H_k)+35)\\ &\equiv 32(3^{p-1}-1)-16(3^{p-1}-1)^2 \pmod{p^3}.\end{aligned} \tag{76}$$

Remark 27 As conjectured by the author [28] and confirmed in [18], we have

$$\sum_{k=1}^{\infty}\frac{(35k-8)81^k}{k^3\binom{2k}{k}^2\binom{4k}{2k}}=12\pi^2.$$

Conjecture 28 (2023-01-14) For any prime $p>3$, we have

$$\begin{aligned}&\sum_{k=1}^{p-1}\frac{\binom{2k}{k}^2\binom{4k}{2k}}{(-144)^k}(4(5k+1)(H_{4k}-H_k)+5)\\ &\equiv\left(\frac{p}{3}\right)(5+2p(2q_p(2)+q_p(3))) \pmod{p^2}.\end{aligned} \tag{77}$$

Remark 28 As conjectured by the author [28] and confirmed in [18], we have

$$\sum_{k=1}^{\infty}\frac{(5k-1)(-144)^k}{k^3\binom{2k}{k}^2\binom{4k}{2k}}=-\frac{45}{2}K.$$

We are unable to find the exact value of the series

$$\sum_{k=1}^{\infty}\frac{(-144)^k}{k^3\binom{2k}{k}^2\binom{4k}{2k}}(4(5k-1)(H_{4k-1}-H_{k-1})-5).$$

The classical rational Ramanujan-type series for $1/\pi$ have the following four forms:

$$\sum_{k=0}^{\infty}(ak+b)\frac{\binom{2k}{k}^3}{m^k}=\frac{c\sqrt{d}}{\pi}, \tag{78}$$

$$\sum_{k=0}^{\infty}(ak+b)\frac{\binom{2k}{k}^2\binom{3k}{k}}{m^k}=\frac{c\sqrt{d}}{\pi}, \tag{79}$$

$$\sum_{k=0}^{\infty}(ak+b)\frac{\binom{2k}{k}^2\binom{4k}{2k}}{m^k}=\frac{c\sqrt{d}}{\pi},\tag{80}$$

$$\sum_{k=0}^{\infty}(ak+b)\frac{\binom{2k}{k}\binom{3k}{k}\binom{6k}{3k}}{m^k}=\frac{c\sqrt{d}}{\pi},\tag{81}$$

where $a, b, m \in \mathbb{Z}$, $am \neq 0$, $c \in \mathbb{Q} \setminus \{0\}$, and d is a positive squarefree integer. It is known that there are totally 36 such series, see, e.g., S. Cooper [11, Chap. 14].

For a positive integer m, can we find similar series for $(\log m)/\pi$? Motivated by Ramanujan-type series of the forms (78)–(81), the author formulated the following general conjecture.

Conjecture 29 (General Conjecture, 2022-12-08)

(i) If we have an identity (78) with $a, b, m \in \mathbb{Z}, am \neq 0, c \in \mathbb{Q} \setminus \{0\}$, and $d \in \mathbb{Z}^+$ squarefree, then

$$\sum_{k=0}^{\infty}\frac{\binom{2k}{k}^3}{m^k}(6(ak+b)(H_{2k}-H_k)+a)=\frac{c\sqrt{d}}{\pi}\log|m|,\tag{82}$$

and

$$\sum_{k=0}^{p-1}\frac{\binom{2k}{k}^3}{m^k}(6(ak+b)(H_{2k}-H_k)+a)\equiv\left(\frac{-d}{p}\right)(a+b(m^{p-1}-1)) \pmod{p^2}\tag{83}$$

for any odd prime $p \nmid dm$.

(ii) If we have an identity (79) with $a, b, m \in \mathbb{Z}, am \neq 0, c \in \mathbb{Q} \setminus \{0\}$, and $d \in \mathbb{Z}^+$ squarefree, then

$$\sum_{k=0}^{\infty}\frac{\binom{2k}{k}^2\binom{3k}{k}}{m^k}((ak+b)(3H_{3k}+2H_{2k}-5H_k)+a)=\frac{c\sqrt{d}}{\pi}\log|m|,\tag{84}$$

and

$$\begin{aligned}&\sum_{k=0}^{p-1}\frac{\binom{2k}{k}^2\binom{3k}{k}}{m^k}((ak+b)(3H_{3k}+2H_{2k}-5H_k)+a)\\&\quad\equiv\left(\frac{-d}{p}\right)(a+b(m^{p-1}-1)) \pmod{p^2}\end{aligned}\tag{85}$$

for any odd prime $p \nmid dm$.

(iii) If we have an identity (80) with $a, b, m \in \mathbb{Z}, am \neq 0, c \in \mathbb{Q} \setminus \{0\}$, and $d \in \mathbb{Z}^+$ squarefree, then

$$\sum_{k=0}^{\infty}\frac{\binom{2k}{k}^2\binom{4k}{2k}}{m^k}(4(ak+b)(H_{4k}-H_k)+a)=\frac{c\sqrt{d}}{\pi}\log|m|, \tag{86}$$

and

$$\begin{aligned}&\sum_{k=0}^{p-1}\frac{\binom{2k}{k}^2\binom{4k}{2k}}{m^k}(4(ak+b)(H_{4k}-H_k)+a)\\&\equiv\left(\frac{-d}{p}\right)\left(a+b(m^{p-1}-1)\right)\pmod{p^2}\end{aligned} \tag{87}$$

for any odd prime $p \nmid dm$.

(iv) If we have an identity (81) with $a, b, m \in \mathbb{Z}, am \neq 0, c \in \mathbb{Q}\setminus\{0\}$, and $d \in \mathbb{Z}^+$ squarefree, then

$$\sum_{k=0}^{\infty}\frac{\binom{2k}{k}\binom{3k}{k}\binom{6k}{3k}}{m^k}(3(ak+b)(2H_{6k}-H_{3k}-H_k)+a)=\frac{c\sqrt{d}}{\pi}\log|m|, \tag{88}$$

and

$$\begin{aligned}&\sum_{k=0}^{p-1}\frac{\binom{2k}{k}\binom{3k}{k}\binom{6k}{3k}}{m^k}(3(ak+b)(2H_{6k}-H_{3k}-H_k)+a)\\&\equiv\left(\frac{-d}{p}\right)\left(a+b(m^{p-1}-1)\right)\pmod{p^2}\end{aligned} \tag{89}$$

for any odd prime $p \nmid dm$.

Remark 29 Ramanujan [25] found the irrational series

$$\sum_{k=0}^{\infty}\left(k+\frac{31}{270+48\sqrt{5}}\right)\frac{\binom{2k}{k}^3}{(2^{20}/(\sqrt{5}-1)^8)^k}=\frac{16}{(15+21\sqrt{5})\pi}.$$

In the spirit of part (i) of our general conjecture, we guess that

$$\begin{aligned}&\sum_{k=0}^{\infty}\frac{\binom{2k}{k}^3}{(2^{20}/(\sqrt{5}-1)^8)^k}\left(6\left(k+\frac{31}{270+48\sqrt{5}}\right)(H_{2k}-H_k)+1\right)\\&=\frac{16}{(15+21\sqrt{5})\pi}\times\log\frac{2^{20}}{(\sqrt{5}-1)^8},\end{aligned}$$

which can be easily checked via `Mathematica`. We believe that similar things happen for all irrational Ramanujan-type series.

Example 1 In view of the Ramanujan series (cf. [11, Chap. 14] and [25])

$$\sum_{k=0}^{\infty}(5k+1)\frac{\binom{2k}{k}^2\binom{3k}{k}}{(-192)^k}=\frac{4\sqrt{3}}{\pi}$$

and the known identity

$$\sum_{k=0}^{\infty}\frac{\binom{2k}{k}^2\binom{3k}{k}}{(-192)^k}=\frac{4\sqrt{\pi}}{3\sqrt{3}\,\Gamma(5/6)^3}$$

(cf. (14.29) of [11, p. 624]), by part (ii) of Conjecture 29 we should have

$$\sum_{k=0}^{\infty}\frac{\binom{2k}{k}^2\binom{3k}{k}}{(-192)^k}(5k+1)(3H_{3k}+2H_{2k}-5H_k)=\frac{4\log 192}{\sqrt{3}\pi}-\frac{20\sqrt{\pi}}{3\sqrt{3}\,\Gamma(5/6)^3}.$$

In 2013, Guillera [16, (32)] proved the identity

$$\sum_{k=0}^{\infty}\frac{\binom{2k}{k}^3}{(-64)^k}\left((4k+1)H_k-\frac{2}{3}\right)=-\frac{4\log 2}{\pi}$$

motivated by the Bauer series (8).

Conjecture 30 (2022-10-16) We have

$$\sum_{k=0}^{\infty}(4k+1)\frac{\binom{2k}{k}^3}{(-64)^k}H_{2k}^{(3)}=\frac{15\zeta(3)}{4\pi}-2G. \tag{90}$$

Remark 30 For any $m,n\in\mathbb{Z}^+$ and odd prime p not dividing $2^{2m-1}-1$, we have $H_{p-1}^{(2m-1)}\equiv 0 \pmod p$ since $\sum_{j=1}^{p-1}1/(2j)^{2m-1}\equiv\sum_{k=1}^{p-1}1/k^{2m-1} \pmod p$, thus

$$\begin{aligned}
&\sum_{k=0}^{(p-1)/2}(4k+1)\binom{(p-1)/2}{k}^n H_{2k}^{(2m-1)}\\
&=\sum_{k=0}^{(p-1)/2}\left(4\left(\frac{p-1}{2}-k\right)+1\right)\binom{(p-1)/2}{k}^n H_{p-1-2k}^{(2m-1)}\\
&\equiv-\sum_{k=0}^{(p-1)/2}(4k+1)\binom{(p-1)/2}{k}^n H_{2k}^{(2m-1)} \pmod p
\end{aligned}$$

and hence
$$\sum_{k=0}^{(p-1)/2}(4k+1)\frac{\binom{2k}{k}^n}{(-4)^{kn}}H_{2k}^{(2m-1)}\equiv 0 \pmod{p}. \tag{91}$$

(Note that $\binom{-1/2}{k}=\binom{2k}{k}/(-4)^k$ for any $k\in\mathbb{N}$.)

Conjecture 31 (2022-12-04)
(i) We have

$$\sum_{k=0}^{\infty}(6k+1)\frac{\binom{2k}{k}^3}{256^k}H_k=\frac{4}{3}\cdot\frac{\sqrt[3]{2}\sqrt{\pi}}{\Gamma(5/6)^3}-\frac{8\log 2}{\pi} \tag{92}$$

and
$$\sum_{k=0}^{\infty}(6k+1)\frac{\binom{2k}{k}^3}{256^k}H_{2k}=\frac{2}{3}\cdot\frac{\sqrt[3]{2}\sqrt{\pi}}{\Gamma(5/6)^3}-\frac{8\log 2}{3\pi}. \tag{93}$$

(ii) Let p be an odd prime. Then

$$\sum_{k=0}^{(p-1)/2}\frac{\binom{2k}{k}^3}{256^k}((6k+1)(3H_{2k}-H_k)-1)\equiv(-1)^{(p+1)/2} \pmod{p^4}. \tag{94}$$

If $p>3$, then

$$\begin{aligned}&\sum_{k=0}^{(p-1)/2}\frac{\binom{2k}{k}^3}{256^k}((6k+1)(H_{2k}-H_k)+1)\\ &\equiv\left(\frac{-1}{p}\right)\left(1+\frac{4}{3}p\,q_p(2)-\frac{2}{3}p^2q_p(2)^2\right) \pmod{p^3}.\end{aligned} \tag{95}$$

Remark 31 It is known (cf. (14.27) of [11, p. 623]) that

$$\sum_{k=0}^{\infty}\frac{\binom{2k}{k}^3}{256^k}=\frac{2}{3}\cdot\frac{\sqrt[3]{2}\sqrt{\pi}}{\Gamma(5/6)^3}.$$

In view of this, (92) and (93) together implies the identities

$$\sum_{k=0}^{\infty}\frac{\binom{2k}{k}^3}{256^k}((6k+1)(3H_{2k}-H_k)-1)=0$$

and
$$\sum_{k=0}^{\infty}\frac{\binom{2k}{k}^3}{256^k}((6k+1)(H_{2k}-H_k)+1)=\frac{16\log 2}{3\pi}.$$

The last identity is also implied by Conjecture 29(i).

Conjecture 32 (2022-10-12) Let $p > 3$ be a prime.

(i) We have

$$\sum_{k=0}^{(p-1)/2} (6k+1)\frac{\binom{2k}{k}^3}{256^k}\left(H_{2k}^{(2)} - \frac{5}{16}H_k^{(2)}\right) \equiv \frac{7}{24}\left(\frac{-1}{p}\right)p^2 B_{p-3} \pmod{p^3}, \quad (96)$$

and

$$\sum_{k=0}^{p-1}(6k+1)\frac{\binom{2k}{k}^3}{256^k}\left(H_{2k}^{(2)} - \frac{5}{16}H_k^{(2)}\right) \equiv -pE_{p-3} \pmod{p^2}. \quad (97)$$

(ii) If $p \equiv 2 \pmod 3$, then

$$\sum_{k=0}^{(p-1)/2} \frac{\binom{2k}{k}^3}{256^k}\left(H_{2k}^{(2)} - \frac{5}{16}H_k^{(2)}\right) \equiv 0 \pmod{p}. \quad (98)$$

Remark 32 For any prime $p > 3$, Guo and Lian [19] proved that the two sides of (96) are congruent modulo p^2.

Conjecture 33 (2022-10-11)

(i) We have the identity

$$\sum_{k=0}^{\infty}(6k+1)\frac{\binom{2k}{k}^3}{256^k}\left(H_{2k}^{(3)} - \frac{7}{64}H_k^{(3)}\right) = \frac{25\zeta(3)}{8\pi} - G. \quad (99)$$

(ii) Let p be an odd prime. Then

$$\sum_{k=0}^{(p-1)/2} (6k+1)\frac{\binom{2k}{k}^3}{256^k}\left(H_{2k}^{(3)} - \frac{7}{64}H_k^{(3)}\right) \equiv -\frac{1}{2}E_{p-3} \pmod{p} \quad (100)$$

and

$$\sum_{k=0}^{p-1}(6k+1)\frac{\binom{2k}{k}^3}{256^k}\left(H_{2k}^{(3)} - \frac{7}{64}H_k^{(3)}\right) \equiv -\frac{3}{2}E_{p-3} \pmod{p}. \quad (101)$$

Remark 33 For any $k \in \mathbb{Z}^+$, it is easy to see that

$$H_{2k}^{(3)} - \frac{7}{64}H_k^{(3)} = \sum_{j=1}^{k}\left(\frac{1}{(2j-1)^3} + \frac{1}{(4j)^3}\right).$$

Conjecture 34 (2022-10-12) Let $p > 3$ be a prime. Then

$$\sum_{k=0}^{(p-1)/2}(6k+1)\frac{\binom{2k}{k}^3}{(-512)^k}\left(H_{2k}^{(2)}-\frac{5}{16}H_k^{(2)}\right)\equiv\left(\frac{2}{p}\right)\frac{p}{4}E_{p-3}\pmod{p^2} \tag{102}$$

and

$$\sum_{k=0}^{p-1}(6k+1)\frac{\binom{2k}{k}^3}{(-512)^k}\left(H_{2k}^{(2)}-\frac{5}{16}H_k^{(2)}\right)\equiv\frac{p}{16}E_{p-3}\left(\frac{1}{4}\right)\pmod{p^2}. \tag{103}$$

Remark 34 Note that (102) is stronger than (7).

Conjecture 35 (2022-10-16)

(i) We have the identity

$$\sum_{k=0}^{\infty}(6k+1)\frac{\binom{2k}{k}^3}{(-512)^k}\left(H_{2k}^{(3)}-\frac{7}{64}H_k^{(3)}\right)=\frac{57}{16}\cdot\frac{\zeta(3)}{\sqrt{2}\pi}-L, \tag{104}$$

where

$$L:=L\left(2,\left(\frac{-8}{\cdot}\right)\right)=\sum_{n=1}^{\infty}\frac{\left(\frac{-8}{n}\right)}{n^2}=\sum_{k=0}^{\infty}\frac{(-1)^{k(k-1)/2}}{(2k+1)^2}$$

with $(\frac{-8}{\cdot})$ the Kronecker symbol.

(ii) Let p be an odd prime. Then

$$\sum_{k=0}^{p-1}(6k+1)\frac{\binom{2k}{k}^3}{(-512)^k}\left(H_{2k}^{(3)}-\frac{7}{64}H_k^{(3)}\right)\equiv 0\pmod{p}. \tag{105}$$

Remark 35 For a general odd prime p, we are unable to find a closed form for the left-hand side of the congruence (105) modulo p^2.

Conjecture 36 (2022-12-09) For any prime $p > 3$, we have

$$\begin{aligned}&\sum_{k=0}^{(p-1)/2}\frac{\binom{2k}{k}^3}{4096^k}((42k+5)(H_{2k}-H_k)+7)\\&\equiv\left(\frac{-1}{p}\right)\left(7+10p\,q_p(2)-5p^2q_p(2)^2\right)\pmod{p^3}.\end{aligned} \tag{106}$$

Remark 36 In view of the Ramanujan series (cf. [25])

$$\sum_{k=0}^{\infty}(42k+5)\frac{\binom{2k}{k}^3}{4096^k}=\frac{16}{\pi},$$

by part (i) of Conjecture 29 we should have

$$\sum_{k=0}^{\infty}\frac{\binom{2k}{k}^3}{4096^k}((42k+5)(H_{2k}-H_k)+7)=\frac{32\log 2}{\pi},$$

and

$$\sum_{k=0}^{p-1}\frac{\binom{2k}{k}^3}{4096^k}(6(42k+5)(H_{2k}-H_k)+42)$$
$$\equiv\left(\frac{-1}{p}\right)\left(42+5p\,q_p(2^{12})\right)\equiv\left(\frac{-1}{p}\right)\left(42+60p\,q_p(2)\right)\pmod{p^2}$$

for any odd prime p.

Conjecture 37 (2022-10-11)
(i) We have the identity

$$\sum_{k=0}^{\infty}(42k+5)\frac{\binom{2k}{k}^3}{4096^k}\left(H_{2k}^{(2)}-\frac{25}{92}H_k^{(2)}\right)=\frac{2\pi}{69}. \tag{107}$$

(ii) Let $p>3$ be a prime. If $p\neq 23$, then

$$\sum_{k=0}^{(p-1)/2}(42k+5)\frac{\binom{2k}{k}^3}{4096^k}\left(H_{2k}^{(2)}-\frac{25}{92}H_k^{(2)}\right)\equiv\left(\frac{-1}{p}\right)\frac{3}{20}\left(p^4B_{p-5}-5H_{p-1}\right)\pmod{p^5}. \tag{108}$$

Also,

$$\sum_{k=0}^{p-1}(42k+5)\frac{\binom{2k}{k}^3}{4096^k}\left(H_{2k}^{(2)}-\frac{25}{92}H_k^{(2)}\right)\equiv -pE_{p-3}\pmod{p^2}. \tag{109}$$

Remark 37 It is interesting to compare this conjecture with Remark 36.

Conjecture 38 (2022-10-12)
(i) We have

$$\sum_{k=0}^{\infty}(42k+5)\frac{\binom{2k}{k}^3}{4096^k}\left(H_{2k}^{(3)}-\frac{43}{352}H_k^{(3)}\right)=\frac{555}{77}\cdot\frac{\zeta(3)}{\pi}-\frac{32}{11}G. \tag{110}$$

(ii) For any prime $p > 7$, we have

$$\sum_{k=0}^{(p-1)/2}(42k+5)\frac{\binom{2k}{k}^3}{4096^k}\left(11H_{2k}^{(3)}-\frac{43}{32}H_k^{(3)}\right)\equiv -16E_{p-3}\pmod p \tag{111}$$

and

$$\sum_{k=0}^{p-1}(42k+5)\frac{\binom{2k}{k}^3}{4096^k}\left(11H_{2k}^{(3)}-\frac{43}{32}H_k^{(3)}\right)\equiv -27E_{p-3}\pmod p. \tag{112}$$

Remark 38 Conjecture 38 looks quite challenging.

Conjecture 39 (2022-12-05)
(i) We have

$$\sum_{k=0}^{\infty}\frac{\binom{2k}{k}^2\binom{3k}{k}}{216^k}((6k+1)(H_{2k}-2H_k)+3)=\frac{9\sqrt{3}\log 3}{2\pi} \tag{113}$$

and

$$\sum_{k=0}^{\infty}\frac{\binom{2k}{k}^2\binom{3k}{k}}{216^k}(6k+1)(3H_{3k}-H_k)=\frac{9\sqrt{3}\log 2}{\pi}. \tag{114}$$

(ii) For any prime $p > 3$, we have

$$\sum_{k=0}^{p-1}\frac{\binom{2k}{k}^2\binom{3k}{k}}{216^k}((6k+1)(H_{2k}-2H_k)+3)\equiv\left(\frac{p}{3}\right)\frac{3^p+3}{2}\pmod{p^2} \tag{115}$$

and

$$\sum_{k=0}^{p-1}\frac{\binom{2k}{k}^2\binom{3k}{k}}{216^k}(6k+1)(3H_{3k}-H_k)\equiv 3\left(\frac{p}{3}\right)p\,q_p(2)\pmod{p^2}. \tag{116}$$

Remark 39 This is motivated by the Ramanujan series (cf. [11, Chap. 14] and [25])

$$\sum_{k=0}^{\infty}(6k+1)\frac{\binom{2k}{k}^2\binom{3k}{k}}{216^k}=\frac{3\sqrt{3}}{\pi}.$$

Conjecture 40 (2022-12-04)
(i) We have

$$\sum_{k=0}^{\infty}\frac{\binom{2k}{k}^2\binom{4k}{2k}}{(-1024)^k}((20k+3)(H_{2k}-3H_k)+12)=\frac{56\log 2}{\pi}. \tag{117}$$

(ii) For any odd prime p, we have

$$\sum_{k=0}^{p-1}\frac{\binom{2k}{k}^2\binom{4k}{2k}}{(-1024)^k}((20k+3)(H_{2k}-3H_k)+12)\equiv\left(\frac{-1}{p}\right)(12+21p\,q_p(2))\pmod{p^2}. \tag{118}$$

Remark 40 This is motivated by the Ramanujan series (cf. [11, Chap. 14] and [25])

$$\sum_{k=0}^{\infty}(20k+3)\frac{\binom{2k}{k}^2\binom{4k}{2k}}{(-1024)^k}=\frac{8}{\pi}.$$

Conjecture 41 (2023-06-16)
(i) We have

$$\sum_{k=0}^{\infty}\frac{\binom{2k}{k}^2\binom{4k}{2k}}{(-1024)^k}\left((20k+3)\left(H_{4k}^{(2)}-\frac{H_{2k}^{(2)}}{4}-\frac{H_k^{(2)}}{16}\right)+\frac{1}{4k+1}\right)=\frac{\pi}{6}. \tag{119}$$

(ii) For any prime $p>3$, we have

$$\begin{aligned}&p\sum_{k=0}^{p-1}\frac{\binom{2k}{k}^2\binom{4k}{2k}}{(-1024)^k}\left((20k+3)\left(H_{4k}^{(2)}-\frac{H_{2k}^{(2)}}{4}-\frac{H_k^{(2)}}{16}\right)+\frac{1}{4k+1}\right)\\&\equiv\left(\frac{-1}{p}\right)+\frac{10}{3}p^2E_{p-3}\pmod{p^3}.\end{aligned} \tag{120}$$

Remark 41 The identity (119) looks curious and challenging.

Conjecture 42 (2022-12-04)
(i) We have

$$\sum_{k=0}^{\infty}\frac{\binom{2k}{k}^2\binom{4k}{2k}}{48^{2k}}((8k+1)(3H_{2k}-4H_k)+6)=\frac{16\sqrt{3}\log 2}{\pi}. \tag{121}$$

(ii) For any prime $p>3$, we have

$$\sum_{k=0}^{p-1}\frac{\binom{2k}{k}^2\binom{4k}{2k}}{48^{2k}}((8k+1)(3H_{2k}-4H_k)+6)\equiv\left(\frac{p}{3}\right)(6+8p\,q_p(2))\pmod{p^2}. \tag{122}$$

Remark 42 This is motivated by the Ramanujan series (9).

Conjecture 43 (2022-10-15) For any prime $p>3$, we have

$$\sum_{k=0}^{(p-1)/2}(8k+1)\frac{\binom{2k}{k}^2\binom{4k}{2k}}{48^{2k}}\left(H_{2k}^{(2)}-\frac{5}{18}H_k^{(2)}\right)\equiv\frac{p}{36}B_{p-2}\left(\frac{1}{3}\right)\pmod{p^2}, \tag{123}$$

$$\sum_{k=0}^{p-1}(8k+1)\frac{\binom{2k}{k}^2\binom{4k}{2k}}{48^{2k}}\left(H_{2k}^{(2)}-\frac{5}{18}H_k^{(2)}\right)\equiv-\frac{5}{24}pB_{p-2}\left(\frac{1}{3}\right)\pmod{p^2}. \tag{124}$$

Remark 43 The congruence (123) is motivated by (11).

Conjecture 44 (2022-10-19)
(i) For any prime $p>3$, we have

$$\begin{aligned}&\sum_{k=0}^{(p-1)/2}(4k+1)\frac{\binom{2k}{k}^4}{256^k}\left(H_{2k}^{(2)}-\frac{1}{2}H_k^{(2)}\right)\\&\equiv\sum_{k=0}^{p-1}(4k+1)\frac{\binom{2k}{k}^4}{256^k}\left(H_{2k}^{(2)}-\frac{1}{2}H_k^{(2)}\right)\equiv\frac{7}{6}p^2B_{p-3}\pmod{p^3}.\end{aligned} \tag{125}$$

(ii) For any odd prime p, we have

$$\sum_{k=0}^{(p-1)/2}(4k+1)\frac{\binom{2k}{k}^4}{256^k}H_{2k}^{(3)}\equiv\sum_{k=0}^{p-1}(4k+1)\frac{\binom{2k}{k}^4}{256^k}H_{2k}^{(3)}\equiv\frac{3}{2}pB_{p-3}\pmod{p^2}. \tag{126}$$

Remark 44 Guo and Lian [19, (1.7)] proved that for any prime $p>3$ we have

$$\sum_{k=0}^{(p-1)/2}(4k+1)\frac{\binom{2k}{k}^4}{256^k}\left(H_{2k}^{(2)}-\frac{1}{2}H_k^{(2)}\right)\equiv 0\pmod{p^2}.$$

4 Series and Congruences with Summands Containing at Least Five Binomial Coefficients

The following two conjectures are motivated by the identity

$$\sum_{k=1}^{\infty}\frac{(-1)^k(205k^2-160k+32)}{k^5\binom{2k}{k}^5}=-2\zeta(3)$$

established by T. Amdeberhan and D. Zeilberger [2] in 1997 via the WZ method.

Conjecture 45 (2022-12-09)
(i) We have

$$\sum_{k=1}^{\infty}\frac{(-1)^{k-1}}{k^5\binom{2k}{k}^5}\left((205k^2-160k+32)(H_{2k-1}-H_{k-1})-41k+16\right)=\frac{\pi^4}{60}. \tag{127}$$

(ii) For any prime $p > 5$, we have

$$\sum_{k=0}^{p-1}(-1)^k\binom{2k}{k}^5\big((205k^2+160k+32)(H_{2k}-H_k)+41k+16\big) \equiv 16p+64p^2H_{p-1} \pmod{p^6}. \tag{128}$$

Conjecture 46 (2022-12-09)
(i) We have

$$\sum_{k=1}^{\infty}\frac{(-1)^k((205k^2-160k+32)(4H_{2k-1}^{(2)}-12H_{k-1}^{(2)})-43)}{k^5\binom{2k}{k}^5} = -8\zeta(5). \tag{129}$$

(ii) Let $p > 3$ be a prime. Then

$$\sum_{k=1}^{(p-1)/2}\frac{(-1)^k((205k^2-160k+32)(4H_{2k-1}^{(2)}-12H_{k-1}^{(2)})-43)}{k^5\binom{2k}{k}^5} \equiv -200B_{p-5} \pmod{p}, \tag{130}$$

and

$$\sum_{k=1}^{p-1}(-1)^k\binom{2k}{k}^5((205k^2+160k+32)(4H_{2k}^{(2)}-12H_k^{(2)})+43) \equiv 256pH_{p-1} \pmod{p^5} \tag{131}$$

if $p > 5$.

The following two conjectures are motivated by the identity

$$\sum_{k=1}^{\infty}\frac{(10k^2-6k+1)(-256)^k}{k^5\binom{2k}{k}^5} = -28\zeta(3)$$

(cf. [14, Identity 8]).

Conjecture 47 (2022-12-09)
(i) We have

$$\sum_{k=1}^{\infty}\frac{(-256)^k}{k^5\binom{2k}{k}^5}\big((10k^2-6k+1)(2H_{2k-1}-H_{k-1})-3k+1\big) = -\frac{\pi^4}{2}. \tag{132}$$

(ii) For any prime $p > 3$, we have

$$\sum_{k=0}^{(p-1)/2}\frac{\binom{2k}{k}^5}{(-256)^k}\big((10k^2+6k+1)(2H_{2k}-H_k)+3k+1\big)\equiv p+\frac{14}{3}p^4B_{p-3}\pmod{p^5}. \tag{133}$$

Remark 45 For any prime $p > 3$, the author [37, Conjecture 31(ii)] conjectured the congruence

$$\sum_{k=(p+1)/2}^{p-1}\frac{\binom{2k}{k}^5}{(-256)^k}(10k^2+6k+1)\equiv-\frac{7}{2}p^5B_{p-3}\pmod{p^6}$$

which implies that

$$\begin{aligned}&\sum_{k=(p+1)/2}^{p-1}\frac{\binom{2k}{k}^5}{(-256)^k}\big((10k^2+6k+1)(2H_{2k}-H_k)+3k+1\big)\\ \equiv&\sum_{k=(p+1)/2}^{p-1}\frac{\binom{2k}{k}^5}{(-256)^k}(10k^2+6k+1)\cdot\frac{2}{p}\equiv-7p^4B_{p-3}\pmod{p^5}.\end{aligned}$$

Conjecture 48 (2022-12-09)
(i) We have

$$\sum_{k=1}^{\infty}\frac{(-256)^k((10k^2-6k+1)(4H^{(2)}_{2k-1}-3H^{(2)}_{k-1})-2)}{k^5\binom{2k}{k}^5}=-124\zeta(5). \tag{134}$$

(ii) For any prime $p > 3$, we have

$$\sum_{k=1}^{(p-1)/2}\frac{(-256)^k}{k^5\binom{2k}{k}^5}((10k^2-6k+1)(4H^{(2)}_{2k-1}-3H^{(2)}_{k-1})-2)\equiv-124B_{p-5}\pmod{p} \tag{135}$$

and

$$\sum_{k=1}^{(p-1)/2}\frac{\binom{2k}{k}^5}{(-256)^k}((10k^2+6k+1)(4H^{(2)}_{2k}-3H^{(2)}_{k})+2)\equiv\frac{28}{3}p^3B_{p-3}\pmod{p^4}. \tag{136}$$

The following two conjectures are motivated by the identity

$$\sum_{k=1}^{\infty}\frac{(28k^2-18k+3)(-64)^k}{k^5\binom{2k}{k}^4\binom{3k}{k}}=-14\zeta(3)$$

conjectured by the author [30] and confirmed recently by Au [5].

Conjecture 49 (2022-12-09)
(i) We have

$$\sum_{k=1}^{\infty}\frac{(-64)^k}{k^5\binom{2k}{k}^4\binom{3k}{k}}\big((28k^2-18k+3)(4H_{2k-1}-3H_{k-1})-20k+6\big)=-\frac{\pi^4}{2}. \quad (137)$$

(ii) For any odd prime p, we have

$$\begin{aligned}&\sum_{k=0}^{p-1}\frac{\binom{2k}{k}^4\binom{3k}{k}}{(-64)^k}\big((28k^2+18k+3)(4H_{2k}-3H_k)+20k+6\big)\\&\equiv 6p-14p^4B_{p-3}\pmod{p^5}.\end{aligned} \quad (138)$$

Conjecture 50 (2022-12-09)
(i) We have

$$\sum_{k=1}^{\infty}\frac{(-64)^k((28k^2-18k+3)(2H_{2k-1}^{(2)}-3H_{k-1}^{(2)})-2)}{k^5\binom{2k}{k}^4\binom{3k}{k}}=-31\zeta(5). \quad (139)$$

(ii) For any odd prime p, we have

$$\begin{aligned}&\sum_{k=1}^{p-1}\frac{\binom{2k}{k}^4\binom{3k}{k}}{(-64)^k}\Big((28k^2+18k+3)(2H_{2k}^{(2)}-3H_k^{(2)})+2\Big)\\&\equiv -7p^3B_{p-3}\pmod{p^4}.\end{aligned} \quad (140)$$

The following two conjectures are motivated by the identity

$$\sum_{k=0}^{\infty}(20k^2+8k+1)\frac{\binom{2k}{k}^5}{(-4096)^k}=\frac{8}{\pi^2}$$

(cf. [14, Identity 8]).

Conjecture 51 (2022-12-09)
(i) We have

$$\sum_{k=0}^{\infty}\frac{\binom{2k}{k}^5}{(-4096)^k}((20k^2+8k+1)H_k-6k-1)=-\frac{16\log 2}{\pi^2} \quad (141)$$

and

$$\sum_{k=0}^{\infty}\frac{\binom{2k}{k}^5}{(-4096)^k}(5(20k^2+8k+1)H_{2k}-10k-1)=-\frac{32\log 2}{\pi^2}. \quad (142)$$

(ii) For any prime $p > 3$, we have

$$\sum_{k=0}^{(p-1)/2}\frac{\binom{2k}{k}^5}{(-4096)^k}((20k^2+8k+1)H_k-6k-1) \equiv -p-2p^2q_p(2)+p^3q_p(2)^2-\frac{2}{3}p^4q_p(2)^3 \pmod{p^5}, \tag{143}$$

and

$$\sum_{k=0}^{(p-1)/2}\frac{\binom{2k}{k}^5}{(-4096)^k}(5(20k^2+8k+1)H_k-10k-1) \equiv -p-4p^2q_p(2)+2p^3q_p(2)^2 \pmod{p^4}. \tag{144}$$

Conjecture 52 (2022-12-09)
(i) We have

$$\sum_{k=1}^{\infty}\frac{\binom{2k}{k}^5}{(-4096)^k}((20k^2+8k+1)(8H_{2k}^{(2)}-3H_k^{(2)})+4) = -\frac{4}{3}. \tag{145}$$

(ii) For any prime $p > 3$, we have

$$\sum_{k=1}^{(p-1)/2}\frac{\binom{2k}{k}^5}{(-4096)^k}((20k^2+8k+1)(8H_{2k}^{(2)}-3H_k^{(2)})+4) \equiv \frac{14}{3}p^3B_{p-3} \pmod{p^4} \tag{146}$$

and

$$\sum_{k=1}^{p-1}\frac{\binom{2k}{k}^5}{(-4096)^k}((20k^2+8k+1)(8H_{2k}^{(2)}-3H_k^{(2)})+4) \equiv -\frac{28}{3}p^3B_{p-3} \pmod{p^4}. \tag{147}$$

The following two conjectures are motivated by the identity

$$\sum_{k=0}^{\infty}(820k^2+180k+13)\frac{\binom{2k}{k}^5}{(-2^{20})^k} = \frac{128}{\pi^2}$$

(cf. [10, Identity 9]).

Conjecture 53 (2022-12-09)
(i) We have

$$\sum_{k=0}^{\infty}\frac{\binom{2k}{k}^5}{(-2^{20})^k}\Big((820k^2+180k+13)(H_{2k}-H_k)+164k+18\Big) = \frac{256\log 2}{\pi^2}. \tag{148}$$

(ii) For any odd prime p, we have

$$\sum_{k=0}^{(p-1)/2}\frac{\binom{2k}{k}^5}{(-2^{20})^k}\big((820k^2+180k+13)(H_{2k}-H_k)+164k+18\big)\equiv 18p+26p^2q_p(2)-13p^3q_p(2)^2\pmod{p^4}. \tag{149}$$

Conjecture 54 (2022-12-09)
(i) We have

$$\sum_{k=1}^{\infty}\frac{\binom{2k}{k}^5}{(-2^{20})^k}((820k^2+180k+13)(11H_{2k}^{(2)}-3H_k^{(2)})+43)=-\frac{1}{3}. \tag{150}$$

(ii) Let $p>3$ be a prime. Then

$$\sum_{k=1}^{p-1}\frac{\binom{2k}{k}^5}{(-2^{20})^k}((820k^2+180k+13)(11H_{2k}^{(2)}-3H_k^{(2)})+43)\equiv -\frac{77}{6}p^3B_{p-3}\pmod{p^4}, \tag{151}$$

and

$$\sum_{k=1}^{(p-1)/2}\frac{\binom{2k}{k}^5}{(-2^{20})^k}((820k^2+180k+13)(11H_{2k}^{(2)}-3H_k^{(2)})+43)\equiv -\frac{11}{4}pH_{p-1}\pmod{p^5} \tag{152}$$

if $p>5$.

The following two conjectures are motivated by the known identity

$$\sum_{k=0}^{\infty}(74k^2+27k+3)\frac{\binom{2k}{k}^4\binom{3k}{k}}{4096^k}=\frac{48}{\pi^2}$$

(cf. [15]).

Conjecture 55 (2022-12-09)
(i) We have

$$\sum_{k=0}^{\infty}\frac{\binom{2k}{k}^4\binom{3k}{k}}{4096^k}\big((74k^2+27k+3)H_{2k}-17k-3\big)=0 \tag{153}$$

and

$$\sum_{k=0}^{\infty}\frac{\binom{2k}{k}^4\binom{3k}{k}}{4096^k}((74k^2+27k+3)(51H_{3k}+250H_{2k}-153H_k)+15) = \frac{9792\log 2}{\pi^2}. \tag{154}$$

(ii) For any odd prime p, we have

$$\sum_{k=0}^{p-1}\frac{\binom{2k}{k}^4\binom{3k}{k}}{4096^k}((74k^2+27k+3)H_{2k}-17k-3) \equiv -3p+7p^4B_{p-3} \pmod{p^5}, \tag{155}$$

and

$$\sum_{k=0}^{p-1}\frac{\binom{2k}{k}^4\binom{3k}{k}}{4096^k}((74k^2+27k+3)(51H_{3k}+250H_{2k}-153H_k)+15) \equiv 15p+612p^2q_p(2)-306p^3q_p(2)^2 \pmod{p^4}. \tag{156}$$

Conjecture 56 (2022-12-09)
(i) We have

$$\sum_{k=1}^{\infty}\frac{\binom{2k}{k}^4\binom{3k}{k}}{4096^k}((74k^2+27k+3)(92H_{2k}^{(2)}-33H_k^{(2)})+112) = 160. \tag{157}$$

(ii) For any odd prime p, we have

$$\sum_{k=1}^{p-1}\frac{\binom{2k}{k}^4\binom{3k}{k}}{4096^k}\left((74k^2+27k+3)(92H_{2k}^{(2)}-33H_k^{(2)})+112\right) \equiv 644p^3B_{p-3} \pmod{p^4}. \tag{158}$$

The following two conjectures are motivated by the identity

$$\sum_{k=0}^{\infty}(120k^2+34k+3)\frac{\binom{2k}{k}^4\binom{4k}{2k}}{2^{16k}} = \frac{32}{\pi^2}$$

(cf. [14, Identity 10]).

Conjecture 57 (2022-12-09)
(i) We have

$$\sum_{k=0}^{\infty}\frac{\binom{2k}{k}^4\binom{4k}{2k}}{2^{16k}}(2(120k^2+34k+3)H_{4k}-16k-1) = 0 \tag{159}$$

and

$$\sum_{k=0}^{\infty}\frac{\binom{2k}{k}^4\binom{4k}{2k}}{2^{16k}}((120k^2+34k+3)(H_{2k}-2H_k)+68k+9)=\frac{128\log 2}{\pi^2}. \quad (160)$$

(ii) Let p be an odd prime. Then

$$\sum_{k=0}^{p-1}\frac{\binom{2k}{k}^4\binom{4k}{2k}}{2^{16k}}((120k^2+34k+3)(H_{2k}-2H_k)+68k+9) \equiv 9p+12p^2q_p(2)-6p^3q_p(2)^2 \pmod{p^4}, \quad (161)$$

and

$$\sum_{k=0}^{p-1}\frac{\binom{2k}{k}^4\binom{4k}{2k}}{2^{16k}}(2(120k^2+34k+3)H_{4k}-16k-1)\equiv -p+\frac{77}{6}p^4B_{p-3} \pmod{p^5} \quad (162)$$

if $p>3$.

Conjecture 58 (2022-12-09)
(i) We have

$$\sum_{k=1}^{\infty}\frac{\binom{2k}{k}^4\binom{4k}{2k}}{2^{16k}}((120k^2+34k+3)(23H_{2k}^{(2)}-7H_k^{(2)})+24)=\frac{16}{3}. \quad (163)$$

(ii) Let p be an odd prime. Then

$$\sum_{k=1}^{p-1}\frac{\binom{2k}{k}^4\binom{4k}{2k}}{2^{16k}}((120k^2+34k+3)(23H_{2k}^{(2)}-7H_k^{(2)})+24)\equiv\frac{161}{2}p^3B_{p-3} \pmod{p^4}, \quad (164)$$

and

$$\sum_{k=1}^{(p-1)/2}\frac{\binom{2k}{k}^4\binom{4k}{2k}}{2^{16k}}((120k^2+34k+3)(23H_{2k}^{(2)}-7H_k^{(2)})+24) \equiv -23pH_{p-1} \pmod{p^5} \quad (165)$$

if $p\neq 5$.

Conjecture 59 (2022-12-09)
(i) We have

$$\sum_{k=0}^{\infty}\frac{\binom{2k}{k}^3\binom{3k}{k}\binom{4k}{2k}}{(-24^4)^k}\big((252k^2+63k+5)(4H_{4k}+3H_{3k}-7H_k)+504k+63\big) = \frac{192\log 24}{\pi^2}. \tag{166}$$

(ii) For any prime $p>3$, we have

$$\sum_{k=0}^{p-1}\frac{\binom{2k}{k}^3\binom{3k}{k}\binom{4k}{2k}}{(-24^4)^k}\big((252k^2+63k+5)(4H_{4k}+3H_{3k}-7H_k)+504k+63\big) \equiv 63p+5p^2q_p(24^4)-\frac{5}{2}p^3q_p(24^4)^2 \pmod{p^4}. \tag{167}$$

Remark 46 Conjecture 59 is motivated by the identity

$$\sum_{k=0}^{\infty}(252k^2+63k+5)\frac{\binom{2k}{k}^3\binom{3k}{k}\binom{4k}{2k}}{(-24^4)^k}=\frac{48}{\pi^2}$$

(cf. [10]).

Conjecture 60 (2023-01-16)
(i) We have

$$\sum_{k=0}^{\infty}\frac{\binom{2k}{k}^2\binom{3k}{k}^2\binom{6k}{3k}}{10^{6k}}\big(3(532k^2+126k+9)(H_{6k}-H_k)+532k+63\big) = \frac{1125\log 10}{4\pi^2}. \tag{168}$$

(ii) For any odd prime $p\neq 5$, we have

$$\sum_{k=0}^{p-1}\frac{\binom{2k}{k}^2\binom{3k}{k}^2\binom{6k}{3k}}{10^{6k}}\big(3(532k^2+126k+9)(H_{6k}-H_k)+532k+63\big) \equiv 63p+\frac{9}{2}p^2q_p(10^6)-\frac{9}{4}p^3q_p(10^6)^2 \pmod{p^4}. \tag{169}$$

Remark 47 Conjecture 60 is motivated by the identity

$$\sum_{k=0}^{\infty}(532k^2+126k+9)\frac{\binom{2k}{k}^2\binom{3k}{k}^2\binom{6k}{3k}}{10^{6k}}=\frac{375}{4\pi^2}$$

(cf. [10]).

Conjecture 61 (2023-01-17)

(i) We have

$$\sum_{k=0}^{\infty}\frac{\binom{2k}{k}^2\binom{3k}{k}^2\binom{6k}{3k}}{(-2^{18})^k}\big(6(1930k^2+549k+45)(H_{6k}-H_k)+3860k+549\big)=\frac{6912\log 2}{\pi^2}. \tag{170}$$

(ii) For any odd prime p, we have

$$\sum_{k=0}^{p-1}\frac{\binom{2k}{k}^2\binom{3k}{k}^2\binom{6k}{3k}}{(-2^{18})^k}\big(6(1930k^2+549k+45)(H_{6k}-H_k)+3860k+549\big)\equiv 549p+45p^2q_p(2^{18})-\frac{45}{2}p^3q_p(2^{18})^2 \pmod{p^4}. \tag{171}$$

Remark 48 Conjecture 61 is motivated by the identity

$$\sum_{k=0}^{\infty}(1930k^2+549k+45)\frac{\binom{2k}{k}^2\binom{3k}{k}^2\binom{6k}{3k}}{(-2^{18})^k}=\frac{384}{\pi^2}$$

(cf. [10]).

Conjecture 62 (2023-01-17)

(i) We have

$$\sum_{k=0}^{\infty}\frac{\binom{2k}{k}^2\binom{3k}{k}^2\binom{6k}{3k}}{(-2^{18}3^65^3)^k}\big(2(5418k^2+693k+29)(H_{6k}-H_k)+3612k+231\big)=\frac{128\sqrt{5}}{\pi^2}\log(2^63^25). \tag{172}$$

(ii) For any prime $p>5$, we have

$$\sum_{k=0}^{p-1}\frac{\binom{2k}{k}^2\binom{3k}{k}^2\binom{6k}{3k}}{(-2^{18}3^65^3)^k}\big(2(5418k^2+693k+29)(H_{6k}-H_k)+3612k+231\big)\equiv\left(\frac{5}{p}\right)\left(231p+\frac{29}{3}p^2q_p(2^{18}3^65^3)-\frac{29}{6}p^3q_p(2^{18}3^65^3)^2\right) \pmod{p^4}. \tag{173}$$

Remark 49 Conjecture 62 is motivated by the identity

$$\sum_{k=0}^{\infty}(5418k^2+693k+29)\frac{\binom{2k}{k}^2\binom{3k}{k}^2\binom{6k}{3k}}{(-2^{18}3^65^3)^k}=\frac{128\sqrt{5}}{\pi^2}$$

(cf. [10]).

Conjecture 63 (2023-01-17) For $k \in \mathbb{N}$, set

$$H(k) := 6H_{6k} + 4H_{4k} - 3H_{3k} - 2H_{2k} - 5H_k.$$

(i) We have

$$\begin{aligned}&\sum_{k=0}^{\infty} \frac{\binom{2k}{k}^2\binom{3k}{k}\binom{4k}{2k}\binom{6k}{3k}}{(-2^{22}3^3)^k}\big((1640k^2+278k+15)H(k)+3280k+278\big)\\ &\quad = \frac{256}{\sqrt{3}\pi^2}\log(2^{22}3^3).\end{aligned} \tag{174}$$

(ii) For any prime $p > 3$, we have

$$\begin{aligned}&\sum_{k=0}^{p-1} \frac{\binom{2k}{k}^2\binom{3k}{k}\binom{4k}{2k}\binom{6k}{3k}}{(-2^{22}3^3)^k}\big((1640k^2+278k+15)H(k)+3280k+278\big)\\ &\equiv \left(\frac{3}{p}\right)\left(278p + 15p^2q_p(2^{22}3^3) - \frac{15}{2}p^3q_p(2^{22}3^3)^2\right) \pmod{p^4}.\end{aligned} \tag{175}$$

Remark 50 Conjecture 63 is motivated by the identity

$$\sum_{k=0}^{\infty}(1640k^2+278k+15)\frac{\binom{2k}{k}^2\binom{3k}{k}\binom{4k}{2k}\binom{6k}{3k}}{(-2^{22}3^3)^k} = \frac{256}{\sqrt{3}\pi^2}$$

(cf. [10]).

Conjecture 64 (2023-01-17)
For $k \in \mathbb{N}$, set

$$\mathcal{H}(k) := 4H_{8k} - 2H_{4k} + H_{2k} - 3H_k.$$

(i) We have

$$\begin{aligned}&\sum_{k=0}^{\infty} \frac{\binom{2k}{k}^3\binom{4k}{2k}\binom{8k}{4k}}{(2^{18}7^4)^k}\big((1920k^2+304k+15)\mathcal{H}(k)+1920k+152\big)\\ &\quad = \frac{56\sqrt{7}}{\pi^2}(9\log 2 + 2\log 7).\end{aligned} \tag{176}$$

(ii) For any odd prime $p \neq 7$, we have

$$\begin{aligned}&\sum_{k=0}^{p-1} \frac{\binom{2k}{k}^3\binom{4k}{2k}\binom{8k}{4k}}{(2^{18}7^4)^k}\big((1920k^2+304k+15)\mathcal{H}(k)+1920k+152\big)\\ &\equiv \left(\frac{7}{p}\right)\left(152p + \frac{15}{2}p^2q_p(2^{18}7^4) - \frac{15}{4}p^3q_p(2^{18}7^4)^2\right) \pmod{p^4}.\end{aligned} \tag{177}$$

Remark 51 Conjecture 64 is motivated by the identity

$$\sum_{k=0}^{\infty}(1920k^2+304k+15)\frac{\binom{2k}{k}^3\binom{4k}{2k}\binom{8k}{4k}}{(2^{18}7^4)^k}=\frac{56\sqrt{7}}{\pi^2}$$

(cf. [10]).

Conjecture 65 (2022-12-09)
(i) We have

$$\sum_{k=1}^{\infty}\frac{256^k}{k^7\binom{2k}{k}^7}\left((21k^3-22k^2+8k-1)(4H_{2k-1}^{(2)}-5H_{k-1}^{(2)})-6k+2\right)=\frac{\pi^6}{24}. \tag{178}$$

(ii) For any odd prime p, we have

$$\sum_{k=0}^{(p-1)/2}\frac{\binom{2k}{k}^7}{256^k}\left((21k^3+22k^2+8k+1)(4H_{2k}^{(2)}-5H_k^{(2)})+6k+2\right) \equiv 2p \pmod{p^5}. \tag{179}$$

Remark 52 This is motivated by the identity

$$\sum_{k=1}^{\infty}\frac{(21k^3-22k^2+8k-1)256^k}{k^7\binom{2k}{k}^7}=\frac{\pi^4}{8}$$

conjectured by Guillera [13].

The following three conjectures are motivated by the identity

$$\sum_{k=0}^{\infty}(168k^3+76k^2+14k+1)\frac{\binom{2k}{k}^7}{2^{20k}}=\frac{32}{\pi^3} \tag{180}$$

conjectured by B. Gourevich (cf. [10]).

Conjecture 66 (2022-12-09)
(i) We have

$$\sum_{k=0}^{\infty}\frac{\binom{2k}{k}^7}{2^{20k}}\left(7(168k^3+76k^2+14k+1)(H_{2k}-H_k)+252k^2+76k+7\right) = \frac{320\log 2}{\pi^3}. \tag{181}$$

(ii) For any prime $p > 5$, we have

$$\sum_{k=0}^{(p-1)/2} \frac{\binom{2k}{k}^7}{2^{20k}} \left(7(168k^3 + 76k^2 + 14k + 1)(H_{2k} - H_k) + 252k^2 + 76k + 7\right)$$
$$\equiv \left(\frac{-1}{p}\right) \left(7p^2 + 10p^3 q_p(2) - 5p^4 q_p(2)^2 + \frac{10}{3} p^5 q_p(2)^3 - \frac{5}{2} p^6 q_p(2)^4\right) \pmod{p^7}. \tag{182}$$

Conjecture 67 (2022-12-09)
(i) We have

$$\sum_{k=0}^{\infty} \frac{\binom{2k}{k}^7}{2^{20k}} \left((168k^3 + 76k^2 + 14k + 1)(16H_{2k}^{(2)} - 5H_k^{(2)}) + 8(6k + 1)\right) = \frac{80}{3\pi}. \tag{183}$$

(ii) For any prime $p > 5$, we have

$$\sum_{k=0}^{(p-1)/2} \frac{\binom{2k}{k}^7}{2^{20k}} \left((168k^3 + 76k^2 + 14k + 1)(16H_{2k}^{(2)} - 5H_k^{(2)}) + 8(6k + 1)\right) \equiv \left(\frac{-1}{p}\right) 8p \pmod{p^6}. \tag{184}$$

Conjecture 68 (2023-06-19)
(i) We have

$$\sum_{k=0}^{\infty} \frac{\binom{2k}{k}^7}{2^{20k}} \left((168k^3 + 76k^2 + 14k + 1)(128H_{2k}^{(4)} - 7H_k^{(4)}) + \frac{64}{2k + 1}\right) = \frac{976}{45}\pi. \tag{185}$$

(ii) For any odd prime $p \neq 5$, we have

$$\sum_{k=0}^{(p-3)/2} \frac{\binom{2k}{k}^7}{2^{20k}} \left((168k^3 + 76k^2 + 14k + 1)\left(128H_{2k}^{(4)} - 7H_k^{(4)}\right) + \frac{64}{2k + 1}\right) \equiv -256 q_p(2) \pmod{p}. \tag{186}$$

Conjecture 69 (2023-06-19) Set

$$P(x) = 4528x^4 + 3180x^3 + 972x^2 + 147x + 9.$$

(i) We have

$$\sum_{k=0}^{\infty} \frac{\binom{2k}{k}^7 \binom{3k}{k} \binom{4k}{2k}}{(-2^{24})^k} \left(\left(H_{4k} + \frac{H_{2k}}{2}\right) P(k) - 484k^3 - 108k^2 + 9k + 3\right) = 0, \tag{187}$$

$$\sum_{k=0}^{\infty}\frac{\binom{2k}{k}^7\binom{3k}{k}\binom{4k}{2k}}{(-2^{24})^k}\left(\left(11H_{2k}^{(2)}-4H_k^{(2)}\right)P(k)+1780k^2+633k+63\right)=\frac{512}{\pi^2}, \tag{188}$$

$$\sum_{k=0}^{\infty}\frac{\binom{2k}{k}^7\binom{3k}{k}\binom{4k}{2k}}{(-2^{24})^k}\left(\left(9H_{2k}^{(3)}-2H_k^{(3)}\right)P(k)-\frac{916k^2+65k-9}{2k+1}\right)=0, \tag{189}$$

and

$$\sum_{k=0}^{\infty}\frac{\binom{2k}{k}^7\binom{3k}{k}\binom{4k}{2k}}{(-2^{24})^k}\left(\left(49H_{2k}^{(4)}-2H_k^{(4)}\right)P(k)+\frac{5996k^2+3071k+329}{(2k+1)^2}\right)=\frac{4096}{15}. \tag{190}$$

(ii) For any prime $p>3$, we have

$$\begin{aligned}&\sum_{k=0}^{p-1}\frac{\binom{2k}{k}^7\binom{3k}{k}\binom{4k}{2k}}{(-2^{24})^k}\left(\left(H_{4k}+\frac{H_{2k}}{2}\right)P(k)-484k^3-108k^2+9k+3\right)\\&\equiv 3p^3-\frac{1953}{10}p^8B_{p-5}\pmod{p^9},\end{aligned} \tag{191}$$

$$\begin{aligned}&\sum_{k=0}^{p-1}\frac{\binom{2k}{k}^7\binom{3k}{k}\binom{4k}{2k}}{(-2^{24})^k}\left(\left(11H_{2k}^{(2)}-4H_k^{(2)}\right)P(k)+1780k^2+633k+63\right)\\&\equiv 63p^2-\frac{9207}{10}p^7B_{p-5}\pmod{p^8},\end{aligned} \tag{192}$$

$$\begin{aligned}&\sum_{k=0}^{(p-3)/2}\frac{\binom{2k}{k}^7\binom{3k}{k}\binom{4k}{2k}}{(-2^{24})^k}\left(\left(9H_{2k}^{(3)}-2H_k^{(3)}\right)P(k)-\frac{916k^2+65k-9}{2k+1}\right)\\&\equiv 9p-\frac{81}{2}p^2H_{p-1}-\frac{243}{8}p^6B_{p-5}\pmod{p^7},\end{aligned} \tag{193}$$

and

$$\begin{aligned}&\sum_{k=1}^{(p-1)/2}\frac{\binom{2k}{k}^7\binom{3k}{k}\binom{4k}{2k}}{(-2^{24})^k}\left(\left(49H_{2k}^{(4)}-2H_k^{(4)}\right)P(k)+\frac{5996k^2+3071k+329}{(2k+1)^2}\right)\\&\equiv -\frac{441}{2}\left(pH_{p-1}+\frac{3}{4}p^5B_{p-5}\right)\pmod{p^6}.\end{aligned} \tag{194}$$

Remark 53 This is motivated by the conjectural identity

$$\sum_{k=0}^{\infty}P(k)\frac{\binom{2k}{k}^7\binom{3k}{k}\binom{4k}{2k}}{(-2^{24})^k}=\frac{768}{\pi^4}$$

(cf. [10]).

Conjecture 70 (2023-06-19) Set

$$Q(x) = 43680k^4 + 20632k^3 + 4340k^2 + 466k + 21$$

and

$$R(x) = 87360x^3 + 30948x^2 + 4340x + 233.$$

(i) We have

$$\sum_{k=0}^{\infty} \frac{\binom{2k}{k}^8\binom{4k}{2k}}{2^{32k}}((2(H_{4k} + 3H_{2k} - 4H_k)Q(k) + R(k)) = \frac{2^{15}\log 2}{\pi^4} \quad (195)$$

and

$$\sum_{k=0}^{\infty} \frac{\binom{2k}{k}^8\binom{4k}{2k}}{2^{32k}}\left(\left(7H_{2k}^{(2)} - 2H_k^{(2)}\right)Q(k) + 3624k^2 + 926k + 69\right) = \frac{2048}{3\pi^2}. \quad (196)$$

(ii) Let p be an odd prime. If $p \neq 5$, then

$$\sum_{k=0}^{p-1} \frac{\binom{2k}{k}^8\binom{4k}{2k}}{2^{32k}}((2(H_{4k} + 3H_{2k} - 4H_k)Q(k) + R(k)) \equiv 233p^3 + 336p^4q_p(2) - 168p^5q_p(2)^2 + 112p^6q_p(2)^3 - 84p^7q_p(2)^4 \pmod{p^8}. \quad (197)$$

Provided $p > 3$, we have

$$\sum_{k=0}^{p-1} \frac{\binom{2k}{k}^8\binom{4k}{2k}}{2^{32k}}\left(\left(7H_{2k}^{(2)} - 2H_k^{(2)}\right)Q(k) + 3624k^2 + 926k + 69\right) \equiv 69p^2 + \frac{1953}{20}p^7B_{p-5} \pmod{p^8}. \quad (198)$$

Remark 54 This is motivated by the conjectural identity

$$\sum_{k=0}^{\infty} Q(k)\frac{\binom{2k}{k}^8\binom{4k}{2k}}{2^{32k}} = \frac{2048}{\pi^4}$$

(cf. [10]).

Acknowledgements The work was supported by the National Natural Science Foundation of China (Grant No. 12371004).

References

1. G. Almkvist, C. Krattenthaler and J. Petersson, Some new series for π, Experiment. Math. **12** (2003), 441–456.
2. T. Amdeberhan and D. Zeilberger, Hypergeometric series acceleration via the WZ method, Electron. J. Combin., **4**(2) (1997), #R3.
3. R. Apéry, Irrationalité de $\zeta(2)$ et $\zeta(3)$, Astérisque **61** (1979), 11–13.
4. K. C. Au, Iterated integrals and multiple polylogarithm at algebraic arguments, arXiv:2201.01676, 2022.
5. K. C. Au, Multiple zeta values, WZ-pairs and infinite sums computations, arXiv:2212.02986, 2022.
6. J. M. Borwein and M. Chamberland, Integer powers of arcsin, Int. J. Math. Math. Sci. **2007** (2007), Article ID 19381, 10pp.
7. K. N. Boyadzhiev, Series with central binomial coefficients, Catalan numbers, and harmonic numbers, J. Integer Seq. **15** (2012), Article 12.1.7, 11pp.
8. H. Chen, Interesting series associated with central binomial coefficients, Catalan numbers and harmonic numbers, J. Integer Seq. **19** (2016), Article 16.1.5, 11pp.
9. W. Chu, Alternating series of Apéry-type for the Riemann zeta function, Contrib. Discrete Math. **15** (2020), 108–116.
10. H. Cohen and J. Guillera, Rational hypergeometric identities for $1/\pi^c$: survey and generalizations, arXiv:2101.12592, 2021.
11. S. Cooper, Ramanujan's Theta Functions, Springer, Cham, 2017.
12. J.W.L. Glaisher, On the residues of the sums of products of the first $p-1$ numbers, and their powers, to modulus p^2 or p^3, Quart. J. Math. **31** (1900), 321–353.
13. J. Guillera, About a new kind of Ramanujan-type series, Experiment. Math. **12** (2003), 507-510.
14. J. Guillera, Hypergeometric identities for 10 extended Ramanujan-type series, Ramanujan J. **15** (2008), 219–234.
15. J. Guillera, A new Ramanujan-like series for $1/\pi^2$, Ramanujan J. **26** (2011), 369–374.
16. J. Guillera, More hypergeometric identities related to Ramanujan-type series, Ramanujan J. **32** (2013), 5–22.
17. J. Guillera, WZ-proofs of "divergent" Ramanujan-type series, in: Advances in Combinatorics (eds., I. Kotsireas and E. Zima), Springer, 2013, 187–195.
18. J. Guillera and M. Rogers, Ramanujan series upside-down, J. Austral. Math. Soc. **97** (2014), 78–106.
19. V.J.W. Guo and X. Lian, Some q-congruences on double basic hypergeometric sums, J. Difference Equ. Appl. **27** (2021), 453–461.
20. Kh. Hessami Pilehrood and T. Hessami Pilehrood, Bivariate identities for values of the Hurwitz zeta function and supercongruences, Electron. J. Combin. **18** (2012), #P35, 30pp.
21. Q.-H. Hou, C. Krattenthaler and Z.-W. Sun, On q-analogues of some series for π and π^2, Proc. Amer. Math. Soc. **147** (2019), 1953–1961.
22. T. Huber, D. Schultz and D. Ye, Series for $1/\pi$ of level 20, J. Number Theory **188** (2018), 121–136.
23. L. Long, Hypergeometric evaluation identities and supercongruences, Pacific J. Math. **249** (2011), 405–418.
24. G.-S. Mao and H. Pan, Congruences corresponding to hypergeometric identities I. ${}_2F_1$ transformations, J. Math. Anal. Appl. **505** (2022), Article ID: 125527.
25. S. Ramanujan, Modular equations and approximations to π, Quart. J. Math. (Oxford) (2) **45** (1914), 350–372.
26. Z.-H. Sun, Congruences concerning Legendre polynomials, Proc. Amer. Math. Soc. **139** (2011), 1915–1929.
27. Z.-W. Sun, On congruences related to central binomial coefficients, J. Number Theory **131** (2011), 2219–2238.
28. Z.-W. Sun, Super congruences and Euler numbers, Sci. China Math. **54** (2011), 2509–2535.

29. Z.-W. Sun, Arithmetic theory of harmonic numbers, Proc. Amer. Math. Soc. **140** (2012), 415–428.
30. Z.-W. Sun, Products and sums divisible by central binomial coefficients, Electron. J. Combin. **20** (2013), no. 1, #P9, 1-14.
31. Z.-W. Sun, Supercongruences involving products of two binomial coefficients, Finite Fields Appl. **22** (2013), 24–44.
32. Z.-W. Sun, Conjectures and results on x^2 mod p^2 with $4p = x^2 + dy^2$, in: Number Theory and Related Area (eds., Y. Ouyang, C. Xing, F. Xu and P. Zhang), Adv. Lect. Math. 27, Higher Education Press and International Press, Beijing-Boston, 2013, pp. 149–197.
33. Z.-W. Sun, p-adic congruences motivated by series, J. Number Theory **134** (2014), 181–196.
34. Z.-W. Sun, New series for some special values of L-functions, Nanjing Univ. J. Math. Biquarterly **32** (2015), no. 2, 189–218.
35. Z.-W. Sun, A new series for π^3 and related congruences, Internat. J. Math. **26** (2015), no. 8, Article ID: 1550055 (23 pages).
36. Z.-W. Sun, Supercongruences motivated by e, J. Number Theory **147** (2015), 326–341.
37. Z.-W. Sun, Open conjectures on congruences, Nanjing Univ. J. Math. Biquarterly **36** (2019), no. 1, 1–99.
38. Z.-W. Sun, New series for powers of π and related congruences, Electron. Res. Arch. **28** (2020), no. 3, 1273–1342.
39. Z.-W. Sun, New type series for powers of π, J. Comb. Number Theory **12** (2020), 157–208.
40. Z.-W. Sun, New Conjectures in Number Theory and Combinatorics (in Chinese), Harbin Institute of Technology Press, 2021.
41. Z.-W. Sun, Some new series for $1/\pi$ motivated by congruences, Colloq. Math. **173** (2023), 89–109.
42. Z.-W. Sun and L.-L. Zhao, Arithmetic theory of harmonic numbers (II), Colloq. Math. **130** (2013), 67–78.
43. H. Swisher, On the supercongruence conjecture of van Hamme, Res. Math. Sci. **2** (2015), Article ID: 18.
44. L. van Hamme, Some conjectures concerning partial sums of generalized hypergeometric series, in: p-adic Functional Analysis, Lecture Notes in Pure and Appl. Math., 192, Dekker, New York, 1997, 223–236.
45. C. Wang and Z.-W. Sun, Proof of some conjectural hypergeometric supercongruences via curious identities, **505** (2022), Article ID: 125575.
46. C. Wei, On two double series for π and their q-analogues, Ramanujan J. **60** (2023), 615–625.
47. C. Wei and G. Ruan, Some double series for π and their q-ananlogues, J. Math. Anal. Appl. **537** (2024), Article ID:128309.
48. J. Wolstenholme, On certain properties of prime numbers, Quart. J. Appl. Math. **5** (1862), 35–39.
49. D. Zeilberger, Closed form (pun intended!), Contemporary Math. **143** (1993), 579–607.

Condensation and Densification for Sets of Large Diameter

Trevor D. Wooley

Abstract Consider a set of integers $\mathscr{A}$ having finite diameter X, and a system of simultaneous polynomial equations to be solved over $\mathscr{A}$. In many circumstances, it is known that the number of solutions of this system is $O(X^{\varepsilon}|\mathscr{A}|^{\theta})$ for a suitable $\theta > 0$ and any $\varepsilon > 0$. These estimates become worse than trivial when the diameter X is very large compared to $|\mathscr{A}|$, or equivalently, when the set $\mathscr{A}$ is very sparse. This motivates the problem of seeking a new set of integers $\mathscr{B}$, in a certain sense isomorphic to $\mathscr{A}$, having the property that the diameter X' of $\mathscr{B}$ is smaller than X, and at the same time the set $\mathscr{B}$ preserves the salient features of the solution set of the system of equations in question. We report on our speculative investigations concerning this problem closely associated with the topic of Freiman homomorphisms.

1 Introduction

Given a system of polynomial equations having integral coefficients, the investigation of solution sets with variables restricted to a given finite set of integers $\mathscr{A}$ is of basic interest in arithmetic combinatorics. Even for a fixed system of equations, comprehensive knowledge concerning such solution sets seems a goal far too ambitious to be realised, for the sets $\mathscr{A}$ to which variables are restricted may contain extraordinarily complicated constellations of arbitrarily large size. In this paper we seek to understand such solution sets in terms of related sets of integers, every element of which is bounded purely in terms of the cardinality of $\mathscr{A}$ and the data associated with the system of polynomials in question. Thus, in a certain sense, our conclusions derive faithful models of solution sets in arithmetic combinatorics. Any model of this type having elements of least size might reasonably be interpreted as a minimal model. The interest in such models lies in the hope that a minimal model might be more easily understood than a non-minimal and potentially very sparse counterpart.

T. D. Wooley (✉)
Department of Mathematics, Purdue University, 150 N. University Street,
West Lafayette, IN 47907-2067, USA
e-mail: twooley@purdue.edu

M. B. Nathanson (ed.), *Combinatorial and Additive Number Theory VI*,
Springer Proceedings in Mathematics & Statistics 464,
https://doi.org/10.1007/978-3-031-65064-2_22

There are close parallels with the concept of Freiman homomorphisms and isomorphisms (see [4, 5] and, for example, [14, Definition 5.21]) in the situation wherein these mappings take one set of integers to another. Although we comment further on such considerations in due course, we emphasise for now the importance for us of remaining within the same ring rather than mapping to a finite field. We express the hope that, despite our investigations on these matters being rudimentary in nature, they may provide a stimulus for further work.

Further discussion requires the introduction of some notation. We are interested primarily in finite sets of integers $\mathscr{A} \subset \mathbb{Z}$. We write A for $\mathrm{card}(\mathscr{A})$. Two notions of the size of the elements of $\mathscr{A}$ play a role in our discussions. First, there is the *diameter* of $\mathscr{A}$, namely

$$\mathrm{diam}(\mathscr{A}) = \max \mathscr{A} - \min \mathscr{A} + 1.$$

Second, there is the *enveloping radius* of $\mathscr{A}$, by which we mean

$$\mathrm{env}(\mathscr{A}) = \max\{|a| : a \in \mathscr{A}\} + 1.$$

It is apparent that $\mathrm{diam}(\mathscr{A})$ and $\mathrm{env}(\mathscr{A})$ provide very crude measures of the complexity of the set $\mathscr{A}$ in wide generality.[1] One focus of interest for us concerns translation-dilation invariant (TDI) systems of equations, such as the familiar linear equation $x_1 + x_2 = x_3 + x_4$. When considering the solutions of such an equation with $\mathbf{x} \in \mathscr{A}^4$, it is apparent that no information concerning the solution set is lost if one translates the elements of $\mathscr{A}$ by a fixed integer b to obtain a new set $\mathscr{A}' = \{a - b : a \in \mathscr{A}\}$. Consequently, there is no loss of generality in assuming that $\min \mathscr{A} = 0$, and in such circumstances it is more natural to measure the complexity of the set $\mathscr{A}$ by means of its diameter rather than its enveloping radius.

The measures $\mathrm{env}(\mathscr{A})$ and $\mathrm{diam}(\mathscr{A})$ play a critical role in the best available upper bounds for certain mean values of additive number theory. For example, when s and k are natural numbers and $\mathscr{A} \subset \mathbb{Z}$ is finite, let $J_{s,k}(\mathscr{A})$ denote the number of solutions of the system of equations

$$x_1^j + \ldots + x_s^j = x_{s+1}^j + \ldots + x_{2s}^j \quad (1 \leqslant j \leqslant k),$$

with $x_i \in \mathscr{A}$ $(1 \leqslant i \leqslant 2s)$. Likewise, when $\varphi_j \in \mathbb{Z}[t]$ $(1 \leqslant j \leqslant k)$, denote by $J_{s,k}(\mathscr{A}; \varphi)$ the number of solutions of the system of equations

$$\varphi_j(x_1) + \ldots + \varphi_j(x_s) = \varphi_j(x_{s+1}) + \ldots + \varphi_j(x_{2s}) \quad (1 \leqslant j \leqslant k),$$

with $x_i \in \mathscr{A}$ $(1 \leqslant i \leqslant 2s)$. We begin by recalling a consequence of recent work resolving the main conjecture in Vinogradov's mean value theorem.

[1] The presence of the additional term 1 in these definitions may seem mysterious, but is designed to align with a subsequent definition appropriate for the situation in algebraic number fields.

Theorem 1 *Let $\mathscr{A} \subset \mathbb{Z}$ be finite.*

(i) Suppose that $\varphi_j \in \mathbb{Z}[t]$ $(1 \leqslant j \leqslant k)$ is a system of polynomials with

$$det\left(\frac{\mathrm{d}^i \varphi_j(t)}{\mathrm{d}t^i}\right)_{1 \leqslant i,j \leqslant k} \neq 0.$$

Also, let s and k be natural numbers with $s \leqslant k(k+1)/2$. Then for each $\varepsilon > 0$, one has

$$J_{s,k}(\mathscr{A}; \boldsymbol{\varphi}) \ll env(\mathscr{A})^{\varepsilon} A^s. \tag{1}$$

(ii) For all natural numbers s and k, and each $\varepsilon > 0$, one has

$$J_{s,k}(\mathscr{A}) \ll diam(\mathscr{A})^{\varepsilon}(A^s + A^{2s-k(k+1)/2}). \tag{2}$$

In each asymptotic bound, the constants implicit in Vinogradov's notation $\ll$ may depend on ε, s, k and the coefficients of $\boldsymbol{\varphi}$.

Both of the conclusions of Theorem 1 are immediate consequences of Wooley [17, Theorem 1.1], as we explain in §9 below, and the conclusion (ii) is also immediate from the work of Bourgain, Demeter and Guth [3]. The motivating observation we wish to highlight here is that both estimates (1) and (2) are worse than trivial when the set $\mathscr{A}$ is extremely sparse. Suppose, for example, that

$$\mathrm{diam}(\mathscr{A}) \asymp \exp(\exp(A)).$$

Then the estimates (1) and (2) are inferior to the trivial bounds $J_{s,k}(\mathscr{A}; \boldsymbol{\varphi}) \leqslant A^{2s}$ and $J_{s,k}(\mathscr{A}) \leqslant A^{2s}$. This observation remains valid for the improved estimates for $J_{3,2}(\mathscr{A})$ and $J_{6,3}(\mathscr{A})$ made available, respectively in the very recent work reported in [8, 9] and [13]. It seems reasonable to speculate that the extremal situation is that in which $\mathscr{A}$ consists of A consecutive integers.

Conjecture 1 Suppose that $\mathscr{A} \subset \mathbb{Z}$ is finite and $s, k \in \mathbb{N}$. Then

$$J_{s,k}(\mathscr{A}) \leqslant J_{s,k}(\{1, 2, \ldots, A\}).$$

Moreover, for each $\varepsilon > 0$, one has

$$J_{s,k}(\mathscr{A}) \ll_{\varepsilon,s,k} A^{s+\varepsilon} + A^{2s-k(k+1)/2}. \tag{3}$$

By elimination, one finds that the bound $J_{s,k}(\mathscr{A}) \ll A^s + A^{2s-k}$ is essentially trivial. When $k \geqslant 2$, this estimate remains very far from that asserted in Conjecture 1. The only other non-trivial bound of which the author is aware is an estimate very slightly stronger than

$$J_{s,2}(\mathscr{A}) \ll A^{2s-3+2^{2-s}} \quad (s \geqslant 3)$$

due to Mudgal [12, Theorem 1.1]. With progress towards Conjecture 1 in mind, it would be desirable to have available a set $\mathscr{C}$ associated with a sparse set $\mathscr{A}$ having the property that

$$J_{s,k}(\mathscr{A}) = J_{s,k}(\mathscr{C}),$$

or even merely

$$J_{s,k}(\mathscr{A}) \ll J_{s,k}(\mathscr{C}),$$

and, moreover, having much smaller diameter than $\mathscr{A}$. Were one to have the upper bound $\mathrm{diam}(\mathscr{C}) \leqslant A^c$, for some fixed $c > 0$, for example, then the conjecture (3) would follow from Theorem 1(ii). Although such cannot be true in general, one is led to the broader problem of determining the extent to which such compressions might be achieved in practice. This problem concerning *condensations* is formalised in §2, and explored in §§3, 4 and 5. We direct the reader to Theorem 5 in Sect. 5 for our most general conclusions concerning condensations associated with systems of polynomial equations. Write c_1 and c_2 for suitable positive constants. Then a very rough idea of these conclusions can be surmised if we note, first, that we are forced to work in a number field of degree as large as $\exp_2(c_1 A)$, and second, that our condensations contain elements roughly of size $\exp_4(c_2 A)$. Here, and throughout, we use $\exp_m(\cdot)$ to denote the m-fold iterated exponential function. Thus

$$\exp_1(x) = e^x, \quad \exp_2(x) = e^{e^x},$$

and so on.

An alternate strategy for obtaining bounds of the shape (3) has a very different flavour. One might surmise that the difficulty in applying Theorem 1 to establish the estimate (3) of Conjecture 1 stems from the awkward nature of ultra-sparse sets $\mathscr{A}$ having very large diameter compared to their cardinality A. One might therefore seek to obtain a much denser set $\mathscr{D}$ associated with $\mathscr{A}$, having the property that for some large integer N one has

$$\mathrm{card}(\mathscr{D}) = (\mathrm{card}(\mathscr{A}))^N \quad \text{and} \quad J_{s,k}(\mathscr{A}) = J_{s,k}(\mathscr{D})^{1/N},$$

while at the same time $\mathrm{diam}(\mathscr{D})$ is not much larger than $\mathrm{diam}(\mathscr{A})$. This set $\mathscr{D}$ would be a much denser analogue of $\mathscr{A}$ with the potential that $\mathrm{diam}(\mathscr{D}) \leqslant (\mathrm{card}(\mathscr{D}))^c$, for some fixed $c > 0$. In these circumstances, the conjectured estimate (3) would again follow from Theorem 1(ii). We formalise this problem of *densification* in §6 and explore it in §7.

It seems worth remarking that the concepts of condensation and densification possess interpretations also in the scenario wherein sets of integers are replaced by finite sets of real numbers, or even finite subsets of a characteristic zero integral domain. We make some remarks in this direction in §8.

We view both the strategies of condensation and densification of sets of large diameter as being of interest in their own right. We emphasise that our conclusions do not achieve the level whereby application to Conjecture 1 can reasonably be envisioned.

Acknowledgements: This work was supported in its initial phases by a European Research Council Advanced Grant under the European Union's Horizon 2020 research and innovation programme via grant agreement No. 695223. The bulk of the work reported here was obtained while the author was supported by the National Science Foundation via Grant No. DMS-1854398 and DMS-2001549. The author wishes to express his gratitude to Julia Wolf for some early discussions on the topic of this paper, and to Ben Barber for discussion concerning an idea that led to Theorem 2.

We write $X \asymp Y$ when, in Vinogradov's notation, we have

$$X \ll Y \ll X.$$

Also, when θ is a real number, we write $\lceil \theta \rceil$ for the least integer m with $m \geqslant \theta$, and likewise $\lfloor \theta \rfloor$ for the largest integer m with $m \leqslant \theta$. In addition, we write $\|\theta\|$ for $\min\{|\theta - m| : m \in \mathbb{Z}\}$. Finally, we make frequent use of vector notation in the form $\mathbf{x} = (x_1, \dots, x_r)$. Here, the dimension r depends on the course of the argument.

2 Condensations of Sets

The informal introduction of condensations in §1 provides a framework insufficient for the more serious discussion on which we now embark. We begin by introducing a notion generalising that of a Freiman isomorphism.

Definition 1 Let $\mathscr{A}$ and $\mathscr{B}$ be finite sets of integers, and suppose that we are given polynomials $P_1, \dots, P_r \in \mathbb{Z}[x_1, \dots, x_s]$. We say that a bijection $\psi : \mathscr{A} \to \mathscr{B}$ is a *Freiman* **P**-*isomorphism* if, whenever one has $(x_1, \dots, x_s) \in \mathscr{A}^s$, then

$$P_i(x_1, \dots, x_s) = 0 \quad (1 \leqslant i \leqslant r)$$

if and only if

$$P_i(\psi(x_1), \dots, \psi(x_s)) = 0 \quad (1 \leqslant i \leqslant r).$$

We emphasise here that a Freiman **P**-isomorphism is specific to a particular polynomial tuple **P**, since our focus will lie on the solution set of a fixed polynomial system. This is in contrast with a similar definition given in work of Grosu (see the preamble to the statement of [7, Theorem 1.3]). Moreover, also in contrast to the latter and indeed other sources concerning Freiman isomorphisms, we shall only be interested in situations wherein both $\mathscr{A}$ and $\mathscr{B}$ lie in the same ring. This restriction permits iterative approaches in which one composes a sequence of Freiman **P**-isomorphisms $\psi_1, \psi_2, \dots, \psi_n$ to obtain a new Freiman **P**-isomorphism $\psi_n \circ \psi_{n-1} \circ \dots \circ \psi_1$.

A few words of explanation seem warranted concerning our interest in Freiman $\mathbf{P}$-isomorphisms. We are interested in the structure of the solutions of the system of polynomials

$$P_i(x_1, \dots, x_s) = 0 \quad (1 \leqslant i \leqslant r), \tag{4}$$

with $\mathbf{x} \in \mathscr{A}^s$. This is described precisely by the hypergraph $\Gamma(\mathscr{A}; \mathbf{P})$ with the elements of $\mathscr{A}$ as vertices, and having hyperedges defined by the s-tuples $\mathbf{x}$ from $\mathscr{A}^s$ satisfying the system of equations (4). With this characterisation of the structure of the solution set of (4) in mind, it is apparent that the mapping

$$\Psi : \Gamma(\mathscr{A}; \mathbf{P}) \to \Gamma(\mathscr{B}; \mathbf{P}),$$

induced by a Freiman $\mathbf{P}$-isomorphism $\psi : \mathscr{A} \to \mathscr{B}$, delivers a bijection that preserves every feature of the solution set of (4) as one replaces $\mathscr{A}$ by $\mathscr{B} = \psi(\mathscr{A})$.

Given a finite set of integers $\mathscr{A}$ and a system of polynomials $\mathbf{P} \in \mathbb{Z}[\mathbf{x}]^r$, our interest lies in finding a set $\mathscr{B}$ Freiman $\mathbf{P}$-isomorphic to $\mathscr{A}$ with $\mathscr{B}$ having elements intrinsically smaller than those of $\mathscr{A}$. Since $\Gamma(\mathscr{B}; \mathbf{P})$ is in bijective correspondence with $\Gamma(\mathscr{A}; \mathbf{P})$, one may expect that the salient features of the solution structure of the system (4) with $\mathbf{x} \in \mathscr{A}^s$ may be more easily determined by instead considering solutions $\mathbf{x} \in \mathscr{B}^s$. This motivates the next definition.

Definition 2 We say that a mapping $\psi : \mathscr{A} \to \mathscr{C}$ is a $\mathbf{P}$-*condenser of* $\mathscr{A}$ if it is a Freiman $\mathbf{P}$-isomorphism for which $\mathrm{env}(\mathscr{C}) \leqslant \mathrm{env}(\mathscr{A})$. When the latter inequality is strict, we refer to ψ as a *strict* $\mathbf{P}$-*condenser of* $\mathscr{A}$. In either case, we refer to $\mathscr{C}$ as being a $\mathbf{P}$-*condensation of* $\mathscr{A}$.

We make an observation here concerning TDI systems of polynomials $\mathbf{P}$. Suppose that $\min \mathscr{A} = b$. Then by considering the mapping $\psi : \mathscr{A} \to \mathbb{Z}$ defined by $a \mapsto a - b$, we see that $\mathscr{A}$ possesses a $\mathbf{P}$-condensation $\mathscr{B}$ with $\mathrm{env}(\mathscr{B}) = \mathrm{diam}(\mathscr{B})$.

Of particular interest are the $\mathbf{P}$-condensations $\mathscr{B}$ of $\mathscr{A}$ distinguished by the property that $\mathrm{env}(\mathscr{B})$ is minimal.

Definition 3 The $\mathbf{P}$-*essential enveloping radius* of $\mathscr{A}$ is

$$\mathrm{env}^*(\mathscr{A}; \mathbf{P}) = \min\{\mathrm{env}(\psi(\mathscr{A})) : \psi \text{ is a } \mathbf{P}\text{-condenser of } \mathscr{A}\},$$

and the $\mathbf{P}$-*essential diameter of* $\mathscr{A}$ is

$$\mathrm{diam}^*(\mathscr{A}; \mathbf{P}) = \min\{\mathrm{diam}(\psi(\mathscr{A})) : \psi \text{ is a } \mathbf{P}\text{-condenser of } \mathscr{A}\}.$$

The notion of the $\mathbf{P}$-essential enveloping radius of a finite set $\mathscr{A} \subset \mathbb{Z}$ provides a measure of the complexity of $\mathscr{A}$ with respect to the system of equations (4), for it describes the minimal footprint of a set $\mathscr{B}$ for which the hypergraph $\Gamma(\mathscr{B}; \mathbf{P})$ associated with the solution set faithfully describes that of interest, namely $\Gamma(\mathscr{A}; \mathbf{P})$.

In general, the sharpest conclusions concerning $\mathrm{env}^*(\mathscr{A}; \mathbf{P})$ of which the author is aware are the trivial ones recorded in the following theorem.

Theorem 2 *Let* $\mathbf{P} \in \mathbb{Z}[x_1, \ldots, x_s]^r$ *be a polynomial system, and let* $\mathscr{A} \subset \mathbb{Z}$ *be a finite set of integers having cardinality* A*. Then one has*

$$\lceil (A+1)/2 \rceil \leqslant env^*(\mathscr{A}; \mathbf{P}) \ll_{A,\mathbf{P}} 1$$

and

$$A \leqslant diam^*(\mathscr{A}; \mathbf{P}) \ll_{A,\mathbf{P}} 1.$$

We emphasise here that the upper bounds recorded in this theorem indicate that the $\mathbf{P}$-essential enveloping radius (respectively, the $\mathbf{P}$-essential diameter) of $\mathscr{A}$ depends at most on $A = \mathrm{card}(\mathscr{A})$ and the polynomials comprising $\mathbf{P}$, but not on the specific identity of the elements of $\mathscr{A}$. A few moments of reflection should disabuse the puzzled reader that this conclusion might be in any sense non-trivial.

Proof (of Theorem 2) The solution set of the polynomial system

$$P_i(x_1, \ldots, x_s) = 0 \quad (1 \leqslant i \leqslant r),$$

with $\mathbf{x} \in \mathscr{A}^s$, defines the hypergraph $\Gamma(\mathscr{A}; \mathbf{P})$. Let $\mathscr{C} \subset \mathbb{Z}$ be any set of integers of cardinality A having smallest enveloping radius for which $\Gamma(\mathscr{C}; \mathbf{P})$ is isomorphic to $\Gamma(\mathscr{A}; \mathbf{P})$. Denote by ψ the mapping from $\mathscr{A}$ to $\mathscr{C}$ induced by this hypergraph isomorphism, and note that one possibility is that ψ is the identity mapping. The definitions of $\Gamma(\mathscr{C}; \mathbf{P})$ and $\Gamma(\mathscr{A}; \mathbf{P})$ ensure that $\psi : \mathscr{A} \to \mathscr{C}$ is a bijection satisfying the property that whenever $(x_1, \ldots, x_s) \in \mathscr{A}^s$, then

$$P_i(x_1, \ldots, x_s) = 0 \quad (1 \leqslant i \leqslant r)$$

if and only if

$$P_i(\psi(x_1), \ldots, \psi(x_s)) = 0 \quad (1 \leqslant i \leqslant r).$$

Hence, we see that ψ is a Freiman $\mathbf{P}$-isomorphism and also a $\mathbf{P}$-condenser of $\mathscr{A}$ with $\mathscr{C} = \psi(\mathscr{A})$.

The set of all hypergraphs on A vertices with hyperedges defined by s-tuples of vertices is finite in number. Indeed, the number of such hypergraphs depends at most on s and A. Thus, since $\mathscr{C}$ depends at most on the hypergraph isomorphism class of $\Gamma(\mathscr{A}; \mathbf{P})$ and the polynomial system $\mathbf{P}$, one sees that $\mathrm{env}(\mathscr{C})$ depends at most on s, $\mathbf{P}$ and A. Since $\mathscr{C} = \psi(\mathscr{A})$ with ψ a $\mathbf{P}$-condenser of $\mathscr{A}$, it follows that $\mathrm{env}^*(\mathscr{A}; \mathbf{P}) \ll_{A,\mathbf{P}} 1$. A similar conclusion is apparent also for $\mathrm{diam}^*(\mathscr{A}; \mathbf{P})$ by arguing mutatis mutandis.

The lower bounds $\mathrm{env}^*(\mathscr{A}; \mathbf{P}) \geqslant \lceil (A+1)/2 \rceil$ and $\mathrm{diam}^*(\mathscr{A}; \mathbf{P}) \geqslant A$ follow by considering sets $\mathscr{A}$ containing A consecutive integers.

3 Condensations for Linear Systems of Equations

There is one class of polynomial systems for which the quantitative aspects of condensations are explicit, and for which the underlying methods possess familiar themes. Thus, the analysis of systems of linear polynomials $\mathbf{P}(\mathbf{x})$ is both simple and instructive, and serves as a warm-up for the analysis of the next two sections concerning polynomial systems. We focus in this section on such linear systems in order to motivate the more general discussion of the next section.

In order to fix ideas, suppose that $s \geqslant 2$, $r \geqslant 1$, and for $1 \leqslant i \leqslant r$ fix $b_i \in \mathbb{Z}$ and $c_{ij} \in \mathbb{Z}$ $(1 \leqslant j \leqslant s)$. We ignore the trivial situation in which for some index i one has $c_{ij} = 0$ for $1 \leqslant j \leqslant s$, since this will correspond to a case in which r is smaller. The system of polynomials of interest to us in this section is

$$P_i(\mathbf{x}) = \sum_{j=1}^{s} c_{ij}x_j - b_i \quad (1 \leqslant i \leqslant r).$$

When $\mathscr{A} \subset \mathbb{Z}$ is a finite set of integers, we write $S(\mathscr{A}; \mathbf{P})$ for the set of solutions of the system of equations $P_i(\mathbf{x}) = 0$ $(1 \leqslant i \leqslant r)$, with $\mathbf{x} \in \mathscr{A}^s$. Also, we define the integer $\Lambda = \Lambda(\mathbf{c}, \mathbf{b})$ by putting

$$\Lambda = \max_{1 \leqslant i \leqslant r} \left(|b_i| + \sum_{j=1}^{s} |c_{ij}| \right).$$

Thus, the quantity Λ provides a measure of the height of the coefficient matrix defining $\mathbf{P}$. Finally, it is convenient both here and elsewhere to introduce an integer that encapsulates both distinctness of the elements of $\mathscr{A}$, and also whether or not an s-tuple $\mathbf{a}$ lies in $S(\mathscr{A}; \mathbf{P})$. Thus, we define

$$\Upsilon = \Bigg(\prod_{\substack{a_1, a_2 \in \mathscr{A} \\ a_1 \neq a_2}} |a_1 - a_2| \Bigg) \Bigg(\prod_{\substack{\mathbf{a} \in \mathscr{A}^s \\ \mathbf{a} \notin S(\mathscr{A}; \mathbf{P})}} \sum_{i=1}^{r} |P_i(\mathbf{a})| \Bigg). \tag{5}$$

Theorem 3 *Consider a system* $\mathbf{P}$ *of linear polynomials as described in the preamble, and consider a finite set of integers* $\mathscr{A}$. *Then provided that* $A = \mathrm{card}(\mathscr{A})$ *is sufficiently large in terms of* r *and* s, *one has*

$$\mathrm{env}^*(\mathscr{A}; \mathbf{P}) < \exp\left(3(\Lambda + 1)^A\right). \tag{6}$$

Proof A moment of reflection reveals that there is no loss of generality in supposing that $\Lambda \geqslant 2$ and $s \geqslant 2$. Write $X = \mathrm{env}(\mathscr{A}) - 1$. Our strategy is to find an integer h with $\Lambda < h < 2X$ having the following three properties:

(i) when $a_1, a_2 \in \mathscr{A}$ satisfy $a_1 \neq a_2$, then $a_1 \not\equiv a_2 \pmod{h}$;
(ii) when $\mathbf{a} \notin S(\mathscr{A}; \mathbf{P})$, then there is an index i with $1 \leqslant i \leqslant r$ for which one has $P_i(\mathbf{a}) \not\equiv 0 \pmod{h}$;
(iii) for every $a \in \mathscr{A}$, one has $\|a/h\| < 1/\Lambda$.

If such an integer h can be found, then we may define the map $\psi : \mathscr{A} \to \mathbb{Z}$ by putting

$$\psi(a) = [a \pmod{h}],$$

where $[a \pmod{h}]$ denotes the numerically least residue of a modulo h. To be clear, the numerically least residue of a modulo h is the integer m with $-h/2 < m \leqslant h/2$ for which $a \equiv m \pmod{h}$. Property (i) then ensures that the set $\mathscr{B} = \psi(\mathscr{A})$ is in bijective correspondence with $\mathscr{A}$. Also, property (ii) ensures that whenever $\mathbf{a} \notin S(\mathscr{A}; \mathbf{P})$, then for some index i with $1 \leqslant i \leqslant r$, one has

$$P_i(\psi(\mathbf{a})) \equiv P_i(\mathbf{a}) \not\equiv 0 \pmod{h},$$

whence $P_i(\psi(\mathbf{a})) \neq 0$. However, when $\mathbf{a} \in S(\mathscr{A}; \mathbf{P})$, one has

$$P_i(\psi(\mathbf{a})) \equiv P_i(\mathbf{a}) \equiv 0 \pmod{h} \quad (1 \leqslant i \leqslant r).$$

At the same time, in view of property (iii), one has

$$|P_i(\psi(\mathbf{a}))| < \left(|b_i| + \sum_{j=1}^{s} |c_{ij}| \right) \frac{h}{\Lambda} \leqslant h,$$

whence $P_i(\psi(\mathbf{a})) = 0$ $(1 \leqslant i \leqslant r)$. We therefore infer that the map ψ is a Freiman $\mathbf{P}$-isomorphism from $\mathscr{A}$ to $\mathscr{B}$. Consequently, since

$$\mathrm{env}(\mathscr{B}) - 1 \leqslant h/2 < X = \mathrm{env}(\mathscr{A}) - 1,$$

we have confirmed the existence of a strict $\mathbf{P}$-condenser of $\mathscr{A}$.

We establish the existence of a suitable integer h by modifying very slightly an argument employed by Baker and Harman (see [1, Proposition 1]). Recall the definition (5) of the integer Υ. A crude estimate delivers the bounds

$$1 \leqslant \Upsilon \leqslant (2X)^{A^2-A}(r\Lambda X)^{A^s} \leqslant \tfrac{1}{3}(r\Lambda X)^{2A^s}. \tag{7}$$

The number of prime divisors of Υ exceeding $\log(3\Upsilon)$ cannot exceed

$$\frac{\log(3\Upsilon)}{\log\log(3\Upsilon)}.$$

Thus, an application of the prime number theorem reveals that whenever Y is large and $Y \geqslant 2\log(3\Upsilon)$, then in any interval $(Y, 2Y)$, there exists a prime π with $\pi \nmid \Upsilon$. It therefore follows from (7) that we may choose a prime π with $\pi \nmid \Upsilon$ for which

$$\pi < 4\log(3\Upsilon) \leqslant 8A^s \log(r\Lambda X). \tag{8}$$

We put

$$L = \mathrm{lcm}[1, 2, \ldots, (\Lambda + 1)^A]$$

and note that an elementary application of the prime number theorem ensures that, when A is large, one has $L \leqslant \exp(2(\Lambda + 1)^A)$. Next, by applying the multidimensional version of Dirichlet's box principle to the real numbers $a/(\pi L)$ $(a \in \mathscr{A})$, it follows that for some $\rho \in \mathbb{N}$ satisfying $1 \leqslant \rho \leqslant (\Lambda + 1)^A$, one has

$$\left\| \frac{\rho a}{\pi L} \right\| \leqslant \frac{1}{\Lambda + 1} \quad (a \in \mathscr{A}).$$

Since $\rho | L$, we may therefore define the integer $h = \pi L/\rho$, and then we see that

$$\|a/h\| \leqslant (\Lambda + 1)^{-1} \quad (a \in \mathscr{A}). \tag{9}$$

By construction, the integer h is divisible by π. We now exploit the fact that $\pi \nmid \Upsilon$ using the definition (5). Thus, when $a_1, a_2 \in \mathscr{A}$ satisfy $a_1 \neq a_2$, one has $a_1 \not\equiv a_2 \pmod{\pi}$. Moreover, when $\mathbf{a} \in \mathscr{A}^s$ one sees that $P_i(\mathbf{a}) \equiv 0 \pmod{\pi}$ $(1 \leqslant i \leqslant r)$ if and only if $\mathbf{a} \in S(\mathscr{A}; \mathbf{P})$. In combination with (9), therefore, it is apparent that the properties (i), (ii) and (iii) above all hold for the integer h that we have constructed. In particular, the map $\psi : \mathscr{A} \to \mathbb{Z}$ defined by putting $\psi(a) = [a \pmod{h}]$ gives a Freiman $\mathbf{P}$-isomorphism from $\mathscr{A}$ to $\mathscr{B} = \psi(\mathscr{A})$ in which, on recalling (8), we see that

$$\mathrm{env}(\mathscr{B}) - 1 \leqslant h/2 \leqslant \pi L/2 < 4A^s \log(r\Lambda X) \exp(2(\Lambda + 1)^A).$$

We may summarise our deliberations thus far in the following form. Whenever $\mathscr{A}$ is a finite subset of $\mathbb{Z}$ with cardinality A and enveloping radius $X + 1$, then $\mathscr{A}$ possesses a $\mathbf{P}$-condensation $\mathscr{A}_1 = \psi(\mathscr{A})$ with enveloping radius at most $X_1 + 1$, where

$$X_1 = 4A^s \log(r\Lambda X) \exp\left(2(\Lambda + 1)^A\right).$$

When A is large in terms of r and s, and $X + 1 \geqslant \exp(3(\Lambda + 1)^A)$, we have

$$4A^s (\log(r\Lambda) + \log X) \leqslant \tfrac{1}{2} X^{1/3}.$$

Thus, under the same conditions on A and X, it follows that

$$X_1 \leqslant \tfrac{1}{2} X^{1/3} (X + 1)^{2/3} < X,$$

and consequently one has $\mathrm{env}(\mathscr{A}_1) < \mathrm{env}(\mathscr{A})$. Provided that

$$\mathrm{env}(\mathscr{A}_1) \geqslant \exp\left(3(\Lambda + 1)^A\right),$$

we may apply this process again, next showing that $\mathscr{A}_1$ has a $\mathbf{P}$-condensation $\mathscr{A}_2$ with $\mathrm{env}(\mathscr{A}_2) < \mathrm{env}(\mathscr{A}_1)$. Since $\mathbf{P}$-condensers may be composed, it follows that $\mathscr{A}$ also has a $\mathbf{P}$-condensation $\mathscr{A}_2$ with enveloping radius smaller than $\mathrm{env}(\mathscr{A}_1)$. By iterating this process repeatedly, with each iteration reducing the enveloping radius of the condensation of $\mathscr{A}$, we ultimately obtain a condensation $\mathscr{A}^*$ of $\mathscr{A}$ for which

$$\mathrm{env}(\mathscr{A}^*) < \exp\left(3(\Lambda+1)^A\right).$$

This establishes the bound (6), and the proof of the theorem is complete.

In the situation in which $\mathbf{b} = \mathbf{0}$, so that all of the linear polynomials are homogeneous, there is more freedom to apply changes of variable to advantage. Here the arguments are reminiscent of those employed in the proof of versions of Freiman's theorem (see, for example, the proof of [2, Theorem 2.1]).

Theorem 4 *Consider a system* $\mathbf{P}$ *of linear polynomials with* $\mathbf{b} = \mathbf{0}$*, as described in the preamble to the statement of Theorem 3. Also, consider a finite set of integers* $\mathscr{A}$*. Then provided that* $A = \mathrm{card}(\mathscr{A})$ *is sufficiently large in terms of* r *and* s*, one has*

$$env^*(\mathscr{A}; \mathbf{P}) \leqslant (\Lambda+1)^A. \tag{10}$$

Proof We proceed much as in the proof of Theorem 3, though with a twist en route. First, writing $X = \mathrm{env}(\mathscr{A}) - 1$ and defining the integer Υ as in (5), we again obtain the bound (7), and conclude that there exists a prime number π with $\pi \nmid \Upsilon$ satisfying the property that

$$(\Lambda+1)^A < \pi < 2\max\{(\Lambda+1)^A, 4A^s \log(r\Lambda X)\}. \tag{11}$$

By applying the multidimensional version of Dirichlet's approximation theorem to the real numbers a/π $(a \in \mathscr{A})$, it follows that for some $\rho \in \mathbb{N}$ with $\rho \leqslant (\Lambda+1)^A$, one has

$$\left\| \frac{\rho a}{\pi} \right\| \leqslant \frac{1}{\Lambda+1} \quad (a \in \mathscr{A}).$$

We fix any such integer ρ, noting that since $\pi > \rho$, one has $(\rho, \pi) = 1$. It follows that:

(i) whenever $a_1, a_2 \in \mathscr{A}$ satisfy $a_1 \neq a_2$, then $\rho a_1 \not\equiv \rho a_2 \pmod{\pi}$;
(ii) whenever $\mathbf{a} \notin S(\mathscr{A}; \mathbf{P})$, then there is an index i with $1 \leqslant i \leqslant r$ for which one has $P_i(\rho\mathbf{a}) \not\equiv 0 \pmod{\pi}$;
(iii) for every $a \in \mathscr{A}$, one has $\|\rho a/\pi\| < 1/\Lambda$.

We now define the map $\psi : \mathscr{A} \to \mathbb{Z}$ by putting

$$\psi(a) = [\rho a \pmod{\pi}].$$

Property (i) then ensures that the set $\mathscr{C} = \psi(\mathscr{A})$ is in bijective correspondence with $\mathscr{A}$. Also, property (ii) ensures that whenever $\mathbf{a} \notin S(\mathscr{A}; \mathbf{P})$, then for some index i with $1 \leqslant i \leqslant r$, one has

$$P_i(\psi(\mathbf{a})) \equiv \rho P_i(\mathbf{a}) \not\equiv 0 \pmod{\pi},$$

whence $P_i(\psi(\mathbf{a})) \neq 0$. However, when $\mathbf{a} \in S(\mathscr{A}; \mathbf{P})$, one has

$$P_i(\psi(\mathbf{a})) \equiv \rho P_i(\mathbf{a}) \equiv 0 \pmod{\pi} \quad (1 \leqslant i \leqslant r).$$

At the same time, in view of property (iii), one has

$$|P_i(\psi(\mathbf{a}))| < \frac{\pi}{\Lambda} \sum_{j=1}^{s} |c_{ij}| \leqslant \pi,$$

whence $P_i(\psi(\mathbf{a})) = 0$ $(1 \leqslant i \leqslant r)$. We therefore infer that ψ is a Freiman **P**-isomorphism from $\mathscr{A}$ to $\mathscr{C}$. Consequently, provided that $\pi < 2X$, we see that

$$\text{env}(\mathscr{C}) - 1 \leqslant \pi/2 < X = \text{env}(\mathscr{A}) - 1,$$

and thus we have established the existence of a strict **P**-condenser of $\mathscr{A}$.

Notice here that in view of (11), one has

$$\text{env}(\mathscr{C}) - 1 \leqslant \pi/2 < \max\{(\Lambda + 1)^A, 4A^s \log(r\Lambda X)\},$$

and we again have available an iterative process for reducing the enveloping radius of **P**-condensations of $\mathscr{A}$ similar to that made available in the concluding stages of the proof of Theorem 3. When A is sufficiently large in terms of r and s, and $X \geqslant (\Lambda + 1)^A$, we have

$$4A^s(\log(r\Lambda) + \log X) < X.$$

In this instance, therefore, under the same conditions on A and X, we discern from (11) that $\pi < 2X$ and hence that $\text{env}(\mathscr{C}) < \text{env}(\mathscr{A})$. Thus, our iteration continues until we obtain a **P**-condensation $\mathscr{A}^*$ of $\mathscr{A}$ for which $\text{env}(\mathscr{A}^*) - 1 < (\Lambda + 1)^A$. This confirms the bound (10), and thus the proof of the theorem is complete.

The problem of obtaining lower bounds on $\text{env}^*(\mathscr{A}; \mathbf{P})$ has been considered in special cases such as that in which the system **P** consists of the single polynomial $x_1 + x_2 = x_3 + x_4$. Here, it is apparent that the set $\mathscr{A} = \{0, 1, 2, 4, \ldots, 2^{A-2}\}$ cannot be condensed into a fundamentally smaller set (see [10, § 5]). Thus, in this special case, one has $\text{env}^*(\mathscr{A}; \mathbf{P}) \gg 2^A$, and it is apparent that the upper bound on $\text{env}^*(\mathscr{A}; \mathbf{P})$ provided by Theorem 4 cannot in general be replaced by a quantity subexponential in A.

4 Condensations for Non-linear Systems of Equations, I

Equipped with the discussion of §3 applicable to linear equations, we move on in this section to consider the corresponding situation for the solubility of polynomial equations of degree exceeding one over a finite subset $\mathscr{A}$ of the integers. Here, any attempt to merely mimic the proofs of Theorems 3 and 4 must plainly be abandoned. Suppose, for example, that we seek to analyse the solubility with $\mathbf{x} \in \mathscr{A}^4$ of the equation $x_1^2 + x_2^2 = x_3^2 + x_4^2$ by utilising the map $\psi : \mathbb{Z} \to \mathbb{Z}/h\mathbb{Z}$ defined by putting $\psi(a) = [a \pmod{h}]$ for a suitable positive integer h. The optimistic notion that the congruence

$$\psi(a_1)^2 + \psi(a_2)^2 \equiv \psi(a_3)^2 + \psi(a_4)^2 \pmod{h}$$

might imply that

$$\psi(a_1)^2 + \psi(a_2)^2 = \psi(a_3)^2 + \psi(a_4)^2$$

would seem to demand a choice for h ensuring that $|\psi(a)| < \frac{1}{4}h^{1/2}$ for all $a \in \mathscr{A}$. Such an eventuality cannot reasonably be countenanced for any but very special sets $\mathscr{A}$. However, a means of mapping subsets of finite fields into subsets of $\mathbb{C}$, while preserving associated solution structures, has been made available in work of Grosu [7]. With care, this approach can be wrought to yield a **P**-condenser of sorts in the non-linear situation currently of interest to us.

The discussion of this section and the next requires the introduction of notions somewhat more flexible than those defined in §2. We have in mind now that the sets of integers under consideration will be replaced by elements of some algebraic number field. For the sake of simplicity, we shall restrict the polynomial equations under consideration to have coefficients lying in $\mathbb{Z}$, though it is straightforward to relax this condition so that the coefficient ring $\mathbb{Z}$ is replaced by the ring of integers from some other number field.

Definition 4 Let $\mathscr{A}$ and $\mathscr{B}$ be finite sets of algebraic numbers. Suppose that $P_1, \dots, P_r \in \mathbb{Z}[x_1, \dots, x_s]$. We say that a bijection $\psi : \mathscr{A} \to \mathscr{B}$ is an *algebraic Freiman* ***P****-isomorphism* if, whenever $(x_1 \dots, x_s) \in \mathscr{A}^s$, then

$$P_i(x_1, \dots, x_s) = 0 \quad (1 \leqslant i \leqslant r)$$

if and only if

$$P_i(\psi(x_1), \dots, \psi(x_s)) = 0 \quad (1 \leqslant i \leqslant r).$$

Notice here that we have not insisted that $\mathscr{A}$ and $\mathscr{B}$ lie in the same number field. Thus, for example, it is possible that one has $\mathscr{A} \subset \mathbb{Z}[\sqrt{-1}]$ and $\mathscr{B} \subset \mathbb{Z}[\sqrt[3]{2}]$. Given this flexibility in the choice of the image set, an appropriate definition of the analogue of a **P**-condenser takes some care. First, when $\mathscr{C} = \{c_1, \dots, c_n\}$ is a finite set of algebraic numbers, we define the number field $K = \mathbb{Q}(\mathscr{C})$ by putting $K = \mathbb{Q}(c_1, \dots, c_n)$. We then put

$$d(\mathscr{C}) = [K : \mathbb{Q}] \quad \text{and} \quad D(\mathscr{C}) = \mathrm{Disc}(K : \mathbb{Q}).$$

Rather than become entangled with a coordinate basis for K over $\mathbb{Q}$, we instead work with minimal polynomials associated with each element $c \in \mathscr{C}$. Here, by the minimal polynomial $m_c \in \mathbb{Z}[x]$ of $c \in \mathscr{C}$, we mean the irreducible polynomial in $\mathbb{Z}[x]$ with content 1 and positive leading coefficient satisfying the condition that $m_c(c) = 0$. Note that if m_c has leading coefficient l, then $l^{-1}m_c$ is the conventional minimal polynomial of c over $\mathbb{Q}$. Given a polynomial $f \in \mathbb{Z}[x]$ with $f(x) = f_0 + f_1 x + \ldots + f_d x^d$, we define

$$\|f\|_q = \left(|f_0|^q + |f_1|^q + \ldots + |f_d|^q\right)^{1/q} \quad (q = 1, 2).$$

Then, as a measure of the enveloping radius of the set $\mathscr{C}$, we work with the *algebraic enveloping radius*

$$\mathrm{Env}(\mathscr{C}) = \max\{\|m_c\|_1 : c \in \mathscr{C}\}.$$

If $\mathscr{C}$ is a set of rational integers, then it is apparent that $\mathrm{Env}(\mathscr{C}) = \mathrm{env}(\mathscr{C})$. Notice that $\mathrm{Env}(\mathscr{C})$ is independent of any particular coordinate basis for $\mathbb{Q}(\mathscr{C})$.

Our goal is now to map a set of algebraic numbers $\mathscr{A}$, having a large algebraic enveloping radius $\mathrm{Env}(\mathscr{A})$, to a new set $\mathscr{B}$ having smaller algebraic enveloping radius $\mathrm{Env}(\mathscr{B})$, via an algebraic Freiman $\mathbf{P}$-isomorphism $\psi : \mathscr{A} \to \mathscr{B}$. In this way, the size of the elements of $\mathscr{B}$ is morally speaking smaller than the corresponding size of the elements of $\mathscr{A}$, and yet $\mathscr{B}$ preserves the salient features of the solubility of the system $\mathbf{P}(\mathbf{x}) = \mathbf{0}$ exhibited by $\mathscr{A}$. This objective motivates the following analogues of Definitions 2 and 3.

Definition 5 We say that a mapping $\psi : \mathscr{A} \to \mathscr{C}$ is a *d-algebraic* $\mathbf{P}$*-condenser* of $\mathscr{A}$ if it is an algebraic Freiman $\mathbf{P}$-isomorphism having the property that

$$\mathrm{Env}(\mathscr{C}) \leqslant \mathrm{Env}(\mathscr{A}) \quad \text{and} \quad d(\mathscr{C}) = d.$$

When the inequality here is strict, we refer to ψ as a *strict d-algebraic* $\mathbf{P}$*-condenser* of $\mathscr{A}$. In either case, we refer to $\mathscr{C}$ as being a *d-algebraic condensation* of $\mathscr{A}$.

Definition 6 Let $\mathscr{A}$ be a finite set of algebraic numbers, and denote by $\Psi_{\mathbf{P}}(\delta)$ the set of all d-algebraic $\mathbf{P}$-condensers of $\mathscr{A}$ with $d \leqslant \delta$. Then the δ*-limited* $\mathbf{P}$*-essential enveloping radius* of a finite set $\mathscr{A}$ of algebraic integers is

$$\mathrm{Env}^*_\delta(\mathscr{A}; \mathbf{P}) = \min\{\mathrm{Env}(\psi(\mathscr{A})) : \psi \in \Psi_{\mathbf{P}}(\delta)\}.$$

We are now equipped to describe, in broad and rough terms, our strategy for condensing algebraic sets $\mathscr{A}$ into sets $\mathscr{C}$ establishing that $\mathrm{Env}^*_\delta(\mathscr{A}; \mathbf{P})$ is bounded purely in terms of $|\mathscr{A}|$ and $\mathbf{P}$, while at the same time maintaining δ to be likewise bounded purely in terms of $|\mathscr{A}|$ and $\mathbf{P}$. The details of this process will be the subject of the next section.

Let $\mathscr{A}$ be a finite set of algebraic integers with $\mathrm{card}(\mathscr{A}) = A$. In fact we shall need to consider finite sets of algebraic numbers, and this generates additional complications relative to the situation where the algebraic numbers are in fact algebraic integers. However, this simplified case allows us to sketch out the necessary argument. Put $X = \mathrm{Env}(\mathscr{A})$ and suppose that $d(\mathscr{A}) = d_0$, with d_0 bounded above by some absolute constant.

Our first step is to seek a rational prime π having the property that $\pi \nmid (a_1 - a_2)$ for any distinct elements $a_1, a_2 \in \mathscr{A}$, and also $\pi \nmid P_i(\mathbf{a})$ for any $\mathbf{a} \in \mathscr{A}^s$ with $P_i(\mathbf{a}) \neq 0$

$(1 \leqslant i \leqslant r)$. For the sake of concreteness, we shall in fact interpret these divisibility relations by taking norms of the algebraic numbers in question. It is apparent that an argument similar to that applied in §3 will deliver such a prime to us with $\pi \ll_{A,\mathbf{P}} \log X$. Unfortunately, we must ensure that this prime number behaves congenially with respect to the number field $K_0 = \mathbb{Q}(\mathscr{A})$, because we intend subsequently to consider the set $\mathscr{A}$ modulo π to be a subset of the finite field $\mathbb{F}_\pi$, and thence consider the associated system of congruences

$$P_i(\mathbf{a}) \equiv 0 \pmod{\pi} \quad (1 \leqslant i \leqslant r).$$

We therefore seek an appropriately sized prime π having the property that a certain minimal polynomial associated with $\mathscr{A}$ splits into linear factors over $\mathbb{F}_\pi$. If we assume a certain Generalised Riemann Hypothesis, then it follows from an appropriate version of the Chebotarev density theorem that such a prime can be shown to exist with $\pi \ll_{A,\mathbf{P}} (\log X)^3$.

Having obtained a prime π with the properties just described, our second step is to apply the argument of Grosu [7] to rectify the set $\mathscr{A} \pmod{\pi}$. Provided that π is chosen large enough in terms of A and $\mathbf{P}$, this argument shows that the set $\mathscr{A} \pmod{\pi}$ can be mapped to a new set $\mathscr{B}$ algebraic Freiman $\mathbf{P}$-isomorphic to $\mathscr{A}$, and having the property that

$$\mathrm{Env}(\mathscr{B}) \ll_{A,\mathbf{P}} (\log X)^{c(A,\mathbf{P})} \quad \text{and} \quad d(\mathscr{B}) \ll_{A,\mathbf{P}} 1.$$

Here, the positive number $c(A, \mathbf{P})$ depends at most on A and $\mathbf{P}$. Notice here that, whilst the set $\mathscr{A}$ has elements of size roughly X, the elements of $\mathscr{B}$ have size roughly a power of $\log X$. This reduction in size is crucial to our condensation argument.

By iterating these two steps sufficiently many times, much as was done in §3 in the simpler linear case in the proofs of Theorems 3 and 4, we ultimately obtain a set $\mathscr{C}$ algebraically Freiman $\mathbf{P}$-isomorphic to $\mathscr{A}$, and satisfying the property that

$$\mathrm{Env}(\mathscr{C}) \ll_{A,\mathbf{P}} 1 \quad \text{and} \quad d(\mathscr{B}) \ll_{A,\mathbf{P}} 1.$$

All that remains is to take care in controlling the behaviour in these results of the implicit constants depending on A and $\mathbf{P}$.

We describe the details of the argument just outlined in the next section. For now, it suffices to say that in the setting countenanced in the above discussion, we are able to show that, subject to the validity of the Generalised Riemann Hypothesis for all Dedekind zeta functions, there is a set of algebraic numbers $\mathscr{C}$ algebraic Freiman $\mathbf{P}$-isomorphic to $\mathscr{A}$ with

$$d(\mathscr{C}) \leqslant \exp_2(\kappa_1 A) \quad \text{and} \quad \mathrm{Env}(\mathscr{C}) \leqslant \exp_4(\kappa_2 A),$$

where κ_1 and κ_2 are positive numbers depending at most on $\mathbf{P}$.

5 Condensations for Non-linear Systems of Equations, II

Let us now put the plan of the previous section into action. It is worth stressing that our bounds will be extraordinarily weak. In consequence, it makes sense to avoid stress on detailed bounds, but instead opt for estimates somewhat weaker than might be achieved with greater attention to detail, but ones nonetheless simple to state in suitable notation.

Let s and r be natural numbers with $s \geqslant 2$, and for $1 \leqslant i \leqslant r$ consider fixed polynomials $P_i = P_i(\mathbf{x}) \in \mathbb{Z}[x_1, \ldots, x_s]$ of respective degrees t_i. We write $\| P_i \|_1$ for the sum of the absolute values of the coefficients of P_i, and we suppose that $t_i \leqslant t$ and $\| P_i \|_1 \leqslant k$ for $1 \leqslant i \leqslant r$. Then, in the sense of Grosu [7], the polynomial system $\mathbf{P}$ is (k, t)-bounded. Note that, in view of our work in §3, there is no loss of generality in supposing that $t \geqslant 2$. Our initial discussion will be focused on establishing the iterative step described in the previous section. Suppose then that $\mathscr{A}$ is a set of algebraic numbers with $\text{card}(\mathscr{A}) = A \geqslant 2$. Our discussion will be simplified by introducing the function

$$\nu(n) = (2t)^{(2t)^{2^n}}. \tag{12}$$

To avoid any potential ambiguity, we note that this is a 3-fold iterated exponential function of n. Equipped with this notation, it is convenient to suppose that

$$d(\mathscr{A}) \leqslant (2t)^{2^A} \quad \text{and} \quad \text{Env}(\mathscr{A}) = X. \tag{13}$$

In most familiar applications of algebraic number theory, analytic number theorists are used to working with a fixed number field K wherein the degree and discriminant are well-controlled. Unfortunately for us, we require discussions of field extensions of $\mathbb{Q}$ with enormous degree and discriminant, and so we are forced to pay attention to details that in normal circumstances would not delay our argument. Our initial focus lies on the non-zero algebraic number

$$\Upsilon(\mathscr{A}) = \Bigg(\prod_{\substack{a_1, a_2 \in \mathscr{A} \\ a_1 \neq a_2}} (a_1 - a_2) \Bigg) \Bigg(\prod_{i=1}^{r} \prod_{\substack{\mathbf{a} \in \mathscr{A}^s \\ P_i(\mathbf{a}) \neq 0}} P_i(\mathbf{a}) \Bigg). \tag{14}$$

We seek a rational prime number π with properties associated to $\Upsilon(\mathscr{A})$ outlined in the discussion of the previous section. In preparation for our application of the Chebotarev density theorem, we discuss the Galois closure K^c of $K = \mathbb{Q}(\mathscr{A})$, and some of its properties.

Lemma 1 *One has* $[K^c : \mathbb{Q}] \leqslant \nu(A + 1)$.

Proof By the primitive element theorem, there is some algebraic number $\theta \in K$ for which $K = \mathbb{Q}(\theta)$. The minimal polynomial m_θ of θ over $\mathbb{Z}$ has degree $[K : \mathbb{Q}] = d(\mathscr{A}) \leqslant (2t)^{2^A}$. Thus, the splitting field of m_θ, which contains K^c, has degree at most $d(\mathscr{A})!$. We therefore conclude that

$$[K^c : \mathbb{Q}] \leqslant d(\mathscr{A})^{d(\mathscr{A})} \leqslant (2t)^B,$$

where

$$B = 2^A (2t)^{2^A} \leqslant (2t)^{A+2^A} \leqslant (2t)^{2^{A+1}}.$$

Thus, on recalling the notation (12), we find that $[K^c : \mathbb{Q}] \leqslant \nu(A+1)$, and the proof of the lemma is complete.

Lemma 2 *One has* $Disc(K^c : \mathbb{Q}) \leqslant (2tX)^{\nu(A+4)}$.

Proof We begin by considering a typical element $a \in \mathscr{A}$. Suppose that $[\mathbb{Q}(a) : \mathbb{Q}] = d$, whence the minimal polynomial m_a of a over $\mathbb{Z}$ has degree d. We note for future reference that $d \leqslant [K : \mathbb{Q}] = d(\mathscr{A})$. Let $S(a)$ denote the splitting field for m_a over $\mathbb{Q}$, so that $S(a) = \mathbb{Q}(\beta_1, \ldots, \beta_d)$ for some distinct algebraic numbers $\beta_1, \ldots, \beta_d$. We put $S_0(a) = \mathbb{Q}$, and when $1 \leqslant j \leqslant d$ we define

$$S_j(a) = \mathbb{Q}(\beta_1, \ldots, \beta_j).$$

For each index j with $1 \leqslant j \leqslant d$, we have

$$\mathrm{Disc}(\mathbb{Q}(\beta_j) : \mathbb{Q}) \leqslant \mathrm{Disc}(m_{\beta_j}) = \mathrm{Disc}(m_a). \tag{15}$$

Recall the upper bounds (13). Then, by considering the resultant of m_a and m_a' in terms of the determinant of the associated Sylvester matrix, noting that the coefficients of m_a are bounded in absolute value by $\mathrm{Env}(\mathscr{A})$, we see that

$$\begin{aligned}\mathrm{Disc}(m_a) &\leqslant (\deg(m_a))!\,(\mathrm{Env}(\mathscr{A}))^{2\deg(m_a)}\\ &\leqslant d(\mathscr{A})^{d(\mathscr{A})} X^{2d(\mathscr{A})}\\ &\leqslant \nu(A+1) X^{2(2t)^{2^A}}.\end{aligned} \tag{16}$$

Now observe that, as a consequence of a simple bound of Tôyoma [15], whenever $E : \mathbb{Q}$ and $F : \mathbb{Q}$ are two field extensions, then

$$\mathrm{Disc}(EF : \mathbb{Q}) \leqslant (\mathrm{Disc}(E : \mathbb{Q})\,\mathrm{Disc}(F : \mathbb{Q}))^{[EF:\mathbb{Q}]}. \tag{17}$$

Thus, for $1 \leqslant j < d$, it follows from (15) that

$$\begin{aligned}\mathrm{Disc}(S_{j+1}(a) : \mathbb{Q}) &\leqslant \big(\mathrm{Disc}(S_j(a) : \mathbb{Q})\,\mathrm{Disc}(\mathbb{Q}(\beta_{j+1}) : \mathbb{Q})\big)^{[K^c:\mathbb{Q}]}\\ &\leqslant \big(\mathrm{Disc}(S_j(a) : \mathbb{Q})\,\mathrm{Disc}(m_a)\big)^{[K^c:\mathbb{Q}]}.\end{aligned} \tag{18}$$

Since it also follows from (15) that $\mathrm{Disc}(S_1(a) : \mathbb{Q}) \leqslant \mathrm{Disc}(m_a)$, we may apply the relation (18) inductively to derive the relation

$$\mathrm{Disc}(S_d(a) : \mathbb{Q}) \leqslant (\mathrm{Disc}(m_a))^{d[K^c:\mathbb{Q}]^{d-1}}. \tag{19}$$

We therefore deduce from (16) and Lemma 1 that

$$\mathrm{Disc}(S(a):\mathbb{Q}) \leqslant \left(\nu(A+1)X^{2(2t)^{2^A}}\right)^{\nu(A+1)^{d(\mathscr{A})}}.$$

A modicum of computation confirms that

$$\nu(A+1)^{d(\mathscr{A})} \leqslant (2t)^{(2t)^{2^{A+1}}(2t)^{2^A}} \leqslant \nu(A+2),$$

whilst

$$\nu(A+1)^{\nu(A+2)} \leqslant (2t)^{\nu(A+3)} \quad \text{and} \quad 2(2t)^{2^A}\nu(A+2) \leqslant \nu(A+3).$$

Consequently, we arrive at the simplified upper bound

$$\mathrm{Disc}(S(a):\mathbb{Q}) \leqslant (2tX)^{\nu(A+3)}. \tag{20}$$

At this point, we have bounded the discriminant associated to only one element of $\mathscr{A}$. The Galois closure K^c of $\mathbb{Q}(\mathscr{A})$, however, is the compositum of all the splitting fields $S(a)$ for $a \in \mathscr{A}$. We therefore apply the relation (17) as in the deduction of (19) to establish the bound

$$\mathrm{Disc}(K^c:\mathbb{Q}) \leqslant \left(\max_{a\in\mathscr{A}} \mathrm{Disc}(S(a):\mathbb{Q})\right)^{A[K^c:\mathbb{Q}]^{A-1}}.$$

By Lemma 1, one has

$$A[K^c:\mathbb{Q}]^{A-1} \leqslant \nu(A+1)^A,$$

so the upper bound (20) yields the estimate

$$\mathrm{Disc}(K^c:\mathbb{Q}) \leqslant (2tX)^B,$$

where

$$B = \nu(A+3)\nu(A+1)^A = (2t)^{A(2t)^{2^{A+1}}+(2t)^{2^{A+3}}} \leqslant (2t)^{(2t)^{2^{A+4}}} = \nu(A+4).$$

Thus we conclude that $\mathrm{Disc}(K^c:\mathbb{Q}) \leqslant (2tX)^{\nu(A+4)}$, completing the proof.

We shall need to identify a rational prime number π having the property that the algebraic number $\Upsilon(\mathscr{A})$ is a unit modulo π. Let $\Upsilon_0(\mathscr{A})$ denote the least positive (rational) integer having the property that the algebraic number

$$\Upsilon_1(\mathscr{A}) = \Upsilon_0(\mathscr{A})\Upsilon(\mathscr{A})$$

is an algebraic integer. Then, by taking norms, it is evident that it suffices to arrange that π does not divide the rational integer

$$\Upsilon^*(\mathscr{A}) = |N_{K^c:\mathbb{Q}}(\Upsilon_0(\mathscr{A}))N_{K^c:\mathbb{Q}}(\Upsilon_1(\mathscr{A}))|.$$

Lemma 3 *One has* $1 \leqslant \Upsilon^*(\mathscr{A}) \leqslant (2kX)^{r(2A)^s\nu(A+2)}$.

Proof We begin by taking a crude approach to bounding $|N_{K^c:\mathbb{Q}}(\Upsilon(\mathscr{A}))|$, applying bounds for the complex absolute values of the conjugates of each element of $\mathscr{A}$. Let a be a typical element of $\mathscr{A}$. Since $\mathrm{Env}(\mathscr{A}) = X$, the minimal polynomial m_a of a over $\mathbb{Z}$ satisfies the relation $\|m_a\|_1 \leqslant X$. Also, since $\deg(m_a) \leqslant d(\mathscr{A})$, we find that m_a takes the form

$$b_0(a)x^d + \ldots + b_{d-1}(a)x + b_d(a), \tag{21}$$

in which $d \leqslant d(\mathscr{A})$ and $|b_0(a)| \geqslant 1$. The (complex) absolute value of a therefore satisfies either the upper bound $|a| \leqslant 1$, or else is constrained by the inequality

$$|b_0(a)a^d| \leqslant |a|^{d-1}\|m_a\|_1 \leqslant |a|^{d-1}\mathrm{Env}(\mathscr{A}) = |a|^{d-1}X,$$

whence $|a| \leqslant X$. Thus, in any case, one has $|a| \leqslant X$. Since the conjugates of a are also roots of the polynomial m_a, one sees in this way that every conjugate of a in K^c has absolute value bounded above by X.

Recall the formula (14). It follows from our discussion thus far that the element $\Upsilon(\mathscr{A})$ of K^c satisfies the bound

$$|\Upsilon(\mathscr{A})| \leqslant (2X)^{A^2}(kX^t)^{rA^s}.$$

Here, we have made use of the observation that, since each polynomial $P_i(\mathbf{x})$ is (k,t)-bounded for $1 \leqslant i \leqslant r$, then for $\mathbf{a} \in \mathscr{A}^s$ one has

$$|P_i(\mathbf{a})| \leqslant \|P_i\|_1\Big(\max_{1\leqslant i\leqslant s}|a_i|\Big)^{t_i} \leqslant kX^t.$$

In order to bound the norm of $\Upsilon(\mathscr{A})$, we must multiply all of the conjugates of $\Upsilon(\mathscr{A})$ together. However, since we assume that $s \geqslant 2$, the concluding remark of the preceding paragraph shows that each of these conjugates is bounded above by

$$(2X)^{A^2}(kX^t)^{rA^s} \leqslant (2kX)^{2rtA^s}.$$

Thus, multiplying all of these conjugates together, we find that

$$|N_{K^c:\mathbb{Q}}(\Upsilon(\mathscr{A}))| \leqslant \left((2kX)^{2rtA^s}\right)^{[K^c:\mathbb{Q}]}. \tag{22}$$

Next, we investigate the denominator $\Upsilon_0(\mathscr{A})$. Referring to the minimal polynomial (21) of a over $\mathbb{Z}$, we see that $b_0(a)a$ is an algebraic integer and $|b_0(a)| \leqslant X$. An inspection of (14) therefore shows that $\Upsilon_0(\mathscr{A})$ is a positive rational integer dividing

$$\Bigg(\prod_{a_1,a_2\in\mathscr{A}} b_0(a_1)b_0(a_2)\Bigg)\Bigg(\prod_{i=1}^{r}\prod_{\mathbf{a}\in\mathscr{A}^s} b_0(a_1)^t\cdots b_0(a_s)^t\Bigg).$$

Thus, we have

$$\Upsilon_0(\mathscr{A}) \leqslant X^{2A^2+rstA^s},$$

whence

$$|N_{K^c:\mathbb{Q}}(\Upsilon_0(\mathscr{A}))| \leqslant \big(X^{(s+2)rtA^s}\big)^{[K^c:\mathbb{Q}]}.$$

By combining this estimate together with (22), we therefore discern that

$$\begin{aligned}\Upsilon^*(\mathscr{A}) &= |N_{K^c:\mathbb{Q}}(\Upsilon_0(\mathscr{A}))^2 N_{K^c:\mathbb{Q}}(\Upsilon(\mathscr{A}))| \\ &\leqslant \big((2kX)^{(2s+6)rtA^s}\big)^{[K^c:\mathbb{Q}]}. \qquad (23)\end{aligned}$$

Finally, we simplify the estimate supplied by (23). Observe first that, since $s \geqslant 2$, we have $2s+6 \leqslant 2^{s+2}$. Thus, on applying Lemma 1, we infer that

$$\Upsilon^*(\mathscr{A}) \leqslant (2kX)^{4rt(2A)^s\nu(A+1)}.$$

However, one has $4t\nu(A+1) \leqslant \nu(A+2)$. We consequently conclude that

$$\Upsilon^*(\mathscr{A}) \leqslant (2kX)^{r(2A)^s\nu(A+2)}.$$

On noting that $\Upsilon^*(\mathscr{A})$ is non-zero, by construction, the proof of the lemma is complete.

It is now time to select the rational prime number π by applying the Chebotarev density theorem. By the primitive element theorem, there exists an element $\Theta \in K^c$ having the property that $K^c = \mathbb{Q}(\Theta)$. It is apparent, moreover, that by making an appropriate choice for Θ, we may assume not only that $\mathscr{A} \subset \mathbb{Z}[\Theta]$, but also that all of the conjugates of the elements of $\mathscr{A}$ within K^c lie in $\mathbb{Z}[\Theta]$. With this choice for Θ now fixed in such a manner, we seek a rational prime number π with $\pi \nmid \Upsilon^*(\mathscr{A})$ having the property that the minimal polynomial m_Θ of Θ over $\mathbb{Z}$ splits into linear factors modulo π. This allows us to bijectively map the set $\mathscr{A}$ into a set of residues modulo π, while preserving the salient features of the solution set associated with the system of polynomial equations $\mathbf{P} = \mathbf{0}$. Throughout, we abbreviate Generalised Riemann Hypothesis to GRH.

Lemma 4 *There is an effectively computable positive absolute constant M_1 with the following property. Suppose that GRH holds for the Dedekind zeta function associated with the field extension $K^c : \mathbb{Q}$. In addition, assume that*

$$Y \geqslant r(2A)^s(2t)^{\nu(A+3)}\log(2kX) \qquad (24)$$

and

$$Y \geqslant M_1(2t)^{2\nu(A+4)}(\log(2tX))^4. \tag{25}$$

Then there exists a rational prime number π, with $Y < \pi \leqslant 16Y$ and $\pi \nmid \Upsilon^*(\mathscr{A})$, having the property that the minimal polynomial m_Θ of Θ over $\mathbb{Z}$ splits into linear factors over $\mathbb{F}_\pi[t]$.

Proof We work under the hypotheses (24) and (25) throughout. An effective version of the Chebotarev density theorem is provided by Lagarias and Odlyzko under the assumption of GRH for the Dedekind zeta function associated with the field extension $K^c : \mathbb{Q}$. Denote by $\pi_\Theta(x)$ the number of rational prime numbers p with $p \leqslant x$ having the property that m_Θ splits into linear factors over $\mathbb{F}_p$. Put

$$G = \mathrm{Gal}(K^c : \mathbb{Q}), \quad n = [K^c : \mathbb{Q}] \quad \text{and} \quad D = \mathrm{Disc}(K^c : \mathbb{Q}).$$

Then it follows from [11, Theorem 1.1] that there exists a positive absolute constant M_0 such that

$$\left| \pi_\Theta(x) - \frac{\mathrm{Li}(x)}{|G|} \right| \leqslant M_0 \left(\frac{x^{1/2} \log(Dx^n)}{|G|} + \log D \right). \tag{26}$$

Here, we have written $\mathrm{Li}(x)$ for the usual logarithmic integral function.

On recalling Lemmata 1 and 2, we find that when $x \leqslant X$, we have

$$\begin{aligned} \log(Dx^n) &\leqslant \nu(A+4)\log(2tX) + \nu(A+1)\log x \\ &\leqslant 2\nu(A+4)\log(2tX). \end{aligned}$$

Moreover, it follows from a trivial upper bound for $|G|$ together with Lemma 1 that

$$|G| \leqslant [K^c : \mathbb{Q}]! \leqslant [K^c : \mathbb{Q}]^{[K^c:\mathbb{Q}]} \leqslant \nu(A+1)^{\nu(A+1)} \leqslant (2t)^{\nu(A+2)}, \tag{27}$$

whence

$$\begin{aligned} |G| \log D &\leqslant (2t)^{\nu(A+2)} \nu(A+4)\log(2tX) \\ &\leqslant (2t)^{\nu(A+3)} \log(2tX). \end{aligned}$$

Thus, we deduce from (26) that

$$\left| \pi_\Theta(x) - \frac{\mathrm{Li}(x)}{|G|} \right| \leqslant \frac{M_0}{|G|} \left(2x^{1/2}\nu(A+4)\log(2tX) + (2t)^{\nu(A+3)}\log(2tX) \right).$$

Suppose that M_1 is sufficiently large in terms of M_0 and $16Y \leqslant X$. Then, under the hypothesis (25), we have

$$\begin{aligned}\pi_\Theta(16Y) &\geqslant \frac{1}{|G|}\Big(\frac{16Y}{\log(16Y)} - \big(8Y^{1/2}\nu(A+4) + (2t)^{\nu(A+3)}\big)M_0\log(2tX)\Big)\\ &\geqslant \frac{1}{|G|}\Big(\frac{16Y}{\log(16Y)} - \frac{4Y}{\log Y}\Big) \geqslant \frac{8Y}{|G|\log Y}.\end{aligned}$$

Meanwhile, in a similar manner one finds that

$$\begin{aligned}\pi_\Theta(Y) &\leqslant \frac{1}{|G|}\Big(\frac{2Y}{\log Y} + \big(2Y^{1/2}\nu(A+4) + (2t)^{\nu(A+3)}\big)M_0\log(2tX)\Big)\\ &\leqslant \frac{1}{|G|}\Big(\frac{2Y}{\log Y} + \frac{2Y}{\log Y}\Big) = \frac{4Y}{|G|\log Y}.\end{aligned}$$

Pedestrian adjustments of this argument confirm that the same conclusions hold when $16Y > X$. Thus, we discern that

$$\pi_\Theta(16Y) - \pi_\Theta(Y) \geqslant \frac{4Y}{|G|\log Y}.$$

Let Π denote the set of rational prime numbers p with $Y < p \leqslant 16Y$ for which m_Θ splits into linear factors over $\mathbb{F}_p$. Then

$$\prod_{p\in\Pi} p \geqslant Y^{4Y/(|G|\log Y)} = \exp\left(4Y/|G|\right).$$

Meanwhile, from Lemma 3 we find that

$$\log \Upsilon^*(\mathscr{A}) \leqslant r(2A)^s\nu(A+2)\log(2kX).$$

Thus, recalling the upper bound (27) for $|G|$, we have

$$\prod_{p\in\Pi} p > \Upsilon^*(\mathscr{A})$$

provided only that

$$\frac{4Y}{(2t)^{\nu(A+2)}} > r(2A)^s\nu(A+2)\log(2kX). \tag{28}$$

But $\nu(A+2)(2t)^{\nu(A+2)} \leqslant (2t)^{\nu(A+3)}$, and so the hypothesis (24) is sufficient to ensure the validity of (28). With this condition now satisfied, we conclude that there exists a rational prime number π with $Y < \pi \leqslant 16Y$ satisfying $\pi \nmid \Upsilon^*(\mathscr{A})$, and such that m_Θ splits into linear factors over $\mathbb{F}_\pi$. The conclusion of the lemma follows.

We are now in a position to move on to the second step in the inductive phase of the argument, applying the method of Grosu [7]. The conclusion of Lemma 4 shows that there is a rational prime number π with

$$\pi \leqslant 16M_1 r(2A)^s(2t)^{2\nu(A+4)}(\log(2tkX))^4$$

having the property that $\pi \nmid \Upsilon^*(\mathscr{A})$, and such that m_Θ splits into linear factors over $\mathbb{F}_\pi$. Let a_0 be any zero of the polynomial m_Θ in $\mathbb{F}_\pi$. Since $\mathscr{A} \subset \mathbb{Z}[\Theta]$, the ring homomorphism $\Phi : \mathbb{Z}[\Theta] \to \mathbb{F}_\pi$ defined by putting $\Phi(\Theta) = a_0$ restricts to a well-defined map $\varphi : \mathbb{Z}[\mathscr{A}] \to \mathbb{F}_\pi$.

We claim that the set $\mathscr{A}$ has image $\mathscr{B} = \varphi(\mathscr{A})$ in which, for pairs of elements $a_1, a_2 \in \mathscr{A}$, one has $\varphi(a_1) = \varphi(a_2)$ if and only if $a_1 = a_2$. This claim will be confirmed by verifying that whenever $a_1 \neq a_2$, then $\varphi(a_1) \neq \varphi(a_2)$. By way of seeking a contradiction, suppose that $a_1 \neq a_2$ and $\varphi(a_1) = \varphi(a_2)$. Then we have $\Phi(a_1) = \Phi(a_2)$, and the homomorphism property of Φ implies that $\Phi(a_1 - a_2) = 0$. But Υ^* is a multiple of $a_1 - a_2$, say $\Upsilon^* = \mu(a_1 - a_2)$ for a suitable element μ of $\mathbb{Z}[\Theta]$. The homomorphism property of Φ thus ensures that $\Phi(\Upsilon^*) = \Phi(\mu)\Phi(a_1 - a_2) = 0$. However, we have $\Upsilon^* \in \mathbb{Z}$, and since $\pi \nmid \Upsilon^*$ we find that $0 = \Phi(\Upsilon^*) = \Upsilon^*\Phi(1)$ and hence $\Phi(1) = 0$. This contradicts the homomorphism property of Φ, confirming our earlier claim.

We also claim that, for $1 \leqslant i \leqslant r$ and $\mathbf{a} \in \mathscr{A}^s$, one has $P_i(\mathbf{a}) = 0$ if and only if $P_i(\varphi(\mathbf{a})) = 0$. In this instance it suffices to show that when $P_i(\mathbf{a}) \neq 0$, then $P_i(\varphi(\mathbf{a})) \neq 0$. We again proceed by seeking a contradiction, assuming that $P_i(\mathbf{a}) \neq 0$ and $P_i(\varphi(\mathbf{a})) = 0$. Then the homomorphism property of Φ ensures that $\Phi(P_i(\mathbf{a})) = 0$. But Υ^* is a multiple of $P_i(\mathbf{a})$, say $\Upsilon^* = \mu' P_i(\mathbf{a})$ for a suitable element μ' of $\mathbb{Z}[\Theta]$. Thus, in a similar manner to that described in the previous paragraph, we find that $\Phi(\Upsilon^*) = \Phi(\mu')\Phi(P_i(\mathbf{a})) = 0$, contradicting the fact that $\pi \nmid \Upsilon^*$. This contradiction again confirms our claim.

Lemma 5 *Suppose that π is a prime number having the property that $\pi \nmid \Upsilon^*(\mathscr{A})$, and such that m_Θ splits into linear factors over $\mathbb{F}_\pi$. Suppose also that*

$$\pi > (2kt)^{(2t)^{2A+1}}. \tag{29}$$

Then there exists an algebraic extension L of $\mathbb{Q}$ of degree at most $(2t)^{2^A}$, and a subset $\mathscr{C} \subset L$ with card$(\mathscr{C}) = A$, having the following properties:

(a) there is an injective map $\omega : \mathscr{C} \to \mathbb{F}_\pi$, with $\omega(\mathscr{C}) = \varphi(\mathscr{A})$, having the property that the canonical induced map $\widetilde{\omega} : \mathbb{Z}[\mathscr{C}] \to \mathbb{F}_\pi$ is a ring homomorphism;
(b) given $\mathbf{a} \in \mathscr{A}^s$, define $c_i = \omega^{-1}(\varphi(a_i))$ $(1 \leqslant i \leqslant s)$. Then one has $P_i(\mathbf{c}) = 0$ $(1 \leqslant i \leqslant r)$ for $\mathbf{c} \in \mathscr{C}^s$ if and only if $P_i(\varphi(\mathbf{a})) = 0$ $(1 \leqslant i \leqslant r)$ for $\mathbf{a} \in \mathscr{A}^s$;
(c) one has Env$(\mathscr{C}) \leqslant \nu(A + 1)\pi$.

In order to avoid ambiguity, we stress that the map $\widetilde{\omega}$ is defined for $f(\mathbf{x}) \in \mathbb{Z}[x_1, \ldots, x_m]$ by taking

$$\widetilde{\omega}(f(c_1, \ldots, c_m)) = f(\omega(c_1), \ldots, \omega(c_m)).$$

Proof (of Lemma 5) The desired conclusion is a consequence of the argument of Grosu [7, Lemma 8.1], though care is required in interpreting the argument underlying the latter proof so as to obtain the desired outcome. Following the general strategy of Grosu, we assign distinct indeterminates x_a to each element $\varphi(a)$ of $\mathscr{B} = \varphi(\mathscr{A})$. Certain equations are then known to have solutions over $\mathbb{F}_\pi$, specifically

$$P_i(x_{a_1}, \ldots, x_{a_s}) = 0 \quad (1 \leqslant i \leqslant r) \tag{30}$$

has a solution $(x_{a_1}, \ldots, x_{a_s}) = (\varphi(a_1), \ldots, \varphi(a_s))$ whenever

$$P_i(a_1, \ldots, a_s) = 0 \quad (1 \leqslant i \leqslant r)$$

for $\mathbf{a} \in \mathscr{A}^s$. In addition, one has certain non-equations. First, of course, one has

$$x_{a_1} - x_{a_2} \neq 0$$

whenever $x_{a_1} = \varphi(a_1)$ and $x_{a_2} = \varphi(a_2)$ for $a_1 \neq a_2$ with $a_1, a_2 \in \mathscr{A}$. Moreover, we have

$$P_i(x_{a_1}, \ldots, x_{a_s}) \neq 0$$

whenever $x_{a_j} = \varphi(a_j)$ $(1 \leqslant j \leqslant s)$ for $(a_1, \ldots, a_s) \in \mathscr{A}^s$ satisfying the condition $P_i(a_1, \ldots, a_s) \neq 0$. Taken together, we now have a list of equations and non-equations in the variables x_a $(a \in \mathscr{A})$, all defined over $\mathbb{F}_\pi$, and with the defining equations all (k, t)-bounded. It is worth emphasising, for the uninitiated, that the number of equations here may be very large. When $r = 1$, for example, the number of equations may be as large, roughly speaking, as A^{s-1}.

The argument of the proof of Grosu [7, Lemma 8.1] now shows via an elimination procedure using resultants that over the algebraic closure $\overline{\mathbb{Q}}$ of $\mathbb{Q}$, the equations (30) possess a solution if and only if certain eliminant polynomials are constant and equal to 0. By applying the same elimination procedure over $\mathbb{F}_\pi$, however, one sees that these constant polynomials must be 0 over $\mathbb{F}_\pi$, since the system (30) possesses the solution $x_a = \varphi(a)$ $(a \in \mathscr{A})$ in that setting. Provided that these eliminant polynomials have small enough coefficients in terms of π, therefore, one finds that in the setting of $\overline{\mathbb{Q}}$, these eliminant polynomials are indeed 0, and hence the system (30) possesses a solution, say $x_a = \psi(a)$ $(a \in \mathscr{A})$, lying in $\overline{\mathbb{Q}}$.

This is not the end of the story. It is shown first by Grosu [7, Lemma 8.1] that the field $L = \mathbb{Q}(\psi(\mathscr{A}))$ has degree at most $\tilde{t} = (2t)^{2^A}$ over $\mathbb{Q}$. Second, all of the eliminant polynomials to which we alluded above have coefficients bounded by

$$\tilde{k} = (2kt)^{(2t)^{2^{A+1}}},$$

and thus the condition (29) suffices for the desired conclusion. Indeed, Grosu shows that the eliminant polynomials are all $(\tilde{k}, \tilde{t})$-bounded. Third, one finds that the elements $\psi(a)$ $(a \in \mathscr{A})$ may be chosen in such a manner that there is a ring homomorphism $\gamma : \mathbb{Z}[\mathscr{C}] \to \mathbb{F}_\pi$ which sends $\psi(a)$ to $\varphi(a)$ for each $a \in \mathscr{A}$. A subtle detail of this last conclusion is that it may be necessary to take $\psi(a)$ to be a rational integer b lying in the set $\{0, 1, \ldots, \pi - 1\}$ with $b \equiv \varphi(a) \pmod{\pi}$ (see the fifth and sixth paragraphs of Step 2 of the proof of [7, Lemma 8.1]).

We are now in possession of sufficient detail to complete the proof of the lemma. We have already confirmed the claims made in the statement of the lemma concerning the existence of L, the subset $\mathscr{C}$, the degree of L over $\mathbb{Q}$, and we have also explained the hypothesis (29). We define $\omega : \mathscr{C} \to \mathbb{F}_\pi$ by taking $\omega(c) = \gamma(c)$ for $c \in \mathscr{C}$, and then $\widetilde{\omega}$ coincides with γ by virtue of the ring homomorphism property of γ. Moreover, if $\omega(c_1) = \omega(c_2)$ for some elements c_1 and c_2 of $\mathscr{C}$, then $\gamma(c_1) = \gamma(c_2)$. When $c_i = \psi(a_i)$ $(i = 1, 2)$ with $a_1, a_2 \in \mathscr{A}$, then we have

$$\varphi(a_1) = \gamma(\psi(a_1)) = \gamma(c_1) = \gamma(c_2) = \gamma(\psi(a_2)) = \varphi(a_2).$$

Thus, from the properties of the mapping φ, we have $a_1 = a_2$. Hence

$$c_1 = \psi(a_1) = \psi(a_2) = c_2.$$

The mapping ω is consequently injective. This confirms the claim (a).

By construction, if $c_i = \psi(a_i)$ for $1 \leqslant i \leqslant s$ and $\mathbf{a} \in \mathscr{A}^s$, then

$$P_i(\mathbf{c}) = 0 \quad (1 \leqslant i \leqslant r)$$

if and only if

$$\widetilde{\omega}(P_i(\mathbf{c})) = 0 \quad (1 \leqslant i \leqslant r).$$

But

$$\widetilde{\omega}(P_i(\mathbf{c})) = P_i(\omega(c_1), \dots, \omega(c_s)) = P_i(\varphi(a_1), \dots, \varphi(a_s)) = 0 \quad (1 \leqslant i \leqslant r).$$

Thus $P_i(\mathbf{c}) = 0$ $(1 \leqslant i \leqslant r)$ for $\mathbf{c} \in \mathscr{C}^s$ if and only if $P_i(\varphi(\mathbf{a})) = 0$ $(1 \leqslant i \leqslant r)$. This confirms the claim (b).

Finally, each element $c \in \mathscr{C}$ is either an integer lying in the set $\{0, 1, \dots, \pi - 1\}$, or else satisfies a $(\tilde{k}, \tilde{t})$-bounded polynomial of degree at most $(2t)^{2^A}$ having integral coefficients of absolute value at most

$$(2kt)^{(2t)^{2^{A+1}}} \leqslant \pi.$$

In the latter case, the minimal polynomial m_c of c over $\mathbb{Z}$ is a divisor of a polynomial $q \in \mathbb{Z}[t]$, with $\|q\|_1 \leqslant \tilde{k} \leqslant \pi$ and $\deg(q) \leqslant (2t)^{2^A}$. If $\deg(q) = d$ and we write $q(x) = q_d x^d + \dots + q_1 x + q_0$ with $q_i \in \mathbb{Z}$ $(0 \leqslant i \leqslant d)$, then we have

$$\sum_{l=0}^{d} |q_l|^2 \leqslant \left(\sum_{l=0}^{d} |q_l| \right)^2 = \|q\|_1^2.$$

It is then a consequence of the corollary to the main theorem of Granville [6] that if $r \in \mathbb{Z}[x]$ is any polynomial divisor of q, then

$$\|r\|_2 \leqslant \Big(\frac{\sqrt{5}+1}{2}\Big)^d \|q\|_2 \leqslant \Big(\frac{\sqrt{5}+1}{2}\Big)^d \|q\|_1 \leqslant \Big(\frac{\sqrt{5}+1}{2}\Big)^d \pi.$$

By Cauchy's inequality, therefore, we have

$$\|r\|_1 \leqslant (\deg(r)+1)^{1/2}\|r\|_2 \leqslant 2d\Big(\frac{\sqrt{5}+1}{2}\Big)^d \pi.$$

Hence, every element $c \in \mathscr{C}$ has minimal polynomial m_c over $\mathbb{Z}$ having degree at most $(2t)^{2^A}$, with

$$\|m_c\|_1 \leqslant 2(2t)^{2^A} 2^{(2t)^{2^A}} \pi \leqslant \nu(A+1)\pi.$$

Thus $\mathrm{Env}(\mathscr{C}) \leqslant \nu(A+1)\pi$, completing the proof of claim (c).

The plan outlined in the previous section may now be applied to good effect. Suppose that $\mathscr{A}$ is a set of algebraic numbers with

$$d(\mathscr{A}) \leqslant (2t)^{2^A} \quad \text{and} \quad \mathrm{Env}(\mathscr{A}) = X.$$

Then, assuming GRH for all Dedekind zeta functions, it follows from Lemma 4 that we can find a rational prime number π with

$$\pi > (2kt)^{(2t)^{2^{A+1}}}$$

and

$$\pi \leqslant \max\{16(2kt)^{(2t)^{2^{A+1}}}, 16M_1 r A^{2s}(2t)^{2\nu(A+4)}(\log(2tkX))^4\} \tag{31}$$

such that $\pi \nmid \Upsilon^*(\mathscr{A})$, and having the property that m_Θ splits into linear factors over $\mathbb{F}_\pi$. As a consequence of Lemma 5, we deduce that there is a set $\mathscr{C}$ of algebraic numbers having the property that

$$d(\mathscr{C}) \leqslant (2t)^{2^A} \quad \text{and} \quad \mathrm{Env}(\mathscr{C}) \leqslant \nu(A+1)\pi,$$

and having the property, moreover, that there is a bijection $\psi : \mathscr{A} \to \mathscr{C}$ which is an algebraic Freiman **P**-isomorphism.

We may now iterate this step, starting with the set of algebraic numbers $\mathscr{C}$, and deriving a new set $\mathscr{C}'$ algebraic Freiman **P**-isomorphic to $\mathscr{C}$, and with

$$d(\mathscr{C}') \leqslant (2t)^{2^A} \quad \text{and} \quad \mathrm{Env}(\mathscr{C}') \leqslant \nu(A+1)\pi',$$

where

$$\pi' \leqslant \max\{16(2kt)^{(2t)^{2^{A+1}}}, 16M_1 r A^{2s}(2t)^{2\nu(A+4)}(\log(2tk\nu(A+1)\pi))^4\}.$$

The composition of the two algebraic Freiman **P**-isomorphisms that we have encountered here provides an algebraic Freiman **P**-isomorphism from $\mathscr{A}$ to $\mathscr{C}'$. Plainly, it makes sense to iterate this process repeatedly so long as the associated algebraic enveloping radius is decreasing.

In order to assess the strength of the ensuing bounds, it makes sense to simplify our estimates so as to make iteration tractable. Observe first that the bound (31) may be simplified by noting that, when it is satisfied, one has

$$\pi \leqslant 16M_1 r A^{2s}(2kt)^{2\nu(A+4)}(\log(2tkX))^4,$$

whence

$$\begin{aligned}\nu(A+1)\pi &\leqslant 16M_1 r A^{2s}(2kt)^{2\nu(A+4)}\nu(A+1)(\log(2tkX))^4\\ &\leqslant M_1 r A^{2s}(2kt)^{\nu(A+5)}(\log(2tkX))^4.\end{aligned}$$

We therefore have

$$\mathrm{Env}(\mathscr{C}) \leqslant \nu(A+1)\pi \leqslant \tfrac{1}{2}X = \tfrac{1}{2}\mathrm{Env}(\mathscr{A})$$

provided that X is large and

$$X \geqslant M_1^2 r^2 A^{4s}(2kt)^{2\nu(A+5)+1}.$$

Indeed, provided that X is large enough, one has

$$(\log(2tkX))^4 \leqslant \tfrac{1}{2}\sqrt{tkX},$$

and hence

$$\begin{aligned}M_1 r A^{2s}(2kt)^{\nu(A+5)}(\log(2tkX))^4 &\leqslant \tfrac{1}{2}M_1 r A^{2s}(2kt)^{\nu(A+5)}(tk)^{1/2}X^{1/2}\\ &\leqslant \tfrac{1}{2}X.\end{aligned}$$

Thus, by iterating this condensation process, we may ensure that $\mathscr{A}$ is algebraic Freiman **P**-isomorphic to a set $\mathscr{B}$ of algebraic numbers with

$$d(\mathscr{B}) \leqslant (2t)^{2^A} \quad \text{and} \quad \mathrm{Env}(\mathscr{B}) \leqslant M_1^2 r^2 A^{4s}(2kt)^{\nu(A+6)}.$$

We summarise these deliberations in the form of a theorem.

Theorem 5 *Assume GRH for all Dedekind zeta functions. Let*

$$P_i(\mathbf{x}) \in \mathbb{Z}[x_1, \dots, x_s] \quad (1 \leqslant i \leqslant r)$$

be polynomials each of degree at most t, and with $\|P_i\|_1 \leqslant k$ $(1 \leqslant i \leqslant r)$. *Suppose that* $\mathscr{A}$ *is a set of algebraic numbers with* $d(\mathscr{A}) \leqslant (2t)^{2^A}$. *Then* $\mathscr{A}$ *is algebraic Freiman* **P**-*isomorphic to a set of algebraic numbers* $\mathscr{B}$ *with*

$$d(\mathscr{B}) \leqslant (2t)^{2^A} \quad \textit{and} \quad \mathit{Env}(\mathscr{B}) \ll r^2 A^{4s} (2kt)^{(2t)^{(2t)^{2^{A+6}}}}.$$

Here, the implicit constant in Vinogradov's notation is absolute.

We remark that, by inflating the parameter t so that $(2t)^{2^A} \geqslant d(\mathscr{A})$, the theorem can be applied so as to accomodate sets $\mathscr{A}$ of algebraic integers of arbitrarily large finite degree $d(\mathscr{A})$. The following corollary may make the conclusion of Theorem 5 more transparent.

Corollary 1 *In the setting of Theorem 5, we have*

$$\mathit{Env}_\delta^*(\mathscr{A}; \mathbf{P}) \leqslant \exp_4(c_1 A) \quad \textit{with} \quad \delta \leqslant (2t)^{2^A},$$

where $c_1 = c_1(r, s, t, k)$ *is a positive number depending at most on* r, s, t *and* k.

In certain situations, one may be interested in working with algebraic integers rather than more general algebraic numbers. For homogeneous polynomials, this is of course easily handled by clearing denominators.

Corollary 2 *Assume GRH for all Dedekind zeta functions. Let*

$$P_i(\mathbf{x}) \in \mathbb{Z}[x_1, \ldots, x_s] \quad (1 \leqslant i \leqslant r)$$

be homogeneous polynomials each of degree at most t, and with $\|P_i\|_1 \leqslant k$ $(1 \leqslant i \leqslant r)$. *Suppose that* $\mathscr{A}$ *is a set of algebraic integers with* $d(\mathscr{A}) \leqslant (2t)^{2^A}$. *Then* $\mathscr{A}$ *is algebraic Freiman* **P**-*isomorphic to a set of algebraic integers* $\mathscr{B}$ *with*

$$d(\mathscr{B}) \leqslant (2t)^{2^A} \quad \textit{and} \quad \mathit{Env}(\mathscr{B}) \ll (r^{2A} A^{4sA})^{(2t)^{2^A}} (2kt)^{(2t)^{(2t)^{2^{A+7}}}}.$$

Here, the implicit constant in Vinogradov's notation is absolute.

Proof Under the hypotheses of the statement of the corollary, it follows from Theorem 5 that $\mathscr{A}$ is algebraic Freiman **P**-isomorphic to a set of algebraic numbers $\mathscr{C}$ with

$$d(\mathscr{C}) \leqslant (2t)^{2^A} \quad \text{and} \quad \mathrm{Env}(\mathscr{C}) \ll r^2 A^{4s} (2kt)^{\nu(A+6)}.$$

Consider a typical element $c \in \mathscr{C}$ and its minimal polynomial m_c over $\mathbb{Z}$. For some integer $d = d_c \leqslant d(\mathscr{C})$, we can write

$$m_c(x) = g_0 x^d + \ldots + g_{d-1} x + g_d,$$

where $g_i = g_i(c) \in \mathbb{Z}$ satisfies $|g_i| \leqslant \mathrm{Env}(\mathscr{C})$ for $0 \leqslant i \leqslant d$. Let G be the least common multiple of all of the integers $g_0(c)$ with $c \in \mathscr{C}$, so that

$$G \leqslant \prod_{c\in\mathscr{C}} g_0(c) \leqslant (\mathrm{Env}(\mathscr{C}))^A$$

and for each $c_0 \in \mathscr{C}$ one has

$$G/g_0(c_0) \leqslant \prod_{c\in\mathscr{C}\setminus\{c_0\}} g_0(c) \leqslant (\mathrm{Env}(\mathscr{C}))^{A-1}.$$

Observe that when $c \in \mathscr{C}$, the polynomial $(G^d/g_0)m_c(x)$ is equal to

$$(Gx)^d + (G/g_0)g_1(Gx)^{d-1} + \ldots + (G^{d-1}/g_0)g_{d-1}(Gx) + (G^d/g_0)g_d,$$

so that Gc is an algebraic integer whose minimal polynomial m_{Gc} satisfies

$$\|m_{Gc}\|_1 \leqslant (G^d/g_0(c))\mathrm{Env}(\mathscr{C}) \leqslant (\mathrm{Env}(\mathscr{C}))^{dA}.$$

We consider the set

$$\mathscr{B} = \{Gc : c \in \mathscr{C}\}.$$

It follows from the above discussion that $\mathscr{B}$ is a set of algebraic integers with $d(\mathscr{B}) = d(\mathscr{C}) \leqslant (2t)^{2^A}$ and

$$\mathrm{Env}(\mathscr{B}) \leqslant (\mathrm{Env}(\mathscr{C}))^{d(\mathscr{B})A} \leqslant \left(r^{2A}A^{4As}(2kt)^{A\nu(A+6)}\right)^{d(\mathscr{B})}.$$

The conclusion of the corollary follows with a modicum of computation.

Corollary 3 *In the setting of Corollary 2, the set of algebraic integers $\mathscr{A}$ is algebraic Freiman* **P***-isomorphic to a set of algebraic integers $\mathscr{B}$ with*

$$d(\mathscr{B}) \leqslant (2t)^{2^A} \quad \text{and} \quad Env(\mathscr{B}) \ll \exp_4(c_2 A),$$

where $c_2 = c_2(r, s, t, k)$ is a positive number depending at most on r, s, t and k.

We finish this section by remarking that, in certain non-linear situations, conclusions significantly stronger than are made available via Theorem 5 can be obtained by making use of underlying linear structure.

Theorem 6 *Let $P_i(\mathbf{x}) \in \mathbb{Z}[x_1, \ldots, x_s]$ $(1 \leqslant i \leqslant r)$ be diagonal polynomials of the shape*

$$P_i(\mathbf{x}) = \sum_{j=1}^{s} c_{ij}x_j^t \quad (1 \leqslant i \leqslant r),$$

where $\|P_i\|_1 \leqslant k$ $(1 \leqslant i \leqslant r)$. Suppose that $\mathscr{A}$ is a finite set of integers. Then, when card($\mathscr{A}$) is large, one has

$$Env_\delta^*(\mathscr{A}; \mathbf{P}) \leqslant t2^{t+1}(k+1)^A \quad with \quad \delta \leqslant t^A.$$

Proof We consider the set of integers $\mathscr{A}_t = \{a^t : a \in \mathscr{A}\}$ and the set of linear polynomials

$$L_i(\mathbf{y}) = \sum_{j=1}^{s} c_{ij} y_j \quad (1 \leqslant i \leqslant r).$$

By Theorem 4, the set $\mathscr{A}_t$ is Freiman **L**-isomorphic to a set of integers $\mathscr{B}_t$ with $\mathrm{env}(\mathscr{B}_t) \leqslant (k+1)^A$. Now consider the set

$$\mathscr{B} = \{b^{1/t} : b \in \mathscr{B}_t\}.$$

One has $d(\mathscr{B}) \leqslant t^{\mathrm{card}(\mathscr{B})} = t^A$. Moreover, given $c \in \mathscr{B}$, one has $c^t \in \mathbb{Z}$ with $|c|^t < \mathrm{env}(\mathscr{B}_t) \leqslant (k+1)^A$. By applying the corollary to the main theorem of Granville [6], much as in the conclusion of the proof of Lemma 5, we find that the minimal polynomial m_c of c over $\mathbb{Z}$ is a divisor of the polynomial $f_{t,l}(x) = x^t - l$, where $l = c^t \in \mathscr{B}_t$. Thus,

$$\begin{aligned} \|m_c\|_1 &\leqslant 2\deg(f_{t,l})\Big(\frac{\sqrt{5}+1}{2}\Big)^{\deg(f_{t,l})} \|f_{t,l}\|_1 \\ &\leqslant t2^{t+1}\mathrm{env}(\mathscr{B}_t) \\ &\leqslant t2^{t+1}(k+1)^A. \end{aligned}$$

Thus $\mathrm{Env}(\mathscr{B}) \leqslant t2^{t+1}(k+1)^A$, and $\mathscr{B}$ is algebraic Freiman **P**-isomorphic to $\mathscr{A}$ with $d(\mathscr{B}) \leqslant t^A$. This completes the proof of the theorem.

6 Densifications of Sets

We turn next to a discussion of the densification idea to which we alluded in the introduction. We begin with an analogue of the Freiman **P**-isomorphism defined in Definition 1 suitable for the discussion of cartesian products. In this context, when $P_1, \dots, P_r \in \mathbb{Z}[x_1, \dots, x_s]$ and $\mathscr{C} \subset \mathbb{Z}$, we again write

$$S(\mathscr{C}; \mathbf{P}) = \{\mathbf{x} \in \mathscr{C}^s : P_i(\mathbf{x}) = 0\,(1 \leqslant i \leqslant r)\}.$$

Also, when $\mathbf{x}_1, \dots, \mathbf{x}_t \in \mathscr{C}^s$, we have in mind the notational convention that

$$\mathbf{x}_i = (x_{i1}, \dots, x_{is}).$$

Then, when $(\mathbf{x}_1, \dots, \mathbf{x}_t) \in \mathscr{C}^s \times \dots \times \mathscr{C}^s$, it is convenient to abbreviate the t-tuple $(x_{1j}, x_{2j}, \dots, x_{tj})$ as $\mathbf{x}^{(j)}$.

Definition 7 Let $t \in \mathbb{N}$, and suppose that $\mathscr{A}$ and $\mathscr{B}$ are finite sets of integers with $|\mathscr{B}| = |\mathscr{A}|^t$. Suppose in addition that the polynomials $P_1, \dots, P_r$ lie in $\mathbb{Z}[x_1, \dots, x_s]$.

We say that a bijection $\omega : \mathscr{A}^t \to \mathscr{B}$ is a *t-fold Freiman* **P**-*isomorphism* (from $\mathscr{A}$ to $\mathscr{B}$) if it is the case that

$$(\mathbf{x}_1, \ldots, \mathbf{x}_t) \in S(\mathscr{A}; \mathbf{P})^t$$

if and only if

$$\big(\omega(\mathbf{x}^{(1)}), \ldots, \omega(\mathbf{x}^{(s)})\big) \in S(\mathscr{B}; \mathbf{P}).$$

As in the discussion of §2, we emphasise that a t-fold Freiman **P**-isomorphism is specific to a particular polynomial tuple **P**, and maps a t-tuple of integers to an integer. This once again permits an iterative approach in which t-fold Freiman **P**-isomorphisms are successively composed in the natural manner.

It may be useful to highlight the utility of such a definition. When $t > 1$, the structure of the solutions of the system of polynomials

$$P_i(\mathbf{x}) = 0 \quad (1 \leqslant i \leqslant r), \tag{32}$$

with $\mathbf{x} \in \mathscr{A}^s$, both determines and is determined by

$$S(\mathscr{A}; \mathbf{P})^t = S(\mathscr{A}; \mathbf{P}) \times \ldots \times S(\mathscr{A}; \mathbf{P}).$$

When $\omega : \mathscr{A}^t \to \mathscr{B}$ is a t-fold Freiman **P**-isomorphism, it follows from Definition 7 that $S(\mathscr{A}; \mathbf{P})^t$ is in bijective correspondence with

$$S(\omega(\mathscr{A}^t); \mathbf{P}) = S(\mathscr{B}; \mathbf{P}).$$

Thus, the structure of the solutions of the system (32) with $\mathbf{x} \in \mathscr{A}^s$ both determines and is determined by the structure of the solutions of the system (32) with $\mathbf{x} \in \mathscr{B}^s$. A particularly simple consequence of this observation is that, just as $|\mathscr{B}| = |\mathscr{A}|^t$, so too one has

$$|S(\mathscr{B}; \mathbf{P})| = |S(\mathscr{A}; \mathbf{P})|^t.$$

Provided that $\mathrm{env}(\mathscr{B})$ is not too much larger than $\mathrm{env}(\mathscr{A})$, then the solution set $S(\mathscr{A}; \mathbf{P})$ of a sparse set $\mathscr{A}$ may be understood precisely in terms of a potentially denser set $\mathscr{B}$ and its solution set $S(\mathscr{B}; \mathbf{P})$. This motivates the next definition.

Definition 8 We say that a map $\omega : \mathscr{A}^t \to \mathscr{D}$ is a *t-fold* **P**-*densifier of* $\mathscr{A}$ if it is a t-fold Freiman **P**-isomorphism having the property that $\mathrm{env}(\mathscr{D}) \leqslant \mathrm{env}(\mathscr{A})^t$. When the latter inequality is strict, we refer to ω as a *strict t-fold* **P**-*densifier* of $\mathscr{A}$. In either case, we refer to $\mathscr{D}$ as being a t-fold **P**-*densification* of $\mathscr{A}$.

Of particular interest are the t-fold $\mathbf{P}$-densifications $\mathscr{D}_t$ of $\mathscr{A}$ distinguished by the property that

$$\frac{\log \operatorname{env}(\mathscr{D}_t)}{\log |\mathscr{D}_t|}$$

is particularly small.

Definition 9 Let $\mathscr{A}$ be a finite set of integers, and suppose that the polynomials $P_1, \ldots, P_r$ lie in $\mathbb{Z}[x_1, \ldots, x_s]$. We say that the set $\mathscr{A}$ has $\mathbf{P}$-*densification exponent* κ when

$$\kappa = \liminf_{t\to\infty} \left\{ \frac{\log \operatorname{env}(\mathscr{D}_t)}{\log |\mathscr{D}_t|} : \mathscr{D}_t \text{ is a } t\text{-fold } \mathbf{P}\text{-densification of } \mathscr{A} \right\}.$$

It follows that when $\mathscr{A}$ has finite $\mathbf{P}$-densification exponent κ, then for each $\varepsilon > 0$ there is a natural number t and a t-fold $\mathbf{P}$-densification $\mathscr{D}_t$ of $\mathscr{A}$ such that

$$\operatorname{env}(\mathscr{D}_t) \leqslant |\mathscr{D}_t|^{\kappa+\varepsilon}.$$

Suppose that, in addition, one has an estimate of the shape

$$|S(\mathscr{B}; \mathbf{P})| \ll \operatorname{env}(\mathscr{B})^{\varepsilon} |\mathscr{B}|^{\theta},$$

valid for all finite sets of integers $\mathscr{B}$. Then we may infer that

$$|S(\mathscr{A}; \mathbf{P})|^t = |S(\mathscr{D}_t; \mathbf{P})| \ll \operatorname{env}(\mathscr{D}_t)^{\varepsilon} |\mathscr{D}_t|^{\theta} < |\mathscr{D}_t|^{\theta+2\kappa\varepsilon} \ll |\mathscr{A}|^{t(\theta+2\kappa\varepsilon)},$$

whence

$$|S(\mathscr{A}; \mathbf{P})| \ll |\mathscr{A}|^{\theta+2\kappa\varepsilon}.$$

In this way, it should be apparent that the existence of finite $\mathbf{P}$-densification exponents would lead from conclusions such as Theorem 1 to the validity of conjectures of the shape of that recorded in Conjecture 1. We shall see in the next section that, while such objectives are attainable for linear systems $\mathbf{P}$, it would seem that for systems $\mathbf{P}$ of higher degree, currently accessible conclusions are necessarily weaker.

7 Densifications for Linear Systems of Equations

The polynomial systems most amenable to densification via the circle of ideas already presented in §3 are systems of homogeneous linear equations. Since the results concerning such systems are both simple and instructive, we expend the bulk of this section on their analysis. In order to fix ideas, suppose that $s \geqslant 2$, $r \geqslant 1$ and for $1 \leqslant i \leqslant r$ one has $c_{ij} \in \mathbb{Z}$ $(1 \leqslant j \leqslant s)$. We again ignore the trivial situation in which for some index i one has $c_{ij} = 0$ for $1 \leqslant j \leqslant s$. The system of polynomials initially of interest to us in this section is

$$P_i(\mathbf{x}) = \sum_{j=1}^{s} c_{ij} x_j \quad (1 \leqslant i \leqslant r).$$

Next, when $\mathscr{A} \subset \mathbb{Z}$ is a finite set of integers, we recall the notation of writing $S(\mathscr{A}; \mathbf{P})$ for the set of solutions of the system of equations $P_i(\mathbf{x}) = 0$ $(1 \leqslant i \leqslant r)$, with $\mathbf{x} \in \mathscr{A}^s$. In accordance with the treatment of §3, we define $\Lambda = \Lambda(\mathbf{c})$ by putting

$$\Lambda = \max_{1 \leqslant i \leqslant r} \sum_{j=1}^{s} |c_{ij}|.$$

Theorem 7 *Consider a system* $\mathbf{P}$ *of linear polynomials as described in the preamble, and consider a finite set of integers* $\mathscr{A}$*. Then provided that* $A = \mathrm{card}(\mathscr{A})$ *is sufficiently large in terms of* r *and* s*, the set* $\mathscr{A}$ *has a finite* $\mathbf{P}$*-densification exponent* κ *satisfying* $\kappa \leqslant s$*. In particular, whenever* $\varepsilon > 0$*, there exists a natural number* t *and a* t*-fold Freiman* $\mathbf{P}$*-isomorphism* $\omega : \mathscr{A}^t \to \mathscr{D}$ *having the property that* $\mathrm{env}(\mathscr{D}) \leqslant \mathrm{card}(\mathscr{D})^{s+\varepsilon}$.

Proof Fix a small positive number ε. We seek to apply an iterative strategy that, given a set $\mathscr{D}$ that is t-fold Freiman $\mathbf{P}$-isomorphic to $\mathscr{A}$, generates a new set $\mathscr{D}'$ that is t'-fold Freiman $\mathbf{P}$-isomorphic to $\mathscr{D}$ and satisfies

$$\frac{\log \mathrm{env}(\mathscr{D}')}{\log \mathrm{card}(\mathscr{D}')} \leqslant (1-\varepsilon)\frac{\log \mathrm{env}(\mathscr{D})}{\log \mathrm{card}(\mathscr{D})}. \tag{33}$$

Notice that the composition of a t-fold Freiman $\mathbf{P}$-isomorphism from $\mathscr{A}$ to $\mathscr{D}$, together with a t'-fold Freiman $\mathbf{P}$-isomorphism from $\mathscr{D}$ to $\mathscr{D}'$, gives a tt'-fold Freiman $\mathbf{P}$-isomorphism from $\mathscr{A}$ to $\mathscr{D}'$. Thus, the relation (33) suggests an improvement in the densification exponent. Provided that we are able to iterate this process sufficiently many times, we find that a $\mathbf{P}$-densification $\mathscr{D}$ of $\mathscr{A}$ exists with

$$\frac{\log \mathrm{env}(\mathscr{D})}{\log \mathrm{card}(\mathscr{D})} \leqslant (1-\varepsilon)^n \frac{\log \mathrm{env}(\mathscr{A})}{\log \mathrm{card}(\mathscr{A})},$$

with n as large as is necessary. It transpires that when

$$\mathrm{env}(\mathscr{D}) > \mathrm{card}(\mathscr{D})^{s+\varepsilon},$$

then further iteration is possible, and in this way we see that the $\mathbf{P}$-densification exponent of $\mathscr{A}$ is at most s.

We now initiate the proof proper. We may suppose without loss of generality that $\Lambda \geqslant 2$ and $s \geqslant 2$. We consider a finite set of integers $\mathscr{D}$ that is t-fold Freiman $\mathbf{P}$-isomorphic to $\mathscr{A}$, so that $|\mathscr{D}| = A^t \geqslant A$. Write $D = |\mathscr{D}|$ and $X = \mathrm{env}(\mathscr{D}) - 1$. If one were to have

$$X + 1 \leqslant D^{s(1+4\varepsilon)},$$

then the desired conclusion would follow, since $\varepsilon > 0$ may be taken arbitrarily small. We may therefore suppose that $X + 1 > D^{s(1+4\varepsilon)}$.

Next, in accordance with (5), we define the natural number Υ by putting

$$\Upsilon = \Bigg(\prod_{\substack{d_1, d_2 \in \mathscr{D} \\ d_1 \neq d_2}} |d_1 - d_2| \Bigg) \Bigg(\prod_{\substack{\mathbf{d} \in \mathscr{D}^s \\ \mathbf{d} \notin S(\mathscr{D}; \mathbf{P})}} \sum_{i=1}^{r} |P_i(\mathbf{d})| \Bigg).$$

Then one finds that

$$1 \leqslant \Upsilon \leqslant (2X)^{D^2} (r\Lambda X)^{D^s} \leqslant \tfrac{1}{3} (r\Lambda X)^{2D^s}.$$

Note that

$$2 \log(3\Upsilon) \leqslant 4D^s \log(r\Lambda X).$$

Then provided that $Y \geqslant 4D^s \log(r\Lambda X)$, it follows from the prime number theorem that in any interval $(Y, 2Y)$, there exist at least D prime numbers π with $\pi \nmid \Upsilon$. Let $\pi_1, \ldots, \pi_D$ be any D such distinct prime numbers.

We next construct a map $\omega : \mathscr{D}^D \to \mathbb{Z}$ as follows. When

$$\mathbf{d} = (d_1, \ldots, d_D) \in \mathscr{D}^D,$$

we define

$$\omega(\mathbf{d}) = \sum_{i=1}^{D} d_i \prod_{\substack{1 \leqslant j \leqslant D \\ j \neq i}} \pi_j. \tag{34}$$

Write $\mathscr{E} = \omega(\mathscr{D}^D)$. Then we claim that the mapping $\omega : \mathscr{D}^D \to \mathscr{E}$ is a D-fold Freiman $\mathbf{P}$-isomorphism from $\mathscr{D}$ to $\mathscr{E}$.

We first verify that $\omega : \mathscr{D}^D \to \mathscr{E}$ is a bijection, and for this it suffices to check that ω is injective. However, if $\mathbf{d}, \mathbf{d}' \in \mathscr{D}^D$ and $\omega(\mathbf{d}) = \omega(\mathbf{d}')$, then it is apparent that

$$\sum_{i=1}^{D} d_i \prod_{\substack{1 \leqslant j \leqslant D \\ j \neq i}} \pi_j \equiv \sum_{i=1}^{D} d_i' \prod_{\substack{1 \leqslant j \leqslant D \\ j \neq i}} \pi_j \pmod{\pi_k} \quad (1 \leqslant k \leqslant D),$$

whence

$$(d_k - d_k') \prod_{\substack{1 \leqslant j \leqslant D \\ j \neq k}} \pi_j \equiv 0 \pmod{\pi_k} \quad (1 \leqslant k \leqslant D).$$

For each index k, however, one has

$$\Bigg(\pi_k, \prod_{\substack{1 \leqslant j \leqslant D \\ j \neq k}} \pi_j \Bigg) = 1,$$

and thus we deduce that $d_k \equiv d_k' \pmod{\pi_k}$ $(1 \leqslant k \leqslant D)$. Recalling the definition of Υ, however, one sees that $\pi_k \nmid (d_k - d_k')$ whenever $d_k \neq d_k'$, and so we must have $d_k = d_k'$ $(1 \leqslant k \leqslant D)$. In this way, we conclude that $\mathbf{d} = \mathbf{d}'$, whence $\omega : \mathscr{D}^D \to \mathscr{E}$ is indeed bijective.

Next, whenever $(\mathbf{d}_1, \ldots, \mathbf{d}_D) \in S(\mathscr{D}; \mathbf{P})^D$, the linearity of the polynomials $\mathbf{P}$ ensures that for $1 \leqslant l \leqslant r$, one has

$$P_l\left(\omega(\mathbf{d}^{(1)}), \ldots, \omega(\mathbf{d}^{(s)})\right) = \sum_{i=1}^{D} P_l(d_{i1}, \ldots, d_{is}) \prod_{\substack{1 \leqslant j \leqslant D \\ j \neq i}} \pi_j = 0.$$

Thus $\left(\omega(\mathbf{d}^{(1)}), \ldots, \omega(\mathbf{d}^{(s)})\right) \in S(\mathscr{E}; \mathbf{P})$. Also, when

$$(\mathbf{d}_1, \ldots, \mathbf{d}_D) \notin S(\mathscr{D}; \mathbf{P})^D,$$

then for some index l with $1 \leqslant l \leqslant r$, and some index k with $1 \leqslant k \leqslant D$, one has

$$P_l(d_{k1}, \ldots, d_{ks}) \neq 0.$$

Meanwhile, if one were to have

$$P_l\left(\omega(\mathbf{d}^{(1)}), \ldots, \omega(\mathbf{d}^{(s)})\right) = 0 \quad (1 \leqslant l \leqslant r), \tag{35}$$

then in particular,

$$\sum_{i=1}^{D} P_l(d_{i1}, \ldots, d_{is}) \prod_{\substack{1 \leqslant j \leqslant D \\ j \neq i}} \pi_j \equiv 0 \pmod{\pi_k} \quad (1 \leqslant l \leqslant r).$$

The latter congruences imply that

$$P_l(d_{k1}, \ldots, d_{ks}) \prod_{\substack{1 \leqslant j \leqslant D \\ j \neq k}} \pi_j \equiv 0 \pmod{\pi_k} \quad (1 \leqslant l \leqslant r),$$

whence

$$P_l(d_{k1}, \ldots, d_{ks}) \equiv 0 \pmod{\pi_k} \quad (1 \leqslant l \leqslant r). \tag{36}$$

But the definition of Υ ensures that when $\mathbf{d}_k \notin S(\mathscr{D}; \mathbf{P})$, as we may assume, then

$$\sum_{l=1}^{r} |P_l(\mathbf{d}_k)| \not\equiv 0 \pmod{\pi_k}.$$

Thus we have $P_l(\mathbf{d}_k) \not\equiv 0 \pmod{\pi_k}$ for some index l with $1 \leqslant l \leqslant r$, and this contradicts the relation (36). We therefore conclude that (35) cannot hold. In consequence, when $(\mathbf{d}_1, \dots, \mathbf{d}_D) \notin S(\mathscr{D}; \mathbf{P})^D$, one must have

$$\left(\omega(\mathbf{d}^{(1)}), \dots, \omega(\mathbf{d}^{(s)})\right) \notin S(\mathscr{E}; \mathbf{P}).$$

We have thus shown that the map $\omega : \mathscr{D}^D \to \mathscr{E}$ is a D-fold Freiman $\mathbf{P}$-isomorphism.

We next investigate the $\mathbf{P}$-densification exponent associated with the mapping $\omega : \mathscr{D}^D \to \mathscr{E}$. Observe first that the definition (34) shows that

$$\operatorname{env}(\mathscr{E}) \leqslant D(2Y)^{D-1} \max_{1 \leqslant i \leqslant D} |d_i| \leqslant D(2Y)^{D-1} \operatorname{env}(\mathscr{D}).$$

We take $Y = 4D^s \log(r\Lambda X)$, in which we recall that $X = \operatorname{env}(\mathscr{D}) - 1$. Thus

$$\begin{aligned} \frac{\log \operatorname{env}(\mathscr{E})}{\log |\mathscr{E}|} &\leqslant \frac{\log \operatorname{env}(\mathscr{D}) + \log D + (D-1)\log(2Y)}{D \log D} \\ &= \frac{1}{D}\left(\frac{\log \operatorname{env}(\mathscr{D})}{\log D}\right) + \left(1 - \frac{1}{D}\right)\left(\frac{\log(2Y)}{\log D}\right) + \frac{1}{D}. \end{aligned}$$

It follows that whenever

$$\frac{\log(2Y)}{\log D} \leqslant (1 - 2\varepsilon)\frac{\log \operatorname{env}(\mathscr{D})}{\log D}, \tag{37}$$

then one has

$$\frac{\log \operatorname{env}(\mathscr{E})}{\log |\mathscr{E}|} \leqslant (1 - \varepsilon)\frac{\log \operatorname{env}(\mathscr{D})}{\log |\mathscr{D}|}. \tag{38}$$

This is the improving $\mathbf{P}$-densification argument outlined in the opening discussion of the proof.

Let us return to examine the condition (37). This condition is satisfied provided that

$$2Y \leqslant (\operatorname{env}(\mathscr{D}))^{1-2\varepsilon} = (X+1)^{1-2\varepsilon},$$

which is to say that

$$8D^s \log(r\Lambda X) \leqslant (X+1)^{1-2\varepsilon}.$$

However, in the opening discussion of the proof, we were at liberty to suppose that $X + 1 > D^{s(1+4\varepsilon)}$. Thus we have

$$\frac{(X+1)^{1-2\varepsilon}}{\log(r\Lambda X)} > D^{s(1+\varepsilon)} > 8D^s,$$

and in consequence the condition (37) is fulfilled. This justifies the conclusion (38).

As we explained in the opening discussion of the proof, the upper bound (38) permits an iterative approach to be employed that delivers a t-fold $\mathbf{P}$-densification $\mathscr{D}$ of $\mathscr{A}$ satisfying the property that

$$\frac{\log \operatorname{env}(\mathscr{D})}{\log \operatorname{card}(\mathscr{D})} \leqslant (1-\varepsilon)^n \frac{\log \operatorname{env}(\mathscr{A})}{\log \operatorname{card}(\mathscr{A})}, \tag{39}$$

with n arbitarily large, provided only that $\operatorname{env}(\mathscr{D}) > (\operatorname{card}(\mathscr{D}))^{s(1+4\varepsilon)}$. Since for sufficiently large n, the bound (39) contradicts the condition $\operatorname{env}(\mathscr{D}) > (\operatorname{card}(\mathscr{D}))^{s(1+4\varepsilon)}$, we are forced to conclude that such a t-fold $\mathbf{P}$-densification $\mathscr{D}$ exists in which $\operatorname{env}(\mathscr{D}) \leqslant (\operatorname{card}(\mathscr{D}))^{s(1+4\varepsilon)}$. By taking $\varepsilon > 0$ arbitrarily small, this shows that

$$\liminf_{t\to\infty} \left\{ \frac{\log \operatorname{env}(\mathscr{D}_t)}{\log |\mathscr{D}_t|} : \mathscr{D}_t \text{ is a } t\text{-fold } \mathbf{P}\text{-densification of } \mathscr{A} \right\} \leqslant s.$$

This completes the proof of the theorem.

The strategy underlying the proof of Theorem 7 can be generalised in some sense both to inhomogeneous systems, and also to systems of equations of degree exceeding 1. In order to illustrate ideas, consider a system of homogeneous polynomials $P_i(\mathbf{x}) \in \mathbb{Z}[x_1, \ldots, x_s]^r$, not necessarily linear. Suppose that these polynomials are of degree at most k, and that the sum of the absolute values of the coefficients in the polynomial $P_i(\mathbf{x})$ is at most Λ for $1 \leqslant i \leqslant r$. Let $\mathscr{D}$ be a finite set of integers that is t-fold Freiman $\mathbf{P}$-isomorphic to $\mathscr{A}$, and write $D = |\mathscr{D}|$ and $X = \operatorname{env}(\mathscr{D})$. Also, define the integer Υ now by putting

$$\Upsilon = \Bigg(\prod_{\substack{d_1, d_2 \in \mathscr{D} \\ d_1 \neq d_2}} |d_1 - d_2| \Bigg) \Bigg(\prod_{i=1}^{r} \prod_{\substack{\mathbf{d} \in \mathscr{D}^s \\ P_i(\mathbf{d}) \neq 0}} |P_i(\mathbf{d})| \Bigg).$$

Then provided that $Y \geqslant 4r D^{2s} \log(\Lambda X^k)$, it follows from the prime number theorem that in any interval $(Y, 2Y)$, there exist at least D prime numbers π with $\pi \nmid \Upsilon$. Let $\pi_1, \ldots, \pi_D$ be any D such distinct prime numbers.

We again define a map $\omega : \mathscr{D}^D \to \mathbb{Z}/(\pi_1 \ldots \pi_D \mathbb{Z})$ via (34), and write $\mathscr{E} = \omega(\mathscr{D}^D)$. The map ω is a bijection from $\mathscr{D}^D$ to $\mathscr{E}$, just as in the analogous argument in the proof of Theorem 7. We observe that for $1 \leqslant l \leqslant r$, one has

$$\begin{aligned} &P_l\left(\omega(\mathbf{d}^{(1)}), \ldots, \omega(\mathbf{d}^{(s)})\right) \\ &\equiv \sum_{i=1}^{D} P_l(d_{i1}, \ldots, d_{is}) \Bigg(\prod_{\substack{1 \leqslant j \leqslant D \\ j \neq i}} \pi_j \Bigg)^{\deg(P_l)} \pmod{\pi_1 \ldots \pi_D}. \end{aligned}$$

If $(\mathbf{d}_1, \ldots, \mathbf{d}_D) \notin S(\mathscr{D}; \mathbf{P})^D$, then for some index l with $1 \leqslant l \leqslant r$, and some index k with $1 \leqslant k \leqslant D$, one has

$$P_l(d_{k1}, \ldots, d_{ks}) \not= 0.$$

Since $\pi_j \nmid \Upsilon$ for $1 \leqslant j \leqslant D$, one cannot have

$$P_l(d_{k1}, \ldots, d_{ks}) \equiv 0 \pmod{\pi_k},$$

and consequently

$$P_l\left(\omega(\mathbf{d}^{(1)}), \ldots, \omega(\mathbf{d}^{(s)})\right) \not\equiv 0 \pmod{\pi_1 \ldots \pi_D} \quad (1 \leqslant l \leqslant r).$$

On the other hand, whenever $(\mathbf{d}_1, \ldots, \mathbf{d}_D) \in S(\mathscr{D}; \mathbf{P})^D$, then for $1 \leqslant l \leqslant r$ one must have

$$P_l\left(\omega(\mathbf{d}^{(1)}), \ldots, \omega(\mathbf{d}^{(s)})\right) \equiv 0 \pmod{\pi_1 \ldots \pi_D}.$$

We thus perceive that the solution structure of $S(\mathscr{D}; \mathbf{P})^D$ is preserved by the map ω in a manner analogous to that in our discussion of densifications. One can now attempt to rectify the set $\omega(\mathscr{D}^D) \subseteq \mathbb{Z}/(\pi_1 \ldots \pi_D\mathbb{Z})$ to obtain a new set $\mathcal{F} \subset \overline{\mathbb{Q}}$ by means of the method of Grosu [7]. In this way one perceives the possibility of a densification process for sets of algebraic numbers. However, in common with the method of Grosu, there is only weak control of the degree and other data associated with the field extension in which the elements of $\mathcal{F}$ are embedded. This level of control would appear to be far too weak to facilitate useful densification conclusions.

We finish this section with some comments concerning the main conclusion of Theorem 7. We are interested in understanding the set of solutions $S(\mathscr{A}; \mathbf{P})$ of a given system of polynomial equations

$$P_i(\mathbf{x}) = 0 \quad (1 \leqslant i \leqslant r),$$

with variables restricted to a set $\mathscr{A}$. The conclusion of Theorem 7 shows that, in circumstances wherein the polynomials $P_i(\mathbf{x})$ are both homogeneous and linear at least, this objective can be achieved by studying instead a related set of integers $\mathscr{D}$ with $\mathrm{env}(\mathscr{D}) \leqslant |\mathscr{D}|^{s+\varepsilon}$. While this polynomial dependence of $\mathrm{env}(\mathscr{D})$ on $|\mathscr{D}|$ may seem significantly superior to the exponential dependence available in the condensation results of §3, one may interpret this nonetheless as a "non-result". If it is the case that $\mathscr{D}$ is a typical set having roughly $X^{1/s-\varepsilon}$ elements in a box of size X, then conventional heuristics suggest nothing more than that the number of solutions of the system $P_i(\mathbf{x}) = 0$ $(1 \leqslant i \leqslant r)$, with $\mathbf{x} \in \mathscr{D}^s$, could be $O(1)$ or even 0. In other words, the exponent $1/s$ is already small enough that in general little or nothing can be learned from the counting function for $|\mathscr{D} \cap [1, X]|$ alone. Perhaps it is more illuminating to point out that more or less any solution behaviour can be encoded in a set $\mathscr{D}$ for which $|\mathscr{D} \cap [1, X]| \ll X^{1/s-\varepsilon}$.

8 Remarks on Sets of Real Points

We now explore some consequences of work of Vu, Wood and Wood [16, Theorem 1.1]. Let D be an integral domain of characteristic zero, such as the field of real numbers $\mathbb{R}$, and let $\mathscr{D}$ be a finite subset of D. Consider a system of polynomials $P_i(\mathbf{x}) \in \mathbb{Z}[x_1, \ldots, x_s]$ $(1 \leqslant i \leqslant r)$. In this section, we are interested in the set $S(\mathscr{D}; \mathbf{P})$ of solutions $\mathbf{x} \in \mathscr{D}^s$ of the simultaneous equations

$$P_i(x_1, \ldots, x_s) = 0 \quad (1 \leqslant i \leqslant r).$$

The structure of the solution set $S(\mathscr{D}; \mathbf{P})$ is determined by the hypergraph $\Gamma(\mathscr{D}; \mathbf{P})$ defined just as in the analogous discussion of §2.

Given a large prime number p, one may seek a ring homomorphism $\varphi_p : \mathbb{Z}[\mathscr{D}] \to \mathbb{F}_p$ with the property that, whenever $(x_1, \ldots, x_s) \in \mathscr{D}^s$, then

$$P_i(x_1, \ldots, x_s) = 0 \quad (1 \leqslant i \leqslant r)$$

if and only if

$$P_i(\varphi_p(x_1), \ldots, \varphi_p(x_s)) = 0 \quad (1 \leqslant i \leqslant r).$$

We emphasise here that the latter system of equations over $\mathbb{F}_p$ amount to a system of congruences. The conclusion of [16, Theorem 1.1] demonstrates that there exists an infinite sequence of primes with positive relative density having the property that such a ring homomorphism exists. This conclusion may not at first sight be obvious from [16, Theorem 1.1]. Of course, any ring homomorphism $\varphi_p : \mathbb{Z}[\mathscr{D}] \to \mathbb{F}_p$ has the property that, whenever $(x_1, \ldots, x_s) \in \mathscr{D}^s$ satisfies $P_i(x_1, \ldots, x_s) = 0$ $(1 \leqslant i \leqslant r)$, then

$$P_i(\varphi_p(x_1), \ldots, \varphi_p(x_s)) = \varphi_p(P_i(x_1, \ldots, x_s)) = \varphi_p(0) = 0 \quad (1 \leqslant i \leqslant r). \tag{40}$$

Thus, the interesting feature for us is that whenever

$$(x_1, \ldots, x_s) \in \mathscr{D}^s \quad \text{and} \quad P_i(x_1, \ldots, x_s) \neq 0$$

for some index i with $1 \leqslant i \leqslant r$, then

$$P_i(\varphi_p(x_1), \ldots, \varphi_p(x_s)) = \varphi_p(P_i(x_1, \ldots, x_s)) \neq 0.$$

The approach here is to define a set L of all elements

$$P_i(x_1, \ldots, x_s) \in \mathbb{Z}[\mathscr{D}],$$

with $\mathbf{x} \in \mathscr{D}^s$, having the property that $P_i(x_1, \ldots, x_s) \neq 0$. The conclusion of [16, Theorem 1.1] guarantees that the ring homomorphisms φ_p, whose existence is asserted, may be constructed in such a manner that $0 \notin \varphi_p(L)$. This last assertion guarantees that the condition (40) holds, and this ensures that the sought after ring homomorphisms φ_p do indeed exist.

Equipped with these ring homomorphisms $\varphi_p : \mathbb{Z}[\mathscr{D}] \to \mathbb{F}_p$, we see that $\Gamma(\mathscr{D}; \mathbf{P})$ is isomorphic as a hypergraph to $\Gamma(\varphi_p(\mathscr{D}); \mathbf{P})$. Thus, the solution structure of $S(\mathscr{D}; \mathbf{P})$ may be faithfully embedded into appropriate finite fields $\mathbb{F}_p$. If the prime number p has been chosen sufficiently large, then one may apply [7, Theorem 1.3] to obtain a faithful model of the finite field solution structure inside a number field K with degree at most $\exp(\exp(c_{\mathbf{P}}|\mathscr{D}|))$, for a suitable real number $c_{\mathbf{P}}$ depending at most on $\mathbf{P}$. For systems of linear equations, moreover, one can restrict to an integer model. In this way, one sees that linear problems involving sets of real points, for example, may be considered instead as linear problems involving sets of integers. For non-linear polynomial problems, we must instead work with sets of algebraic numbers of bounded algebraic enveloping radius. In both settings, the condensation and densification ideas of this paper become applicable.

9 The Conclusion of Theorem 1

As promised in the introduction, we briefly justify the conclusion of Theorem 1. Suppose that $\mathscr{A} \subset \mathbb{Z}$ is finite with $A = \text{card}(\mathscr{A})$, and define

$$\mathfrak{a}_n = \begin{cases} 1, & \text{when } n \in \mathscr{A}, \\ 0, & \text{when } n \notin \mathscr{A}. \end{cases}$$

Suppose first that $\varphi_j \in \mathbb{Z}[t]$ $(1 \leqslant j \leqslant k)$ is a system of polynomials with

$$\det\left(\frac{\mathrm{d}^i \varphi_j(t)}{\mathrm{d}t^i}\right)_{1 \leqslant i,j \leqslant k} \neq 0.$$

Let s and k be natural numbers with $s \leqslant k(k+1)/2$. Then for each $\varepsilon > 0$, the conclusion of [17, Theorem 1.1] shows that

$$\begin{aligned} \int_{[0,1)^k} \left| \sum_{|n| \leqslant X} \mathfrak{a}_n e(\alpha_1 \varphi_1(n) + \ldots + \alpha_k \varphi_k(n)) \right|^{2s} \mathrm{d}\boldsymbol{\alpha} &\ll X^{\varepsilon} \Big(\sum_{|n| \leqslant X} |\mathfrak{a}_n|^2 \Big)^s \\ &\ll X^{\varepsilon} A^s. \end{aligned}$$

Since for each $n \in \mathscr{A}$, one has $|n| \leqslant \text{env}(\mathscr{A})$, the first conclusion of Theorem 1 follows on setting $X = \text{env}(\mathscr{A})$.

The second conclusion of Theorem 1 follows on making use of the translation invariance property of the system of equations

$$x_1^j + \ldots + x_s^j = x_{s+1}^j + \ldots + x_{2s}^j \quad (1 \leqslant j \leqslant k).$$

Put $m = \min \mathscr{A}$, and observe that whenever $\mathbf{x} \in \mathscr{A}^{2s}$ satisfies this system of equations, then as a consequence of the binomial theorem, one has

$$(x_1 - m)^j + \ldots + (x_s - m)^j = (x_{s+1} - m)^j + \ldots + (x_{2s} - m)^j \quad (1 \leqslant j \leqslant k).$$

Thus, if we put $\mathscr{B} = \{a - m : a \in \mathscr{A}\}$, then we have $J_{s,k}(\mathscr{A}) = J_{s,k}(\mathscr{B})$. We therefore deduce from the special case $\varphi_j(t) = t^j$ $(1 \leqslant j \leqslant k)$ of the first part of the theorem that

$$J_{s,k}(\mathscr{A}) \leqslant (\mathrm{env}(\mathscr{B}))^{\varepsilon} A^s = (\max(\mathscr{A}) - \min(\mathscr{A}) + 1)^{\varepsilon} A^s = (\mathrm{diam}(\mathscr{A}))^{\varepsilon} A^s.$$

The second conclusion of Theorem 1 follows when $s \leqslant k(k+1)/2$. When instead $s > k(k+1)/2$, we observe that a trivial estimate combines with orthogonality to show that

$$\begin{aligned} J_{s,k}(\mathscr{A}) &\leqslant A^{2s-k(k+1)} \int_{[0,1)^k} \Big| \sum_{|n| \leqslant X} \mathfrak{a}_n e(\alpha_1 n + \ldots + \alpha_k n^k) \Big|^{k(k+1)} \mathrm{d}\boldsymbol{\alpha} \\ &\ll A^{2s-k(k+1)} \cdot (\mathrm{diam}(\mathscr{A}))^{\varepsilon} A^{k(k+1)/2}. \end{aligned}$$

The desired conclusion is now immediate in this case, since

$$A^{2s-k(k+1)/2} > A^s.$$

References

1. R. C. Baker and G. Harman, *Small remainder of a vector to a suitable modulus*, Math. Z. **221** (1996), no. 1, 59–71.
2. Y. F. Bilu, V. F. Lev and I. Z. Ruzsa, *Rectification principles in additive number theory*, Discrete Comput. Geom. **19** (1998), no. 3, 343–353.
3. J. Bourgain, C. Demeter and L. Guth, *Proof of the main conjecture in Vinogradov's mean value theorem for degrees higher than three*, Ann. of Math. (2) **184** (2016), no. 2, 633–682.
4. G. A. Freiman, *Addition of finite sets*, Dokl. Akad. Nauk SSSR **158** (1964), 1038–1041.
5. G. A. Freiman, *Foundations of a structure theory of set addition*, Translations of Math. Monographs **37** (1973), American Math. Soc., Providence, RI.
6. A. Granville, *Bounding the coefficients of a divisor of a given polynomial*, Monatsh. Math. **109** (1990), no. 4, 271–277.
7. C. Grosu, $\mathbb{F}_p$ *is locally like* $\mathbb{C}$, J. London Math. Soc. (2) **89** (2014), no. 3, 724–744.
8. S. Guo, Z. K. Li and P.-L. Yung, *Improved discrete restriction for the parabola*, Math. Res. Letters, **30** (2023), no. 5, 1375–1409.
9. L. Guth, D. Maldague and H. Wang, *Improved decoupling for the parabola*, J. Eur. Math. Soc., **26** (2024), no. 3, 875–917.
10. S. V. Konyagin and V. F. Lev, *Combinatorics and linear algebra of Freiman's isomorphism*, Mathematika **47** (2000), no. 1-2, 39–51.
11. J. C. Lagarias and A. M. Odlyzko, *Effective versions of the Chebotarev density theorem*, in: Algebraic number fields: L-functions and Galois properties (Proc. Symposia in Math., Univ. Durham, Durham, 1975), pp. 409–464, Academic Press, London, 1977.

12. A. Mudgal, *Diameter free estimates for the quadratic Vinogradov mean value theorem*, Proc. London Math. Soc. (3) **126** (2023), no. 1, 76–128.
13. R. Schippa, *Improved decoupling for the moment curve in three dimensions*, arXiv:2302.10884v1.
14. T. Tao and V. Vu, *Additive combinatorics*, Cambridge University Press, Cambridge, 2006.
15. H. Tôyama, *A note on the different of the composed field*, Kodai Math. Sem. Rep. **7** (1955), no. 2, 43–44.
16. V. H. Vu, M. M. Wood and P. M. Wood, *Mapping incidences*, J. London Math. Soc. (2) **84** (2011), no. 2, 433–445.
17. T. D. Wooley, *Nested efficient congruencing and relatives of Vinogradov's mean value theorem*, Proc. London Math. Soc. (3) **118** (2019), no. 4, 942–1016.

The manufacturer's authorised representative in the EU is Springer Nature Customer Service Centre GmbH, Europaplatz 3, 69115 Heidelberg, Germany. If you have any concerns regarding our products, please contact ProductSafety@springernature.com

Printed and bound by CPI Group (UK) Ltd, Croydon, CR0 4YY
01/10/2026
02225932-0018